Smart Innovation, Systems and Technologies

Volume 225

Series Editors

Robert J. Howlett, Bournemouth University and KES International, Shoreham-by-Sea, UK

Lakhmi C. Jain, KES International, Shoreham-by-Sea, UK

The Smart Innovation, Systems and Technologies book series encompasses the topics of knowledge, intelligence, innovation and sustainability. The aim of the series is to make available a platform for the publication of books on all aspects of single and multi-disciplinary research on these themes in order to make the latest results available in a readily-accessible form. Volumes on interdisciplinary research combining two or more of these areas is particularly sought.

The series covers systems and paradigms that employ knowledge and intelligence in a broad sense. Its scope is systems having embedded knowledge and intelligence, which may be applied to the solution of world problems in industry, the environment and the community. It also focusses on the knowledge-transfer methodologies and innovation strategies employed to make this happen effectively. The combination of intelligent systems tools and a broad range of applications introduces a need for a synergy of disciplines from science, technology, business and the humanities. The series will include conference proceedings, edited collections, monographs, handbooks, reference books, and other relevant types of book in areas of science and technology where smart systems and technologies can offer innovative solutions.

High quality content is an essential feature for all book proposals accepted for the series. It is expected that editors of all accepted volumes will ensure that contributions are subjected to an appropriate level of reviewing process and adhere to KES quality principles.

Indexed by SCOPUS, EI Compendex, INSPEC, WTI Frankfurt eG, zbMATH, Japanese Science and Technology Agency (JST), SCImago, DBLP.

All books published in the series are submitted for consideration in Web of Science.

More information about this series at http://www.springer.com/series/8767

Suresh Chandra Satapathy · Vikrant Bhateja ·
Margarita N. Favorskaya · T. Adilakshmi
Editors

Smart Computing Techniques and Applications

Proceedings of the Fourth International
Conference on Smart Computing
and Informatics, Volume 1

 Springer

Editors
Suresh Chandra Satapathy
School of Computer Engineering
KIIT University
Bhubaneswar, Odisha, India

Margarita N. Favorskaya
Reshetnev Siberian State University
of Science and Technology
Krasnoyarsk, Russia

Vikrant Bhateja
Department of Electronics
and Communication Engineering
Shri Ramswaroop Memorial Group
of Professional Colleges (SRMGPC)
Lucknow, Uttar Pradesh, India

Dr. A.P.J. Abdul Kalam Technical
University, Lucknow, Uttar Pradesh, India

T. Adilakshmi
Department of Computer Science
and Engineering, Vasavi College
of Engineering, Hyderabad
Telangana, India

ISSN 2190-3018 ISSN 2190-3026 (electronic)
Smart Innovation, Systems and Technologies
ISBN 978-981-16-0880-3 ISBN 978-981-16-0878-0 (eBook)
https://doi.org/10.1007/978-981-16-0878-0

Conference Committees

Chief Patrons

Sri M. Krishna Murthy, Secretary, VAE
Sri P. Balaji, CEO, VCE

Patron

Dr. S. V. Ramana, Principal, VCE

Honorary Chair

Dr. Lakhmi C. Jain, Australia

General Chair

Dr. Suresh Chandra Satapathy, KIIT DU, Bhubaneswar

Organizing Chair

Dr. T. Adilakshmi, Professor and HOD, CSE, VCE

Publication Chairs

Dr. Nagaratna P. Hegde, Professor, CSE, VCE
Dr. Vikrant Bhateja, SRMGPC, Lucknow, UP, India

Program Committee

Dr. S. Ramachandram, Former Vice Chancellor, OU
Dr. Banshidhar Majhi, Director, IITDM Kancheepuram
Dr. Siba K. Udgata, Professor, HCU
Dr. Sourav Mukhopadhyay, Associate Professor, IIT Kharagpur
Dr. P. Radha Krishna, Professor, CSE, NIT Warangal
Dr. M. M. Gore, Professor, MNNIT, Allahabad
Dr. S. M. Hegde, Professor, NIT Surathkal
Dr. S. Bapi Raju, Professor, IIIT Hyderabad
Dr. Rajendra Hegadi, Associate Professor, IIIT Dharwad
Dr. S. Sameen Fatima, Former Principal, OU
Dr. K. Shyamala, Professor, OU
Dr. Naveen Sivadasan, TCS Innovation Labs, Hyderabad
Dr. Badrinath G. Srinivas, Research Scientist—III, Amazon Development Centre, Hyderabad
Dr. Ravindra S. Hegadi, Professor, PAH Solapur University
Dr. S. P. Algur, Professor and Chairman, CSE, Rani Channamma University, Belagavi
Dr. R. Sridevi, HOD, Department of CSE, JNTUH

International Advisory Committee/Technical Program Committee

Dr. Rammohan, South Korea
Dr. Kailash C. Patidar, South Africa
Dr. Naeem Hanoon, Malaysia
Dr. Vimal Kumar, The University of Waikato, New Zealand
Dr. Akshay Sadananda Uppinakudru Pai, University of Copenhagen, Denmark
Dr. K. C. Santosh, The University of South Dakota
Dr. Ayush Goyal, Texas A&M University, Kingsville
Dr. Sobhan Babu, Associate Professor, IIT Hyderabad
Dr. D. V. L. N. Somayajulu, Director, IIIT Kurnool
Dr. Siba Udgata, Professor, HCU
Dr. R. B. V. Subramaanyam, Professor, NITW
Dr. S. G. Sanjeevi, Professor, NITW

Dr. Sanjay Sengupta, CSIR, New Delhi
Dr. A. Govardhan, Rector, JNTU Hyderabad
Prof. Chintan Bhatt, Chandubhai Patel Institute of Technology, Gujarat
Dr. Munesh Chandra Trivedi, ABES Engineering College, Ghaziabad
Dr. Alok Aggarwal, Professor
Dr. Anuja Arora, Jaypee Institute of Information Technology, Noida, India
Dr. Divakar Yadav, Associate Professor, MMMUT, Gorakhpur, India
Dr. Kuda Nageswar Rao, Andhra University, Visakhapatnam
Dr. M. Ramakrishna Murthy, ANITS, Visakhapatnam
Dr. Suberna Kumar, MVGR, Vizianagaram
Dr. J. V. R. Murthy, Director Incubation and IPR, JNTU Kakinada
Dr. D. Ravi, IDRBT, Hyderabad
Dr. Badrinath G. Srinivas, Research Scientist—III, Amazon Development Center, Hyderabad
Dr. K. Shyamala, Professor, OU
Dr. P. V. Sudha, Professor, OU
Dr. M. A. Hameed, Assistant Professor, OU
Dr. B. Sujatha, Assistant Professor, OU
Dr. T. Adilakshmi, Professor and HOD, CSE, VCE
Dr. Nagaratna P. Hegde, Professor, CSE, VCE
Dr. V. Sireesha, Assistant Professor, CSE, VCE

Organizing Committee

Dr. D. Baswaraj, Professor, CSE, VCE
Dr. K. Srinivas, Associate Professor, CSE, VCE
Dr. V. Sireesha, Assistant Professor, CSE, VCE
Mr. S. Vinay Kumar, Assistant Professor, CSE, VCE
Mr. M. Sashi Kumar, Assistant Professor, CSE, VCE
M. Sunitha Reddy, Assistant Professor, CSE, VCE
R. Sateesh Kumar, Assistant Professor, CSE, VCE
Mr. T. Nishitha, Assistant Professor, CSE, VCE

Publicity Committee

Dr. M. Shanmukhi, Professor, CSE, VCE
Mr. C. Gireesh, Assistant Professor, CSE, VCE
Ms. T. Jalaja, Assistant Professor, CSE, VCE
Mr. I. Navakanth, Assistant Professor, CSE, VCE
Ms. S. Komal Kaur, Assistant Professor, CSE, VCE
Mr. T. Saikanth, Assistant Professor, CSE, VCE

Ms. K. Mamatha, Assistant Professor, CSE, VCE
Mr. P. Narasiah, Assistant Professor, CSE, VCE

Website Committee

Mr. S. Vinay Kumar, Assistant Professor, CSE, VCE
Mr. M. S. V. Sashi Kumar, Assistant Professor, CSE, VCE

Preface

This volume contains the selected papers presented at the 4th International Conference on Smart Computing and Informatics (SCI-2020) organized by the Department of Computer Science and Engineering, Vasavi College of Engineering (Autonomous), Ibrahimbagh, Hyderabad, Telangana, during October 9–10, 2020. It provided a great platform for researchers from across the world to report, deliberate, and review the latest progress in the cutting-edge research pertaining to smart computing and its applications to various engineering fields. The response to SCI 2020 was overwhelming with a good number of submissions from different areas relating to artificial intelligence, machine learning, cognitive computing, computational intelligence, and its applications in main tracks. After a rigorous peer review with the help of Technical Program Committee members and external reviewers, only quality papers were accepted for presentation and subsequent publication in this volume of SIST series of Springer.

Several special sessions were floated by eminent professors in cutting-edge technologies such as blockchain, AI, ML, data engineering, computational intelligence, big data analytics and business analytics, and intelligent systems. Eminent researchers and academicians delivered talks addressing the participants in their respective field of proficiency. Our thanks are due to Prof. Roman Senkerik, Head of AI Lab, Tomas Bata University in Zlin, Czech Republic; Shri. Shankarnarayan Bhat, Director Engineering, Intel Technologies India Pvt. Ltd.; Ms. Krupa Rajendran, Associate VP, HCL Technologies; and Mr. Aninda Bose, Springer India, for delivering keynote addresses for the benefit of the participants. We would like to express our appreciation to the members of the Technical Program Committee for their support and cooperation in this publication. We are also thankful to team from Springer for providing a meticulous service for the timely production of this volume.

Our heartfelt thanks to Shri. M. Krishna Murthy, Secretary, VAE; Sri. P. Balaji, CEO, VCE; and Dr. S. V. Ramana, Principal, VCE, for extending support to conduct this conference in Vasavi College of Engineering. Profound thanks to Prof. Lakhmi C. Jain, Australia, for his continuous guidance and support from the beginning of the conference. Without his support, we could never have executed such a mega-event. We are grateful to all the eminent guests, special chairs, track managers, and reviewers for their excellent support. Special vote of thanks to numerous authors across the country as well as abroad for their valued submissions and to all the delegates for their fruitful discussions that made this conference a great success.

Bhubaneswar, India   Suresh Chandra Satapathy
Lucknow, India   Vikrant Bhateja
Krasnoyarsk, Russia   Margarita N. Favorskaya
Hyderabad, India   T. Adilakshmi

List of Special Sessions Collocated with SCI-2020

SS_01: Next Generation Data Engineering and Communication Technology
Dr. Suresh Limkar, AISSMS Institute of Information Technology, Pune

SS_02: Artificial Intelligence and Machine Learning Applications (AIML)
Dr. V. Sowmya, CEN, Amrita Vishwa Vidyapeetham, Coimbatore
Dr. M. Anand Kumar, NIT Karnataka
Dr. M. Venkatesan, NIT Karnataka
Prof. K. P. Soman, Amrita Vishwa Vidyapeetham, Coimbatore

SS_03: Advances in Computational Intelligence and Its Applications
Dr. C. Kishor Kumar Reddy, Stanley College of Engineering and Technology for Women, Hyderabad
P. R Anisha, Stanley College of Engineering and Technology for Women, Hyderabad

SS_04: Blockchain Technology: Foundations, Challenges and Applications
Prof. Sandeep Kumar Panda, Faculty of Science and Technology, ICFAI Foundation for Higher Education, Hyderabad
Prof. Santosh Kumar Swain, School of Computer Engineering, KIIT (Deemed to be) University, Bhubaneswar

SS_05: Application of Machine Learning for Intelligent System Design
Dr. Minakhi Rout, KIIT (Deemed to be) University, Bhubaneswar

SS_06: Advances in Big Data Analytics and Business Intelligence
Dr. Vijay B. Gadicha, G. H. Raisoni Academy of Engineering and Technology Nagpur
Dr. Ajay B. Gadicha, P. R. Pote College of Engineering and Management, Amravati

SS_07: Recent Advances in Artificial Intelligence—Applications, Challenges and Future Trends

Dr. S. Velliangiri, CMR Institute of Technology, Hyderabad
Dr. P. Karthikeyan, Presidency University, Bengaluru
Dr. Iwin Thanakumar Joseph, KITS, Coimbatore

Contents

About the Editors

Suresh Chandra Satapathy is currently working as Professor, KIIT Deemed to be University, Odisha, India. He obtained his Ph.D. in Computer Science Engineering from JNTUH, Hyderabad, and master's degree in Computer Science and Engineering from National Institute of Technology (NIT), Rourkela, Odisha. He has more than 27 years of teaching and research experience. His research interest includes machine learning, data mining, swarm intelligence studies and their applications to engineering. He has more than 98 publications to his credit in various reputed international journals and conference proceedings. He has edited many volumes from Springer AISC, LNEE, SIST and LNCS in past, and he is also the editorial board member in few international journals. He is a senior member of IEEE and a life member of Computer Society of India. Currently, he is National Chairman of Division-V (Education and Research) of Computer Society of India.

Vikrant Bhateja is Associate Professor, Department of ECE in SRMGPC, Lucknow. His areas of research include digital image and video processing, computer vision, medical imaging, machine learning, pattern analysis and recognition. He has around 160 quality publications in various international journals and conference proceedings. He is associate editor of IJSE and IJACI. He has edited more than 30 volumes of conference proceedings with Springer Nature and is presently EiC of IGI Global: IJNCR journal.

Dr. Margarita N. Favorskaya is Professor and Head of the Department of Informatics and Computer Techniques at Reshetnev Siberian State University of Science and Technology, Russian Federation. Professor Favorskaya is a member of KES organization since 2010, the IPC member and Chair of invited sessions of over 30 international conferences. She serves as a reviewer in international journals (Neurocomputing, Knowledge Engineering and Soft Data Paradigms, Pattern Recognition Letters, Engineering Applications of Artificial Intelligence), an associate editor of Intelligent Decision Technologies Journal, International Journal of Knowledge-Based and Intelligent Engineering Systems, International Journal of Reasoning-based Intelligent Systems, a honorary editor of the International Journal of Knowledge Engineering and Soft Data Paradigms, reviewer, a guest editor, and a book editor (Springer). She is the author or the co-author of 200 publications and 20 educational manuals in computer science. She co-authored/co-edited seven books for Springer recently. She supervised nine Ph.D. candidates and presently supervising four Ph.D. students. Her main research interests are digital image and videos processing, remote sensing, pattern recognition, fractal image processing, artificial intelligence, and information technologies.

Dr. T. Adilakshmi is currently working as Professor and Head of the Department, Vasavi College of Engineering. She completed her Bachelor of Engineering from Vasavi College of Engineering, Osmania University in the year 1986 and did her Master of Technology in CSE from Manipal Institute of Technology, Mangalore, in 1993. She received Ph.D. from Hyderabad Central University (HCU) in 2006 in the area of Artificial Intelligence. Her research interests include data mining, image processing, artificial intelligence, machine learning, computer networks and cloud computing. She has 23 journal publications to her credit and presented 28 papers at international and national conferences. She has been recognized as a research supervisor by Osmania University (OU) and Jawaharlal Nehru Technological University (JNTU). Two research scholars were awarded Ph.D. under her supervision, and she is currently supervising 11 Ph.D. students.

Chapter 1
Pad Vending Machine with Cashless Payment

Anjali Prajapati and Anandu M. Dharan

Abstract A vending machine aims to provide required product or the service to the customer with certain ease, wherein not much effort is required. This research work aims to design a pad vending machine with an option of payment using QR code which is implemented using blockchain to make the system much more efficient and reliable than the existing systems present in the Indian market. The system is divided into two parts, first being the working of the machine and second being the mode of payment which is implemented using a blockchain. It is noticed many times that due to unpredictable menstrual flow women tend to face a lot problems. To overcome this problem, a pad vending machine is proposed with certain advancements through which women can help themselves in the stated circumstances.

1.1 Introduction

There are different types of networking systems through which a technology or an application can be implemented. There are majorly three types of networking systems which are majorly used to implement any application and are classified into three, namely centralized, distributed, and decentralized. Each type of networking system serves in a different and unique way to address the problem, and almost every application can be implemented using each of the network system; however, we should always look for the most efficient way of implementing any application in a network. In the current scenario, most of our applications are implemented on a centralized network which follows a three-tier architecture which includes a server, a client, and a receiver. Since there are three layers involved in the architecture, the process consumes a lot of time in sending and receiving requests. But if the server which is present in the middle of the receiver and the sender is removed, it would save a lot of time in the networking process by making the whole process a decentralized one, which makes the process easy as it has only two parties which are

A. Prajapati (✉) · A. M. Dharan
Christ (Deemed to be University), Faculty of Engineering, Bangalore, India
e-mail: anjali.prajapati@btech.christuniversity.in

© The Author(s), under exclusive license to Springer Nature Singapore Pte Ltd. 2021
S. C. Satapathy et al. (eds.), *Smart Computing Techniques and Applications*,
Smart Innovation, Systems and Technologies 225,
https://doi.org/10.1007/978-981-16-0878-0_1

communicating to each other directly. Everything around us which is implemented through centralized system can be switched to a decentralized one. For example, a simple money transaction; consider A wants to send Rs. 100 to B; in a centralized network, A will send a request to the bank for the transaction of Rs. 100, and once the bank approves the transaction, the money would be credited to B's account. Consider the same situation in a decentralized network; since there is no middle layer involved, there would not be any requirement of the approval from the bank, A can directly transfer the money to B's account, and the bank would only record that on a cloud or in a database. Hence, a decentralized system is introduced to make any process simpler, easier and hassle-free with full security.

1.2   Literature Review

The concept of vending machine is to provide the required product to the customer through a machine with just a push button. There are different vending machines available in the market, for example coffee vending machine, food and beverages vending machine, pad vending machine, etc. Most of these machines available in the Indian market requires their customers to have hard cash for the payment to be done. However, there are some food and beverages vending machine available in the Indian market which provides an option of payment using Paytm. It is majorly required and essential as not every time the customer would have a hard cash to do the payment and purchase the product.

In [1], the author has defined vending machine as a terminal device that interacts with the ultimate customer for the direct selling the proposed product or service and aims to design a system that makes the customer service faster and efficient. The author also aims to reduce the maintenance cost of the whole system and making it more interactive.

In [2], the author informs the readers about the success rate of vending machine in Japan and USA because of cashless payment and ease of access to the products required. It also helped these countries in reducing the crime rates because of cashless payment methods.

In [3], the author puts up a thought about the security of the payment using mobile phones, and it requires authentication as it can be hacked. The payment can be done using mobile phones through QR code. QR code can store more information than a traditional barcode, and the security of the system can be maintained by using Public Key Infrastructure (PKI) which adds an additional layer of security to the system.

A lot of research has been made to make sure any vending machine available in the market should be interactive enough for the customers and should have fast and secure mode of payment with an ease of use. Specifically, in the case of a pad vending machines which are used by women in case of emergency of unpredicted menstrual flow. There are different researcher who worked on the different parts of the vending machine to make it efficient and fast for its customers. The major problems faced by the customer while using the vending machine was the machine being less interactive and the mode of payment. As said earlier, majorly the mode of payment in a vending machine is via coin system or cash payment, and it is not necessary that every time a user would have a coin or cash; hence, an addition of smart and easy mode of payment would really help the customers, especially in case of a pad vending machine.

1.3 Proposed System

The proposed system is a pad vending machine which has an interactive UI for the customer, and the customers can scan the QR code at the end of the process for completing the payment process. An architecture of the model is given below (Fig. 1.1):

The proposed system aims to develop a decentralized system using blockchain and remove the third-party involvement; hence, the system has no administration. The architecture of the Pad Vending Machine (PVM) includes only two parties, namely the user and the vendor and two status in two different variable stack.

The description of each and every and every part of the system is given below:

A.   **User X**: A random customer X would come to the PVM and will select the size of the sanitary napkin required via a push button; the system would check the availability of the same, and if it is available, it will be delivered to the customer once the payment is done. In case, the select size of the napkin is not available, the user would be asked by the system to opt another size of the pad.

B.   **Vendor**: The second involvement in the system would be of a vendor who would be responsible for refilling the PVM with the napkins of different sizes as and when required, and the payment of each napkins would be done by the

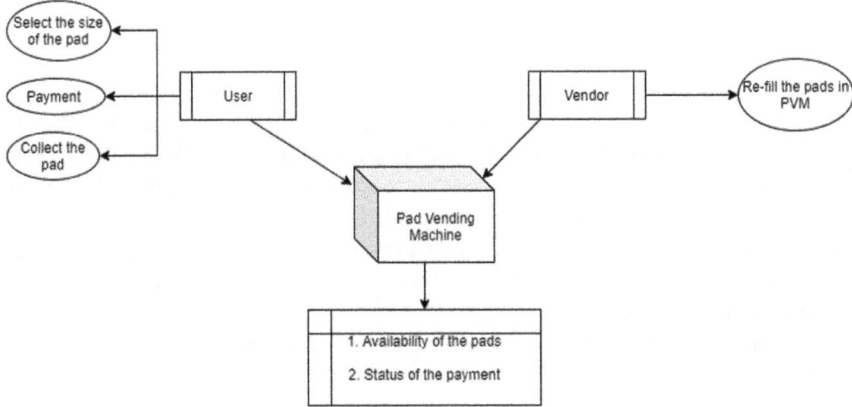

Fig. 1.1 Architecture of the model

customers via QR which would be directly send to the vendor's account without any approval from third party as it uses blockchain for the payment to be done.

C. **Availability of the pad**: In PVM, there is a stack which keeps a check on the availability of the sanitary pads of each size. For each size of the sanitary napkin, there are two values to be updated, and those are 0 and 1. '1' defines that there are napkins available in the machine, and '0' defines that there are no napkins left in the machine of that size.

D. **Status of the payment**: It is a single variable which is defined to check the payment status, and it contains only Boolean values 0 and 1. '1' defines the payment is done, and '0' defines the payment is pending.

1.4   Implementation

The sequence of the steps involved in the proposed system is shown in the flow chart given in Fig. 1.2. The proposed system is a pad vending machine which uses QR scanning method for payment which is implemented using blockchain. The flow of the events in the system is quite simple and easy. The process starts with a user standing in front of the machine in the need of the sanitary napkin. In the first step, the user is expected to use the push button present in the machine to select the size of the required sanitary napkin.

The second step is to scan the QR code available on the screen of the machine, through which the payment is supposed to be done. Since the mode of payment is through a QR code and it is implemented using a blockchain mechanism, each time the user takes a pad, the machine generates a new QR code for the payment to be done to the same vendor, which ensures the safety of the system as it is anonymous and cannot be hacked. The mode of the payment is so designed using blockchain which makes the system decentralized wherein the system is governed by a network of nodes and all nodes works according to the predefined protocol. Also, all the funds transfer are controlled by the users and not a centralized party like a bank.

As soon as the machine generates a new QR code to be scanned, it actually makes a request for the transaction to be done. The transaction is represented as a 'block.' Thereafter, the block is broadcasted to every party in the network. Once the network of nodes receives the transaction, they validate it. Now, a new block of data is added to the existing blockchain and the transaction is completed. The existing blockchain is a data structure that keeps the record of all the transaction made to the vendor, and a new block is added every time a new payment is done. The transaction would be completed in less than a minute as the transaction does not required any sort of approval from the bank.

Once the payment is done by the user and the vendor receives it, the payment variable would be turned to '1' from '0' automatically stating that the payment is done, and now, the sanitary napkin can be delivered to the user, and thereafter, the sanitary napkin would be delivered.

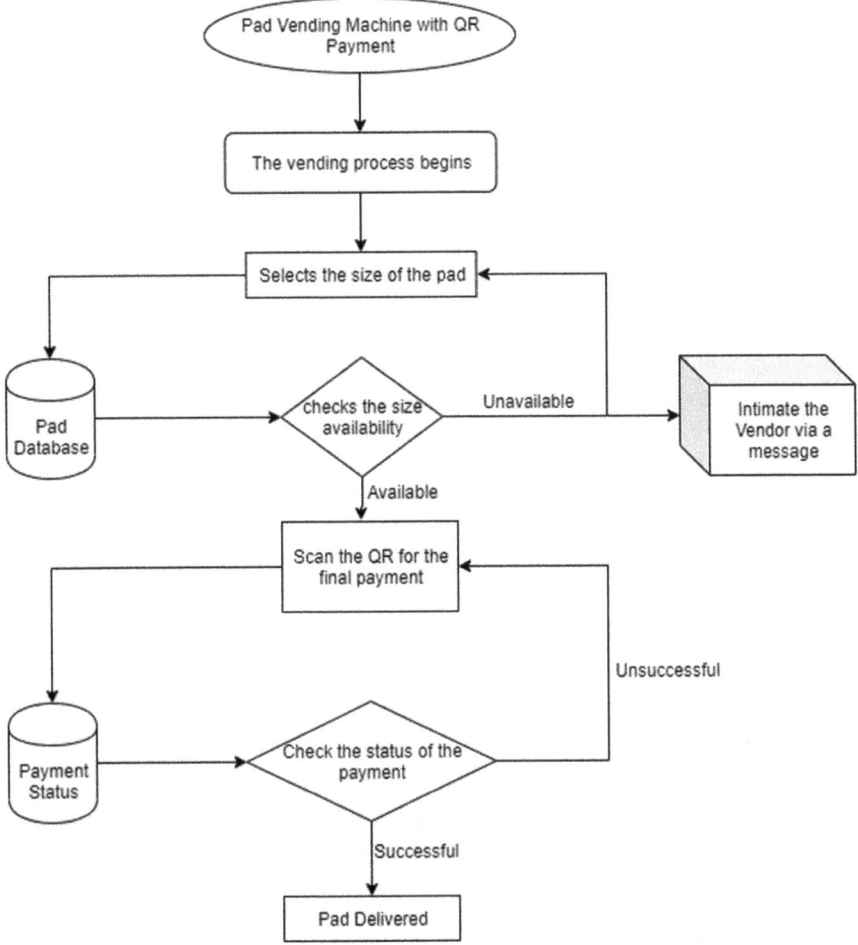

Fig. 1.2 Flow chart of the process

On the other hand, if a situation arises wherein the selected size of the pad is not available, the value in the pad availability stack for that particular size of the pad would be turned to '0' and a message would be sent to the vendor immediately for the re-fill of the pads.

1.5 Results and Comparative Study

The system provides the sanitary napkin of the required size on the basis of the customer's request. Once the request is made to the machine, it checks for the availability of the size of the napkin and generates a QR code for the payment to be done

Fig. 1.3 Comparative study
of the proposed system with
existing systems

to the vendor. As soon as the payment is done and verified by the system, the sanitary napkin is delivered to the user. However, if there a situation wherein the selected size of the napkin is not available, the machine asks the user to select a different size and procedure with it.

The system is compared with different existing systems present in the Indian market and stands out different because of its fast processing and efficient working, and a graphical representation of the same is shown below in Fig. 1.3. With the introduction of blockchain in the payment, the system becomes really efficient and secure, uses less time to process the request and the density of providing the service increases.

1.6 Future Scope

The stack which is being used for storing the information about the availability of the size of the sanitary napkins and the variable which stores the information regarding the status of the payment can be replaced with two other blockchain structures. This would again reduce the cost of the whole system as there would not be any requirement of storage. Also, the efficiency of the system would increase exponentially.

1.7 Conclusion

The system proposed above is a pad vending machine in which the mode of payment is the major advancement as it is implemented using a blockchain mechanism which makes the system faster in terms of processing the request and serving the customer in less than a minute. Also, it makes the system a decentralized peer-to-peer network, and if the data is entered once, it cannot be altered by any other unreliable source.

References

1. Rifa, H., Budi, R.: Blockchain Based E-Voting Recording System Design. IEEE (2017). 978-1-5386-3546-9/17/$31.00
2. Koç, A.K., Yavuz, E., Çabuk, U.C., Dalklç, G.: Towards Secure E-Voting Using Ethereum Blockchain. IEEE (2018) 978-1-5386-3449-3/18/$31.00
3. Kshetri, N., Voas, J.: Blockchain-Enabled E-Voting. IEEE (2018). 0740-7459/18/$33.00
4. Christian.: Desain Dan Implementasi Visual Cryptography Pada Sistem E-Voting Untuk Meningkatkan Anonymity. InstituteTeknologi Bandung (2017)
5. Dougherty, C.: Vote Chain: Secure Democratic Voting (2016)

Chapter 2
Design of Wrung Order Layout Using Advanced Optimization Techniques with Integrated Variable Batch Scheduling

K. Mallikarjuna and Y. Hariprasad Reddy

Abstract Universally, specialists and scientists accept that flexibility assumes a elementary play in modern factory segment. Only associated with modest parcel size generation since agility adaptable is an indispensable part to be incorporate into course of action of racks in format plan among the assembling fragment. In view of such conditions, considering NP hard double target issues is, regularly, a lumbering responsibility. In this work, researchers tended to about a populace-based high end search techniques like differential development (DE) and sheep run technique (SRT) for making wrung structure configuration matters in lithe system of manufacturing location. The instigators focused on twofold aim headway connected with fundamental objective is stressed over the versatile slot (FJSP) arranging issue, the accompanying objective focused on wrung order layout matters where expelling the interest of machineries within lead-ins of wrung steps to control rigid transference cost and hoarding working time of employments on machineries. The accomplishment of the estimation (SRT and TS) is crisscross by standard issues. At long last, it is pondered that SRT yields better outcomes at the point on par with TS.

2.1 Introduction

In the present situation, mechanized assembling ventures are under prodigious stress which brought about by the increasing expense of vitality, materials, works, capital, and strengthening overall challenge. While these patterns will stay for quite a while, the issue fronting producing today runs much cavernous. By and large, they come

K. Mallikarjuna (✉)
Department of Mechanical Engineering, G. Pullaiah College of Engineering and Technology, Kurnool, India

Y. Hariprasad Reddy
Rayalaseema University College of Engineering, Kurnool, India

© The Author(s), under exclusive license to Springer Nature Singapore Pte Ltd. 2021   9
S. C. Satapathy et al. (eds.), *Smart Computing Techniques and Applications*,
Smart Innovation, Systems and Technologies 225,
https://doi.org/10.1007/978-981-16-0878-0_2

from the very idea of the assembling procedure itself. So as to beat that, adaptable manufacturing frameworks (@FMSs) are viewed as one of the most productive strategies to use in lessening or taking out assembling issues. FMS is in excess of a specialized arrangement [1]; it is a factory-based driven arrangement prompting to improve gainfulness via decreasing process times and stock levels and improved assembling viability through expanded operational adaptability, consistency, and control.

The FMS design includes assigning different hold for achieving full skill. The plan has an impact eager for elapsed time and cost [2] which ought to be resolved in the beginning of the FMS [3]. By and by, the most generally utilized sort of FMS designs [4] is as per the following:

1.   Line or single column design
2.   Circle format or oval shape design
3.   Stepladder or pecking order or wrung structure format
4.   U-shaped format.

Amongst the above designs, this task addressed about wrung order layout with scheduling as restrictions utilizing sheep run technique (SRT) and Tabu search method (TS).

2.2   Problematic Depiction

The work plan got from Abraham et al. [5] cited for this work. Analyst's complement on construction of wrung position in versatile course of action of collecting with [FJSP] versatile job shop booking as impediment.

2.2.1   Multi-goal Mathematical Representations

Here, maker familiarizes the twofold aim condition with the versatile variable bunch arranging problems which are united with wrung design arrangement primes to lessen the goal of work such as throughput, make run,
minimize MAKSP, F (S_{maxi})

$$\text{Minimize, } F(S_{\text{max}}) = S_{n,m} \tag{2.1}$$

Connected to

$$S_{i,j,k} \leq S_{i,j+1,k} - T_{i,j+1}, \quad \text{for} \quad j \ldots = 1, 2, 3 \ldots p$$

$$S_{i,j,k} \geq 0, \quad \text{for} \quad j \ldots = 1, 2, 3 \ldots p$$

2.3   Proposed Advanced Optimization

This work is proposed with two advanced non-traditional optimization techniques such as

(a)   Sheep run technique
(b)   Tabu search method

The above two methods are applied on to multiobjective function of wrung order layout design with scheduling constraints, and necessary throughputs, make run have been obtained.

2.3.1   Sheep Run Technique

In the inception, sheep run techniques (SRT) was developed by Hyunchul and Byungchul [6]. This calculation was started to explain huge scale issues on planning over a period of a few continuous years. And it is competing with performance of another evolutionary algorithm called genetic algorithm calculation, and it is alluded as staggered hereditary activities can get great arrangements. This calculation was commonly founded on the common difference in sheep in the group.

Let us consider two runs of sheep in a structure with two persons for vigilance who being monitoring and observing those sheep runs. At the moment when these two were busy in chitchat [7], there is a probability of unification of herds been taken place, and in such condition, both the persons arrive at the mixed gathering and endeavor to segregate the sheep from mixed surges and keep the groups as of now [8]. In actuality, the effect of inheritance will cause imperfection on specific surges. In this, necessary steps which describe the multilevel genetic operation for scheduling problem which is a constraint in layout design in sheep run techniques are given below.

Multilevel Genetic Operation

```
int b1Makespan, b2Makespan;
        int b1=1, b2=2;
        b1Makespan = chsMakespan[1];
        b2Makespan = chsMakespan[2];

        //Best 2 chromosomes
        for (int p = 2; p <= no_populations; ++p) {
            if(b1Makespan > chsMakespan[p]){
                b1Makespan = chsMakespan[p];
            }   b1 = p;
        }

cross overing the strings using subchromosomal
        for (inte p = 1; p <= no__ populations;++ p)
{
            inte *ttt;
            ttt = subbChromosomeCrossOvers(chss[p],
sizee*subChssCrossOverRatios, size);
            for (inte b = 1; b <= sizee; b ++ ) {
                chsSUB[pp][b] == ttt[b];
            }
        }
```

2.3.2 *Tabu Search*

Tabu search (TS) is a heuristic strategy initially proposed by Glover in 1986 to different combination problems have noted up in the activities explore writing. In a few cases, the strategies portrayed give arrangements near optimality [9] and are among the best, if not the best, to handle the troublesome issues within reach. These victories have made TS incredibly well known among those keen on discovering great answers for the huge combinatorial issues experienced in numerous down to earth settings. Here, a necessary procedural step for scheduling using Tabu search [10] is furnished through pseudocode.

Pseudocode

```
int
makespan_OdjectWithWaitingTimes(Schedulable_operations
representatives1[],int sequence[]){
            Schedulable_operations
S[MAX],J[MAX],Job_order[MAX],RR[MAX],JJ[MAX];
            int
R[no_machines+1],RT[no_machines+1],SMPT[no_batchs+1],OP[n
o_batchs+1],
            nn=no_batchs*no_operations,order=1,n=0;
            int num=0;
            int OP1[MAX];
            int Morder[MAX_MC][MAX];
            int Morder1[MAX_MC][MAX];
            initialize();

        //display_ReprestativesSequence(representatives1,se
quence);
            for(int k=1;k<=no_machines;k++){
                for (int l = 1; l<=
no_batchs*no_operations; l++) {
                        Morder[k][l] = 0;
                        Morder1[k][l] = 0;
                }
                OP1[k] = 1;
            }
```

2.4   Results and Discussion

Six different machine slots with different processing of jobs on machines to be allocated in machine slots in wrung order layout for effective design of layout are considered, along with necessary process times, interslot distance between machines and load and unloading cost is also considered. Different combinations of these six slots for machine allocations, six jobs, and varied operations of jobs for different cluster combinations were used, and four layouts are used to generate five example problems [1]. Here, Table 2.1 shows production summary of wrung order layout, and Table 2.2 shows batch variabilities with batch magnitudes of wrung order layout.

Table 2.1 Framework of manufacture system

Layout pattern	Number of machines	Number of batches	Number of processes	Load/unload stations
Wrung order	6	6	6	2

Table 2.2 Batch variabilities with batch magnitudes of the pecking layout with six machines with six batches

Batch No.		B1	B2	B3	B4	B5	B6
Batch variabilities	VBS	100	70	55	60	80	40

Table 2.3 Assessment of make run of the proposed evolutionary algorithms (for VBS with 100 generations)

Problem Instance	SFHA	TS
	MAKRN (min)	MAKRN (min)
KMK 1—(6 × 6 × 6)	5020	5520
KMK 2—(6 × 5 × 5)	4200	4800
KMK 3—(6 × 5 × 4)	3445	3950
KMK 4—(6 × 4 × 4)	3025	3300
KMK 5—(6 × 4 × 3)	2554	2995

Table 2.4 Assessment of overall conveyance cost of the projected evolutionary algorithms (VBS through 100 generations)

Problem Instance	SFHA	TS
	OCC	OCC
KMK 1—(6 × 6 × 6)	8400	9200
KMK 2—(6 × 5 × 5)	7652	8560
KMK 3—(6 × 5 × 4)	7025	8020
KMK 4—(6 × 4 × 4)	6654	7885
KMK 5—(6 × 4 × 3)	5951	6995

2.4.1 Inferences

By thorough analysis From Tables 2.3 and 2.4 and Figs. 2.1 and 2.2 it is revealed that accomplishment of SRT and TS for figuring Overall Conveyance penalty and Makerun (MAKRN) is weakening as issue ration is littler according to the issue scope. By relative investigation, it is seen that OCC are progressed for SRT.

2.5 Conclusion

This work addressed on the presenting of wrung order layout configuration with incorporated planning for which the recurrence of excursions of jobs amid machines, the gap among the machines with stacking and emptying good ways from stacking/emptying place to all machineries, and component material taking care of

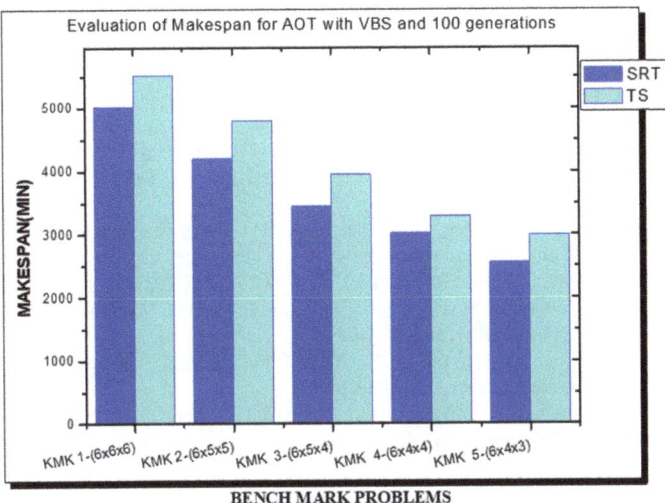

Fig. 2.1 Plot of make run to evaluate performance of projected algorithm

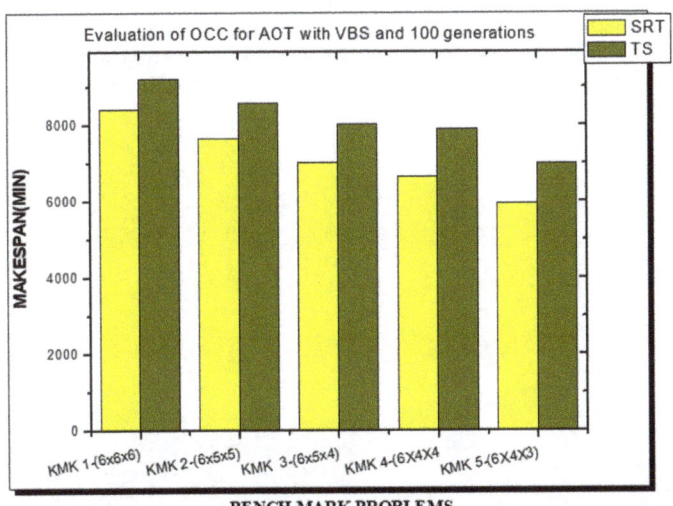

Fig. 2.2 Plot of overall conveyance charge to evaluate performance of projected algorithm

expense (MHD) are evaluated in an unexpected way. The issue is encircled as the quadratic project issues (QPI) proposal of capacity layout problems. This is inferable from the point that in the QPI models, the separation between the places of allocation is recognized well ahead of time yet it is structure subordinate for the problems considered in this paper. From the outcomes, we infer that wrung order design is upgraded utilizing SRT and is superior to TS with varible batch sizes and constant MHD cost and recurrence of outings between machines.

References

1. Mallikarjuna, K., Veeranna, V., Reddy, K.H.: A new meta-heuristics for optimum design of loop layout in flexible manufacturing system with integrated scheduling. Int. J. Adv. Manuf. Technol. **84**(9–12), 1841–1860 (2016). https://doi.org/10.1007/s00170-015-7715-9
2. Apple, J.M.: Plant Layout and Material Handling, 3rd edn. The Ronald Press Co, New York (1977)
3. Satheesh Kumar, R.M., Asokan, P., Kumanan, S.: Design of loop layout in flexible manufacturing system using non-traditional optimization technique. Int. J. Adv. Manuf. Technol. **38**(5–6), 594–599 (2008). https://doi.org/10.1007/s00170-007-1032-x
4. Groover, M.P.: Production Systems and Computer Integrated Manufacturing. Automation Prentice-Hall Inc. (1992). ISBN 0-87692-618-9
5. Liu, H., Abraham, A., Grosan, C., Li, N.: ICDIM 2007. LNCS, vol. 1, pp 138–145. https://www.springer.com/lncs. Accessed 21 Nov 2016
6. Kim, H., Ahn, B.: A new evolutionary algorithm based on sheep flocks heredity model. IEEE Pacific RIM Conf. Commun. Comput. Signal Process. Proc. (2001). https://doi.org/10.1109/pacrim.2001.953683
7. Subbaiah, K.V., Nageswara Rao, M., Narayana Rao, K.:. Scheduling of AGVs and machines in FMS with makespan criteria using sheep flock heredity algorithm. Int. J. Phys. Sci. (2009)
8. Nara, K., Takeyama, T., Kim, H.: New evolutionary algorithm based on sheep flocks heredity model and its application to scheduling problem. Proc. IEEE Int. Conf. Syst. Man Cybernet. (1999). https://doi.org/10.1109/icsmc.1999.816603
9. Glover, F.: Tabu search—Part I. ORSA J. Comput. (1989). https://doi.org/10.1287/ijoc.1.3.190
10. Ekşioğlu, B., Ekşioğlu, S.D., Jain, P.: A tabu search algorithm for the flowshop scheduling problem with changing neighborhoods. Comput. Ind. Eng. (2008). https://doi.org/10.1016/j.cie.2007.04.004

Chapter 3
Adaptive Beamforming Using Radial Basis Function

Gowtham Narayan, Sharath Kumar, Abhinav Bharadwaj, Srikar Vangala, and R. Lavanya

Abstract With the advancements in 5G mobile communication, the need for improvements in spatial signal processing has become indispensable. Beamforming is one such technique used to transmit or receive wireless signals in a particular direction in space and suppress interfering signals. This is achieved by making use of a uniform antenna array. This paper focuses on the receiver antenna. MUSIC algorithm is highly popular for calculating the direction of the received signal. However, it is computationally expensive and hence not suitable for real-time applications. The idea proposed in this paper is to develop a neural network model based on MUSIC, to cater to real-time need of modern communication systems.

3.1 Introduction

Antenna arrays provide an effective way to track and process signals arriving in various directions in wireless communication. An array of sensors can arbitrarily alter its beampattern by altering the weights of its array elements compared to a single antenna that is constrained in directionality and bandwidth [1]. Ideally, maximum signal reception has to be towards the direction of arrival (DoA) of the desired signal, and minimal signal reception has to be towards the directions of the interfering signals. Among the algorithms for DoA estimation, MUSIC provides more accurate results at the cost of comparatively higher computational power [2, 3] (Fig. 3.1).

In real-world scenarios, the desired and interfering signals change their directions continuously. A fast beamforming technique is required to keep track of these changes in directions. Though the MUSIC algorithm is known for its accuracy, the computational complexity makes it unsuitable for real-time applications such as communication-on-the-move applications. To alleviate this problem, a neural

G. Narayan · S. Kumar (✉) · A. Bharadwaj · S. Vangala · R. Lavanya
Department of Electronics and Communication Engineering, Amrita School of Engineering, Amrita Vishwa Vidyapeetham, Coimbatore, India

R. Lavanya
e-mail: r_lavanya@cb.amrita.edu

S. C. Satapathy et al. (eds.), *Smart Computing Techniques and Applications*,
Smart Innovation, Systems and Technologies 225,
https://doi.org/10.1007/978-981-16-0878-0_3

Fig. 3.1 MUSIC algorithm
flowchart

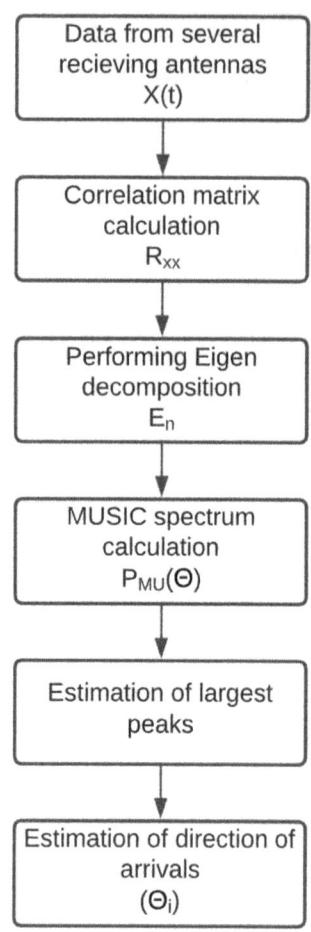

network is employed to learn the function underlying the MUSIC algorithm such
that the latter's accuracy is retained and at the same time the processing speed is
high. Neural networks are very efficient in pattern recognition and function approximation and are becoming increasingly popular in diverse areas and applications [4,
5]. A neural network with radial basis as the activation function is employed in this
work. The network output is a linear combination of inputs' radial base functions
and neuron parameters [6].

3.2 Methodology

The first step for developing the neural network model is to generate dataset with
received signals as input and DoA as output. This is done by varying the angle between

[+90° −90°] and sampling the outputs from each element. The signal received by the ULA with N elements is modelled as

$$X(t) = A(\theta)S(t) + N(t) \tag{3.1}$$

where

$$X(t) = \left[x_1(t) \; x_2(t) \; \ldots \; x_N(t) \right]^{\mathrm{T}} \tag{3.2}$$

$$N(t) = \left[n_1(t) \; n_2(t) \; \ldots \; n_N(t) \right]^{\mathrm{T}} \tag{3.3}$$

and

$$S(t) = \left[s_1(t) \; s_2(t) \; \ldots \; s_M(t) \right]^{\mathrm{T}} \tag{3.4}$$

$X(t)$ is a complex vector of antenna array outputs. $N(t)$ is the noise vector caused by the channel and is modelled as White Gaussian noise. $S(t)$ is the actual source signals vector transmitted with M individual signals. The $N \times M$ matrix A of the incident source signals can be defined as

$$A(\theta) = \left[a_1(\theta) \; a_2(\theta) \; \ldots \; a_M(\theta) \right]^{\mathrm{T}} \tag{3.5}$$

where

$$a(\theta_m) = \left[e^{-j(n-1)\Psi_m} \right]^{\mathrm{T}} : 1 \leq n \leq N \tag{3.6}$$

Ψ_{m} represents the phase shift from element to element along the array which is defined by

$$\Psi_m = 2\pi \left(\frac{d}{\lambda} \right) \sin(\theta_m) \tag{3.7}$$

where d is the element spacing and λ is the wavelength of the received signal. After varying the DoA and generating the dataset for the neural network, the dataset is split in the ratio of 7:3. 70% of the dataset is used for training with multiple epochs till $R = 1$ is achieved. The remaining 30% of the dataset is used to test the neural network model.

The neural network architecture chosen was RBFNN due to its short learning time, easy design and strong tolerance to noise. RBFNN uses the activation functions of radial basis functions. The input to the neural network is the spatial covariance matrix R of signals received by the elements of the ULA.

The spatial covariance matrix is given by

$$R = E\{X(t)X(t)^H\}$$
$$= AE\{S(t)S(t)^H\} + E\{N(t)N(t)^H\}$$
$$= APA^H + \sigma^2 I$$
$$= \sum_{i=1}^{M} \lambda_i e_i e_i^H \tag{3.8}$$

P denotes the power matrix of the obtained signals by the array antenna. λ_i denotes the eigenvalues of matrix R, σ^2 is the variance of independent White noise received by the array antenna.

This spatial covariance matrix is used as input to MUSIC algorithm to obtain DoA's. The spatial covariance matrix along with the DoAs is used as ground truths for the neural network.

3.3 RBFNN Overview

A three-layer feed forward neural network is a radial basis function (RBF). Radial base function networks are commonly used for approximation problems as a type of artificial neural network. They are differentiated from other neural network methods because of their ability to perform universal approximation and have a faster learning speed. As stated earlier, the input layer, hidden layer and output layer are the three layers that are associated with this particular neural network system. They have their own set of tasks to be performed. A brief explanation on the RBF network is presented as follows. The process of RBF training is completed once the computed error reaches the desired values or the total number of iterations of training is terminated. Commonly, an RBF network is chosen based on a specific number of nodes in the hidden layer. Computational units for transfer functions are generally handled by a Gaussian function. In comparison with multi-layer perceptron, the amount of time needed for completion of training is usually less for RBF. The neural network of radial basis function (RBFNN) consists of many elementary processing units called neurons, organised into layers. The subsequent adjacent layer is related to each neuron in a layer. There are no connections between neurons of the same layer. Neurons connect with each other by weight, and according to the received signals, they can be triggered accordingly. There are three layers in the RBFNN, namely input, output and one hidden layer, which are interconnected in a feed forward manner. The mapping is nonlinear from the input to the hidden layer, while the mapping from the hidden layer to the output layer is linear (Fig. 3.2).

The mapping functionality relies on the distance between the vector of the input and the vector of the centre. The RBFNN with input x of dimension N and output y of dimension M can be expressed as a weighted sum of a finite number of radial basis functions given by

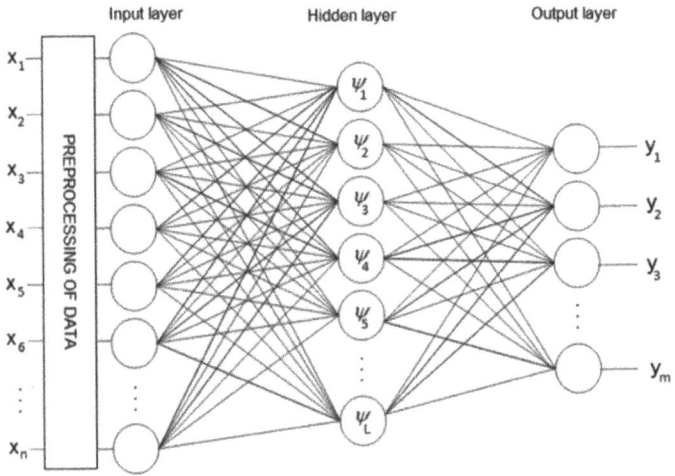

Fig. 3.2 Architecture of RBFNN

$$y = F(x) = \sum_{i=1}^{L} w_i \Psi(\|x - x_i\|) \tag{3.9}$$

$\Psi(\|x - x_i\|)$ is the radial basis function calculated by shifting $\Psi(x_i)$. L is the set of arbitrary functions and x_i is the centre of RBF $\Psi(\|x - x_i\|)$. $\Psi_s(x_i)$ is given by

$$\Psi\left(e^{-x^2}\right) \tag{3.10}$$

RBF is a strongly nonlinear function which offers optimal functionality for progressive learning. As a transition function, the neurons in the secret layer of RBFNN have RBF. By integrating the weighted inputs and biases, RBF determines its weighted inputs with the Euclidean distance weight function and obtains its net input. During the training phase, neurons are eventually introduced to the secret layer.

Many learning strategies exist for training RBFNN. One technique is to use the K-means algorithm to classify the initial centres of the hidden layer's Gaussian functions. The standard deviation of the Gaussian function is the mean distance of the other Gaussian functions to the first hundred closest neighbours. This method allows the weights from the input to the hidden layer to be defined. The weights from the hidden layer to the output layer, implemented on single-layer networks, are calculated by supervised learning known as the delta law.

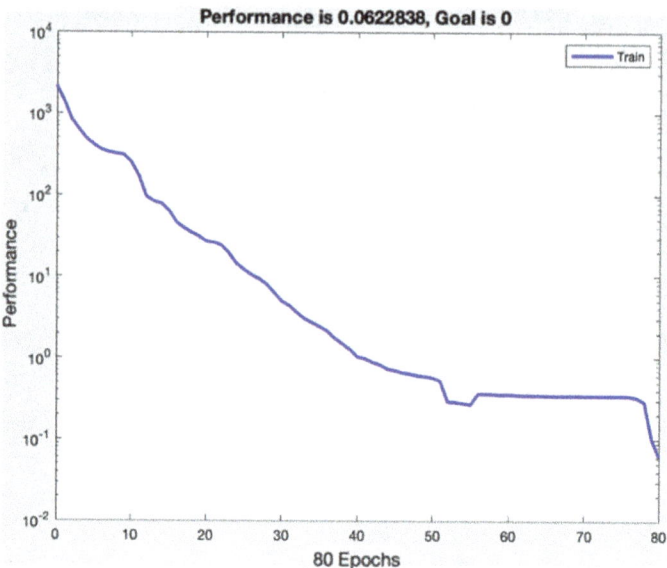

Fig. 3.3 MSE plot

3.4 Results

A 10 element ULA with helical antenna elements and element spacing given by $d = \lambda/2$ is considered for the study. The input to the neural network is given as the spatial correlation matrix of the array output vector. Before feeding the spatial correlation matrix to the neural network, pre-processing of the data has to be performed. When the matrix is separated into its real and imaginary parts, they are both symmetrical matrices. Also, the diagonal elements of the imaginary matrix are all zeros. Thus, only the elements of the upper triangular matrix are taken as input to the neural network to decrease the complexity, thus providing us with $N \times (N-1)$ inputs rather than N^2 inputs. Training, validation and testing were performed on the developed neural network model. The output of the neural network is the DoA of the incoming signals. The mean squared error (MSE) of the developed neural network model was found to be 0.0622838 (Figs. 3.3, 3.4, 3.5 and 3.6).

3.5 Conclusion

In this work, a neural network-based DoA estimation algorithm is proposed. The proposed scheme employs radial basis network which is trained based on MUSIC algorithm, combining the merits of fast convergence of RBFNN and high performance of MUSIC. The algorithm is highly suitable for deployment in real-time

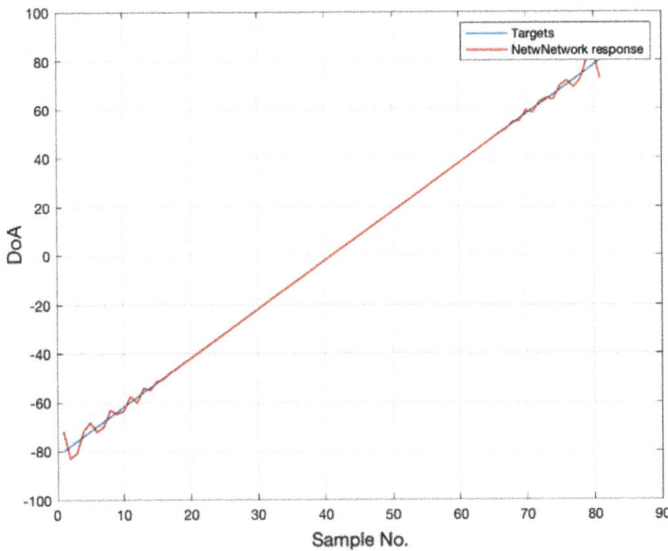

Fig. 3.4 Comparison of target DoA and RBFNN-estimated DoA

Fig. 3.5 Training regression plot

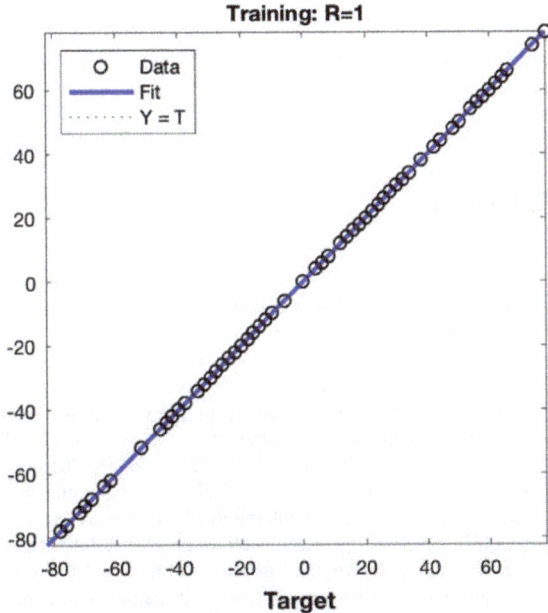

Fig. 3.6 Testing regression plot

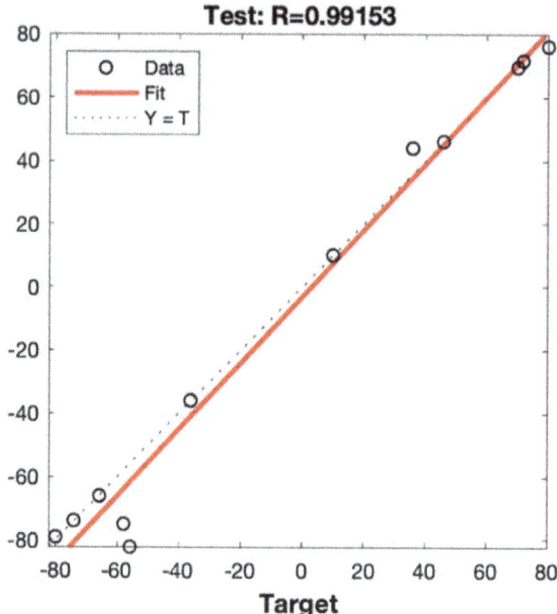

communication applications. The work can be further extended for beamforming module as well.

References

1. Pei, B., Han, H., Sheng, Y., Qiu, B.: Research on smart antenna beamforming by generalized regression neural network. In: 2013 IEEE International Conference on Signal Processing, Communication and Computing (ICSPCC 2013), pp. 1–4. KunMing (2013)
2. Vikas, B., Vakula, D.: Performance comparision of MUSIC and ESPRIT algorithms in presence of coherent signals for DoA estimation. In: International conference of Electronics, Communication and Aerospace Technology (ICECA) Coimbatore, pp. 403–405 (2017)
3. Qian, R., Sellathurai, M., Wilcox, D.: A study on MVDR Beamforming applied to an ESPAR antenna. IEEE Signal Process. Lett. **22**(1), 67–70 (2015)
4. Praveena, S., Lavanya, R,: Superpixel based segmentation for multilesion detection in diabetic retinopathy. In: 2019 3rd International Conference on Trends in Electronics and Informatics (ICOEI), pp. 314–319. Tirunelveli, India (2019). https://doi.org/10.1109/ICOEI.2019.8862636
5. Rajendiran, G., Kumar, C., Vishnuprasad K., Ramachandran, K.I.: Feature mapping techniques for improving the performance of fault diagnosis of synchronous generator. Int. J. Prognost. Health Manage. **6**, 1–12 (2015)
6. Ping, Z.: DOA Estimation method based on neural network. In: 2015 10th International Conference on P2P, Parallel, Grid, Cloud and Internet Computing (3PGCIC), pp. 828–831. Krakow (2015)

Chapter 4
A Cloud-Based Centralized Smart Electricity Billing and Payment Management System

P. Pramod Kumar, Nethaji Achha, G. Ranadheer Reddy, and V. Pranathi

Abstract Energy sources are very important in our day-to-day life. One of the most important energy sources is electricity; we pay for the electricity which we use. As we are living in and around the space of artificial intelligence, we need to develop ourselves to live with artificial intelligence. In past five years, there are many changes took place in the electricity department to provide better services to the consumers. Many of the people suspect that many malpractices are going on in the electricity department while issuing a bill for consumed units and at payment time. To avoid this type of malpractices, we can go with the atomized billing system with artificial intelligence. This artificial intelligence can reduce the usage of paper and manpower too. As it is automatized, it can maintain the accurate time, which follows perfect rules, exact bill and many more flexibilities. In this article, we discuss artificial intelligence, and we propose an idea on how to use this artificial intelligence to avoid the abovementioned malpractices and problems.

4.1 Introduction

In this busy world, everyone is bothering about their time, nowadays, time is very precious, and people are forgetting about their regular bill payments (like electricity). Presently, the electricity billing is manual and time-consuming process. With the proposed cloud-based billing system, we can overcome the disadvantages which are

P. Pramod Kumar (✉)
Department of Computer Science and Artificial Intelligence, SR University, Warangal, Telangana, India

N. Achha
Department of Electronics and Communication Engineering, S R Engineering College, Warangal, Telangana, India

G. Ranadheer Reddy · V. Pranathi
Department of Computer Science and Engineering, Sumathi Reddy Institute of Technology for Women, Warangal, Telangana, India

© The Author(s), under exclusive license to Springer Nature Singapore Pte Ltd. 2021
S. C. Satapathy et al. (eds.), *Smart Computing Techniques and Applications*,
Smart Innovation, Systems and Technologies 225,
https://doi.org/10.1007/978-981-16-0878-0_4

occurring due to the present manual billing system [1]. In this cloud-based system, there is no need of producing the bill and distributing the bill manually [2].

The main idea of this paper is to give the best and well-automated electricity billing and payment management system; this makes our lives easy in this faster generation. This system reduces or eliminates the paper wastage and manpower; to pay our bills using this system, we are required to have a fourth-generation (4G) subscriber identity module (SIM) [3]; this 4G SIM will be linked with the Aadhar number of the person on whose name the electric meter number was issued. With the help of the Aadhar number, details of the person will be accessed. The SIM which is fixed to the meter reads the number of units consumed by the consumer, and then it automatically calculates the charges. The consumer can easily pay the bill amount online through various e-payment gateways and Web sites. With this kind of system, no one can escape from the fines or punishments. First of every month consumers will automatically get the amount to be paid through the SMS, e-mail and departmental applications which are provided. This service perfectly follows the government rules like charges per unit, taxes and punishments (like disconnecting the power supply if consumer fails to pay after 15 days of the bill issued). The consumer will be allowed to pay the amount more than the bill generated, but they cannot pay the less amount than the bill generated, and extra paid amount will be added to the adjustment charges to reduce from the next month bill.

4.2 Related Work

Many people have worked on similar kind of systems in different ways and proposed versatile methodologies to avoid the problems faced in the manual billing process. In the present traditional system of bill generation, one employee from electricity department visits to each and every house in that area with a billing machine provided by the department [4]. He places that billing machine in front of the meter, then it automatically scans, counts the number units consumed and prints the automatically calculated bill including different taxes and charges, and finally, a receipt of paper will be issued to the consumer. However, the bill is calculated automatically but here still the manpower is required and paper wastage is done. To overcome this type of problems, we are making automatized billing, where the manpower is not required and paper wastage is reduced. However, raw details stored in the cloud storage and bill can be printed as on the request [5]. The below given figure is one such kind of smart billing system published in July 2015 (Fig. 4.1) [6].

The below figure is one such kind of a proficient and smart electricity billing published on July 13, 2019 (Fig. 4.2).

Here, a work is done that displays the warning to pay the bill on the light-emitting diodes (LED) fixed to it. The block diagram of prepaid energy meter is shown in Fig. (i). It consists of microcontroller LPC2148, buzzer, relay, single-phase energy meter, MAX232, infrared (IR) sensor circuit and liquid crystal display (LCD) [2] (Figs. 4.2 and 4.3).

Fig. 4.1 Smart billing
system published in July
2015

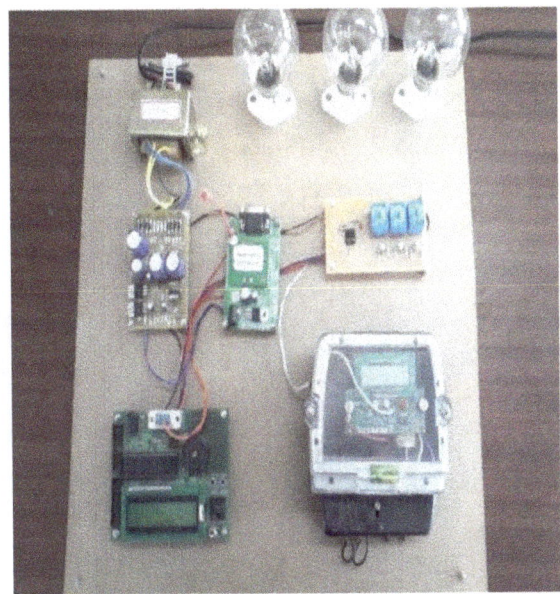

Fig. 4.2 Proficient and
smart electricity billing
published on July 13, 2019

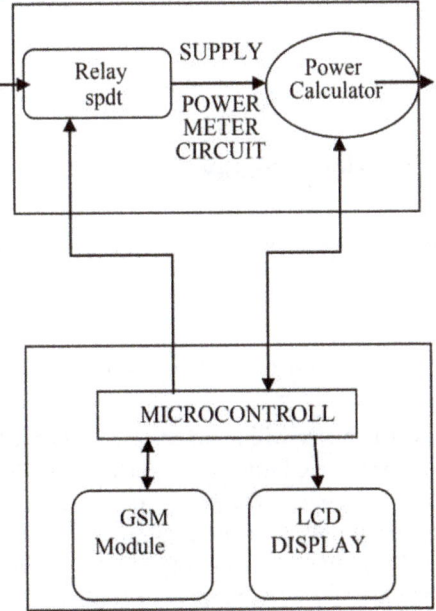

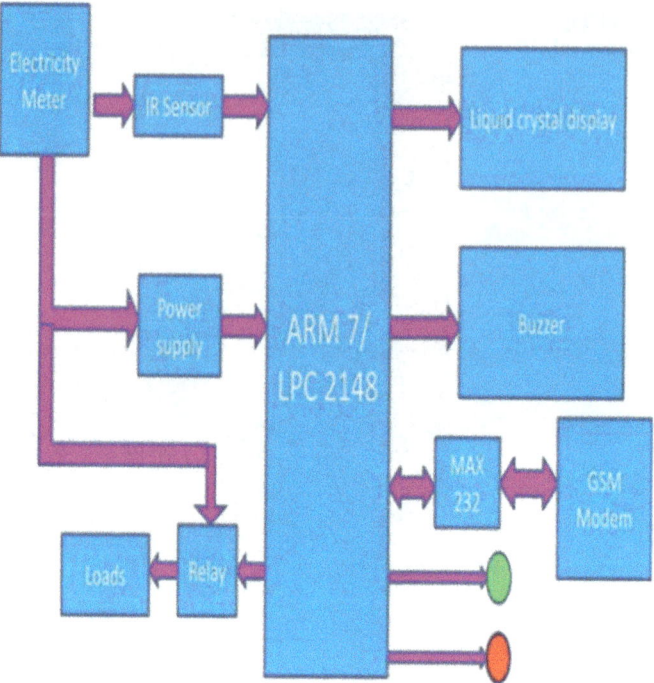

Fig. 4.3 Smart electricity billing system from volume-6, issue-3, May–June 2016

4.3   Methodology

We discuss the proposed methodology in three different sections, as follows.

4.3.1   *Automatic or Calculation Section*

Electric meter will be fixed with a 4G SIM [3, 7], which is linked with the Aadhar number of a consumer on whose name the meter is registered. If the house owner wants to change the details or wants to update any thing, they have to show the perfect proofs as documents, so that details of consumer will be updated or changed [8, 9]. In the beginning of the every month, the details of the consumer including the meter reading will be sent automatically to the cloud space with the help of the SIM and the data will be stored in the cloud [10, 11, 12]. The meter number, meter reading and other required details will be sent to the automatic calculator, and then the calculator will generate the bill which includes all taxes and charges. Then, the final bill amount will be sent back to the cloud storage (Fig. 4.4).

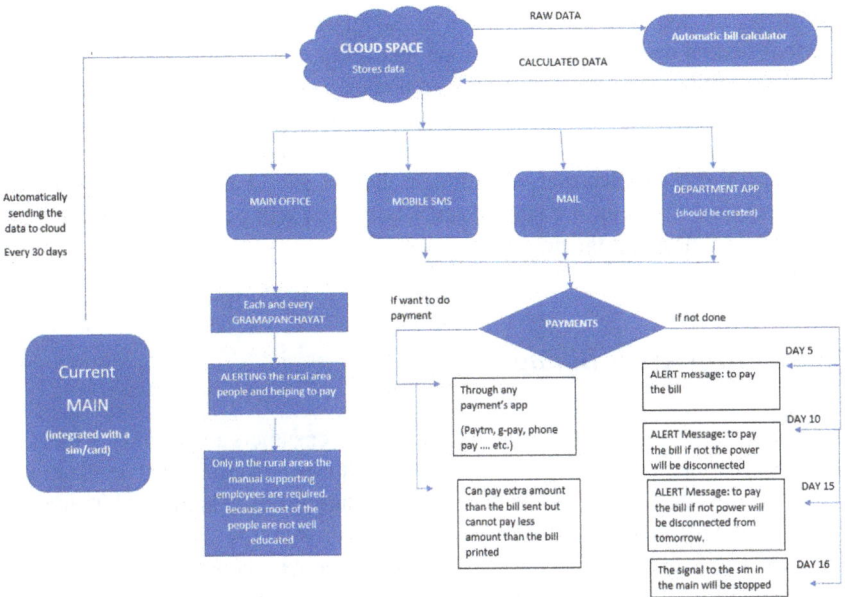

Fig. 4.4 Cloud-based centralized smart electricity billing and payment management system

4.3.2 Payment Section

From the cloud storage, bill data will be sent to the electricity department head office database and to the consumer through the mobile SMS and e-mail which are linked to Aadhar number with a payment link added to it [10, 11, 13]. The app published by the electricity department will be notifying the consumers with payment details. When the consumer wants to pay, then the consumer will be directed to the online payment Web site. If consumer fails to pay the bill, then alert messages will be sent, the first alert message will be on 5th day after the bill issuing date to remind to pay the bill before the last date mentioned, the second alert message will be sent on 10th day after the bill issuing date to remind to pay the bill and to warn about the power supply disconnections, and the third alert message will be sent on the 15th day after the bill issuing date, that is to remind bill payment. After three alert messages, if no payment is made by the consumer, then by the next day automatically the electric supply will be disconnected by disabling the SIM inside the meter. If the consumer wants to regain the power connection, they must have to pay the previous pending bills and have to put a request for connection resume [6, 14, 15].

During the payment of bill, the consumer is directed to the secured payment Web sites, and there the consumer will have an option to edit the amount to be paid (adjustment charges) [16, 17]. They can pay the amount more than the bill generated, and that extra paid amount will be adjusted for the future consumptions. But, they cannot pay the amount less than the bill amount generated. If the payment is done before due date, the connection is stable; otherwise, the connection will be disconnected by deactivating the SIM automatically.

This is a fully automatic process where no one can escape from the punishments and the late payments. Since this is a fully automatic system with all functions, the urban area consumers can do their payments through the e-payment services as they are aware. From the electricity department side, if any charges are changed (like charge per unit, taxes, etc.), each and every consumer will be intimated through the SMS and e-mails about the new regulations.

4.3.3 Rural Areas Billing

When it comes to the scenario of rural areas, all the consumers are not mostly literates and everyone may not be aware of using smart phones and e-payment applications. So, the manual process should be done by local authorities to help the illiterate and old people on the first of every month. As the details from the cloud storage are sent to the electricity central office, from there the village wise consumer list along with generated bill amount for that current month will be sent to respective local electricity offices and that list will be made available on notice board. Hence, everyone can see that list, and by taking the office employee help, rural consumers also can pay bills in time. Those employees should take care about the alerts that to be given manually by moving to each and every house, whose bills are not paid on every 5th, 10th and 15th day after the bill issued [6, 18, 19].

However, the most of the work is done automatically with the help of the SIM and cloud storage. It can be used in the rural areas also. By this we can reduce the manpower utility and usage of the paper.

4.4 Performance Analysis

4.4.1 Comparison of Smart Meter with Traditional Meters

The below table gives the differences between smart meter and traditional meters, and as per the properties listed we can say that the performance of smart meter is much more better than traditional and helpful to both electricity department and consumers (Table 4.1).

Table 4.1 Comparison of smart meter with traditional meters

Property	Electromechanical meter	Digital meter	Smart meter
Accuracy	Low	High	Very high
Theft detection	Low	Possible at node level	Possible at network level
Communication	Manual reading	One-way communication	Two-way communication
Control	No	Limited	Full
Consumer participation	Nil	Less	High
Time wise reading	No	After a fix interval	According to requirement
Day-to-day billing info	Not possible	Not possible	Possible

4.4.2   Research Limitations and Practical Implications

If this automatic system is brought to existence, there will be a lot of changes in our activities. It reduces the paper wastage and manpower utility. Many miscalculations can be avoided, and strict punishments will be executed. However, in rural areas more than the 50% of the work is done automatically. As this is useful to electricity department and to the consumers, it can be implemented as soon as possible. There will be a helpline services for 24*7 to rectify the problems and to provide help [20].

4.4.3   Originality and Outcomes

In this way, we can use the artificial intelligence to decrease the manpower utility and to decrease the malpractices. After using this artificial mechanism, these are some outputs we get.

- We can do these things with less a cost.
- It decreases the usage of the paper.
- This mechanism can be used in anywhere irrespective of the area (village, town/city)
- This does not require the heavy manpower.
- Malpractices cannot be done.
- Strict rules will be followed by the program where no one can miss the rules regarding punishments or anything.
- Changes in the charges can be done directly through the server.
- Changed charges and details will be informed and notified through the SMS and e-mail services.

- Through SMS, e-mails and app notifications, the bill and payment details will be intimated.
- While paying the amount, we can pay the amount more than the bill issued but cannot pay the bill less than the bill issued, and those extra money will be adjusted for future balance.
- There will be the app which is more helpful.
- It is developed with a user-friendly environment.

4.5 Conclusion and Future Scope

In this well-turned out society, we need smart techniques to get rid of the problem. This kind of technology of artificial intelligence is the solution for many problems such as, wastage of manpower, incompetent and inaccurate billing, profusely increasing dereliction of duty and asymmetrical payments in electricity billing departments. Proper execution of tax and bills will amplify nation's economy.

In this article, we have given a solution to the problems which we have mentioned in the abstract. If we use this type of artificial intelligent mechanism, we can make our day-to-day life easy. The human power which we are using now can be reduced. We can make better understanding of the power consumption. We can provide the accurate billing and those customers who have these meters installed and want to have strict control over monthly bills can also go for prepayment using smart billing tools. From the side of government, they can have the control over the revenue, since this also works in the rural areas.

In future, if government wants to the setup to be easy and user friendly, they should develop an app or update the version of present available mobile app to make easy access to online payments and in that app we can also be able to file complaints, apply for new meter connections, etc.

Funding No funding sources.

Conflict of Interest Authors have declared that no conflict of interest.

Ethical Approval This article does not contain any studies with human participants or animals performed by any of the authors.

References

1. Senavirathna, A.M.N.N., Wijesinghe, W.A.S., Premachandra, C.: Smart home energy management system to reduce monthly electricity bill. In: 2019 IEEE 8th Global Conference on Consumer Electronics (GCCE), pp. 243–245. Osaka, Japan (2019). https://doi.org/10.1109/GCCE46687.2019.9015221
2. The Bible of electric meters, continuously updated since electricity was discovered. In: Handbook for Electricity Metering. The Edison Electric Institute

3. Jain, A., Kumar, D., Kedia, J.: Smart and intelligent GSM based automatic meter reading system. Int. J. Eng. Res. Technol. (IJERT) **2**(3),1–6. ISSN: 2278–0181
4. Leung, K.K., Yung, N.H.C., Cheung, P.Y.S.: Novel neighborhood search for multiprocessor scheduling with pipelining. In: Proceedings Fourth International Conference/Exhibition on High Performance Computing in the Asia-Pacific Region, vol. 1, pp. 296–301. Beijing, China (2000). https://doi.org/10.1109/HPC.2000.846565
5. Kanthimathi, K., et al.: GSM based automatic electricity billing system. Int. J. Adv. Res. Trends Eng. Technol. (IJARTET) **2**(7), 16–21 (2015)
6. Through SMS, e-mails and app notifications the bill and payment details will be intimated.
7. Pramod Kumar, P., Sagar, K.: A proficient and smart electricity billing management system. In: 1st International Conference on Emerging Trends in Engineering (ICETE 2019), held in Hyderabad, pp. 156–160 (2019)
8. Kumar, K.S., Bhavana, J.: A study on data mining towards cloud computing. Indian J. Public Health Res. Develop. **9**(11) (2018)
9. Pramod Kumar, P., Thirupathi, V., Monica, D.: Enhancements in mobility management forfuture wireless networks. Int. J. Adv. Res. Comput. Commun. Eng. **2**(2) (2013)
10. Kundeti, K., Pallagani, S.: Smart electricity billing system. Int. J. Eng. Manage. Res. 526–529
11. Alam, M.J., Shahriar, F.M.: Electricity billing systems at residential level in sylhet city: is pre-paid system perceived as a better option by the subscribers? Indust. Eng. Lett. **2**(3) (2012). ISSN 2224–6096 (print) ISSN 2225–0581 (online)
12. P. Pramod Kumar, S. Naresh Kumar, V. Thirupathi, Ch. Sandeep, "QOS AND SECURITY PROBLEMS IN 4G NETWORKS AND QOS MECHANISMS OFFERED BY 4G", International Journal of Advanced Science and Technology, Vol. 28, No. 20, (2019), pp. 600–606.
13. Rastogi, S., Sharma, M., Varshney, P.: Internet of Things based smart electricity meters. Int. J. Comput. Appl. (0975–8887) **133**(8):13–164 (2016)
14. Rathnayaka, M.R.M.S.B., Jayasinghe, I.D.S., et al.: Mobile based electricity billing system (MoBEBIS). Int. J. Sci. Res. Publ. **3**(4):1–5 (2013). ISSN: 2250–3153
15. Sandeep, C.H., Naresh Kumar, S., Pramod Kumar, P.: Security challenges and issues of the IoT system. Indian J. Public Health Res. Develop. **11**, 748–753 (2018)
16. Shinde, M.V., Kulkarni, P.W.: Automation of electricity billing process. In: 2014 International Conference on Power, Automation and Communication (INPAC), pp. 18–22. Amravati (2014). https://doi.org/10.1109/INPAC.2014.6981128
17. Shoults, R.R., Chen, M.S., Domijan, A.: The energy systems research center electric power system simulation laboratory and energy management system control center. IEEE Power Eng. Rev. PER **7**(2), 49–50
18. Tamarkin, T.D.: Automatic Meter Reading. Public. Power Mag. **50**(5), 934–937 (1992)
19. Tariq J.: Design and implementation of a wireless automatic meter reading system. In: Proceedings of the World Congress on Engineering (2008)
20. Mohale, V.P., Hingmire, A.G., Babar, D.G.: Ingenious energy monitoring, control, and management of electrical supply. In: 2015 international conference on energy systems and applications, pp. 254–257. Pune (2015)

Chapter 5
Real-Time Driver Distraction Detection Using OpenCV and Machine Learning Algorithms

V. N. V. L. S. Swathi, D. Akhilesh, G. Senthil Kumar, and A. Vani Vathsala

Abstract Continuous driver interruption identification is the center to numerous interruption countermeasures and major for building a driver-focused driver help framework. As more remote correspondence, diversion and driver help frameworks multiply the vehicle advertise, and the rate of interruption-related accidents is relied upon height. This article presents a diagram promising methodology which is to grow continuous driver interruption countermeasures, including three categories: distraction prevention before distraction occurs; distraction mitigation after distraction occurs; collision system adjustment when a potential collision is estimated.

5.1 Introduction

Nowadays, many accidents occurring due to driver distraction play important role. According to National Highway Traffic Safety Administration (NHTSA) and World Health Organization (WHO), consistently more than 1.24 million individuals kick the bucket in car crashes far and wide, which are among children, heavy vehicle driver, due to proper sleep. In this venture, we plan to build up a driver interruption discovery framework which is utilized to distinguish the driver from threat. In this system, we are going detect whether the driver is feeling sleepy or not.

In the proposed system, the vehicle is tracked continuously and monitored using the system which contains Raspberry Pi 3, pi camera. If the driver distraction is due to sleepy, wrong path, the alert system will active, then status of the vehicle is sent to the user as alert message via mail (or) message.

The paper is organized as follows: Section 1 contains the introduction, Sect. 2 contains the background, Sect. 3 contains the related study, Sect. 4 contains the

---

V. N. V. L. S. Swathi (✉) · D. Akhilesh · A. V. Vathsala
CVR College of Engineering, Hyderabad, India

G. Senthil Kumar
SRM Institute of Technology, Chennai, India

© The Author(s), under exclusive license to Springer Nature Singapore Pte Ltd. 2021   35
S. C. Satapathy et al. (eds.), *Smart Computing Techniques and Applications*,
Smart Innovation, Systems and Technologies 225,
https://doi.org/10.1007/978-981-16-0878-0_5

methodology, Sect. 5 contains the Implementation, Sect. 6 explains the results, and Sect. 7 describes conclusion and future work.

5.2 Background

5.2.1 Real-Time System [1]

Real time is the very short amount needed for computer system to receive data and information and then communicate it or make it available.

The continuous working framework utilized for a constant application implies for those application where information preparing ought to be done in the fixed and little quantum of time. It is not quite the same as broadly useful PC where time idea isn't considered as much critical as progressively working framework.

5.2.2 Driver Distraction Detection System [2]

Mainly driver interruption frequently brought about by following elements: sleep, work period, work pressure and actual individuals regularly attempt to accomplish work much in a day and they lose valuable rest because of this. They take coffee, tea or different simulant individuals keep on remaining alert that happens the body at long last implodes and the individual falls rest. The variables caused the driver interruption, and human mind is prepared to think there are times the body should be sleeping; basically this happens during the hour of sunrise and sunset. Between the long periods of 1–6 am, this reason truly ill suited, by being either under or overweight will cause weakness. In this location, framework was clarified in methodology.

5.2.3 OpenCV

OpenCV (open-source computer vision library) supports a lot of computer vision and machine learning application. Mainly, OpenCV supports programming languages like Python, C++ , Java. In Python, OpenCV Python API and other libraries like SciPy, Matplotlib and NumPy can be used with this.

5.3   Related Study

Related study is also known as literature survey; it is most important step in development of the software process. Before developing the software, it is necessary to determine which platform can be developed for our project (driver distraction detection system). In the previous long term, numerous explorers have been chipping away at the advancement of driver distraction detection system to forestall street mishap by utilizing various strategies. The ongoing methods depend on physiological measures like brain waves, pulse, heart rate… … and so on,

We have referred many papers; from those papers we understood what are the limitation of the existing system. One of the existing system [3]. According to the experts, it has been observed that when the driver does not take break, they tend to run a high risk of becoming drowsy; this study shows that accidents occur due to sleepy driver in need of a rest, which implies that street mishaps happen more because of sluggishness as opposed to drink driving. Consideration help can warm of in mindfulness and laziness their present status of exhaustion and the driving time since in the last break.

The most reinforcement and recuperation procedures that have been created in driving discovery strategies, for example,

ECG and EEG [4]
Electrocardiogram (ECG), electroencephalogram (EEG) are the physiological sign most usually to gauge tiredness. The ECG and EEG signals have different frequency bands and range. Its disadvantages are implementation and complexity.

Steering Wheel Movement (SWM) [5]
It is estimated utilizing directing point sensor, and it is broadly utilized for vehicle-based measure for the recognizing the degree of driver tiredness. To eliminate the effects of lane changes, the researches thought about which are expected to change the horizontal situation inside the path. Model a vehicle organizations, for example, Nissan and Renault have received SWMs yet it works in exceptionally restricted circumstances. Its disadvantages are costly technique, redundancy issue, and complexity.

5.4   Methodology

In this work, we have characterized the Driver Distraction Detection Algorithm. We utilize the camera-based framework to notice his visual conduct and the vehicle sensor to quantify the driving conduct. According to development of real-time driver distraction countermeasure includes three categories by using it.

5.4.1 Distraction Prevention Before Distraction Occurs

In this method, we are using to prevent distraction before distraction occurs, the vehicle real-time identification measure by recording the video arrangement of the drivers. The video transfers got from web camera are driven the framework, starting doesn't known the underlying situation of the face in light of the fact that the eye squint is a fast closing and continuing of a characteristic eye. Each individual has fairly one of a kind case of squint, the model shifts in the speed of closing and opening, a degree of pounding the eye and in a glimmer range the eye squint props up approximately 100–400 ms.

The framework gets the main picture and attempts to discover the face district in the picture utilizing the histogram of oriented gradients and linear SVM object indicator algorithm explicitly for the errand of face location.

To build our face detection, the author used Viola–Jones algorithm. By comparing both algorithms, the HOG [6] accuracy is more. It is used to prepare higher-precise item classifier in their specific investigation, human identifiers.

HOG

A HOG is a component descriptor by and largely utilized for object location. A HOG depends on the property of items inside in a picture to have the conveyance of power angles or edge headings. Slopes are determined inside a picture for each squares. A square is considered as a pixel lattice wherein angles are established from the size and heading of progress in the forces of the pixel with in the square.

HOG figures the classifier for each picture in and size of the picture, applied the sliding windows, removed HOG descriptor at every window and applied classifier. All the face test photos of an individual are dealt with to the part descriptor extraction figuring, i.e., a HOG. The descriptors are angle vectors produced per pixel of the picture. The slope for every pixel comprises of size and heading, determined utilizing the accompanying formulae:

$$g = \sqrt{g_x^2 + g_y^2}$$
$$\theta = \arctan\frac{g_y}{g_x}, \tag{5.1}$$

In the current formulae, G_x and G_y are separately the level and vertical segments of the adjustment in the pixel power. A window size of 96×144 is utilized for face pictures since it coordinates the overall viewpoint proportion of human countenances. The descriptors are resolved over squares of pixels with 8×8 estimations. These descriptor esteems for every pixel more than 8×8 square are quantized into 9 canisters, where each receptacle speaks to a directional point of slope and incentive in that container, which is the summation of the extents of all pixels with a similar point. Further, the histogram is then standardized over a 16×16 square size, which implies four squares of 8×8 are standardized together to limit light conditions. This instrument mitigates the precision drop because of an adjustment in light. On the off chance that the classifier identified an item with enough likelihood that takes after a face, the classifier recording the bouncing box of the window and applied non-greatest concealment to make the precision expanded (Fig. 5.1).

Linear SVM [7]

Support vector machines (SVMs) or backing vector machines (BVMs) are managed AI models that partition and order information. The SVM model is readied using different HOG vectors for various appearances (Fig. 5.2).

Facial Landmarks Detection with Dlib

The Dlib C++ library is a cross-stage bundle for stringing, organizing, numerical activities, AI, PC vision, and pressure, setting a solid accentuation on very and compact code. From a PC vision point of view, Dlib has various best in class executions, including facial milestone discovery, correlation following, deep measurement learning. Facial landmarks they are used in computer vision applications. Recognizing facial places of interest is a subset of the shape conjecture issue. Given an info

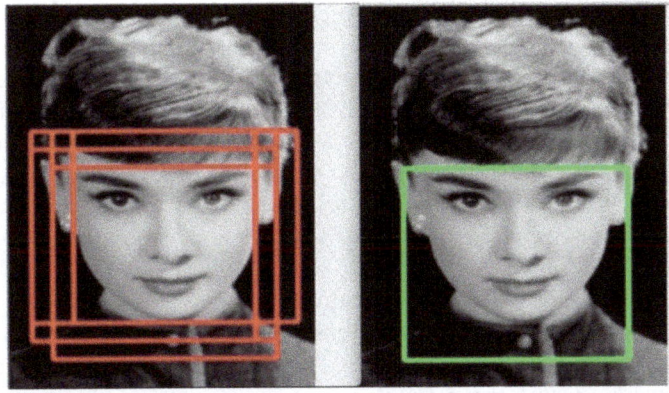

Fig. 5.1 (Left) Detecting numerous covering bouncing boxes around the face we need to identify. (Right) Applying non-greatest concealment to eliminate the excess bouncing boxes

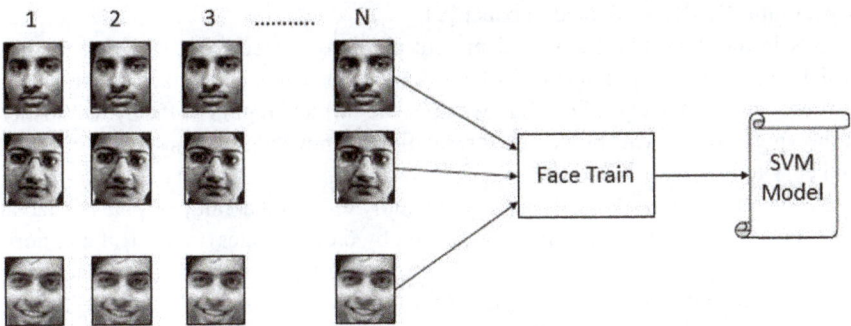

Fig. 5.2 Face training on linear SVM

picture for (HOG + VSVM) and regularly a ROI that determines the object of interest a shape predicator endeavors to neighborhood key focal points along the shape. With respect to facial achievements, our goal is recognize critical facial structures on the face using shape desire methods.

Recognizing facial tourist spots is subsequently a two-stage measure:

Stage 1: Point the part of the face in the image.

Stage 2: Finding key facial structures on the face region. We may apply a pre-arranged HOG + Linear SVM object identifier unequivocally for the endeavor of face acknowledgment. In one or the other case, the certifiable figuring used to perceive the face in the image doesn't have any kind of effect. Or maybe, what's huge is that through some methodology we get the face ricocheting box (i.e., the (x, y)—bearings of the face in the image). Given the face locale we would then be able to apply recognizing key facial structures in the face area. There are a grouping of facial achievement pointers, yet all strategies fundamentally endeavor to limit and name the going with facial locale: right eye, left eye, nose, mouth.

The facial landmark detector and Dlib library of the paper [8].

This strategy begins by utilizing: A preparation set of named facial tourist spots on a picture. These pictures are physically marked, indicating explicit (x, y)—directions of locales encompassing facial structure. Priors, of even more expressly, the probability on partition between sets of information pixels. Given this readiness data, a social occasion of backslide trees is set up to survey the facial achievement positions genuinely from the pixel powers themselves (i.e., no "feature extraction" is happening). The end result is a facial achievement identifier that can be used to recognize facial places of interest dynamically with first rate conjectures.

Advance Dlib's Facial Milestone Indicator

The advance facial milestone indicator and Dlib library are used for 68 (x, y)—coordinates on the face. The below image describes the 68 coordinates of Dlib's facial landmark detector (Fig. 5.3).

Eye Blink Detection

To fabricate our blink identifier, we'll be registering a measurement called the eye aspect ratio (EAR), presented in paper [9].

Not in any way like traditional picture planning systems for enrolling squints which regularly incorporate a mix of eye repression. Threshold value tells about the whites region of the eyes. Deciding whether the "white" region of the eyes vanishes it showing a squint. This procedure for eye glint acknowledgment is speedy, capable, and easy to execute.

As far as flicker identification, we are simply enthused about two plans of facial structures the eyes. Each eye point is spoken by 6 coordinates, in vertical and horizontal axis. From the point p1 of the eye, we make point of the location which are working clockwise region around the eye.

In the above Fig. 5.4, we should eliminate on key point. There is a connection between the width and the stature of these directions. In view of the work [5], we

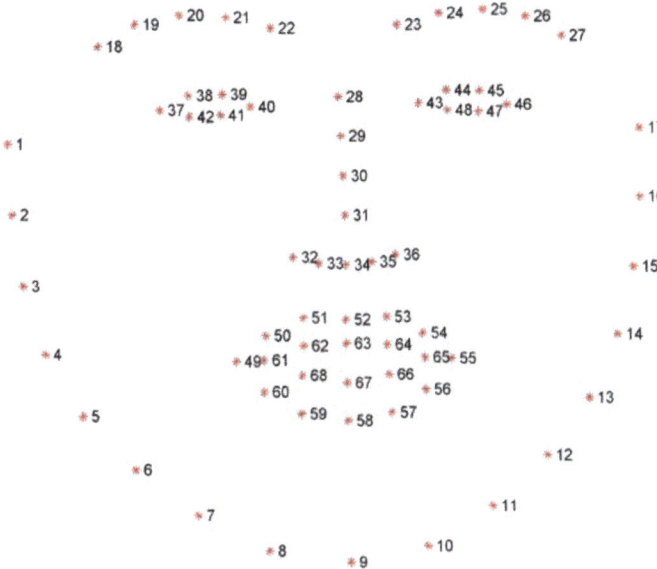

Fig. 5.3 68 coordinates of Dlib's facial landmark

would then be able to infer a condition (2) that mirrors this connection called the eye aspect ratio (EAR):

$$\text{EAR} = \frac{||p_2 - p_6|| + ||p_3 - p_5||}{2||p_1 - p_4||}, \tag{5.2}$$

In the above Eq. (5.2) $p_1, p_2, p_3, p_4, p_5, p_6$ are 2D facial milestone areas. The numerator of this condition calculates the detachment between the vertical eye places of interest while the denominator enrolls the partition between level eye achievements,

Fig. 5.4 6-point location of the eye

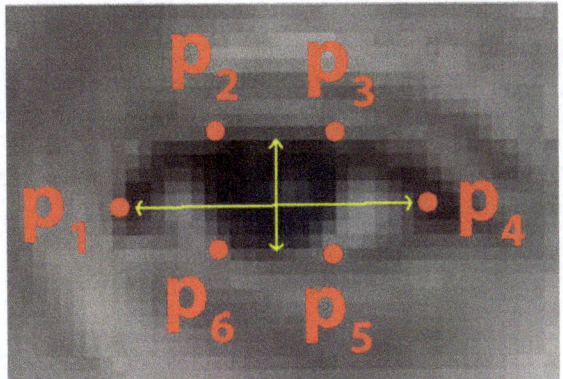

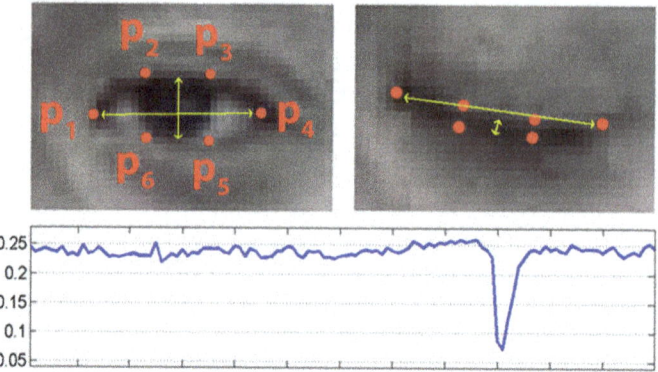

Fig. 5.5 6-point location of the eye is open. Eye indicator when the eye is closed. Graph of the eye aspect ratio over time

weighting the denominator fittingly since there is only one parcel of even concentrates yet two plans of vertical core interests. For sure, as we'll find, the eye viewpoint extent is around consistent while the eye is open, anyway will rapidly tumble to zero when a squint is happening. Utilizing this basic condition, we can dodge picture preparing procedures and just depend on the proportion of eye milestone separations to decide whether an individual is flickering.

To make this more clear, consider the following Fig. 5.5.

In Fig. 5.5 top-left, we have an eye that is totally open the eye perspective extent here would be large and decently reliable after some time. Be that as it may, when the individual squints (upper right) the eye perspective proportion diminishes drastically, moving toward zero. The down graph tells about the eye angle proportion after some time for a video cut. As should be obvious, the eye viewpoint proportion is steady, at that point quickly drops near zero, at that point increments once more, showing a solitary flicker has occurred.

In our next segment, we'll figure out how to execute the eye angle proportion for flicker location utilizing facial tourist spots, OpenCV, Python, and Dlib.

Drowiness Detection Results

To run the above detection, we are writing a program code on Python, to run this program on your own Raspberry Pi, as shown in system architecture Fig. 5.6. The detection system checks for driver drowsy or not. If drive is drowsy an alarm is generated to alert the driver who is in a drowsy state, otherwise the process is terminated.

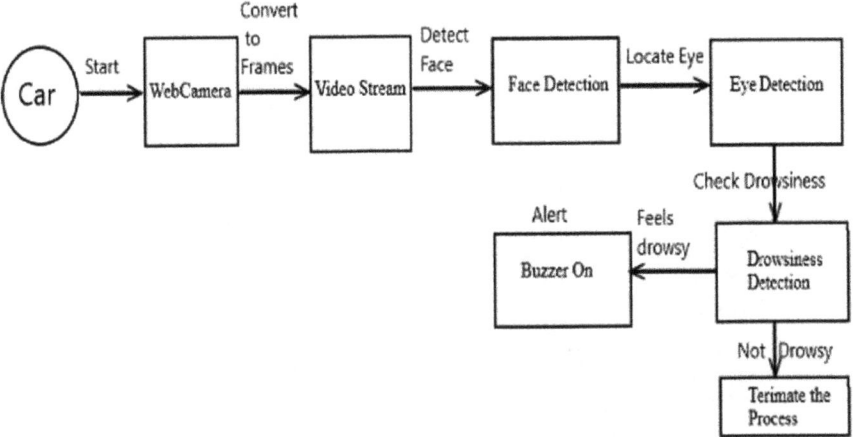

Fig. 5.6 System architecture

5.4.2   *Distraction Mitigation After Distraction Occurs*

In this case were going implement the driver alert system through send message and location of the car, in the case of continuous drowsy of a person.

5.4.3   *Collision System Adjustment When a Potential Collision*

The system can be enchased with more additional sensors to improve the safety of the users in the vehicle like speed limit, 360 degree camera and automatic sensors.

5.5   Implementation

The tools are connected in this way, so that we can collect the data. The tools used are Raspberry Pi 3, pi-camera, GPS, and USB (Fig. 5.7).

The process of collecting the data from the system that contains Raspberry Pi 3, pi-camera, GPS is done in the following way:

1. The webcam is used mainly to capture the static image or the video stream. This captured image or video stream is given as contribution to the framework.
2. Captured image processing module will process the obtained image or video input fed into the system. It processes the input and converts it into frames for further processing.

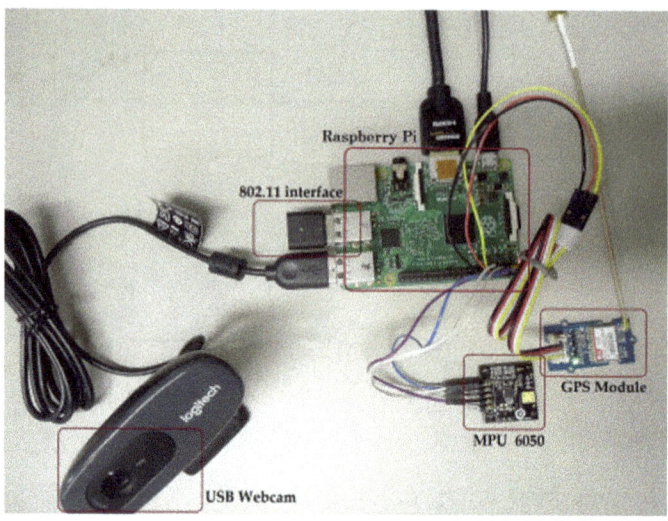

Fig. 5.7 Proposed system architecture kit

3. The face detection technique is used to locate the face from the image. The video stream which is given as input is converted into frames, and the face is detected from those frames. Haar classifier and HOG + SVM algorithm is used to detect the face from a given image.
4. The situation of the driver's eye is controlled by utilizing suitable limit. In this work, edge area of the eyes region is considered.
5. After locating the eye region in the frame, the system finds out whether the eye is in a closed state or in open state. In the event that the eye stays open, at that point the framework gives the message that the driver isn't feeling languid. On the off chance that the eyes stay shut, at that point the framework gives the message that the driver is feeling sluggish.
6. The output from the system includes: Alert Message to the user via buzzer and terimate the process. The indication is given in order to prevent road accidents which occur due to the drowsiness of the driver.

5.6 Result

Our framework created the ready when the driver gets weakness and on the caution, gets ready message, area of vehicle to the client precisely when driver gets occupied.
    The system is tested, and the results obtained are furnished below:

1. Checking for the driver distraction as shown in Fig. 5.8.

2. Giving alert when driver gets distracted as given below Fig. 5.9.

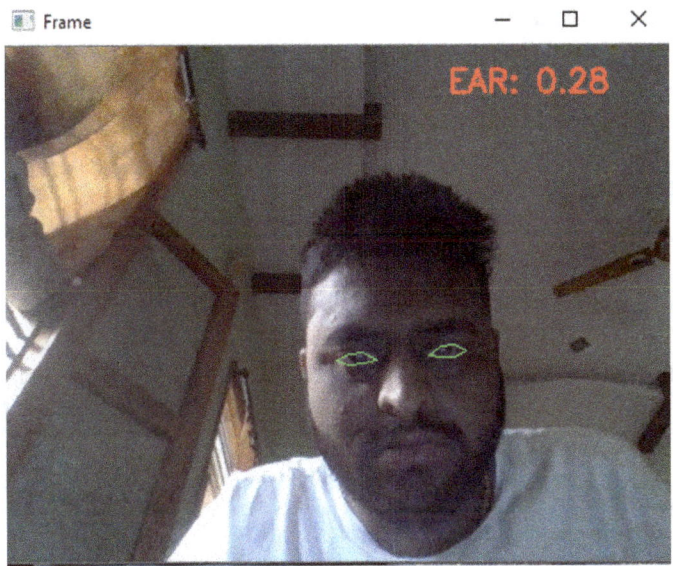

Fig. 5.8 Checking for drowsiness

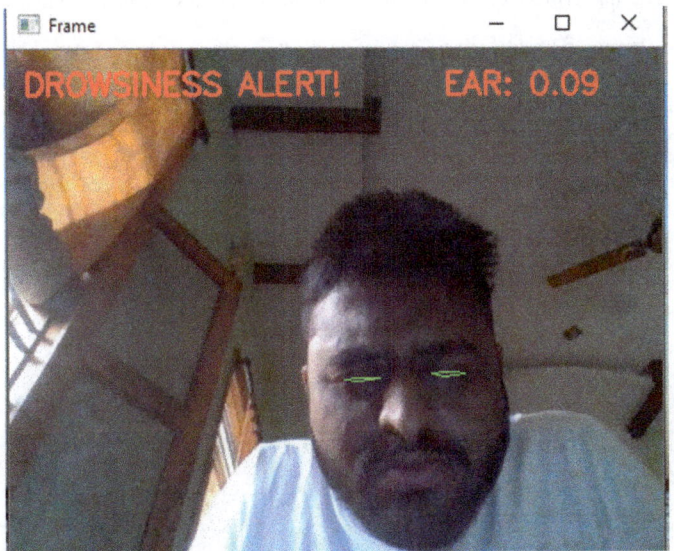

Fig. 5.9 Alert

5.7 Conclusion and Future Work

In this paper, driver distraction detection system will efficiently find the drive fatigue and sends alert message, location of vehicle to the user. The system can be enchased with more additional sensors to improve the safety of the users in the vehicle. The invoice generation can also be implemented as additional feature in future.

References

1. Zhang, H., Smith, M.R.H., Witt, G.J.: Identification of real-time diagnostic measures of visual distraction with an automatic eye-tracking system. Human Factors. **48**(4), 805–821 (2006)
2. Liu, T., Yang, Y., Huang, G.-B., Yeo, Y.K., Lin, Z.: Driver Distraction Detection Using Semi-Supervised Machine Learning. IEEE Trans. Intell. Transp. Syst. **17**(4), 1108–1120 (2016). https://doi.org/10.1109/tits.2015.2496157
3. VandnaSaini, R.S.: Driver Drowsiness Detection System and Techniques: A Review. Int. J. Comput. Sci. Informat. Technol. **5**(3), 4245–4424 (2014)
4. Singh, K., Kaur, R.: Physical and physiological drowsiness detection methods. IJIEASR, **2**, 35–43 (2013)
5. Krajewski, J., Sommer, D.: Steering wheel behavior based estimation of fatigue. In: Fifth International Driving Symposium on Human Factors in Driver Assessment, Training and Vehicle Design, Germany
6. Dalal, N., Triggs, B.: Histograms of oriented gradients for human detection. In: 2005 IEEE Computer Society Conference on Computer Vision and Pattern Recognition (CVPR'05), vol. 1, pp. 886–883. San Diego, CA, USA (2005). https://doi.org/10.1109/CVPR.2005.177
7. Tomasz, M., Gupta, A., Efros, A.A.: Ensemble of exemplar-SVMs for object detection and beyond. In: 2011 International Conference on Computer Vision, 89–96 (2011)
8. Kazemi, V., Sullivan, J.: One millisecond face alignment with an ensemble of regression trees. In: 2014 IEEE Conference on Computer Vision and Pattern Recognition, pp. 1867–1874. Columbus, OH (2014). https://doi.org/10.1109/CVPR.2014.241
9. Tereza, S., Cech, J.: Real-Time Eye Blink Detection using Facial Landmarks (2016)

Chapter 6
Data-Driven Prediction Model for Crime Patterns

Y. Gayathri, Y. Sri Lalitha, M. V. Aditya Nag, and Sk. Althaf Hussain Basha

Abstract With the ever-increasing technology widgets and media, there is enough awareness of good and bad in the surroundings. Unemployment or abundant money is making people to indulge in crime. Throughout the world, crime prediction and prevention are the two major issues of all governments for the safety and security of their citizens. With machine learning and data mining techniques to predict unknown events, crime data analysis and prediction has attracted the attention of many researchers in recent times. There is still generalized, accurate and optimized requirement of crime data analysis and prediction models. This work proposes to address crime data analysis and prediction using visual analytics and machine learning algorithms on Kaggle crime dataset. The proposed models provide better insights on the crime zones and frequency of occurrences of certain crimes in a zone. We have compared Naïve Bayesian and Random Forest Machine Learning Models to predict the crime, and we have noticed that our models exhibited better accuracy.

6.1 Introduction

Crime is an un-lawful act punishable by state or authority. Crime is an act harmful not only to some individual but also to a community, society or the state. Therefore it is socio-economic problem. The nature of the crime varies depending on society, education, unemployment and poverty levels. The types of crimes such as assault, aggravated assault, burglary, murder, rape, kidnapping and stabbing are common and occur periodically. Knowing in advance the occurrence of crime and preventing it is a great help to society and man-kind. Predicting a crime type or its reoccurrence

Y. Gayathri (✉) · Y. Sri Lalitha
Gokaraju Rangaraju Institute of Engineering and Technology, Hyderabad, India

M. V. Aditya Nag
Institute of Aeronautical Engineering, Dundigal, Hyderabad, India

Sk. Althaf Hussain Basha
A1, Global Institute of Engineering and Technology, Markapur, India

© The Author(s), under exclusive license to Springer Nature Singapore Pte Ltd. 2021   47
S. C. Satapathy et al. (eds.), *Smart Computing Techniques and Applications*,
Smart Innovation, Systems and Technologies 225,
https://doi.org/10.1007/978-981-16-0878-0_6

is not trivial. The advancements in machine learning techniques and open-source tools to implement the same is making it possible to provide solutions to many real-time problems. There is a vibrant research community contributing improved models in crime data analysis and prediction. The generalized, accurate models are yet to be generated. The accurate prediction models use data and statistical analysis for predicting crime types and are useful for law-enforcement departments.

In this work, we explored visual analytics for crime data analysis and machine learning models to predict crime categories. In the process, we used R language an open-source statistical tool to build visualization and prediction models. With the high volumes of data storage mechanism, we can record all the crimes and the criminal activities in structured and un-structured forms. To derive the required knowledge from this voluminous data, visualization tools are most helpful. It is impossible to know the characteristics of data without visual analytics. Visualization helps in performing exploratory data analysis and identifies required features for building a prediction model. R language provides extensive visualization functionalities; we explored this here to determine high-crime-prone locations, crime types, days of a week of high crime activity and the time of its occurrence. The machine learning algorithms such as Naïve Bayesian and Random Forest Prediction Models are implemented for predicting a crime categories.

6.2 Literature Survey

Literature study on crime data analysis and crime prediction significant work has taken place. The researchers proposed solutions ranging from application of statistics, visual analytics, machine learning, data mining, GIS-based analysis to artificial neural networks [13]. Still this area requires *generalized, accurate and optimized solutions constituting* various aspects of crime data recorded. This section briefs certain previous works proposed.

In many research papers, statistics is applied to analyze crime data for frequent crime types, crime zones and applied visual analytics to get insights of crime data [12]. In [1], human social data obtained from social media via mobile activity combined with demographic information using real crime data to predict crime hotspots in London, UK. Machine Learning methods such as Naïve Bayesian, Decision Tree using WEKA an open-source data mining technique applied in [2]. Considering characteristics such as weather, road, driver and car conditions, the occurrences of road accidents were explored in [3] using K-NN, Naïve Bayesian and decision trees algorithms of data mining with 79–81% accuracy. At University of Arizona, the data mining techniques were applied to identify the crime zones [5, 8, 9]. Reviews different crime-prediction techniques and suggested that statistics, ML, AI and Data Mining are the effective mechanisms in crime prediction. In [10], Gradient Boost and Page Rank techniques were applied to determine and destroy the hidden links in criminal networks. GIS-based analytics is applied in [11] to identify areas of crime and alert cops for necessary actions. A comparative study of linear regression, additive regression and Decision stump algorithms using WEKA is made for crime prediction in [4]. In [7] drug-related crime data and emerging crime zones were identified using visual and spatial analytics. Mobile network activity and demographic information together used to predict crime zones in UK. This work uses visual analytics with R to analyze the crime data and predict crime type using naïve Bayes and random forest approaches.

6.3   Our Approach

This work involves the steps like dataset collection, preprocessing, analyzing nature of dataset using visualization for feature selection [6], building a prediction model and evaluating the models with unknown criminal event.

Our process of building a prediction model is depicted in Fig. 6.1.

Dataset Collection
We have considered Kaggle Crime dataset which constitutes data from 2001 till date. It is a collection of criminal and traffic offenses in the country. The work here considered data from 2012 to 2017. It constitutes around 457,000 crime records. From this collection after preprocessing, we have considered 333,830 records of different offense categories for experimenting.

Data Preprocessing
The dataset is composed of 19 attributes. Not all of these attributes are relevant for our analysis. Only the relevant attributes are considered for efficient performance. The dataset is cleaned by removing null valued rows and invalid dates. Selected records randomly for each offense_category thereby forming a preprocessed dataset. The dataset consists of 15 different offense categories. Each category is classified into suboffense category ranging between 3 and 58 subcrime categories. We have to consider the offense categories that have less than 20 subcategories.

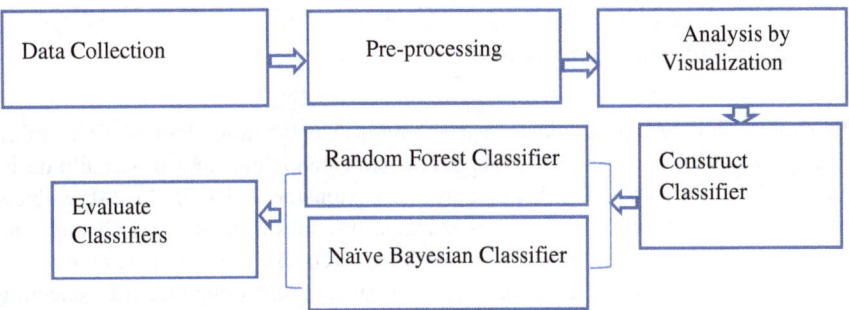

Fig. 6.1 Crime data visualization and prediction process

Analysis by Visualization

One important aspect of this study is to design methods of analysis and interpretation of crime data through visual analytics. To determine frequent prone areas of crime, offense category, period of occurrences, neighborhood locations that are victims of similar type of crimes, data visualizations are applied. Visualization provides a quick summary of huge data.

Construct Classifier

The work considers two popular machine learning approaches *probabilistic* and *ensemble methods. Probabilistic:* Given an observation of input, a probability distribution over a set of classes the **probabilistic classifier** can predict the output class. Bayes theorem $P(A|B) = (P(B|A)P(A))/P(B)$, the probability of A is predicted given the occurrence of B in dataset with strong independence among the features of dataset, the Naive Bayes approach is designed. This work uses Naive Bayes model to predict the offense category for Crime dataset. Naive Bayes is simple and fast classifier, highly scalable, requires less training data, can be used for binary, multi-class classification problems. Naive bayes is often "good enough" in a lot of real-world applications. When data is dynamic and keeps changing, Naïve Bayes can adapt quickly to the changes. *Ensemble*: These methods uses multiple learning algorithms to obtain better prediction than obtained by any single learning model. We employed random forest in this work. Random forest method is a decision forest method applied to Classification and Regression tasks. The method constructs multiple Decision trees at learning stage and outputs a model that has accurate many a times. With random forest approach, overfitting is reduced by which the prediction accuracy improves. Multiple trees reduce the chance of stumbling across a classifier that doesn't perform well. In case of unbalanced datasets, random forest has balancing error in class population. It has capabilities to compute similarities in the data and identify outliers. Thus, it can be extended to unlabeled data, leading to unsupervised learning, data views and outlier detection.

Section 6.4 deals with experimental approach that discusses the analysis of dataset using visualization and crime patterns using prediction models. Section 6.5 discusses evaluation of the prediction models and in Sect. 6.6 conclusions.

6.4 Experimental Approach

We explore R language for data preprocessing, visualization, classification. Shiny Package was used for design of Webpage, embed standalone apps or to build dashboards. The significance of Shiny is it provides extension with CSS, HTML widgets and JavaScript Actions. Shinythemes package provides a flexibility to change the overall appearance of a webpage. This work explored date attribute for knowing the frequent days of crimes or traffic activities, removal of null valued records, selecting significant attributes for classification. Lubridate package is used extensively for manipulating date attribute to better understand the crime occurrence patterns based on day, month, time in terms of hours of the day. In Data Visualization phase, we extensively used Plotly and ggplot2 and ggthemes packages for crime data analysis. Ggplot2 is a system for declaratively creating graphics, based on the grammar of graphics. Just provide the data, tell ggplot2 how to map variables to aesthetics, what graphical primitives to use, and it takes care of the details. It embodies a deep philosophy of visualization. For interactive data visualization with high-quality graphs online, Plotly R graphing library makes interactive, publication-quality graphs online.

6.4.1   Result Analysis by Visualization

Crime Data Analysis and Visualization is a systematic approach for identifying and analyzing patterns and trends in crime. The work here experimented by analyzing crimes through visualizing the prominent offense categories based on month in an year recorded more offenses of a particular category of offense, on which day of the week such offenses are more to happen and so on. Using the ggplot2 of R, we can visualize through map the offense category and the regions where it has occurred. Our observations from the data are similar type of offense categories take place in the neighborhood areas. We explored neighborhood regions and their crime nature.

Table 6.1 shows different offense category types and the no. of cases recorded during 2012–17. It can be seen that 13 different categories of crime types with different sub-category types in each category. For our work, we considered first five crime types for analysis using visualization.

Table 6.1 Offense category-wise crimes recorded

Sl. No.	Offense_category_id	No. of cases
1	Traffic-accident	104,739
2	Larceny	43,334
3	Public-disorder	41,269
4	Drug-alcohol	33,384
5	Burglary	28,354
6	Theft-from-motor-vehicle	23,085
7	Auto-theft	21,679
8	Other-crimes-against-persons	17,701
9	Aggravated-assault	8680
10	White-collar-crime	5944
11	Robbery	4994
12	Arson	498
13	Murder	169

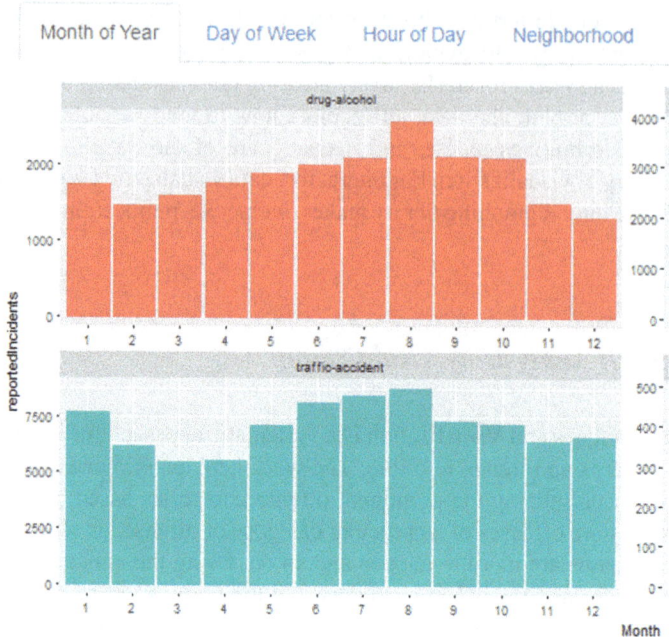

Fig. 6.2 Histogram of traffic accidents, drug crimes month wise

From Fig. 6.2, It is observed that in the month of August, drug-alcohol crimes were high and consistent throughout the year the crimes are above 1000 excluding the month of December which is slightly less. Figure 6.3 shows that these cases occurred in the mid of the week and are very less on weekends. It can be interpreted as people indulge in drug-alcohol crimes in the weekdays under the influence of friends or colleagues. Similarly, Fig. 6.4 explores public disorder, alcohol, burglary and larceny cases. Public disorder which constitutes gambling, criminal mischief, threaten to injure, harassment, disturbing peace is some subtypes of public disorder occurred consistently in all weekdays, but were at peek during June to August months, traffic accidents were at peek between June–August and quiet low in March and April. Clear weather may be one of the reasons for less accident cases in March–April. Larceny cases are consistent in weekdays, and more on weekends. Burglary crimes are more in April and consistent on all weekdays. Knowing the month and weekdays of peek crime types will help the authorities to take appropriate measures to address them.

Figure 6.5 depicts the hour of the day the crime taken place. It is observed that drug-alcohol cases are more in the evening hours and slowly reduced in the nights. Similarly, traffic accident in 3rd quadrant of the figure is during the office hours, slowly reduced toward night. Public disorder cases are more or less consistent throughout the day from 9 am onward. White-collar cases are during the office hours, and they are note minimum in the morning hours of the day.

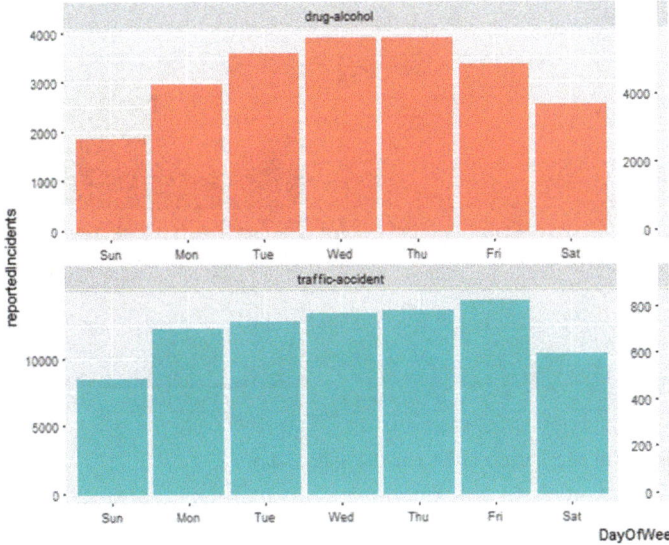

Fig. 6.3 Histogram of traffic accidents, drugs day of the week

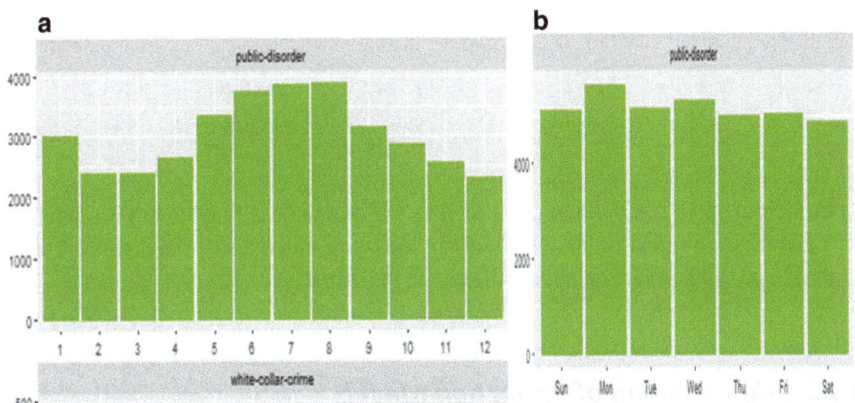

Fig. 6.4 **a** Public disorder crimes month of the year, **b** day of the week

Another type of analysis is identifying the neighborhood cities that are victims of similar type of crimes. In total, 80 neighborhood cities exist in dataset. Table 6.2 shows maximum and minimum count of crime for five neighborhood ids are listed.

It should also be noted for the five offense categories that this paper study shows that five-points is a place where public disorder, burglary and drug alcohol cases are very high and Wellshire, DIA and Indian-creek are the places of least occurred cases. Traffic accidents and Larceny cases were noticed many in Stapleton and less in Wellshire. This may end up in interpreting that Wellshire, Indian-Creek are the

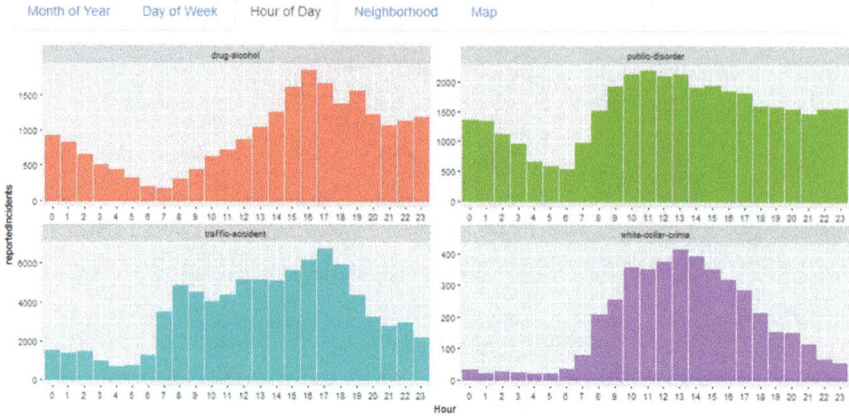

Fig. 6.5 Histogram of crime occurrence in the hour of day

Table 6.2 Neighborhoods maximum and minimum crime occurrences

Neighborhood Id	Maximum no. of crimes	Neighborhood Id	Minimum no. of crimes
Five-points	18,761	Rosedale	1300
Stapleton	14,463	Skyland	1178
Capitol-hill	12,636	Ry-club	928
Cbd	11,842	Wellshire	495
Baker	9981	Indian-creek	444

peaceful and safe places when compared to five-points and Stapleton. Visualization provides a broad range of analysis without knowing much about the dataset. Quick analysis and collective measures will reduce the crime.

6.4.2   Construction of Prediction Models

We have applied Naive Bayes and Random Forest algorithms on the dataset to predict the crime category. The test set contained 25 records. Train set contains 9 categories. The category 'others crimes against persons' has 58 subcategories; therefore, this category is excluded from the dataset. The dataset for prediction has 240,562 records out of which 168,393 were training data which is 70% of the dataset and 72,169 is the testing data.

Naïve Bayes model details of test and train data, construction and prediction are seen in Figs. 6.6, 6.7, 6.8 and 6.9.

Fig. 6.6 Levels in offense category in train set

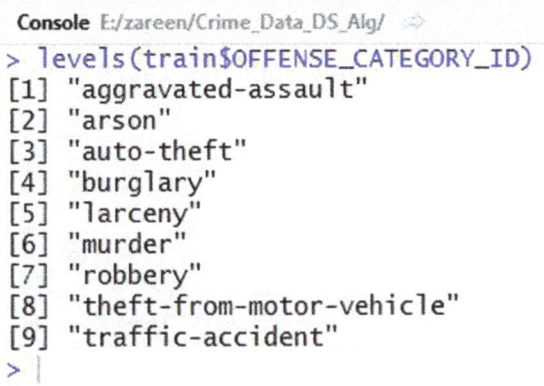

```
Console E:/zareen/Crime_Data_DS_Alg/
> levels(train$OFFENSE_CATEGORY_ID)
[1] "aggravated-assault"
[2] "arson"
[3] "auto-theft"
[4] "burglary"
[5] "larceny"
[6] "murder"
[7] "robbery"
[8] "theft-from-motor-vehicle"
[9] "traffic-accident"
>
```

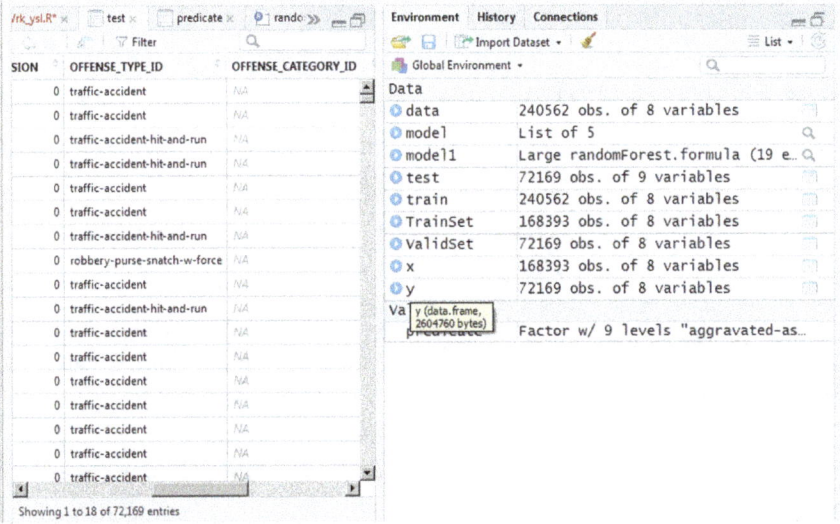

Fig. 6.7 View of environment and test set

Figure 6.6 depicts the offense categories in the dataset. Figure 6.7 depicts the environment variables and test set details. Figure 6.8 shows the Naïve Bayes model construction and Fig. 6.9 prediction on test set.

Random Forest model details of test and train data, construction and prediction are seen in Figs. 6.9, 6.10 and 6.11. Random forest algorithm works when the no. of sub categories are less than 53. Due to this reason, we removed other crimes from people category records from the dataset. Figure 6.9 depicts the Random forest model, train set details. Figure 6.10 shows the confusion matrix of random forest offense categories predictions and Fig. 6.11 prediction on test set and compares with that of Naïve Bayes model.

```
> model<-naiveBayes(train$OFFENSE_CATEGORY_ID~.,data = train)
> model

Naive Bayes Classifier for Discrete Predictors

Call:
naiveBayes.default(x = X, y = Y, laplace = laplace)

A-priori probabilities:
Y
        aggravated-assault                         arson                     auto-theft
            0.0360821742                    0.0020701524                   0.0959627871
                burglary                       larceny                         murder
            0.0901181400                    0.1801365137                   0.0007025216
                 robbery theft-from-motor-vehicle             traffic-accident
            0.0207597210                    0.1387750351                   0.4353929548

Conditional probabilities:
                          offense
Y                             [,1]      [,2]
  aggravated-assault     239534.4 108734.11
  arson                  250242.7  98530.74
```

Fig. 6.8 Naïve Bayes model training

```
> model1 <- randomForest(x$OFFENSE_CATEGORY_ID ~ ., data = x, importance = TRUE)
> model1

Call:
 randomForest(formula = x$OFFENSE_CATEGORY_ID ~ ., data = x, importance = TRUE)
               Type of random forest: classification
                     Number of trees: 500
No. of variables tried at each split: 2

        OOB estimate of  error rate: 0%
```

Fig. 6.9 Random forest model construction

```
Confusion matrix:
                        aggravated-assault arson auto-theft burglary larceny murder robbery
aggravated-assault                    6040     0          0        0       0      0       0
arson                                    0   343          0        0       0      0       0
auto-theft                               0     0      16196        0       0      0       0
burglary                                 0     0          0    15179       0      0       0
larceny                                  0     0          0        0   30386      0       0
murder                                   0     0          0        0       0    108       0
robbery                                  0     0          0        0       0      0    3485
theft-from-motor-vehicle                 0     0          0        0       0      0       0
traffic-accident                         0     0          0        0       0      0       0
                        theft-from-motor-vehicle traffic-accident class.error
aggravated-assault                             0                0           0
arson                                          0                0           0
auto-theft                                     0                0           0
burglary                                       0                0           0
larceny                                        0                0           0
murder                                         0                0           0
robbery                                        0                0           0
theft-from-motor-vehicle                   23371                0           0
traffic-accident                               0            73285           0
> predicate <- predict(model1, ValidSet)
```

Fig. 6.10 Confusion matrix of random forest offense category prediction

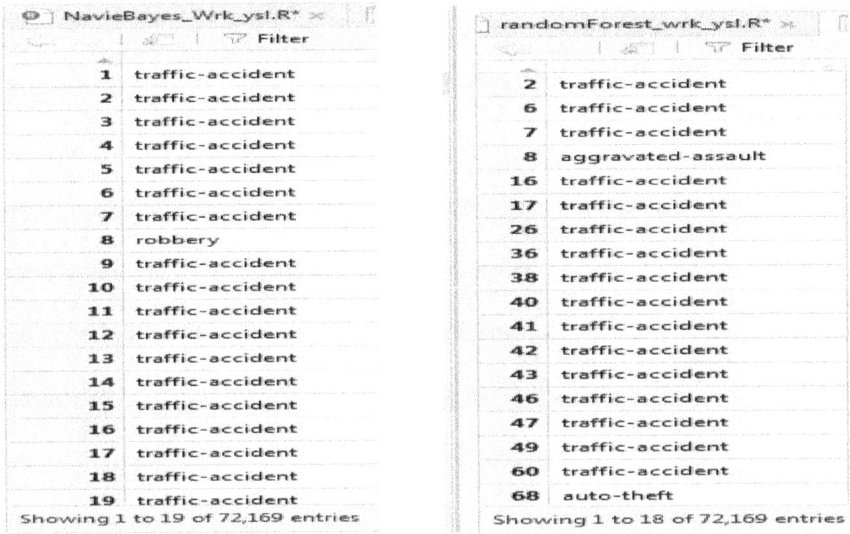

NavieBayes_Wrk_ysl.R* ×		randomForest_wrk_ysl.R* ×	
	Filter		Filter
1	traffic-accident	2	traffic-accident
2	traffic-accident	6	traffic-accident
3	traffic-accident	7	traffic-accident
4	traffic-accident	8	aggravated-assault
5	traffic-accident	16	traffic-accident
6	traffic-accident	17	traffic-accident
7	traffic-accident	26	traffic-accident
8	robbery	36	traffic-accident
9	traffic-accident	38	traffic-accident
10	traffic-accident	40	traffic-accident
11	traffic-accident	41	traffic-accident
12	traffic-accident	42	traffic-accident
13	traffic-accident	43	traffic-accident
14	traffic-accident	46	traffic-accident
15	traffic-accident	47	traffic-accident
16	traffic-accident	49	traffic-accident
17	traffic-accident	60	traffic-accident
18	traffic-accident	68	auto-theft
19	traffic-accident		
Showing 1 to 19 of 72,169 entries		Showing 1 to 18 of 72,169 entries	

Fig. 6.11 Predicted values using naïve bayes and random forest

Table 6.3 Naive Bayes wrongly classified classes

Correct class	Wrongly classified
Auto theft	Theft-from-motor-vehicle
Burglary	Theft-from-motor-vehicle, aggravated assault, auto-theft
Larceny	Aggravated assault, robbery
Murder	Theft-from-motor-vehicle

6.5   Evaluation

The observations from prediction models are random forest generated 500 trees, and its accuracy is 100%. It could correctly predict the classes of Offense Category for the dataset of about 72,169 records. The problem with the model is it took more time to build the model. Prediction was fast. Whereas Naïve Bayes algorithm took less time to build the model, but its accuracy is given 98.42%. It wrongly classified 1134 records. Its Prediction time is slow (Table 6.3).

6.6   Conclusions

The objective of this work is to study the impact of data visualization in analysis and interpretation of huge data. In this process, we demonstrated how data and visual analytics together will assess the characteristics of Crime Data. Using visual analytics, we could summarize that traffic accidents are more common offenses, apart from burglary and public disorders. We have also noticed that there are peaceful areas and crime-prone areas by identifying commonly occurring neighborhood areas. Visualization presents a quick summary of crime types and its network. Neighborhood analysis of offense category summarizes the crime-prone areas. Crime data analysis can help law enforcement officers to take collective measures increase level of crime prevention. Our prediction models are working quite well. But random forest has a limitation that more than 53 subcategories are not accepted. In our future work, we want to look in this direction. We wish to work with more advanced versions of machine learning algorithms, with huge data. Using Spark, we want to study this data more precisely.

References

1. Bogomolov, B., Lepri, J., Staiano, N., Oliver, F., Pianesi, Pentland, A.: Once upon a crime: towards crime prediction from demographics and mobile data. In: Proceedings of the 16th International Conference on Multimodal Interaction, pp. 427–434 (2014)
2. Iqbal, R., Murad, M.A.A., Mustapha, A., Shariat Panahy, P.H., Khanahmadliravi, N.: An experimental study of classification algorithms for crime prediction. Indian J. Sci. Technol. 6(3) 4219–4225 (2013)
3. Chen, H., Chung, W., Xu, J.J., Wang, G., Qin, Y., Chau, M.: Crime data mining: a general framework and some examples. IEEE Comput. 37(4), 50–56 (2004)
4. Beshah, T., Hill, S.: Mining road traffic accident data to improve safety: role of road-related factors on accident severity in Ethiopia. In: Proceeding of Artificial Intelligence for Developments (AID 2010), pp. 14–19 (2010)
5. Zhang, Q., Yuan, P., Zhou, Q., Yang, Z.: Mixed spatial-temporal characteristics based crime hot spots prediction. In: IEEE 20th International Conference on Computer Supported Cooperative Work in Design (CSCWD), Nanchang, China (2016)
6. Sri Lalitha, Y., Govardhan, A.: Semantic framework for text clustering with neighbors. In: ICT and Critical Infrastructure: Proceedings of the 48th Annual Convention of CSI, Volume II, Advances in Intelligent Systems and Computing 249, pp. 261–271. © Springer International Publishing, Switzerland (2013). ISBN: 978-3-319-03095-1
7. Mahmud, N., Ibn Zinnah, K., Ar Rahman, Y., Ahmed, N.: CRIMECAST: a crime prediction and strategy direction service. IEEE 19th International Conference on Computer and Information Technology. Dhaka, Bangladesh (2016)
8. Lin, Y.L., Yu, L.C., Chen, T.Y.: Using machine learning to assist crime prevention. In: IEEE 6th International Congress on Advanced Applied Informatics (IIAI-AAI). Hamamatsu, Japan (2017)
9. Grover, V., Adderley, R., Bramer, M.: Review of current crime prediction techniques. In: International Conference on Innovative Techniques and Applications of Artificial Intelligence, pp. 233–237. Springer, London (2007)
10. McClendon, L., Meghanathan, N.: Using machine learning algorithms to analyze crime data. Mach. Learn. Appl. Intl. J. (MLAIJ) 2(1) (2015)
11. Budur, E., Lee, S., Kong, V.S.: Structural analysis of criminal network and predicting hidden links using machine learning. arXiv:1507.05739 (2015)
12. Chainey, S., Ratcliffe, J.: GIS and Crime Mapping. Wiley, USA (2015)
13. Kim, S., Joshi, P., Kalsi, P.S., Taheri, P.: Crime analysis using machine learning. In: IEEE 9th Annual Information Technology, Electronics and Mobile Communication Conference (2018)

Chapter 7
Automatic Reminder List Generation to Assist People with Alzheimer's

Shireesha Gundeti and Bhageshwari Ratkal

Abstract No matter how well people set up to-do lists and calendars, they usually can't get things done unless they have a reliable way of reminding themselves to do them. However hard they try to remember them all, it is very common for people to forget an appointment with a doctor, or to wish a friend on birthday. And moreover, as people grow older, their brains change and they may have problems related to memory, every once in a while. These details may include simple things like not remembering when to take medicines or important details like forgetting their own house address. An application that can automatically generate a reminder list from people's conversation and remind them the appropriate task on appropriate time would definitely help people in avoiding such problems. The proposed idea would avoid the need of people themselves typing in the task and setting the time for the reminder, manually, as this application would take the person's casual conversation as input and would generate the reminder list automatically based on the related information extracted from the conversation. And based on the automatically generated reminder list which would be connected to one's calendar, the user would receive a notification when the clock ticks the time on a particular date as set in the reminder, thus helping the user to remember and do all the tasks at the required time.

7.1 Introduction

People always forget to get their things done, no matter how well they set up their to-do lists or calendars. It's very natural for even a young person to forget to catch up on a business meeting or to attend a family get-together. Usually, people tend to forget those events or tasks that are not their habits. If we suppose that all the events or tasks in a typical human life be classified into two, then they would be

S. Gundeti (✉) · B. Ratkal
Department of Computer Science and Engineering, G. Narayanamma Institute of Technology and Science, Shaikpet, Hyderabad, India

B. Ratkal
e-mail: bhagya@ratkal.com

© The Author(s), under exclusive license to Springer Nature Singapore Pte Ltd. 2021
S. C. Satapathy et al. (eds.), *Smart Computing Techniques and Applications*,
Smart Innovation, Systems and Technologies 225,
https://doi.org/10.1007/978-981-16-0878-0_7

habitual events/tasks and occasional or random events/tasks [2]. Habitual events include events such as brushing teeth, bathing, having food, and a regular person would never forget or misdo such tasks. Whereas occasional or random events include meetings, appointments, birthdays, etc. And forgetting a couple of events of this kind is a tendency that is often seen even in young and healthy human beings. Also, this tendency is observed to grow, as people grow older. And, elderly people are more vulnerable to being attacked by Alzheimer's. Alzheimer's disease causes problems with memory, thinking, and behavior [1]. As the disease progresses, a person with Alzheimer's disease will develop severe memory impairment and lose the ability to carry out everyday tasks.

So, forgetting things is an absolute problem that almost every human being on earth is facing. And a reliable way of reminding people, about their events or tasks is all they need. To address this issue, many of the reminder setting apps have been introduced into the market. But the major drawback in almost all of the existing apps is that they require the users to set the reminders manually. This itself is a tedious and boring task; it also encompasses another risk of the user forgetting to set the reminder.

Now, to address this drawback, an application named Automatic Generation of Reminder List (AGRL), based on Semantic Analysis is developed that can automatically set reminders based on the conversations that people make, with each other, regarding any event or task [1]. That is, if the user is discussing, with someone, about some event that is going to be held in future, then this conversation would be recorded by the application and based on the conversation, the user would be reminded of the event on a particular date at a particular time as mentioned by the user in the conversation. This solution avoids the necessity of manual setting of reminders and thus helps people to not miss out on events. Especially people affected with Alzheimer's can get greater help from this application and can lead an easier routine life.

7.2 Existing Systems

Most of the available reminder apps have good reviews from their users and are used by a good portion of the population. Almost all of these apps have an easy-to-use interface, voice-enabled interfaces for setting reminders, and many more options for the users so that they won't miss out on their events [3].

To discuss in detail, if a user using any of these apps wants to set a reminder, the user should type the title or something about the event that can remind the user of the event, the date of the event or the task, and time to be specific. Or there is also another facility in a few of the apps available now, which provides a voice recording option so that the user can record the title or name of the event instead of typing it in. This is obviously a great help to people who are busy and don't have time to type in the title or to people who are not that comfortable with typing on the phone. But the point to be noted here is that even though entering the title is made easier, the task of setting the reminder which includes selecting the date, time if it should repeat when to repeat and the number of times it should repeat, is still left to the user to do.

For the whole task of setting the reminder to be automated, it requires the automation of the following subtasks:

- Entering the title or name of the event
- Choosing the date
- Setting the time

So, the main objective of the proposed application is to automate the process of setting the reminders which include the above-mentioned subtasks.

7.3 Proposed System—Automatic Generation of Reminder List (AGRL)

The working of the proposed system is depicted in Fig. 7.1.

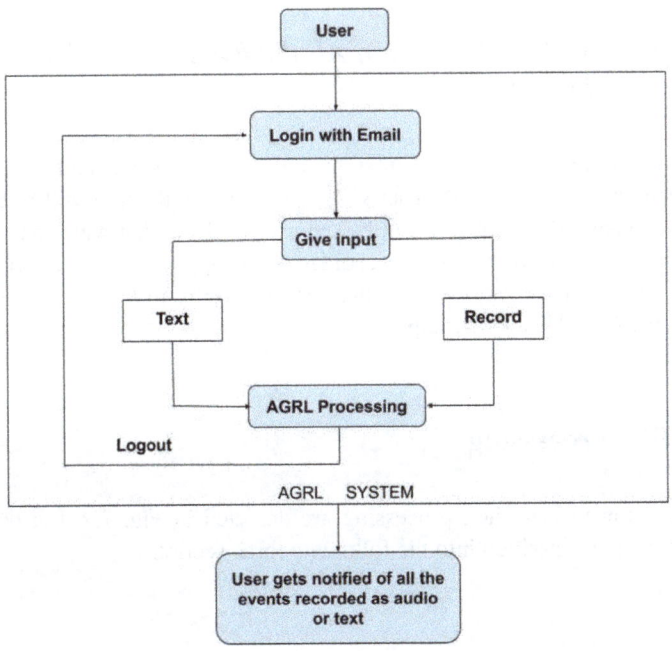

Fig. 7.1 AGRL Architecture

7.3.1 User Input

Once the user has logged into the application using a valid email-id, the user can give the input in either of the following two ways.

Text: The user can simply type in the event, time, and date, maybe, in the form of a simple sentence.

Record: Or the user can choose to record the conversation which may include information about some events and the dates when they might occur in the future.

7.3.2 AGRL Processing

The input might be in the form of text or audio, either way, it requires a lot of processing to extract the required task or event name and the associated time when it might take place. The methodology involved in this processing is elaborated in the section IV.

7.3.3 The User Gets Notified of All the Events Recorded as Audio or Text

Once the processing of the given text or audio is completed and the extracted reminders are connected with the user's Google calendar, the user will be reminded of all those events that were typed in by the user or those that were mentioned by the user in the conversation that was recorded in the application. At the time of the reminder, the user may receive a notification, according to the user's notification settings in the Google calendar app.

7.4 AGRL Processing

The steps followed in AGRL processing are depicted in Fig. 7.2. Let the AGRL processing part be classified into the following three sections.

7.4.1 Text—Preprocessing

This step comes into action, only when the user gives the input in the form of text. The text preprocessing includes only one step namely sentence tokenization [6]. And this step assumes that the user would stick to the following grammar conventions:

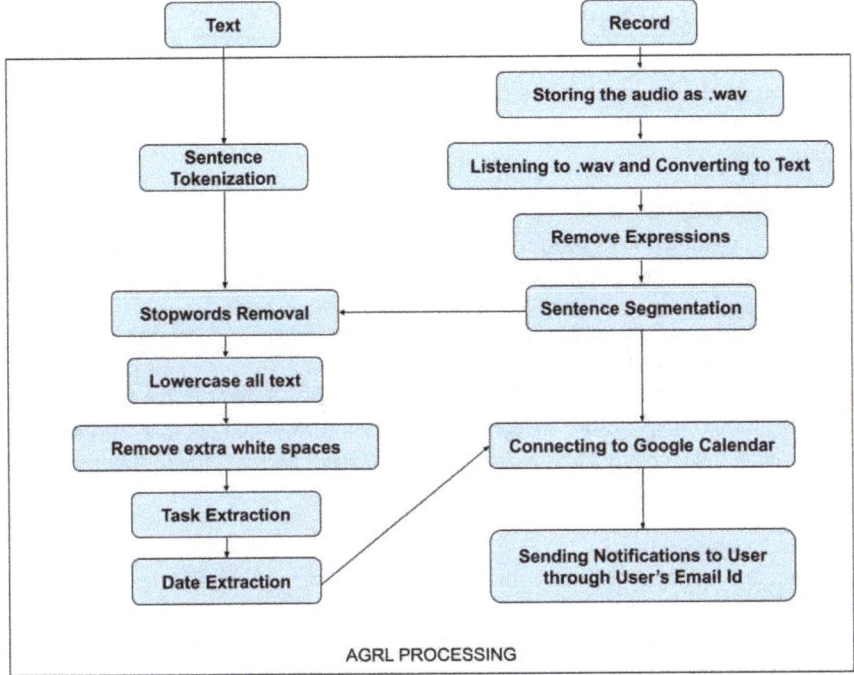

Fig. 7.2 AGRL processing

- Use clear and correctly formed sentences in English while typing in the text
- Use correct punctuation such as full stops(.), question marks(?), to indicate the end of the sentences, in case of more than one sentences
- Do not use chat language or short-cuts

  The process involved in this step is as follows.

Sentence Tokenization

The text given by the user can be a single sentence or a paragraph. If the input is a single sentence, it goes unchanged in this step of preprocessing. But if the input contains more than one sentence or a paragraph, then it needs to be sentence tokenized, i.e., the given paragraph needs to be split into sentences at the specified sentence separators, which in most cases is a full stop(.).

7.4.2   Record—Preprocessing

If the user gives the input in the form of audio, only then the processing steps to be mentioned under this section would come into play. The steps involved are as follows.

Storing the Audio as .wav

The speech or the conversation recorded by the user would be initially stored as an audio file with the.wav extension. The audio content in this file will be processed in further steps.

Listening to .wav and Converting to Text

Here a speech recognition is used to listen to the audio content in the .wav file and convert it to text. Only if the speech or conversation recorded is clear, audible, and loud enough, the converted text would be correct as per user and the following results would be as expected.

Remove Expressions

While recording the speech or conversation, if the user or the other speakers in the conversation show some kind of expressions such as laugh or smile or cry or grin, they will also get recorded in the audio and would get converted to text as well. But these expressions would add no value to the text related to the reminders and hence these should be removed or neglected before proceeding to extract reminders related text. To do so, when some of the speech-to-text conversions containing expressions are noticed, it is observed that when the sound associated with the expressions is converted to text, they are enclosed within some of the special expressions such as [] or {} or **, etc. For example, '[laughing]', '{crying}', *grins*, etc. Using the Re module in Python, the expressions are identified in the text and eliminated.

Sentence Segmentation

Unlike the case when the input is given as text, when the input is given as audio and later converted to text, there would be no sentence separators specifying the end of each sentence and hence the process used in sentence tokenization would not be appropriate to use here. In order to address this situation, sentence segmenter is used.

7.4.3   Text and Record—Processing

After the input is processed as per the process mentioned in the above sections, the resulting text is further processed through the following steps. Note that the steps to be mentioned are commonly performed in both cases of input, text, or audio.

Stopwords Removal

All the not so important words in each sentence need to be removed as they sometimes might add inaccuracy to the final output obtained after processing the input.

Lowercase all Text

Lower casing all the text is done to avoid having multiple copies of the same words. For example, while analyzing the words or pattern matching, 'Analytics' and 'analytics' will be taken as different words, if not lowercased before processing.

Remove Extra White Spaces

Extra white spaces, if present in the given text, also should be removed, so that all the words in the text are separated by single spaces. This step is needed because sometimes extra spaces might also cause problems while pattern matching.

Task Extraction

The task extraction is carried out using Parts-of-Speech Tagging and Chunking [5]. The sub-steps involved in this step are as follows:

- Tokenize text: Each sentence is tokenized into words.
- POS_Tag is applied to each of the words obtained in the above step.
- POS Tagging is combined with Regular Expressions to form and extract chunks of noun and verb phrases.

Basically, a chunk grammar needs to be defined to specify the rules that indicate how the sentences should be chunked [4]. And the grammar that was defined for the task at hand is as follows.

grammar = "NP:{<VB.*>?<VBP>?<RB>?<PRP.*>?<IN>?<DT>?<VBZ>? <VBN>?<POS> ?<NN>?<VBZ>?<JJ.*>*<NN.*>+<VBZ>?<TO>?<VB>?}"

The above grammar specifies that the NP chunk should be formed whenever the chunker finds an optional verb type, followed by an optional verb in present tense without third person singular, followed by an optional adverb, followed by an optional personal pronoun, followed by an optional preposition, followed by an optional determiner, followed by an optional verb in the present tense with third-person singular, followed by an optional verb past participle, followed by an optional possessive ending, followed by an optional noun, followed by an optional verb in the present tense with third-person singular, followed by any number of adjectives, followed by one or more number of nouns, followed by an optional verb in the present tense with third-person singular, followed by an optional infinite marker, followed by an optional verb.

The chunks that are formed in each sentence are ranked based on the summation of their individual POS ranks set as follows.

POS_rank = {"DT":2, "IN":2, "RB":3, "PR":3, "VB":6, "JJ":5, "NN":5, "PO":5,"TO":1}

Among the chunks formed, one with the maximum chunk rank would be considered as the required task name that can remind the user of the event.

Date Extraction

To extract the date and time relevant to the reminder that needs to be reminded, different cases in which the date and time can be specified in a sentence are all analyzed. And for all the cases that were analyzed, a regular expression pattern was constructed, so that when a date-time string matching the corresponding pattern is found, it would trigger the required processing needed in order to extract the date and time from the given sentence.

The regular expression in the python programming language is a method used for matching text patterns. The 're' module in Python provides regular expression support.

7.5 Discussions on Results

The main aim of the application, which was to assist the people affected with Alzheimer's to remember things that can be reminded, is achieved. The performance of the application is satisfactory in terms of accuracy, security, speed, usability, etc.

To test the accuracy of the model used in the application, more than 1000 sentences, that may contain information related to reminders, were collected from some random online websites [7]. The labels and the datetime information, if present, were manually extracted from each of those sentences. And all this information is organized in the form of a dataset containing 3 attributes; sent, label, datetime. This dataset is shown in Fig. 7.3.

The values that these attributes can take are as follows.

sent: The sentence that may contain information related to reminders. In other words, it can be any sentence that may or may not indicate a reminder.

label: The task or event name related to the reminder, if present in the sentence. The values that this attribute would take can be categorized into two.

- if task name or event name appears in the sentence → task or event name
- else → Not Found

	sent	label	datetime
0	Set a reminder on date 23rd November 2016	Not Found	23-11-2016 00:00
1	remind me 6 pm today evening	Not Found	06-04-2020 18:00
2	Remind me at 28 December for recharge	recharge	28-12-2020 00:00
3	remind me at 7pm on 8 Jan	on 8 Jan	08-01-2020 19:00
4	Set a reminder on 4th Dec of going to meet son...	meet sonal miss	04-12-2020 00:00
...	...	...	...
1177	Remind me to go to Eat Panipuri at 5:00 PM today	go to Eat Panipuri	04-06-2020 17:00
1178	Remind me to order garnier roll on 1 January at 1	order garnier roll	01-01-2020 13:00
1179	Please remind the same on 13th November	Not Found	NaN
1180	Hi I want alarm with tone saying time is 3.30a...	3.30am...it's time booja	04-06-2020 15:30
1181	Hey remind me to mail the scanned copy of canc...	mail the scanned copy of cancelled check to Vi...	04-06-2020 16:00

1182 rows × 3 columns

Fig. 7.3 Testing dataset

	sent	label	datetime	label_testing	datetime_testing
0	Set a reminder on date 23rd November 2016	Not Found	23-11-2016 00:00	Set a	23-11-2016 00:00
1	remind me 6 pm today evening	Not Found	06-04-2020 18:00	Not Found	06-04-2020 18:00
2	Remind me at 28 December for recharge	recharge	28-12-2020 00:00	for recharge	28-12-2020 00:00
3	remind me at 7pm on 8 Jan	on 8 Jan	08-01-2020 19:00	Not Found	08-01-2020 19:00
4	Set a reminder on 4th Dec of going to meet son…	meet sonal miss	04-12-2020 00:00	meet sonal miss	04-12-2020 00:00
…	…	…	…	…	…
1177	Remind me to go to Eat Panipuri at 5:00 PM today	go to Eat Panipuri	04-06-2020 17:00	Eat Panipuri	06-04-2020 17:00
1178	Remind me to order garnier roll on 1 January at 1	order garnier roll	01-01-2020 13:00	order garnier roll	01-01-2001 00:00
1179	Please remind the same on 13th November	Not Found	NaN	Not Found	13-11-2020 00:00
1180	Hi I want alarm with tone saying time is 3.30a…	3.30am…it's time booja	04-06-2020 15:30	saying time is	06-04-2020 03:00
1181	Hey remind me to mail the scanned copy of canc…	mail the scanned copy of cancelled check to Vi…	04-06-2020 16:00	of cancelled check to	06-04-2020 16:00

1182 rows × 5 columns

Fig. 7.4 After adding the columns generated by the AGRL model

datetime: The date and time information related to the reminder, if present in the sentence. The values that this attribute would take can be categorized into two.

- if the date and time information appear in the sentence → datetime
- else → Nan (empty)

Once this dataset was ready with around 1100 records, the model used in the application was executed giving the sentences, from this dataset, as input to give the label_testing and datetime_testing values as output. After adding this output as another two columns in the previous dataset, it looked as in Fig. 7.4.

Here, the extra attributes indicate the following:

label_testing: The task name or event name extracted from the sentence, by the model constructed.

datetime_testing: The datetime information extracted from the sentence by the model used in the application.

Now, the accuracies of individual results of extraction of labels and datetime were estimated, and then the accuracy of the combined output was calculated using the following formulas.

$$\textbf{Label\_Accuracy} = \frac{\text{Number of Correctly Estimated Labels}}{\text{Total number of records}} \times 100$$

$$\text{where, Correctly Estimated Labels} = \begin{cases} 1; \text{if label} = \text{label\_testing} \\ 0; \text{otherwise} \end{cases} \quad (7.1)$$

$$\textbf{Datetime\_Accuracy} = \frac{\text{Number of Correctly Estimated Datetime}}{\text{Total number of records}} \times 100$$

$$\text{where, Correctly Estimated Datetime} = \begin{cases} 1; \text{if datetime} = \text{datetime\_testing} \\ 0; \text{otherwise} \end{cases} \quad (7.2)$$

$$\textbf{Label\_Datetime\_Accuracy} = \frac{\text{Number of Correctly Estimated Labels and Datetime}}{\text{Total number of records}} \times 100$$

$$\text{where, Correctly Estimated Labels and Datetime} = \begin{cases} 1; \text{if label} + \text{datetime} = \text{label\_testing} + \text{datetime\_testing} \\ 0; \text{otherwise} \end{cases}$$

$$(7.3)$$

Using the Eqs. (7.1), (7.2), and (7.3), the accuracies were estimated which were found to be 71% for label extraction, 84% for Datetime extraction, and 60% for the combined extraction. The accuracies obtained are pictorially depicted in Figs. 7.5, 7.6, and 7.7.

In Fig. 7.5, it is clearly shown that 71% of the labels were correctly extracted. And this accuracy was calculated based on the following assumptions.

- It is assumed that extracting something about the actual reminder is always better than not extracting anything. So, even if the label value in the dataset is 'Not

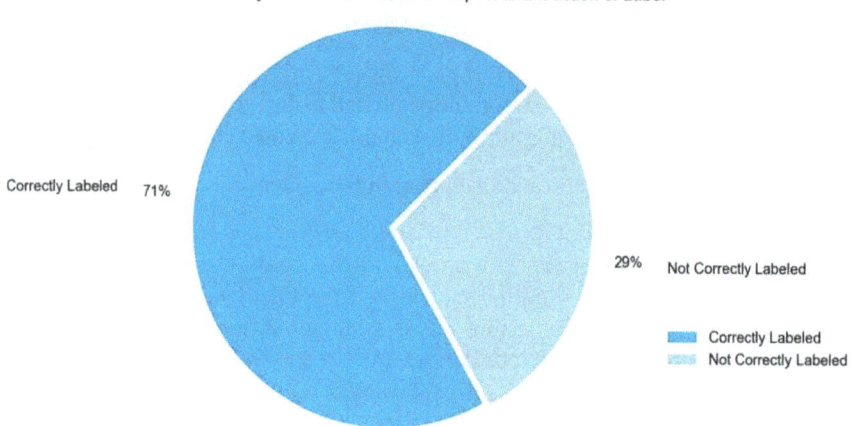

Fig. 7.5 Accuracy with regard to extraction of labels

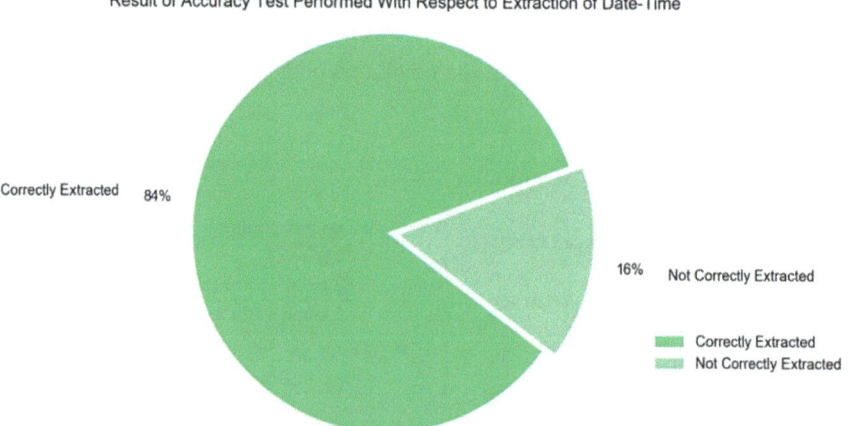

Fig. 7.6 Accuracy with regard to extraction of date–time

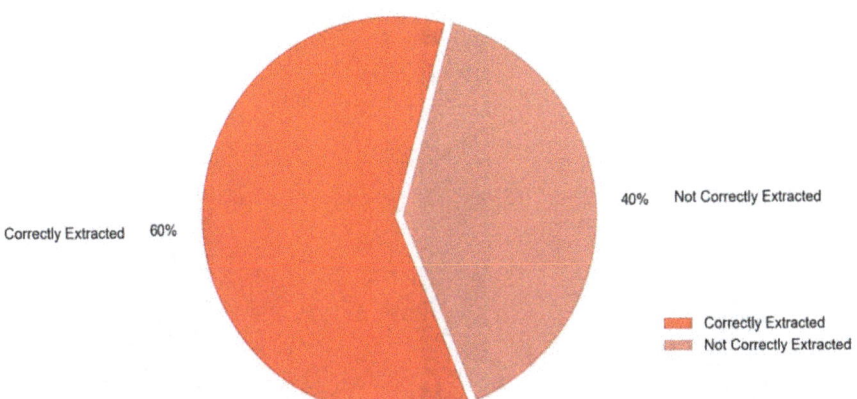

Fig. 7.7 Accuracy with regard to extraction of label and date–time

Found' and the label extracted using the application found to be some word from the input sentence, it is still considered as 'Correctly Labeled'.

- It is assumed that even a single word related to the event, to be reminded, is always enough to remind the user of the event. So, even if a single word in the label extracted using the application matches with that in the dataset, it is considered as 'Correctly Labeled'.

Figure 7.6 depicts the pie diagram, showing the percentages of both correctly and wrongly estimated dates and times related to the reminders. Unlike in the extraction of labels, no assumptions or exemptions are made in the extraction of dates and times, because a word predicted wrong can still remind the user of the event, but if the user is reminded of the event late, by even 5 min, is not admissible.

Here, while calculating the accuracy, the datetime from the dataset is exactly compared to that extracted using the application.

Figure 7.7, shows that about 60% of the extracted labels and datetime values are correct while 40% of them are wrongly estimated. Here, instead of separately extracting and comparing the results with those values in the dataset, this is to test the accuracy of the combined result extracted from the given input sentences. The steps followed in calculating this accuracy are based on the below algorithm.

- check if datetime from dataset matches exactly to the extracted datetime_testing
- If yes:
- check if label == 'Not Found' or label_testing ! = 'Not Found'
- If yes: consider as Correctly Extracted
- If no: consider as Not Correctly Extracted

The bar graph in Fig. 7.8, has the three different Accuracy Tests Conducted on Extraction of Labels and Datetime on the X-axis and the Accuracy, obtained in each

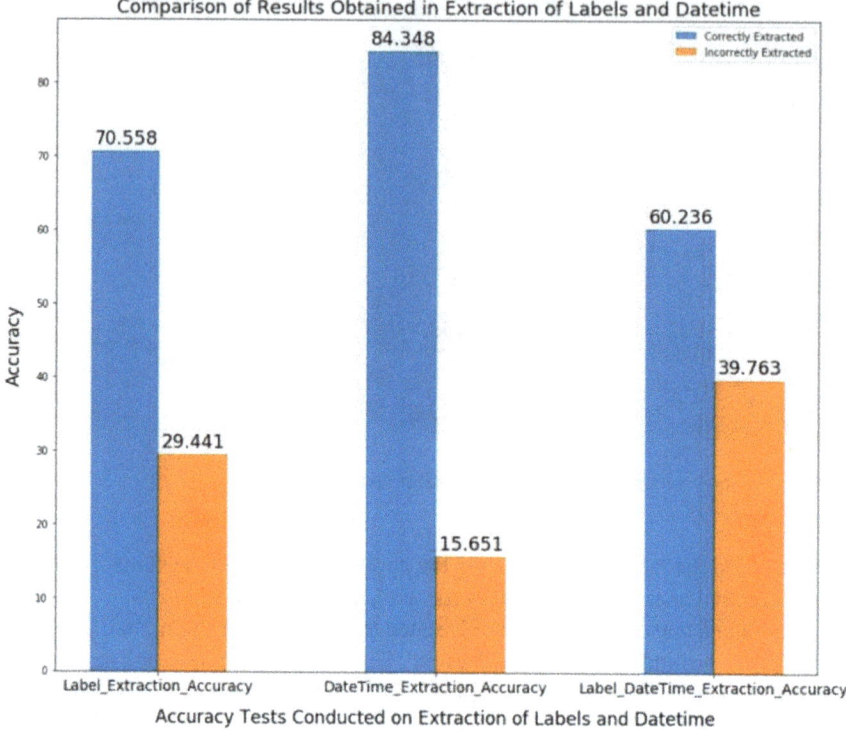

Fig. 7.8 Comparison of results obtained in the extraction of labels and datetime-I

test, on the Y-axis. It shows the comparison of percentages of the correctly extracted and incorrectly extracted results in each of the three test cases.

7.6 Conclusion

Using the proposed idea, any person suffering with Alzheimer's would now be able to attend almost all their events or tasks without fail. The users would now be able to avoid all the repetitious work involved in setting reminders.

References

1. https://www.mayoclinic.org/diseases-conditions/alzheimers-disease/symptoms-causes/syc-203 50447
2. https://www.lifehack.org/articles/featured/back-to-basics-reminders.html

3. https://www.gadgetsnow.com/apps/6-best-reminder-apps-for-android-and-ios/articleshow/646 84028.cms
4. https://www.guru99.com/pos-tagging-chunking-nltk.html#:~:text=Chunking%20is%20used% 20to%20add,of%20more%20than%20one%20level
5. Varghese, N.,Punithavalli, M.: Lexical and semantic analysis of sacred texts using machine learning and natural language processing. Int J Sci. Technol. Res. 8(12) (2019)
6. https://www.nltk.org
7. https://www.researchgate.net/publication/330901752_Latent_Semantic_Analysis_Methods/ link/5c5ab75245851582c3d1bdb2/download

Chapter 8
Recognizing the Faces from Surveillance Video

T. Shreekumar, K. Karunakara, H. Manjunath, B. Ravinarayana, and D. R. Annappa Swami

Abstract The face recognition (FR) is getting extremely well known in the field of biometric due to its unique nature. In spite of the fact that the progression of FR advancement has accomplished a moderate level of improvement with deep learning strategies, still there exists some disadvantage, for example, illumination variation, change in pose angle, occlusion, masked face, and so on. The deep learning strategies will be productive only if the learning database contains a large number of images for training purpose. It is a known fact that, a large number of images are very difficult to get always, especially from village people. In order to overcome all the discussed problems, this paper proposes two different FR approaches which functions admirably with the database even with few training images. The two methods proposed are (i) face recognition with local linear regression (LLR) using a trained artificial neural network (ANN), (ii) FR with support vector machine (SVM) by utilizing the well-known particle swarm optimization (PSO) technique for optimizing the SVM kernel parameters. The performance of both the techniques is measured in terms of statistical measures and the computation speed. The performance of the proposed methods is also compared with a very recent and similar active appearance model (AAM)-based FR. From the experiment and result analysis, it is clear that the proposed PSO-SVM-based FR yields a maximum accuracy of 95%, an average accuracy of 91.5% with a minimum time requirement of 0.03 s.

T. Shreekumar (✉) · H. Manjunath · B. Ravinarayana · D. R. A. Swami
Mangalore Institute of Technology and Engineering, Moodabidre, Mangalore, Karnataka, India

D. R. A. Swami
e-mail: annappaswamy@mite.ac.in

K. Karunakara
Sri Siddhartha Institute of Technology, SAHE, Tumakuru, Karnataka, India
e-mail: karunakarak@ssit.edu.in

© The Author(s), under exclusive license to Springer Nature Singapore Pte Ltd. 2021
S. C. Satapathy et al. (eds.), *Smart Computing Techniques and Applications*,
Smart Innovation, Systems and Technologies 225,
https://doi.org/10.1007/978-981-16-0878-0_8

8.1 Introduction

The requirement of accurate person identification system is very essential nowadays, in most of the government and commercial applications. The demand for the reliable user authorization methods is increasing because of the rapid developments in the fields of social networking, communication, and mobility. Biometric is a method of identifying the genuineness of the user, based on the physiological or behavioral traits. Biometric systems are now used in various applications like forensics, commercial, and other civilian applications. Face, signature, hand geometry, fingerprints, retinal scan, iris scan, hand vein, facial thermogram, voice, etc., are prominent ones in the biometric space. Biometric is getting popular starting from issuing an Adhaar card to gaining permission to enter a nation.

Face recognition forms the most trending technology nowadays. It forms the major way for identifying a person and also it becomes a better way for providing the much needed security for many kinds of applications. It forms the most reliable biometric due to its potential, when it is compared with other biometrics. This system has a long range of applications, both in the commercial as well as personal fields. The face recognition system uses many features from the face image which are used for recognizing the people. Although in the olden days, the face recognition equipments were using only geometrical features derived from the face directly, and they were not as accurate as they are now. Of late, due to the technological advancements, various methods which are used to identify the person are readily available. Nowadays, the face identification technology is used even for scam revealing and finding criminals too.

Face acknowledgment by utilizing an existing face image set verifies and confirms the individuals in the specified video stream or in the still image. [1–3]. It has an amorous application in security and authentication-related applications [4]. The entire face recognition methods can be divided into feature-based and appearance-based.

Feature-based method uses the geometric parameters from the face image for facial matching [7, 8] and appearance-based method uses intensity parameter [1, 17]. The face acknowledgment should be possible in the given picture also in the given video. The procedure of face recognition method is performed on the still image or even in a video streaming. Face detection and the face identification are the two stages present in a face recognition system [4]. In the face detection stage, the face is located in the video stream or in the still image. In the face detection stage, the face is located in the image or in the video stream. Then, the face identification algorithm utilizes the located face for identifying the individual from the image. Hence, both face detection and recognition algorithms are reasonably important in facial identification [13].

As it has been proved that motion enables identification of (well-known) faces when negation, inversion, or threshold operation is performed on images, video-based face recognition is preferred over using still image-based face recognition [3]. It has also been proved that animated faces compared to randomly reordered images from the same collection are better recognized by humans. The video-based facial identi-fication methods that consistently use both spatial and temporal information began only some years before and still require more investigation [9]. But the difficulty in video-based recognition is low frame quality [3, 11].

A good face recognition algorithm should overcome the bottlenecks like pose change, variations in illumination, aging, partial occlusion, sunglass, make-up, facial expression variation, change in hairstyle, scale variation, etc. Recognizing an image under severe illumination change and pose change is an extremely challenging problem in face recognition [12]. Illumination change can be handled by feature-based method easily and when the problem is with the pose, and the appearance-based method is predominantly used to recognize the faces. In appearance-based methods, the image of the individual is recognized from the image captured with a diverse in view angles [15, 16].

The face recognition has plenty of applications which include securing the mobile and computer applications, Internet communication, computer entertainment, veri-fying a person, securing a system using CCTV, and such others. Lately, the convo-lutional neural network-based face recognition has shown significant progress in the efficiency of face recognition task, but still the face recognition remains a chal-lenging problem when the question is of recognizing a person from a poor quality video sequence or a noisy image or even from a group photos.

8.2   Literature Survey

The [17] introduces a face recognition method to identify the faces from variety of pose angles. This work is mainly focused on pose normalization and face recognition. Pose normalization is carried out with the help of local linear regression (LLR). The normalized face is then subjected to dimensionality reduction using principal component analysis (PCA). Then, the face image is recognized using Fisher's linear discriminant analysis (FLDA) from the reduced dimension. This approach has shown a good performance on Yale B Face Database, with 93.4% of accuracy.

The author [18] proposed a model for face identification using local binary pattern (LBP) and convolutional neural network (CNN). Initially, LBP features are extracted from the image and stored in a feature set. CNN training is carried out using this stored LBP features. The trained CNN is then used to identify the query image. From the experiment, author concluded that the input to CNN is an image then pixel level knowledge is extracted and the recognition rate is less. A substantial improvement in the recognition performance is observed when the LBP feature map is used as input to the CNN.

The work [19] has proposed a face acknowledgment framework to detect and identify the faces in still images and low resolution video sequence, by identifying the face images from the image gallery of known individuals, utilizing genetic algorithm and ant colony optimization techniques. In this method, initially, the pre-processing step is carried out on the input images, feature extraction is done using the ant colony optimization, and finally, the recognition is carried out with genetic algorithm. This method is suitable in single and multi-threaded processing, real-time video processing. The proposed technique is robust to expression change and illumination change.

The face recognition method [20] has proposed a new method to identify the faces from a video sequence known as the 'analysis-by-synthesis framework' to overcome the variation in pose and illumination changes. This method works by tracking the entire video sequence to integrate the motion, lighting, and shape and to generate model from the gathered information. The information gathered from video tracking is used to estimate the illumination condition, pose alignment, and the structure of the face in each frame with the help of extended light source and multiple camera points. And then, corresponding to the illumination and pose conditions estimated from the video sequence, a 3D face model is generated. After that, the similarity measure is obtained by comparing the synthesized images and the probe video. This strategy can deal with circumstances where the face posture and lighting conditions in the training and testing information are totally disjointed.

In [21], an algorithm is proposed for recognizing the faces from a video sequence using the adaptive multiple face matchers. This method briefly populates the databases in order to capture the intraclass variations. The method uses active appearance model (AAM) to extract dynamic information about the facial poses in each frame and 3D face reconstruction model in order to factorize the same from the video sequence. The motion blur is estimated using discrete cosine transform (DCT). The experiment is carried on CMU's face-in-action (FIA) database and the result showed a good improvement when experimented with 204 subjects and with different face matchers.

In [22], a method is proposed for pose invariant face acknowledgment. Their face acknowledgment framework was insensitive to the changes in view direction and requires just one sample per person. The method used here is to find similarities of a face image against a set of face images of the same view, from the training set to establish a view invariant face acknowledgment.

The method [23] proposed a regression-based virtual frontal view generation in order to compensate pose variation. This frontal virtual view was then used to identify a person. They used local linear regression for virtual frontal view generation. They also tried with global linear regression (GLR) for virtual frontal view generation. The result indicated that the accuracy of LLR is maximum compared to the GLR-based virtual view generation. LLR uses a linear mapping function between the current view and the virtual frontal view. GLR uses the entire image and generates a virtual frontal image. The LLR divides the image into smaller blocks called, 'patches' with a fixed patch size of 20×20, or 10×10. They used dense sampling-based overlapped patch creation. Each patch is then subjected to LLR to get the linear mapping to the virtual frontal view. After the entire iteration of mapping for each patch, all the newly generated frontal patches are combined to form the virtual frontal face image. The experiment was carried out with CMU PIE face set and compared with eigen light-field method. The result indicates better compared to eigen light-field method.

The work [24] proposed a novel face recognition method for efficient face matching. Within this framework, the unprocessed greyscale face images of two individuals are compared using simple image processing filters. They successfully estimated the discrepancy in the illumination condition between the test image and the training set implicitly and used to weigh the contribution of query and the gallery representation. This filter-specific matching algorithm works offline by minimizing the online overhead. The experiment recorded the best performance, by correctly recognizing 97% of the individuals.

Face detection and identification from surveillance videos are quiet difficult because of image blur, illumination, pose variations, occlusion, and noise. A convolution neural network (CNN)-based recognition system is proposed in [25], which estimates the blur function from the difference between the artificially blurred images and the original image of the real-world video frames. Initially, the training data is formed using the original still images and the artificially blurred images of the same counterpart. The CNN is used to learn the blur function during the training phase, from the training data, made out of both the artificially blurred data and the still images. Secondly, in order to overcome the pose problem and occlusion, a Trunk-Branch Ensemble CNN (TBE-CNN) is proposed, which adequately separates the correlative data from the face images. Third, a triplet loss function is estimated to improve the effectiveness of feature representation of the extracted feature.

A facial recognition technique to extract the core region of face analysis, besides, all the while alter the facial posture and accomplish face alignment has been proposed in [26]. It can maintain a scheme distance from "mis-alignment disaster" amid the component matching process. The video face recognition system of little examples capacity had been built up to test the technique. The technique can create the face tests in the uncontrolled environment and upgrade the recognition rate of the face recognition system.

8.3 Face Recognition with LLR and ANN

In this approach, the face image is recognized with two different stages: (1) training section and (2) testing section. In the first section, an artificial neural network (ANN) is trained using the combined score [27] obtained from principal component analysis (PCA) and Fischer linear discriminant analysis (FLDA). In this method, the dimension reduction is done using PCA, feature vector is obtained using the FLDA, and then a combined score of PAC and FLDA is stored in a vector. The combined score of PCA and FLDA is then used to train the ANN. In the second section, the video of the person is obtained and normalized for illumination and pose variation problems. For illumination normalization, popular discrete cosine transform (DCT) and the pose problem are addressed with local linear regression method. The normalized frame is then subjected to get the combined score of PCA and LDA. The combined score is then used to recognize the face using already trained ANN. Figure 8.1 shows the structure of face recognition using local linear regression.

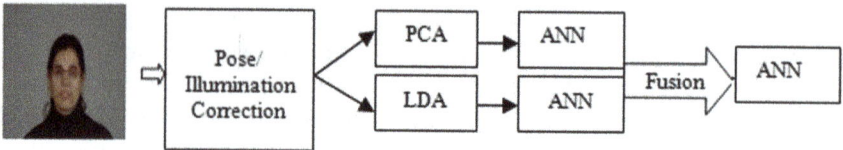

Fig. 8.1 PCA and LDA fusion using NN

8.4 Face Recognition with PSO and SVM

This method uses a popular and effective SVM to recognize the person from video frame. In this method, the SVM is trained with training database, and during the training phase, the SVM parameters are optimized with an optimization technique, particle swam optimization (PSO). Initially, the pre-processing of the image is carried out with adaptive median filter. In the next stage, the feature vector is obtained with active appearance model. Then, the SVM is used to recognize the face image. Figure 8.2 shows the structure of recognition system (Fig. 8.3).

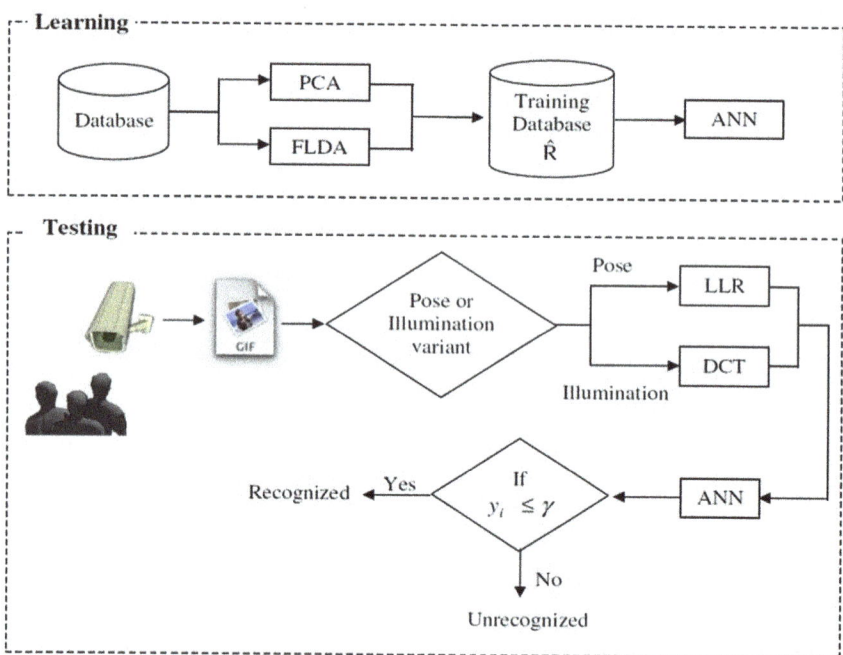

Fig. 8.2 LLR-based face recognition system

Video Face Recognition System

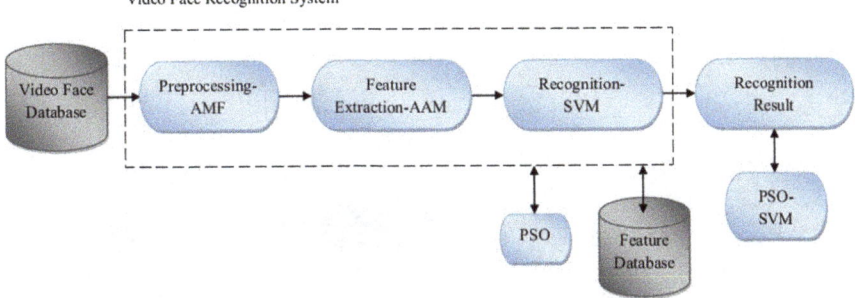

Fig. 8.3 Face recognition system based on PSO-SVM

8.5 Experiment and Result

In this section, the experimental outcome is described in detail, which is implemented in PC with 3.2 GHz Pentium Core i5 processor using MATLAB (version 7.2). To evaluate the effectiveness of the proposed method, the performance of proposed algorithm is evaluated on the UPC face database. This database has been made with the primary reason for testing the robustness of the face acknowledgment method against strong face posture variation and illumination variations. This database incorporates an aggregate of 44 people with 27 pictures for each individual which relate to various face posture views (0°, 30°, 45°, 60°, and 90°) under three distinct illumination conditions. Besides, this database incorporates 10 more extra frontal view of the face images with various occlusion and expression variations. The resolution of the pictures is 240 × 320 and they are in BMP format. Figure 8.4 shows the sample from UPC dataset.

This paper proposes LLR-based face recognition system [28] and PSO-SVM-based face recognition [29]. For evaluating the algorithms, we use UPC database. We divide the database into four datasets. The experiment is conducted to recognize the face images using all the four datasets.

From the analysis of results, it is clear that the proposed methods recognize the face images accurately. The proposed LLR-ANN method records an accuracy of 75%, and the second method PSO-SVM-based FR shows an accuracy of 95%.

Next, for comparison purpose, a recent AAM-based face recognition algorithm [30] is selected. This algorithm attains an accuracy of 90% in the experiment. The recognition performance of all the three methods is compared and tabulated in Table 8.1

Figure 8.5 shows the recognition performance of selected FR algorithms in terms of statistical measures.

As shown in Table 8.1 and Fig. 8.5, the experimental results indicate that PSO-SVM has higher face recognition performance compared to LLR-ANN-based method and AAM-based FR in terms of statistical measures. The performance

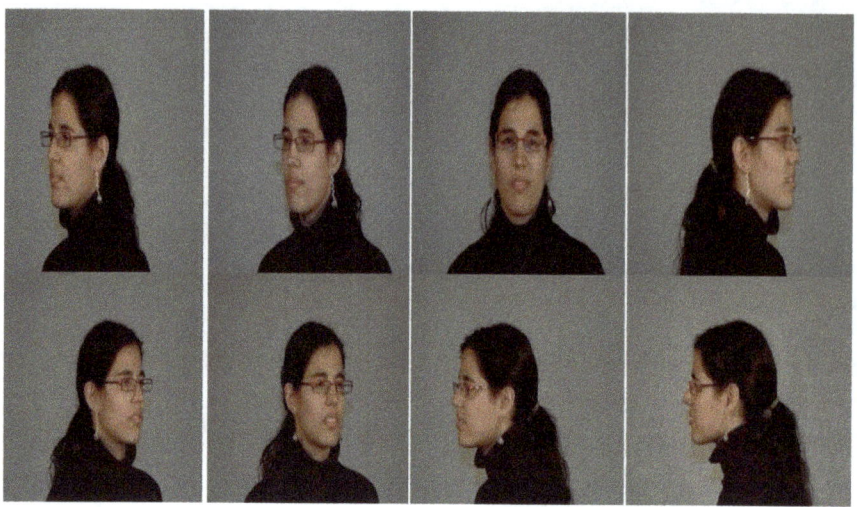

Fig. 8.4 Sample video frames from the UPC database

Table 8.1 Performance result of PSO-SVM and LLR method

Performance	PSO-SVM	AAM-based FR	LLR-ANN
Sensitivity	96.67	80.00	70.00
Specificity	93.33	100.00	80.00
Precision	93.55	100.00	77.78
NPV	96.55	83.00	72.73
FPR	06.67	0.00	20.00
FDR	06.45	0.00	22.22
FNR	03.33	20.00	30.00
Accuracy	95.00	90.00	75.00
F1-score	95.08	88.89	73.68
MCC	90.05	81.65	50.25

metrics show that the proposed PSO-SVM video face recognition technique is more accurately recognizes the face images than other two methods.

The computation time is also an important factor in face recognition. The time taken by LLR-based face recognition system [28], PSO-SVM-based methods [29], and AAM-based FR [30] are also compared, and the result is tabulated in Table 8.2. As can be seen from Table 8.2 that the PSO-SVM face recognition technique needs very less time for recognizing the faces compared to other two methods.

The main reason for selecting the UPC database is that it contains the face images of the individuals with different degrees of pose variation. The experiment is conducted on a dataset formed by selecting ten individuals. Table 8.3 tabulates the performance accuracy of all the three methods.

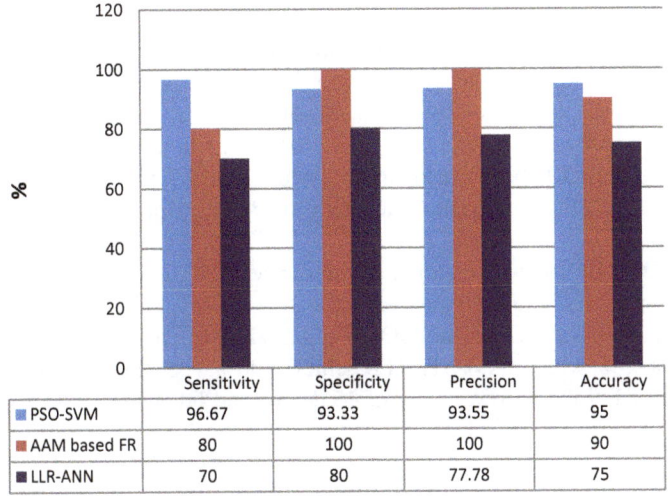

Fig. 8.5 Analysis of sensitivity, specificity, precision, and accuracy

	Sensitivity	Specificity	Precision	Accuracy
PSO-SVM	96.67	93.33	93.55	95
AAM based FR	80	100	100	90
LLR-ANN	70	80	77.78	75

Table 8.2 Comparison of computational time in terms of seconds

Experiment no	PSO-SVM	LLR-ANN approach	AAM-based approach
1	0.036011	08.1322	0.04679
2	0.047519	09.7772	0.05780
3	0.056591	10.5866	0.06501
4	0.054959	10.7222	0.06829
5	0.058237	10.1094	0.07184
6	0.049964	10.6153	0.06712
7	0.048922	09.4988	0.05793
8	0.050033	11.1353	0.07518
9	0.036495	08.6391	0.05745
10	0.055194	11.4592	0.07790

Table 8.3 Accuracy with different pose angle (%)

Pose angle/methods	0°	+30°	−30°	+45°	−45°	Avg
PSO-SVM	95.00	93.33	93.33	92.5	92.5	91.55
AAM-based FR	90.00	87.00	86.5	82.00	82.5	85.6
LLR-ANN	80.00	75.00	75.00	72.5	71.5	71.64

From Table 8.3, it is clear that when the pose variation is 0°, the recognition accuracy is high. The average accuracy of all the accuracy in each method is recorded separately for the comparison purpose. From Table 8.3, it is clear that the PSO-SVM performs better compared to other methods with an average accuracy of 91.55%.

The result analysis is given in Fig. 8.6. From Fig. 8.6, it is clear that PSO-SVM outperforms the LLR-based recognition system and AAM-based face acknowledgment by yielding a maximum accuracy of 95% and an average accuracy of 91.55%.

Table 8.4 shows the recognition performance of PSO-SVM and AAM-based methods in terms of statistical measures.

From Table 8.4, it is clear that PSO-SVM-based system has achieved a maximum accuracy of 95%, while AAM-based FR achieved 90% when the pose angle variation of 30 degree. For the pose variation of 45 degree and 60 degree, PSO-SVM-based

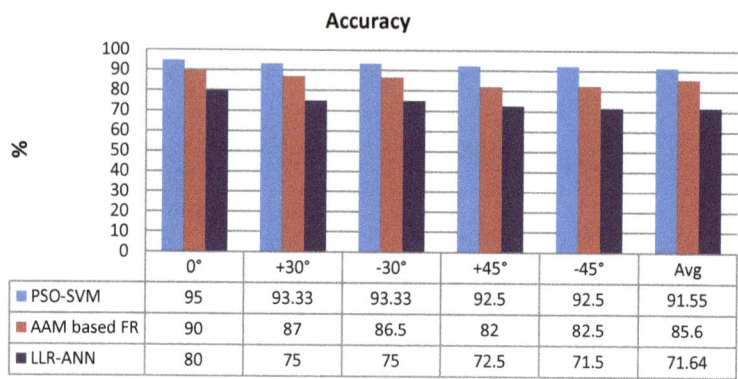

Fig. 8.6 Analysis of accuracy on different pose angles

Table 8.4 Statistical measures on different degrees of pose variations on UPC database

Statistical measures	PSO-SVM			AAM-based FR		
	30°	45°	60°	30°	45°	60°
Sensitivity	96.67	93.33	90.00	80.00	80.00	60.00
Specificity	90.00	90.00	80.00	100.00	80.00	80.00
Precision	96.67	96.55	81.82	100.00	80.00	75.00
NPV	90.00	81.82	88.89	83.00	80.00	66.67
FPR	10.00	10.00	20.00	0.00	20.00	20.00
FDR	03.33	03.45	18.18	0.00	20.00	25.00
FNR	03.33	06.67	10.00	20.00	20.00	40.00
Accuracy	95.00	92.50	85.00	90.00	80.00	70.00
F1-score	96.67	94.92	85.71	88.89	80.00	66.67
MCC	86.67	80.81	70.35	81.65	60.00	40.82

strategy shows an accuracy of 92.5% and 80%, respectively, which is comparatively higher rate than AAM-based method.

The above figures show the analysis of the proposed PSO-SVM-based FR [29] and AAM-based method [30] on different statistical measures. Figure 8.7 shows the accuracy, Fig. 8.8 shows recall, Fig. 8.9 shows specificity, and Fig. 8.10 shows recall. From the above figures, it is clear that the proposed PSO-SVM performs better compared to AAM-based face recognition.

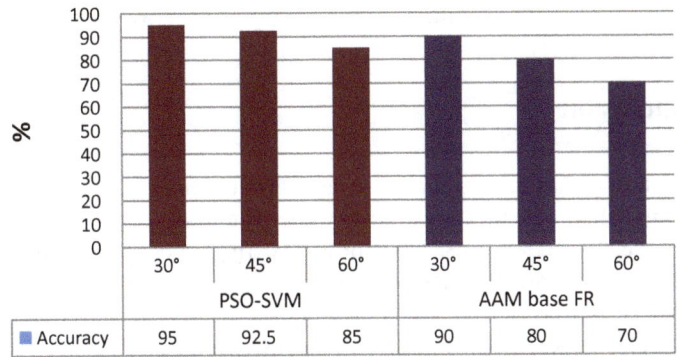

Fig. 8.7 Accuracy

Fig. 8.8 Sensitivity

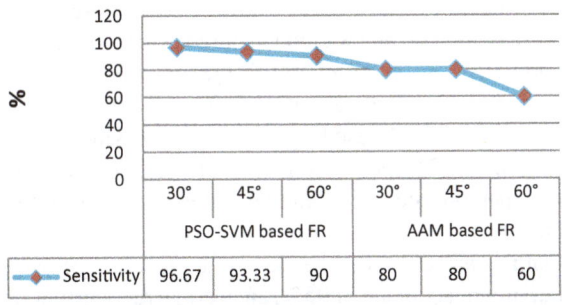

Fig. 8.9 Specificity

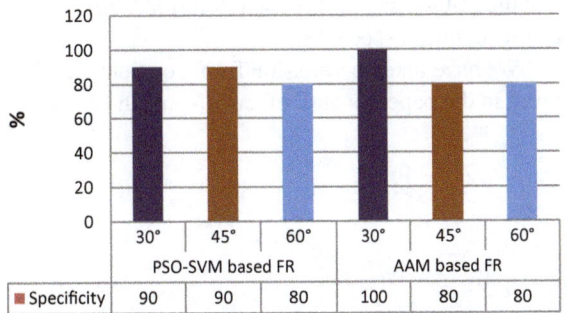

Fig. 8.10 Precision

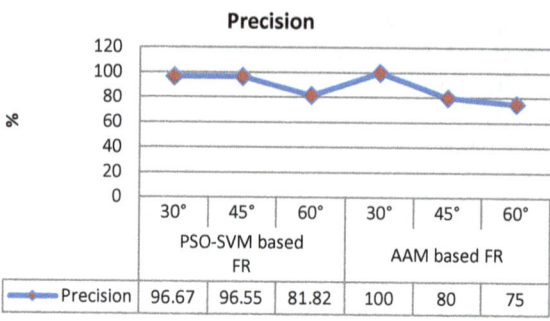

8.6 Conclusion

This research work has proposed a basic face recognition technique to perform the face recognition task with the least process value. The work is expected to illuminate the disadvantage of high process value of the existing recognition techniques. The exploratory outcomes are incontestable on the grounds that the projected technique outperforms the past work concerning both, the recognition accuracy and the other computational complexities including the computation time.

In the integrative approach (LLR-ANN), initially, the PCA and LDA features of each image sample are combined and represented in feature space efficiently. During the testing phase, the image is tested for the illumination problem and pose variation. Illumination is normalized with DCT and pose is normalized with LLR. In any of these two cases, any way the feature vector is obtained using the combination of PCA and LDA. The test result of this method reported a maximum accuracy of 93.5%, an average recognition rate of 71.64%, and a recognition speed of 8.13 s.

In PSO-SVM-based approach, a new video face recognition system has been introduced. The proposed PSO is utilized to optimize the parameters of the SVM. In this method, the facial features are extracted from the face image using both AAM and shape model, and then the same is represented in feature space. These stored features are utilized to recognize the face image using SVM. This method yields a maximum accuracy of 95%, an average recognition rate of 91.55% with a minimum processing speed of 0.03second. By comparing the three selected methods, it is clear that the PSO-based face recognition stands first both in terms of statistical measures and recognition speed.

"We have taken permission from competent authorities to use the images/data as given in the paper. In case of any dispute in future, we shall be wholly responsible."

References

1. Jain, A.K., Ross, A., Prabhakar, S.: An introduction to biometric recognition. In: Appeared in IEEE Transactions on Circuits and Systems for Video Technology, Special Issue on Image- and Video-Based Biometrics, vol. 14(1) (2004)
2. Delac, K.,Grgic, M.: A survey of biometric recognition. In: 46th International Symposium Electronics in Marine, ELMAR-2004, 16–18. Zadar, Croatia (2004)
3. Prabhakar, S., Pankanti, S., Jain, A.K.: Biometric recognition: security and privacy concerns. IEEE Sec. Privacy Magz. 1(2), 33–42 (2003)
4. Maltoni, D., Maio, D., Jain, A.K., Prabhakar, S.: Handbook of Fingerprint Recognition. Springer, NY (2003)
5. Jain, A.K., Bolle, R., Pankanti, S.: Biometrics: Personal Identification in Networked Society. Kluwer Academic Publishers (1999)
6. Turk, M., Pentland, A.: Eigenfaces for recognition. Cognitive Neurosci. 3, 72–86 (1991)
7. Kirby, M., Sirovich, L.: Application of the Karhunen-Loeve procedure for the characterization of human faces. IEEE Trans. Pattern Anal. Mach. Intell. 12(1), 103–108 (1990)
8. Zhao, W., Chellappa, R.: Robust face recognition using symmetric shapefrom-shading. Technical Report, Center for Automation Research, University of Maryland (1999)
9. Zhao, W., Chellappa, R., Phillips, P.J., Rosenfeld, A.: Face recognition: a literature survey. ACM Comput. Surv. (CSUR) 35(4), 399–458 (2003)
10. Wu, Y.-L., Jiao, L., Wu, G., Chang, E.Y., Wang, Y.-F.: Invariant feature extraction and biased statistical inference for video surveillance. In: Proceedings of IEEE Conference on Advanced Video and Signal Based Surveillance, pp. 284–289 (2003)
11. Ramanathan, N., Chellappa, R.: Face verification across age progression. In: Proceedings of IEEE Conference on Computer Vision and Pattern Recognition, vol. 2, pp. 462–469 (2005)
12. Chai, X., Shan, S., Chen, X., Gao, W.: Local linear regression (LLR) for pose invariant face recognition. EEE Trans. Image Process 16(7), 1716–25. https://doi.org/10.1109/tip.2007.899195
13. Portera, G., Doran, G.: An anatomical and photographic technique for forensic facial identification. Forensic Sci. Int. 114, 97–105 (2000)
14. Aggarwal, G., Roy-Chowdhury, A.K., Chellappa, R.: A system identification approach for video-based face recognition. Proc. Int. Conf. Pattern Recognit. 4, 175–178 (2004)
15. Zhou, S., Krueger, V., Chellappa, R.: Probabilistic recognition of human faces from video. Comput. Vis. Image Underst. 91, 214–245 (2003)
16. Stillman, S., Tanawongsuwan, R., Essa, I.: A system for tracking and recognizing multiple people with multiple cameras. In: Proceedings of International Conference on Audio and Video-Based Biometric Person Authentication, pp. 96–101 (1999)
17. Shermina, J.: Impact of locally linear regression and fisher linear discriminant analysis in pose invariant face recognition. Int. J. Comput. Sci. Netw. Sec. 10(10) 106–110 (2010)
18. Zhang, H., Qu, Z., Yuan, L., Li, G.: A face recognition method based on LBP feature for CNN. In: 2017 IEEE 2nd advanced information technology, electronic and automation control conference (IAEAC), pp. 544–547 (2017)
19. Venkatesan, S., Srinivasa Rao Madane, S.: Face recognition system with genetic algorithm and ANT colony optimization. Int. J. Innov. Manage. Technol. 1(5) (2010)
20. Xu, Y., Roy-Chowdhury, A., Patel, K.: Integrating illumination, motion, and shape models for robust face recognition in video. EURASIP J. Adv. Signal Process. p. 13 (2007)
21. Connolly, J.F., Granger, É., Sabourin, R.: An adaptive classification system for video-based face recognition, information sciences (In Press). https://doi.org/10.1016/j.ins.2010.02.026,march2010
22. Ng, H.U.: Pose-invariant face recognition security system. Asian J. Health Informat. Sci. 1(1), 101–111 (2006)
23. Chai, X., Shan, S.: Locally linear regression for pose-invariantface recognition. IEEE Trans. Image Process. 16(7), 1716–1725 (2007)

24. Arandjelovic, O., Cipolla, R.: A Methodology for Rapid Illumination-Invariant Face Recognition Using Image Processing Filters, pp. 159–171. Elsevier (2009)

25. Ding, C., Tao, D.: Trunk-branch ensemble convolutional neural networks for video-based face recognition. IEEE Trans. Pattern Anal. Mach. Intell. 1–14 (2017). https://doi.org/10.1109/TPAMI.2017.2700390

26. Wang, K., et al.: Facial Standardization Method in Video Face Recognition System (2016)

27. Liu, N.: Multiple Instance Learning with Deep Instance Selection for Video-based Face Recognition (2016). In: Dhiren, P., Jayesh, D. (eds.) PCA and LDA Method with Neural Network for Primary Diagnosis of Genetic Syndrome. International Advanced Research Journal in Science, Engineering and Technology, vol. 2, issue 10 (2015). https://doi.org/10.17148/IARJSET.201 5.210. ISSN (Online) 2393–8021

28. Shreekumar, T., Karunakara, K.: Face pose and illumination normalization for unconstraint face recognition from direct interview videos. Int. J. Recent Technol. Eng. (TM), **7**(6S4) 59–68 (2019). ISSN: 2277- 3878 (Online)

29. Shreekumar, T., Karunakara, K.: A video face recognition system with aid of support vector machine and particle swarm optimization (PSO-SVM). J. Adv. Res. Dyn. Control Syst (JARDCS) **10**, 496–507 (2018)

30. Prasanna, K.M., Rai, C.S.: A new approach for face recognition from video sequence. In: 2018 2nd International Conference on Inventive Systems and Control (ICISC), pp. 89–95. Coimbatore (2018)

31. Shreekumar, T., Karunakara, K.: Active appearance model based for face recognition from surveillance video. Test Eng. Manage. **83**, 6969–6981 (2020). ISSN:0193–410

32. Shreekumar, T.,Karunakara, K.: Face pose and blur normalization for unconstraint face recognition from video/still images. Int. J. Innov. Comput. Appl. Indersci (2020). ISSN:1751–648X

33. Shreekumar, T., Karunakara, K.: Identifying the faces from poor quality image/video. Int. J. Innov. Technol. Exploring Eng. (IJITEE). **8**(12), 1346–1353 (2019). ISSN: 2278–3075.

34. Amrutha, K., Shreekumar, T.: Instant warning system to detect drivers in fatigue. Int. J. Sci. Res. (IJSR), **4**(2), 791–79 (2015)

Chapter 9
Application of Bat Optimization Algorithm for Power System Loss Reduction Using DSSSC

M. Thirumalai and T. Yuvaraj

Abstract A novel method is proposed for finding the optimal location along with the sizing of DSSSC in order to mitigate the power losses of the system. DSSSC sitting and sizing are determined by bat algorithm. Testing of the pro-posed work is done on the IEEE 69-bus which is standard system of radial bus distribution. The results obtained shows that the placement of optimal point of DSSSC with its sizing has reduced the power losses of the total radial distribution network system.

9.1 Introduction

The power system consist of three sections namely generation of power, transmission of power and distribution of power, among which resistance to reactance ratio is higher in the case of distribution which results in instability of the voltage and more power loss. The conventional methods for the maintenance of acceptable voltage stabilization in the distribution system use series voltage regulator and shunt capacitor [1–3]. But there are some demerits with these conventional methods as reactive power generation is not possible by the series voltage regulator and operates step by step leading to slower response [4]. Also variable reactive power is not generated by the shunt capacitors continuously. The behavior of the distribution capacitors are oscillatory when used along with inductive components within the same circuit is also a problem to be focused. Introduction of devices based on the concept of FACTS resolves the defects due to series voltage regulators and the shunted capacitors. The FACTS concept was initially introduced for transmission systems but later applied also for distribution systems to mitigate the system losses, power factor correction and for improved voltage of the power system.

M. Thirumalai · T. Yuvaraj (✉)
Department of Electrical and Electronics Engineering, Saveetha School of Engineering, Saveetha Institute of Medical and Technical Sciences, Chennai, India

M. Thirumalai
Department of Electronics and Communication Engineering, Saveetha Engineering College, Chennai, India

S. C. Satapathy et al. (eds.), *Smart Computing Techniques and Applications*,
Smart Innovation, Systems and Technologies 225,
https://doi.org/10.1007/978-981-16-0878-0_9

A synchronous generator of static type namely static synchronous series compensator (SSSC) operates without electrical energy as a controllable voltage is injected by the series compensator using a VSC where the inoculated voltage is in quadrature with the power network's route current. Hence rapid impedance compensation is provided both capacitive and inductive by the device without depending on the line current of the power network. DSSSC is comparatively better than conventional compensation devices for reactive power as it has various features namely modification of power factor by inoculation of voltage continuously together with organized regulator, balancing of load in distributed organized systems, covering the power demand both reactive & capacitive, control of power flow, reduced harmonic distortion by the process of active filtering [5, 6]. Designing SSSC based controller based on the technique of GA for improving the transient performance using the approach of multi-objective optimization in power system is done in [7]. Sizing of SSSC controller is done by optimized method in the network of transmission [8].

Reduction of transmission losses is done by the PSO technique which modifies the algorithm of Newton—Raphson load flow improving the stability using SSSC damped controller for minimizing the objection function based on the time domain. Oscillation damping to power system is done efficiently to develop the bus voltage [9]. In the system of radial distribution, locating the optimum point of DSTATCOM has greater influence. If the placement is done improperly then the benefits can be reduced which will still endanger the operation of the entire system [10].

So far only one researcher used DSSSC in distribution system for loss mitigation. The approach of particle swarm algorithm is implemented in [12] which identify the appropriate allocation of DSSSC with reduced power loss as objective function was investigated.

The proposed method implements bat optimization algorithm for identifying the best location of DSSSC along with its sizing in order to mitigate the total system losses in the RDS. Testing of the present method is done on IEEE 69-bus on RDS. Evaluation of the outcomes acquired from the present approach with that of the already available approaches is done and found to be better.

9.2 Formulation of the Problem

9.2.1 Analysis of Load Flow

The conventional methods for studying the load flow such as GS method, NR technique and the method of Decoupled load flow are not suited for obtaining the line flow & voltages in the system of RDS as the ratio of R/X is higher. A straight methodology for the power flow for the distribution system is discussed in [13]. A simple distribution system is represented by the single line diagram (see Fig. 9.1).

At node t, the current injected is obtained from Fig. 9.1 as follows

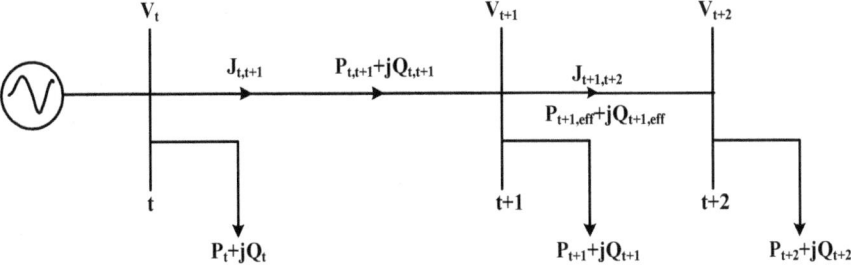

Fig. 9.1 One line diagram

$$I_t = \left(\frac{P_t + jQ_t}{V_t} \right)^*$$ (9.1)

The branch current is determined using KCL between the nodes t & $t + 1$ from Fig. 9.1 which is obtained as,

$$J_{t,t+1} = (I_{t+1}) + (I_{t+2})$$ (9.2)

By utilizing the Bus Injected to Branch Current matrix, the Eq. (9.2) is signified in matrix form as below.

$$[J] = [I][\text{BIBC}]$$ (9.3)

The bus voltage at $t + 1$ is determined using KVL from the Fig. 9.1 which is obtained as,

$$V_{t+1} = V_t - \left[J_{t,t+1}(R_{t,t+1} + jX_{t,t+1}) \right]$$ (9.4)

The real power loss and imaginary power loss between the nodes t & $t + 1$ in the line section can be obtained as

$$P_{\text{Loss}(t,t+1)} = \left(\frac{P_{t,t+1}^2 + Q_{t,t+1}^2}{|V_{t,t+1}|^2} \right) * R_{t,t+1}$$ (9.5)

$$Q_{\text{Loss}(t,t+1)} = \left(\frac{P_{t,t+1}^2 + Q_{t,t+1}^2}{|V_{t,t+1}|^2} \right) * X_{t,t+1}$$ (9.6)

In the RDS, the average system loss is obtained by adding the losses occur-ring in all the line section as follows,

$$P_{T\,\text{Loss}} = \sum_{t=1}^{b} P_{(\text{Loss},t,t+1)}$$ (9.7)

9.2.2 Power Loss Mitigation

The aim of allocation of DSSSC in the RDS is to mitigate the losses while satisfies the constraints of equality and in equality. The main aim of the work is given by

$$\text{Minimize}(F) = \text{Min}(P_{T\text{ Loss}}) \tag{9.8}$$

9.3 Proposed Bat Algorithm

9.3.1 Overview of Bat Algorithm

The most power algorithm for the optimization problems of tough power system are the nature inspired algorithms proposed by Yang [11]. Bats are the only mammals with wings have the special feature of finding their prey with advanced capability of echo location. The objects in the surrounding are detected by the radiation of the sound signal known as echo location so that way is found even when it is completely dark.

Development of Bat algorithm is idealized by the appearances of bats. The three rules romanticized are assumed as:

1. The characteristic of echo location is utilized by each bat to sense the distance with the knowledge of differentiating prey and other obstacles by nature.
2. The motion of bat is random with velocity u_i, position y_i, frequency $g_{\min}$, variable wavelength λ & loudness B_0 for seeking the prey. Bats have the ability of regulating the frequency of the emitted pulse also to regulate the pulse emission rate r within the array of [0, 1].
3. Though there is possibility for the variation in loudness, assumption is done that the loudness differs from positive great value B_0 to a constant least value $B_{\min}$.

9.3.2 Population Initialization

Initially for the BA population is produced arbitrarily hence the no of virtual bats are randomly selected. The population of the simulated bats can be between 10 & 40. Once after the fitness done initially, updating of the values of pulse rate, loudness, movement is performed for a given function.

9.3.3 Motion of Virtual Bats

The rules are defined in bat algorithm for updating the velocity u_i and position y_i of the simulated bats which are assumed as follows:

$$g_i = g_{min} + (g_{max} - g_{min})\beta \tag{9.9}$$

$$u_i^t = u_i^{t-1} + (y_i^{t-1} - y^*)g_i \tag{9.10}$$

$$y_i^t = y_i^{t-1} + y_i^t \tag{9.11}$$

where the random vector $\beta \in [0, 1]$ obtained from uniform distribution where y^* is the best location indicating the current solution out of the "n bats".

The innovative result generated locally for all the bats with arbitrary walk is specified by the Eq. (9.12).

$$y_{new} = y_{old} + \varepsilon A^t \tag{9.12}$$

where ε is the arbitrary no in the assortment $[0, 1]$, while $B^t = (B_i^t)$ is the middling loudness of all the bats at this time step.

9.3.4 Emission of Pulse and Loudness

The loudness B_i is proceeded based on the iteration with pulse emission rate r_i is updated when a bat is nearing its prey, the volume reductions with increase in the rate of pulse production hence the convergence equation is given as

$$B_i^{t+1} = \alpha B_i^t \tag{9.13}$$

$$R_i^{t+1} = r_i^0[1 - \exp(-\gamma t)] \tag{9.14}$$

where α and γ are continual values.

For any instant of "$0 < \alpha < 1$" and $\gamma > 0$ we have.

$B_i^t \to 0, r_i^t \to r_i^0$ as $t \to \infty$.

The starting value of loudness B_0 can characteristically be in the sort of $[0, 1]$, while the starting value of emission rate r_i can be in the sort of $[0, 1]$.

Bat Algorithm is implemented step by step based on the approximation discussed above as follows:

Step 1: The input data namely the bus data and load data are read by the system.

Step 2: The load flow distribution of the base case is run to determine to the reactive power loss, real power loss along with the voltages.

Step 3: The applicant buses are identified for the location of DSSSC.

Step 4: The control parameters of the bat algorithm such as loudness, pulse rate, pulse frequency, maximum number of iterations are set along with the inferior and higher bound of the constraints.

Step 5: The population of bat initially is generated in a random manner. The optimal size of each bat is encouraging (p.u) for the devices of DSSC in the distributed network.

Step 6: The suitability function is evaluated. Calculation of the expected reactive power losses, active losses and the objective function voltage deviance is done for each bat or solution using the method of direct load flow.

Step 7: The bat with minimum value of power loss is selected as the best bat among the population.

Step 8: The population of bat is updated.

Step 9: The load flow is run for determining the unknown true and imaginary power loss values.

Step 10: The criteria for closure is checked which can be extreme no of repetitions for updating the populace of the BA or the value for which the OF will be least. Once the condition is satisfied the algorithm ends else returned to step 5.

Step 11: Optimal solutions are displayed.

The above steps in the bat algorithm are followed for minimizing the objective function.

9.4   Discussion on Results

9.4.1   Test System for IEEE 69-Bus

The present method shows performance proving its effectiveness by testing on standard IEEE 69-bus system. The test system involves of 69 buses & 68 divisions hence a large scale system with a total reactive load of 2.69 MVAr and total real load of 3.80 MW respectively. The IEEE 69-bus radial distributed system is displayed with its one line illustration in Fig. 9.2. The proposed method is simulated using MATLAB which runs the MATLAB code of the RDS to calculate the reactive power loss and the real power loss for identifying the optimum sitting of DSSSC and the sizing of DSSSC. In this work, the optimal installation of DSSSC is done by selecting the 61st bus system.

The comparison of the locations, real power losses, voltage values of the existing methods (p.u) & the present approach is done in Table 9.1. The proposed method has reduced the real power loss to 24 kW which indicates 89.39% of mitigation of power loss after the installation of DSSSC. This mitigation in power loss is high compared to that of the existing methods. The IEEE 69-bus bus voltage diagram with DSSSC

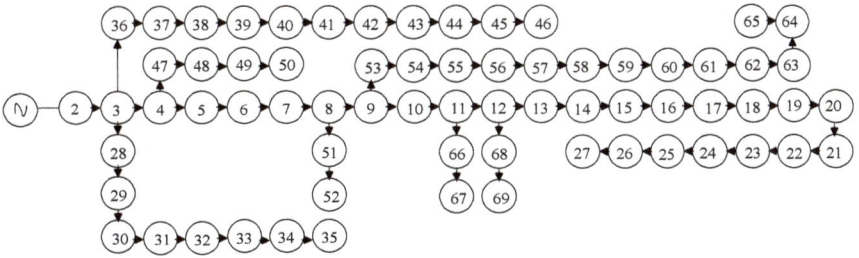

Fig. 9.2 One line diagram of IEEE 69-bus system

Table 9.1 The results of DSSSC allocation for a 69-bus system

	Uncompensated	Analytical method	PSO[12]	BA
KVAr (p.u)	–	1.9229	2.1377	2.0737
Location	–	60	60	61
Power loss (kW)	225.4	125.5	37	24
% of power loss reduction	–	44.32	83.58	89.35
Voltage (p.u)	0.9196	1.0	1.0065	1.0

& without DSSSC is shown in Fig. 9.3. Effective improvement is viewed in the bus voltage diagram of the RDS as implemented with DSSSC. Hence optimization based on the present BA approach is efficient related to the prevailing approaches.

Fig. 9.3 Bus voltage diagram improvement for 69 bus system

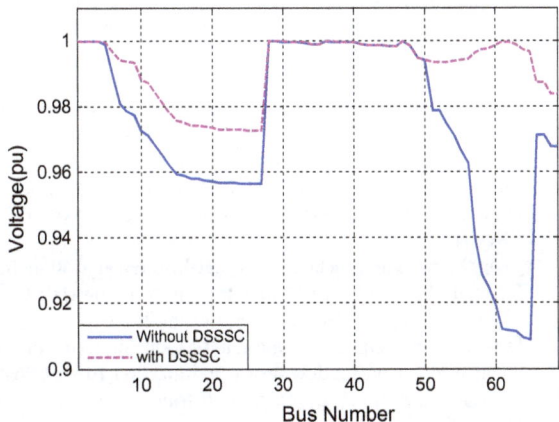

9.5 Conclusion

A new BA based approach is presented in this paper which finds the optimal location of DSSSC along with its sizing of the RDS. Testing of the present approach is done on standard IEEE 69—bus system. Proper location and sizing of the DSSSC by present approach ensures the greater loss mitigation. The simulated results show that the implementation of the proposed method with DSSSC has reduced the total real and imaginary power losses better than the existing methods. Bat algorithm is accurate in finding the solution for optimal location which can be applied for n no of buses.

References

1. Casillas, G.I., et al.: Voltage regulators, capacitor banks and distributed resources allocation in a distribution network system. In: 2017 IEEE PES Innovative Smart Grid Technologies Conference-Latin America (ISGT Latin America). IEEE (2017)
2. Sarfi, V., Livani, H.: Optimal Volt/VAR control in distribution systems with prosumer DERs. Electr. Power Syst. Res. **188**, 106520 (2020)
3. Xie, Q., et al.: Use of demand response for voltage regulation in power distribution systems with flexible resources. IET Generat. Trans. Distribut. **14**(5), 883–892 (2020)
4. Yuvaraj, T., Devabalaji, K.R., Thanikanti, S.B.: Simultaneous allocation of DG and DSTATCOM using whale optimization algorithm. Iranian J. Sci. Technol. Trans. Electri. Eng. **44**(2), 879–896 (2020)
5. Hingorani, N.G., Gyugyi, L.: Understanding FACTS; Concepts and Technology of Flexible AC Transmission Systems. IEEE Press book (2000)
6. El-Zonkoly, A.: Optimal sizing of SSSC controllers to minimize transmission loss and a novel model of SSSC to study transient response. Electr. Power Syst. Res. **78**(11), 1856–1864 (2008)
7. Sidhartha, P.: Multi-objective evolutionary algorithm for SSSC-based controller design. Electr. Power Syst. Res. **79**, 937–944 (2009)
8. El-Zonkoly, A.: Optimal sizing of SSSC controllers to minimize transmission loss and a novel model of SSSC to study transient response. Electr. Power Syst. Res. **78**, 1856–64 (2008)
9. Panda, S., Padhy, N.P., Patel, R.N.: Power-system stability improvement by PSO optmized SSSC-based damping controller. Electr. Power Compon. Syst. **36**, 468–90 (2008)
10. Yuvaraj, T., Ravi, K., Devabalaji, K.R.: DSTATCOM allocation in distribution networks considering load variations using bat algorithm. Accepted on 26 September 2015. Ain Shams Eng. J. Elsevier
11. Yang, X.-S.: A new meta heuristic bat-Inspired algorithm. In: Gonzalez, J.R. et al. (eds.) Nature Inspired Cooperative Strategies for Optimization (NISCO 2010). Studies in Computational Intelligence, pp. 65–74. Springer, Berlin (2010)
12. Devi, S., Geethanjali, M.: Optimal location and sizing of distribution static synchr nous series compensator using particle swarm optimization. Electri. Power Energy Syst **62**, 646–653 (2014)
13. Jen-Hao, T.: A direct approach for distribution system load flow solutions. IEEE Trans. Power Deliv. **18**(3), 882–887 (2003)

Chapter 10
Malaria Cell Detection Model

Prasannajeet Bajpai and Sanchita Chourawar

Abstract Malaria is the most ubiquitous endemic and a deadly disease which is caused by parasites which are transmitted to people through bites of an infected female Anopheles mosquito. This disease is present in several continents, which is a matter of concern for several bodies and associations. All the governments around the world are trying to reduce this morbidity by taking various preventive measures. Model will help us to detect five species of Plasmodium (Malaria Parasites) that have long been recognized to infect humans. Now with our model which is a pre-trained CNN model with 13,775 images of infected cells and 13,775 images of uninfected cells with valid efficiency more than 93% and then all it is needed is just upload image or images that we want to test which then will be fed to our mode and then the model will give the output with concerning images in the same sequence in which it was uploaded. By using this model, we are primarily promoting reusability, accessibility, cost-efficient, eco-friendly as we are not using any kits to create more plastic waste.

10.1 Introduction

Malaria is caused by Plasmodium parasites. These parasites spread to people by the bite of infected female Anopheles mosquitoes also called to as "malaria vector." Five parasites species which are found in humans, and a couple of those species are— *P. falciparum* and *P. vivax* which pose a great threat. With over quite 200 million cases per annum and most of its cases are from remote or rural areas of sub-Saharan Africa South-East Asia, Eastern Mediterranean, Western Pacific, and therefore, the Americas where immediate testing is not always possible but by an invention of 20% "hand-powered blood centrifuge" by Manu Prakash which can reach up to 125,000 rpm enough to separate plasma from blood after which we can directly capture an image of an individual cell and then feed it to our model after which the model will give us the result. There are only three essential steps for the implementation of this model:

P. Bajpai · S. Chourawar (✉)
Government Engineering College Bilaspur, Bilaspur, India

© The Author(s), under exclusive license to Springer Nature Singapore Pte Ltd. 2021   95
S. C. Satapathy et al. (eds.), *Smart Computing Techniques and Applications*,
Smart Innovation, Systems and Technologies 225,
https://doi.org/10.1007/978-981-16-0878-0_10

(a) Collect and Arrange Data:

- First, we will gather multiple images of parasitic and uninfected cells.
- Now we classify them into two categories "train" and "test."
- Then inside both test and train, we will keep parasitic and uninfected cells images which will be used to rain our model properly.

(b) Training the Model:

- Before doing anything, we will need to import all important libraries like shutil, TensorFlow, matplotlib.
- Now after having all the libraries imported, we will have to make our model which will be "keras.models.Sequential" in this case.
- And then we will optimize it with RMSprop as an optimizer.
- Then the final step is to fit this model which gives use of valid-accuracy more than 94.

(c) Using the Model:

- Now, all we need to do is just upload the image.
- Then we will get results corresponding to our images in the same order in which we uploaded.

10.2 Literature Review

ImageNet classification which is a set of a large number of images and with the help of deep convolutional neural network this is where we can say that it all started [1]. Albeit deep learning has been around since the 70s with AI heavyweights Geoff Hinton, Yann LeCun, and Yoshua Bengio giving all their time and efforts on convolutional neural network until AlexNet (2012) brought deep learning into the mainstream where he builds an easy model to differentiate between cat, dog, and duck [2, 3]. Usage and hindrance of malaria rapid diagnostic testing (2009) within the war-torn The Democratic Republic of Congo made a search paper where they concluded that CHWs will be allowed to use RDTs for the management of febrile children safely and effectively; but, the cost-efficiency of RDTs is restricted in the area of high malaria prevalence [4, 5]. Multimodal Semantic Division: The fusion of RGB and Depth Data in CNN research by Zoltan Kopanyi, Dorota Iwaszczuuk, Bing Zhao, Can JozefSaul, Charles K. Toth, Alper Yilmaz (2019) where they showed how an RGB image goes through a number of phases it goes through where it shows Feature Extraction, Thresholding, Edge Detection, and many more which are important for classification of an image [6]. From the research paper to deep learning model in Keras by using python for image segmentation by Peterfitchcsiro (2020) where we use convolutional neural network to take input and then pass it through multiple layers of Conv2D and MaxPooling2D to create a proper sequential model [7, 8]. Malaria: The Past and the Present by Jasminka, Talapko, Ivana Skrlec, Tamara Alembić, Melita,

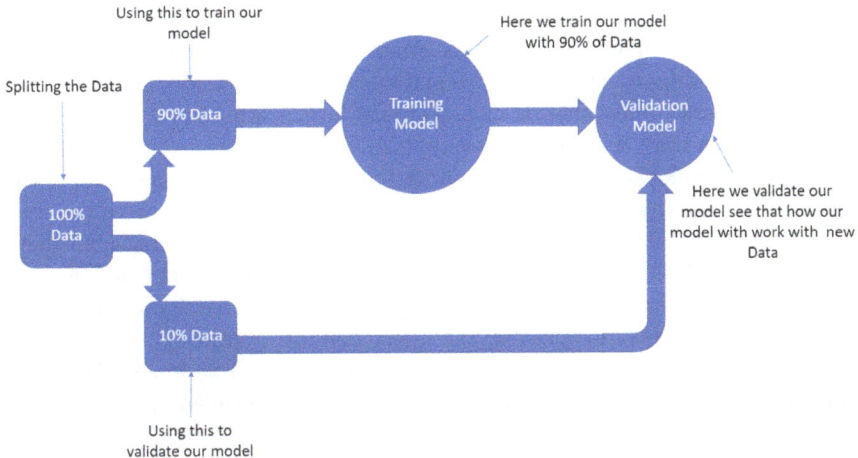

Fig. 10.1 Flowchart of "Training and Validating Model"

Jukić, Aleksandar, Vćev by (2019) which escalated our knowledge about this disease [9]. Having an alternative solution like using a pre-trained CNN model will help us overcome many odds like reducing the use of plastic which is a promising step to a future generation, providing easy access to testing as we need only a microscope and a camera to capture the image of a cell and the camera could also, be of a smartphone which would make it easier for anyone to access it an as large number of the population has access to smartphones and the government also making huge efforts to make them available to everyone, with such a high efficiency we can easily rely on models like this and in near future, we can gather more data and then we won't just be using it to detect malaria in fact when can use it detect multiple diseases [10] (Fig. 10.1).

10.3   Architecture and Working

Now moving any further, we want to show how we trained our data and how exactly it works initially we collected a set of data where we have thousands of images of parasitic cells and uninfected cells. Now we use a simple split function where we will be splitting the data into two parts where one will be a training dataset and another one will be for testing or validating our model which gives a proper picture of how our model will perform with new data.

Further when it comes to the model building which here in the case will be a "Keras.models" and hyperparameter is not given meticulous attention in most of the frameworks in machine learning, but intuitively they govern the underlying system on a "High Level" than the primary parameters of interest and our CNN model consists of 3 Conv2D Layers, 3 2DMaxPooling Layers, 2 Dense Layers, and a Flatten Layer.

Conv2Dlayer produces a convolution kernel that's involved in the layer input to build a tensor of outputs. If use bias is True, a learning vector is formed and is attached to the outputs. The primary requirement within the Conv2D parameter is the sum of filters from which the convolutional layer learns. Layers early within the specification learn lesser from convolutional filters while from the layers deeper within the network (i.e., near to the output predictions), it will study better with filters. Conv2D layers in middle of this will grasp more information from this filter than the first Conv2D layers but fewer filters than the layers closer to the output that's why we would like to use multiple layers during a model for correct working.

2D MaxPooling reduces the dimensions of the data, the quantity of parameters, the quantity of computation needed, and it also controls overfitting. You'll remark that 2D MaxPooling is analogous to lower the dimension of an image. This layer outputs a smaller tensor than its input, which suggests downstream layers will need fewer parameters and amount of computation; it also serves to regulate overfitting. It has five parameters which are used to:

- Horizontal Pooling Factor
- Vertical Pooling Factor
- Horizontal Stride
- Vertical Stride
- Padding.

Flatten Layer is used to flatten the input as we applied for an input batch size of (2,2) which will be converted to a batch size of 4 what a flatten layer is that it converts our CNN to DNN for efficient training (Fig. 10.2).

Dense layer is one of the most frequently used layers and is also known as "regular deeply connected neural network layer."

The optimizer is one of the most important hyperparameters for tweaking weight of the model and gets pretty optimized results for our model. We have a variety of optimizers for optimizing but here we have used RMSprop with lr = 0.001 but in options, we also have Adagrad, SGD, Nesterov Momentum, Adam, BFGS, etc.

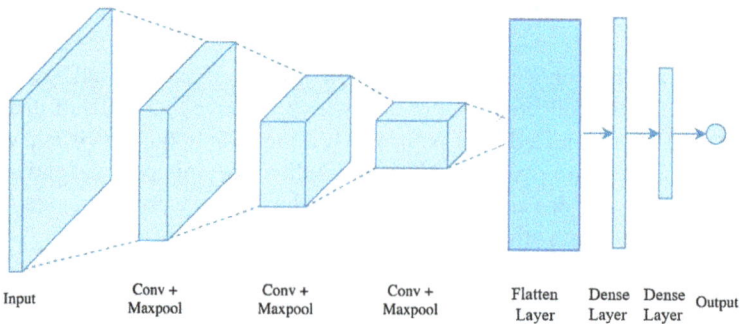

Fig. 10.2 Flowchart of sequential model

```
[ ]  history = model.fit(train_generator,
                         validation_data=validation_generator,
                         steps_per_epoch=100,
                         epochs=5,
                         validation_steps=50,
                         verbose=2)

     Epoch 1/5
     100/100 - 8s - loss: 0.7438 - accuracy: 0.5365 - val_loss: 0.6422 - val_accuracy: 0.6460
     Epoch 2/5
     100/100 - 8s - loss: 0.6597 - accuracy: 0.6070 - val_loss: 0.6695 - val_accuracy: 0.5550
     Epoch 3/5
     100/100 - 8s - loss: 0.5551 - accuracy: 0.7285 - val_loss: 0.3889 - val_accuracy: 0.8470
     Epoch 4/5
     100/100 - 9s - loss: 0.3342 - accuracy: 0.8750 - val_loss: 0.2639 - val_accuracy: 0.8980
     Epoch 5/5
     100/100 - 8s - loss: 0.2487 - accuracy: 0.9200 - val_loss: 0.2039 - val_accuracy: 0.9340
```

Fig. 10.3 Fitting the model

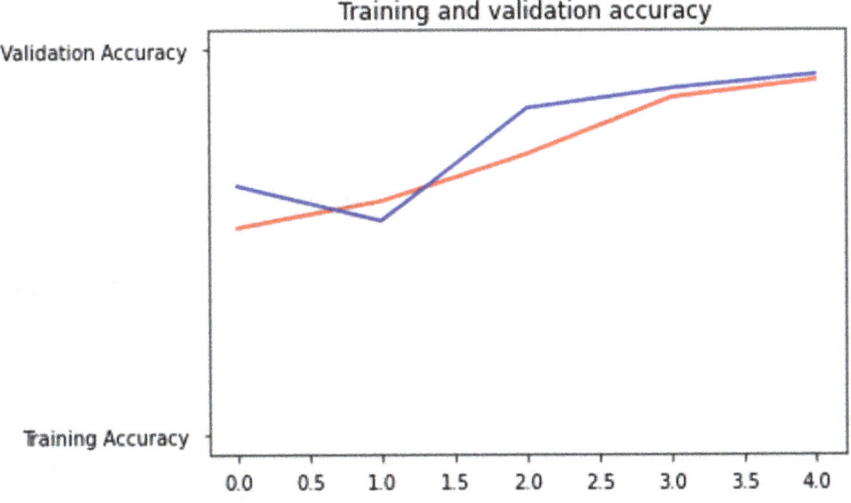

Fig. 10.4 Graphical representation of "Training and Validation Accuracy" throughout epoch cycles

Then we will have to fit the model in which steps per epoch is been set as 100 with epochs as 5 and 50 validation steps with 2 verbose, after 5 epochs cycles, we got a model with + 94% in its validation (Fig. 10.3).

Here to accomplish such a high accuracy at such a less time is possible due to a high GPU utilization and as we build this model on an online platform so we were gifted with a high-end GPU card "GTX TITAN X" which has more than 3000 cores which is a lot to work with these number of data (Figs. 10.4, 10.5, 10.6 and 10.7).

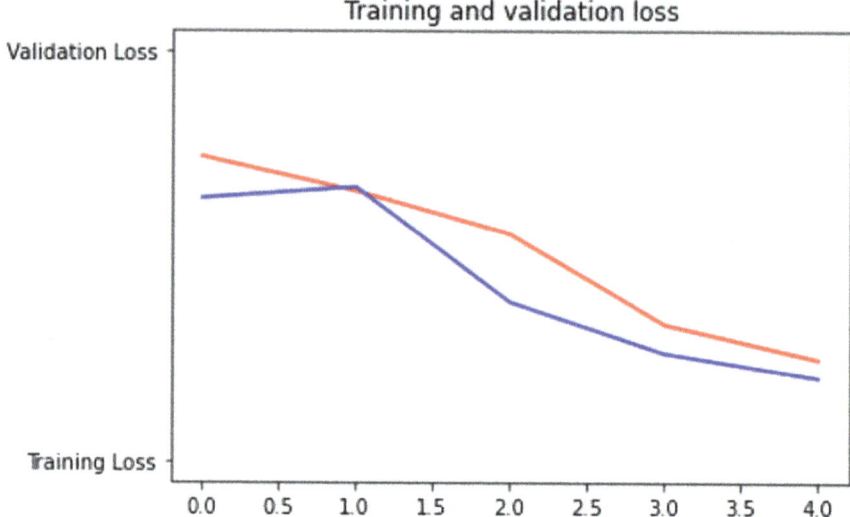

Fig. 10.5 Graphical representation of "Training and Validation Loss" throughout epoch cycles

Fig. 10.6 Snap of cell images

```
Choose Files  4 files
• C4.png(image/png) - 14451 bytes, last modified: 01/07/2020 - 100% done
• C3.png(image/png) - 13727 bytes, last modified: 01/07/2020 - 100% done
• C2.png(image/png) - 8568 bytes, last modified: 01/07/2020 - 100% done
• C1.png(image/png) - 8536 bytes, last modified: 01/07/2020 - 100% done
Saving C4.png to C4 (1).png
Saving C3.png to C3 (1).png
Saving C2.png to C2 (1).png
Saving C1.png to C1 (1).png
[0.]
C4.png is a parasitic
[0.]
C3.png is a parasitic
[1.]
C2.png is uninfected
[1.]
C1.png is uninfected
```

Fig. 10.7 Output of our model after prediction

10.4 Result Analysis

Now we come to the final step where we will put some new untrained images.

Now after successfully uploading images, our model will use its pre-trained data and will display us the output as shown below.

Now deploying this model is very easy all we have to do is just convert it into either .h5 or tflite model. Tflite is always a better option if we are trying to make a model which is to be made pretty portable because of the size difference between .h5 model and tflite model, and tflite model be implemented in android applications also make our model more effortless and convenient for use.

10.5 Conclusion

By implementation of this model, we are able to make a new tech for the detection of malaria disease and also an alternative for the kits which are traditionally used. Our model has lots of merits because with our we are be able to achieve more accurate results as some of the kits are faulty or damaged. It is also reusable and eco-friendly as we do not have to litter kit's waste. By implementing our model in an android device or any other micro-controller, we can increase its compatibility and mobility. Our model is just one-time investment as we do not have to buy any equipment. It is also expandable because currently, we are working with just malaria cells but in future we might be able to detect multiple diseases making our model more versatile with some more dataset.

Acknowledgements In the present world of completion, there is a contest of living in which those that have the desire to come forward will only succeed, and with a model like this, we construct a bridge between theoretical and practical working. Like there is a saying "you never know your limits unless you push them," and this project made me realize my limits about which I am pretty sure that I will be able to push it further later. A teacher is a compass that activates the magnets of curiosity, knowledge and wisdom in their students and I am so fortunate that I have Professor Sanchita Chourawar (Assistance Professor, Department of Computer Science and Engineering). A special thanks to our Head of Department Professor Sourabh Yadav (Department of Computer Science and Engineering) for his constant cheer up and wholehearted support. I am deeply indebted and that I would really like to manifest my sincere gratitude to our honored Head of Institute Dr. B. S. Chawla (Principal, Government Engineering College, Bilaspur) for providing me with an opportunity to do this project.

References

1. AlexNet.: ImageNet classification with the help of CNN **115**, 11–97 (2012)
2. Community health employee of democratic republic of congo for usage and hindrance of malaria rapid diagnostic testing. Malaria J. **8**(308) (2019)
3. Hinton, G., LeCun, Y., Bengio, Y.: Deep learning. Nature **521**(7553), 436–44 (2019)

4. Tseng, C., Lee, S.: A supervised learning method for the design of linear phase FIR digital filter using Keras, pp. 1–2 (2019)
5. Talapko, J., Skrlec, I., Alebić, T., Jukic, M., Véev, A.: Malaria: the past and the present **7**(6):179 (2019)
6. Hinton, G.E., Krizhevsaky, A., Srivastava, N., Sutskever, I., Salakhuttdinov, P.R.: Dropout an easy method to avoid neural networks from overfitting **15**(56), 1929–1958 (2014)
7. Peer, J.: Image processing in python **19**(2), e453 (2014)
8. Jai, X.: Image recognition method based on deep learning. In: CCDC (2017)
9. Lina, B.: Deep Learning with Keras (2017)
10. Gajurel, A., Louis, S.J.: GPU acceleration in sparse neural networks (2020). arXiv:2005.04347 [cs.DC]

Chapter 11
Design and Analysis of Fractal Monopole Antennas for Multiband Wireless Applications

Siddhi Oja and P. M. Menghal

Abstract A design of fractal monopole patch antenna using metamaterial-based substrate and Minkowski fractal has been presented in this paper. The antenna has been designed for a frequency range from 1 to 6 GHz which is mainly covering the wireless fidelity and Wi-Max ranges. Antenna parameters like VSWR, return loss, and bandwidth have been analyzed for the proposed antenna. The overall results of proposed antenna in design after carrying out parametric study are found: gain above 4 dB, return loss $(S-11)$ -37.11 dB $(f_L = 0.3)$ and -59 dB $(f_L = 0.2)$, VSWR -1.02, and bandwidth -77.01%. The simulated characteristics of the antenna along with the 3D radiation patterns and gain are presented and discussed by using CST simulation software.

11.1 Introduction

An antenna is an indispensable device for any wireless communication system. Swift development in satellite and wireless communication mainly in commercial as well as government communication systems in the last few years has compelled more and more requirement of developing low value, low profile, minimal weight, and broadband antennas that are capable of maintaining high performance over an extensive range of frequencies. The most common version of a printed antenna is a microstrip antenna or patch antenna. In this type of antenna, a metallic patch is designed in coordination with ground and its dimensions are appropriately chosen as per the requirement of operability in the particular frequency domain. These antennae are more promoted because of their prostrate cost, simple fabrication, and effortless integration with circuit components [1]. The most exclusive attribute of these antennas is that they are conformal to a planer and non-planer surfaces, hence can be easily mounted on curvy surfaces as of aircraft, missiles, space crafts, satellite systems

S. Oja · P. M. Menghal (✉)
Faculty of Electronics, Military College of Electronics and Mechanical Engineering, Secunderabad 500015, Telangana, India
e-mail: prashant_menghal@ieee.org

© The Author(s), under exclusive license to Springer Nature Singapore Pte Ltd. 2021
S. C. Satapathy et al. (eds.), *Smart Computing Techniques and Applications*,
Smart Innovation, Systems and Technologies 225,
https://doi.org/10.1007/978-981-16-0878-0_11

radars, and mobile devices [2–4]. Also, it is highly versatile in terms of polarization, pattern, and impedance, once an operable mode of patch and form is chosen as per various requirements. The versatile designs that can be achieved with microstrip antenna in all probability exceed that of the other kind of antenna elements [4–8]. In this paper, a flexible monopole fractal antenna is designed using flexible Rogers RO4003C as a substrate. Minkowski fractal geometry is used as a base, which is modified to design the new geometry of the antenna.

11.2   Fractal Monopole Antennas

Fractal is a technology that can lead to antenna miniaturization, multifrequency, and ultra-wideband applications, and this paper is talking about the same. The existing communication system is evolved to broadband and integration, but the people's need for portable mobile communication has increased. This needs antenna development depending upon the broadband technology, multifrequency, and miniaturization technology. Multifrequency antenna is gauging attention, owing to its small volume, lighter in weight, easy, and active circuitry integration advantages, especially due to the emerging wireless communications [2–4]. For antenna operating on different frequencies, with small size, operable in wideband and has a circular polarization, such technology will be the domestic and foreign research key areas [5, 6]. In times to come, communications will lead to some serious technical challenges in the field of antenna design, demanding a miniaturized structure to execute its tasks. The fractal antenna can cater to the need for antenna requirements necessary for modern communication devices with a very thin section, small size, and is easy to manufacture and cheap price [7, 8]. Fractal is produced with the self-similarity of fractal dimension structure through the number of iterations. It will not only lead to reducing the dimensions of the antenna, but also strengthened the characteristics of a directional antenna. Minkowski fractals are given by Hermann Minkowski in the form of representation and definition of geometries in the year 1885 [9, 10]. The square patch geometry on being further compressed into five smaller squares leads to the formation of the Minkowski fractal geometry given in Fig. 11.1. This is explained in terms of affine transformation here.

11.3   Antena Design

The design has been evolved after a series of steps starting from designing a simple patch antenna to a monopole patch antenna, followed by applying Minkowski fractal and finally modifying the Minkowski geometry to achieve the final design. The antenna parameters taken into consideration to study the results of the various designs are return loss, VSWR, and gain. The antenna design equation is given in Appendix

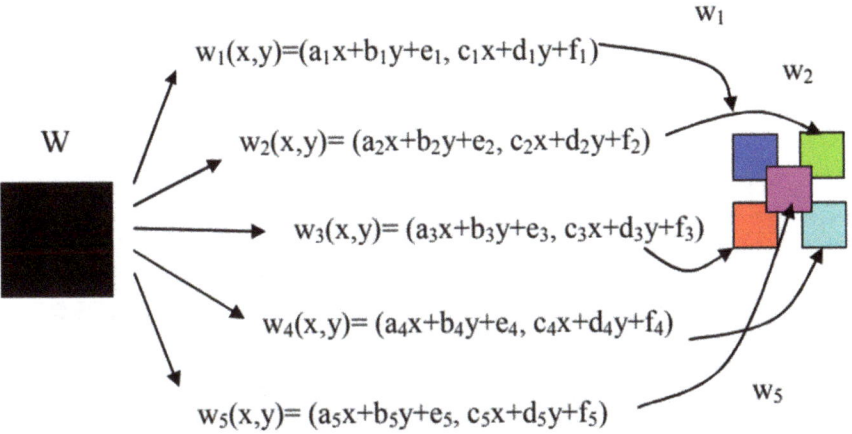

Fig. 11.1 Generation of Minkowski fractal geometry

'A' The design specification of the proposed antenna based on the calculation as mentioned above is given in Table 11.1.

The process of design evaluation of antenna is shown in Fig. 11.2.

11.4 Simulation Results and Comparative Analysis

In the proposed antenna design, different substrate techniques and multilayer substrates are used to enhance the bandwidth, return loss (S−11), and other properties of the antenna. The proposed multilayer monopole patch antenna is designed for gigahertz (GHz) frequency range up to 6 GHz. The proposed frequency range accommodates the various bands between 1 and 6 GHz which lies between the wireless fidelity and Wi-Max range. The multilayer monopole patch-based microstrip antennas are gaining importance in the applications of wireless local area networks (WLAN) and wireless fidelity. The simulation of various parameters such as return loss (S11), VSWR, and radiation pattern and Bandwidth are carried out using CST software, and the screenshots of the results are given out in this chapter. The new proposed microstrip antenna has enhanced gain and bandwidth. The optimized dimensions of the geometric parameters are given in Table 11.1. There is a good agreement with the simulated results of CST software. All the comparable results of the software are achieved by simulation and approximation for the proposed design. There are different parameters of antenna which are used to examine the efficient functioning of the antenna. These are return loss (S−11), voltage standing wave ratio (VSWR), gain, bandwidth, and number of bands.

Table 11.1 Design parameters of fractal monopole antenna

S. No	Design	Dimensions (mm)³				
		Substrate	Ground	Patch		
				Patch-1	Patch-2	Upper patch
1.	Design-I monopole patch	40 × 45 × 1.6	40 × 45 × 0.0635	21 × 23 × 0.0635	30 × 28 × 0.035	20 × 16 × 0.035
2.	Design-II monopole Minkowski fractal	40 × 45 × 0.0126	40 × 13 × 0.0635	21 × 23 × 0.0635	30 × 28 × 0.035	20 × 16 × 0.035

Design—III Monopole Modified Minkowski Fractal

Antenna dimension	Dimension length (L) × Width (W) × height(h)mm³
Substrate (s) RT/duroid	40 × 45 × 0.125
Ground (Gl) copper	4D × 12 × 0.0635
P1[Rectangular pole copper]	21 × 23 × 0.0635
P2[Rectangular pole SU8]	30 × 25 × 0.12
P1 feed line (f_L) P2 feed line (f_L)	**(0.6:0.2) × 10 × 0.0635** 1 × 10 × 0.12

Design—IV Monopole Modified Minkowski Fractal

Antenna dimension	Dimension length (L) × Width (W) × height(h)mm³
Substrate (s) RT/duroid	40 × 45 × 0.125
Ground (Gl) copper	40 × 2 × 0.0635
P1 [Rectangular pole copper]	21 × 23 × 0.0635
P2 [Rectangular pole SU8]	30 × 28 × 0.12
FL [SU8]	1 × 28 × 0.12

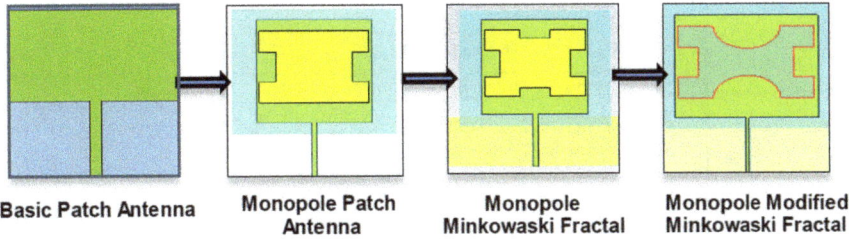

Basic Patch Antenna   Monopole Patch   Monopole   Monopole Modified
                       Antenna   Minkowaski Fractal   Minkowaski Fractal

Fig. 11.2 Design of evaluation of different antenna

(a)   Design-I Monopole Patch

Figure 11.3 shows the complete result with all three bands. The three different return loss (S-11) are obtained in between range 1 and 6 GHz. The S-parameter shows the RF energy propagated in the multi-port network. The X-axis shows the frequency range, and the Y-axis shows the return loss in dB. The result shows that there are two different bands between 1 and 6 GHz. First discuss a band that is obtained in the range of 2.475–3.017 GHz; in this band the obtained resonating frequency is 2.68 GHz; the obtained return loss is −14.244 dB; second one is at 5.2 GHz; the obtained return loss is −17.95. Totally, there are two narrow bands obtained in this design, bandwidth 22% and 23.52%. Since the bandwidth of both bands are under 50% that is why both are narrowband and proposed design 1 lies in the category of narrowband antenna (Figs. 11.4, 11.5, 11.6 and Table 11.2).

The gain of the proposed design at 5.5 GHz is near to 3.65 dB.

The performance of the above design is not up to the mark, in terms of return loss (S−11) and VSWR. The design requires enhancement and changes in the geometry of the fractal antenna (Fig. 11.7).

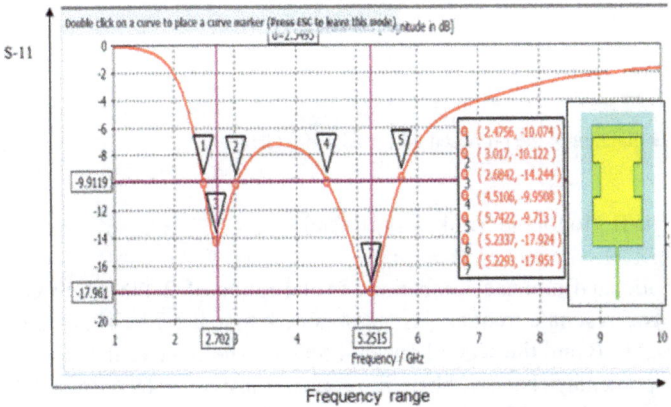

Fig. 11.3 Return loss (S−11) of proposed antenna combined two bands

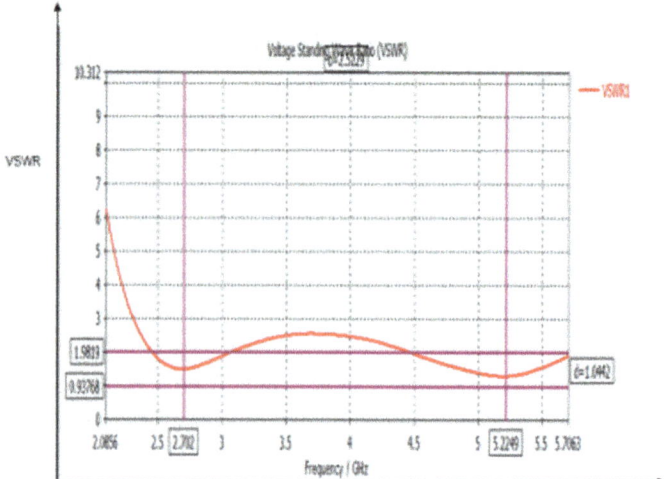

Fig. 11.4 VSWR

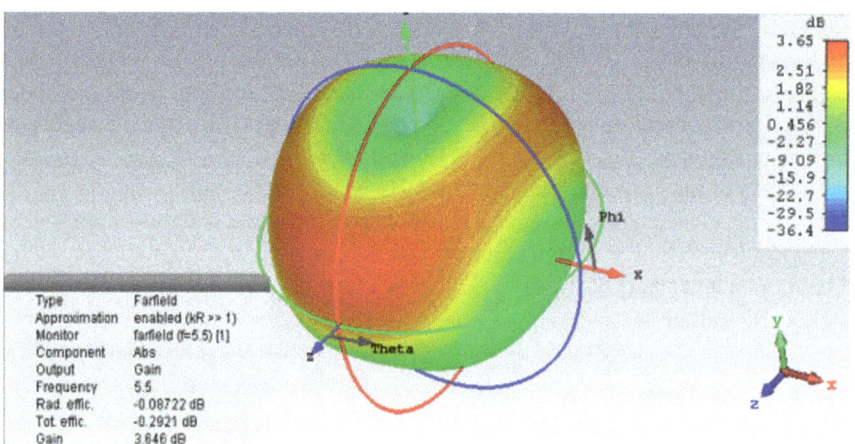

Fig. 11.5 3D diagram of the proposed design ($f = 5.5$ GHz)

(b)   Design-II Monopole Minkowski fractal

A single wideband antenna is obtained in the range of 2.475–4.99 GHz; in this wideband, two resonate frequencies are obtained, first at 2.75 GHz where the return loss is −20.23 dB and the second one at 4.5 GHz where the return loss is −15.404. The overall percentage bandwidth of the proposed design 2 is 69.61%. The bandwidth of both bands is above 50% that is why design 2 exist in the wideband region. The output of the proposed design is shown in Fig. 11.8 wherein the VSWR at 2.7 GHz

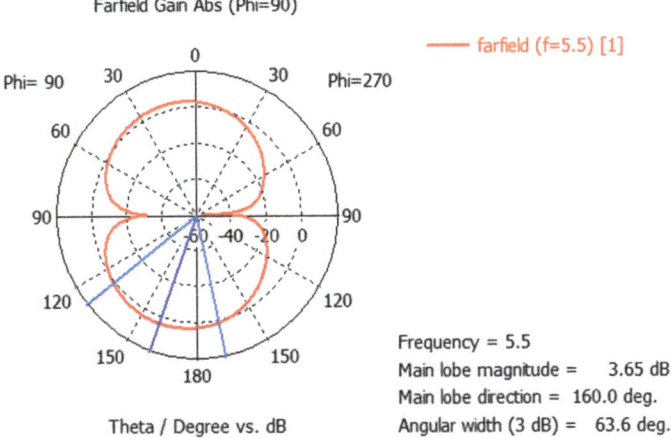

Fig. 11.6 Far-field pattern of the ($f = 5.5$ GHz)

is 1.2, and at 4.5 GHz, the VSWR is 1.4. At both resonating frequencies, the VSWR always lies under 2 (Figs. 11.9, 11.10, 11.11 and 11.12 and Table 11.3).

(c)   Design—III Monopole Modified Minkowski Fractal

In design III, a parametric study is carried out of the different antenna parameters. In this design, we have analyzed the microstrip feed line in the range from 0.2 to 0.5 mm. In this range, the optimum result is obtained at a 0.2 mm feed line (f_l). Table 11.4 shows the different outcomes at different frequencies.

From the above results, it is observed that when the width of the feed line deceases, the return loss incenses and a similar effect on bandwidth also reduces with reducing the bandwidth feed line.

The output of proposed antenna design III for VSWR is good in overall bands. The VSWR at different frequencies from 2.3 to 5.9 GHz always lies under 2.5 (Figs. 11.13 and 11.14).

The gain of the proposed design at 5.5 GHz is near to 4.07 dB.

In the above Table 11.5, the comparison of all three designs having different advantages is carried out; in the case of design I, it contains two separate bands, design II contains a single band with 69.61% bandwidth and two resonant frequencies. But the design III is better as compared to all three designs (Table 11.6).

11.5   Conclusion

This paper represents the design and analysis of microstrip patch antenna by modifying Minkowski geometry for a frequency band of 1 to 6 GHz. The simulation is done by using CST studio suite 2016. The return loss is −37.11 dB for fL = 0.3 and

Table 11.2 Analytical results for antenna—design-I

Method	Shape	Feed technique	Range	S-parameter (dB)	VSWR	Bandwidth (%)	No. of bands
Proposed Method—1	Multilayer proposed—1	Microstrip feed	1-6	2.68 GHz= −14.35dB 5.2 GHz= −17.98dB	1.45 1.26	22 23.52	Two Narrow Bands

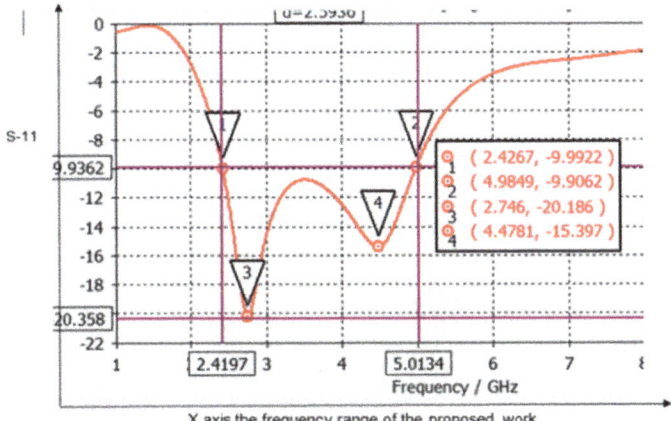

X axis the frequency range of the proposed work

Fig. 11.7 Return loss (S−11) of proposed design II (Single wideband)

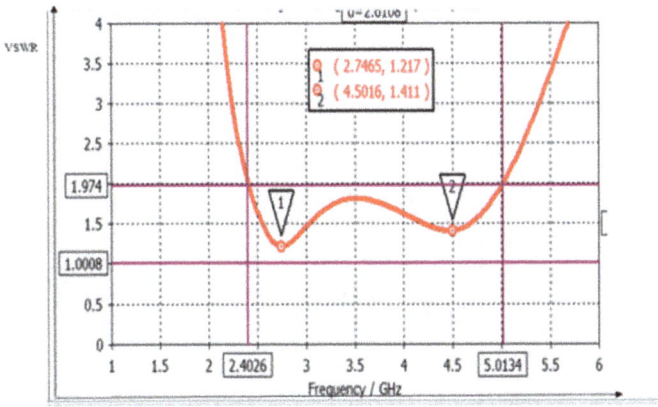

Fig. 11.8 VSWR

−59 dB for *f*L = 0.2. The VSWR is 1.02 (*f*L = 0.3), and bandwidth is 77.1% (*f*L = 0.3). This presented work shows the design of a multilayer monopole patch antenna with a Minkowski fractal. The proposed design shows good results in terms of basic antenna parameters such as VSWR, gain, return loss, and bandwidth. The proposed antenna shows a wideband and covers Wi-Fi and Wi-Max ranges whose frequencies are between 1 and 6 GHz. The overall gain of the proposed antenna is above 4db. Also, it shows the good result in terms of return loss that is (S−11) −37.11 dB as well as VSWR that is 1.02, and important parameter is percentage bandwidth which is 77.01%.

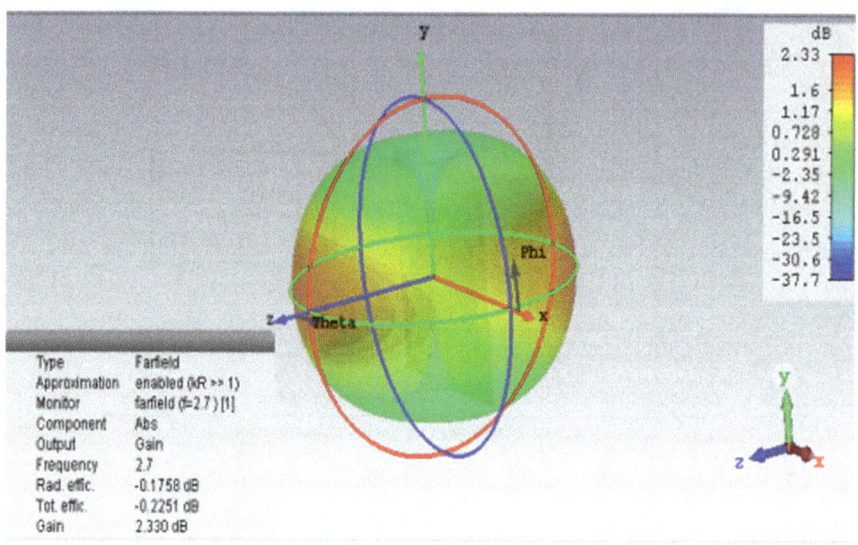

Fig. 11.9 3D diagram of the proposed design ($f = 2.7$ GHz)

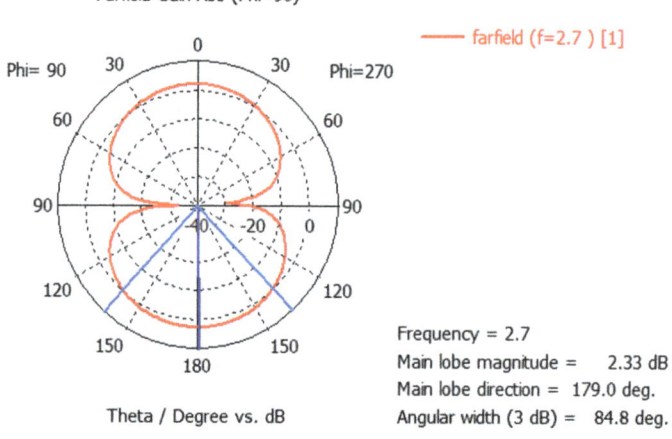

Fig. 11.10 Far-field pattern

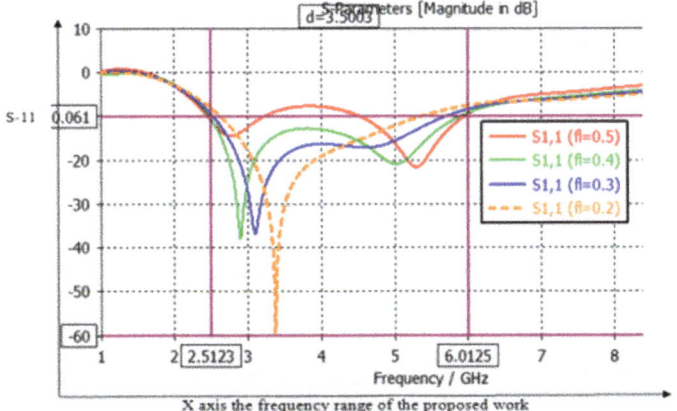

Fig. 11.11 Return loss (S-11) antenna design III (Single wideband)

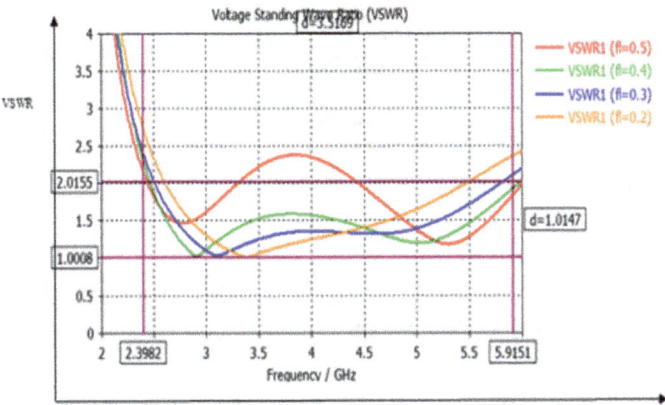

Fig. 11.12 VSWR

Table 11.3 Analytical results for antenna—design-II

Method	Shape	Feed technique	Range	S-Parameter (dB)	VSWR	Bandwidth	No. of band
Proposed Method-2	Multilayer proposed -2	Microstrip feed	1–6	2.7 GHz = −20.37dB 4.5 GHz = −15.58dB	1.2 1.4	69.61%	Single Wide Band

Table 11.4 Analytical results for antenna—design-III

Parameters feed line (f_l).	Resultant return loss (S-11) dB	Bandwidth (B.W.) %	VSWR
Fl = 0.4	2.9 GHz = −38.56	82.35% (2.48−5.91)	1.01
Fl = 0.3	3.1 GHz = −37.11	77.10% (2.55−5.75)	1.02
Fl = 02	3.7 GHz = −59.36	67.8% (2.63−5.4)	1.01

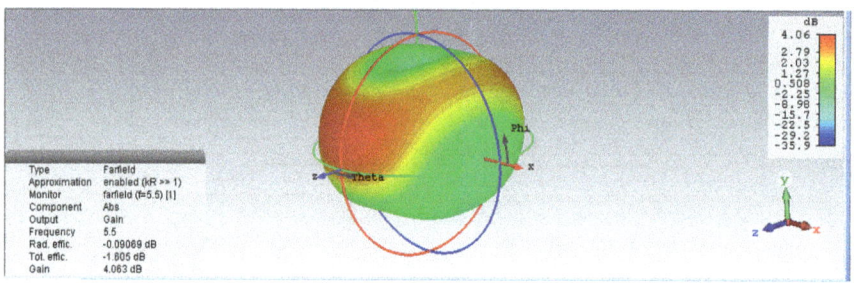

Fig. 11.13 3D pattern ($f = 5.5$ GHz)

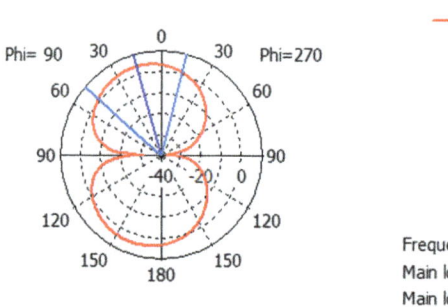

Farfield Gain Abs (Phi=90)

—— farfield (f=5.5) [1]

Frequency = 5.5
Main lobe magnitude = 4.07 dB
Main lobe direction = 16.0 deg.
Angular width (3 dB) = 63.2 deg.

Theta / Degree vs. dB

Fig. 11.14 2D pattern ($f = 5.5$ GHz)

Table 11.5 Comparative analysis between different design of antenna

Shape	Feed technique	Range	S—parameter (dB)	VSWR	Band width (%)	No. of band
Multilayer proposed—1	Microstrip feed	1–6	2.68 GHz = −14.35dB 5.2 GHz=−17.98dB	1.45 1.26	22 23.52	2 narrow
Multilayer proposed—2	Microstrip feed	1–6	2.7 GHz =−20.37dB 4.5 GHz = −15.58dB	1.2 1.4	**69.61**	Single wideband
Multilayer proposed design—3	Microstrip feed line	1–6	**3.3 GHz =−59.57dB**	1.1	67.8	Single wideband
Multilayer proposed design—3	Microstrip feed line	1–6	**3.1 GHz =−37.11**	1.02	**77.1**	Single wideband

Table 11.6 Comparison on the basis of S−11 and number of bands

S. No	Year	Antenna (Shape/ modification)	Feed technique	Range	S-Parameter	No. of bands
(a)	2018	Proposed multilayer patch antenna	Microstrip feed	1–6 GHz	**3.3 GHz = −59.57dB**	1 (Wideband)
(b)	2017	Flexible bow Tie antenna	CPW feed	1–6 GHz	−35 dB (Wideband bandwidth 1.79 GHz)	1

Appendix 'A'

Antena Design and Parameters

For design purposes, the Rogers RO4003C substrate is used which has a dielectric constant of 3.38. Frequency −1 to 6 GHz.

$$W = \frac{C}{2fo\sqrt{\frac{\varepsilon r+1}{2}}} \qquad \varepsilon_{\text{reff}} = \frac{\varepsilon r + 1}{2} + \frac{\varepsilon r - 1}{2}\left[1 + 12\frac{h}{w}\right]^{-1/2}$$

$$L_{\text{eff}} = \frac{c}{2\,fo\sqrt{\varepsilon_{\text{reff}}}} \qquad \frac{\Delta L}{h} = \frac{0.412(\varepsilon_{\text{reff}} + 0.3)\left(\frac{w}{h} + 0.264\right)}{(\varepsilon_{\text{reff}} - 0.258)\left(\frac{w}{h} + 0.8\right)}$$

$$L = L_{\text{eff}} - 2\Delta L \quad Wg = 6h + L \quad Z_o = 50\ \Omega \quad h = 1.6\ \text{mm}. \quad \lambda_o = c\,/\,f$$

$$\frac{w}{h} = \left(\frac{\exp(H')}{8} - \frac{1}{4\exp(H')}\right)^{-1}$$

$$H' = \frac{Z_0\sqrt{2(\varepsilon_r+1)}}{119.9} + \frac{1}{2}\left(\frac{\varepsilon_r-1}{\varepsilon_r+1}\right)\left(l_n\frac{\pi}{2} + \frac{1}{\varepsilon_r}l_n\frac{4}{\pi}\right)$$

$$\varepsilon_{\text{eff}} = \frac{\varepsilon_r+1}{2}\left[1 - \frac{1}{2H'}\left(\frac{\varepsilon_r-1}{\varepsilon_r+1}\right)\left(l_n\frac{\pi}{2} + \frac{1}{\varepsilon_r}l_n\frac{4}{\pi}\right)\right]^{-2}$$

References

1. Liu, P., Zou, Y., Xie, B., Liu, X., Sun, B.: Compact CPW-fed tri-band printed antenna with meandering split-ring slot for WLAN/WiMAX applications. IEEE Antennas Wireless Propag. Lett. **11**, 1242–1244 (2012)
2. Ramarao, B.: CPW-fed monopole antenna with l shaped and stair shape slot for dual-band WLAN/WiMAX applications. In: International Conference on Innovations in Engineering and Technology, December 2013, vol. 13. pp. 65–75. (2013)
3. Moosazadeh, M., Kharkovsky, S.: Compact and small planar monopole antenna with symmetrical L-and U-shaped slots for WLAN/WiMAX applications. IEEE Antennas Wireless Propag. Lett. **13**, 1536–1225 (2014)
4. Shao, S., Kiourti, A., Burkholder, R.J., Volakis, J.L.: Broadband textile-based passive UHF RFID tag antenna for elastic material. IEEE Antennas Wirel. Propag. Lett. **14**, 1385–1388 (2015)
5. Hamouda, Z., Wojkiewicz, J.-L., Pud, A.A., Kone, L., Belaabed, B., Bergheul, S., Lasri, T.: Dual-band elliptical planar conductive polymer antenna printed on a flexible substrate. IEEE Trans. Antennas Propag. **63**(12), 5864–5867 (2015)
6. Quarfoth, R., Zhou, Y., Sievenpiper, D.: Flexible patch antennas using patterned metal sheets on silicone. IEEE Antennas Wirel. Propag. Lett. **14**, 1354–1357 (2015)
7. McKerricher, G., Titterington, D., Shamim, A.: A fully inkjet-printed 3-D honeycomb-inspired patch antenna. IEEE Antennas Wirel. Propag. Lett. **15**, 544–547 (2016)
8. Higashi, H., Rutkowski, T.M., Tanaka, T., Tanaka, Y.: Multilinear discriminant analysis with subspace constraints for single-trial classification of event-related potentials. IEEE J. Selected Topics Signal Process. **10**(7), 1295–1305 (2016)
9. Ali, T., Khaleeq, M.M., Biradar, R.C.: A multiband reconfigurable slot antenna for wireless applications. Elsevier Int. J. Electron. Commun. **84**, 273–280 (2018)
10. Singh, H.K., Paik, H.: A compact single layer differentially-fed monopole antenna for UWB applications. In: International Conference on Communication and Signal Processing, July 28–30, pp 1–4. (2020)

Chapter 12
Fake News Detection Using Text Analytics

Uma Maheshwar Amanchi, Nithesh Badam, and Rama Lakshmi Elaganti

Abstract Fake news is a form of news consisting of false statements from the real ones spread via news media or online social media. In this paper, we aim for the fake news detection model which is capable of detecting the fake news from large amounts of data that are daily produced on online platforms. The approach for our model is a machine learning technique which is text analysis and for classifying fake news we have used k means clustering. Using the data preprocessing, classification, and topic modeling we get topics from the article, and they are compared with legitimate news. We modeled a framework named Fake News Detection (FND) which is used to classify the news articles. By streaming detection of fake information, we can control false or inaccurate content.

12.1 Introduction

Fake news is considered as the dangerous thing happening in today's world and previous fake news detection techniques are inadequate in classifying the fake news from the vast amount of data that daily publishing on online platforms. So we have decided to create a fake news detection system that deliberately classifies the fake news from a large amount of data with better credibility evaluation and accuracy. In our paper, we decided the text analytics approach to detecting fake news. This approach is used in various applications and gives better results in classification projects. The text analytics approach is the machine learning model which is used to analysis the meaning of the text. (Manzoor et al. 2019) In the first phase, the data is formed into topic clusters using classification module and formed clusters based on topics using topic modelling. Next the topic clusters from the testing are formed into a list and compared with the real news topics. According to the credibility score of

U. M. Amanchi (✉) · N. Badam · R. L. Elaganti
Chaitanya Bharathi Institute of Technology, Hyderabad, India

R. L. Elaganti
e-mail: eramalakshmi_it@cbit.ac.in

© The Author(s), under exclusive license to Springer Nature Singapore Pte Ltd. 2021
S. C. Satapathy et al. (eds.), *Smart Computing Techniques and Applications*,
Smart Innovation, Systems and Technologies 225,
https://doi.org/10.1007/978-981-16-0878-0_12

117

an article we classify the article as real or fake (Hiramath & Deshpande 2019). This approach is able to give better credibility and classification accuracy than previous researches we have searched.

12.2   Proposed System

In our proposed model the model training framework has several stages of the following: We proposed a topic-based approach to detecting fake news. In first module, from the legitimate news we extract topics, then the topics are classified into various clusters. Every cluster is formed based on similar news which are common in topic. Each topic clusters contain large number of topic which are legitimate news.

12.2.1   Module 1

- At first, we collect the legitimate news and we extract topics from collected training dataset, next the news is categorized into selected number of clusters based on topics.
- Each cluster which is formed using clustering are centered on common topics of a news article.

We collected the training dataset from Kaggle website for model training framework, which are regarded as legitimate news because it is a trustworthy website by large audience. In detecting the fake news we created a framework fake news detection system (FND) undergo three modules and data from testing datasets are formed as topic clusters. Then the topic clusters are which are from the testing data is compared against the topic article from training data and using the threshold we calculate the credibility of testing articles.

12.2.2   Module 2

- In this module, the article which is to be checked is converted into clusters according to topics, the we validate this clusters with the topic clusters formed from legitimate news.

To calculate the credibility of news article we compare the topic and clusters and based on the threshold values we will estimate whether the news is real or deceptive news.

12.3   Implementation

In this section, first, we describe the model training framework which tells the detailed information about how the legitimate news datasets undergo various stages to form a well-defined topic-based cluster which is legitimate to classify from the fake ones. Then the stages of modern training framework describe data preprocessing where the data is cleaned. Second feature extraction to obtain the features from data to form clusters. Finally, the topic modeling approach is used to form clusters that are surrounded by common news topics. The required evaluation metrics calculations are described in detail with formulas.

By The design and development of the model training framework are as shown in Fig. 12.1. The model training framework is driven by a ground-truth knowledge base comprised of legitimate-news topic clusters. The model-training framework drives the functioning of how it will create clusters based on topics such that news articles in the same cluster share a set of similar topics. Topics classified in separate clusters are different from each other cluster.

This framework guides the design and development of detecting fake news. The model training framework describes the detailed procedure of how legitimate news is classified according to similar news topics. This will act as a ground truth model for converting any news which is to be verified. The legitimate news is free from stemming and stop words when it undergoes through the framework. All the datasets are first undergone through model training framework and comparison is done to detecting the news which is fake.

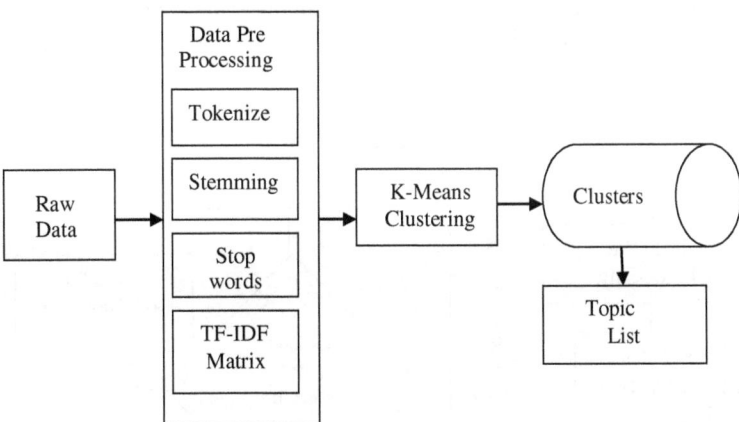

Fig. 12.1   Model training framework

12.3.1 FND (Fake News Detection System)

They are various stages involved in the FND system of detecting fake news from various sources. The input from any source undergoes many operations as described above in the model training framework which are data preprocessing, feature extraction, clustering, and topic modeling. The FND system checks the news which should be verified as fake or real news from the dataset. It compares the news which should be verified and the news will undergo the model training framework and compared with the legitimate news clusters from the model training framework.

In the fake news detection system, the input from any source is converted into the CSV file format. The data will undergoes data preprocessing to free from all data errors. Then the data will undergo TF-IDF vectorization to behave as a feature to form as a cluster with similar topics. Similar topics are under a single cluster. Each cluster consists of various large numbers of topics. The comparison of a large number of topics is not easy to handle by the clustering technique. The topics are formed as a cluster based on topics using topic modeling Similar topics are surrounded by common shared topics using topic modeling (Ibrishimova 2019). We have used the LDA model to separate the topics and the clusters are compared against the legitimate news topic cluster to verify whether the news is fake or real.

Figure 12.2 describes FND system consists of three modules each module i.e., data preprocessing, classification module, verification module. We collected the two datasets which are training and testing from Kaggle for model training, which are real and fake news articles respectively. In our training dataset there are 2000 articles and 100 articles in testing dataset. They also provide fake news dataset where the news should be verified. The data from the training dataset will undergo various techniques to be free and using the clustering algorithm the similar news formed as

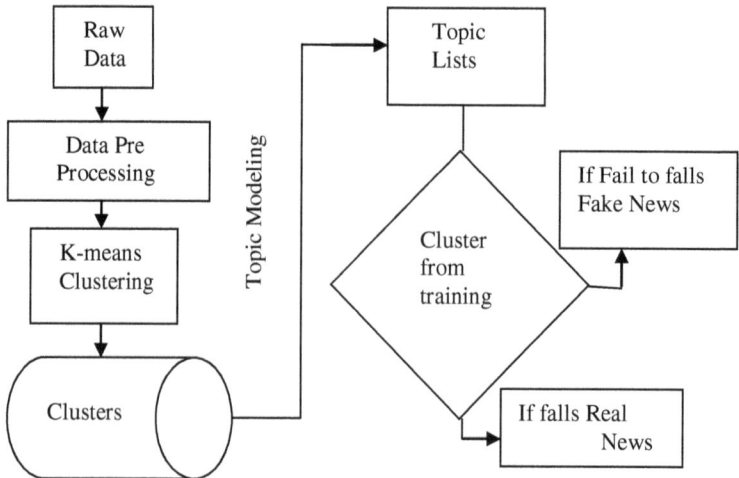

Fig. 12.2 Fake news detection system

a single cluster. Various clusters are formed from the clustering algorithm and the clusters are converted to topic clusters where the article with similar topic information into single topic cluster. Topics from the testing dataset compared against the topics of training dataset based on the threshold we calculate the credibility of the news article.

12.3.2   Data Preprocessing

Data Preprocessing is the process of analyzing the data that has not been screened for such problems that can produce misleading results. The raw data from the various sources are saved as ground-truth and fake news database. Data undergoes various stages of data cleaning which can give the data free from errors. The various techniques involved in data preprocessing are the following:

Tokenization: It is a very common task in NLP while dealing with the characters, the process of chopping a character into pieces (which are known as tokens) called tokenization.

Stemming: The process of reducing the word into its root word is called stemming.

Lemmatization: It is a process of grouping the inflected form of words so they can be analyzed as a single term.

Stop Words: Those are the word which is filtered on and or before data preprocessing techniques applied on data.

12.3.3   TF-IDF Vectorization

TF: Term Frequency
IDF: Inverse Document Frequency
TF-IDF = TF*IDF
TF = (Number of times term the particular word appears an article)/(Number of words in an article)
IDF = log(S/s)
S is the total number of articles in the datasets, and s is the number of articles the particular word has appeared in the dataset. Clustering can be formed by using only numerical values, hence TF-IDF values for each word is calculated and clustering is formed by using the values.

12.3.4   Clustering

In classification module, we have used an unsupervised clustering algorithm to classify similar news from large datasets. We have used the k means clustering algorithm to classify the fake news from the real news. In our model, we have formed 50 clusters each cluster has centers and surrounded by similar news. The clusters formed from the algorithm may contain a large number of news and handling a large number of news in a single cluster difficult and the articles undergo topic modeling technique to form topic-based clusters.

In the implementation of k means clustering first (Mahid 2018), we assume that we get an output of k clusters. Then we randomly select the k number of items and using euclidean distance each k item is surrounded with similar news with a centroid. Then the process of euclidean distance is repeated until the appropriate news is fallen into the particular cluster.

12.3.5 Topic Modeling

In this paper, the topic modeling is used to create the topic based clusters. We have used LDA topic modeling technique for the topic-based representation of clusters. The clusters are formed based on topics and each topic contain similar information. The topics from the training and testing are stored in the list and the topics of article which is to be checked is compared against the topics of training and calculate the credibility of article.

12.3.6 Evaluation Credibility

The information of articles from testing data are used to check the credibility of system. The credibility of test news article is calculated through a function which described whether the news is fake topic and real topic, using the threshold we display the false credibility of an article.

$$g(\alpha) = \frac{\sum_{j=1}^{w_i} (f_E(e_i^j))}{w_i} \tag{12.1}$$

Thus, we have where α is the news article which is to be tested, w_i is the number of topics generated from the test news, f_E (e^{i^j}) is the Boolean valued function.

12.3.7 Performance Evaluation

- $d_{r \to r}$ be the number of news which are true and truly detected as true information.
- $d_{r \to f}$ is the number of news which are true and falsely detected as false information.

- $d_{f \to r}$ is the number of news which are false and falsely detected as true information.
- $d_{f \to f}$ is the number of news which are false and truly detected as false information.

Accuracy Rate (A)

$$A = \frac{d_{r \to r} + d_{f \to f}}{d_{r \to r} + d_{r \to f} + d_{f \to r} + d_{f \to f}} \tag{12.2}$$

Precision (P)

$$P = \frac{d_{f \to f}}{d_{r \to f} + d_{f \to f}} \tag{12.3}$$

Recall (R)

$$R = \frac{d_{f \to f}}{d_{f \to r} + d_{f \to f}} \tag{12.4}$$

F-Score (F)

$$F = \frac{2 \times P \times R}{P + R} \tag{12.5}$$

12.4   Results

12.4.1   Cluster Evaluation

Topics from the trained datasets are compared against the training dataset based on threshold valued the testing article is classified as legitimate or deceptive news respectively is shown in Fig. 12.3.

12.4.2   Accuracy of FND System

Figure 12.4 Illustrates the proposed approach for detecting the fake news from large amount of data which is generated from social media using the machine learning techniques. Our model got better classification accuracy than various other approaches to detecting fake news and accuracy about 88%.

```
gsum=0
for i in test:
    if i in out:
        gsum=gsum+1
gfinal=gsum/10
print(gfinal)
```

```
gsum=0
for i in test:
    if i in outtopics:
        gsum=gsum+1
gfinal=gsum/10
print(gfinal)
```

0.4   0.0

```
threshold=0.4
if gfinal<threshold:
    print("Fake")
else:
    print("Legitimate")
```

```
threshold=0.4
if gfinal<threshold:
    print("Fake")
else:
    print("Legitimate")
```

Legitimate   Fake

Fig. 12.3 Evaluation of fake news

Fig. 12.4 Accuracy of FND systems

```
c=data['Stance']
ct=0
for i in range(0,904):
    if l[i]==1 and c[i]=='Fake':
        ct+=1
    if l[i]==0 and c[i]=='Legitimate':
        ct+=1
print('Accuracy:',ct*100/904)
```

Accuracy: 87.83185840707965

12.5   Conclusion

Detecting fake news has become a crucial role in everyday life, various techniques are lack in detecting the deceptive news from large data produced every day from various social platforms. In this paper, we have proposed a model that is used to detect fake news continuously produced on online media. We have created a text analytics-based approach system mainly focused on representing the news as topic-based representation and the system is used to classify the news as deceptive or real. We have used three modules in classifying the news which is to be verified those three modules make various changes to the data and made a topic-based representation of news. Topics from the testing data is compared against the topics from training data which are also derived using three modules used to classify the news. Each topic is surrounded by similar news and based on threshold value the news is classified as fake news. Our model got better classification accuracy than various other approaches to detecting fake news.

References

1. Hiramath, C.K., Deshpande, G.C.: Fake news detection using deep learning techniques. In: Science Direct on Multimedia, Chikmagalur, India, pp. 411–415 (2019)
2. Ibrishimova, M.D., Li, K.F.: A machine learning approach to fake news detection using knowledge verification and natural language processing. In: Barolli, L., Nishino, H., Miwa, H. (eds.) Advances in Intelligent Networking and Collaborative Systems. INCoS 2019. Advances in Intelligent Systems and Computing, vol 1035. Springer, Cham (2020)
3. Manzoor, S.I., Singla, J., Nikita: Fake news detection using machine learning approaches: a systematic review. In: 2019 3rd International Conference on Trends in Electronics and Informatics (ICOEI), Tirunelveli, India, pp. 230–234 (2019)
4. Mahid, Z.I., Manickam, S., Karuppayah, S.: Fake news on social media: brief review on detection techniques. In: Science Direct, Subang Jaya, Malaysia, pp. 1–5 (2018)

Chapter 13
An Extreme Learning Machine-Based Model for Cryptocurrencies Prediction

Sarat Chandra Nayak, B. Satyanarayana, Bimal Prasad Kar, and J. Karthik

Abstract Cryptocurrency, in a short period of time has got wide popularity and considered as an investment asset. Prediction of cryptocurrency is a recent area of research interest and budding fast. The price trend of cryptocurrency behaves arbitrarily and fluctuates like other stock markets due to inherent volatility. Though few computational intelligence methods are available, sophisticated methodologies for accurate prediction of cryptocurrency are still lacking and need to be explored. Extreme learning machine (ELM) is a faster and better learning method for neural networks with solitary hidden layer and has enhanced generalization performance. This study proposes an ELM based approach for prediction of four emerging cryptocurrencies such as Litecoin, Ethereum, Ripple, and Bitcoin. The prediction ability of the proposed approach is compared with few similar methods such as RBFN, SVM, MLP, ARIMA, and LSE. From exhaustive simulation studies and comparative result analysis it is found that the ELM method performed better than others and hence can be suggested as an efficient tool for cryptocurrencies prediction.

13.1 Introduction

Cryptocurrency is an open source and decentralized digital currency based on principle of cryptography. The relevance of cryptocurrencies as an emerging market in the financial world is increasing significantly. Now days more and more financial organizations are getting concerned about cryptocurrency trading. The trading opportunities and profitability in the cryptocurrency market are studied in [1]. A survey on systems of cryptocurrency is elaborated in [2]. The characteristics of this

S. C. Nayak (✉) · J. Karthik
CMR College of Engineering and Technology, Hyderabad 501401, India

B. Satyanarayana
CMR Institute of Technology, Hyderabad 501401, India

B. P. Kar
Gandhi Institute for Technological Advancement, Bhubaneswar 752054, India

S. C. Satapathy et al. (eds.), *Smart Computing Techniques and Applications*,
Smart Innovation, Systems and Technologies 225,
https://doi.org/10.1007/978-981-16-0878-0_13

market such as high availability of market data, high volatility, smaller capitalization, and decentralized control are studied in [3]. This market fluctuates like other stock markets due to inherent volatility and the confidence of investors have been reflected on it [4]. The feasibility of this market in relation to other financial markets is documented in the literature [5]. Like other stock market, cryptocurrency market prices behave arbitrarily and coupled with high nonlinearity and dynamics. Though few computational intelligence methods are available, sophisticated methodologies for accurate prediction of cryptocurrency are still lacking and need to be explored.

As the cryptocurrencies market prices are growing fast and behaves similarly as other stock market price movement, speculators as well as researchers are anxious about prediction of their market values. During last few years machine learning and statistical methodologies are used for prediction of Bitcoin [6, 7]. Forecasting accuracy of a neural model is greatly subjective to the network magnitude and learning method. Gradient descent learning is a common method for neural network training. However, suffering from sluggish convergence rate, imprecise learning and prone to local minima are the general drawbacks of this technique. It adds computational overhead to the model [8]. To overcome these, ELM was proposed in [9]. It chooses the input-hidden weights at random. The hidden-output weights are determined analytically in contrast to iterative fine-tuning. Numerous researches and experimentations are done for diverse applications using ELM [10]. It is a faster and better learning method for networks with solitary hidden layer and has enhanced generalization performance.

With objective of achieving improved prediction accuracy, here we use ELM for training of a neural network having single hidden layer. The trained network is then used to extrapolate the historical prices of the four cryptocurrency and predict the next day price. Contrast to conventional method of segregating the dataset into two parts (i.e. training and testing), we followed a moving window method to train and test pattern generation. To access the performance of ELM based approach, five other models such as RBFN, SVM, MLP, ARIMA, and LSE are developed in similar fashion. Prediction accuracies of all the methods are accessed in terms of MAPE, ARV, and computation times from four cryptocurrency datasets.

The remaining parts such as methods and materials are described in Sect. 13.2, the ELM-based forecasting method is explained in Sect. 13.3, and simulation results are summarized in Sect. 13.4 followed by concluding remarks.

13.2 Methods and Materials

The methods such as ELM, and proposed ELM-based forecasting and materials such as cryptocurrency datasets collected from online sources are presented in this section.

13.2.1   ELM

As stated earlier, to overcome the limitations of gradient descent learning ELM was proposed [9]. It was proposed for training networks with one hidden layer. It considers random weight and bias for hidden nodes. As an alternative to iterative tuning the output weights are determined through generalized inverse function on the outputs from the hidden layer. For an input vector x_j, weight vector between input neuron and ith hidden neuron $w_i = [w_{i1}, w_{i2}, \cdots, w_{iN}]^T$, $(i = 1, 2, \cdots, N_h)$ and output weights linking the ith hidden neuron and the output neurons $\beta_i = [\beta_{i1}, \beta_{i2}, \cdots, \beta_{iN}]^T$, the output vector O_j is computed as:

$$O_j = \sum_{i=1}^{N_h} \beta_i * f(w_i x_j + \text{bias}_i), \ j = 1, 2, \ldots, N \tag{13.1}$$

Here N_h is the hidden layer size. The output weight vector β_i obtained by solving $H\beta = Y$, where:

$$H(w_i, x_i, b_i) = \begin{bmatrix} f(w_1 x_1 + b_1) & \cdots & f(w_{N_h} x_1 + b_{N_h}) \\ \vdots & \ddots & \vdots \\ f(w_1 x_N + b_1) & \cdots & f(w_{N_h} x_N + b_{N_h}) \end{bmatrix}_{N \times N_h} \tag{13.2}$$

$$\beta = \begin{bmatrix} \beta_1^T \\ \vdots \\ \beta_{N_h}^T \end{bmatrix}_{N_h \times m} \quad Y = \begin{bmatrix} y_1^T \\ \vdots \\ y_N^T \end{bmatrix}_{N \times m}$$

In real cases, the inequalities $N_h \ll N$ holds true. So, H is a non-square matrix and may be non-singular in the majority of cases. Therefore, a combination (w_i, b_i, β_i) satisfying Eq. (13.2) may not exist. So, the network can be trained by finding the least square minimum norm solution $\hat{\beta}$ of (13.2) as follows:

$$\left\| H\hat{\beta} - Y \right\| = \min_\beta \| H\beta - Y \| \tag{13.3}$$

The minimum norm least square solution of Eq. (13.2) is calculated as in Eq. (13.4). Here H^+ is the pseudo inverse or Moore–Penrose inverse of H.

$$\hat{\beta} = H^+ Y \tag{13.4}$$

13.2.2 Proposed ELM-Based Prediction

A pictorial view of ELM-based forecasting of cryptocurrency is depicted in Fig. 13.1. Here, the base model is a neural network having single hidden layer. The network is trained by ELM. The input-hidden weights are assigned with random values. The output weights are computed as per Eq. 13.4. We created the input patterns for the model using moving window from the original series. For example one train/test set is formed as follows.

$$
\begin{array}{c}
\left.\begin{array}{ccc}
x(i) & x(i+1)\ x(i+2) & \vdots\ x(i+3) \\
x(i+1)\ x(i+2)\ x(i+3) & \vdots\ x(i+4) \\
x(i+2)\ x(i+3)\ x(i+4) & \vdots \\
\hline
\underbrace{\qquad\qquad\qquad\qquad}_{\text{Training data}} & \vdots\ x(i+5)
\end{array}\right. \\
\underbrace{\text{Target}}
\end{array}
$$

$$
\underbrace{x(i+3)\ x(i+4)\ x(i+5)}_{\text{Test data}} \ \vdots\ \underbrace{x(i+6)}_{\text{Target}}
$$

The patterns are then normalized to scale the data into the same range for each input feature in order to minimize bias [11]. We used tanh estimator method as in Eq. 13.5 which is an efficient scheme to standardize time series data. The mean and standard deviation of the moving window are represented as μ and σ respectively.

$$
\hat{x} = 0.5 * \left(\tanh\left(\frac{0.01 * (x - \mu)}{\sigma} \right) + 1 \right) \tag{13.5}
$$

After normalization the patterns are inputted sequentially to the network. The hidden neurons calculate the product of input and associated weight adds a bias and

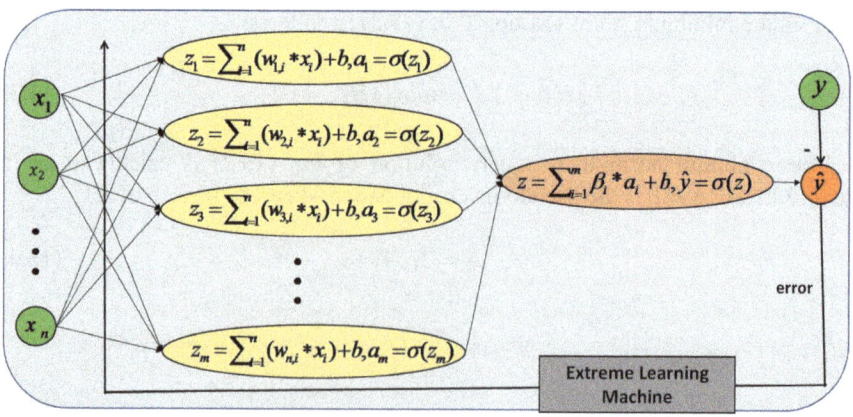

Fig. 13.1 ELM-based forecasting

Table 13.1 Statistical summary of four cryptocurrencies

Statistic	Bitcoin	Litecoin	Ethereum	Ripple
Minimum	68.4300	1.1600	0.4348	0.0028
Mean	$1.4857e^{+03}$	20.4966	147.7843	0.0984
Median	482.8100	3.9100	12.0200	0.0079
Variance	$8.7573e^{+06}$	$2.2240e^{+03}$	$6.9765e^{+04}$	0.1028
Maximum	$1.9497e^{+04}$	358.3400	$1.3964e^{+03}$	3.3800
Std.dev	$2.9593e^{+03}$	47.1594	264.1308	0.3206
Skewness	3.5394	4.1627	2.38440	5.8039
Kurtosis	15.9671	21.5925	8.51220	42.9469
Corr. Co.eff	0.00030	0.00211	−0.00520	0.00027
No. of data	1760	1760	929	1662

applies sigmoid activation to compute the net output. The output unit computes the product of weights and activations from the hidden units adds a bias and applied activation. The model estimated an output ($\hat{y}$) at the output layer as in Eq. 13.6. The amount of deviation of the estimation from the target (y) is calculated as error, i.e. Equation 13.7. The minimum is the error; more closer the model estimation towards the actual, hence better is the prediction accuracy of the model.

$$\hat{y} = \sum_{i=1}^{m} \sigma\left(\beta_i * a_i + b\right) \tag{13.6}$$

$$error = \text{abs}\left(y - \hat{y}\right) \tag{13.7}$$

13.2.3   Cryptocurrency Data

The models are evaluated on experimenting cryptocurrencies historical closing prices collected from kaggle.com. The currencies are Bitcoin, Litecoin, Ethereum, and Ripple. A summary of statistics of the four datasets are given in Table 13.1. The price series are plotted in Fig. 13.3. The total number of data points on the Bitcoin, Litecoin, Ethereum, and Ripple are 1760, 1760, 929, and 1662 respectively.

13.3   Experimental Results and Analysis

To evaluate the models, we used them to forecast the prices of four cryptocurrency datasets separately. To access the capacity of the ELM method, five other methods

Table 13.2 MAPE and ARV from all models

MODEL	MAPE				ARV			
	Bitcoin	Litecoin	Ehtereum	Ripple	Bitcoin	Litecoin	Ehtereum	Ripple
LSE	0.084185	0.172655	0.078047	0.088747	0.015777	0.024671	0.019375	0.038572
ARIMA	0.060645	0.094002	0.075884	0.082855	0.008352	0.009558	0.046743	0.016345
MLP	0.047281	0.083777	0.059358	0.049708	0.007763	0.007994	0.005905	0.017437
SVM	0.039462	0.055652	0.046352	0.047638	0.006584	0.006042	0.006355	0.006854
RBFN	0.045470	0.053754	0.049535	**0.045733**	0.007563	0.007962	0.005756	0.014362
ELM	**0.032283**	**0.051376**	**0.043901**	0.047335	**0.006137**	**0.005815**	**0.003982**	**0.005333**

The best MAPE and ARV statistics values are shown in bold face.

such as SVM, MLP, RBFN, ARIMA, and LSE are developed in this study and a comparison is done. Two performance metrics such as mean absolute percentage of error (MAPE), and average relative variance (ARV) are used to measure the forecast accuracy as follows:

$$\text{MAPE} = \frac{1}{N}\sum_{i=1}^{N}\frac{|x_i - \hat{x}_i|}{x_i} \times 100\% \tag{13.8}$$

$$\text{ARV} = \frac{\sum_{i=1}^{N}(\hat{x}_i - x_i)^2}{\sum_{i=1}^{N}(\hat{x}_i - \overline{X})^2} \tag{13.9}$$

The mean values from twenty simulations are considered as the performance of a model. The MAPE and ARV values are summarized in Table 13.2. The best statistics are highlighted in bold. It may be seen that the MAPE values from the ELM are lower than others expect Ripple dataset. For example, MAPE of ELM is 0.032283 from Bitcoin, 0.051376 from Litecoin, and 0.043901 from Ethereum. For Ripple the RBFN produced lowest, i.e. 0.047335. However, the ELM achieved lowest MAPE thrice. It achieved lowest ARV values for all four datasets. The overall performance of ELM is better than others. It also may be observed that the SVM and RBFN are placed in second and third position. Further, to show the goodness of the ELM method the predicted prices are plotted against the actual and depicted in Figs. 13.2, 13.3, 13.4, 13.5 respectively. It can be seen that the ELM predicted prices are very closer to the actual prices and following the trend accurately.

13.4 Conclusions

Prediction of cryptocurrency is a recent area of research interest and budding fast. The trend it follows behaves arbitrarily and fluctuates like other stock markets due

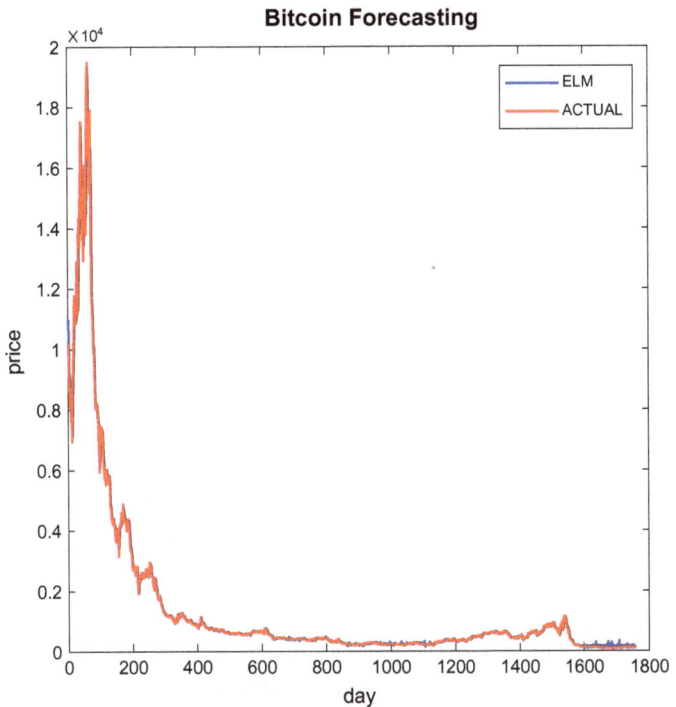

Fig. 13.2 Forecasted prices versus actual from bitcoin

to inherent volatility. Therefore, capturing the under laying dynamics of cryptocurrency prices movement is challenging. This article proposed an ELM-based forecasting approach for predicting the future price of four popular currencies such as Bitcoin, Litecoin, Ethereum, and Ripple. ELM has faster convergence rate and better generalization ability compared to gradient descent learning. Here, we used ELM for training a single hidden layer neural network. The ELM predictability is compared with that of SVM, RBFN, MLP, ARIMA, and LSE in terms of MAPE and ARV. From comparative analysis it is concluded that the ELM-based forecasting is good enough to forecast the cryptocurrencies than others. The future plan may explore other neural networks to be trained with ELM and the current approach may be applied for other financial forecasting problems.

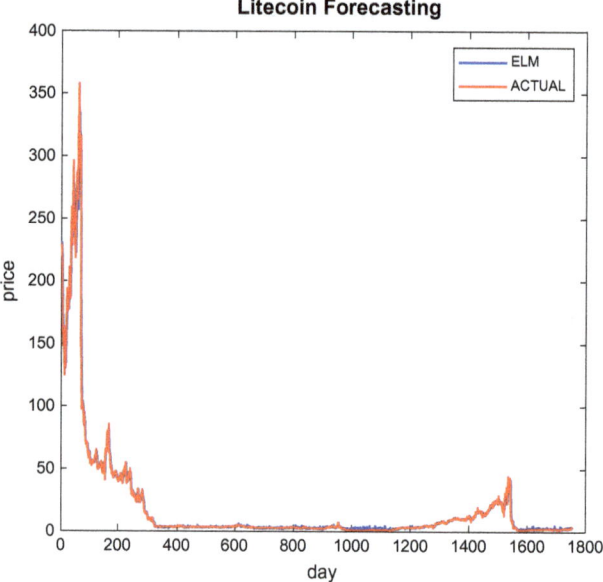

Fig. 13.3 Forecasted prices versus actual from litecoin

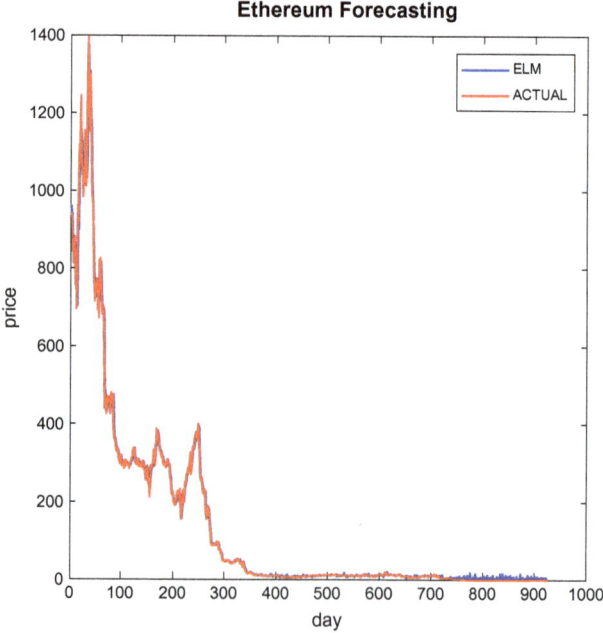

Fig. 13.4 Forecasted prices versus actual from ethereum

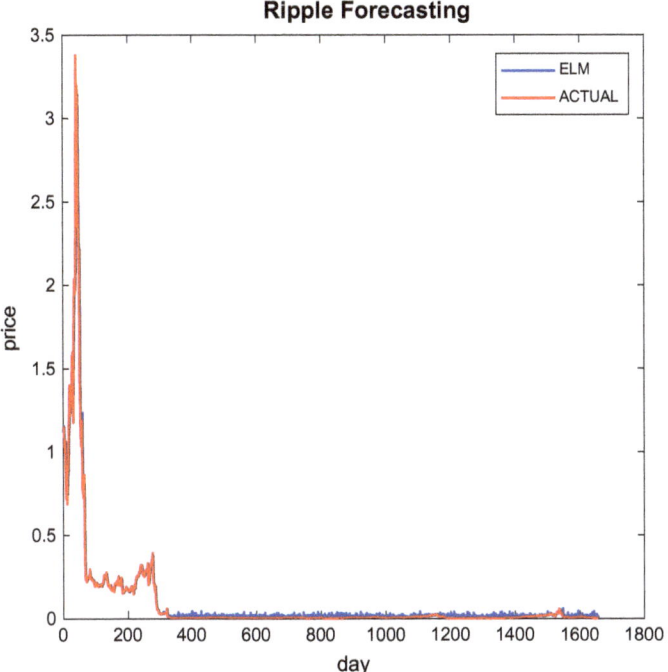

Fig. 13.5 Forecasted prices versus actual from ripple

References

1. Kyriazis, N.A.: A survey on efficiency and profitable trading opportunities in cryptocurrency markets. J. Risk Finan. Manage. **12**, 67 (2019)
2. Mukhopadhyay, U., Skjellum, A., Hambolu, O., Oakley, J., Yu, L., Brooks, R.: A brief survey of cryptocurrency systems. In: 2016 14th Annual Conference on Privacy, Security and Trust (PST), IEEE. pp. 745–752. (2016)
3. Ferreira, M., Rodrigues, S., Reis, C.I., Maximiano, M.: Blockchain: a tale of two applications. Appl. Sci. **8**, 1506 (2018)
4. Grinberg, R.: Bitcoin: an innovative alternative digital currency. Hastings Sci. Technol. Law J. **4**, 160–207 (2011)
5. Corelli, A.: Cryptocurrencies and exchange rates: a relationship and causality analysis. Risks **6**, 111 (2018)
6. Chu, J., Nadarajah, S., Chan, S.: Statistical analysis of the exchange rate of bitcoin. PloS One **10**(7) (2015)
7. Núñez, J.A., Contreras-Valdez, M.I., Franco-Ruiz, C.A.: Statistical analysis of bitcoin during explosive behavior periods. PloS One, **14**(3) (2019)
8. Fernández-Navarro, F., Hervás-Martínez, C., Ruiz, R., Riquelme, J.C.: Evolutionary generalized radial basis function neural networks for improving prediction accuracy in gene classification using feature selection. Appl. Soft Comput. **12**(6), 1787–1800 (2012)
9. Huang, G.B., Zhu, Q.Y., Siew, C.K.: Extreme learning machine: theory and applications. Neurocomputing **70**(1–3), 489–501 (2006)

10. Nayak, S.C., Misra, B.B.: Extreme learning with chemical reaction optimization for stock volatility prediction. Finan. Innov. **6**(1), 1–23 (2020)
11. Nayak, S.C., Misra, B.B., Behera, H.S.: Impact of data normalization on stock index forecasting. Int. J. Comput. Info. Syst. Indus. Manage. Appl. **6**(2014), 257–269 (2014)

Chapter 14
An Ensemble Approach for Intrusion Detection in Collaborative Attack Environment

Shraddha R. Khonde, G. Kulanthaivel, and V. Ulagamuthalvi

Abstract Data confidentiality and integrity are the main issue in modern era of Internet. Network security is compromised by intruders with help of attacks. Collaborative attack is a new attack used by intruder which is to be monitored by intrusion detection system (IDS). In this paper, we are proposing IDS which can handle collaborative attack with the help of ensemble classifiers. Proposed IDS makes use of various feature selection techniques to reduce number of feature required for attack detection. Various classifiers are used in ensemble to increase attack detection rate and reduce false alarm rate (FAR). Dataset used for experiments is UNSW NB15. Comparison of individual classifier is done with ensemble approach. Results show reduction of 5% in FAR and improvement of 2% in accuracy for ensemble approach. Ensemble approach and feature selection provide improved performance of IDS in terms of precision and recall.

14.1 Introduction

New era of Internet makes use of various technologies and architectures to provide ease and efficient data handling through network and Internet. To provide security to data transferred in Internet, network security policies should be strong enough. To maintain security of the network, various tools are used by administrators to handle outside threats. When data is compromised by any insider, it becomes difficult to identify and handle it. To handle such type of attack that is inside as well as outside attacks, an intelligent system is used known as IDS. This system monitors all activities of the network whether data is coming from outside sender or from inside sender; it is analysed and checked for security. While monitoring activities in network if any abnormal activity is observed by IDS, an alarm is generated for

S. R. Khonde (✉) · V. Ulagamuthalvi
Sathyabama Institute of Science and Technology, Chennai, Tamilnadu, India

G. Kulanthaivel
National Institute of Technical Teachers Training and Research, Chennai, Tamilnadu, India

© The Author(s), under exclusive license to Springer Nature Singapore Pte Ltd. 2021   137
S. C. Satapathy et al. (eds.), *Smart Computing Techniques and Applications*,
Smart Innovation, Systems and Technologies 225,
https://doi.org/10.1007/978-981-16-0878-0_14

administrator for taking further action on source and destination. Various types of attacks are available in this modern era which not only compromise the system, but takes whole control of system and violates the network such as *active, passive and collaborative*. Active attacks modifies data captured from the network and penetrate back modified data in the network. This type of attack is dangerous to the system as modified data can have different or severe impact on receiver. Passive attack does not modify data of network it only monitors network activities. Using this type of attack, intruders try to pass confidential information outside the network. Collaborative attack is a new type of attack used by intruders, nowadays, in this multimedia modern era of data transmission. This type of attack is created with the help of multiple attacks. Combinations of multiple attacks are used to generate new attacks to get entry into the system. As the signatures of most well-known single attacks are stored in dataset, it becomes cumbersome for intruder to get entry into network using single attack. This is the main reason that intruders are trying to create and use collaborative attacks to get access of network. To detect attacks, IDS makes use of various machine learning algorithms. These algorithms help IDS in classifying network traffic in normal and attack category. Attack detection by IDS is categorised as *signature-based IDS* and *anomaly-based IDS*. In signature detection, signatures stored in the dataset are used to identify malicious activities happening in the network. Signatures are nothing but footprints stored as a signature in a dataset. Each packet entering in network is matched with signature for attack detection [1]. Limitation of this detection method is datasets that are not up-to-date for modern attacks [2]. Anomaly-based IDS focussed on the behaviour of node or the system or the network to find abnormal activities. Any packet of data entering in the system is analysed according to normal behaviour. If it is deviated from normal behaviour, it is considered as an abnormal behaviour and marked as novel or anonymous attack. Little deviation in normal behaviour is considered as attack becomes limitation of this detection method which can be handled by various machine learning algorithms [3]. In this paper, we are proposing better IDS which will be able to identify or detect collaborative attack along with single attack at a time. Collaborative attack is defined as multiple attacks are happening on the network at the same time or with the difference of fraction of seconds. It can also be any new attack which is created by combining two or more existing attacks. To implement this IDS, use of various machine learning algorithms is done. To improve detection accuracy and reducing false alarm rate, ensemble of classifier is done. Experiments were done on UNSW NB15 dataset to handle modern day attacks. Collaborative attack signatures are created and updated in dataset. Feature reduction is used to reduce features for training classifiers.

14.2 Literature Survey

Network security is prime area on research nowadays because of huge usage of data sharing and Internet. This takes focus of many researchers towards various machine learning, deep learning and data mining algorithms used to improve performance of IDS. Various datasets are available like KDD99, NSL-KDD, DARPA, UNSW-NB15, CIC-IDS2017 and many more. These all datasets are used by researchers to implement IDS for attack detection. In [4], importance and performance of supervised machine learning algorithm are explained with the help of multilayer perceptron neural network model for improving performance of IDS. Importance of ensemble method over single classifier is explained in [5]. An ensemble approach of genetic algorithm and support vector machine to improve performance of IDS is elaborated in [6] which make use of fitness function to reduce error rate. Approaches like ensemble and hybrid classifiers are used to implement IDS [7]. IDS with ensemble classifiers random forest, naïve Bayes and average one-dependence estimator is used by increasing accuracy in [8]. Another ensemble approach is presented in [9] where core vector machine is used to identify all categories of attacks of KDD99 dataset. For each category, one CVM is implemented and tested for attack detection. Results prove that this ensemble approach with CVM provides good efficiency and less computational time. Feature selection is better explored in [10, 11] as chi square, gain ratio and information gain on ensemble and single classifiers. Comparison of various machine learning algorithms according to their performance is explained in [12]. Ensemble with best classifier along with feature selection technique improves IDS performance [13]. Use of feature selection based on genetic algorithms to obtained 99.71% accuracy is elaborated in [14] over NSL-KDD dataset with 15 features. Wrapper feature selection technique can also use to receive better accuracy from classifiers [15]. Performance parameters are elaborated in [16] which help to evaluate overall performance of IDS. Hybrid architecture using both detection methods provides good performance in attack detection [17]. It helps in identifying both known and unknown attacks.

A novel framework is proposed in [18], which use SVM ensemble with feature augmentation. Dataset is transformed into new supervised dataset using marginal density function. Feature selection technique to improve performance of IDS in binary and multiclass classifiers along with hybrid approach is used to test results on NSL-KDD dataset [19]. This approach provides accuracy of 99.77% for binary and 99.63% for multiclass detection. In [20], author focusses on reducing false alarm rate in IDS to improve its performance. Performance of IDS is not only depending on the classifiers used or a feature selection technique used. Ensemble shows better performance as compared to individual classifier [21]. Survey of standard datasets is presented in [22] used for network IDS. Customization of dataset is required for maintaining performance of IDS and increasing detection rate. UNSW NB15 dataset shows better performance in terms of modern attacks [23]. This dataset consists of signatures for all modern attacks which helps IDS to detect various up-to-date attacks.

As per observations in literature survey presented above, we can say that IDS provides improved performance if classifiers are used in ensemble approach rather than individually. Feature selection is also important to reduce computation and training time of classifier. In proposed IDS, multiple supervised and unsupervised machine learning classifiers are used in ensemble to implement IDS. Experiments are conducted on UNSW NB15 dataset. To reduce number of features from dataset, feature selection techniques are used. As none of the benchmarked dataset includes signature of collaborative attack, UNSW NB15 dataset is updated by adding signatures of collaborative attacks. For creation of signature, attack is generated using scripts and penetrated in network so that packets can be captured and signatures can be created. Performance evaluation is done by considering various parameters such as precision, recall, FAR, and accuracy of classifier. Comparison of individual classifier is observed with ensemble approach to check performance of IDS.

14.3 Methodology of Proposed System

In proposed IDS, an ensemble of supervised and unsupervised classifiers is done to improve performance. Classifiers used for ensembling are decision tree (DT), support vector machine (SVM) and XGboost. UNSW NB15 dataset is used for training all classifiers. Testing is performed in real-time environment to check robustness of IDS. To reduce training time of classifier, feature reduction techniques are used.

14.3.1 Signature Creation for Collaborative Attack Over UNSW NB15

UNSW NB15 dataset is generated in Australian Centre for cyber security (ACCS) in real-time environment consists of nine types of attack. Size of dataset is 100 GB with 49 features having last label for attack or normal data. The instances available for this dataset are two million and 540,044. The number of instances in the training set is 175,341, and the testing set is 82,332 instances [24]. This dataset provides improved performance over other dataset [25]. To generate signature of collaborative attack, scripts are written in scapy so that packets of attacks are created and sent in the network. Scapy allows creating packet for new attacks as per the features of dataset. Packets for some collaborative attacks are created which represents combination of multiple attacks. To generate attacks, probe and DoS attacks were combined together to create new attack. Signatures are created update with class label collaborative attack in dataset. Packets for collaborative attacks are captured in real-time environment where scripts are used to generate it. After merging all signatures to dataset, all classifiers are trained using dataset to test in real-time environment. Process for signature creation is represented in Fig. 14.1. Packets of collaborative attacks are randomly generated and penetrated into the network along with the normal traffic. Packets are captured using ethereal and saved in.pcap file. This file is passed for pre-processing and features are extracted from it.

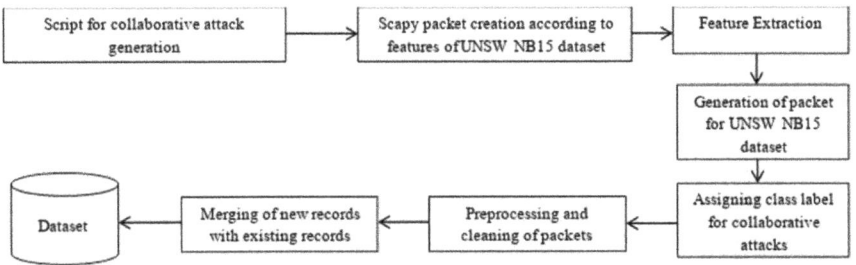

Fig. 14.1 Process for signature creation for collaborative attack

14.3.2   Feature Selection

Feature selection is used for reducing number of features from the dataset. It will help to reduce computational cost as well as training time for classifiers. Features should be reduced in manner that it will not affect performance or accuracy or detection rate of the IDS. Proposed IDS use information gain (IG) for feature selection which is used to calculate quality of each split done for finding the accuracy of the classifier. As decision tree, XGboost and SVM classifiers are used in proposed system, information gain provides better performance as compared to other techniques. Information gain values for each feature are calculated with formula given in Eq. 14.1. As shown in Eq. 14.1, information gain for all 49 features is calculated, and only features having good gain for information are selected. Number of features obtained after reducing it using information gain is 35 which will be used to train all classifiers used in proposed IDS.

$$IG = -\sum_{i}^{c} p_i \log_2 pi \qquad (14.1)$$

where c—number of classes for attacks.

pi—probability of random occurrence of selecting record of class.

14.3.3   Proposed System Architecture

As shown in Fig. 14.2, IDS is implemented using ensemble of DT, SVM and XGboost.
All classifiers were trained using reduced UNSW NB15 dataset with 35 features. As each classifier has its own limitation, it can happen that sometimes classifier results in biased that is wrong output. Ensemble technique is used to improve performance of the IDS. In ensemble, all the four classifier output is passed as input to the majority voting algorithm. This algorithm collects input from all four classifiers, and then according to the input, the final prediction is made. Final decision is made according

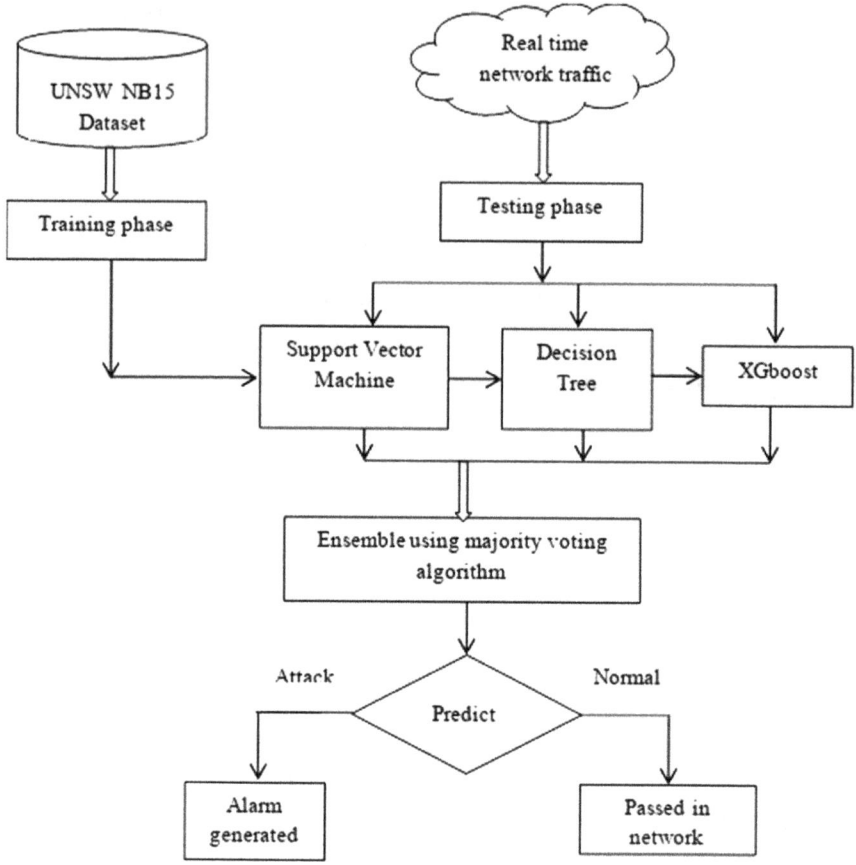

Fig. 14.2 Architecture of proposed ensemble-based IDS

to majority predictions received from all four classifiers. Using ensembling, biased
and incorrect predictions can be reduced. If final prediction is attack, then alarm
is generated for administrator. If prediction is normal, then packet is passed in the
network. All classifiers performance along with ensembling is recorded in real-time
environment with various performance parameters. IDS work efficiently and provides
very good detection rate for both single and collaborative attack detection. This
system provides reduced FAR and increases in accuracy.

14.4 Results and Discussions

Proposed IDS with ensemble of three classifiers is tested for single and collaborative
attack detection with help of new updated dataset. Performance of the classifiers is

evaluated using precision, recall, accuracy and false alarm rate parameters. All classifiers are tested individually on UNSW NB15 dataset along with ensemble technique. Equations from 14.2 to 14.4 show Formula to calculate parameters precision, recall and accuracy.

$$precision = \frac{TP}{TP + FP} \tag{14.2}$$

$$accuracy = \frac{TP + TN}{total} \tag{14.3}$$

$$recall = \frac{TP}{TP + FN} \tag{14.4}$$

where TP—Prediction is positive value and the prediction is correct.

FP—Prediction is positive value, and the prediction is incorrect.

TN—Prediction is negative value, and the prediction is correct.

FN—Prediction is negative value, and the prediction is incorrect.

All these values are used to calculate precision, recall, false alarm rate and accuracy. Comparison of it is shown in Table 14.1.

Precision, recall and accuracy are used to show the robustness and completeness of classifier. Another parameter false positive rate is also used for evaluation. FAR is the total number of wrong predictions made out of actual value. All these parameters are tested for three classifiers individually and in ensemble approach. As per observations from Table 14.1, we can conclude that ensemble approach shows better performance in detection of single attack and collaborative attack. Graphical representation for all parameters with classifier is shown in Fig. 14.3. Figure shows reduction in false alarm rate for ensemble approach and improvement in precision, recall and accuracy of IDS.

As per shown in Fig. 14.3, we can observe that ensemble classifier provides better performance in precision, recall and accuracy as compared to individual classifier. Ensemble approach shows reduced FAR as compared to individual classifier. We can summarise that ensemble approach provides good exactness and completeness in terms of attack detection, increase of 2% accuracy value and 5% reduction in FAR. Ensemble approach removes prediction incorrectness and limitations of biased predication.

Table 14.1 Comparison of individual classifiers with ensemble approach for all parameters

Classifier	Precision	Recall	FAR	Accuracy (%)
DT	0.92	0.78	0.14	91.22
SVM	0.89	0.72	0.19	87.55
XGboost	0.98	0.75	0.15	95.79
Ensemble approach	0.99	0.85	0.07	97.96

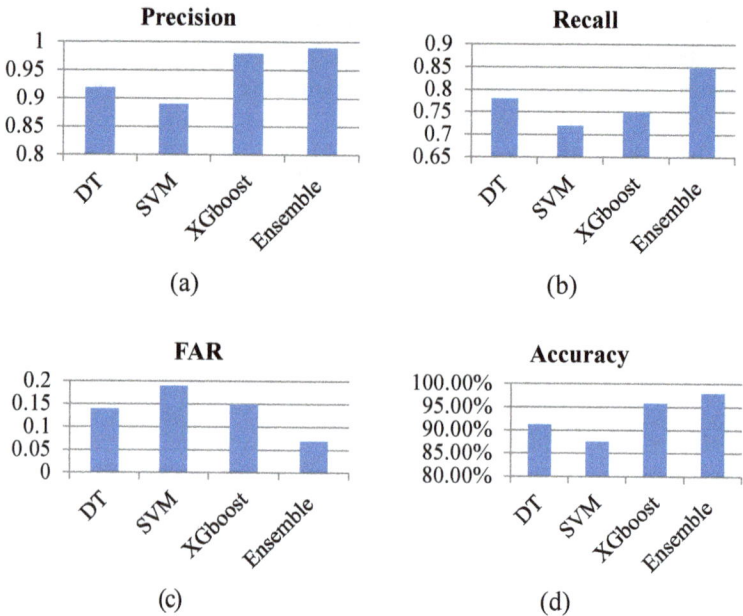

Fig. 14.3 a, b, c, d Graphical representation of comparison of individual classifier with ensemble approach for all parameters

14.5 Conclusion

An ensemble approach is presented in the paper which shows improved performance in terms of precision, recall, false alarm rate and accuracy. Most of the IDS are handling single attack at a time but this IDS will able to handle single as well as collaborative attacks. Signatures for some collaborative attacks are created and updated in existing dataset. Experiments are conducted on three classifiers as DT, SVM, and XGboost. All classifiers are trained using UNSW NB15 reduced 35 dataset features. To avoid biased and incorrect predictions, ensemble technique is used. Ensemble of classifier is done using majority voting algorithm which takes input from all three classifiers gives final prediction according to inputs. Experiments are conducted in real-time traffic, and ensemble approach shows better accuracy than single classifiers. Results prove the improvement of 2% in accuracy and reduction of more than 5% in false alarm rate. In future, proposed IDS can be modified for anomaly detection of unknown attacks. As proposed IDS use only signature-based detection, new or novel attacks cannot be identified and can harm a network. In future, signature and anomaly-based IDS can be combined together to form a new hybrid framework for attack detection.

References

1. Depren, O., Topallar, M., Anarim, E., Kemal, C.: An intelligent intrusion detection system (IDS) for anomaly and misuse detection in computer networks. Expert Syst Appl. **29**(4), 713–722 (2005)
2. Thakkar, A., Lohiya, R.: A review of the advancement in intrusion detection datasets. Proc. Comput. Sci. **167**, 636–645 (2020)
3. Saranya, T., Sridevi, S., Deisy, C., DucChung, T., AhamedKhan, M.: Performance analysis of machine learning algorithms in intrusion detection system: a review. Proc. Comput. Sci. **171**, 1251–1260 (2020)
4. Mebawondu, J., Alowolodu, O., Mebawondu, J., Adetunmbi, A.: Network intrusion detection system using supervised learning paradigm. Sci. African **9**, e00497 (2020)
5. Folino, G., Sabatino, P.: Ensemble based collaborative and distributed intrusion detection systems: aa survey. J. Netw. Comput. Appl. **66**, 1–16 (2016)
6. Tao, P., Sun, Z., Sun, Z.: An improved intrusion detection algorithm based on GA and SVM. IEEE Access **6**, 13624–13631 (2018)
7. Aburomman, A., Reaz, M.: A survey of intrusion detection systems based on ensemble and hybrid classifiers. Comput. Secur. **65**, 135–152 (2017)
8. Jabbar, M., Aluvalu, R., Satyanarayana, S.: RFAODE: a novel ensemble intrusion detection system. Proc. Comput. Sci. **115**, 226–234 (2017)
9. Divyasree, T.H., Sherly, K.K.: A network using intrusion detection system based approach on ensemble cvm efficient feature selection using efficient feature selection approach. Proc. Comput. Sci. **143**, 442–449 (2018)
10. Vinutha, H.P, Poornima, B.: An ensemble classifier approach on different feature selection methods for intrusion detection, Inf. Syst. Des. Intell. Appl. **672**, 442–451 (2018)
11. Gupta, G., Kulariya, M.A.: Framework for fast and efficient cyber security network intrusion detection using apache spark. Proc. Comput. Sci. **93**, 824–831 (2016)
12. Saranya, T., Sridevi, S., Deisyc, C., Chungd, T., Khane, M.: Performance analysis of machine learning algorithms in intrusion detection system: a review. Proc. Comput. Sci. **171**, 1251–1260 (2020)
13. Kunal, Dua, M.: Attribute selection and ensemble classifier based novel approach to intrusion detection system. Proc. Comput. Sci. **167**, 2191–2199 (2020)
14. Gaikwad, D.P., Thool, R.: Intrusion detection system using bagging with partial decision tree base classifier. Proc. Comput. Sci. **49**, 92–98 (2015)
15. Almasoudy, F., Al-Yaseen, W., Idrees, A.: Differential evolution wrapper feature selection for intrusion detection system. Proced. Comput. Sci. **167**, 1230–1239 (2020)
16. Niksefat, S., Kaghazgaran, P., Sadeghiyan, B.: Privacy issues in intrusion detection systems: a taxonomy survey and future directions. Comput. Sci. Rev. **25**, 69–78 (2017)
17. Khonde, S.R., Ulagamuthalvi, V.: Hybrid framework for intrusion detection system using ensemble approach. Int. J. Adv. Trends Comput. Sci. Eng. **9**(4), 4881–4890 (2020)
18. Gu, J., Wang, L., Wang, H., Wang, S.: A novel approach to intrusion detection using SVM ensemble with feature augmentation. Comput. Secur. **86**, 53–62 (2019)
19. Aljawarneh, S., Aldwairi, M., Yassein, M.: Anomaly-based intrusion detection system through feature selection analysis and building hybrid efficient model. J. Comput. Sci. **25**, 152–160 (2018)
20. Hubballi, N., Suryanarayanan, V.: False alarm minimization techniques in signature-based intrusion detection systems: a survey. Comput. Commun. **49**, 1–17 (2014)
21. Li, Y., Wang, J., Tian, Z., Lu, T., Young, C.: Building lightweight intrusion detection system using wrapper-based feature selection mechanisms. Comput. Secur. **28**(6), 466–475 (2009)
22. Ring, M., Wunderlich, S., Scheuring, D., Landes, D., Hotho, A.: A survey of network-based intrusion detection data sets. Comput. Secur. **86**, 147–167 (2019)
23. Moustafa, N., Slay, J.: UNSW-NB15: a comprehensive data set for network intrusion detection systems (UNSW-NB15 network data set). In: Military Communications and Information Systems Conference (MilCIS), IEEE Explore (2015)

24. Moustafa, N., Slay, J.: The evaluation of network anomaly detection systems: statistical analysis of the UNSW-NB15 dataset and the comparison with the KDD99 dataset. Info. Secur. J.: A Global Persp. 1–14 (2016)
25. Sonule, A., Kalla, M., Jain, A., Chouhan, D.: Unsw-Nb15 dataset and machine learning based intrusion detection systems. Int. J. Eng. Adv. Technol. (IJEAT) 9(3), 2638–2648 (2020)

Chapter 15
A Decision Based Asymmetrically Trimmed Modified Geometric Mean Algorithm for the Removal of High Density Salt and Pepper Noise in Images and Videos

K. Vasanth, C. N. Ravi, S. Nagaraj, M. Vadivel, K. Gopi, S. Pradeep Kumar Reddy, and Thulasi Prasad

Abstract A decision based asymmetrically trimmed modified geometric mean algorithm (DBATMGMA) for the removal of high density salt and pepper noise in images and videos is proposed. The algorithm uses fixed 3×3 windows for increasing noise densities. The algorithm initially checks for the presence of outliers (0 or 255) in the processed pixel. If it holds the outliers, the processed pixel is termed faulty. If the processed pixel is faulty, check for the four neighbors are noisy or not (0 or 255). If all the four neighbors are noisy, then check for all the pixels is noisy in the confined neighborhood. If the entire window is noisy, then the corrupted pixel is replaced by mean of all the elements else the corrupted pixel is replaced by mean of the four neighbors. If all the four neighbors are not noisy, then the faulty pixels are replaced with asymmetrically trimmed modified geometric mean. If the pixels does not hold the outlier, then the pixel is considered undamaged and left unaltered. The proposed algorithm is compared with standard algorithms on an image database. The noise suppression along with information preservation capability of the proposed algorithm was found to be very good both in terms of qualitative and quantitative measures.

K. Vasanth (✉) · M. Vadivel · S. P. K. Reddy · T. Prasad
Department of ECE, Vidya Jyothi Institute of Technology, Hyderabad, Telangana, India

C. N. Ravi
Department of EEE, Vidya Jyothi Institute of Technology, Hyderabad, Telangana, India

S. Nagaraj
Department of ECE, Sri Venkateswara College of Engineering and Technology, Chitoor, Andhra pradesh, India

K. Gopi
Department of ECE, Sreenivasa Institute of Technology and Management Studies, Chitoor, Andhra pradesh, India

© The Author(s), under exclusive license to Springer Nature Singapore Pte Ltd. 2021   147
S. C. Satapathy et al. (eds.), *Smart Computing Techniques and Applications*,
Smart Innovation, Systems and Technologies 225,
https://doi.org/10.1007/978-981-16-0878-0_15

15.1 Introduction

Fixed valued impulse noise also called as salt and pepper noise (SPN) distorts image due to transmission errors. Fixed valued impulse noise with equal noise probability for an 8 bit image for very pixel is given by the noise model in Eq. 15.1. If $K(i, j)$ takes pixel values between 0 to 255 for all (i, j) for an 8 bit image, then noise model is given as

$$O(i, j) = \begin{array}{ll} 0 \text{ for probability } p \\ K(i, j) \text{ for probability } 1 - p - q \\ 255 \text{ for probability } q \end{array} \qquad (15.1)$$

where, $p+q$ gives the noise level

A nonlinear operation such as median filter is best suited to eliminate fixed valued impulse noise [1]. Over the years, many nonlinear filters have been proposed and few are listed here below. An adaptive median filter is good in elimination of corrupted pixels but a large window size blurs the image and a smaller window lacks information to restore the image [2]. A decision-based median filter (DBA) [3] used median only for corrupted pixels. At high noise densities a pre-processed neighbor. This induced streaking of edges in images. An asymmetrically trimmed decision-based median filter [4] was proposed. This gave very good noise elimination capability but induced fading effect at very high noise densities. This fading was controlled by an improved algorithm [5] which used global mean instead of local mean if all nine pixels of processing window are corrupted. A new class of trimmed modified Winsorized filter was proposed. These algorithms used modified Winsorized mean [6] or median [7] or variants [8] for the removal of high density salt and pepper noise. All the algorithms work well in eliminating the high density salt and pepper noise. But these algorithms induce ambiguity while eliminating the high density noise. An algorithm has to be formulated for the elimination of high density salt and pepper noise without inducing ambiguity. The attempt in this paper is to achieve the above is by keeping a fixed 3 × 3 window. The proposed algorithm will eliminate the noise proposed in Eq. 15.1. Section 15.2 gives the flowchart of the proposed algorithm. Section 15.3 gives the simulation results and discussions. Section 15.4 gives the conclusion of the work.

15.2 Decision-Based Neighborhood Referred Unsymmetrical Trimmed Modified Geometric Mean (DBNRUTMGMA)

The decision-based neighborhood referred asymmetrical trimmed modified geometric mean algorithm (DBNRATMGMF) is introduced for the removal of high density salt and pepper noise. The DBNRATMGMF filter will also be referred as proposed algorithm. The conventional geometric mean is not stable for increasing noise densities. In order to make it stable, few additional conditions were included as part of the algorithm. Hence, the instability found in conventional geometric mean is eluded in modified geometric mean. In case of videos, the frames are converted into images, and the proposed algorithm is applied on each image and appended later to for a restored video. The flowchart of the proposed algorithm is given in Fig. 15.1.

15.3   Simulation Results

The images used in the paper was used from website of USC Viterbi School of Engineering SIPI database [9]. The noise was added on images from 10 to 90% in increments of 10%. Different algorithms were tested on the noisy images, and performance were tabulated using the quantitative measures such as peak signal to noise ratio (PSNR) and structural similarity index metric (SSIM). The noise removal capability of the algorithm is calculated using the metrics given in Eqs. 15.2, 15.3, and 15.4, respectively.

$$PSNR = 10 \log 10 \left(\frac{255^2}{MSE} \right) \tag{15.2}$$

$$MSE = \frac{\sum_i \sum_j (I_{ij} - R_{ij})2}{\text{size of processed Image}} \tag{15.3}$$

I indicates reference image, and R indicates the filtered image. The SSIM metric is evaluated on image using many 3×3 windows maneuvered in an image. In any area of the image arbitrarily, the evaluation of common window size between two equal sized windows is given by Eq. 15.3.

$$SSIM(x, y) = \frac{(2\mu_I \mu_R + D1)(2\sigma_I \sigma_R + D2)}{(\mu_I^2 + \mu_R^2 + D1)(\sigma_I^2 + \sigma_R^2 + D2)} \tag{15.4}$$

where μ refers to average, and σ indicates the standard deviation of small windows. [10]. The value of the PSNR should lie between 20–24db for subsequent processing of images. A value closer to 1 indicates the structural preservation property which is close to the original image. To quantify the good result of the algorithms additional metrics such as image enhancement factor (IEF), Pratt's figure of merit (Pratt's FOM), normalized cross-correlation (NCC), blur metric (BM) and universal quality index (UQI). Tables 15.1 and 15.2 give the peak signal to noise ratio (PSNR) and structural similarity index metric (SSIM) performance of different algorithms on bridge image corrupted by fixed valued impulse noise. It was inferred from Table 15.1 that high value of PSNR indicates very good noise elimination capabilities of the proposed

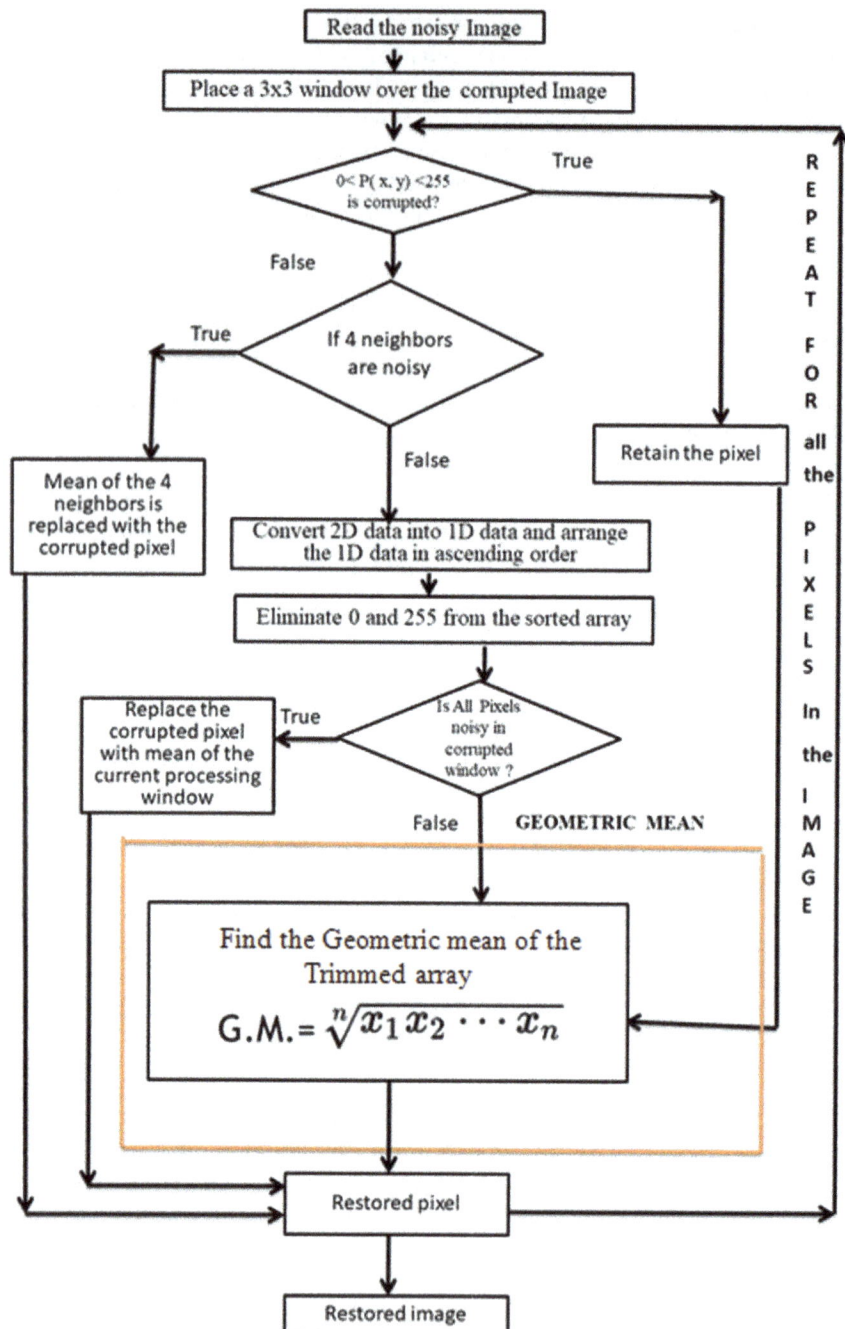

Fig. 15.1 Flowchart of decision-based neighborhood referred asymmetrical trimmed modified geometric mean

Table 15.1 Measure of PSNR of various algorithms on bridge image corrupted by SPN

ND	DBA [3]	MDBUTMF [4]	MDBUTMF_GM [5]	Win mean [6]	Win median [7]	PA
10	28.75	34.68	34.67	34.31	34.44	34.68
20	28.04	31.28	31.21	31.31	31.02	31.61
30	26.56	29.04	28.98	29.33	28.88	29.46
40	25.24	27.16	27.27	27.78	26.98	27.83
50	24.18	25.53	26.68	26.52	25.49	26.51
60	23.08	23.98	23.68	25.17	24.05	25.25
70	21.77	22.23	22.14	23.83	22.20	23.68
80	20.37	20.72	20.34	22.20	20.13	22.06
90	NAN	18.42	18.34	20.04	NAN	19.52

Table 15.2 Measure of SSIM of various algorithms on bridge image corrupted by SPN

ND	DBA [3]	MDBUTMF [4]	MDBUTMF_GM [5]	Win mean [6]	Win median [7]	PA
10	0.928	0.978	0.979	0.977	0.978	0.979
20	0.911	0.955	0.935	0.954	0.953	0.956
30	0.879	0.924	0.925	0.927	0.923	0.929
40	0.836	0.884	0.885	0.895	0.882	0.896
50	0.784	0.832	0.834	0.855	0.831	0.857
60	0.720	0.763	0.763	0.805	0.764	0.805
70	0.644	0.666	0.662	0.737	0.664	0.733
80	0.543	0.529	0.529	0.638	0.523	0.628
90	0	0.342	0.340	0.484	0	0.458

algorithm. It was also noted that structure of the image was well preserved even at very high noise densities. Table 15.3 illustrates the performance of proposed algorithm on different images of the database corrupted by 90% SPN for different image measures. In order to evaluate the information preservation performance of the algorithms, a synthetic image was created using eight visually differentiable gray levels. The image was corrupted by fixed valued impulse noise, and the algorithms were applied on it. It was vivid from image 2 that the proposed algorithm preserved ramp edge but attenuates the roof and step edge, respectively. Ramp edge corresponds to different gray levels of the image. The attenuated similar look alike gray levels are referred to as step image, and they are attenuated by the proposed algorithm. In case of videos as shown in Fig. 15.3, the performance of proposed algorithm on different videos such as viptraffic.avi, rhino.avi, shuttle.avi is corrupted by 50% SPN. It was found that proposed algorithm is very good in noise suppression and information preservation. Hence, this algorithm will be suitable for low-level processing of images corrupted by high density salt and pepper noise. The good performance of the proposed algorithm

Table 15.3 Measures of the proposed filter on various database images with 90% of SPN

Images	PSNR	IEF	MSE	SSIM	FOM	Blur metric	NCC	UQI
aero.tiff	30.43	396	58	0.729	0.352	0.321	0.982	0.255
barbara.tif	21.31	35	480	0.600	0.449	0.366	0.958	0.413
bird.tif	24.61	73	224	0.812	0.480	0.432	0.991	0.428
elaine.png	23.86	62	267	0.645	0.490	0.412	0.982	0.435
fingerprint.bmp	17.15	13	1252	0.443	0.506	0.310	0.919	0.440
Hill.jpg	23.75	62	272	0.594	0.422	0.352	0.985	0.397
Butterfly.bmp	19.92	25	661	0.656	0.476	0.375	0.923	0.396
peppers.gif	18.92	20	833	0.663	0.455	0.425	0.945	0.496
two.bmp	18.25	16	971	0.623	0.467	0.327	0.926	0.390
Pentogon.tiff	23.81	56	269	0.563	0.468	0.334	0.981	0.355
Zelda.png	26.35	114	150	0.768	0.399	0.438	0.979	0.563
House.bmp	22.93	50	331	0.713	0.444	0.366	0.986	0.331
Crater.png	26.48	104	146	0.650	0.445	0.367	0.989	0.358
Cameraman.bmp	16.35	12	1503	0.580	0.448	0.258	0.981	0.272

is mainly due the fact that the asymmetrical trimming of outliers from the data set for which geometric mean is evaluated. The values computed do not use outliers. Hence, the computed modified geometric mean is outlier free. Hence, the performance of the algorithm was found good at very high noise densities (Figs. 15.2, 15.3 and 15.4).

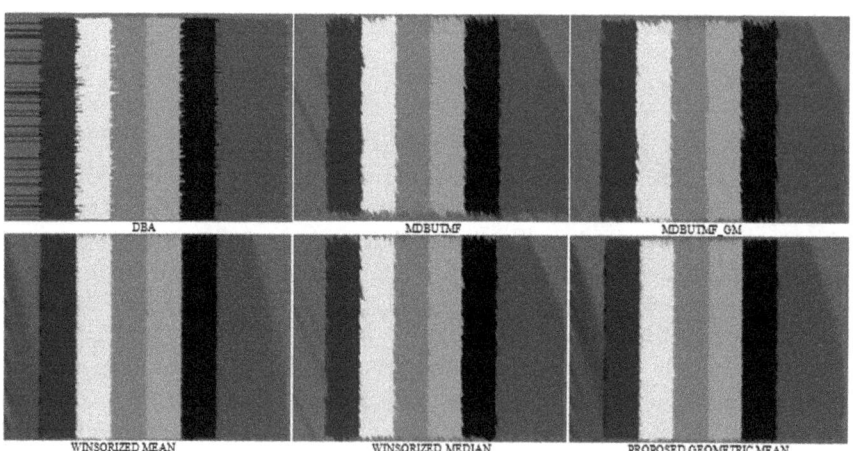

Fig. 15.2 Visual results of different filters/algorithms on ramp synthetic image

Fig. 15.3 Visual results of proposed filter on various images of the database corrupted by 90% SPN

Fig. 15.4 Visual results of decision-based neighborhood referred asymmetrical trimmed modified geometric mean on videos (viptraffic, rhinos, shuttle) corrupted by 50% SPN

15.4 Conclusion

A decision-based neighborhood referred asymmetrically trimmed modified geometric mean filter is proposed for the removal of high density salt and pepper noise in images and videos. The proposed algorithm used a fixed 3 × 3 kernel at high noise densities. The proposed algorithm does not induce any artifacts during the noise removal. Hence, this algorithm was a suitable for high density SPN removal for both images and videos.

References

1. Astola, J., Kusmanen, P.: In: Fundamentals of Non Linear Digital Filtering, CRC press (1997)
2. Hwang, H., Haddad, R.A.: Adaptive median filters: new algorithms and results. IEEE Trans. Image Process. **4**, 499–502 (1995)
3. Srinivasan, K.S., Ebenezer, D.: A new fast and efficient decision-based algorithm for removal of high-density Impulse noises. IEEE Signal Process. Lett. **14**(3), 189–192 (2007)
4. Esakkirajan, S., Veerakumar, T., Subramanyam, A.N., Prem Chand, C.H. : Removal of high density Salt and pepper noise through modified decision based unsymmetrical trimmed median filter. IEEE Signal Process. Lett. **18**(5), 287–290 (2011)
5. Veerakumar, T., Esakkirajan, S., Vennila, I.: An approach to minimize very high density salt and pepper noise through trimmed global mean. Int. J. Comput. Appl. **39**–12, 29–33 (2012)
6. Vasanth, K., Manjunath, T.G., NirmalRaj, S.: A decision based unsymmetrical trimmed modified winsorized mean filter for the removal of high density salt and pepper noise in images and videos. Proc. Comput. Sci. **54**, 595–604 (2015)
7. Christo, M.S., Vasanth, K., Varatharajan, R.: A decision based asymmetrically trimmed modified winsorized median filter for the removal of salt and pepper noise in images and videos. Multimed. Tools Appl. **79**, 415–432 (2020)
8. Vasanth, K., Varatharajan, R.: An adaptive content based closer proximity pixel replacement algorithm for high density salt and pepper noise removal in images. J. Ambient Intell. Human Comput. (2020). https://doi.org/10.1007/s12652-020-02376-2
9. Signal and image processing institute, University of South California https://sipi.usc.edu/dat abase/ Accessed on 10 March (2020)
10. Wang, Z., Bovik, A.C., Sheikh, H.R., Simoncelli, E.P.: Image quality assessment: from error measurement to structural similarity. IEEE Trans. Image Process. **13**(4), 102–109 (2004)

Chapter 16
Anti-poaching System Using Wireless Sensor Networking

T. Asha, H. S. Jahanavi, Sony S. Rao, S. Sinchana, and K. A. Priyanka

Abstract Our life is dependent upon trees. There is a deep relationship between man and trees. Nowadays, there are many incidents about smuggling of trees like sandal and sagwan. These trees are very costly and meager. They are used in the medical sciences, cosmetics and many more. To restrict their smuggling and to save forests around the globe, some preventive measures need to be deployed. A system has been proposed that uses RFID tags and four sensors along with Wi-Fi module which can be used to restrict smuggling. The proposed system is cost effective and efficiently signals the forest officer or any authorities regarding any illegal smuggling activities.

16.1 Introduction

Forests represent more or less half hour of the worldwide acreage. They provide environment for both humans and a few species which share the product of the precious ecosystem. Amerciable research represents one amongst the largest challenges of forests property. Preparation is not associated exclusively with Bharat; China, Australia and African countries are also fighting the same issue. Indian sandalwood prices 12,000 to 13,000 agencies per kilogram while red sandalwood prices INR 10 crore per ton in the international market. The Indian tree has become scarce in recent years. The Indian Government is seeking to restrict the exportation of wood in an attempt to control its attainable loss. Forests provide clean water and air, wood product timber, living environments, fertile soil and leisure opportunities, and these trees make the environment beautified. In addition, they are an important economic resource, which produces highly marketable timber. Despite this amount, ample forestland is being eliminated from forest development, largely due to urban area pretending to be wooded areas, which will remain replaceable.

Wood smuggling has generated problems of social, cultural, legal, law and order in areas bordering Bharat. Trees, which are primarily influenced by the square measure,

T. Asha (✉) · H. S. Jahanavi · S. S. Rao · S. Sinchana · K. A. Priyanka
Department of ECE, GM Institute of Technology, Davangere, India

© The Author(s), under exclusive license to Springer Nature Singapore Pte Ltd. 2021   155
S. C. Satapathy et al. (eds.), *Smart Computing Techniques and Applications*,
Smart Innovation, Systems and Technologies 225,
https://doi.org/10.1007/978-981-16-0878-0_16

represent timber, teak timber and rosewood. There are several measures undertaken by entirely different organizations, and particularly Bharat's Government, to ease this drawback. It involves the housing, coaching and forest-wide training of anti-poaching seekers. Nevertheless, some of the precautionary steps remain ineffective. The highest hopeful resolution is—"Protecting valuable trees from smuggling using RFID and sensors" which can be a strong, reliable and practical technology for watching. In addition, a machine-controlled smart device is developed to fix these issues.

These modules are supplied by the combination of the new wireless communication systems and embedded solutions. Each tree should have one very small system unit embedded with: sensors and RFID tags. The distance of elements higher than the aforementioned will submit this tree status to the server, using Wi-Fi module. The information is sent via twitter, and IoT organize is formed here from now on. The identified data is deciphered by the BLYNK APP, which maintains the information of each of those trees. In cloud, information is investigated to see whether the tree is being covered or pushed down. Nevertheless, the information continues to be constantly informative about the condition of the trees, and the forest specialists should discuss the circumstances of the crisis. Section 16.1 briefly explains about the introduction, problem statement and objective of the project. Chapter 16.2 briefly explains the different papers referred for the literature survey. Sections 16.3 and 16.4 give the brief description about the system design and implementation. Results and conclusion are discussed in the Sects. 16.5 and 16.6.

16.2 Literature Survey

In this section, some of the relevant papers have been studied to gain insight into the existing systems.

Raghavendra et al. [1] proposed IoT-based anti-smuggling system for trees in forest where the main objective of the system is to restrict the smuggling and save the valuable trees so to maintain balanced ecosystem by preventing deforestation. The system uses the GPS technology from which they can find the location of the tree where the poaching is done. It also uses a chip (board) on which various sensors are embedded which are controlled through IoT. These sensors supervise and determine the parameters like tilting, burning and cutting of the trees, and these are accessed on android app installed in android smartphone. Pushpalatha et al. [2] proposed an IOT-based GSM-based real-time forest anti-smuggling monitoring system where the proposed work foiled the plan of an enemy of pirating framework which is valuable in protected forest regions. The thieving and illegal business trees production, such as sandalwood, teak, sagwan and so on, is an important concern. It is a more burglary to flora and fauna of the forests. As residence involved, this belief system is built to combat such sneaking exercise by using the most recent advances.

Naveenraj et al. proposed an IoT-based anti-poaching alarm system for trees in forest to prevent sneaking [3]. The structural frame uses three sensor tilt sensors to recognize the tree's propensity once it is cut, temperature-sensing element to identify timberland fires and sound sensing elements to successfully discover illegal logging. Information obtained from those sensors is calculated consistently. Ringer is enacted for the tilt sensing and sound sensing function, and a water syphon is actuated for temperature sensing portion. Using the Wi-Fi module, generated information is stored in the cloud server. When any occasion arises, forest authorities are suggested, so that correct step can be made. Suguvanam et al [4] presented an idea for innovative protection of valuable trees from smuggling using RFID and sensors. This paper presents a pattern of preventing the importation into the forest of valuable trees [4] such as sandalwood tress, red sandal trees by using GSM module and RFID. In the proposed device, vibration sensors and continuity sensor are used to find scaling down of valuable trees instantly. Here, the vibration sensor is used to detect the disturbance produced when the trees are being cut and the continuity sensor is used to test the continuity between the neighboring trees. RFID is used within the current device to find the missing tags in the tree.

Kotkar et al [5] put forward an anti-smuggling system for trees in forest using flex sensor and Zigbee. They proposed a system which able to restrict smuggling of tree in the forest system using GSM module, Zigbee module, microcontroller and flex sensor. The program also essentially aims to reduce global warming. The species natural environment is also partially conserved. Through this device the trees can also be covered from forest fire. It is in effect tends to reduce the deforestation. Sonwane et al. [6] proposed microcontroller-based design and development of wireless sensor node for anti-poaching capable of exposing theft by keeping track of vibrations produced during the cutting trees using a MEMS accelerometer.

16.3 System Design

The block diagram of anti-poaching system which consists of ESP8266, flame sensor, metal detector sensor, vibration sensor and RFID is shown in Fig. 16.1. The main idea is to design a transportable wireless detector node, which will be a part of wireless detector network. This method includes two modules: One is the sensor module, and the other is the controller module which will be at the tree spot. The information can be received from the detector continuously in the cloud. This is an IOT-based project where information about the detector is continuously uploaded to the cloud server over a Wi-Fi module.

In case of temperature or fire sensor, buzzer is turned on at the time of forest fire. Fire sensor is used to detect the flame or fire. When fire burns, it emits a small amount of infrared light; this light will be received by the photodiode on the sensor module. Fire detector consists of photodiode, resistor, capacitor, potentiometer and LM393 comparator in an IC.

In case of metal sensor, if any metal like axe is came in its range, it will sense and signals are sent to cloud. The operation of metal sensor is based upon the principles of electromagnetic induction. Metal detectors contain one or more inductor coils

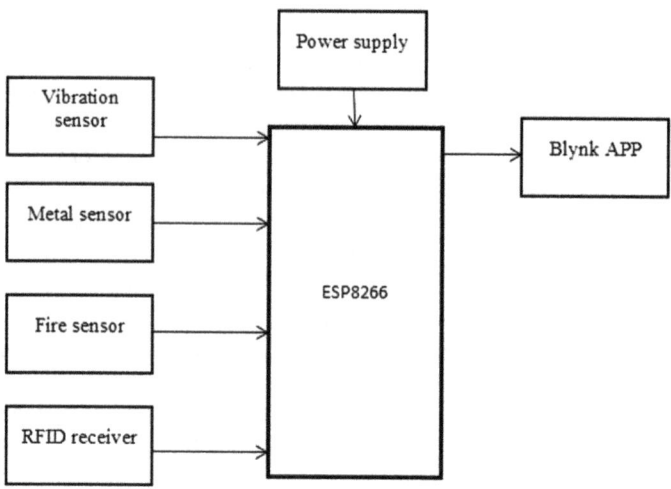

Fig. 16.1 Block diagram of anti-poaching system using wireless sensor networking

that are used to interact with metallic elements. A pulsing current is applied to the coil, which then induces a magnetic field. When the magnetic field of the coil moves across metal, the field induces electric currents (called eddy currents). The eddy currents induce their own magnetic field, which generates an opposite current in the coil, which induces a signal indicating the presence of metal.

RFID is an identification tool for frequency. A simple, low-cost and automatic identification technology uses RF frequency to transfer information between an RFID reader and an RFID tag. It carries a small amount of distinctive information—a serial range or specific characteristic of the object. The data is scanned from a distance, no contact or maybe line of sight required. RFID technology uses an RFID tag, which is a small item, like a paste sticker, which can be hooked up to or inserted into a device. RFID tags include antennas to adjust them from an RFID transceiver to receive and respond to radio-frequency queries.

The vibration sensor type used is SW-18015P/20p, where series are unit spring form, no directional vibration detector trigger switch, and no angle can activate. The switch is an OFF-state electrical circuit, as it is still, when an external force comes into contact and equal vibration, or movement speed provides centrifugal (partial) power, semiconducting system feet can generate instant physical phenomenon is ON-state instant, once the external force is gone, turn back to OFF-state electric circuit. Switch has the identification P at the bottom of the rock to be completely airtight; the lifetime of two lakhs cycles can be reached within the conventional switch period. This series switches on high sensitivity control.

16.4   System Implementation

In these section, implementation details of system design in previous section are explained. Figure 16.2 shows the circuit diagram of anti-poaching system.

16.4.1   Flame Sensor

Flame sensor is used to detect fire flames. The fire sensor module consists of fire sensor and comparator to detect fire up to a range of 1 m. The flame sensor consists of D0 pin, ground pin and VCC pin. D0 pin of flame sensor is connected to the D2 pin of NodeMCU ESP8266, GND pin of flame sensor is connected to GND pin of ESP8266, and VCC pin is connected to 3V3 pin of ESP8266. Input to this module is flame or smoke, which is detected while burning. When the flame is detected, the system sends notification through Blynk application that fire is detected.

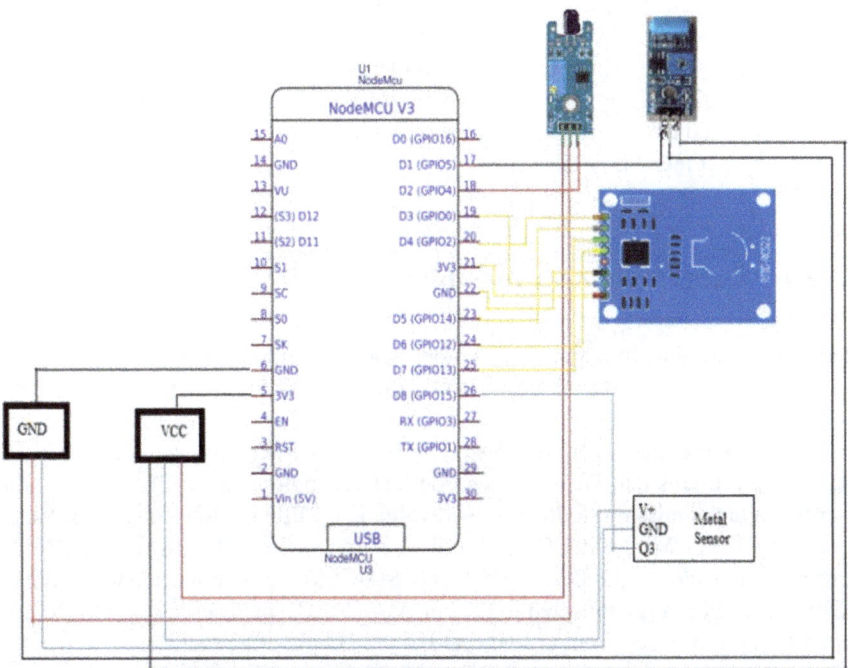

Fig. 16.2 Circuit diagram of anti-poaching system using wireless sensor networking

16.4.2   Vibration Sensor

The vibration-sensing element is also called as piezoelectric sensor. This sensor is used to measure the intensity of vibration. Input to this system is the vibration which is detected while a tree is being cut. When this sensor detects continuous vibration, a notification will be sent to the android phone through Blynk application showing the degree of intensity of vibration. This sensor consists of three pins, namely D0 pin, GND pin and 5 V pin. D0 pin of vibration sensor is connected to the general-purpose D1 pin of ESP8266, GND pin is connected to GND pin, and 5 V pin is connected to 3V3 pin of ESP8266.

16.4.3   Metal Detector Sensor

Metal detector sensor is used to detect the presence of metal. The metal detector sensor module operates by the currents caused in metal objects and responding once it happens. The buzzer rings when the metal is detected. The metal detector sensor consists of three pins, namely D0 pin, GND pin and VCC pin. D0 pin of the metal detector sensor is connected to the D8 pin of ESP8266, GND pin to GND pin and VCC to 3V3 pin of ESP8266. Input to this sensor is the presence of metal. When metal is detected by the sensor, a notification will be sent to the user through Blynk application that metal is detected.

16.4.4   Radio-Frequency Identification (RFID)

Radio-frequency identification (RFID) uses magnetic fields to mechanically establish and track tags connected to things. RFID is one methodology of automatic identification and data capture. The tags contain electronically hold on data. Input to this module is the presence of RFID tags or card. Trees will be provided with an RFID tag. When a tree is cut, RFID tag will be lost and notification will be sent to the user that a tag is missing. RC522 sensor which is a RFID module consists of VCC, reset, GND, IRQ, MISO/SCL/TX, MOSI, SCK and SS/SDA/Rx. RST pin of RC522 sensor is connected to the D3 pin ESP8266, SDA(SS) pin is connected to D4 pin of ESP8266, SCK pin is connected to D5 pin, MISO(SCL) pin is connected to D6 pin, and MOSI pin of RC522 pin is connected to the D7 pin of ESP8266.

16.4.5   Bylnk App

Blynk could be a new platform allowing to quickly create interfaces for managing and track IOS and android hardware ventures. Afterward, a project dashboard is created by installing the Blynk software and organize buttons, sliders, diagrams and other on-screen widgets. Pins can be switched on using the widgets and turn off or view sensor data. Whatever the project is, probably there are many tutorials that render the hardware component relatively straightforward, but software interface construction is very tough. However, the software side is much simpler than the hardware, with Blynk. Blynk is suitable for interfacing with basic tasks such as the temperature control of tank or remotely switch lights on and off. Blynk actually supports most of the Arduino modules, versions Raspberry Pi, ESP8266, ESP32, particle core and a couple of microcontrollers and single-panel computers. Wi-Fi and Ethernet shields are powered by Arduino but can also monitor devices that are often attached to the USB port of a device.

16.4.6   Arduino Ide

Arduino has integrated IDE development environment, a cross-platform application written in the Java programming language. A program for Arduino which is written with the IDE is called a sketch. Sketches are stored on the production machine as text files with the.info file extension, pre-1.0 sketches stored by Arduino software (IDE) with the.pde extension. The Arduino IDE supports the languages C and C++ using special rules of code structuring.

16.5   Results

The snapshots of different steps involved in designing and development of anti-poaching system using wireless sensor networking are shown Figs. 16.3, 16.4, 16.5 and 16.6.

16.6   Conclusion

The theory was completed in order to abstain from bringing valuable trees in protected woodland region. There are various approaches to securing trees, and here, an excellent technique has been done to connect a few sensors around trees with an IoT system. The idea was to recognize each tree as a smart tree (with a controller device and sensors) and to put several such trees under a network (Internet of Things). This

Fig. 16.3 Snap shot of Anti-poaching system

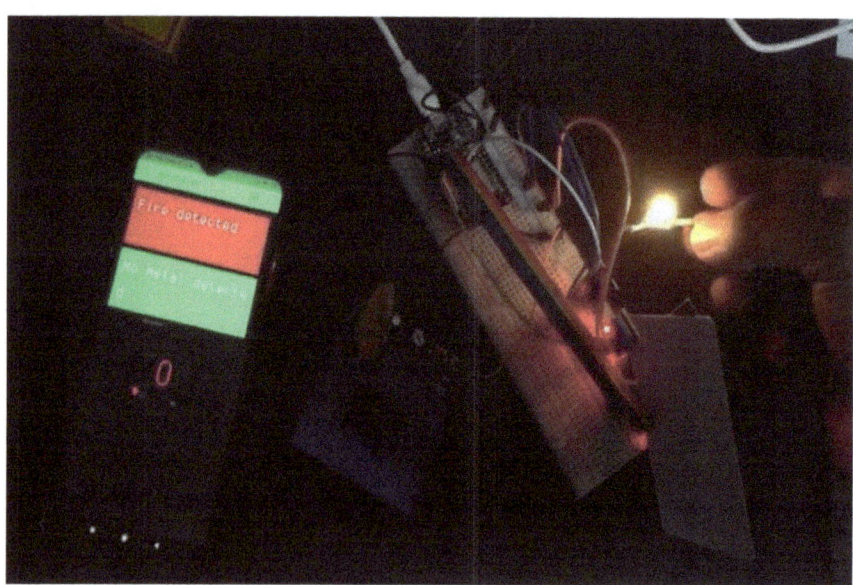

Fig. 16.4 Snapshot of metal detected shown in Blynk app

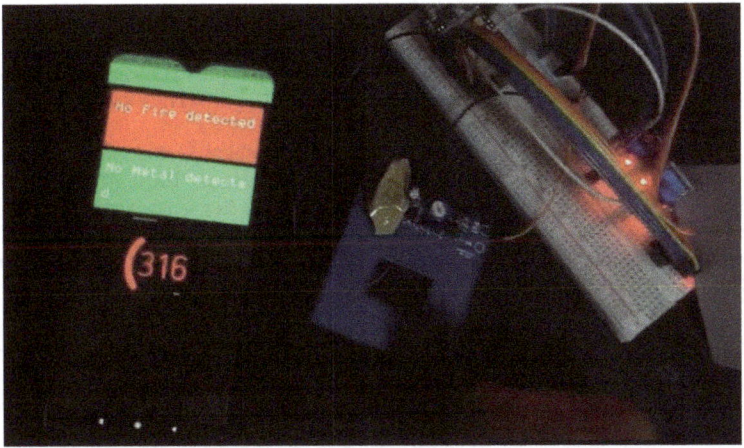

Fig. 16.5 Snapshot of vibration detected shown in Blynk app

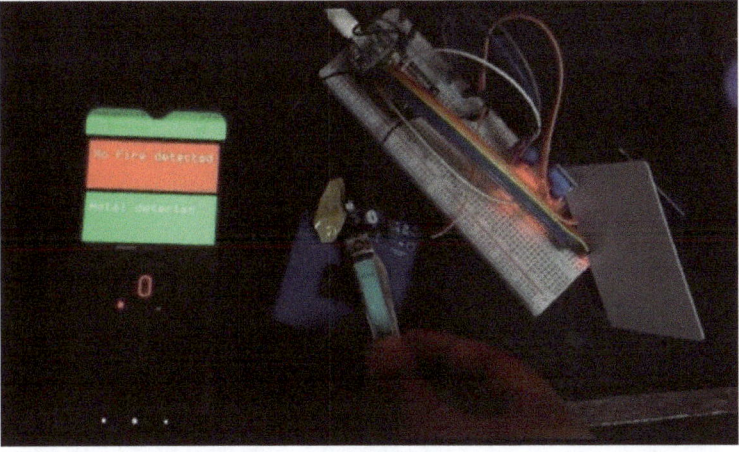

Fig. 16.6 Snapshot of fire detected shown in the Blynk app

system is designed to limit the sneaking of trees in the backwoods where individuals are not present to provide security. Such a system is designed in the backwoods where trees are exorbitant and where their protection is fundamental reality.

References

1. Hegde, R.I.: IoT Based anti-smuggling system for trees in forest. Int. J. Sci. Res. Rev. **8**(6), (2019)

2. Pushpalatha, R., Darshini, M.S.: Real time forest anti-smuggling monitoring system based on IOT using GSM. Int. J. Adv. Res. Comput. Commun. Eng. **8**(2), (2019)
3. Naveenraj, M., Arunprasath, Jeevabarathi, C.T., Srinivasan, R.: IoT Based anti-poaching alarm system for trees in forest. Int. J. Innov. Technol. Explor. Eng. (IJITEE) **8** (6S), (2019). ISSN: 2278–3075
4. Suguvanam, K.R., Senthil Kumar, R., Partha Sarathy, S., Karthick, K., Raj Kumar, S.: Innovative protection of valuable trees from smuggling using rfid and sensors. JIRSET (2017)
5. Kotkar, N.R.: Anti smuggling system for trees in forest using flex sensor and zigbee. IJARCET (2014)
6. Sonwane, A.D., Bhonge, V.N., Khandare, A.: Design and development of wireless sensor node for anti-poaching. IEEE Xplore (2016)

Chapter 17
Performance Issues and Monitoring Mechanisms for Serverless IoT Applications—An Exploratory Study

Shajulin Benedict

Abstract Performance monitoring and analysis of applications are imperative tasks for application developers, including serverless application developers. Application developers, in general, focus on developing their application logic rather than looking into the performance bottlenecks of serverless functions. In this review article, available performance analysis tools and monitoring mechanisms for IoT-enabled serverless applications are explored. Efforts are taken to investigate the existing performance analysis tools, their merits, demerits, and the performance metrics for IoT-enabled serverless applications. Further, various challenges, highlighting the need for the attributes for implementing sophisticated performance analysis tools of IoT-enabled serverless applications, are discussed. This article will be beneficial for the performance analysis tool developers or the serverless cloud application developers.

17.1 Introduction

IoT-enabled application development has gained a wide attraction in various domains such as transportation, agriculture, industry, smart cities, health care, smart buildings, and so forth. The application of serverless functions, empowered by IoT, cloud, and edge computing, has emerged as a profiting computing paradigm in the recent past. Developers have utilized serverless functions to enable an event-driven IoT programming aspect; to niche functions on the pathway of service flows; to trigger stream processing elements; and to bridge an integrated edge-cloud learning framework.

Although serverless computing had exhorted the minds of cloud and IoT application developers, it appears to have been subdued in the utilization expectations owing to the performance concerns and poor practical knowledge of the users. A few

This work is partially funded by IIITKottayam Faculty Research fund and AIC-IIITKottayam project.

S. Benedict (✉)
Indian Institute of Information Technology Kottayam, Kottayam, Kerala 686635, India
e-mail: shajulin@iiitkottayam.ac.in
URL: https://www.sbenedictglobal.com

© The Author(s), under exclusive license to Springer Nature Singapore Pte Ltd. 2021   165
S. C. Satapathy et al. (eds.), *Smart Computing Techniques and Applications*,
Smart Innovation, Systems and Technologies 225,
https://doi.org/10.1007/978-981-16-0878-0_17

notable performance concerns of serverless applications include (i) resource mapping issues, (ii) non-availability of resources, (iii) slow start and warm start issues, (iv) memory allocation issues, (v) security concerns, and so forth. Obviously, tools need to be assisted for pinpointing the performance bottlenecks of these applications.

This paper reveals the multi-cornered performance issues that underscore IoT-enabled serverless applications; explores a generic IoT-enabled serverless architecture for implementing serverless applications; and, lists the merits and demerits of the available performance analysis tools for serverless-oriented IoT-enabled applications. Besides, the performance metrics that are specific to serverless applications and the future research directions are highlighted in the article.

The rest of the paper is discussed as follows: Sect. 17.2 highlights the related works, Sect. 17.3 introduces the generic architecture of IoT-enabled serverless computing environment; Section 17.4 delves into the key performance concerns of IoT-enabled serverless applications; Section 17.5 discloses the tools and their challenges; and, Sect. 17.6 provides a few conclusions and outlooks.

17.2 Related Works

Research works on developing IoT cloud applications have much evidenced the utilization of serverless functions in the recent past owing to the resource efficiency or scalability features. For instances, Gusev et al. [1] have proposed the application of serverless functions on IoT devices to increase the scalability of streaming applications; Razin et al. [2] have applied serverless computing for green oil gas industry. Apart from IoT cloud-specific applications, serverless functions have also become a pristine concept for blockchains [3] and programmers [4–6].

There exist a few platforms for implementing serverless functions and hosting them. Such platforms are available in two dimensions: Opensource and commercial facets–platforms such as Kubeless, OpenWhisk, Fission, OpenFaaS, and so forth are opensource; and, Platforms such as AWS Lambda, Google Cloud Functions, Microsoft Azure Functions, and so forth are commercial platforms for executing serverless functions. Mohanty et al. [7] have analyzed the efficiency of such opensource serverless platforms in terms of the scalability and the response time of handling multiple client requests; similarly, Lee et al. [8] have revealed the efficiency of different serverless platforms while executing the dynamic workload functions on them.

In general, performance monitoring tools could pinpoint the performance problems of applications [9, 10]. Although there are a few performance analysis tools that discuss on the performance concerns of serverless functions, a comprehensive study that exposes the intricacies of those performance monitoring tools and those tools which highlight the future research directions are very rarely available in the literature. Standalone discussions on analyzing the impact of executing serverless functions on the available computing platforms exist. For instance, Eric et al. [11] have manifested the utilization of suitable communication patterns for serverless environments. This paper disclosed the inner details of the available performance tools of serverless functions when executed on serverless platforms.

17.3   Serverless IoT-Working Model

Serverless computing is an event-driven-based execution model that advances in cost ownerships; i.e., the functions are executed on servers only when they need to be executed. Long-serving server applications could benefit from this computing model as the server functions seek available resources only during the executions. Such functions are often individually managed functions. Most commonly utilized serverless functions are single-task functions such as accessing a database, registering user credentials into the database, functioning actuators, invoking a periodic task, scheduling a task, and so forth.

A pictorial representation of the IoT-enabled serverless architecture, where serverless functions could be augmented at four levels of IoT-enabled applications, is shown in Fig. 17.1. The modes of operations or the possible functionalities of serverless functions at each level of the compute nodes are discussed below:

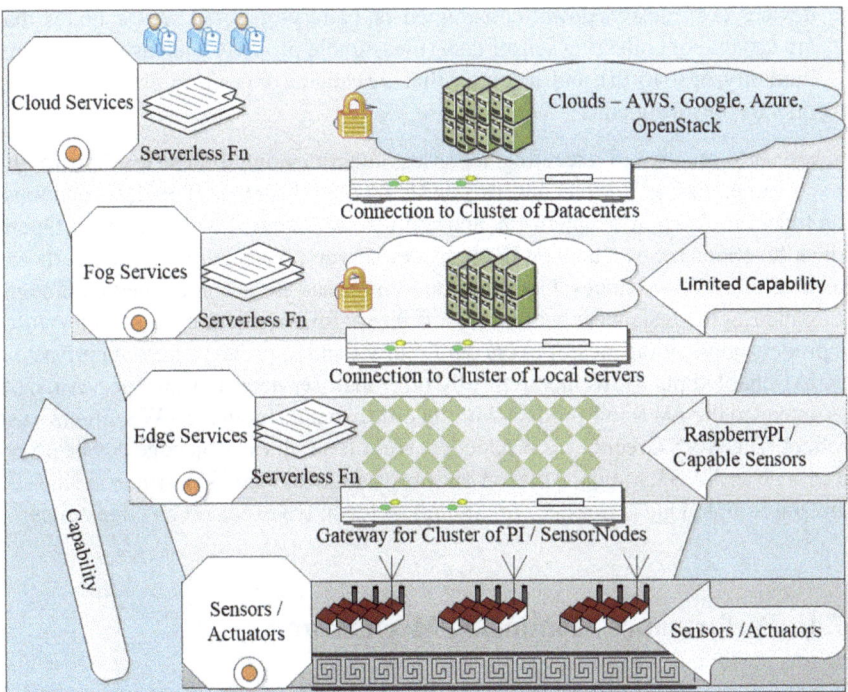

Fig. 17.1   Generic architecture of IoT-enabled applications—the suitability of serverless functions

1. *Cloud level*: Serverless functions are more predominately applied at the cloud level computing machines for various purposes: (a) invoking security flavors, (b) executing data analytic algorithms, (c) performing performance monitoring of cloud resources and applications, (d) dealing with user interfaces, (e) enriching the business capabilities of applications, (f) enabling logging services, and so forth.

2. *Fog Level*: Understanding the performance concerns of transferring all sensor data to the cloud environments and processing them at the cloud, a few compute-capable devices such as servers or VMs are utilized to assist the IoT applications at the fog level. The responsibilities of these fog nodes are (i) to contextually filter sensor data (if required), (ii) to perform inter-operable services between hetero-geneous sensor networks, (iii) to analyze the sensor data based on algorithms, including Machine Learning (ML) algorithms, and so forth.

3. *Edge Level*: Typically, IoT sensors or actuators are connected to edge nodes. The edge nodes are power constraint nodes such as raspberry pi nodes or IoT gateways or pi clusters. The edge-level serverless functions are utilized to trigger sensor nodes or actuator nodes based on the tasks levied in the functions; and, to deal with the fault-tolerance issues of sub-tasks of IoT applications.

4. *Sensor Level*: It is an arduous task to launch serverless functions at the sensor level of IoT applications. IoT sensors or actuators are placed at this level. These devices are, mostly, power-constrained or battery-operated sensor nodes that are capable of collecting sensor data (measurable property such as temperature, humidity, or so forth); and, actuators that are capable of performing actions based on the control signals.

A general practice of executing serverless functions on all these levels is to uti-lize containers/orchestration platforms on servers. For instances, kubeless functions, iron functions, Fission, Fn platform, and so forth are executed on Kubernetes; Open-Whisk functions are executed on IBM Apache servers; AWS Lambda and Sparta are executed on AWS instances; Google Cloud Functions are accomplished on Google Compute Engine instances; and, Microsoft Azure functions are executed on Azure. To process applications nearer to the data sources and to preserve the data privacy, a few IoT-enabled platforms hierarchically offer their services. For instance, Amazon has extended the AWS IoT product with specific capabilities in the AWS GreenGrass product. The AWS GreenGrass product enables cloud services on edge devices. Ser-vices such as AWS Lambda, Amazon Simple Storage Service (S3), Amazon Kinesis, Amazon Simple Queue Service (SQS), and so forth, are executed on edge nodes.

17.4 Performance Monitoring Mechanisms

This section describes the performance monitoring mechanisms and the existing performance issues or challenges of involving serverless functions on IoT-enabled cloud-based applications.

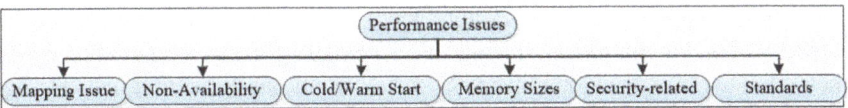

Fig. 17.2 Performance concerns of serverless functions

Performance Issues—Serverless Functions A list of performance issues that influence the efficient working of serverless functions on cloud/edge-level resources is discussed below (see Fig. 17.2).

Mapping issue: Often, the serverless functions are not exactly mapped to the available heterogeneous IoT sensors or edge/fog/cloud resources. The protocols and rules of different sensors belonging to various companies should be compatible so that the sensors are connected each other [4, 12].

Non-Availability: The execution-related resources such as containers, memory, and network should be readily available for keeping serverless functions active in real-time irrespective of their mobility. However, the solidity of availability is beleaguered owing to the provider issues or the issues relating to creating VMs. To improve the availability of functions, retrospection of previous states is provided in some architectures.

Cold or Warm Start Issues: The first time instantiation of VMs or containers for executing serverless functions, also named as `Cold Start` of serverless functions, will certainly draw down the execution time of serverless functions when they are executed on edge environments. Similarly, if the containers are not bootstrapping for the first time, then the image is assumed to be powered ON which is commonly termed as `Warm Start`. In fact, the `Cold Start` leads to a high overhead and the `Warm Start` arrives at high energy consumptions [13, 14].

Impact of Memory Sizes and BigData: Most of the serverless functions offered by providers, such as AWS, Google, or Microsoft, have options to specify the memory sizes—for instance, AWS lamba offers one million requests for function invocations at the free of cost with memory sizes ranging from 128 to 3008 MB. Increasing the memory sizes would extensively increase the CPU instantiation time of applications [15].

Security Concerns: IoT applications, in general, are prone to security issues such as device hacking, authentication issues, cross site scripting, injection attacks, file manipulation, and so forth.

Standards, Protocols, Programming Languages: Lack of proper standards or protocols for serverless functions at the IoT domain has also degraded the performance of IoT-enabled serverless applications. Besides, the utilization of apt programming languages for serverless functions is also an important parameter to improve the performance of IoT applications—for instance, services or functions written in golang or nodejs suffers from a hefty start time due to their inherent dependencies.

Monitoring Mechanisms The possible ways of monitoring serverless functions include: (i) *Tracing Approach:* Tracing based monitoring tracks serverless functions when executed on edge/cloud machines. It is carried out by sniffing into the log files or error messages of serverless functions in a distributed fashion—often, named as *distributed tracing*; (ii) *Sandboxing Approach*: In the sandboxing approach, an isolated and secure environment will be created to study the performance impact of the newer services or functions on computing resources. IoT applications, especially automated vehicles or industry-oriented IIoT applications, require continuous operations or predictive maintenance/allotment [16, 17] to fulfill the automated functioning of machines to enrich the supply-value chain of the upcoming innovative products. Adopting a newer service or a beta-version of data analytic algorithms might lead to a disfunction of the whole machinery. In order to avoid the performance impact of adding such services directly at the production time, the sandboxing approach of performance analysis is recommended; (iii) *Modeling/Prediction Approach*: To avoid performance monitoring overheads, modeling of the end-to-end execution of serverless functions could be a suitable approach for getting the performance insights of functions. For instance, authors of [18] have studied the performance of serverless applications by modeling the dependencies between serverless functions and their associated memories, storage, other computing units; and, (iv) *Language-Inherent or Instrumentation Approach*: The language-inherent or instrumentation approach of monitoring the performance of serverless functions will have the capability of injecting a few predefined functions or notations into the serverless functions before their executions.

17.5 Monitoring Tools—A Comparative Study

The nine most commonly utilized performance analysis tools for monitoring serverless functions in IoT-enabled edge/fog/cloud computing levels are Amazon X-Ray (T1), Google StackDriver (T2), ApplicationInsights (T3), CloudWatch (T4), Prometheus (T5), Thundra (T6), AppDynamics (T7), Wavefront (T8), and Epsagon (T9). A synopsis of these tools are discussed below:

1. T1 -- Amazon X-ray applies the instrumentation features to monitor Lambda functions or services. It is a distributed tracing tool to analyze the performance issues of serverless functions, such as, average latency, utilization chart for databases, function completion ratio (Failure rate), and so forth, when executed on the Amazon Lambda platform;

2. T2 -- Google-Stackdriver offers several preferences such as *Stackdriver Debugger*, *Stackdriver Trace*, and *Stackdriver Profiler*, in the form of the performance monitoring API to the Google Cloud users;

3. T3 -- ApplicationInsights service is most commonly utilized for analyzing the serverless functions of Microsoft Azure namely *Azure Functions*. It requires minimal configurations for delving into the performance details of functions;

4. T4 -- CloudWatch is event-based monitoring tool for AWS Lambda functions;
5. T5 -- Prometheus, combined with the Grafana tool, is a combinatory open-source tool utilized for measuring the performance issues of serverless functions namely (i) Kubeless functions on Kubernetes or (ii) Fx functions of Apache OpenWhisk on container engines;
6. T6 -- Thundra tool provides an asynchronous mode of monitoring of the serverless functions with a provision for including the automated instrumentation feature. It offers an end-to-end monitoring support for serverless applications when executed on heterogeneous computing nodes. The tool is claimed to provide a continuous configuration, enforcement, and verification of security and controls;
7. T7 -- AppDynamics is an application performance management system that analyzes the entire application performance of serverless-enabled applications;
8. T8 -- WaveFront and Delta-WaveFront tools are the products of VMWare that monitor the performance of AWS Lambda serverless functions with a special feature to sum up all metrics together;
9. T9 -- Epsagon tool, an offering from a company with different pricing models, focusses on vividly highlighting the issues relating to the data pipelines of Lambda functions emerging from various IoT sensors. Most preferably, this tool deals with the devices belonging to one organizational domain or docker images.

A comparison of these nine performance analysis tools of serverless functions is discussed in different perspectives: (i) commercial versus open source, (ii) monitoring approach, (iii) expressiveness, (iv) measurement granularity, and (v) solution availability (See Table 17.1). In addition, the capabilities of these tools are given in Table 17.2.

17.5.1 Challenges of Tools—Research Directions

Although performance analysis tools are available in the market, there are several pitfalls which could be further improved in various directions as discussed below: (i) The serverless functions are often short-lived. Naturally, it becomes an arduous task for any performance analysis tools to reproduce the same fault or to trace the same behavior of serverless functions; (ii) Delving into the large volume of performance data available in the log files is a challenging task for the existing tools. Although a few tools visit the log files, they are unable to find the root cause of the performance problem within a short span of time or to depict the faults using sophisticated *visualization* options; (iii) In line with the problems faced by serverless functions of IoT applications, the associated performance analysis tools still experience the issue of vendor lock-in issues. This is a serious concern while considering the realization of a federated IoT-enabled cloud application where the performance analysis tool has to deal with the pros and cons of multiple cloud providers; (iv) Most

Table 17.1 Comparison of performance analysis tools for serverless applications

Description	T1	T2	T3	T4	T5	T6	T7	T8	T9
Commercial	✓	✓	✓	✓		✓	✓	✓	✓
Opensource					✓				
FreeTrial Version	✓	✓	✓	✓	✓		✓	✓	✓
Logs		✓		✓			✓		
Plugin			✓						
Tracing	✓				✓	✓	✓	✓	✓
Library		✓			✓	✓			✓
Daemon/Agent			✓				✓		
Service	✓			✓		✓		✓	✓
Function	✓	✓	✓		✓			✓	
Task		✓	✓						
Application				✓		✓	✓		✓

Table 17.2 Performance monitoring tools—important features

Tools	Measuring devices	Features
T1	Amazon LambdaFunctions, EC2	Allows data annotation
T2	GCE, gateways	Profile approach, distributed tracing
T3	Microsoft Azure Functions	Custom metrics
T4	EC2	Specific to hosts
T5	Containers, nodes, and clusters	Kubernetes cluster based support
T6	Amazon LambdaFunctions, EC2	Asynchronous tracing, low overhead
T7	Nodes, containers	Agent based monitoring
T8	Amazon LambdaFunctions, EC2	Less overhead
T9	Amazon LambdaFunctions, EC2	One organization focus on data pipelines

of the available tools endeavor the integration issues of different sensors or nodes belonging to various vendors. It is a mandatory task for the tools to ensure that they are capable enough to survive such environments while measuring the performance data of the functions; and (v) The existing tools are not capable to precisely identify the energy consumption of serverless functions when executed on federated cloud environments.

17.6   Conclusion

The emergence of IoT-enabled applications has increased in the recent past in various sectors such as smart transportation, smart cities, industry, health care, and so forth. The application of serverless functions for executing such applications improved the performance efficiency of them in terms of metrics such as scalability, latency, and the number of requests handled by services. Although practitioners have continually urged for the performance monitoring tools that could assist them to pinpoint the performance bottlenecks of serverless functions, there exists a very few tools that illustrate the relevant performance metrics. This paper explored the emerging performance concerns of serverless functions that are specific to IoT-enabled serverless applications. In addition, it has given deeper insights into the tools for monitoring serverless functions. The challenges and research directions were also provided for the researchers to focus on in the near future while developing IoT-enabled serverless-based applications or the associated performance monitoring tools.

References

1. Gusev, M., Koteska, B., Kostoska, M., Jakimovski, B., Dustdar, S., Scekic, O., Fahringer, T.: A deviceless edge computing approach for streaming IoT applications. IEEE Internet Comput. **23**(1), 37–45 (2019)
2. Hussain, R.F., Salehi, M.A., Semiari, O.: Serverless Edge Computing for Green Oil and Gas Industry (2019). arXiv:1905.04460v1
3. Chen, H., Zhang, L.: FBaaS: Functional Blockchain as a Service. In: Lecture Notes in Computer Science, pp. 243–250 (2018). https://doi.org/10.1007/978-3-319-94478-4_17
4. Bhattacharjee, A., Barve, Y., Khare, S., Bao, S., Gokhale, A.: Stratum: A Serverless Framework for the Lifecycle Management of Machine Learning-based Data Analytics Tasks. arXiv:1904.01727v1 (2019)
5. Hung, L., Kumanov, D., Niu, X., Lloyd, W., Yeung, K.Y.: Rapid RNA sequencing data analysis using serverless computing (2019). https://doi.org/10.1101/576199
6. Werner, S., Kuhlenkamp, J., Klems, M., Muller, J., Tai, S.: Serverless big data processing using matrix multiplication as example. In: 2018 IEEE International Conference on Big Data (Big Data) (2018). https://doi.org/10.1109/bigdata.2018.8622362
7. Mohanty, S.K., Premsankar, G., Di Francesco, M.: An evaluation of open source serverless computing frameworks. IEEE CloudCom **2018**, (2018). https://doi.org/10.1109/cloudcom2018.2018.00033
8. Lee, H., Satyam, K., Fox, G.: Evaluation of production serverless computing environments. In: IEEE 11th CLOUD'18 (2018). https://doi.org/10.1109/cloud.2018.00062
9. Benedict, S., Rejitha, R.S., Bright, C.: Energy consumption analysis of HPC applications using NoSQL database feature of energyanalyzer. In: ICC 2014, vol. 8993. Springer, LNCS (2014). https://doi.org/10.1007/978-3-319-19848-4_7
10. Benedict, S., Gerndt, M.: Automatic performance analysis of OpenMP codes on a scalable shared memory system using periscope. In: PARA 2010, vol 7134. Springer LNCS (2010). https://doi.org/10.1007/978-3-642-28145-7_44
11. Eric et al. (2019) https://www2.eecs.berkeley.edu/Pubs/TechRpts/2019/EECS-2019-3.html. Accessed in June 2019
12. Aumala, G., Boza, E.F., Ortiz-Aviles, L., Totoy, G., Abad, C.L.: Beyond load balancing: package-aware scheduling for serverless platforms. In: 19th IEEE/ACM International Symposium on CCGRID, pp. 282–291 (2019)
13. Lin, P.-M., Glikson, A.: Mitigating Cold Starts in Serverless Platforms A Pool-Based Approach (2019). arXiv:1903.12221v1
14. Wan, J., Han, S., Zhang, J., Zhu, B., Zhou, L.: An image management system implemented on open-source cloud platform. IEEE IPDPS **2013**, (2013). https://doi.org/10.1109/ipdpsw.2013.176
15. Klimovic, A., Wang, Y., Kozyrakis, C., Stuedi, P., Pfefferle, J.: A Trivedi understanding ephemeral storage for serverless analytics. In: Proceedings of the 2018 USENIX Conference, pp. 789–794 (2018)
16. March, S.T., Scudder, G.D.: Predictive maintenance: strategic use of IT in manufacturing organizations. Inform. Syst. Front. **21**(2), 327–341 (2017). https://doi.org/10.1007/s10796-017-9749-z
17. Sezer, E., Romero, D., Guedea, F., Macchi, M., Emmanouilidis, C.: An Industry 4.0-enabled low cost predictive maintenance approach for SMEs. In: 2018 IEEE ICE/ITMC (2018). https://doi.org/10.1109/ice.2018.8436307
18. Winzinger, S., Wirtz, G.: Model-based analysis of serverless applications. In: 2019 IEEE/ACM 11th International Workshop on Modelling in Software Engineering (MiSE), pp. 82–88 (2019). https://doi.org/10.1109/MiSE.2019.00020

Chapter 18
Detection of Pemphigus Vulgaris in Development Stage of Skin Erosion

Mayuri Vagh and Dipali Kasat

Abstract Pemphigus vulgaris (PV) is an autoimmune blister disease which is a common type of pemphigus. It affects the oral mucosa and skin area. PV blister looks the same as the other blister disease. Thus, diagnosis of PV disease is difficult because of the fact that only one characteristic is identifying the PV blister that is Nikolsky's sign (bulla spread sign). Ruptures of the blisters make erosion on the surface of the skin area. The clinical diagnosis of PV is more painful. Therefore, selecting an area of relevant diagnosis through image processing is important not just for dermatologists but for non-expert person as well. The proposed work is an attempt to provide a solution for the detection of PV in the developmental stages of skin erosion. We developed a systematic and effective framework for identifying PV from relevant information, image data, and the description. Our system works in two sections, the first is interrogation about symptoms and the second to detect blister images. The accuracy rate of the system to successfully detect PV blister is 88.67%.

18.1 Introduction

Blister (Bulla) skin diseases are a group of autoimmune disorders. Pemphigus vulgaris (PV) is one of the blister diseases. The life-threatening PV diseases affect 1–5 patients per 1,000,000 populations per year [1]. PV can be proven to be a serious fatal disease if it is untreated. The majority of occurrences are found between 40 and 60 years of age [2]. The occurrence ratio in women is higher compared to men and the ratio range from 1:2 [3]. In 70–90% of the cases, it presents itself with painful and long-persisting erosion and blisters of the mucous membranes, especially of the oral mucosa that spread to other parts of the body area in its developmental stages [4]. It has an important symptom which is bulla spread sign. In the active phase of PV, it is characterized by the epidermal detachment caused by the pressure at the edge

M. Vagh · D. Kasat (✉)
Computer Engineering Department, Sarvajanik College of Engineering and Technology, Surat, Gujarat, India
e-mail: dipali.kasat@scet.ac.in

© The Author(s), under exclusive license to Springer Nature Singapore Pte Ltd. 2021   175
S. C. Satapathy et al. (eds.), *Smart Computing Techniques and Applications*,
Smart Innovation, Systems and Technologies 225,
https://doi.org/10.1007/978-981-16-0878-0_18

Fig. 18.1 PV blister image
Source by Dermatologist

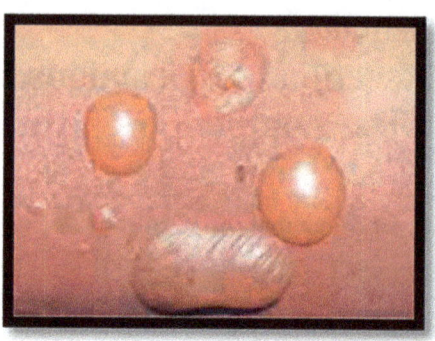

of a blister, and it is also known as Asboe—Hansen sign or Nikolsky's sign [5]. The proposed system provides a non-probing image-based approach for the detection of PV blisters (Fig. 18.1).

18.2 Literature Survey

Schlesinger et al. in their study, "Nail involvement in pemphigus Vulgaris [6]," on 64 patients with PV disease, found that nail changes in 30 out of 64 PV patients, with 14 patients having biopsy-proven PV of the nail. Habibi et al. discussed "Nail changes in pemphigus vulgaris [7]," found that 25 out of 79 PV patients had nail changes during the occurrence of PV disease.

In 2017, Ansari and Sarode, in their work "Skin cancer detection using image processing [8]" used features index information of data present in an image. In 2016, Kumar et al. in their work "Dermatological disease detection using image processing and machine learning [9]" did mapping with the ROIs for skin analysis. It is compared with different classification techniques.

In 2015, Amarathunga, et al. showed "Expert system for diagnosis of skin disease [10]" and performed diagnosis of skin disease by the expert system. The author used the fuzzy logic controller to generate the result with symptoms. In 2018, Zomnrylü, et al., present "The quantitative assessment of progressive dermatologic manifestations in selected ROI (Region of Interest) [11]" developed the algorithm for detecting the blister lesions. It represents a unique attempt to develop a technique of detecting targeted blister lesions in affected skin areas. From the literature survey, we have identified some challenges:

The color after the burst of PV blister is identical to the burnt color and visually the same as the PV blister and the erosion. PV blisters are identical to other autoimmune skin blisters [7]. All autoimmune blisters are visually the same as another blister and covered by the fluid-filled bubble. The identification of significant symptoms of a unique key factor is called Nikolsky's sign [5].

18.3 System Overview

Figure 18.2 depicts a system to decide PV or Not-PV disease based on a flow diagram. The proposed system is divided into two parts: The primary section is an expert system, and the second section is image analysis. In the primary section, we designed the questions for interrogation of symptoms based on some criteria. The next question is determined based on the answers given by the patient to the required questions. In the secondary section, the image analysis consists of preprocessing, segmentation, extract ROI, and feature extraction followed by classification of the blister as "PV" or "Not-PV."

18.3.1 Primary Section/ Interrogate About Symptoms

The primary section is "Interrogate About symptoms" with the expert system. All questions are based on PV symptoms. The questions depend on the answer to the previous question. The system performs diagnosis for PV patients by asking questions related to the conditions that require objective answers.

Figure 18.3 depicts the weightage assigned to answers of the questions asked based on study of symptoms observance in 100 patients. If symptoms are high chance in PV, then the option is more prevalent and carries more weightage. If symptoms show less chance of PV, then the option is less prevalent and carries a lower weightage. The system calculates the weightage at the end of Q/sessions based on the weightage

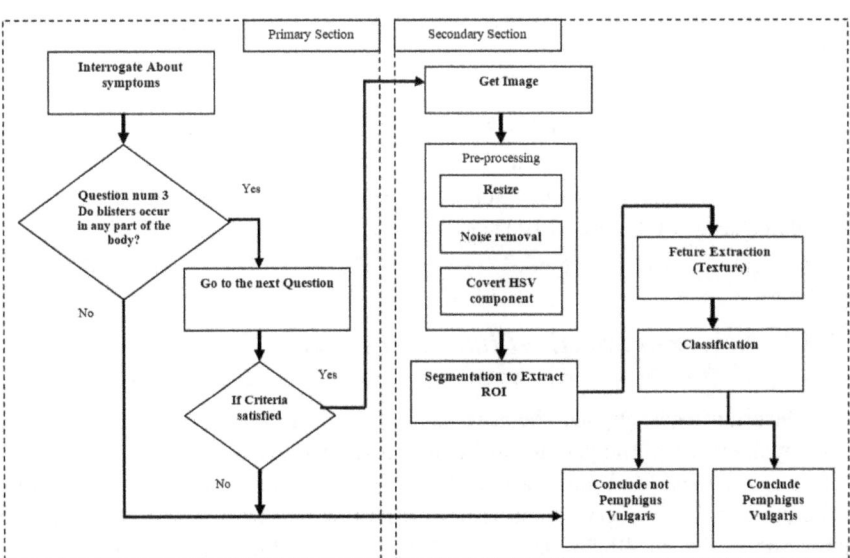

Fig. 18.2 Flow diagram

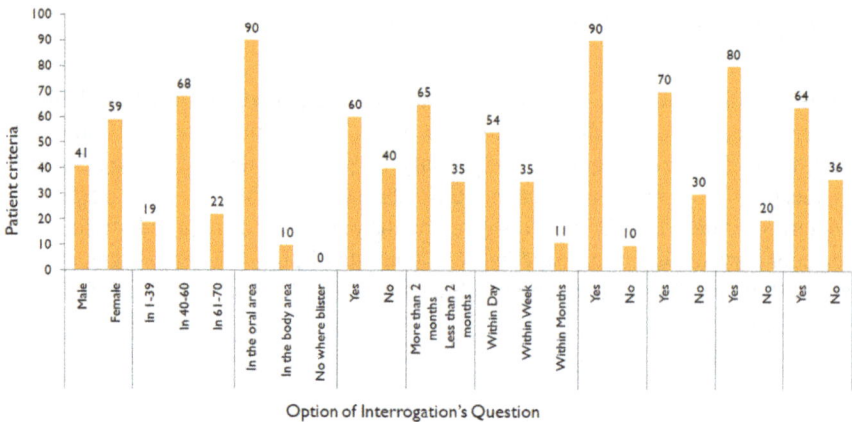

Fig. 18.3 Representation of patient ratio

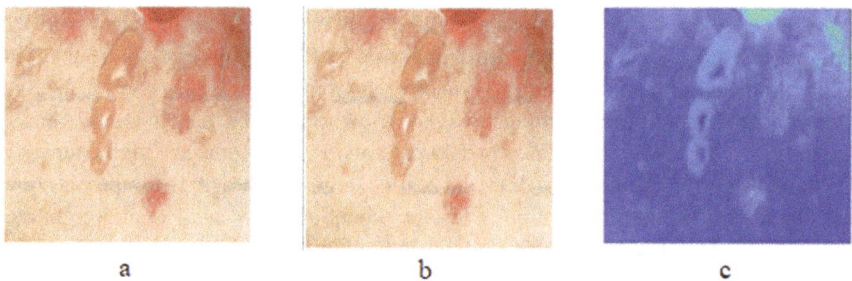

Fig. 18.4 **a** Original image, **b** use resize and noise removal operation, and **c** convert into components of HSV components that are hue, saturation, and intensity

of each answer and deducts the symptoms match result. If the system match is not satisfied with PV disease criteria, then there is no point in going further into the image processing section, but if the criteria are satisfied, then the image analysis part is executed (Fig. 18.4; Table 18.1).

18.3.2 Secondary Section/ Image Analysis

After the criteria get satisfied, the next step is the image analysis of the blister image given by the patient. The patient can submit the image in two ways, capture images using a camera and from the database. After the "get image" process, preprocessing is the primary step. Every image is not in the precise size so, we resize into the standardized size of 400*400. We use Gaussian blur filtering for noise removal in image preprocessing. For the color spaces differentiate the image between hue,

Table 18.1 Questionnaire of interrogation about symptoms

No.	Question	Options	Weight
1	The gender of the patient	Male	8
		Female	10
2	What is the age of the patient?	Age 1–39	8
		Age 40–60	10
		Age 61–70	2
3	Where blisters first appeared?	In the oral area	10
		In the body area	2
		No blister	0
4	Whether blisters are present in the stomatitis area	Yes	10
		No	8
5	A period of blisters in the oral area	More than 2 months	10
		Less than 2 months	2
6	How much time taken for rupturing of blisters	Within day	10
		Within week	8
		Within months	2
7	Whether flaccid arose on healthy skin while rupturing blisters?	Yes	10
		No	2
8	Having itching problem or not?	Yes	2
		No	10
9	Nails are affected?	Yes	10
		No	8
10	Is there any maceration in skin folds?	Yes	10
		No	8

saturation, and intensity components. The saturation component gives the best result. After the saturation color component uses the segmentation method. It is simple and effectively isolates objects in the image. With the help of a threshold, the blister will appear in the foreground. Then, we apply the contour. It will join the line to all the points along the boundary of the image that are having the same intensity. The contour draws the boundary line of the blister shape. We set the region of interest (ROI); it does samples identified for a selected purpose. That blister will be detected with the ROI. Figure 18.5 shows the final results generated by the application with the extracted blister highlighted in green.

The next step is extracting the useful feature which can classify the input image as PV or Not-PV. Important features are color, shape, and texture for the detection of PV blister. We have extracted texture features from the image for classifying the input image as PV or Not-PV. The next step is to classify it as PV or Not-PV. We propose an improvement of classification with the random forest method.

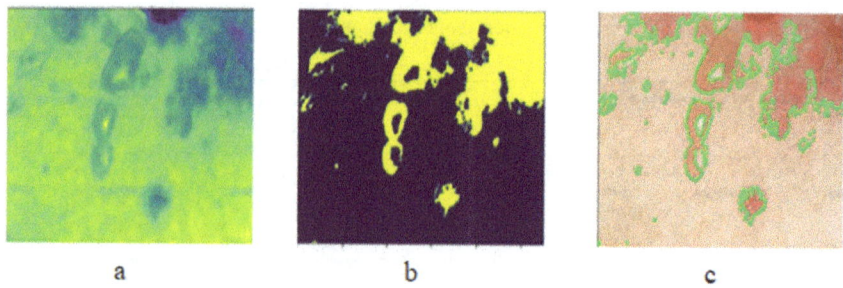

Fig. 18.5 **a** Select saturation component, **b** segmentation, and **c** extracted blisters (ROI)

18.4　Result Analysis

We collected images from different websites of specific blister diseases and the medical expert. The database has 330 images (164 PV blister images, 185 Non-PV). We worked in two different sections: The first section is interrogation about symptoms as the primary section, and the second section is image analysis. In the primary section, the patient will answer the questions from experiences about the blister period. If the criteria of the patient's selected options are calculated to be 50% or less than 50%, then the patient does not have symptoms of PV, so we will suggest a biopsy. And if the criteria of the patient's selected options are calculated to be more than 50%, then the patient has symptoms of PV, so we will suggest uploading the image of blister for the image analysis. After the primary section, the second section is the image analysis. These are the steps: get image (input image), preprocessing, segmentation, feature extraction blister analysis, and the classification of PV blister. For the right accuracy and types of errors, we calculate a confusion matrix. It contains information on the actual and predicted value of the classifier. We get 133 images correctly classify with RF classifier out of 150 images (Table 18.2).

$$\text{Accuracy} = \frac{\text{True Positive} + \text{True Negative}}{\text{True Positives} + \text{False Positive} + \text{True Negative} + \text{False Negative}} \tag{18.1}$$

We classify the image as PV or Not-PV using the RF, SVM, and DT. SVM model sets the clear margin, suitable for a small data set and separation between classes. DT model is a versatile, specific value to each problem, easy to use. Figure 18.6

Table 18.2 Confusion matrix RF, SVM, and DT

	Actual					
	RF		SVM		DT	
Prediction	64	10	45	29	62	12
	7	69	14	62	11	65

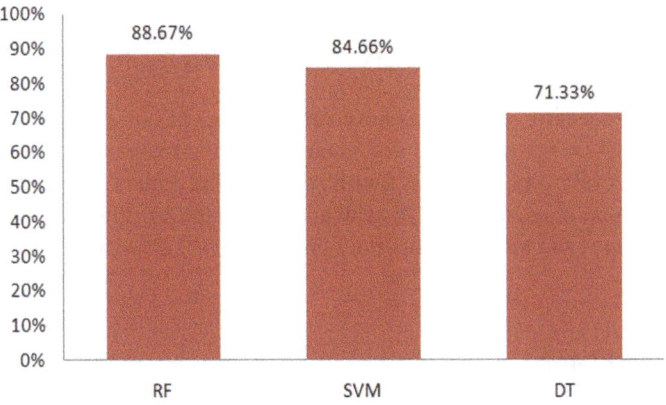

Fig. 18.6 Overall classifier's performance

embarks that RF classifier will handle the missing values and maintain the accuracy of a large amount of data. It would not allow overfitting in the model. RF has a maximum overall accuracy of 88.67% and proved to be better among all for our work. Our system has been tested with different scenarios. The first scenario is if the system does not get any true skin color in the image, then the system not going into the further image processing step. For example, a user uploads a random image without any skin color. The second scenario is if criteria are satisfied but with the user uploading the wrong image (the image blisters are absent). For example, the image is human skin. But the image has no blister. For the PV detection, blister is compulsory. The system will start the further process, but the system will identify the image at the ROI step. The blisters are not available in the image (Table 18.3).

$$Recall = \frac{True\ Positives}{True\ Positives + False\ Negative} \qquad (18.2)$$

Table 18.3 Recall and precision of incorrect input; recall and precision of submit various quality of image

Scenario	Total number of Image	Recall	Precision	Accuracy
Recall and precision of incorrect input				
Not submit the human skin image	20	0.88	0.88	0.9
Submit wrong image after satisfying the criteria	25	0.9	0.83	0.88
Recall and precision of submit various quality of image				
Submit various types of noise image	21	0.94	0.1	0.95
variation in image resolution	25	0.9	0.83	0.88

$$\text{Precision} = \frac{\text{True Positives}}{\text{True Positives} + \text{False Negative}} \qquad (18.3)$$

In the third scenario, the images have various types of noise. By noise in image pixel values that do not reflect the true intensities of the real pixels. But our system classifies the correct image. In the fourth scenario, the patient gives a variation in image resolution. Image pixels seem to blend together to form the image. The system gets glitch images and loses the important pixels that still classify the correct image.

18.5　Conclusion

PV is one of the rare autoimmune blister diseases. PV blisters are painful and develop on the skin, mouth, nose, throat, and genitals. In this paper, we propose a system for the detection of PV. The system works in two sections, the primary section interrogates about symptoms, and the secondary section performs an analysis of the image to classify it as a PV or non-PV image. In classification, phase one is training and the second is the testing phase. Out of 330 total images, 180 images are considered in the training phase, and the remaining 150 images are considered for the testing phase. The resulting analysis proves the 88.67% accuracy obtained. The proposed system can assist non-technical and technical persons to detect the PV blister. The system can be extended for the detection of the other blister disease in the future.

References

1. Shamim, T., Varghese, V.I., Shameena, P.M., Sudha, S.: Pemphigus vulgaris in oral cavity: Clinical analysis of 71 cases. Med. Oral Pathol. Oral Cir. Buccal. **13**(10), 622–666 (2008)
2. Iamaroon, A., Boonyawong, P., Klanrit, P., Prasongtunskul, S., Thongprasom, K.: Characterization of oral pemphigus vulgaris in Thai patients. J. Oral Sci. **48**(1), 43–46 (2006)
3. Tamgadge, S., Tamgadge, A., Bhatt, D.M., Bhalerao, S., Pereira, T.: Pemphigus vulgaris. Contemp. Clin. Dent. **2**(2), 134–137 (2011)
4. Baum, S., Sakka, N., Artsi, O., Trau, H., Barzilai, A.: Diagnosis and classification of autoimmune blistering diseases. Autoimmun. Rev. **13**, 482–489 (2014)
5. Porro, A.M., Seque, C.A., Ferreira, M.C.C., Enokihara, M.M.S.E.S.: Pemphigus vulgaris. Anais brasileiros de dermatologia (2019)
6. Schlesinger, N., Katz, M., Ingber, A.: Nail involvement in pemphigus vulgaris. Br. J. Dermatol. **146**(5), 836–839 (2002)
7. Habibi, M., Mortazavi, H., Shadianloo, S., Balighi, K., Ghodsi, S.Z., Daneshpazhooh, M.: Nail changes in pemphigus vulgaris. Int. J. Dermatol. **47**, 1141–1144 (2008)
8. Bano Ansari, U., Sarode, T.: Skin cancer detection using image processing. Int. Res. J. Eng. Technol. **04** (2017)
9. Kumar, V.B., Kumar, S.S., Saboo, V.: Dermatological disease detection using image processing and machine learning. In: Third International Conference on Artificial Intelligence and Pattern Recognition (AIPR), Lodz, Poland, 19–21 (2016)

10. Amarathunga, A.A.L.C., Ellawala, E.P.W.C., Abeysekara, G.N., Amalraj, C.R.J.: Expert system for diagnosis of skin diseases. Int. J. Sci. Technol. Res **4**(1), 174–178 (2015)
11. Zeljkovic, V., Druzgalski, C., Mayorga, P., Tameze, C.: Quantitative assessment of progressive dermatologie manifestations in selected ROI (Region of Interest), Global Medical Engineering Physics Exchanges/Pan American Health Care Exchanges (GMEPE/PAHCE), Mexico, 20–25 (2018)

Chapter 19
Anomaly-Based Intrusion Detection System in Two Benchmark Datasets Using Various Learning Algorithms

Thongam Jayaluxmi Devi and Khundrakpam Johnson Singh

Abstract The research in network intrusion detection has escalated since past few years. There are various methods and systems being proposed related to intrusion detection. But the changing nature of attack leads to the need to deal with every possible aspect to increase the detection efficiency. The anomaly-based intrusion detection has always been in the limelight because of its detection capability of the unknown pattern. In the study, feature reduction is performed using gain ratio, correlation, information gain and symmetrical uncertainty, and the selected features are used to train the machine learning techniques such as Naive Bayes classifier, sequential minimal optimization (SMO), J48 classifier and random forest. The KDD cup 99 and NSL-KDD datasets were used as a benchmark reference in building the models. The experiment was carried out with an aim to compare the performance of various classifiers between the two datasets. Results show that the feature reduction improves the detection efficiency and can obtain accuracy as high as 99.91%.

19.1 Introduction

The evolution in Internet technologies with concurrent advancement in the number of network attacks has led to the escalation of the importance of research in network intrusion detection. Despite the phenomenal improvements in the area of research, still there are various ways to improve and enhance detection and prevent the network-based attacks. According to [1], "threat is a potential possibility of a deliberate unauthorized attempt to (i) access information, (ii) manipulate information or (iii) render a system unreliable or unusable." The present world solely depends on technology. And with time almost all data are stored in the cloud, and every work is performed online. As such, it becomes a necessity to secure our data or system from intrusions. "The term anomaly-based intrusion detection in networks refers to the problem of finding exceptional patterns in network traffic that do not conform to the expected normal behavior"[2]. "The fundamental concept behind this technique is to define

T. J. Devi (✉) · K. J. Singh
Department of CSE, NIT Manipur, Imphal, India

© The Author(s), under exclusive license to Springer Nature Singapore Pte Ltd. 2021
S. C. Satapathy et al. (eds.), *Smart Computing Techniques and Applications*,
Smart Innovation, Systems and Technologies 225,
https://doi.org/10.1007/978-981-16-0878-0_19

the behavior of the network and/or system, where this predefined behavior is then compared with the normal behavior. The result will be either to accept it or to trigger alarm management system for further investigation"[3].

According to [4], we have two main classifications of intrusion detection system: signature-based and anomaly-based. Signature-based includes exploring network traffic for any malicious bytes or packet sequences. It is easy to establish and acknowledge if we know the behavior of the incoming network. The problem arises when we do not have the pattern of the incoming traffic. In order to increase its validity, we need to update and upgrade our model for any new attack pattern. The anomaly-based intrusion detection system need not know the pattern of the attack in order to alert the system. By defining a normal behavior, the anomaly-based IDS is ready to operate. Any observation of deviation from the normal behavior or the pattern will give an alarm to the system for intrusions.

The rest of the paper consists of related works, datasets, proposed methodology, results and discussion and conclusion.

19.2   Related Works

Table 19.1 shows some of the advantages and limitations of various surveys.

19.3   Datasets

Two datasets have been used in the study: KDD cup 99 and NSL-KDD.

19.3.1   KDD Cup 99

The KDD cup 99 dataset was developed at the MIT Lincoln Laboratories. It is classified into four main attacks: denial of service (DOS), remote to local (R2L), user to root (U2R) and probing. KDD dataset consists of 41 features and one label which specifies the attack name. Table 19.2 shows the features of KDD cup 99.

19.3.2   NSL-KDD

In the study of KDD, two major problems arise, i.e., redundant and duplicate records in training and testing datasets, respectively, which tremendously affect the performance of evaluated systems that leads to the introduction of NSL-KDD.

Table 19.1 List of surveys

Authors	Datasets	Methods	Advantages	Limitation
Veeramreddy et al. [5]	NSL-KDD	Metaheuristic technique, canonical correlation analysis	FCAAIS minimized the process complexity of designing the scale using FAIS	Less detection accuracy
Hossein [6]	KDD cup 99	K-nearest neighbor (K-NN) and K-means clustering (KMC) algorithms, information theory	Good approach for detecting U2R and probe attacks	Fails to detect "phf" and "mailbomb" attacks
Yang et al. [7]	KDD cup 99, NSL-KDD	New deep neural network (NDNN)	Use of adaptive moment estimator makes the parameter more stable and gives an increased and more effective convergence rate	The increase in network structure will lead to exponential progression of computing time and other issues
Mohammed et al. [8]	KDD cup 99, NSL-KDD and Kyoto 2006 +	Least square support vector machine-based IDS (LSSVM-IDS)	The proposed feature selection method eliminates the redundancy parameter required in MIFS and MMIFS	The overall detection rate is low
Yihan et al. [9]	KDD cup 99	Convolutional neural network (CNN), principal component analysis (PCA), auto-encoder (AE)	The proposed method shares same convolutional kernels, which reduces the number of parameters and calculation amount of training and hence can quickly identify attack type of traffic data	Low detection rates of U2R and R2L

Table 19.2 Features of KDD cup 99

Feature No	Feature Label	Feature No	Feature Label
1	duration	22	is_guest_login
2	protocol_type	23	count
3	service	24	srv_count
4	flag	25	serror_rate
5	src_bytes	26	srv_serror_rate
6	dst_bytes	27	rerror_rate
7	land	28	srv_rerror_rate
8	wrong_fragment	29	same_srv_rate
9	urgent	30	diff_srv_rate
10	hot	31	srv_diff_host_rate
11	num_failed_logins	32	dst_host_count
12	logged_in	33	dst_host_srv_count
13	num_compromised	34	host_same_srv_rate
14	root_shell	35	dst_host_diff_srv_rate
15	su-attempted	36	dst_host_same_src_port_rate
16	num_root	37	dst_host_srv_diff_host_rate
17	num_file_creations	38	dst_host_serror_rate
18	num_shells	39	dst_host_srv_serror_rate
19	num_access_files	40	dst_host_rerror_rate
20	num_outbound_cmds	41	dst_host_srv_rerror_rate
21	is_host_login		

19.4   Proposed Methodology

The workflow of the proposed model is shown in Fig. 19.1. It consists of data collection, preprocessing, feature selection and a classifier. The work is implemented in Weka which is a workbench for machine learning.

The data collected from the Web are first converted into comma-separated value (CSV) format. The class labels are given in different attack names, and since it is anomaly-based detection, all the attacks are assigned as anomaly and remaining as normal.

After preprocessing, the next step is feature selection. "Feature selection methods are used to reduce the dimensionality of the data by removing the redundant and

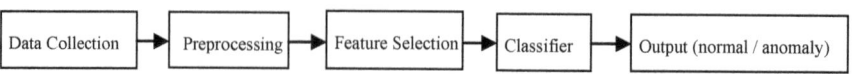

Fig. 19.1 Workflow of the proposed model

irrelevant attributes in a dataset" [10]. In the study, four attributes evaluation, viz. gain ratio, correlation, information gain and symmetrical uncertainty are used. The training data of each dataset run through one attribute evaluator at a time using ranker as the search method which produces some scores for all attributes. And the mean, median and mode of the scores of each attribute evaluator are calculated.

Only those features are selected whose scores are more than or equal to the calculated mean, median and mode separately. Thereafter, selection of features is performed in such a way that the occurrence of the feature is more than or equal to three (since four attribute evaluators are used). Equations (19.1), (19.2) and (19.3) are used for calculating mean, median and mode, respectively.

$$\text{Mean} = \frac{\text{Sum of all feature scores}}{\text{Number of features}} \tag{19.1}$$

$$\text{Median} = \begin{cases} \frac{n}{2}, & \text{when } n \text{ is odd} \\ \frac{(n+1)}{2}, & \text{when } n \text{ is even} \end{cases} \tag{19.2}$$

where n is the number of features

$$\text{Mode} = l + \left(\frac{f_1 - f_0}{2f_1 - f_0 - f_2}\right) \times h \tag{19.3}$$

where l defines the lower limit, h represents the size of the class interval, and f_1, f_0 and f_2 are the frequency of the modal class, class preceding the modal class and class succeeding the modal class, respectively.

In order to get the best features, we have taken the mean and median (as in case of mode, the value comes out to be zero and since the scores of all the features were greater than zero, we need to select all the features and hence it was discarded) as the selection criteria. It was found that the features selected based on median have higher accuracy for both KDD cup 99 and NSL-KDD datasets. These median-based features were used to train the classifiers. The selected features for KDD cup 99 and NSL-KDD datasets are given in Tables 19.3 and19.4, respectively.

The performances can be calculated from the resultant confusion matrix (Table 19.5). For calculating the TP rate and recall, Eq. (19.4) has been used. Equations (19.5), (19.6), (19.7) and (19.8) were used for calculating the FP rate, precision, F-measure and accuracy, respectively.

$$\text{True positive rate (TPR)} = \text{Recall} = \text{TP}/(\text{TP} + \text{FN}) \tag{19.4}$$

$$\text{False positive rate (FPR)} = \text{FP}/(\text{FP} + \text{TN}) \tag{19.5}$$

$$\text{Precision} = \text{TP}/(\text{TP} + \text{FP}) \tag{19.6}$$

Table 19.3 Selected features of KDD cup 99

Feature no.	Feature label	Feature no	Feature label
1	Duration	29	Same_srv_rate
2	Protocol_type	20	Diff_srv_rate
3	Service	31	Srv_diff_host_rate
4	Flag	32	Dst_host_count
5	Src_bytes	34	Host_same_srv_rate
6	Dst_bytes	35	Dst_host_diff_srv_rate
12	Logged_in	36	Dst_host_same_src_port_rate
23	Count	37	Dst_host_srv_diff_host_rate
24	Srv_count	38	Dst_host_serror_rate
25	Serror_rate	39	Dst_host_srv_serror_rate

Table 19.4 Selected features of NSL-KDD

Feature no	Feature label	Feature no	Feature label
3	Service	31	Srv_diff_host_rate
4	Flag	32	Dst_host_count
5	Src_bytes	33	Dst_host_srv_count
6	Dst_bytes	34	Host_same_srv_rate
12	Logged_in	35	Dst_host_diff_srv_rate
23	Count	36	Dst_host_same_src_port_rate
25	Serror_rate	37	Dst_host_srv_diff_host_rate
26	Srv_serror_rate	38	Dst_host_serror_rate
29	Same_srv_rate	39	Dst_host_srv_serror_rate
30	Diff_srv_rate	40	Dst_host_rerror_rate

Table 19.5 Confusion matrix

Predicted			
a	b		
TP	FP	a = normal	Actual
FN	TN	b = anomaly	

$$F - measure = 2/\,(1/Precision) + (1/Recall) \tag{19.7}$$

$$Accuracy = (TP + TN)/\,(TP + TN + FP + FN) \tag{19.8}$$

Table 19.6 Algorithms of classifiers

Algorithm for naive bayes [12]	Algorithm for SMO [13]
Step 1: Read the training dataset T; Step 2: Calculate the mean and standard deviation of the predictor variables in each class; Step 3: Repeat Calculate the probability of f_i using the Gauss density equation in each class; Until the probability of all predictor variables (f_1, f_2, f_3,…, f_n) has been calculated Step 4: Calculate the likelihood for each class; Step 5: Get the greatest likelihood;	Step 1: Input C, kernel, kernel parameter and epsilon Step 2: Initialize $a_i \leftarrow 0$ and $b \leftarrow 0$ Step 3: Let $f_a(x) = b + \sum_{i=1}^{m} y_i a_i k(x, x_i)$ and τ the tolerance Step 4: Find Langrange multiplier a_i which violates KKT optimization Step 5: Choose second multiplier and optimize pair. Repeat steps 4 and 5 till convergence Step 6: Update a_1 and a_2 in one step a_1 can be changed to increase $f_a(x_1)$ a_2 can be changed to decrease $f_a(x_2)$ Step 7: Compute new bias weight b
Algorithm for J48 Classifier [14]	**Algorithm for Random Forest** [15]
Step 1: If all cases are of the same class, the tree is a leaf and so the leaf is returned labeled with this class; Step 2: For each attribute, calculate the potential information provided by a test on the attribute (based on the probabilities of each case having a particular value for the attribute). Also calculate the gain in information that would result from a test on the attribute (based on the probabilities of each case with a particular value for the attribute being of a particular class); Step 3: Depending on the current selection criterion, find the best attribute to branch on	Step 1: The algorithm selects random samples from the dataset provided Step 2: The algorithm will create a decision tree for each sample selected. Then it will get a prediction result from each decision tree created Step 3: Voting will then be performed for every predicted result. For a classification problem, it will use mode, and for a regression problem, it will use mean Step 4: And finally, the algorithm will select the most voted prediction result as the final prediction

19.4.1   The Classifiers

"A classifier is any algorithm that sorts data into labeled classes, or categories of information"[11]. There are four classifiers, namely Naive Bayes, sequential minimal optimization, J48 and random forest which are used in the study and are run separately for the comparison of performance. Table 19.6 shows the algorithms of classifiers used.

19.5   Results and Discussion

In KDD cup 99, the Naive Bayes takes the least time with 2.26 s with an accuracy of 98.37% (See Table 19.7). The SMO takes 3249.26 s, 54 min approx. with an accuracy of 99.18%. The J48 takes 22.49 s, giving an accuracy of 99.97%. RF takes

Table 19.7 Performance of KDD cup 99 and NSL-KDD datasets using various classifiers

Dataset	Classifier	Metrics	Time taken (in sec)	TPR	FPR	Precision	F-Measure	Accuracy (%)
KDD cup 99	Naive Bayes		2.26	0.984	0.012	0.985	0.984	98.37
	SMO		3249.26	0.992	0.004	0.992	0.992	99.18
	J48		22.49	1	0	1	1	99.97
	RF		163.24	1	0	1	1	99.98
NSL-KDD	Naive Bayes		0.78	0.904	0.104	0.908	0.903	90.39
	SMO		1196.11	0.973	0.029	0.973	0.973	97.28
	J48		12.21	0.998	0.002	0.998	0.998	99.76
	RF		56.08	0.999	0.001	0.999	0.999	99.91

163.24 s, approximately 2 min, giving the maximum accuracy of 99.98%. If we go on with the amount of accuracy, RF can be chosen as the best classifier among the rest. However, considering the time taken in building the model as another factor along with the accuracy, the J48 can be a better choice.

In NSL-KDD, the Naive Bayes takes 0.78 s, giving an accuracy of 90.39%. SMO takes 1196.11 s approximately equal to 19 min with an accuracy of 97.28%. The J48 takes 12.21 s in building the model achieving an accuracy of 99.76%. The RF takes 56.08 s and gives the highest accuracy of 99.91%. Though SMO yields a strong accuracy rate, it is not preferable since the amount of time it takes in building the model is high. The Naive Bayes cannot be a good option because of its low accuracy rate, though it is a fast-building model. We can opt for J48 or RF according to our needs and requirements about how fast we want to detect and how accurate we want to get. The graph representation of the four learning algorithms based on accuracy is shown in Fig. 19.2.

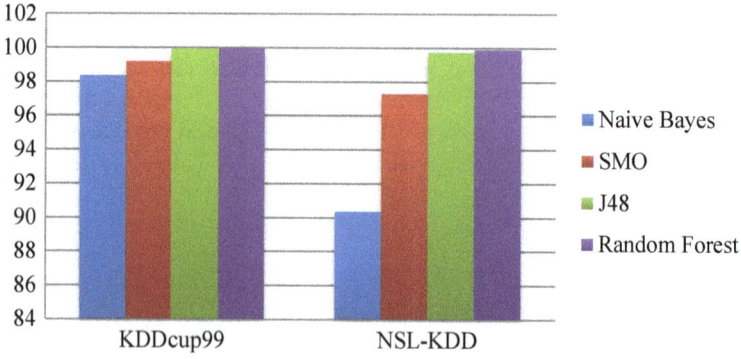

Fig. 19.2 Performance comparison of the four learning algorithms based on accuracy

19.6 Conclusion

From the above analysis, we can see that the accuracy rate based on the KDD cup 99 is greater than the NSL-KDD dataset on every classifier model. This is due to the fact that the KDD cup 99 dataset contains many redundant records which results in the classifier to fall bias toward the more frequent records, hence giving more accuracy rate. The experiment results show that the J48 and random forest classifier effectively detects the intrusion detection using feature reduction methods.

Acknowledgements The authors would like to thank the reviewers whose comments and feedbacks would help in the improvement of the paper.

References

1. Anderson, J.P.: Computer Security Threat Monitoring and Surveillance. James P. Anderson Co. Box 42 Fort Washington, Pa. 19034 (1980)
2. Bhuyan, M.H., Bhattacharyya, D.K., Kalita, J.K.: Network anomaly detection: methods, systems and tools. IEEE Commun. Surv. Tutor. **16**(1), 303–336. First Quarter (2014). https://doi.org/10.1109/SURV.2013.052213.00046
3. Singh, K.J., De, T.: DDOS Attack detection and mitigation technique based on http count and verification using CAPTCHA. In: 2015 International Conference on Computational Intelligence and Networks, Bhubaneshwar, pp. 196–197 (2015). https://doi.org/10.1109/CINE.2015.47
4. Veeramreddy, J., Prasad, V., Koneti, P.: A review of anomaly based intrusion detection systems. Int. J. Comput. Appl. **28**, 26–35 (2011). https://doi.org/10.5120/3399-4730
5. Jyothsna, V., Prasad, K.M.: Anomaly-based intrusion detection system. In: Computer and Network Security, Jaydip Sen, IntechOpen (2019). https://doi.org/10.5772/intechopen.82287
6. Shirazi, H.M.: Anomaly intrusion detection system using information theory, K-NN and KMC algorithms. Austral. J. Basic Appl. Sci. **3**(3), 2581–2597 (2009)
7. Jia, Y., Wang, M., Wang, Y.: Network intrusion detection algorithm based on deep neural network. IET Info. Secur. **13**(1), 48–53 (2018). https://doi.org/10.1049/iet-ifs.2018.5258
8. Ambusaidi, M.A., He, X., Nanda, P., Tan, Z.: Building an intrusion detection system using a filter-based feature selection algorithm. IEEE Trans. Comput. **65**(10), 2986–2998 (2016). https://doi.org/10.1109/TC.2016.2519914
9. Xiao, Y., Xing, C., Zhang, T., Zhao, Z.: An intrusion detection model based on feature reduction and convolutional neural networks. IEEE Access **7**, 42210–42219 (2019). https://doi.org/10.1109/ACCESS.2019.2904620
10. Al Janabi, K.B.S., Kadhim, R.: Data reduction techniques: a comparative study for attribute selection methods. Int. J. Adv. Comput. Sci. Technol. **8**(1), 1–13 (2018)
11. Johnson Singh, K., De, T.: Efficient classification of DDoS attacks using an ensemble feature selection algorithm. J. Intell. Syst. **29**(1), 71–83 (2020)
12. Saputra, M.F.A., Widiyaningtyas, T., Wibawa, A.: Illiteracy classification using K means-naive bayes algorithm. JOIV : Int. J. Inf. Visual. (2018). https://doi.org/10.30630/joiv.2.3.129
13. Deepa, S.N., Devi, B.A.: Neural networks and SMO based classification for brain tumor. In: 2011 World Congress on Information and Communication Technologies, pp. 1032–1037, Mumbai (2011). https://doi.org/10.1109/WICT.2011.6141390
14. Korting, T.S.: C4.5 Algorithm and multivariate decision trees. Image Processing Division, National Institute for Space Research–INPE (2006)

15. David, D.: Random forest classifier tutorial: how to use tree-based algorithms for machine learning. Free code camp. https://www.freecodecamp.org/news/how-to-use-the-tree-based-algorithm-for-machine-learning

Chapter 20
A Machine Learning Approach to Intrusion Detection System Using UNSW-NB-15 and CICDDoS2019 Datasets

Akoijam Priya Devi and Khundrakpam Johnson Singh

Abstract IDS has gained a lot of importance in the recent years for protecting systems and networks from various kinds of attacks. Unfortunately, as new attacks arise and old attacks are evolving, it cannot detect all kinds of attacks. So there comes a need to optimize the efficiency of detecting an intrusion by choosing best characteristics features and classification methods. In this paper, modern attack-based dataset like UNSW-NB-15 and CICDDoS2019 datasets are used. Main objective of the research work is classification of multi-class attacks for anomaly detection system using machine learning techniques. Using Ranker as a search method and Information Gain, Gain Ratio, One Rule, and Correlation Attribute Evaluators, feature selection is done. Analysis of performance of different machine learning algorithms is also done.

20.1 Introduction

Due to the development in Internet and technology, hacking incidents are increasingly reported by various companies every year. New attack techniques and exploits are continuously being developed to compromise the defense mechanisms. Hence, there is a need for more promising second lines of defense after firewalls. Intrusion detection has become one of the most important cybersecurity measures to be taken to prevent from various kinds of cyberattack. An IDS is a security tool which monitors the network traffic and alerts the system or the network administrator of any suspicious activities.

The thesis overview is listed as Sect. 20.2 discusses the literature review of some of the papers related to IDS. Section 20.3 deals with the datasets. Section 20.4 deals with proposed methodology. Section 20.5 deals with the results and conclusion. Section 20.6 contains conclusion and future works and lastly the references.

A. Priya Devi (✉) · K. Johnson Singh
Department of CSE, National Institute of Technology, Manipur, Imphal, India

195

Table 20.1 Literature survey

Authors	Title	Classification	Datasets	Performance
Johnson et al. [1]	Detection and differentiation of application layer DDoS attack from flash events using fuzzy-GA computation	GA-Fuzzy	CAIDA 2007, EPAHTTP, 1998 FIFA World Cup	98.40%
Moustafa et al. [2]	A hybrid feature selection for network intrusion detection systems	EM, NB LR	UNSW-NB-15	Accuracy(LR) 83.0% FAR(EM) 13.1%
Preeti et al. [3]	A detailed investigation and analysis of using machine learning techniques for intrusion detection	ML	KDD'99, UNSW-NB-15	N/A
Moustafa et al. [4]	USWN-NB15: A comprehensive dataset for network intrusion detections systems (UNSW-NB15 network dataset)	ML	UNSW-NB-15	N/A

20.2 Related Works

Table 20.1 given below shows the literature survey on some IDS using machine learning.

Singh et al. [5] proposed a clustering-based model with two clustering methods, namely Gaussian mixture and K-means clustering. The similarity of two different samples is evaluated with distance measure. Shorter the distance between two samples, higher is the similarity. This method gives 99.83% accuracy rate and a FAR of 0.04%.

20.3 Dataset

There are many benchmark datasets [6] available publicly for training and testing of various anomaly detection methods. However, some of these datasets have drawbacks. Datasets like DARPA (1998–1999) and KDD'99 Cup consist of outdated attacks and are less effective on modern networks and infrastructures. Although NSL-KDD dataset removes redundant instances, still some drawbacks of KDD Cup

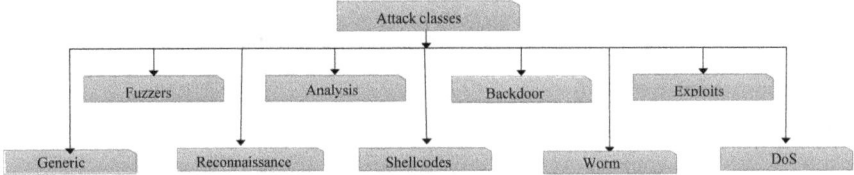

Fig. 20.1 Attack classes of UNSW-NB-15 dataset

remains. CAIDA (2002–2016) dataset lacks traffic feature relevant for identifying individual attacks and also lacks most of the common attacks. UNIBS has only raw traces of traffic and lacks relevant traffic features.

Considering the above drawbacks, we search for a dataset that has current network scenario, less redundant records, one which is properly labeled, and is non-anonymized. Thus, two datasets are considered for my study: UNSW-NB-15 and CICDDoS2019.

20.3.1   UNSW-NB-15 Dataset

The UNSW-NB-15 dataset was created by IXIA Perfect Storm tool in Cyber Range Laboratory of Australian Centre for Cyber Security (ACCS) [7].

Figure 20.1 shows nine attack classes of UNSW-NB-15 dataset. In Fuzzers, an attacker attempts to find the security vulnerabilities in the operating system, program, or in the network due to which resource becomes suspended for some time period and in some case crashing them. Analysis uses combination of various attacks in order to penetrate in a Web site or application. Backdoor are attackers attempt to discover information by bypassing normal authentication process and obtain illegal remote access to the system.

In Exploit, attackers exploits vulnerabilities, glitch or bug in the OS or software, and launch exploits to the system with an intent to cause damage. In Generic, the attacker tries to break the security key of a system. In Reconnaissance, attackers gather information about the victim network to bypass the security control. Shellcode is a code that runs in the victim's machine to gain remote access on the local machine. Worms are malicious programs which can replicate itself to spread to other systems. A DoS attack causes service unavailability to the intended users.

20.3.2   CICDDoS2019 Dataset

CICDDoS2019 dataset [8] contains many new attacks which can be carried out in the application layer. Figure 20.2 shows the output classes of DDoS attack. A DDoS attack is a DoS attack launched from several compromised device that are

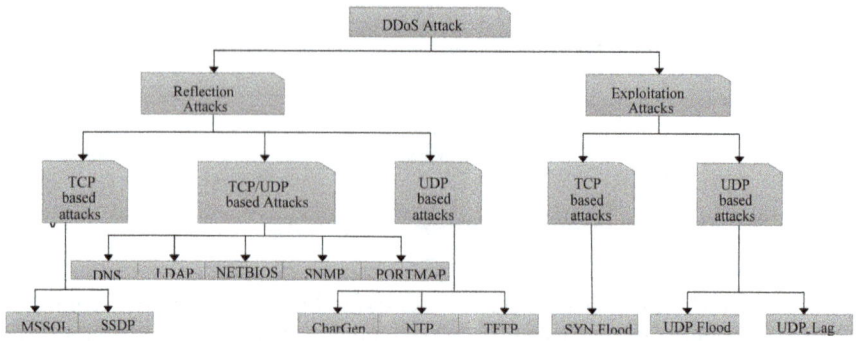

Fig. 20.2 Graphical representation of the output classes of CICDDoS2019 dataset

Table 20.2 Dataset description of UNSW-NB-15 and CICDDoS2019 datasets

Dataset	No. of training records	No. of features	No. of output class
UNSW-NB-15	175,341	49	9
CICDDoS2019	120,848	88	12

distributed globally [9]. Reflection-based DDoS attack can be exploited through the application layer protocols using the TCP, UDP, or combination of both TCP and UDP. MSSQL and SSDP belong to TCP based attack, whereas CharGen, TFTP, and NTP belong to UDP based attacks. Attacks like DNS, NETBIOS, SNMP, and LDAP can be exploited by using either UDP or TCP. Exploitation-based-DDoS attack can be executed through the application layer protocol using TCP and UDP. SYN Flood belong to TCP based exploitation attack category. UDP Flood and UDP_Lag belong to the UDP based exploitation attack category. A dataset description of the two benchmark datasets is given in Table 20.2.

20.4 Proposed Methodology

The workflow diagram of the proposed methodology is shown in Fig. 20.3. It is a model derived from the basic architecture of intrusion detection system [10]. In the first step, the UNSW-NB-15 dataset and CICDDoS2019 dataset are selected. Since the original datasets consists of large amount of data traffic, a part of dataset

Fig. 20.3 Workflow diagram of the proposed model

Table 20.3 UNSW-NB-15 dataset selected features list

Sl	Feature	Sl	Feature name	Sl	Feature name	Sl	Feature name
1	Sbytes	7	Proto	13	ct_dst_sport_ltm	19	Sload
2	Smean	8	ct_srv_dst	14	Sttl	20	Sinpkt
3	Id	9	ct_state_ttl	15	Dload	21	attack_cat
4	Label	10	Dttl	16	ct_src_dport_ltm		
5	Dmean	11	Service	17	ct_dst_src_ltm		
6	Rate	12	ct_srv_src	18	ct_dst_ltm		

Table 20.4 CICDDoS2019 dataset selected feature list

Sl	Feature Name	Sl	Feature Name	Sl	Feature Name
1	Timestamp	12	Source Port	23	Fwd IAT Std
2	Average Packet Size	13	Fwd HeaderLength	24	Subflow Fwd Packets
3	Subflow Fwd Bytes	14	act_data_pkt_fwd	25	Total Fwd Packets
4	Total Length of Fwd Packets	15	Init_Win_bytes_backward	26	Init_Win_bytes_forward
5	Packet length mean	16	Fwd Packets/s	27	ACK Flag Count
6	Avg Fwd Segment Size	17	Flow Packets/s	28	Protocol
7	Fwd Packet Length Mean	18	Flow IAT Mean	29	Min Packet Length
8	Max Packet Length	19	Fwd IAT Total	30	Fwd IAT Max
9	Fwd Packet Length Min	20	Flow IAT Max	31	Flow Duration
10	Fwd Packet Length Max	21	Fwd IAT Mean	32	Flow Bytes
11	Fwd Packet Length Std	22	Flow IAT Std	33	Label

is then extracted and preprocessed in the second step. After using Ranker [11] as a search method and InfoGain, GainRatio, OneR, and Correlation Attribute Evaluators [12], feature selection is done. Twenty-one attributes are selected for UNSW-NB-15 dataset and 33 attributes from the CICDDoS2019 dataset. Tables 20.3 and 20.4 show the selected feature list of UNSW-NB-15 and CICDDoS09 datasets, correspondingly.

20.4.1   J48 Classifier

J48 algorithm is the C4.5 [13] open source implementation in Weka. It accepts discrete and continuous value attributes. Firstly, entropy measure is computed for each feature. Then, set of examples are partitioned in accordance to the possible outcome values of those features which has the lowest entropy. Probabilities are then estimated.

20.4.2   Naïve Bayes Classifier

NB is based on the Bayes' theorem as shown in Eq. (20.1), where posterior probability of class c is calculated given a predictor x.

$$P(c/x) = (P(x/c) \times P(c))/(P(x)) \tag{20.1}$$

Given a training dataset, first convert it into a frequency table. Next, create a likelihood table by computing the probabilities of each feature. Using naïve Bayesain equation, the posterior probability of each of the class are computed. The class having the highest posterior probability becomes the prediction output.

20.4.3   Decision Table Classifier

A decision table (DT) has a more ordered set of "if–then" rules which is more compact, understandable, and simpler than a decision tree [14]. It is usually done by the tables' cross-validation performance measurement for all the different attributes' subsets and then chose the subset with the best performance.

20.4.4   Simple Logistic Based Classifier

Simple logistic regression (SL) is comparable to the linear regression except in the fact that the dependent variable is considered as nominal and not as a measurement. It finds out the equation which predicts best value of a variable Y for each variable X value.

20.5   Results and Discussions

Generation of a performance metrics of a model is a crucial part in understanding the performance of a classification model. Table 20.5 shows the TP Rate, FP Rate, Precision, Recall, F-Measure, and ROC Area [15] of each of the output classes of UNSW-NB-15 using J48, NB, DT, and SL. Table 20.6 gives the overall performance

Table 20.5 Performance metrics for each classes of UNSW-NB-15 using J48, NB, DT, and SL

	Classifier	TP Rate	FP Rate	Precision	Recall	F-Measure	ROC Area
Backdoor	J48	0.241	0.002	0.577	0.241	0.340	0.857
	NB	0.356	0.066	0.051	0.356	0.090	0.863
	DT	0.088	0.000	0.705	0.088	0.156	0.832
	SL	0.006	0.000	0.688	0.006	0.012	0.895
Analysis	J48	0.241	0.002	0.577	0.241	0.327	0.872
	NB	0.168	0.011	0.146	0.168	0.156	0.892
	DT	0.123	0.001	0.704	0.123	0.209	0.866
	SL	0.014	0.000	0.903	0.014	0.028	0.931
Fuzzers	J48	0.887	0.005	0.953	0.887	0.919	0.974
	NB	0.283	0.035	0.483	0.283	0.357	0.877
	DT	0.823	0.010	0.902	0.823	0.861	0.977
	SL	0.796	0.037	0.716	0.796	0.754	0.967
Shellcode	J48	0.742	0.002	0.712	0.74	0.727	0.938
	NB	0.907	0.128	0.044	0.907	0.084	0.939
	DT	0.542	0.002	0.597	0.542	0.568	0.976
	SL	0.000	0.019	0.000	0.000	0.000	0.951
Reconnaisance	J48	0.757	0.005	0.906	0.757	0.825	0.963
	NB	0.225	0.138	0.094	0.225	0.133	0.800
	DT	0.759	0.005	0.915	0.759	0.829	0.967
	SL	0.561	0.017	0.675	0.561	0.613	0.952
Exploit	J48	0.852	0.096	0.675	0.85	0.753	0.943
	NB	0.379	0.016	0.851	0.379	0.524	0.839
	DT	0.898	0.136	0.608	0.898	0.725	0.942
	SL	0.863	0.135	0.600	0.863	0.708	0.942
DoS	J48	0.298	0.034	0.394	0.298	0.340	0.901
	NB	0.004	0.001	0.242	0.004	0.007	0.861
	DT	0.041	0.003	0.475	0.041	0.076	0.916
	SL	0.081	0.010	0.375	0.081	0.134	0.913
Worms	J48	0.531	0.000	0.627	0.531	0.575	0.924
	NB	0.177	0.011	0.012	0.177	0.023	0.911
	DT	0.508	0.000	0.574	0.508	0.539	0.958
	SL	0.000	0.000	0.000	0.000	0.000	0.975
Generic	J48	0.985	0.001	0.995	0.985	0.990	0.996
	NB	0.973	0.008	0.974	0.973	0.974	0.989
	DT	0.979	0.000	0.998	0.979	0.989	0.995
	SL	0.978	0.002	0.992	0.978	0.985	0.998

Table 20.6 Overall performance metrics for UNSW-NB-15 dataset

	Accuracy (%)	Kappa statistic	MAE	RMSE	RAE	RSE (%)
J48	87.45	0.8399	0.0307	0.1311	19.405%	46.6264
NB	60.2323	0.5227	0.0801	0.2642	50.6788	93.9938
DT	85.3485	0.811	0.0539	0.1502	34.1321%	53.4351
SL	81.1231	0.7592	0.0512	0.1635	32.3922%	58.1529

metrics for UNSW-NB-15 by calculating their Accuracy, Kappa statistic, Mean absolute error (MAE), Root mean squared error (RMSE), Relative absolute error (RAE), and Relative squared error (RSE) using the classifier algorithms. Table 20.7 compares the performance metrics for each output classes for CICDDoS2019 using J48, DT, NB, and SL. Table 20.8 gives the overall performance metrics of CICDDoS2019.

Figure 20.4 shows a chart table comparing the accuracy of the two datasets w.r.t. each classifiers used. It is observed that the classifiers have better performance in case of CICDDoS2019 dataset.

20.6   Conclusions and Future Scope

Comparison of performance metrics is done for the two benchmark datasets. J48 gives highest accuracy of 87.45% for UNSW-NB-15 and 99.65% for CICDDoS2019. DT gives 60.23% for UNSW-NB-15 and 99.08% for CICDDoS2019. NB gives 85.35%

Table 20.7 Performance metrics for each classes of CICDDoS2019 using J48, DT, SL, and NB

	Classifier	TP Rate	FP Rate	Precision	Recall	F-Measure	ROC
DrDoS_	J48	0.982	0.001	0.995	0.982	0.988	0.999
DNS	DT	0.978	0.004	0.713	0.978	0.975	1.000
	SL	0.968	0.004	0.971	0.968	0.970	0.984
	NB	0.793	0.012	0.900	0.789	0.846	0.976
Benign	J48	0.956	0.000	0.994	0.968	0.981	0.988
	DT	0.728	0.001	0.941	0.728	0.821	0.993
	SL	0.983	0.000	0.986	0.983	0.985	0.992
	NB	0.991	0.003	0.839	0.991	0.909	1.000
DrDoS_	J48	1.000	0.001	0.995	1.000	0.997	1.000
LDAP	DT	1.000	0.000	1.000	1.000	1.000	1.000

(continued)

Table 20.7   (continued)

	Classifier	TP Rate	FP Rate	Precision	Recall	F-Measure	ROC
	SL	0.985	0.007	0.945	0.985	0.965	0.991
	NB	0.956	0.047	0.724	0.956	0.824	0.988
DrDoS_ MSSQL	J48	1.000	0.001	0.996	1.000	0.998	1.000
	DT	1.000	0.006	0.977	1.000	0.988	1.000
	SL	0.982	0.017	0.933	0.982	0.957	0.984
	NB	0.847	0.016	0.928	0.847	0.886	0.987
DrDoS_ NetBIOS	J48	0.999	0.000	0.998	0.999	0.998	0.999
	DT	1.000	0.000	0.998	1.000	0.999	1.000
	SL	0.986	0.000	1.000	0.986	0.993	0.994
	NB	0.923	0.000	0.997	0.923	0.959	0.997
DrDoS_ NTP	J48	0.999	0.001	0.974	0.987	0.980	0.998
	DT	0.929	0.000	0.996	0.929	0.962	0.999
	SL	0.987	0.000	0.989	0.987	0.988	0.994
	NB	0.892	0.010	0.764	0.892	0.823	0.998
DrDoS_ SSDP	J48	1.000	0.000	1.000	1.000	1.000	1.000
	DT	1.000	0.000	0.999	1.000	1.000	1.000
	SL	0.984	0.000	0.999	0.984	0.992	0.993
	NB	0.866	0.008	0.921	0.866	0.892	0.983
DrDoS_ UDP	J48	0.999	0.000	1.000	0.999	0.999	1.000
	DT	1.000	0.000	1.000	1.000	1.000	1.000
	SL	0.982	0.004	0.968	0.982	0.975	0.990
	NB	0.943	0.017	0.870	0.943	0.905	0.995
Syn	J48	1.000	0.000	1.000	1.000	1.000	1.000
	DT	1.000	0.000	1.000	1.000	1.000	1.000
	SL	0.903	0.003	0.954	0.903	0.928	0.959
	NB	0.996	0.009	0.887	0.996	0.939	1.000
TFTP	J48	0.999	0.000	1.000	0.999	0.999	1.000
	DT	1.000	0.000	1.000	1.000	1.000	1.000
	SL	0.970	0.000	1.000	0.970	0.984	0.987
	NB	0.861	0.001	0.992	0.861	0.922	1.000
UDP-lag	J48	1.000	0.000	0.997	1.000	0.9998	1.000
	DT	1.000	0.000	1.000	1.000	1.000	1.000
	SL	0.910	0.000	1.000	0.910	0.953	0.961
	NB	0.837	0.002	0.971	0.837	0.899	0.982
WebDDoS	J48	0.000	0.000	0.000	0.000	0.000	0.500
	DT	0.000	0.000	0.000	0.991	0.000	0.514
	SL	0.000	0.000	0.000	0.000	0.000	0.647
	NB	1.000	0.003	0.003	1.000	0.007	0.998

Table 20.8 Overall performance metrics of CICDDoS2019 using the classifier models

	Accuracy	Kappa statistic	MAE	RMSE	RAE	RSE
J48	99.6475%	0.996	0.0007	0.0209	0.04732%	7.6939%
NB	88.6742%	0.8725	0.0196	0.1332	13.2728%	49.0389%
DT	99.084%	0.9896	0.0084	0.00447	5.6857%	16.4605%
SL	96.9284%	0.9653	0.0051	0.0715	3.4808%	26.3234%

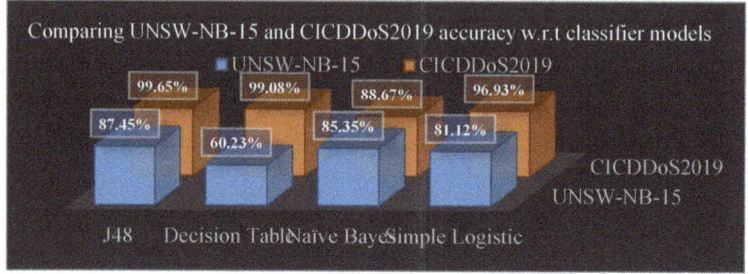

Fig. 20.4 Accuracy of the two datasets with respect to the classifier models

for UNW-NB-15 and 88.67% for CICDDoS2019. SL gives 81.12% for UNSW-NB-15 and 96.93% for CICDDoS2019. This research work also compares the performance metrics of each class such that for a given class output, which algorithm gives the highest score.

Future research will be on in-depth analysis of intrusion detection system models. Hybrid machine learning techniques or deep neural network may be used.

References

1. Singh, K.J., Thongam, K., De, T.: Detection and differentiation of application layer DDoS attack from flash events using fuzzy-GA computation. IET Info. Secur. **12**(6), 502–512 11 (2018)
2. Moustafa, N., Slay, J.: A hybrid feature selection for network intrusion detection systems: central points. In: Proceedings 16th Australian Information Warfare Conference 2015, pp. 1–10. (2015)
3. Mishra, P., Varadharajan, V., Tupakula, U., Pilli, E.S.: A detailed investigation and analysis of using machine learning techniques for intrusion detection. IEEE Commun. Surv. Tutor. **21**(1), 686–728, Firstquarter (2019). https://doi.org/10.1109/COMST.2018.2847722
4. Moustafa, N., Slay, J.: USWN-NB15: a comprehensive data set for network intrusion detections systems (UNSW-NB15 Network Data Set). School of Engineering and Information Technology, Canberra, Australia (2015)
5. Singh, P., Venkatesan, M.: Hybrid Approach for intrusion detection system. In: 2018 International Conference on Current Trends towards Converging Technologies (ICCTCT), Coimbatore, pp. 1–5 (2018)

6. Bhuyan, M.H., Bhattacharyya, D.K., Kalita, J.K.: Network anomaly detection: methods, systems and tools. IEEE Commun. Surv. Tutor. **16**(1), 303–336, First Quarter (2014)
7. The UNSW-NB15 dataset Available at: https://www.unsw.adfa.edu.au/unsw-canberra-cyber/cybersecurity/ADFA-NB15-Datasets/ (Accessed on 12 Feb 2020)
8. CICDDoS2019 dataset Available at: https://www.unb.ca/cic/datasets/ddos-2019.html (Accessed on 25 January 2020)
9. Johnson Singh, K., Thongam, K., De, T.: Entropy-based application layer DDoS attack detection using artificial neural networks. Entropy **18**, 350 (2016)
10. Heady, R., Luger, G., Maccabe, A., Servilla, M.: The architecture of a network level intrusion detection system. Tech. Rep., Computer Science Department, University of New Mexico, New Mexico (1990)
11. Blum, A.I., Langley, P.: Selection of relevant features and examples in machine learning. Artif. Intell. **97**, 245–271 (1997)
12. Mukherjee, S., Sharma, N.: Intrusion detection using naive bayes classifier with feature reduction. Proc. Technol. **4**, 119–128 (2012)
13. Quinlan, J.R.: In: C4.5: Programs for Machine Learning, San Mateo, Morgan Kaufman (1993)
14. Kalmegh, S.R.: Comparative analysis of the WEKA classifiers rules conjunctiverule and decisiontable on indian news dataset by using different test mode. Department of Computer Science, Sant Gadge Baba Amarvati University, Amaravati(M.S.), India (2018)
15. Shone, N., Ngoc, T.N., Phai, V.D., Shi, Q.: A deep learning approach to network intrusion detection. IEEE Trans. Emerg. Topics Comput. Intell. **2**(1), 41–50 (2018)

Chapter 21
Discovery of Localized Malicious Attack in Wireless Networks

S. Gowri, Senduru Srinivasulu, J. Jabez, J. S. Vimali, and A. Sivasangari

Abstract Wireless communication is merely a boon to mankind. With wireless communication, data is transferred between the nodes at a very high rate in recent days. With emerging 5G technology, it is estimated that the devices are going to receive a bandwidth which is as high as 1GBPS which is the usual rate delivered by broadband services. All these are possible with packet data transfer. Packet transfer occurs in wireless networks in two ways which are flooding and routing. However, one challenge in communication is security. We need to ensure that our packets are sent to the receiver securely without any third-party access to our data packets. We propose a system which makes use of the spatial information property of a node and tries to calculate received signal strength (RSS), which is the main parameter for plotting the attacker nodes in the network and eliminating the attacker nodes by using clustering techniques against a radar grid.

21.1 Introduction

Spoofing attack is the most common attack which the attackers use to gain authorization to the devices in a vulnerable way, such that the attacker nodes exist in the system as if they have been authenticated and recognized by the administrator who manages the devices connected on the same network on the receiver side. The most important thing the user of the service thinks about is his data. In a system which does not authenticate its users in a proper way, the data is never secure. If the data is not secure, the usability of the system by its users will be reduced in a significant way. In broad sense, spoofing is nothing but a technique used to gain access to the computer in a vulnerable way by imitating its identification with the available nodes which have already been authenticated by the administrator. Vulnerabilities are the main reason which enable the attackers to enter the system in a vulnerable way. In wired communication systems, the data is kept secure free from vulnerabilities by

S. Gowri (✉) · S. Srinivasulu · J. Jabez · J. S. Vimali · A. Sivasangari
Sathyabama Institute of Science and Technology Jeppiaar Nagar, Rajiv Gandhi Salai, Chennai 600 119, Tamilnadu, India

S. C. Satapathy et al. (eds.), *Smart Computing Techniques and Applications*,
Smart Innovation, Systems and Technologies 225,
https://doi.org/10.1007/978-981-16-0878-0_21

using certain selected encryption algorithms by using several techniques which are sole part of crypt arithmetic which comes under a branch of cryptography whose main motive is to encrypt the data. However, historical studies prove that 90% of the times wired communication is very secure where it will be quite cumbersome for the attackers to enter the system vulnerably.

At present, we have several technologies which run on client side to ensure security. This attack brings us to the most popular point of using API's in computing where we do not have access to the database of the authenticator, and instead, we have several application programming interfaces which enable us to hit them and retrieve the data in the most secured way by generating authentication tokens. The concepts on which this token authentication relies on is BASICAUTH, APIKEY, OAUTH1, and OAUTH2 whose primary purpose is authentication in a more secured and a controlled way. Providing access to the data in a controlled way is quite a decent job to be done before transmitting the data through a network. This can be done by using access specifiers and access modifiers which secure data by acting as their companions. Several strategies have been proposed promote security for identifying and removing the spoofing attackers [1]. One such strategy is crypt arithmetic where we generate private key and send it as a secret key to the receiver [2, 3].

21.2 Literature Survey

One instance of computer security is wireless security, which is nothing but the collection of security threats and security concerns which are most prevalent in wireless networks. Organizations that follow a local area wireless network are specifically concerned about the vulnerabilities to their security breaches which are caused by rogue access points. These countermeasures secure both the network and the valuable materials that are held by the network. But there is one restriction for applying such actions, and the restriction is that it has to be applied to the network devices uniformly. Wireless communication technologies have been increased in domains beyond the original intended, original areas due to its availability, and low cost. Several industrial applications have security requirements which are pretty specific to the application under consideration.

Hence, industrial application characteristics are important to understand, and the vulnerabilities have to be evaluated bearing the highest risk in the context. Wireless networks are easy to break, so the hackers are looking forward to breaking the wireless networks [4]. Usability of wireless networks has grown enormously to an extent that every individual or any enterprise relies mostly on this type of communication to accomplish most of its networking services. A spoofing attack [5, 6] is the most common attack which the attackers use to gain authorization to the devices in a vulnerable way, such that the attacker nodes exist in the system as if they have been authenticated and recognized by the administrator who manages the devices connected on the same network on the receiver side. Vulnerabilities are the main reason which enables the attackers to enter the system in a vulnerable way. With current 802.11 protection strategies, like wired equivalent privacy (WEP), Wi-Fi Safe Access (WPA), or 802.11i, such methods can only secure data frames. Spoofing attacks can further enable a range of traffic intrusion attacks such as access control list attacks, rogue access point (AP) attacks, and ultimately denial of service (DoS) attacks [7, 8]. Multiple opponents can mask the same identity in a large network and cooperate to conduct malicious attacks such as attacks on network resources and denial of service.

21.3 Proposed Methodology

Methodology is quietly the routine model view controller (MVC) where we have several view controllers and they are modeled in such a way that they follow a schematic representation where on procedure calls another and the other call the later and it goes on until the entire system has performed the tasks assigned to them. Our proposed system is going to use a specific property which is known as spatial property which holds the information of a particular node which is concerned with its physical identification and uniqueness which are hard to imitate. Later, it turns out that this specific property makes the node potentially secure by overthrowing the hackers who tries to imitate that node. Usage of this property is often considered as a privilege because the cost of state of the architecture is very low because all these properties associated with the node comes as a package without ourselves having to explicitly provide the functionality to the network nodes. By this way, we are actually implementing model which allows us to not only detect the node hackers but also provide us with an opportunity to eliminate them and swipe them out completely from the existing network by showing the number of hacker nodes which tried to enter.

The architecture which we follow is quite simple where we have a server for management, a database for storing information, and users who are authenticated by the server or an outsourcing community. When the users try to access information from the database by communicating with the server, there might be several spoofing attackers who try to enter into the system by keeping track of vulnerabilities which are associated with the existing system. So, we use several algorithms to identify them by plotting and removing them from the system. This might be enough as of now to recognize and remove the attackers from the network and to promote security. Later, it turns out that this specific property makes the node potentially secure by overthrowing the hackers who tries to imitate that node (Fig. 21.1).

The traditional method to manage hinder disparaging ambushes is to use cryptographic-based approval. We have introduced a secure and proficient key management (SEKM) framework [9]. SEKM creates a public key infrastructure (PKI) by applying a secret sharing arrangement and a basic multicast server gathering. We actualized a key management instrument with irregular key resuscitate and host disavowal to prevent the compromise of affirmation keys. An affirmation framework for different leveled, extraordinarily designated sensor frameworks is proposed.

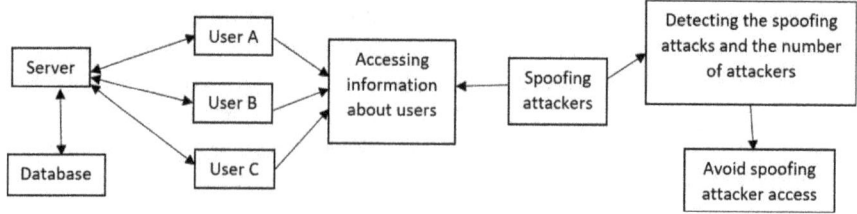

Fig. 21.1 System architecture

In any case, the crypto-sensible approval may not be reliably relevant considering the compelled resources on far off devices and lacking of a fixed key management structure in the distant frameworks.

21.3.1 Detection Model of Generalized Attack

In this segment, we portray our detection model of generalized attack, which comprises of two stages: assault discovery, which recognizes the nearness of an assault, and number assurance, which decides the quantity of foes. The number assurance stage will be introduced.

Framework

The customary restriction approaches are found based on the middle value of RSS from every hub personality contribution to evaluate the situation of a hub. Be that as it may, in remote mocking assaults, the RSS stream of a hub character might be blended in with RSS readings of both the first hub just as satirizing hubs from various physical areas. The conventional strategy for averaging RSS readings cannot separate RSS readings from various areas and consequently is not practical for confining foes. Not quite the same as conventional confinement draws near, our incorporated recognition and limitation framework uses the RSS medoids came back from SILENCE as contributions to restriction calculations to evaluate the places of enemies. The arrival positions from our framework incorporates the area gage of the first hub and the aggressors in the physical space. Taking care of foes utilizing diverse transmission power levels, a foe may change the transmission power levels when performing satirizing assaults with the goal that the limitation framework cannot assess its area precisely. We look at the complex rule models that theys got influence as an element of the separation to the milestone speaks to the transmitting influence of a hub at the reference separation d0, d is the separation between the transmitting hub and the milestone, and is the way misfortune type. Further, we can communicate the distinction of the got power between two tourist spots, I and j, as based on which we found that the distinction of the relating got power between two unique milestones which is free of the transmission power levels. Along these lines, when an enemy dwelling at a physical area fluctuates its transmission capacity to play out a parodying assault, the distinction of the RSS readings between two unique tourist spots from the foe is consistent since the RSS readings are acquired from a solitary physical area. We would then be able to use the distinction of the medoids vectors in signal space got from SILENCE to restrict foes.

Algorithms *Radar-gridded module.* The radar-gridded calculation is a scene-matching restriction calculation. Radar-gridded utilizes an inserted signal guide, which is worked from a lot arrived at the midpoint of RSS readings with known areas. Given a watched RSS perusing with an obscure area, radar restores the x, y of the closest neighbor in the sign guide to the one to limit, where "closest" is characterized as the euclidean separation of RSS that focuses in a N-dimensional sign space, where N is the quantity of milestones.

Area-based probability module. ABP additionally uses an introduced signal guide. Further, the test zone is separated into an ordinary lattice of equivalent estimated tiles. ABP accepts the conveyance of RSS for every milestone and follows a Gaussian appropriation with mean as the normal estimation of RSS perusing vector s. ABP then registers the likelihood of the remote gadget being at each tile Li, with I ¼ 1...L..., so on.

Bayesian networks. BN limitation is a multi-emphasis calculation that encodes the sign to separate proliferation model into the Bayesian graphical model for confinement of the essential Bayesian network utilized for our examination. The vertices X and Y speak to area; the vertex is the RSS perusing from the milestone; and the vertex Di speaks to the euclidean separation between the area indicated by X and Y and the milestone. The estimation follows a sign proliferation model. Foes utilized a similar transmission power levels as the first hub and the returned medoids are utilized. Enemies changed their transmission power level from 15 to 10 dB, and the returned medoids are utilized; and foes changed their transmission power level from 15 to 10 dB, and the distinction of returned medoids are utilized. The key perception from the presentation of utilizing the distinction of returned medoids in dealing with enemies utilizing diverse transmission power levels is similar to the outcomes when foes utilized a similar transmission power levels as the first hub. Further, the confinement execution is a lot of more regrettable than the customary methodologies if the distinction of returned medoids is not utilized while restricting foes utilizing diverse transmission power levels, appeared as the case 2 above. Specifically, when utilizing our methodology, we can achieve the middle mistake of 13 feet for both radar-gridded and ABP on the off chance that 3, a 40–50% execution improvement, contrasting with the middle blunders of 20 and 19 feet for radar-gridded and ABP, individually, on the off chance that 2. Accordingly, it is exceptionally compelling in restricting various foes with or without changing their transmission power levels.

Fig. 21.2 Database
connection

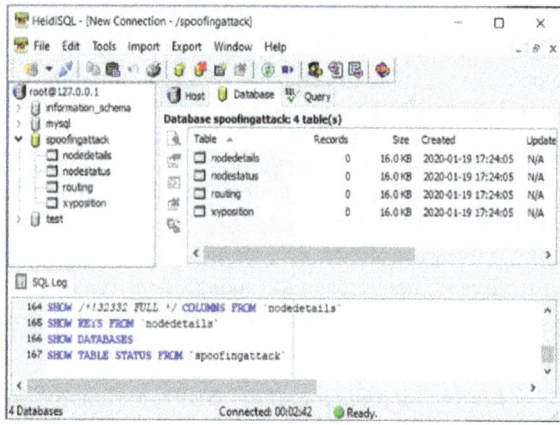

21.4   Results and Discussion

A secured network is free from spoofing attackers and unnecessary traffic which
enables the network to transmit the data in a safe, secure, and faster way. Even
the efficiency and reliability of the network is enhanced as a result of removing
unnecessary traffic. Vulnerabilities are the main reason which enable the attackers to
enter the system in a vulnerable way. In wired communication systems, the data is
kept secure free from vulnerabilities by using certain selected encryption algorithms
by using several techniques which are sole part of crypt arithmetic which comes
under a branch of cryptography whose main motive is to encrypt the data. However,
historical studies prove that 90% of the times wired communication is very secure
where it will be quite cumbersome for the attackers to enter the system vulnerably.
Wireless networks have grown enormously to an extent that every individual or any
enterprise relies mostly on this type of communication to accomplish most of its
networking services.

21.5   Screenshots

See Figs. 21.2, 21.3, 21.4, 21.5, 21.6, 21.7, 21.8, 21.9 and 21.10.

21.6   Conclusion and Future Work

Usability of wireless networks has grown enormously to an extent that every individual or any enterprise relies mostly on this type of communication to accomplish most of its networking services. Therefore, it is highly important to promote

Fig. 21.3 Node creation

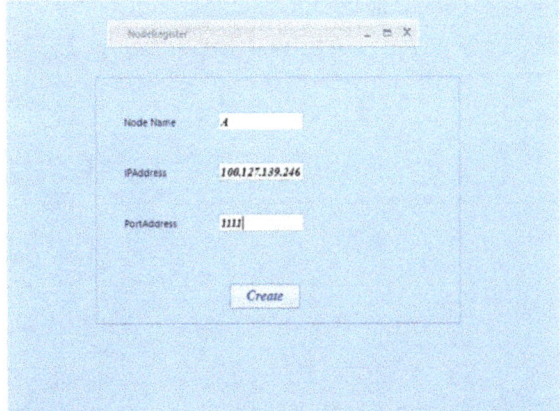

Fig. 21.4 Hacker node

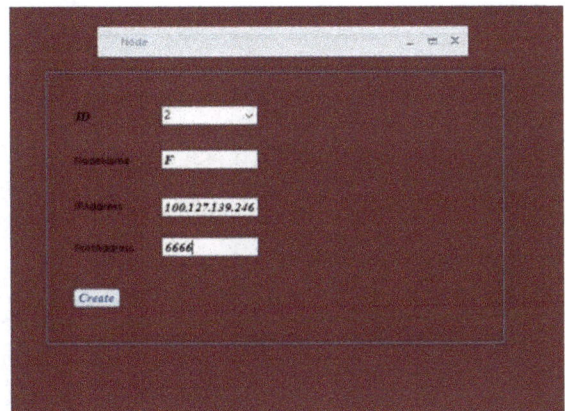

Fig. 21.5 Node details and euclidean dist. calc

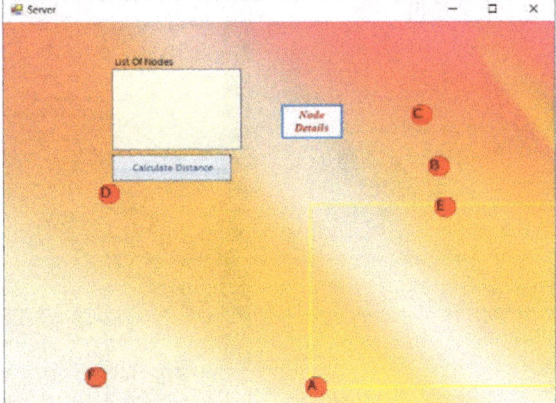

Fig. 21.6 Calculating RSS

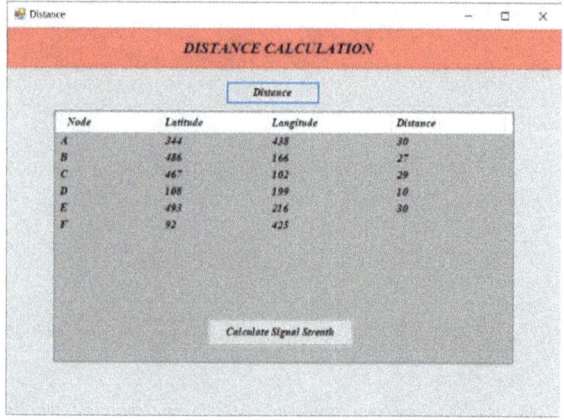

Fig. 21.7 RSS value

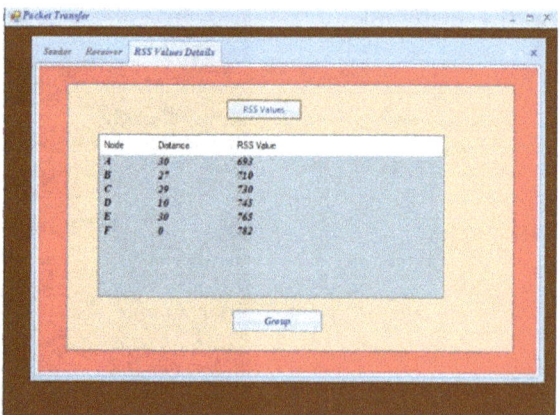

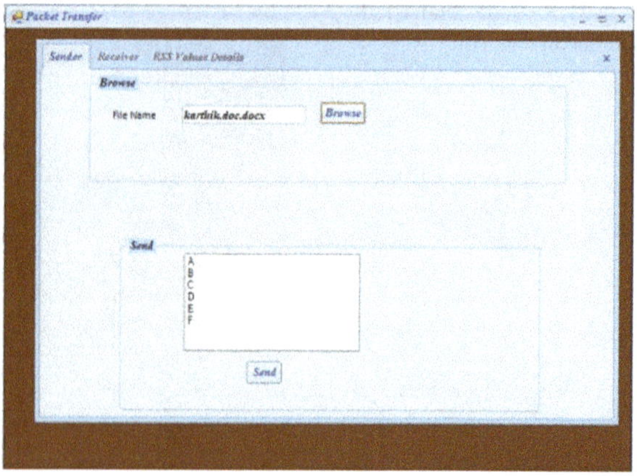

Fig. 21.8 Sending data

Fig. 21.9 Selecting cluster to prevent attack

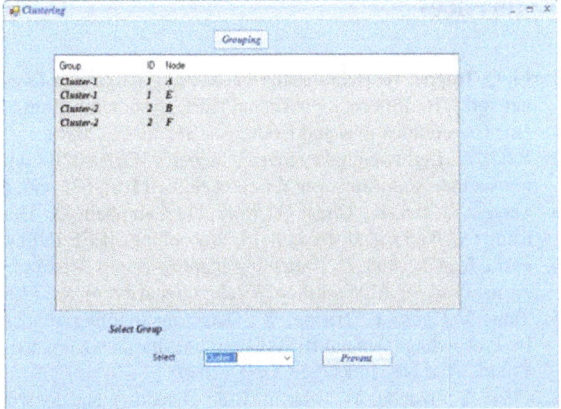

Fig. 21.10 Remove hacker node in network

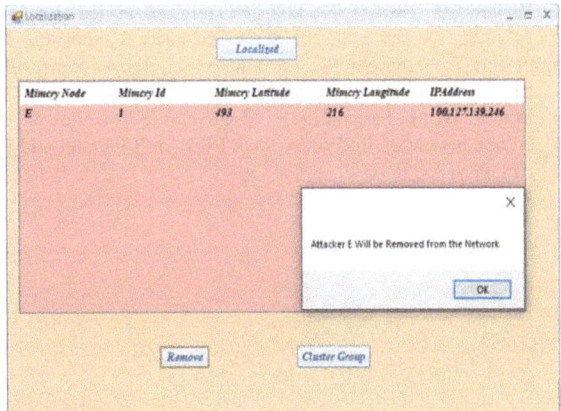

security, flexibility, consistency, reliability in the wireless networks. As a token of development, it is our responsibility to do enough for promoting this security by identifying the attackers and removing them from the networks. It is important to know the properties associated with a network node because it is the primary entity within the network system to do any kind of operations on the network. We found that our location components are exceptionally powerful in both recognizing the nearness of assaults with recognition rates more than 98% and deciding the quantity of foes, achieving more than 90% hit rates and accuracy all the while when utilizing silence and SVM-based instrument. Further, based on the quantity of assailants controlled by our instruments, our incorporated recognition and confinement framework can restrict any number of foes in any event, when aggressors utilizing distinctive transmission power levels.

References

1. Li, Q, Trappe, W: Relationship-based detection of spoofing- related anomalous traffic in ad hoc networks. In: Proceedings Annual IEEE Communications Society on IEEE and Sensor and Ad Hoc Communications and Networks (SECON), (2006)
2. Wool, A.: Lightweight key management for IEEE 802.11 wireless lans with key refresh and host revocation. ACM/Springer Wireless Netw. 11(6), 677–686 (2005)
3. Sheng, Y., Tan, K., Chen, G., Kotz, D., Campbell, D.: Detecting 802.11 MAC layerspoofing using received signal strength. In: Proceedings IEEE INFOCOM, April (2008)
4. Faria, D., Cheriton, D.: Detecting Identity-Based Attacks in Wireless Networks Using Signal-prints. Proc, ACM Workshop Wireless Security (WiSe) (2006)
5. Yang, J., Chen, Y., Trappe, W.: Detecting spoofing attacks in mobile wireless environments. In: Proceedings Annual IEEE Communications Society Conference Sensor, Mesh and Ad Hoc Comm. And Networks (SECON), (2009)
6. Chen, Y., Trappe, W., Martin, R.P.: Detecting and localizing wireless spoofing attacks. In: Proceedings Annual IEEE Communications Society Conference Sensor Mesh and Ad Hoc Communications and Networks (SECON), May (2007)
7. Bellardo, J., Savage, S.: 802.11 Denial-of-service attacks: real vulnerabilities and practical solutions. In: Proceedings USENIX Security Symposium, pp. 15–28. (2003)
8. Ferreri, F., Bernaschi, M., Valcamonici, L.: Access points vulnerabilities to dos attacks in 802.11 networks. In: Proceedings IEEE Wireless Communications and Networking Conference (2004)
9. Wu, B., Wu, J., Fernandez, E., Magliveras, S.: Secure and efficient key management in mobile ad hoc networks. In: Proceedings IEEE Int'l Parallel and Distributed Processing Symposium (IPDPS) (2005)

Chapter 22
An Adapted Approach of Image Steganography Using Pixel Mutation and Bit Augmentation

Ravi Saini, Kamaldeep Joshi, and Rainu Nandal

Abstract This paper presents a new technique of data hiding in spatial domain using digital image as cover file. It uses two logical modules, binary converter and bit extractor in its processing. Binary converter module converts the identified pixel value into binary sequence of eight bits. Bit extractor module extracts the seven bits of pixel value except LSB. The proposed method hides the message in the seven bits of pixel value except LSB as LSB is the most sensitive bit and can be altered on the network by some error. It disburses the message evenly on the pixel intensity. The preliminary results show that it expels various anomalies associated with some state-of-the-art methods. It also provides better imperceptibility and hiding capacity.

22.1 Introduction

The term steganography is consolidation of two Greek words "Steganos" and "Graph". Steganos means "hidden or covered" and graph means "to write" [1]. So, we can say that steganaography means hidden or covered writing. Steganography is not the new field; it has been used from long interval of time. Its origin has been found in the history at about 440 Before Christ in the book "The Histories" written by Herodotus who was a Greek historian. He was known as "Father of the History". The book "The Histories" contained the detail of the series wars between Greek and Persia from 499 BC to 449 BC. It contains two examples of steganography. Histiaeus was the son of Greek King Lysagoras. His subordinate Aristagoras was in prison. Histiaeus sent a message to Aristagoras by using Steganography. He called his most loyal slave. He trimmed the head of his captive and writes the epistle on the scalp of his servant. He kept his slave with himself until his hair has regrown. After that, he sent his slave to prison to meet Aristagoras. Arsitagoras shaved the head of slave and read the message written on the scalp of slave. The second example

R. Saini (✉)
GCW Gurawara, Rewari, India

K. Joshi · R. Nandal
UIET, Maharshi Dayanand University, Rohtak, India

© The Author(s), under exclusive license to Springer Nature Singapore Pte Ltd. 2021   217
S. C. Satapathy et al. (eds.), *Smart Computing Techniques and Applications*,
Smart Innovation, Systems and Technologies 225,
https://doi.org/10.1007/978-981-16-0878-0_22

of steganography was found from the side of Persia. Persia King Demaratus sent a message to his subordinates by writing it to wax tablet. Wax tablet is the tablet made up of wood and covered with wax. It provides reusable writing surface. Another example of steganogrpahy was given by Johannes Trithemius who developed Ave-Maria-Cipher. By using this cipher, information can be hiden within Latin praise of God. Invisible ink has also been used as media for secret communication in the history. The message written by invisible ink can be recouped at the receiver end by putting the media in the light or by some other means [2, 3]. The example of steganography is given below: Suppose the following message is transferred from A to B. SIVANSH AND ADMESH DINED HAPPILY. If B holds the second letter of each word of the above message, then he will get the secret message "INDIA". This type of steganography is called null cipher steganography. In this era of digital media, many steganography techniques have been designed and developed so far. These steganography techniques use digital media like text file, image file, audio file, video file, etc., as cover media for secret information transfer [4]. We can divide the steganography techniques into basically four types reckoning on the type of cover media. Text steganography, image steganography, audio steganography, and video steganography are categories of steganography. Image steganography is very popular as compared to other types of steganography techniques in the era of digital world. Various tools have also been developed which provide insertion and retrieval of message automatically [5]. The examples of steganography tools are Steg, STools, Stegmail, Openpuff, Openstego, Outguess, QuickStego, StegoShare, StegFS, Image-Spyer G2, etc. Many image steganography techniques have been developed in recent times like LSB method [6], parity checker method [7], GLM method [8], etc. In this paper, we will propose a new method for image steganography which will expel the anomalies attached with some traditional methods.

22.2 Proposed Method

A new method of image steganography using pixel mutation and bit augmentation is proposed in this section. It uses the any image file as cover file for concealing the secret data. There are two logical modules, binary converter and bit extractor which are used by this approach. At first, the pixel location is selected by using random number generator function where the message bit is to be inserted. The secret message is also translated into bit sequence using the binary converter module. Every selected pixel can be used for hiding only one bit of secret data. The selected pixel is also converted into binary sequence of eight bits using binary converter. The first seven bits of binary pixel value except LSB are extracted using bit extractor module. The message bit is distributed into these seven bits on the basis of insertion process. The insertion process is shown by Fig. 22.1. Sender engenders the stego image at the sender side and sent it to the accepter. The accepter generates the same locations of stego image where message bits are hidden and extract the pixel value of location.

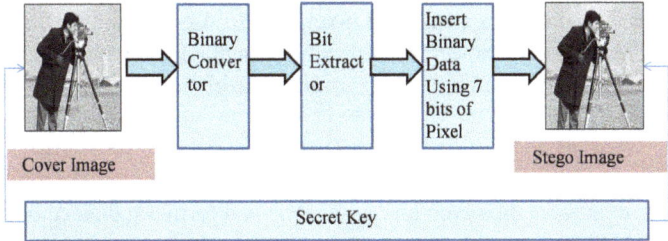

Fig. 22.1 Insertion process

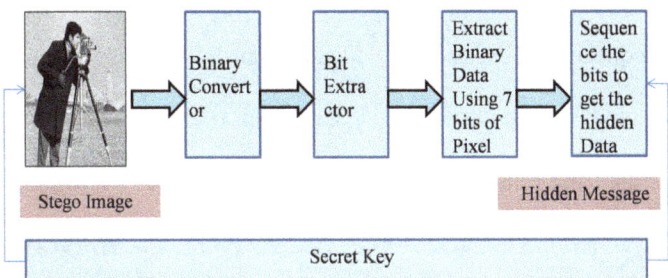

Fig. 22.2 Retrieval process

The message bits are extracted by reverse process at the receiver end and sequenced to form the required message. Secret key is experienced between sender and receiver. The extraction process is shown by Fig. 22.2.

22.3   Algorithm

In this section, we will discuss about the insertion and retrieval algorithm.

22.3.1   Insertion Algorithm

1. Select the cover image in which you want to insert the secret data. Let the cover image is $C(r, c)$ where r symbolizes total rows in the cover image and c symbolizes total columns in the cover image.
2. Convert the secret message M into binary sequence of bits using binary converter.
3. Generate the random location of the pixels using to the Eq. 22.1:

$$Loc(n + 1) = (x * Loc(n) + y)Mod\ M \tag{22.1}$$

$Loc(n + 1)$ = Location of consequent pixel, x = Multiplication Factor, $Loc(n)$ = Location of Previous Pixel Selected and Y = Incremental Factor M = Modulas Factor

4. Bit converter converts the selected pixel value into binary sequence $Bin(P)$ as per Eq. 22.2:

$$Bin(p) = \{b_{ij}|0 < i <= r, 0 < j <= c \; and \; b_{ij}\epsilon(0, 1)\} \qquad (22.2)$$

5. Bit extractor extracts the seven bits of $Bin(p)$ and form a sequence of bits as per Eq. 22.3 given below:

$$Bit(S, 7) = \{S(i)|0 < i <= 7 \; and \; S(i)\epsilon(0, 1)\} \qquad (22.3)$$

6. The bits sequence got in step 4 are added in augmented manner as per Eq. 22.4.

$$Aug(S) = \{S(i) + S(j)|i = 1, 1 < j <= 7 \; and \; 0 <= Aug(S) <= 7\} \quad (22.4)$$

7. Calculate the main value using Eq. 22.5:

$$Main(S) = \{Aug(S)\%2|0 <= Aug(S) <= 7 \; and \; 0 <= Main(S) <= 1\}$$
$$(22.5)$$

8. Insert bit b got in step 2 in sequence of bits compiled is step 4. If $b = 0$ and $Main(S) = 0$, then no alteration in pixel intensity is needed. If $b = 0$ and $Main(S) = 1$, then add or subtract 1 to the pixel selected such that $Main(S)$ becomes 0. If $b = 1$ and $Main(S) = 1$, then no alteration in pixel intensity is needed. If $b = 1$ and $Main(S) = 0$, then add or subtract 1 to the pixel value so that $Main(S)$ becomes 1.

9. Repeat all the steps until all the bits are inserted in the image. Generate the stego image $S(r, c)$ and exit.

22.3.2 Retrieval Algorithm

The seven steps of insertion algorithm are repeated at the receiver side on stego image $S(r, c)$. If $Main(S) = 0$ at the selected pixel value, then message bit is 0; otherwise, message bit is 1. The complete message will be retrieved by sequencing the all bits and translate them into alphanumeric characters.

22.4 Results and Analysis

The imperceptibility of the technique is tested by using practical checking by persons from different age groups. We have sent the cover image and stego image to many persons. We asked them about the difference between cover file and file image into five numeric values as per the following detail:

Fig. 22.3 Response
submitted by different
persons

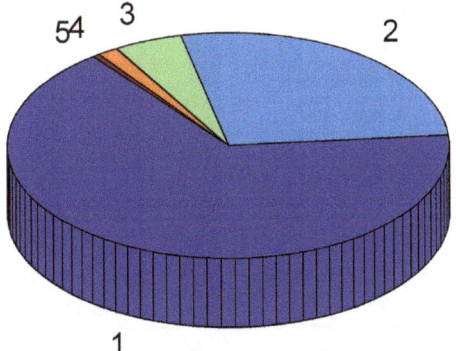

1: No divergence amid cover file and stego file
2: Very minute divergence amid cover file and stego file
3: Average divergence amid cover file and stego file
4: Slightly high divergence amid cover file and stego file
5: Very high divergence amid cover file and stego file

We have got the response of 200 persons by using above numerical values on four different test images. The response submitted by different persons shown by pie graph on one test image in Fig. 22.3.

Mean value of responses is also calculated by using the following equation:

$$\text{Mean Value} = \sum_{k=0}^{200} \text{response}(k) \div 200 \qquad (22.6)$$

Imperceptibility level is also classified into different levels as per different mean values which is given by Table 22.1.

The result of mean value on four different test images after insertion of 1 KB message is given below in Table 22.2. We have analysed the results from Table 22.2 and found that mean value lies between 1 and 2. So, we can say that the proposed method is highly imperceptible. We have also done the PSNR analysis on the above images for different size of messages. The calculation of PSNR is done by using the following equation:

$$\text{PSNR} = 10 \log_{10}(I^2 * \text{MSE}) \qquad (22.7)$$

PSNR denotes peak signal-to-noise ratio, I denotes peak value of pixel, i.e. 255 in this case, and MSE is the mean square error given by Eq. 22.8. PSNR denotes peak signal-to-noise ratio, I denotes peak value of pixel, i.e. 255 in this case, and MSE is the mean square error given by the following equation:

$$\text{MSE} = \sum_{i=1}^{r} \sum_{j=1}^{c} (x_{ij} - y_{ij})^2 \qquad (22.8)$$

Table 22.1 Imperceptibility level as per mean value

Mean value	Imperceptibility level
$1 \leq$ Mean Value < 2	High imperceptibility
$2 \leq$ Mean Value < 3	Average imperceptibility
$3 \leq$ Mean Value < 4	Low imperceptibility
$4 \leq$ Mean Value < 5	Very low imperceptibility

Table 22.2 Mean value of different test images

Image	Image size in pixels	Message size (KB)	Mean value
Building image	50,625	1	1.435
Cameraman image	16,384	1	1.625
Monalisa image	50,625	1	1.475
Ship image	50,625	1	1.730

PSNR on different images for different techniques is given by Table 22.3. The result shows that proposed method produces finer PSNR value than some state-of-the-art techniques. It provides acceptable value of PSNR. The comparison of PSNR values of different images for various size of message is also shown by Figs. 22.4 and 22.5. After analysing various graphs, we have found that proposed method provides better value of PSNR than some traditional methods.

22.5 Conclusion and Future Scope

The proposed method shows improvement over some state-of-the-art methods like LSB method, parity checker method, and GLM method. It provides improved value of PSNR. The subjective test also shows that the proposed method provides greater imperceptibility. Only one bit can be inserted in one pixel value of cover image. Our method also does not depend on LSB for insertion and extraction of the message. So, all the anomalies that are associated with LSB involved approaches are resolved by proposed method. LSB can be changed easily on the network by network noise. Intruder can also alter the LSB for destroying the message. All these types of problems are resolved by our method. In the future, we will try to develop some technique by increasing embedding capacity while retaining robustness and imperceptibility of the cover file.

Table 22.3 Mean value of different test images

Image	Message size (KB)	LSB method	GLM method	Parity checker method	Proposed method
Building image	1	35.51	37.54	38.30	45.20
Cameraman image	1	35.65	37.53	40.30	46.45
Monalisa image	1	35.89	37.02	41.39	46.46
Ship image	1	35.56	36.30	42.87	47.54
Building image	2	34.62	37.90	40.40	45.30
Cameraman image	2	34.65	37.45	40.98	45.29
Monalisa image	2	33.77	35.50	41.03	44.90
Ship image	2	33.31	34.20	39.07	45.67
Building image	3	32.10	36.30	38.70	45.09
Cameraman image	3	32.15	37.92	40.54	44.99
Monalisa image	3	33.17	37.80	40.34	44.78
Ship image	3	34.16	36.40	39.78	44.34

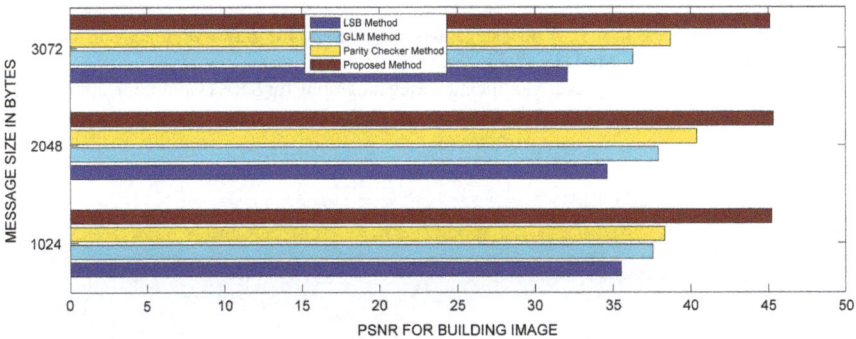

Fig. 22.4 PSNR value comparison of different methods for building image

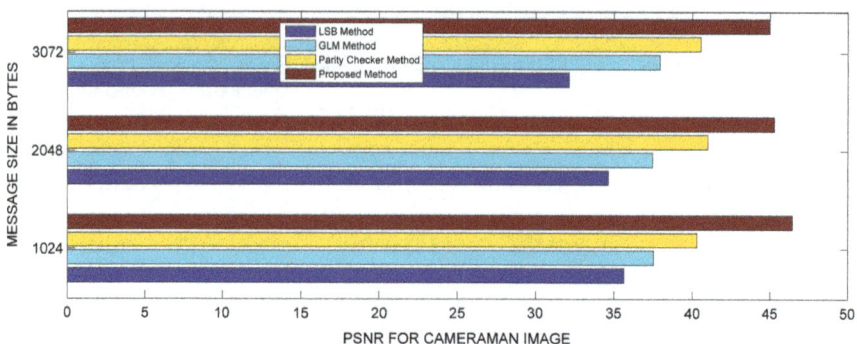

Fig. 22.5 PSNR value comparison of different methods for cameraman image

References

1. Gutub, A., Fattani, M.: A novel Arabic text steganography method using letter points and extensions. World Acad. of Sci. Eng. Technol. **1**, 502–505 (2007)
2. Krenn, R.: Steganography and Steganalysis. Internet Publication. Available at: http://www. krenn.nl/univ/cry/steg/article (2004)
3. Brainos, II, A.C.: A Study of Steganography and The Art of Hiding Information. Internet Publication. Available at www.infosecwriters.com (2007)
4. Provos, N., Honeyman, P.: Hide and Seek: introduction to steganography. IEEE **1**, 32–44 (2003)
5. Ahn, L.V., Hopper, N.J.: Public-Key Steganography. Lecture Notes in Computer Science. Springer, Berlin (2005)
6. Johnson, N.F., Jajodia, S.: Exploring steganography: seeing the unseen. IEEE Comput. **31**, 26–34 (1998)
7. Yadav, R., Rishi, R., Batra, S.: A new steganography method for gray level images using parity checker. Int. J. Comput. Appl. **11**, 0975–8887 (2010)
8. Potdar, V., Chang, E.: Gray Level Modification Steganography for Secret Communication. IEEE, New York (2004)

Chapter 23
CNN Architectures to Recognize Handwritten Telugu Characters

K. Dhana Sree Devi and C. Shoba Bindu

Abstract Artificial neural nets are backbones of many AIs predictive analogies. As common models in many of advancing domains: machine learning, artificial intelligence, and deep learning, they have led their own cognitive ways in developing highly expressive variants that work on image recognitions. From text mining to natural language processing, many of these domains have found neural networks in producing the expected accuracy. This work focuses on recognizing handwritten Telugu character vowels using convolution neural nets. We experimentally demonstrated the performance of CNN on varied architectures in predicting the handwritten Telugu vowels.

23.1 Introduction

The revival of neural networks for pattern recognition has made them to be widely used in image processing, today's most cognitive domain. Before neural networks, pattern recognition problems were most often addressed by KNN classifiers [1], quadratic discriminates [2]. Later with the advent of neural networks, image classification is more facilitated with many of its sophisticated variants. Neural networks with their activation functions [3] and back-propagation algorithm are the beacons in artificial intelligence and deep learning [4] and are almost successful in delivering new insights in image processing. Fourier transformation and linear discrimination approaches are early choices of image recognition and reconstruction. Despite their strong mathematical theories, few weaknesses still persist in LDA [5]. Many other versions of discriminate analysis are developed in pattern recognition [6].

State of the art on neural network predicted them as underlying image recognizers and their prediction was so accurate, that the field started using them in solving many

K. Dhana Sree Devi (✉)
Department of CSE, CVR College of Engineering, Hyderabad, Telangana, India

C. Shoba Bindu
Department of Computer Science Engineering, JNTUA College of Engineering, Anantapur, Andhra Pradesh, India

real-time problems. Their success in image processing was truly because of their well-defined architecture and well-trained on the taxonomy of image processing. Neural network back-propagation is the key contributor to the image compression technique [7]. Their ability to reconstruct the original image [8] in linear time is also a point of conquer. With the advent of machine learning and deep learning, the predictive usage of neural nets grew in a way that led to the development of many of its variants for image classifications. The convolution neural nets, recurrent neural nets, LSTMS, and GANs are popular variants [9–11]. All of these variants have in common neural network architecture of I/O, hidden layers wherein their approaches and purposes are different.

Deep neural networks (DNNs) difference themselves with more hidden layers. Convolution neural networks (CNNs) are a kind of DNNs that are feed-forward [12]. Ordinary neural network and CNN mend a difference in image classification. CNNs are widely used in image processing because of their adaptability and usage of few training parameters. Image classification renders another special kind of neural networks the recurrent neural nets, which showcase a special sequencing operation on inputs. These sequences are used to learn the patterns over time. RNNs architecture uses a connection from output to input which enables to gain a temporary memory [13] of the past data. RNNs are successful in keeping the track of long-term dependencies of the Input sequences. The core of RNN is the back-propagation algorithm due to which the drawback of vanishing gradients do persist in RNNs.

Long short-term memory (LSTM) is improved version of RNNs that have more prominent usage in deep learning. LSTMs were developed to deal with the RNNs drawback of exploding and vanishing gradients [14]. More often LSTMs are used in image labeling, speech recognition by many of major technology companies Apple, Google, and Microsoft.

23.2   Architectural Study of CNN

In this section, we are going to present the architectural components of CNN and their major computational units rendering to performance raise in image recognition.

23.2.1   General Artificial Neural Network

General neural networks which are the backbones for the development of major variants are networks with various layers: the input, output, and a middle hidden layer. The layers include the computational units called the neurons, and these neurons of one layer extend a connection to other layered neurons. Each neuron acts as a multiplier: multiplying the incoming input with the network weights and sums these.

The architectural key aspects are:

- Role of activation function
- Role of bias
- The learning process
- Parameters to improve the image recognition performance.

23.2.1.1 Role of Activation Function

The type of activation function used defines the predictive power of any NN. Simulation using nonlinear activation is practiced by many complex neural networks. To identify the erroneous behavior of the data, the choice is always nonlinear activation functions [15] which can adapt to the nonlinear characteristics of the inputs. Some of the activation functions used are:

- Sigmoid
- Tanh
- ReLU
- Softmax.

Linear functions always do exhibit a constant slope. Slope as we know pictures the rate of change of the predictor variable (y) with respect to the independent variable (x). Real-world problems emphasize on finding the points of minimal or maximal change: may be minimization of loss or maximization of gain. If we take linear activation functions, these changes are constant and cannot be distinguishable between points of minimal and maximal changes. Nonlinear activation functions have gradients that vary between various points. Based on these gradients descent or accent, we can address all the minimization and maximization problems. The nonlinear exponential function e^X has a range between $[0, \infty]$ and has an infinite range. The gradients for the function e^X exist between $[0, \infty]$. To find the optimal gradients, it is quite difficult within this infinite range. Hence, such functions though nonlinear cannot be taken as activation functions. Hence, the needs for activation functions which show active gradients within shorter intervals are needed. One such is the sigmoid activation function which will analyze the gradients between a small $[0, 1]$ interval.

23.2.1.2 Role of Bias

Bias is one more important parameter in an NN. The output from a neuron is of the form: $O/P = \sum(\text{weights}*\text{Inputs}) + \text{Bias}$.

This is more similar to a linear function: $Y = mX + C$; here a positive or a negative constant C shifts Y right or left; even the bias does the same. Weights of the NN are used to steer the steepness of activation function; bias will shift this steepness to the right or left of the curve [16]. These shifts are indications of the delayed triggering of the activation function. Thus, bias is used to preserve the trigging active nature of the activation cell even for zero inputs. If the NN is trained under zero bias, it does not mean that bias cannot be zero. There are NNs which work for zero bias too. But in reality all NNs use bias. Bias is critical for successful learning of the networks; with them the networks can learn different outputs for inputs. A look through Fig. 23.1 reflecting triggering of sigmoid with only weights and without bias. The plot shows steepness in the curves as weights are varying. This is a clear indication that the model is learning fast under zero bias. Neural network fast learning [17] may surpass some of the converging points.

A close look through Fig. 23.2 where a bias term is varied shows the slow triggering of the sigmoid. The moderate learning by the model may analyze various true converging points. Hence, bias terms play a critical role in analyzing the converging points of the model.

Fig. 23.1 NN learning under zero bias

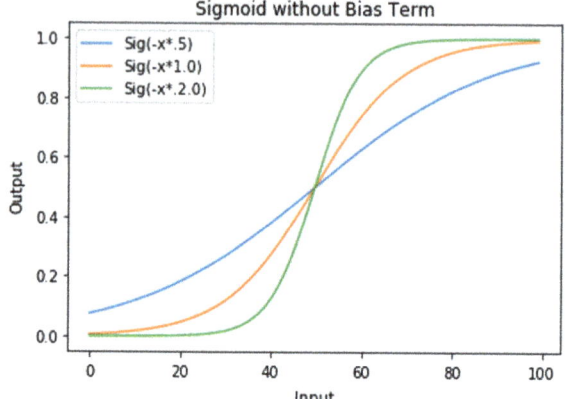

Fig. 23.2 NN learning under some bias

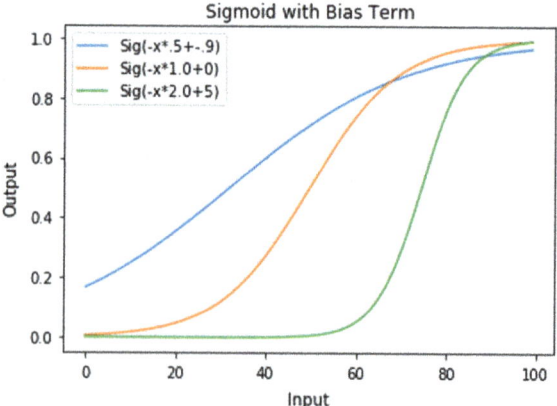

23.2.2   Convolution Neural Networks

They are completely different from other NN variants. Falling into the class of deep neural nets with more hidden layers is most commonly used for image classification. They are different by using high-layered structure with convolution layers, pooling layers, and the classification layer. In usual neural networks, few hand engineered filters are used to accomplish the preprocessing stage; ConvNet need less preprocessing as they have the ability to learn these filter characteristics. The first practically used ConvNet is LeNet-5 [18, 19]. Figure 23.3 shows the architecture of CNN.

The architectural key divisions are:

1. Feature extraction in multiple hidden layers
2. Classification in the output layer.

Feature extraction from CNN involves the following operations:

- Convolution
- Nonlinearity (ReLU)
- Pooling or Sub-sampling.

Whereas classification in the output layer includes:

- Classification using Fully Connected Layer.

23.2.2.1   Feature Extraction in CNN

Before feature extraction little bit of preprocessing is done by normalizing and sizing of image data. Image feature extraction is a kind of dimensionality reduction. Generally, an image is represented as pixels each containing 8-bit information. If the image is too large, this pixel information size is also large and is too much for a model to be trained with. So they initially undergo a feature extraction stage where distinct key parts of the image like corners, edges, and intense patches are identified.

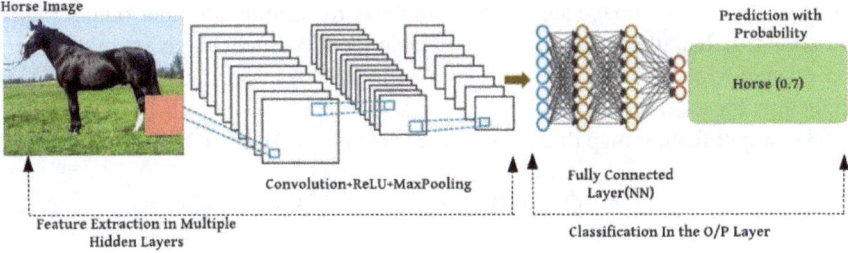

Fig. 23.3 ConvNet architecture

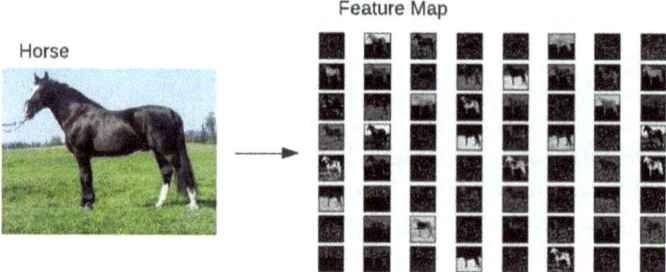

Fig. 23.4 Image feature map

Convolution

The convolution operation anneals in passing the preprocessed data through a convolution layer [20]. Here input image is condensed by extracting key features of interest so as to produce various feature maps. The convolution layer uses various filters in sequence to extract the stack of feature maps. Each filter is convolved with the image to extract the sequence of low level, middle level, and finally the high level feature maps. Figure 23.4 shows feature map extracted from an image.

Each convolution layer is stacked with various parameters; I/0 sizes, depth of feature map, kernel size, and stride. A filter which is a vector of learnable weights is used to convolve the input. These weights are learned using a back-propagation algorithm. Later these learned filters are applied to the input images to create feature map; feature map which summarizes the presence of those features in the image input. Figure 23.5 shows 5×5 filter applied to convolute an image of size 7×7.

Figure 23.6 shows 5×5 image convoluted with 3×3 filter to obtain 3×3 vector feature map. Here the size of the vector image is 5×5 and the size of the filter is 3×3, as the image vector is larger than the filter vector the filter is slide on the image vector as shown in Fig. 23.5. This generates a new output pixel. The slide length, which is the number of pixels we slide, is called as the stride. Figure 23.6 shows a one stride output vector where the pixel value 6 is obtained by convolving the filter with 3×3 image sub vector. Similarly by striding all other output pixels are obtained. The filter used in Fig. 23.6 is an identity filter. There are other filters too. Once the feature maps are obtained, nonlinearity transformation using ReLU is applied on these feature maps.

The output feature map dimensions after convolving is given by:

$$O = ((W - K + 2P)/S) + 1 \tag{23.1}$$

where
    W is the width of the input feature map.
    K is the width of the filter.
    P is the padding layer dimension.

Fig. 23.5 5 × 5 filter to convolute 7 × 7 image

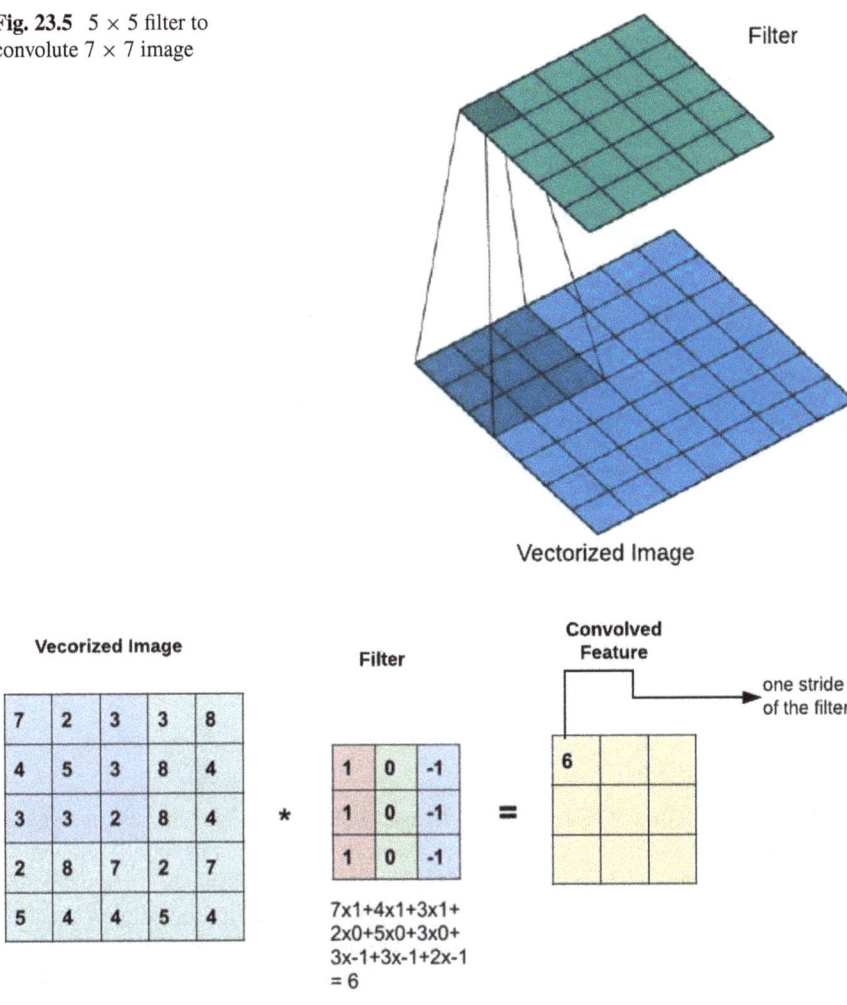

Fig. 23.6 Convolute to extract feature map

S is the stride.

Pooling Layers: The major drawback with the convolution layers is that they mark and record, the exact positions of the features in the input. An indication that small shifts in the feature position of the original input image results in a different feature map. Down sampling [21] is a common approach to address this problem and is achieved by adding pooling layers. These pooling layers are typically added after the convolution layers and after ReLU nonlinearity transformation. Pooling layers does not need a training back-propagation as their operations are fixed. Pooling involves

Fig. 23.7 Applying max pooling using 2 × 2 pooling filter

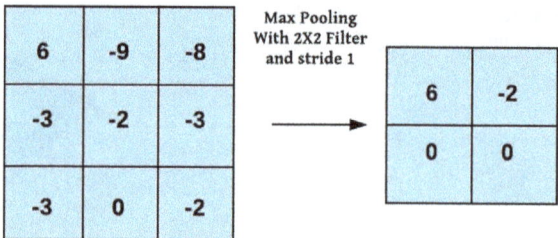

Fig. 23.8 CNN-LeNet-5 structure with Telugu vowel input

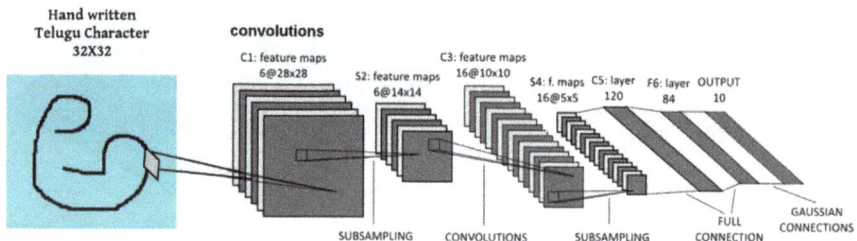

in choosing a suitable pooling operation, and two common pooling operations are given by:

1. Average pooling
2. Maximum polling

For each patch on the feature map, average pooling calculates the average value, whereas the max pooling calculates the maximum value. Thus using either average or maximum approaches produces a pooled feature map with the image feature constrained to the same location. This might be otherwise called as a dimensionality reduction approach by the pooling layers in feature maps of CNNs. The 3 × 3 output feature map in Fig. 23.8 is pooled using Max pooling approach with a 2 × 2 filter and using stride 1 and is shown in Fig. 23.7.

23.2.2.2 Classification in the Output Layer

Classification layer is normal fully connect NN network as discussed in Sect. 23.2.1. It collects the final convoluted feature from the convolution layer and returns a column vector of classes, and with class estimations to which the image belongs as shown in Fig. 23.3.

23.2.2.3   Parameters to Improve the Performance of Image Recognition in CNN

When it comes to deep learning models, performance is a key aspect to be considered. Model accuracy is an important parameter to be brooded over. Once the model is built and when its performance is not meeting the desired, then tuning the performance comes into picture. What we can do to improve the performance of a model is to use various performance improvement techniques. Accuracy of CNNs can be improved in several ways:

- Using a hyper-parameter tuning
- Using improved data.
- Use regularization techniques.

Improving Accuracy with Hyper-Parameter Tuning

In general, the hyper-parameters are specific to the model what we have built and that we want to optimize. These can also be discussed as the settings that are designed to control the model behavior. Some classic hyper-parameters to think of in boosting the performance are:

Number of Hidden Neurons: Number of hidden neurons affects the model performance. With few hidden neurons, the model may fail to learn underlying patterns of the image. So as a first aspect to boost the performance we change: either decrease or increase the hidden neuron count. But this may sometime outperform in test data and may cause to under-fitting or over-fitting, a model with different biased errors. We have to tune this parameter to have an appropriate capacity of fitting the data. Accurate predictions do not always tag a good model; but the one which gives accurate predictions on new data is needed.

Cross-validation: A method is used to estimate the model performance on new data. Model improvement prefers a threefold or a fivefold cross-validation where the entire data is divided into 3 or 5 sub-parts in counter to 70–30 train–test split on the whole data. For example in a threefold cross-validation, train data is divided into 3 subsets T1, T2, T3. In the first run, we fit the model using T1, T2 and use T3 for validation and so as for the next runs and the final accuracy is the mean of the accuracies in each run. When k-fold cross-validation is implemented at one point each and every subset of k-fold is considered new validation data on which the performance is measured. Many works, experimentally showed k-fold cross-validation produced less biased estimate. Here the hyper-parameter is k the number of folds, which can be chosen manually or can be tuned using various parameter tuning approaches, which we will discuss in the later sections.

Using Improved Data

A great challenge is prepare the best data to train the CNN. When we are working on image data set, truly all the images may not be of same kind and size. As to raise the performance, a common approach is preprocessing the data. Various image data preprocessing approaches are: Progressive resizing, Batch Normalization. More often data can be very small then Image augmentation can be used to increase the training data size and can also be discussed as one of the regularization approach to address over-fitting.

Progressive resizing is one of the performance boosters where the images are iteratively resized to analyze the performance. Initially the model is trained with a smaller image size may be 64×64. A transfer learning is then used where the weights obtained in the first training phase with 64×64 image are used to train the model on image sizes 128×128 under the same layered architecture.

Normalization is often done before the model is built in many of the cases; in CNNs normalization takes place inside the network. Under normalization, all the image pixel values are made to fall between 0 and 1 or other ranges. Normalization can scale the data to smaller ranges so that it is easy to tune an optimal learning rate.

One among widely used techniques to improve the CNN performance is Image augmentation. As one of a kind of data augmentation approach, image augmentation generates more of new training samples from the existing and the CNN can be trained with more of newer data. The approach proved to be very useful when we have small training data. Augmentation is where small sub-sampled images are considered from the original image. More advanced Image augmentations like Copyout, CopyPairing showed an improvement in test accuracy. The size of sub-sampled image is one of the hyper-parameter which can be tuned.

Using Regularization Techniques

Most common approach to boost the CNN performance is to avoid over-fitting problem. Regularization is an approach for performance tuning where the learning algorithm is modified so as to generalize better. CNNs use two popular regularization techniques: dropout and early stopping.

As CNN with neuron nodes are concerned, dropout is the most efficient regularization approach, producing optimal results. As known a fully connected NN is where every neuron is connected to every other layered neuron; for a deep network, the structure may be even too large. Also all the neurons of the network may not contribute in meeting the loss function. Such neurons may drop performance. Dropout is where such neurons are identified using a threshold, and these neurons are dropped from the network. On dropping these neurons, the neuron incoming and outgoing connections are also removed and the network becomes smaller. So dropout has become a general choice of regularization for deep networks. The number of neurons to be dropped can be taken as a hyper-parameter.

23.3 Handwritten Telugu Character Recognition

Our work presents the simulation of classical CNN, the LeNet-5, and its two varied architectures, in recognizing the handwritten Telugu vowels and demonstration of experimental performance comparison by tuning various parameters. To add understandings on CNN layered parameters, this section discusses the classical LeNet-5 architecture in detail.

23.3.1 Classical LeNet-5 Architecture

Figure 23.8 shows a classical LeNet-5 CNN learning architecture and is a widely used CNN architecture for handwritten character recognitions. Trained with back-propagation algorithm, the network led for the advent of many advanced CNN architectures. LeNet-5 architecture includes a group of 7 layers, other than the input layer. Three convolution layers C1, C3, C5, two sub-sampling layers S2, S4 and one fully connected layer, one output layer. A 32 × 32 normalized input of a Telugu character image is provided to the network. The images are resized to guarantee better recognition of the features at the center.

23.3.2 Proposed ConvNet Architecture

A follow up of 3 key changes is done in our proposed architectures, shown in Fig. 23.9:

1. Changing the LeNET layers
2. Using ReLU and ELU activations
3. Improving the model training with improved dataset.

We experimented on two datasets: the normal data set and the improved dataset. As Telugu language vowels are more different from other language characters and

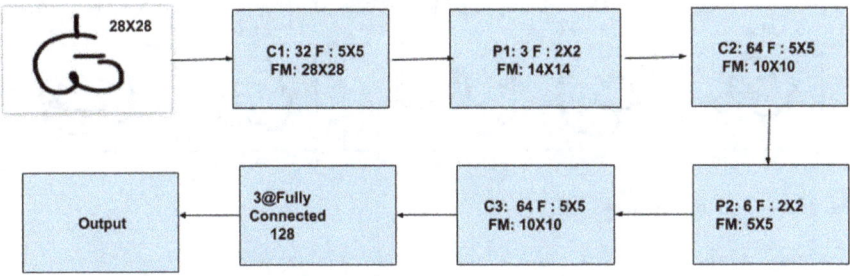

Fig. 23.9 Proposed model architecture

we used two models where we changed the architectural layers and the activation functions. Our model 1 comprised of 3 convolution layers using ReLU activation and a kernel size of 5 × 5. After the last convolution layer, we used a maximum pooling layer with size 3 × 3, the output of which is fed to fully connected layers for image flattening. The final layer is the output layer with 15 nodes and using Softmax and will classify 15 labels. Model 1 is build using normal data what we have initially.

Our model 2 architecture is build to reduce the misclassification error. We changed the model architecture with different layers and output nodes. Here we took 4 convolution layers using ELU activation and with 3 × 3 kernel. We took the same max pooling of 3X3. When experimenting on model 2, we took improved dataset.

We have chosen dropout as regularization approach, a varying dropout from 20 to 30% is what we want to experiment with for both the models.

23.3.3 Data Preparation

Telugu characters are huge in number an actual count of 56 and only 52 of them are more frequently used in language building. Our work included handwritten character classification of 15 vowels. A sample of which is shown in Fig. 23.10. We used English labeling for these 15 vowels accordingly on how they are pronounced. For each character, more than 500 handwritten images are created each of size 200 × 200, to include our normal dataset size of 7000 images.

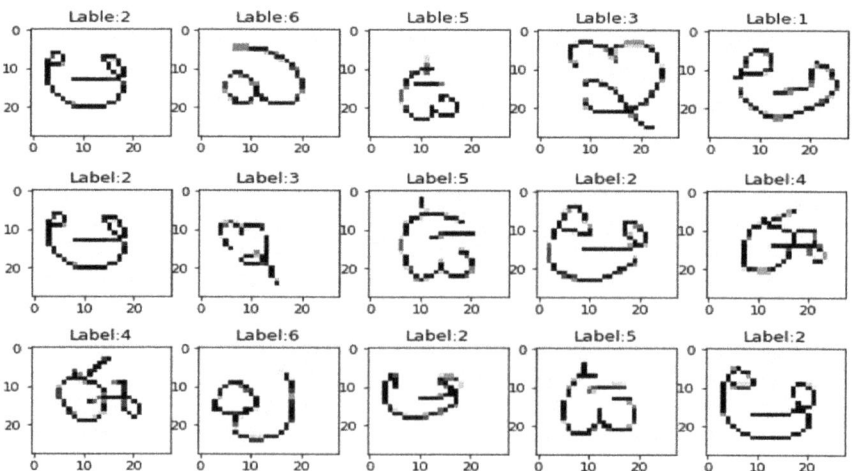

Fig. 23.10 Training samples

23.3.3.1 Improved Data Set

We used data augmentation to raise our sample size. Augmentation added new images to the base dataset using various transformations: image rotation, color change, adding noise. We used small rotation on about 5° as Telugu characters are scripted always with small inclinations from the zero line. Even the image shifts what we used are small of about 5% both vertical and horizontal. We applied ZCA whitening technique.

23.4 Experimental Results

The classical LeNET CNN is the reference for many experimental works, giving a scope to many research ideas. As a part of our proposed work, we modeled Telugu vowel classification on these three models: the LeNET and our proposed architectures model 1 and model 2. We first trained the LeNET to classify the vowels. We trained our model with small data first; the data what we have initially, 7000 images. We observed the accuracy rate at various dropouts; but the misclassification error is too large with 40%.

We then took our proposed model 1, trained with the same 7000 images. We used ReLU activation here. Here also we observed the misclassification error is too large with 40%. Here we observed that two of the vowel classes 'Ee' and 'A' are totally misclassified; as our trained data in these two classes is very less. We switched to our proposed model 2 using ELU activation, and with improved dataset. Using ELU decreased misclassification rate to some extent 34.33%. Here we observed that some of the vowels from 'Ee' and 'A' are correctly classified. We observed that improved dataset with augmentation performed on the data has reduced the misclassification error. We observed the misclassification error on varied dropouts.

In the second stage of experiment, we used our larger dataset which is improved using data augmentation as discussed in Sect. 23.3.3.1. We ran all the three models again by varying the dropout. Model 2 which used ELU activation performed well at dropout 25%; the misclassification error of 30.62%, which is less compared to other dropout observations on other models. Table 23.1 shows the misclassification rates and performance comparison of our two models with classical LeNET. Figure 23.11 shows some misclassified images.

23.5 Conclusions

Native language processing is emerging with new techniques and models with which the linguistic characters are being trained for further recognition. Among such models, the convolution networks are predominantly in use showcasing their accuracy in correctly recognizing the handwritten characters. Convolution neural

Table 23.1 Misclassification error %

Model	Test misclassification error % (Normal dataset)			Test misclassification error % (Improved dataset)		
	Dropout 20%	Dropout 25%	Dropout 30%	Dropout 20%	Dropout 25%	Dropout 30%
LeNET	40.80	38.72	42.78	37.68	35.21	39.55
Proposed model-1	40.01	39.41	43.11	36.91	35.00	38.97
Proposed model-2	37.33	34.76	38.24	33.33	30.62	36.60

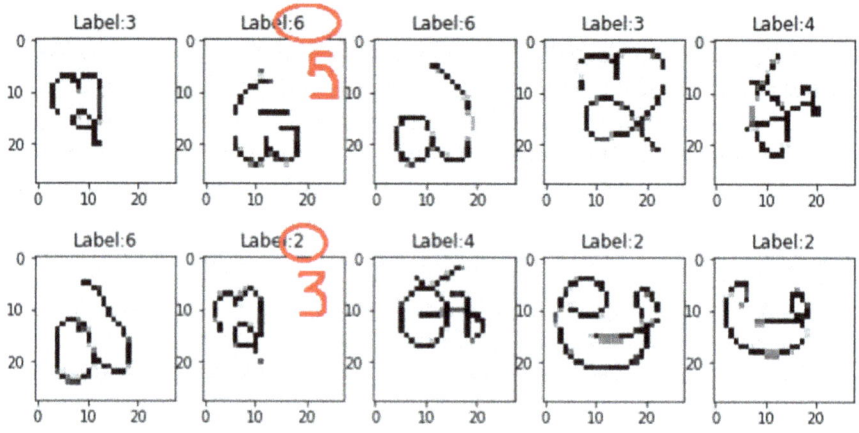

Fig. 23.11 Some misclassified images

networks are deep learning models trained with more of image data. With convolution filters, these models are able to extract recognizable features in images with more predictive powers. Natural language processing is much benefitted using this model. Our work showed one such case of experimental study of pre-trained CNNs in recognizing handwritten characters from Telugu language.

References

1. Vajda, S., Santosh, K.C.: A fast k-Nearest Neighbor classifier using unsupervised clustering. In: International Conference on Recent Trends in Image Processing and Pattern Recognition (Apr. 2017)
2. Bose, S., Pal, A., SahaRay, R., Nayak, J.: Generalized quadratic discriminant analysis. Pattern Recogn. **48**(8), (2015)
3. Dhana Sree, K.: Data analytics: role of activation function in Neural Net. IJITEE **8**(5), (2019)
4. Ongsulee, P.: Artificial intelligence, machine learning and deep learning. In: International Conference on ICT and Knowledge Engineering. IEEE (Nov. 2017)

5. Liang, Y., Gong, W., Pan, Y., Li, W.: Generalizing relevance weighted LDA. Pattern Recogn. **38**(11), (2005)
6. Wen, Y., Shi, Y.: An approach to numeral recognition based on improved LDA and Bhattacharyya distance. In: International Symposium on Communications, Control and Signal Processing. IEEE (2008)
7. Jiang, J.: Image compression with neural networks-a survey. Signal Process.: Image Commun. **14**(9), (1999)
8. Zhou, Y.T., Chellappa, R., Vaid, A., Jenkins, B.K.: Image restoration using neural network. IEEE Trans. Acoust. Speech Signal Process. **36**(7), (1988)
9. Angeline, P.J., Saunders, G.M., Pollack, J.B.: An evolutionary algorithm that constructs recurrent neural networks. IEEE Trans. Neural Netw. **5**(1), (1994)
10. Gers, F.A., Schmidhuber, J., Cummins, F.: Learning to forget: continual prediction with LSTM. ICANN, (1999)
11. Zhang, Y., Zhou, Y., Huang, C.: Generating artificial images by generative adversary network. In: International Conference on Pattern Recognition and Artificial Intelligence (2019)
12. Nebauer, C.: Evaluation of convolutional neural networks for visual recognition. IEEE Trans. Neural Netw. **9**(4), (1998)
13. Sukhbaatar, S., Weston, J., Fergus, R.: End to end memory networks. Adv. Neural Inf. Process. Syst. **28**, (2015)
14. Pascanu, R., Mikolov, T., Bengio, Y.: On the difficulty of training recurrent neural networks. J. Mach. Learn. Res. (2013)
15. Li, H., Ouyang, W., Wang, X.: Multi bias nonlinear activation in deep neural networks. J. Mach. Learn. (2016)
16. Tsujitani, M., Koshimizu, T.: Neural discriminant analysis. IEEE Trans. Neural Netw. (2000)
17. Fu, L.M., Hsu, H.H., Principe, J.C.: Incremental backpropagation learning networks. IEEE Trans. Neural Netw. **7**(3), (1996)
18. Han, S., Pool, J., Tran, J., William D.: Learning both weights and connections for efficient neural network. Neural Inf. Process. Syst. (NIPS) (2015)
19. AI-Jaw, R.: Handwriting Arabic character recognition LeNet using neural network. Int. Arab J. Inf. Technol. **6**(3), (2009)
20. Jeon, Y., Kim, J.: Active convolution: learning the shape of convolution for image classification. In: IEEE Conference on Computer Vision and Pattern Recognition (2017)
21. Sun, M., Song, Z., Jiang, X., Pan, J., Pang, Y.: Learning pooling for convolutional neural network. Neurocomputing **22**

Chapter 24
Rider-Deep Belief Network-Based MapReduce Framework for Big Data Classification

Sridhar Gujjeti and Suresh Pabboju

Abstract Big data is an emerging domain with huge amounts of data and is increasing rapidly in an exponential manner. There exist different eras in the massive datasets, but the analysis of huge data in the medical field is a challenging task. This paper presents a Rider-based Deep belief network (Rider-DBN) for classifying big data in the medical domain to predict diseases. The goal is to provide a MapReduce framework with a deep learning classifier that can categorize normal and abnormal classes with improved accuracy. Here, the information gain is employed for selecting the imperative features using inputted data considering mappers. Thus, the dimension of the features is reduced in such a way that only the selected features are given for the classification to handle the big data. The selected features are employed by the Deep belief network (DBN) for categorizing the normal or abnormal classes in disease prediction. The Rider optimization algorithm (ROA) is employed to train DBN that finds optimum weights of the DBN for disease prediction. The Rider-DBN provides superior performance with the accuracy of 89.3%, sensitivity of 78.6%, and specificity of 92%.

24.1 Introduction

The evolution attained in computer technologies, digital data storage, and scientific encroachment in the communication domain has facilitated the cohort of massive datasets in medical applications. The extraction of patterns using medical data assists the clinicians to diagnose the patients [1]. Many of the hospitals utilize the information of hospital patients and their histories in a storage repository for diagnosis, but these methods may maximize the size of storage and degrade the performance of the system [2]. Healthcare is suffering from several issues for detecting the patient's

S. Gujjeti (✉)
Kakatiya Institute of Technology and Science, Warangal, Telangana, India

S. Pabboju
Chaitanya Bharathi Institute of Technology, Hyderabad, Telangana, India

© The Author(s), under exclusive license to Springer Nature Singapore Pte Ltd. 2021
S. C. Satapathy et al. (eds.), *Smart Computing Techniques and Applications*,
Smart Innovation, Systems and Technologies 225,
https://doi.org/10.1007/978-981-16-0878-0_24

syndrome [3]. The classification of disease using the attributes is a noteworthy research area from a diagnosis standpoint. The effective classification is attained using a suitable mechanism which capitulates improved accuracy [4]. In [5], a rough-fuzzy hybrid classifier is devised for predicting cardiac disease. In [6], SVM was utilized for categorizing tumors based on the Factorizing of Symmetry-Non Negative Matrix. In [7], a genetic-fuzzy system is devised using ECG signals for heart disease diagnosis. In [8], a model is devised for diagnosing cardiac ailment using an optimization algorithm. A neural network [9] is devised for classifying the human brain tumor using magnetic resonance images. In [10], the method was devised for monitoring type 2 diabetes mellitus using patient information.

The major contribution of the paper:

- **Proposed Rider-DBN for disease prediction using massive-sized data**: The classification of big data is carried out using DBN for predicting normal and abnormal classes using input disease data. The training of DBN is done by ROA, which tunes the optimal weights for discovering the classes.

24.2 Literature Review

The four classical disease prediction strategies using big data are illustrated along with their merits and demerits. Lakshmanaprabu et al. [11] devised an IoT-based healthcare system based on Random Forest Classifier (RFC) with MapReduce for big data classification to predict the disease. The method showed effective performance, but the algorithm was computationally slow due to massive datasets. ALzubi et al. [12] devised Weight Optimized Neural Network with Maximum Likelihood Boosting (WONN-MLB) by analyzing huge data for disease prediction. The method showed high diagnosis accuracy, but the method did not apply to other datasets. Nalluri et al. [13] devised a hybrid model by combining SVM and monarch butterfly optimization (MBO) for disease prediction. This method attained effective tradeoffs amongst cataloging three goals, which involve specificity, accuracy, and sensitivity. Meanwhile, the technique was not capable to incorporate multi-objective bi-clustering with parameter evolution. Reddy et al. [14] devised an adaptive genetic algorithm with fuzzy logic (AGAFL) for predicting heart disease and helping the clinicians to spot heart disease in an earlier stage. The method was effective in handling noisy data and poses the potential to deal with a large number of attributes.

24.3 Proposed Rider-DBN for Predicting Disease Using Big Data

Figure 24.1 portrays the architecture of the proposed Rider-DBN for disease prediction using big data. Here, the input data is fed to the MapReduce framework wherein

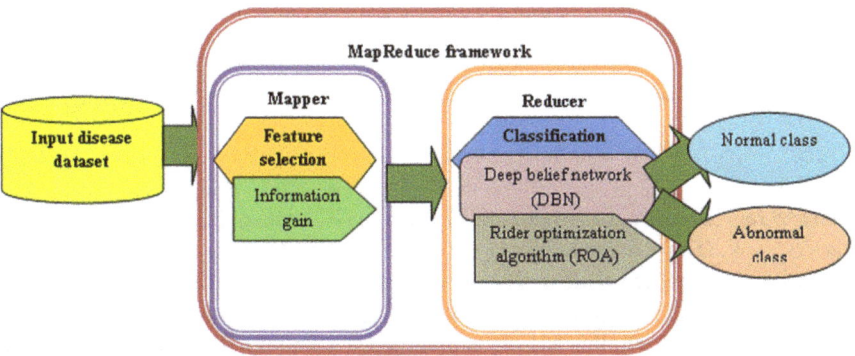

Fig. 24.1 Schematic view of proposed Rider-DBN for predicting disease in big data

the selection of imperative features and classification is performed to predict disease. The feature selection is performed in the mapper function using Information gain and the classification is performed on reducer with DBN, where the training is performed using the ROA that determines the weights of the DBN. Thus, the DBN classifier in the reducer performs the disease prediction and generates two classes, normal or abnormal classes.

24.3.1 Selection of Imperative Feature Using Information Gain in Mapper Phase

The map-reduce model has elevated power for processing since different servers are processed in parallel and poses the capability to perform parallel processing. Consider the input big data is expressed as D with various attributes as given in Eq. (24.1),

$$D = \{d_{e,f}\};\ (1 \le e \le P);\ (1 \le f \le S) \tag{24.1}$$

where $d_{e,f}$ signifies data in big data D with eth data of fth attribute, P denotes total data points, and S represents total attributes. The mapper adapts the mapper function, and ROA is used to select the best features. The importance of selecting features is that imperative features are selected for ensuring dimensionality reduction and to elevate the classification accuracy. Initially, big data is splitted into various subsets, which is represented in Eq. (24.2).

$$d_{e,f} = \{t_u\};\ (1 \le u \le F) \tag{24.2}$$

where, F denotes total subsets. Besides, total subsets are equivalent to total mappers and expressed in Eq. (24.3).

$$M = \{M_1, M_2, ...M_u, ..., M_F\} \tag{24.3}$$

The input to the uth mapper is represented in Eq. (24.4).

$$a_u = \{B_{i,j}\}; \ (1 \le i \le K); \ (1 \le j \le S) \tag{24.4}$$

where, S represents total attributes and K denote total data points in uth subset such that $K < P$. The mapper employs information gain [15] for choosing optimum features and output from all mapper is combined to produce chosen features. The information gain is a measure employed for selecting optimum features. Information gain is also termed as Kullback–Leibler divergence which can be utilized as a measure to rank each feature. For computing information gain, firstly the entropy of the dataset must be computed. The entropy of the dataset D is computed using Eq. (24.5).

$$E(D) = - \sum_{m \in J} P(m) \log P(m) \tag{24.5}$$

where, $P(m)$ represent the probability of instance $d_{e,f}$ to be labeled as m and it is represented in Eq. (24.6).

$$P(m) = \frac{|\{d_{e,f} \in D | q_{e,f} = q\}|}{|D|} \tag{24.6}$$

where, $d_{e,f}$ denote the data in big data D representing eth attribute of the fth data, D is total subset of data, and $q_{e,f}$ indicate label of eth attribute of the fth data. The information gain of lth feature of the dataset D is computed using Eq. (24.7), where the first part indicates the sum of probabilities of the instance $d_{e,f}$ to have a value n of the feature. The second part represents entropy of a subset of instances from D that has the value n of lth feature and is formulated as,

$$IG(D, l) = E(D) - \sum_{n \in N^l} \frac{|\{d_{e,f} \in D | d_{e,f}^l = n\}|}{|D|} E(\{d_{e,f} \in D | d_{e,f}^l = n\}) \tag{24.7}$$

After the application of information gain, the obtained selected features are represented as a feature vector A. The chosen features are subjected as input to reducer, which process to produce needed output with the proposed Rider-DBN model.

24.3.2 Classification of Massive Data with Proposed Rider-DBN in Reducer Phase

The proposed Rider-DBN is employed in reducers to achieve classification by offering optimal classification results to guarantee the effectual extraction of the pattern using big data and to reduce the time taken for processing. Moreover, the reducer is expressed in Eq. (24.8).

$$V = \left\{ V_1, \ V_2, \ ..., \ V_y, \ ...V_z \right\} \tag{24.8}$$

where, z represents total reducers where ($1 \leq y \leq z$). The evaluation of classification results in reducer is adapted for performing effectual classification.

24.3.2.1 Training of DBN Using ROA Algorithm

The DBN [16] is a subset of Deep Network and comprises various layers of Restricted Boltzmann Machines (RBMs) and Multilayer Perceptrons (MLPs). The training of DBN is carried out using the ROA [17] that aims to discover the optimum weights to tune the DBN classifier. The classification of big data employs proposed Rider-DBN to classify data by obtaining the best classification and is capable to handle new data attributes. The steps of the ROA algorithm are depicted in this section.

Step (1) Initialization: At first, the riders are randomly initialized and expressed in Eq. (24.9).

$$Y = \{Y_k\} \ ; (1 \leq k \leq l) \tag{24.9}$$

where, l represents the total number of riders, and Y_k indicates kth rider from the total riders.

Step (2) Evaluate the error: The optimum solution is discovered using fitness function and is considered as a minimization function and hence, the solution with the least Mean Square Error (MSE) is selected as an optimal solution and MSE is formulated in Eq. (24.10).

$$MS_{err} = \frac{1}{P} \sum_{e=1}^{P} \left[O_e - O_e^* \right]^2 \tag{24.10}$$

where, O_e is expected output and O_e^* is the predicted output, P indicate total input data where $1 < e \leq P$.

Step (3) Update the solution based on the ROA algorithm: The solutions are updated after measuring the error and the solution update is performed with ROA using four-rider parameters, namely follower, attacker, overtaker, and bypass rider.

The follower updates position by the leading rider location for reaching the target in rapid mode and is formulated in Eq. (24.11).

$$Y_{s+1}^{f}(k, l) = Y^N(N, l) + [\text{Cos}(W_{k,l}^s * Y^N(N, l) * D_k^s)] \tag{24.11}$$

where, l is coordinate selector, Y^N signifies leading rider position, N denote leading rider index, $W^s_{k,l}$ represent steering angle of kth rider in lth coordinate, and D^s_k is distance.

The updated position of overtaker is employed to maximize the rate of success by discovering the overtaker position and expressed in Eq. (24.12).

$$Y^o_{s+1}(k, l) = Y_s(k, l) + \left[D^*_s(k) * Y^N(N, l)\right] \qquad (24.12)$$

where, $D^*_s(k)$ represent the direction indicator.

The attacker grabs the leader's position by the following leader and the attacker update position is formulated in Eq. (24.13).

$$Y^a_{s+1}(k, b) = Y^N(N, l) + [\text{Cos } W^s_{k,b} * Y^N(N, l)] + D^s_k \qquad (24.13)$$

The bypass riders undergo a common path without following the leading rider and are represented in Eq. (24.14).

$$Y_{s+1}(k, b) = \alpha[Y_s(\mu, b) \times \eta(b) + Y_s(\vartheta, b) * [1 - \eta(b)]] \qquad (24.14)$$

where, α denote a random number between [1, 0], μ is a random number in the interval [1, P], ϑ is a random number between [1, P], η is a random value, b is the coordinate, and k denote the position of the rider, and $Y_s(\mu, b)$ represents the position of μth rider in bth coordinate.

Step (4) Re-computation of error: The error is calculated using Eq. (24.10) and the solution with minimal error is used to train DBN to detect disease.

Step (5) Terminate: The optimum weights are obtained in repeated mode till utmost iterations are attained.

24.4 Results and Discussion

The assessment of the proposed strategy with classical strategies using sensitivity, accuracy, and specificity is illustrated. The dataset employed for experimentation is Heart Disease Data Set [18] donated by David W. Aha. The dataset comprises four databases, namely Cleveland, Hungary, Switzerland, and the VA Long Beach. For this research, two databases namely Hungary and Switzerland are adapted for analyzing disease. The execution of the proposed strategy is performed in JAVA with PC having Windows 10 OS, 2 GB RAM, and Intel i3 core processor.

24.4.1   Comparative Analysis

The methods taken for the assessment include: NN [19], SVNN [20], SVM [13], RideNN [17] and the proposed Rider-DBN.

24.4.1.1   Analysis Considering Hungarian Dataset

Figure 24.2 depicts the comparative analysis of techniques using the Hungarian dataset with accuracy, sensitivity, and specificity. The analysis of techniques with accuracy is deliberated in Fig. 24.2a. When 50% of data is used for training, the accuracy obtained by classical NN, SVNN, SVM, RideNN, and the proposed Rider-DBN are 0.531, 0.605, 0.759, 0.820, and 0.843. The analysis based on sensitivity is deliberated in Fig. 24.2b. The comparative methods, such as NN, SVNN, SVM, RideNN, and the proposed Rider-DBN have the sensitivity of 0.531, 0.605, 0.825,

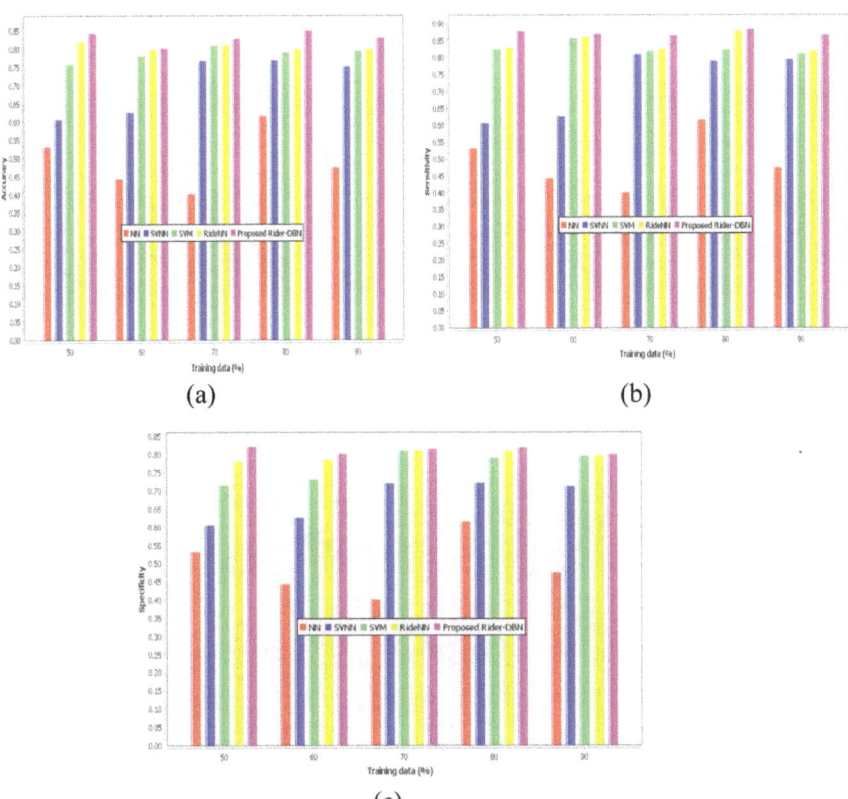

Fig. 24.2 Analysis of methods with Hungarian dataset considering **a** Accuracy **b** Sensitivity **c** Specificity

0.830, and 0.878, when 50% of data is used for training. The analysis using the specificity is deliberated in Fig. 24.2c. When 50% of data is used for training, the specificity of the existing methods, such as NN, SVNN, SVM, RideNN, and the proposed Rider-DBN are 0.531, 0.605, 0.715, 0.780, and 0.820.

24.4.1.2   Analysis Considering Switzerland Dataset

Figure 24.3 displays the comparative analysis using Switzerland dataset. The analysis of techniques using the accuracy is deliberated in Fig. 24.3a. For 50% training data, the accuracy of the existing methods, such as NN, SVNN, SVM, RideNN, and the proposed Rider-DBN are 0.616, 0.830, 0.872, 0.875, and 0.882. The analysis using the sensitivity is deliberated in Fig. 24.3b. For 50% training data, the sensitivity of the comparative methods, such as NN, SVNN, SVM, RideNN, and the proposed Rider-DBN are 0.541, 0.574, 0.683, 0.758, and 0.806. The analysis based on specificity parameter is deliberated in Fig. 24.3c. The methods, such as NN, SVNN, SVM, RideNN, and the proposed Rider-DBN have the specificity of 0.635, 0.893, 0.917, 0.922, and 0.926 when the classifier is trained with 50% of data.

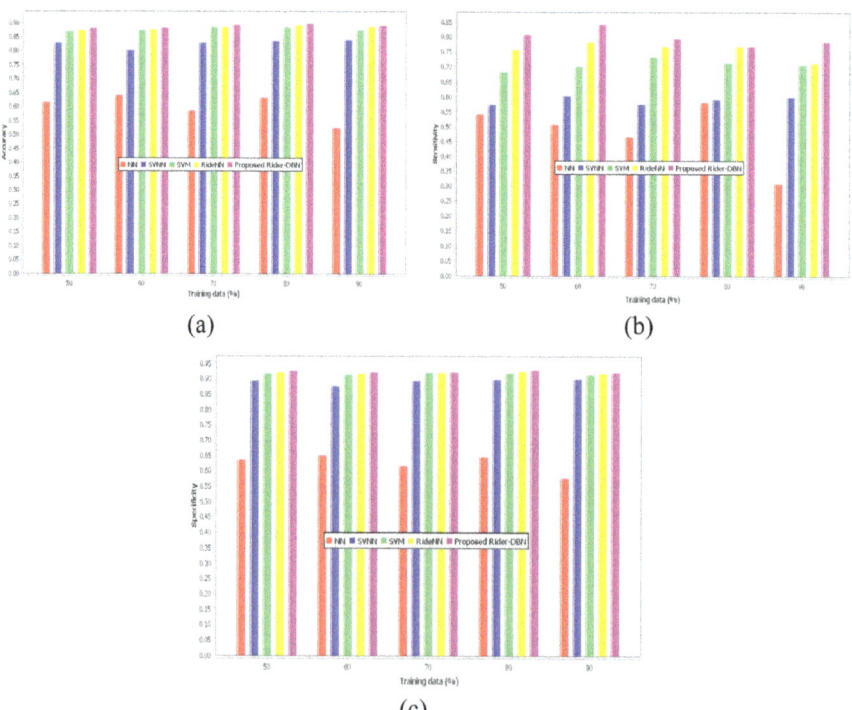

(a)   (b)

(c)

Fig. 24.3 Analysis of methods with Switzerland dataset considering **a** Accuracy **b** Sensitivity **c** Specificity

24.5 Conclusion

This paper presents a Rider-DBN for performing big data classification to predict diseases. The purpose is to provide a MapReduce framework with a deep learning classifier that can classify normal and abnormal classes with better accuracy. The ROA is employed for training the DBN that finds the optimum weights of the DBN for disease prediction. The Rider-DBN provides superior performance with the accuracy of 89.3%, sensitivity of 78.6, and specificity of 92%. In the future, the method will be extended by analyzing the performance of the method with streaming data.

References

1. Tsymbal, A., Bolshakova, N.: Guest editorial introduction to the special section on mining biomedical data. IEEE Trans. Inf. Technol. Biomed. **10**(3), 425–428 (2006)
2. Mane, T.U.: Smart heart disease prediction system using improved K-means and ID3 on big data. In: Proceedings of 2017 International Conference on Data Management, Analytics and Innovation (ICDMAI), pp. 239–245. IEEE (2017)
3. Ephzibah, E.P., Sujatha, R.: Big data management with machine learning inscribed by domain knowledge for health care. Int. J. Eng. Technol. **6**(4), 98–102 (2017)
4. Polat, K., Güne, S.: An expert system approach based on principal component analysis and adaptive neuro-fuzzy inference system to diagnosis of diabetes disease. Digit. Signal Process. **17**, 702–710 (2007)
5. Srinivas, K., Rao, G.R., Govardhan, A.: Rough-fuzzy classifer: a system to predict the heart disease by blending two different set theories. Arab J. Sci. Eng. **39**(4), 2857–2868 (2014)
6. Yuvaraj, N., Vivekanandan, P.: An efficient SVM based tumor classification with symmetry non-negative matrix factorization using gene expression data. In: Proceedings of International Conference on Information Communication and Embedded Systems (ICICES), pp. 761–768. IEEE (2013)
7. Vafaie, M., Ataei, M., Koofgar, H.R.: Heart diseases prediction based on ECG signals' classification using a genetic-fuzzy system and dynamical model of ECG signals. Biomed. Signal Process. Control **14**, 291–296 (2014)
8. Long, N.C., Meesad, P., Unger, H.: A highly accurate firefly based algorithm for heart disease prediction. Expert Syst. Appl. **42**(21), 8221–8231 (2015)
9. Kharat, K.D., Kulkarni, P.P., Nagori, M.: Brain tumor classification using neural network based methods. Int. J. Comput. Sci. Inf. **1**(4), 2231–5292 (2012)
10. Han, J., Rodriguez, J.C., Beheshti, M.: Diabetes data analysis and prediction model discovery using rapidminer. In: Proceedings of Second International Conference on Future Generation Communication and Networking, vol. 3. pp. 96–99. IEEE (2008)
11. Lakshmanaprabu, S.K., Shankar, K., Ilayaraja, M., Nasir, A.W., Vijayakumar, V., Chilamkurti, N.: Random forest for big data classification in the internet of things using optimal features. Int. J. Mach. Learn. Cybernet. **10**(10), 2609–2618 (2019)
12 ALzubi, J.A., Bharathikannan, B., Tanwar, S., Manikandan, R., Khanna, A., Thaventhiran, C.: Boosted neural network ensemble classification for lung cancer disease diagnosis. Appl. Soft Comput. **80**, 579–591 (2019)
13. Nalluri, M.R., Kannan, K., Gao, X.Z., Roy, D.S.: Multiobjective hybrid monarch butterfly optimization for imbalanced disease classification problem. Int. J. Mach. Learn. Cybernet. 1–29 (2019)
14. Reddy, G.T., Reddy, M.P.K., Lakshmanna, K., Rajput, D.S., Kaluri, R., Srivastava, G.: Hybrid genetic algorithm and a fuzzy logic classifier for heart disease diagnosis. Evol. Intell. 1–12 (2019)

15. Zdravevski, E., Lameski, P., Kulakov, A., Jakimovski, B., Filiposka, S., Trajanov, D.: Feature ranking based on information gain for large classification problems with mapreduce. In: Proceedings of IEEE Trustcom/BigDataSE/ISPA, vol. 2, pp.186-191 (2015).
16. Hinton, G.E., Osindero, S., Teh, Y.: A fast learning algorithm for deep belief nets. Neural Comput. **18**, 1527–1554 (2006)
17. Binu, D., Kariyappa, B.S.: RideNN: a new rider optimization algorithm-based neural network for fault diagnosis in analog circuits. IEEE Trans. Instrum. Meas. 1–25 (2018)
18. Heart Disease Data Set, https://archive.ics.uci.edu/ml/datasets/Heart+Disease. Accessed on April 2020
19. Mathan, K., Kumar, P.M., Panchatcharam, P., Manogaran, G., Varadharajan, R.: A novel Gini index decision tree data mining method with neural network classifiers for prediction of heart disease. Des. Autom. Embed. Syst. **22**(3), 225–242 (2018)
20. Gomathi, N., Karlekar, N.P.: Ontology and hybrid optimization based SVNN for privacy preserved medical data classification in cloud. Int. J. Artif. Intell. Tools **28**(3), 1950009 (2019)

Chapter 25
Modified Compressive Sensing and Differential Operation Based Channel Feedback Scheme for Massive MIMO Systems for 5G Applications

V. Baranidharan, R. Shalini, N. Nithya Shree, S. Sasithra, and R. Jaya Murugan

Abstract In 5G communications, the massive Multi-Input Multi-Output (MIMO) systems are one of the promising key technology for future applications. In the massive MIMO systems, modified compressive sensing and differential operation-based channel feedback scheme is widely used to increase the channel matrix dimension. In this paper, a modified compressive sensing and differential operation-based channel feedback scheme has been proposed to reduce the channel feedback overhead of the massive MIMO systems. In order to achieve stronger sparsity, the temporal correlation-based time-varying channels is widely used and exploited in order to generate the Channel Impulse Response (CIR) between the neighboring time slots. The simulation results show that this modified compressive sensing and differential operation-based channel feedback scheme is reducing the overhead arise due to feedback scheme and achieves the stronger sparsity than the existing systems.

V. Baranidharan · R. Shalini (✉) · N. Nithya Shree · S. Sasithra · R. Jaya Murugan
Department of Electronics and Communication Engineering, Bannari Amman Institute of
Technology, Sathy, India
e-mail: shalini.ec16@bitsathy.ac.in

N. Nithya Shree
e-mail: nithyashree.ec16@bitsathy.ac.in

S. Sasithra
e-mail: sasithra.ec16@bitsathy.ac.in

R. Jaya Murugan
e-mail: jayamurugan.ec18@bitsathy.ac.in

© The Author(s), under exclusive license to Springer Nature Singapore Pte Ltd. 2021   251
S. C. Satapathy et al. (eds.), *Smart Computing Techniques and Applications*,
Smart Innovation, Systems and Technologies 225,
https://doi.org/10.1007/978-981-16-0878-0_25

25.1 Introduction

Wireless communication systems will undergo a revolution about once every decade. Now there is a huge demand in the market for a new technology called 5G that should enhance the capacity of the system than 4G technology. The migration towards the higher frequencies leads to increase in the system capacity. For example, millimeter wave (mm wave ranging 30–300 GHz) will release larger bandwidth. The massive Multiple-Input Multiple-Output (MIMO) which contains massive number of antennas (i.e. >= 100 antennas) significantly enhances the spectrum efficiency [1]. The mm-waves are suitable for massive MIMO system because of its higher frequency, very short wavelength, and the physical and enhanced design of the massive antenna arrangement as an array can also be altered.

This method significantly enhances the gain of the MIMO system, and also overcomes the severe path loss of mmWave signals. Therefore, by combining both mm-wave communications and massive MIMO [2], the data rate speed and spectrum efficiency need to be increased. The acquired knowledge about the accurate Channel State Information (CSI) is important in mm wave massive MIMO systems, to improve the link reliability by pre-coding in the uplink and combining in the downlink, because both the Base Station (BS) and Mobile Station (MS) are always equipped with large number of antennas. The CSI channel information acquired is widely used in uplink by leveraging the reciprocity values of the channel. This will be feedback to all the other mobile stations. In comparison with the conventional mmWave MIMO communication systems working at 306 GHz, the channel estimation in the mmWave MIMO systems will be very challenging task due to the large number of antennas, less sparsity, and hardware constraints.

The acquisition of Channel State Information at Transmitter (CSIT) can be done by exploiting the channel reciprocity in Time Division Duplexing (TDD). The problems in TDD like delay in transmitting the information to the receiver due to traffic in the channels are rectified in Frequency Division Duplexing. But still, it fails to hold channel reciprocity which is more important to meet the required system performance of massive MIMO. Conventional codebook based channel feedback method does not suit the massive MIMO systems, because the codebook size needs to be extended in order to acquire a reliable CSIT accuracy [3]. The modified compressive sensing and differential operation-based channel feedback methods based on non-codebook, uses the compressive sensing (CS) to reduce the feedback overhead, since non-codebook possess large channel feedback overhead. CS-based channel feedback methods decrease the overhead by compressing MIMO channel matrix where the spatial correlations are exploited. In practice, when MIMO is distributed, the channels are not spatially correlated. Therefore, the sparsity of time-domain channel is widely exploited to compress the channel impulse response (CIR) directly which leads to the reduction of feedback overhead. But for CIR which is not sufficiently sparse, the overhead is very high.

25.2   Literature Survey

This section explores the many existing techniques which greatly overcome the computational complexity and feedback overhead changes.

Dai et.al. [4] proposed a codebook named 2D space of AoA's and AoD's values, an adaptive subspace to reduce the feedback overhead in FDD multiple access technique-based massive MIMO systems. By using the constant AoD's information, this codebook quantizes the channel vector within the given angle coherence time. But when number of base station antennas is less than 120, the theoretical rate gap is comparatively high.

Sidiropoulos et.al. [5] proposed the Novel Limited Feedback algorithms that lifted the problem of feedback overhead. This algorithm is more suitable for Double Directional Massive MIMO channels where the inherent sparsity is exploited and low computational complexity, particularly on the user side. But, this method is not suitable for conventional uniform dictionaries and the performance gain is poor. Khan et al. [6] proposed a new algorithm called Block Sparsity Adaptive Matching Pursuit (BSAMP). Due to large dimensions of channel vector and unknown value of the channel sparsity, this will lead to computational complexity. Hyukmin and Cho [7] proposed a compressive sensing-based CSI channel feedback method which gains sparsity by exploiting the channels that possess temporal correlation. That is, the sparsity between the channel impulse responses is obtained by temporarily coagulated channels. However, this method results in feedback distortion that further leads to data loss. After analyzing numerically, simulation results show that the sparsity obtained is widely based on the temporally correlated channels values based on the compressed CSI feedbacks.

Kulsoom et al. [8] have proposed a method called Quantized Partially Joint Orthogonal Matching Pursuit (Q-PJOMP). It is also known as Q-PJIHT-based Channel State based algorithm to retrive the various values of CSI from the quantized feedback. In the second stage, Compressed sensing algorithms are used to recover the CSI. The two-step quantization technique is having higher complexity. It also possesses feedback overhead problem.

25.3   System Model

In wireless communication, the information content is sparse due to the random projections. When a massive number of antenna arrays are used, the channel exhibits the sparsity in spatial frequency domain which results in strong spatial correlations and feedback overhead. To overcome this problem, modified compressive sensing (CS) and differential operation based channel feedback scheme is used. Initially, the CS-based method exploits the sparsity by efficiently encoding it without knowing the information about the location. Further, it develops a compression mechanism

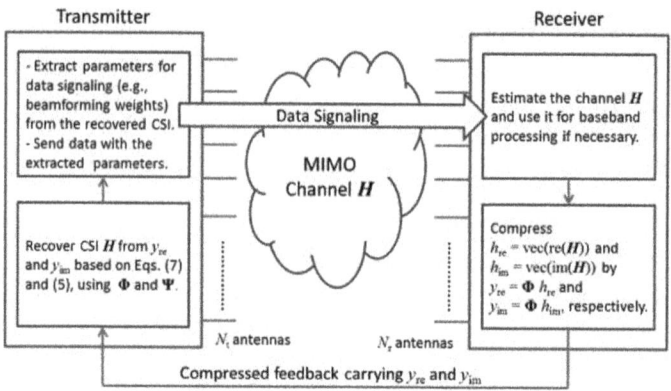

Fig. 25.1 Flow of compressive sensing method

to reduce the Channel State Information by using the small compression ratios (Fig. 25.1).

Consider H as a spatially correlated massive MIMO channel matrix for $N(t) \gg 1$ antennas for transmission and $N(r) >= 1$ antennas used for reception where $H = N(r) * N(t)$. In conventional channel feedback methods, the matrix H is directly fed into the channel feedback whereas the CS-based channel feedback method applies a compression mechanism on H before sending it into the channel feedback. It captures the spatially arranged X non zero elements of X sparse with $N*1$ signal vector, where $X \ll N$ by signal will be compressed into $M \ll N$. The compressed feedback content is represented as $Y = \phi * h$.

25.3.1 Temporal Correlation of Time-Varying MIMO Channels

The lth element of CIR's support vector $p_n^{(t)}$ is varied over T consecutive time slots and it is given by $\{p_n^{(t)}(l)\}_{t=1}^T$. This is known as Markov First Order Process where $p_n^{(t)}(l) \in \{0, 1\}$. The two transition possibilities are represented as:

$$P_{10} = P\{p_n^{(t+1)}(l) = 1 \big| p_n^{(t)}(l) = 0\} \tag{25.1}$$

$$P_{01} = \Pr\{p_n^{(t+1)}(l) = 0 \big| p_n^{(t)}(l) = 1\} \tag{25.2}$$

And the distribution series at time $t = 1$ is denoted as $\mu_n^{(1)} = \Pr\{p_n^{(1)}(l) = 1\}$. If $\Pr\{p_n^{(1)}(l) = 1\} = \mu$ for all time t and n, then $P_{10} = \mu P_{01}/(1 - \mu)$. Similarly, the amplitude vector $a_n^{(t)}$ of CIR is varied over time by using the Autoregressive First-Order and it is given as $a_n^{(t)} = Pa_n^{(t-1)} + (1 - P^2 w^{(t)})^{1/2}$ where $P = J_0(2\pi f_d \tau)$ and is known as correlation coefficient of Bessel function, f_d represents the maximum

Doppler frequency, τ represents the duration of the time slot and $w^{(t)}$ represents the noise vector which is independent of $a_n^{(t-1)}$. All these parameters are independent and are similarly distributed with zero mean Gaussian distribution CN $(0, \sigma_\omega^2)$.

25.3.2 Proposed Modified Differential Channel Feedback Based Compressive Sensing

From the literature survey, many channel feedback reduction methods have been proposed to provide accurate CSI at the transmitter and the receiver side. But the mm-wave massive MIMO antennas are very sensitive to the accuracy of channel state information. A negligible change in the channel state leads to a significant loss in downlink performance, efficiency, and SNR rates. This error level creates a very strong correlation between the successive channel impulse responses (CIR). A good approach to overcome this problem is quantizing the channel impulse response in a differential way because, the time-varying massive MIMO channels have coherence time greater than the time difference between two consecutive channel response.

Differential channel feedback method is a natural method in which the high-level correlation or two-dimensional correlations like temporal or spatial correlation can be exploited to reduce feedback overhead. It uses only the changes in the channel response relative to the previous channel response. That is, it computes the difference between the current and previous channel impulse response by limiting the magnitude of each angle but it maintains the same quantization resolution. Differential quantization is an iterative process. For each iteration, both transmitter and receiver know the accurate information about the current channel response and the last channel estimation. Generally, when conventional feedback methods are used, the receiver feeds back the temporally correlated responses to the transmitter. These responses require greater feedback rate and larger bandwidth. The rate of feedback bit increases when the temporal correlation increases. But the differential channel feedback method reduces the feedback bits to a great extent by exploiting the temporal correlation. Initially, the transmitter predicts the current CIR depending upon the previous CIR. Then, it performs the differential quantization on the channel responses and feeds into the channel feedback. At the receiver side, the quantized responses are captured and recovered. In return, it sends the differential CIR to the transmitter in a compressed way.

The basic idea behind modified compressive sensing and differential operation based channel feedback scheme is that when there is temporal correlation, instead of quantizing the whole channel response which leads to the wastage of feedback resources. The performance of proposed differential channel feedback method is much better than the other conventional methods in low mobility channels. The modified compressive sensing and differential operation based channel feedback scheme differentiates the CIR with stronger sparsity by exploiting the temporal correlation.

The sensing matrix compresses the differential CIR and it is fed into the channel feedback. The received CIR is known to both BS and users in order to recover the original differential CIR by using larger compression ratio. This helps in the avoidance of feedback error propagation because the users execute initialization after certain time slots. The conventional channel feedback method gives feedback about the channel after the estimation in the current time slot estimated values. But the proposed modified compressive sensing and differential operation will be used to generates the sparsity CIR by computing or differentiating two $h(t-1)$ and $h(t)$ estimates where $(t-1)$ denotes the $(n-1)$ which is the previous time slot and t represents the current time slots. Mathematically, the differential CIR $h(t)$ is computed as,

$$\Delta h_n^{(t)} = h_n^{(t)} - h_n^{(t-1)} = p_n^{(t)} \cdot (a_n^{(t)} - a_n^{(t-1)}) + (p_n^{(t)} - p_n^{(t-1)}) \cdot a_n^{(t-1)} \quad (25.3)$$

$$= p_n^{(t)} \cdot [\sqrt{1 - \rho^2}w - (1 - \rho)a_n^{(t-1)}] + (p_n^{(t)} - p_n^{(t-1)}) \cdot a_n^{(t-1)} \quad (25.4)$$

Consider ϕ as the Gaussian matrix (sensing matrix) where $\phi M * L$ with $M \ll L$. This sensing matrix ϕ compresses the sparse signal $\Delta h(t)$ into low dimensional vector Y. The low dimensional measurement vector Y is given by $\phi \Delta h(t)$ which is fed into the base station. When it is fed into the base station, the low dimensional vector Y is added with the noise in the channel and it is given by the equation $Y = \phi \Delta h(t) + n$ where n refers to the noise in the channel feedback. The compressed sparse signal Y is accurately recovered from the noisy base station channel by the most popularly used subspace pursuit (SP) algorithm. This algorithm is chosen for its lesser complexity and noise resistance.

25.4 Simulated Results

The performance metrics are investigated by using the simulation results using MATLAB is explained in this section. The proposed modified CS and Differential operation based channel feedback for massive MIMO systems is simulated and the resulted performance metrics are compared with the existing systems at different channel compression ratio (η). At initially, some parameters are considered for massive MIMO systems for simulations. The simulation parameters used for simulating the system model of massive MIMO systems are tabulated in the Table 25.1.

25.4.1 Channel Impulse Response Comparison

In this method, the channel compression ratio $\eta = M/L$ where, L is the maximum delay spread. Figures 25.2 and 25.3 represents the snapshot of previous Channel

Table 25.1 Simulation parameters

Parameters	Values
Number of the antenna	32
Amplitude vector	$a_n^{(1)}$
Support vector	$p_n^{(1)}$
Channel compression ratio	η

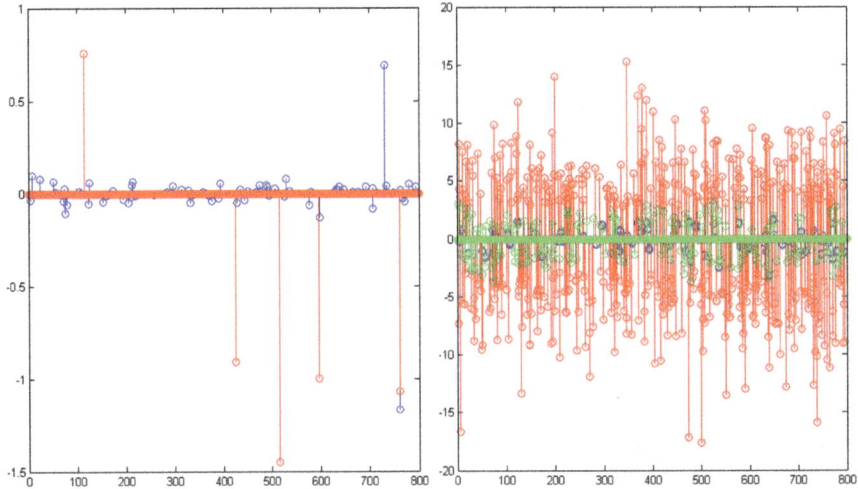

Fig. 25.2 Original CIR

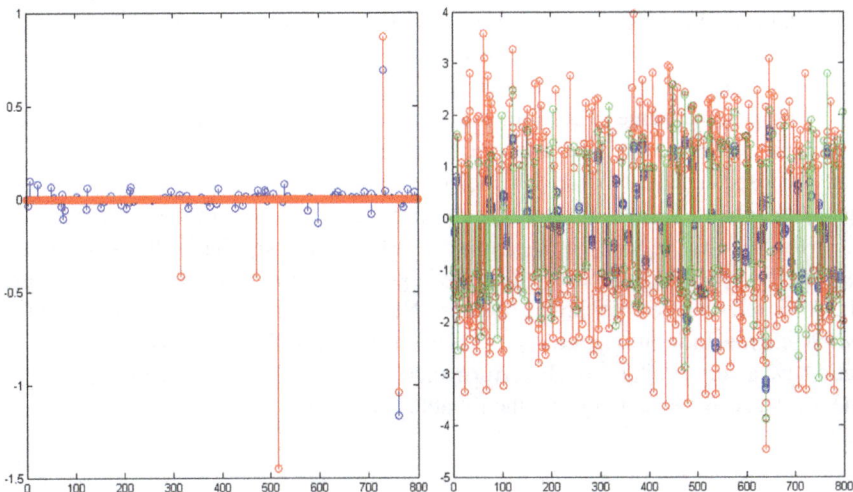

Fig. 25.3 Modified CIR

Table 25.2 CIR simulation parameters

Parameters	Values
L	200
P_{01}	0.05
M	0.1
F_d	10 Hz
T	1 ms
σ_w	1

Impulse Response, current CIR, and differential CIR. The following parameters are considered for simulation and are tabulated in the Table 25.2.

At final, initially the amplitude vector value $a_n^{(1)}$ is considered as random vector. The different entries of independent identically distributed (I.I.D) varied complex Gaussian variables with mean values as zero and unit variance has been considered.

From Fig. 25.3, we observed that the modified differential CIR will give stronger sparsity than the original CIR, which will indicate that it is compressed more but the feedback overhead is reduced comparatively with existing systems.

25.4.2 NMSE Validation

For NMSE validation, initially consider the system with simulation parameters considered for CIR generation. Here, the two different cases are considered $\eta = 25\%$ and the M/L value is changed to $\eta = 45\%$. In the proposed modified differential feedback band channel state information state, η value represents the average range of compression of the CIR over the p feedback slots. The compression ratio of the proposed scheme is 30% for the first case and 40% for the second case. The p-value is considered as 3 for adaptive to the channel condition. Figure 25.4 shows that proposed scheme—based scheme. The direct CS and the proposed differential feedback scheme results statistics are compared and tabulated in the Table 25.3.

In the first case, the existing system work fails to overcome the feedback overhead. But for this case, the proposed system gives good performance. For the second case, the proposed scheme is achieving a better SNR compared with the existing direct CS based scheme. In addition, the proposed scheme for the different channel compression ratio have a very similar NMSE performance, and this scheme can be used to reduce the feedback overhead to ensure the reliable channel feedback.

25.5   Conclusion

In this paper, we analyzed and investigated the issues and challenges in the design of channel feedback with the differential operation in the massive MIMO systems. The

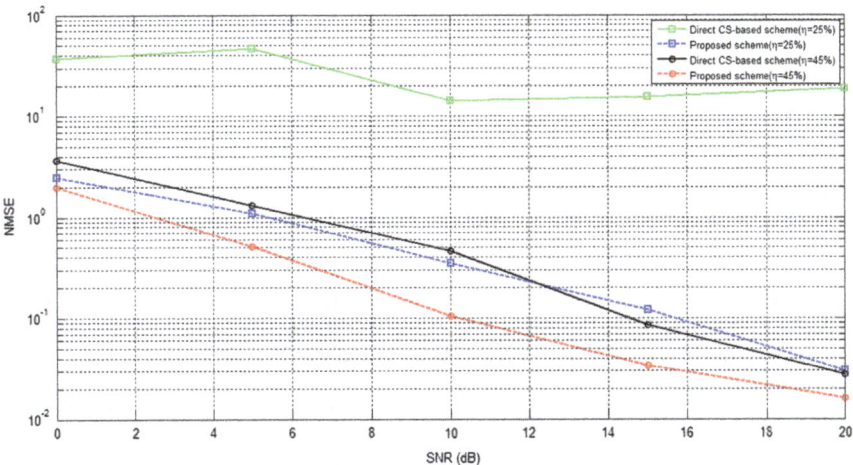

Fig. 25.4 NMSE comparison of the proposed modified differential feedback systems

Table 25.3 Comparison of the proposed differential feedback scheme (CS-DO) with existing systems

Values	$\eta = 25\%$		$\eta = 40\%$	
	Direct CS scheme	CS–DO scheme	Direct CS scheme	CS–DO scheme
Min	14.06	0.03016	0.02743	0.01617
Max	47.31	2.507	3.652	1.999
Mean	26.62	0.8214	1.106	0.5331
Median	18.77	0.3521	0.4642	0.1047
Mode	14.06	0.03016	0.02743	0.01617
Standard deviation	14.93	1.031	1.511	0.844
Angle	33.25	2.477	3.625	1.983

proposed work exploits the modified temporal correlation of the massive MIMO based time-varying channels CIR values and the differential CIR in slots of its neighboring time slots. This modified differential channel feedback scheme gives a stronger sparsity than the existing CIR methods. The modified Compressed sensing and differential operation based channel feedback scheme gives a better Normalized Mean Square Error (NMSE) values than all the other existing methods and it also reduces the feedback overhead for the massive MIMO systems.

References

1. Larsson, E.G., Edfors, O., Tufvesson, F., Marzetta, T.L.: Massive MIMO for next generation wireless systems. IEEE Commun. Mag. **52**(2), (2014)
2. Rappaport, T.S., Xing, Y., MacCartney, G.R., Molisch, A.F., Mellios, E., Zhang, J.: Overview of millimeter wave communications for fifth-generation (5G) wireless networks—with a focus on propagation models. IEEE Trans. Antennas Propag. **65**(12), (2017)
3. Mumtaz, S., Rodriguez, J., Dai, L.: mmWave Massive MIMO, A Paradigm for 5G. Elsevier (2017)
4. Dai, L., Shen, W., Shim, B., Wang, Z., Heath, R.W.: Channel feedback based on AoD-adaptive subspace codebook in FDD massive MIMO systems. IEEE Trans. Commun. **66**(11), (2018)
5. Alevizos, P.N., Fu, X., Sidiropoulos, N.D., Yang, Y., Bletsas, A.: Limited feedback channel estimation in massive MIMO with non-uniform directional dictionaries. IEEE Trans. Signal Process. 66(19), 1 (2018)
6. Khan, I., Singh, M., Singh, D.: Compressive sensing-based sparsity adaptive channel estimation for 5G massive MIMO systems. IEEE Appl. Sci. (2018)
7. Son, H., Cho, Y.-H.: Analysis of compressed CSI feedback in MISO systems. IEEE Wirel. Commun. Lett. 8(6), (2019)
8. Kulsoom, F., Vizziello, A., Chaudhry, H.N., Savazzi, P.: Joint sparse channel recovery with quantized feedback for multi-user massive MIMO system. IEEE Trans. Signal Process. 66(19), 1 (2018)

Chapter 26
Smart Shopping Bag Using IoT

M. Suguna, R. V. Kanimozhi, D. Naveen Kumar, S. Indhushree, and D. Prakash

Abstract Shopping mall is a place where thousands of customers visit every day to purchase many products. Shopping becomes a tedious process. It consumes lots of time as customers have to wait in a long queue. To make shopping easier we have proposed a system titled "SMART SHOPPING BAG USING IOT" where board which will help blind people to verify their purchased product. It will say the details that are displayed in the Liquid Crystal Display (LCD) which will be audible only to the customer carrying the bag. After the final completion, the customer can scan only the bag instead of scanning the product one by one. Once they are done with the scanning, they can pay their bill through mobile phone integrated with Global System for Mobile communication (GSM) module or can pay manually. Products are scanned through Radio Frequency Identification (RFID) tag reader which helps to add up the price of the product purchased. They can switch the mode and remove their product if not needed. This will help to calculate the total amount of their purchase. There is also an encoder-decoder control board which will intimate the customer if they miss their bag or the bags have been replaced. The total value of their purchase is added to their final bill. The bag also features a voice customer are provided with bag which readily displays the product name and details of the item purchased.

M. Suguna (✉) · R. V. Kanimozhi · D. Naveen Kumar · S. Indhushree
Department of Computer Science and Engineering, Kumaraguru College of Technology, Coimbatore, India
e-mail: suguna.m.cse@kct.ac.in

R. V. Kanimozhi
e-mail: kanimozhi.16cs@kct.ac.in

D. Naveen Kumar
e-mail: naveen.16cs@kct.ac.in

S. Indhushree
e-mail: indhushree.16cs@kct.ac.in

D. Prakash
Electrical and Electronics Engineering, VTMT College, Chennai, India

© The Author(s), under exclusive license to Springer Nature Singapore Pte Ltd. 2021
S. C. Satapathy et al. (eds.), *Smart Computing Techniques and Applications*,
Smart Innovation, Systems and Technologies 225,
https://doi.org/10.1007/978-981-16-0878-0_26

26.1   Introduction

IOT has been emerged as an important technology in our day-to-day life. It is a system of interrelated computing devices, mechanical and digital devices. It has been used to transfer data over network without help of humans. Over the last few years IOT has been used in every field and it make our work easier and faster. Using these new technologies in developing variety of systems led to various edges such as doing away with inconsistency and therefore dependency on human labor, raising accuracy and labor speed. Over the last years, several studies have been carried out based on these premises led to adjustment of both these technologies. IoT has many in-built sensors which has the ability to collect the information and transfer them through the network. This will be helpful in real-life decision-making process. There were a lot of IoT applications which has changed the whole technology revolution. It has its emergence in every field because of its lightweight methodology. One such important field is shopping. Shopping is considered as the main source of economy and the rate at which shopping by people is always increasing at a steady rate. Today shopping malls are considered to be crowded area. During weekend or festival season, the amount of purchasing will be increasing at the peak. Though purchasing things get easier, the main problem arises while billing our purchase. Waiting in a billing section is a time-consuming process. IoT has provided a solution by which the time taken for waiting in a queue has been reduced to 1/3rd of the time spent before. It can make shopping easier by allocating more time in shopping than to billing.

26.2   Literature Review

RFID-based smart shopping kart has an idea of using LIFI technology which is most effective and can minimize the queues. When we enter a shopping mall, its regular to take shopping cart and stop wherever we need the products need to buy and waiting in a long queue for billing the products. Hence it is much time-consuming process. The technology of RFID is used for transmission and photodiode for reception. The product details are collected through RFID. When the products are being collected, we use RFID to transmit the products for billing session. When the shopping cart is taken near the billing part then the products in shopping cart is automatically billed and bill is given to the consumer [1].

Sahare et al. [2] proposed outline of using raspberry PI. It is easier to interface and no need for extra modules to interface with RFID and makes the system size more approachable. The key objective is to deliver low-cost, easily accessible, and a system for supporting shopping. The first step of the customer entering shopping is to take the trolley. Every trolley is attached with a RFID reader. The idea focuses on which the customer inspects the details of the things he wants to buy. While inspecting the RFID tag of the item, a cost of the item is taken and secure in the framework's memory. When the customer scans the RFID tag the information of the product is passed to the Raspberry pi through serial communication. Raspberry pi will collect the details of ID from the database. LCD will show product details along with the total amount. As soon as purchase gets over key is pressed to transfer the data to the billing counter [2].

Sivagurunathan et al. [3] has explored the idea of using Zigbee technology which directly sends the information to the billing counter. The trolley can automatically guide the customer to the location of the product once they enter the product name. When a customer enters into the mall, they first take the trolley. Every trolley is joined with a scanner tag and a RFID per user. When they purchase an item, the product is scanned through the RFID tag reader and the details of their purchase will be displayed in the LCD placed in the trolley. At the same time, Arduino will pass the information to the PC with the assistance of RS232 extension. Assuming that every single product checked will give a bell sound making the customer ensure that the product is perused. It will be helpful for security reasons. However, tallying is performed as there is no extension of cost particular item in receipt. On extraction of the unwanted product it will automatically get subtracted on their removal. The product information will be shared via Wireless information process and Zigbee transmitter is used for transferring the final billing amount to the billing system. This will improve the security performance and also the speed [3].

IoT applied smart shopping system has visualized the use of RFID-based production data analysis. Thus, the system uses trolley number for the creation of automatic bill. The smart cart attached with RFID reader automatically reads the items. For data processing microcontroller has been installed and for user interface LCD touch screen has been used. Zigbee technology has been used to communicate smart cart with the server because of its low-power consumption and inexpensiveness. Weight scanner is used for weighing items. RFID reader installed before door is used to check that all the items have been paid. Security and privacy-related issues have been considered. Product quality and quantity, customer satisfaction is automatically monitored in this system. Trolley movement has been monitored using Android application which has been implemented in this system. The application is completely based on the Trolley number and total amount of purchased items [4].

Thangavel, Karthick, Karthikeyan, Karunakaran, Prasanth R has proposed as IoT framework where RFID tag is connected to every product, is read by the RFID reader attached to the bag, given per user. Smart racking system can be included, outfitted with RFID per users, which is used to check screen stock, additionally refreshing a focal server. This system uses RFID to check the product and its price. Once the product is detected the system will process and upload the product price and name. If the stock is in demand or the stock is not available, the system will update it in the IOT [5, 6]. This system also detects the expire date of the product and alert the user through buzzer and update in the IOT website. The load sensor will indicate the overload in the trolley. If the system detects overload it will alert the user through buzzer. The LCD display will display the product name, price, and overload [7].

26.3 Purpose and Objective

At present, we are using barcode scanner to scan the products. There is a LCD display which displays the product name and price. The transfer of data from trolley or basket to PC is handled by Zigbee technology. The amount will get added along with the addition of the product. On the completion, the final amount will be displayed in the LCD. It is much a time-consuming process as it more time to get scanned. Customers have to wait in a long queue for paying their bills which is considered to be a tedious process.

The main objective is to build a smart shopping bag which uses RFID tag to scan the product. The details like product name, amount and the expiry date will be displayed. Along with the LCD display speaker is used to say the details displayed. It will be useful for the blind people and senior citizen. Arduino has been used where the codes are embedded.PIC microcontroller has been replaced by Arduino where we can change our codes easily and in an efficient way. Once we add a product the total will get added along with the total. Once the shopping gets over the final total will be displayed and customer can pay the bill either digitally or manually. The digital payment can be made by using customer card using GSM module. Manual payment will be made in the cash counter where customer can show their bag and receive their receipt of shopping.

26.4 Proposed System

We have proposed a system based on recent technology, IOT. Smart shopping bag will be useful for customer in all shopping malls, supermarkets, and dress shop. The bag usually made of RFID tag reader, LCD display unit, speaker and controller board, encoder decoder module. The initial step of the purchase starts with the scanning of RFID tag of the product scanned. The tag will have a unique code which will be displayed in the LCD display along with the details of the product. Ardunio has been used to store the details of the product. The details of the product will be heard through speaker which will be helpful for blind people and senior citizen. The price of the product is added up after every scan of the product. If customer did not need the product anymore, he can usually pull the switch and scan the product to remove that item from the list. The price of that particular product will be reduced automatically. Once the purchase got completed, he/she can press the button which indicates the purchase completion. Customer can pay their bill through the card of their own which is provided by the shop. The total amount and the balance will be pop up to the mobile. During the time of purchase if they miss their bag the encoder and decoder module present in the bag will automatically warn the owner of that bag (Fig. 26.1).

Fig. 26.1 Block diagram of overall structure of smart shopping bag using IoT

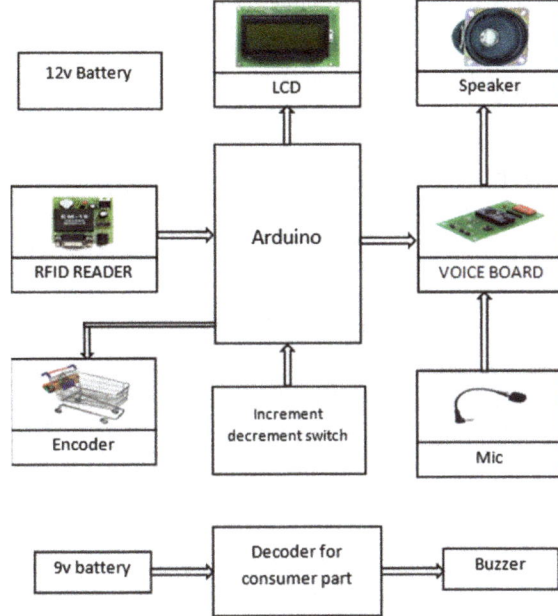

26.5   Module Description

26.5.1   RFID Reader Module

This module is usually composed of RFID tag and a reader. The RFID is scanned when the product is placed in the bag using reader which is placed along the bag side. This will be used to calculate the item scanned into the bill. The ID scanned will be unique to every product. This will get matched to the item details stored in the database. Hytertrm software can be used to verify if the scanned RFID is unique or not.

26.5.2   LCD Display Module

The items scanned will be displayed in this module. When sufficient voltage is applied to the electrodes the liquid crystal molecules will be arranged in a way that displays the required information according to the tag they read. The voltage passed should be of +5 V.

26.5.3   Switch Module

This module can be used either for adding or subtracting the cost. The cost of the item that is scanned through bag is automatically get added to their total amount. This will be reflected in the final amount of purchase. If we switch the mode and scan the product the amount of the respective product will get subtracted from the total.

26.5.4   Blind People Module

According to recent survey, 70% of blind people face difficulty while doing shopping. Even though they can identify the product but the price and details to understand are always challenging. The feedback collected from 3–4 departmental stores nearby shows that it will be useful for blind if they had a speaker attached with the basket, so that they can manage on their own without others help. To sort out the problem faced by them, speaker module is developed. WTV-SR comes under the group of recording serial products. WTV-SR module is used for recording and fixing voice playback, recording of the content uploaded and a variety of control modes can be chosen. WTV-SR module is used for various occasions. On the bottom of WTV-SR, I/O can be adjusted by using different control modes. The voice that is recorded already is played while the respective product is scanned. This will be helpful for the blind people who can know the details of the product by voice that are recorded.

26.5.5   Band Module

This module usually composed and encoder and decoder board. This will intimate the customer if they miss their bag or the bag has been replaced. It produces a buzzer like sound when they are moved 5–10 m away from their bag.

26.5.6   Billing Unit

GSM module is used for sending the bill to the respective customer through their registered mobile number. After the purchase has been completed the total will be displayed. If customer pays their bill through card, then the total amount and balance will be displayed in the mobile. The information is transferred through SIM (Subscriber Identity Module) in GSM module to the registered mobile number.

26.6 Results and Discussion

Figure 26.2 displays the Ardunio software where the code is feeded. It is used to check whether the RFID tag is mapped with the correct tag number. It is earlier stage of verification process where the tags are mapped with their respective number.

Figure 26.3 shows the LCD display unit. It displays the name and price of the product which when scanned through RFID reader module attached to the basket. Whenever the item is scanned using reader the details are collected from the database. As soon as the details are retrieved LCD displays the details of the scanned product.

Figure 26.4 displays the integration of various modules. The board usually contains Ardunio, speaker module, voice module, LCD display unit, and a switch. This board is fitted to the basket at the bottom. The main working of smart shopping

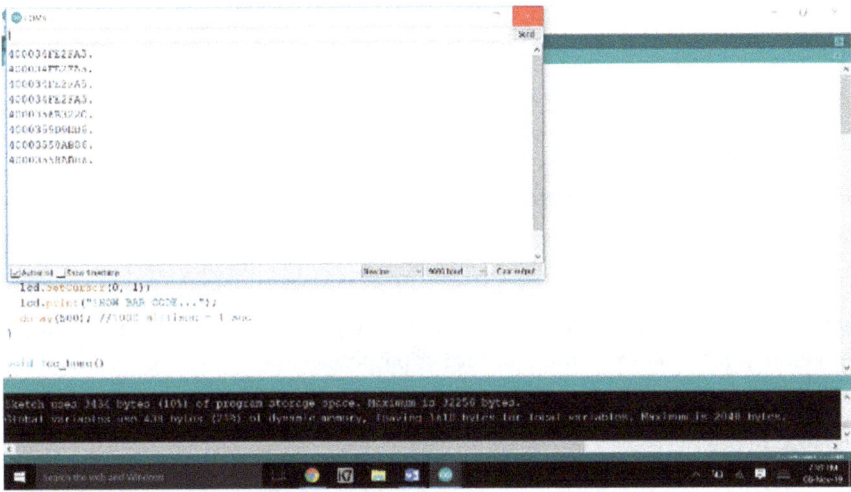

Fig. 26.2 Ardunio verification

Fig. 26.3 LCD display

Fig. 26.4 Smart board

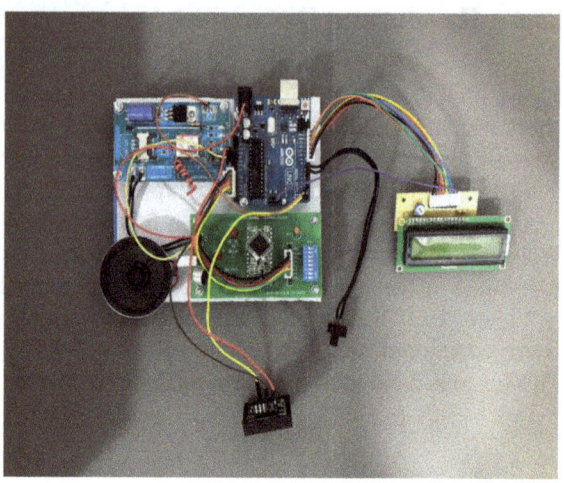

takes place in this board which has an interface to various modules. Speaker acts as a mode for blind people to know the details of the product. The speaker in this board is placed to help the blind people for making their shopping better and can be more confident about the items which they buy.

Figure 26.5 displays the page of billing unit. After shopping gets completed customer can use their card to pay their bills. Once the bill is paid then the card balance and the total cost of their purchase will be sent as a SMS to the registered mobile number. The card mentioned is provided by the shop where we can fill the amount in prior and can be used for further purposes.

Fig. 26.5 Billing unit

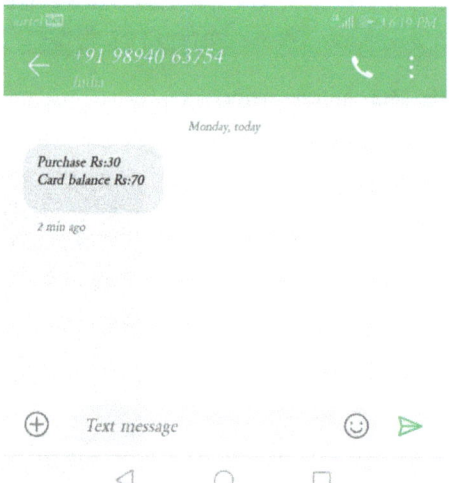

26.7 Conclusion

Thus, the Smart Shopping Bag is made using the recent technologies in IOT. Customer uses this bag for purchase in shopping areas which is kept in the mart so this makes customer easy to purchase products and the total amount will be displayed in the LCD unit. They can scan their bags at the cash counter to pay bills. This will avoid waiting in long queues for billing. Customer can save their time and concentrate more on shopping than waiting in queues for billing. Before billing, they will come to know the total amount of their purchase.

Acknowledgements We affirm that the project titled "SMART SHOPPING USING IOT WITH CUSTOMER ORIENTED SERVICE" is done in focus to help the people who are struggling to do shopping and for waiting in a queue to bill their products. The projects work is carried in our college laboratory and has been developed from the feedbacks received from nearby shopping mart. The information provided is true to the best of our knowledge.

References

1. Sahare, P.S., Gade, A., Rohankar, J.: A review on automated billing for smart shopping system using IOT. Rev. Comput. Eng. Stud. **6**(1), 1–5 (2019). Journal homepage: https://iieta.org/Journals/RCES
2. Sivagurunathan, P.T., Seema, P., Shalini, M., Sindhu, R.: Smart shopping trolley using RFID. Int. J. Pure Appl. Math. **118**(20), 3783–3786 (2018)
3. Vishwas, B., Apoorva, S., Raidurg, S.V., Pawar, A.R., Rananavare, L.B.: IOT application on secure smart shopping system. Int. J. Adv. Res. Comput. Sci. **9**(3), 197 (2018)
4. Devi, R., Pavithira, T., Monisha, E., Likitha, M.: IoT applications on secure smart shopping system. Int. Res. J. Eng. Technol. (IRJET) **05**(03), 1881–1885 (2018)
5. Li, R., Song, T., Capurso, N., Yu, J., Couture, J., Cheng, X.: IoT applications on secure smart shopping system. IEEE Internet Things J. **4**(6), 1945–1954 (2017)
6. Karjol, S., Holla, A.K., Abhilash, C.B.: An IOT based smart shopping cart for smart shopping. In: International Conference on Cognitive Computing and Information Processing, pp. 373–385. Springer, Singapore. Accessed December 2017
7. Shirsath-Nalavade, K.A., Jaiswal, A., Nair, S., Sonawane, G., Suchita, B.A.: IOT based smart shopping cart (SSC) with automated billing and customer relationship management (CRM). Int. J. Res. Appl. Sci. Eng. Technol. (IJRASET) **5**(X), (2017). IC Value: 45.98
8. Shraddha, D.D., Trupti, J.D., Priyanka, S.S.: IOT based intelligent trolley for shopping mall. Int. J. Eng. Dev. Res. **4**(2), 1283–1285 (2016)
9. Li, R., Song, T., Capurso, N., Yu, J., Cheng, X.: Iot applications on secure smart shopping. In: 2016 International Conference on Identification, Information and Knowledge in the Internet of Things (IIKI), pp. 238–243. IEEE. Accessed October 2016
10. Gubbi, J., Buyya, R., Marusic, S., Palaniswami, M.: Internet of things (IoT): a vision, architectural elements, and future directions. Future Gener. Comput. Syst. **29**(7), 1645–1660 (2013)

Chapter 27
An Efficient VM Live Migration Through VM I/O Requests Using Live Scheduling Scheme (LSS)

M. N. Faruk, K. Lakshmi Prasad, and G. Lakshmi Vara Prasad

Abstract Virtual machine (VM) live migration gets essential to execution in the open cloud platform, in order to permit snappy as well as straightforward work deferment scheduling between erratically located hubs, focused on superior and vitality productive use of server framework resources. Regardless, VM Input/Output (I/O) can altogether influence the presentation of VM live storage capacity migration because of conflict aimed at the common disk transfer speed and framework resources. In light of this perception, this paper recommends the Live Scheduling Scheme (LSS) to progress instantaneously the VM execution and migration effectiveness. Live Scheduling Scheme (LSS) lightens the I/O unsettling influence in the middle of VM I/O requests and relocation I/O appeals, by organizing the movement grouping as indicated by the process example of VM I/O demands. Moreover, it advances the store the board productivity by abusing the semantic subtleties from the VMs and relocation threads. The outcomes acquired as of following put together analyses concerning a lightweight model execution of Live Scheduling Scheme (LSS) show that this framework is amazingly robust, enlightening both the VM productivity and migration effectiveness considerably contrasted with prevailing schemes.

M. N. Faruk (✉)
Department of Computer Science and Engineering, Navodaya Institute of Technology, Raichur, India
e-mail: hodcse.nit@navodaya.edu.in

K. L. Prasad
Department of Computer Science and Engineering, Chalapathi Institute of Engineering and Technology, Guntur, India

G. L. V. Prasad
Department of Information Technology, QIS College of Engineering and Technology, Ongole, India

S. C. Satapathy et al. (eds.), *Smart Computing Techniques and Applications*,
Smart Innovation, Systems and Technologies 225,
https://doi.org/10.1007/978-981-16-0878-0_27

27.1 Introduction

In recent years, the expanding selection of distributed computing has changed the IT business. Alongside all-inclusive accessible open cloud arrangements, endeavors likewise exploit comparative arrangements on hybrid and private clouds. Large, virtualized server (Data Centers) is serving the regularly developing interest for storage, networking, and systems administration. The proficient migration of cloud server is progressively significant and complex in the upcoming trends.

Perhaps, web-based administrations, mail management, data source management, and lithography management [1]. The essential hypervisor aimed at the virtualization issue legitimizes the organization of physical devices similarly as the game plan of a collection of cutting edge contraptions to VMs. All the while, hypervisors need to make certain the detachment similarly as reasonableness among diverse VMs on the head of a lone customary physical server, regardless of the way that, improving the basic execution for entire VMs with detectable quality to physical sources [2]. The VM live migration is an action in the part in contemporary hypervisors, that offer assistance to the moving of running VMs starting with one physical agent and continuation within the cloud, either inside the comparable form or all through web server focuses in different regions overall [3].

The key objective of VM real-time migration is to satisfy the boosting essential for lots of coordinating and besides web server commitment mix, system upkeep and update, VM versatility similarly as sensibility in cloud information. With the progression similarly as the colossal use of the VM-based cloud frameworks, in the shared-storage arrangement, VM real-time movement just comprises harmonization of CPU conditions, storage conditions in addition to organize UIs of the objective VM in the middle of resources, and the area of migration [4]. In typical cloud offices, data extra space might be one or the other shared or scattered relying on whether potential data is kept in a focal arrangement, whichever web servers share the physical extra space or scattered (No-Sharing) arrangement, just as each web server consumes its private special submitted extra space [5]. The VM live extra space migration has come to be a basic segment of VM real-time migration in contemporary hypervisors [6]. By way of the executing VM is moving from one hub to multiple hubs, together with the VM's in memory conditions just as computerized disk images, move from the resources web server toward the remote web server. VM online migration, as a crucial down to basic component of the hypervisors, as, QEMU-KVM [7], HyperV [8], and ESX [9], is a much more noticeable advancement. To be point by point, any sort of commotion just as ideal VM live extra space migration framework need to have the holding fast to private or business properties.

(A) *High Performance of VM I/O*: The cloud organization should give precisely the same effectiveness confirmation for each one of those co-found and synchronous executing VMs in the physical web server. All through the VM live extra space migration, solicitations inside the moving VM are as yet executing just as they should be uninformed of the migration methodology. Offered extra space source is radically reached out due to included resources hungry migration threads. The I/O proficiency inside running VMs needs to keep the Service-Level Agreement (SLA) in any capacity times.

(B) *Proficiency to Migrate Several VMs Concurrently*: It is significantly harder to achieve the suitable VM I/O effectiveness of all running VMs as opposed to relocating various VMs to their areas at a reasonable migration rate. Given the huge arrival of VMs in data offices, it wins to move a few VMs further than a multiple web server or move a few VMs directly obsessed by one singular web server.

(C) *Less time utilization on migration*: The component of a VM's extra space picture is much of the time various to 10 s of GBs just as it would take mins or additionally hrs to comprehensive a VM live extra space migration that is well on the way to diminish the limits of framework checking in the cloud data. In contemporary data offices, VMs are running every minute of an everyday premise to offer purchasers around the world.

An assortment of strategies has been prescribed to support the real-time extra space migration proficiency [10], comprising of enhancement on the data impede transmission arrangement and the migration procedure, limiting tedious data transmission, and utilizing heterogeneous capacity gadgets. Because of these checking, this paper suggests the Live Scheduling Scheme (LSS), that timetables the block migration arrangement taking into consideration of VM I/O requirements, to guarantee the burdening disk head exercises could be brought down considerably [11]. On the above of that, by interestingly storing data chunks after movement threads, inbound VM I/O requirements would be offered in the memory, along these lines reducing the assortment of memory gets to all the while. By doing this, both VM I/O productivity and migration proficiency can be supported extensively.

27.1.1 Definitions

Sequential I/O attribute: The productivity of Sequential I/Os is far superior to that of Random I/Os with standard disk searches for and revolutions [26]. The continuous I/O private property is recognized by loads of measurements from both spatial just as temporal dimensions [27] Consecutive I/O building has been only one of the best central standards in framework exploration study area [28], by and large, because of the proficiency variety in the middle of Sequential I/Os just as Random I/Os away storage space frameworks. *Temporal Dimension*: This is the middle of progressive I/O requests. When there is an extended holding up period over the middle of two I/O requests, a couple of different accounts I/O requests could be given in the middle of, which will unquestionably impact the successive I/O private or business property of the underlying I/O stream. Interleaved Streams. This will resemble a blend of I/O requests commencing a few threads, solicitations, or virtual environments. *Spatial Dimension*: The component of statistics to look at or form in the I/O request commonly. Progressive Addresses. Nevertheless, the qualification among LBA of progressive I/O requests remains inside a previously classified limit. Perhaps more classified as simply progressive just as stride access.

A few streamlining techniques have been made to improve the back to back I/O private or business stuff by utilizing the relative tips from the applications. The proficiency of the Random Write I/O stream processed within and it will positively besides be weakened impressively. Different examples of productivity decay for various interleaving I/O streams are situated in [29]. In the log-based records framework [30], enough assortment of modernizes is cushioned in memory previously they remain conveyed to disks as an immense back to back part request to ensure that the disk throughput is enhanced drastically contrasted with private minimal subjective make requests. Look at requests are offered in a practically identical manner, which will positively audit an all-out area from disks simultaneously. Offered the idea of disk openness characteristics and back to back I/O origin of interleaving I/O streams, current information frameworks, both local reports frameworks [31] just as scattered information frameworks [32], support the I/O effectiveness by emphatically conveying sequential I/O requests to the hidden hard disk drive.

27.2 Related Works

The virtual machine (VM) live migration is extremely famous through mutual storage condition, where the two sources just as goal servers protect access to the basic extra space framework. Other than the exchange of a determination of VM, for example, CPU, system, memory and physical devices, moving VM's memory conditions regularly proceeds the greatest bit of relocation time just as an assortment of procedures are anticipated to build the memory migration management represented by [12].

A remarkable procedure CR/TR-Motion [13] is recommended to move checkpoints just as execution follow records instead of memory pages at the pre-copy stage. Furthermore, the author [14] utilizes remote direct memory access (RDMA) to move VM relocation traffic. The post-duplicate strategy, some time ago investigated with regard to process relocation abstract works, has been acquainted with move VMs [15]. The memory "copy" period of the live development is conceded until the VM's CPU conditions have been moved to the objective hub. Pre-duplicate is the commanding ongoing relocation procedure to do the live movement of VMs. These comprise of hypervisor-based methodologies, for example, VMware [16], Xen [17], and KVM [18], working framework degree procedures, perhaps OpenVZ framework [19], just as development over the wide-zone organize [20] Every one of the above frameworks utilizes the pre-duplicate calculation for live VM relocation in a memory-to-memory strategy.

Author [21] identified the downtime of pre-copy migration by the writable operational collection (WWS) of the solicitation. The WWS is a collection of web pages being regularly upgraded to a scope that it is risky to transmit them earlier the last iteration. Subsequently, it will certainly be upgraded once again in every single short duration, as well as any previous transmissions would be lost. Nonetheless, the author did not examine the association between WWS as well as the network condition. A well ahead piece of work by [22] offers simulation designs which can forecast the complete movement time and also interruption (downtime) depend on the pre-copy arrangement. There are two or three examinations considers that have tended to the capacity development of VMs. Authors [23] and [24] applied for Xen VMs through backend disk driver, which can move disk images of a running VM to an area have, in corresponding with memory replicating. Until the third migration of movement depicted in Area II, a moving VM does all I/Os through a source has, making fresh out of the box new squares to the disk images. All through the live development, the backend driver copies all disk squares and more than once moves refreshed squares during disk relocation. An advantage of this component is that there is no exhibition pulverization after development is done. It is once in a while hard to change having hubs quickly; enormous disk development needs long exchange time, in spite of the fact that that must be finished before VM reactivate on an area have. Author [25] uses a duplicate on-compose disk gadget made out of topic images and overhauled squares. A layout disk image, that incorporates standard framework programs that are basic to all VMs, is duplicated to all host hubs ahead of time.

27.3   Proposed Method

This segment focuses on Live Scheduling Scheme (LSS) and its work principle. The second segment explains the block migration process and its scheduling using LSS cache manager. The design objectives as follows. With the intention of progress the enactment of VM live migration, the time constraints play an important role in disk-level operations. The VM I/O performance can be improved via block cache hit ratio and significantly decreasing the disk seek setups. Offering sufficient flexibility with our proposed model, the proposed model flexible enough with respect to diverse applications can be achieved through cache replacement procedures and flexible movement of chunk sizes.

Document-based computerized disk images have been altogether taken on in the virtualized setting [5]. Commencing the VM's perspective, the web-based virtual images concur as to the logical disk that continues all courses of the block layer's application program interface, for example, iSCSI procedural instructions. Like-wise, it works with the mainstream remote OS for instance Windows and Linux OS's. Commencing the extra storage framework's perspective, web-based virtual disk images remain considered as tremendous just as expected reports that can be kept in a great deal of records frameworks. The whole thing of I/O requests since client uses of in succession VM will change the I/O requests aimed at hidden enormous reports that embrace advanced disk-based ISO images. The development stream remains positively audit these colossal records to move the VM's extra storage conditions subtleties. In the direction of support the productivity of online disk images, various submitted document/storage frameworks have been made for advanced ISO images only, for instance, VMWare's VMFS and VM Store. LSS is a fundamental

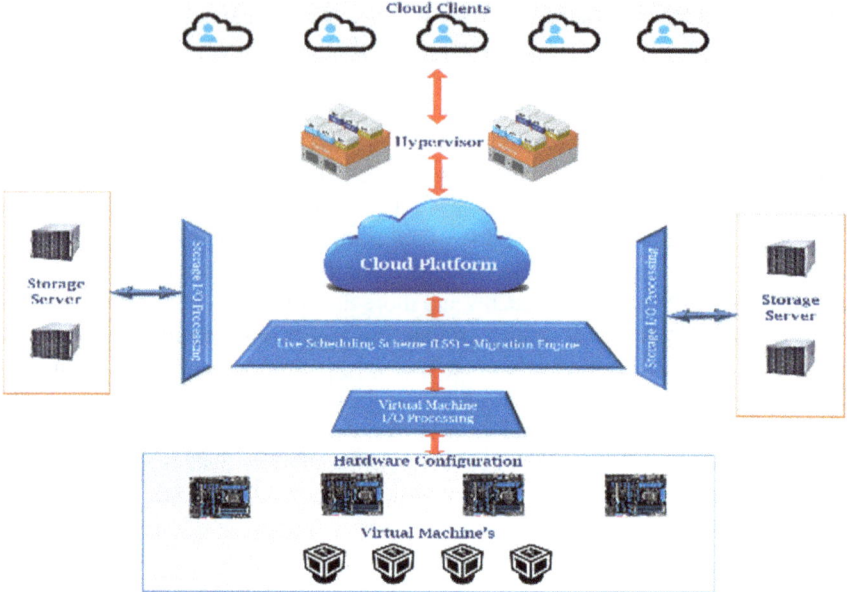

Fig. 27.1 Live scheduling scheme (LSS) system model

yet manageable part that can be integrated directly into present-day hypervisors, for example, XEN [8] hypervisors, and ESX [9]. The aforementioned determinations as development piece measurement and chunks reserve substitute could be tweaked to diverse application work. Aimed at the VM live extra storage development work, merely the web server in the resources side requires to combine the LSS segment, although the webserver on the area side remains by no alterations. LSS is an effectiveness increment layer that could be joined with a standard online development technique, comprising of Dirty Block Tracking also furthermore IO Mirroring, toward extra upgrade both VM IO proficiency just as development productivity. LSS can Fig. 27.1. The structure of the LSS System is put on live to move VMs in the middle of web servers inside exactly the same assortment or all through data offices in diverse territories.

27.3.1 I/O Threads Level Interference

At the point when it concerns the VM live extra space development, one passing thread will surely be allocated for every single moving VM, and also furthermore, it would unquestionably migrate overall conditions evidence of the migrating VMs since the resources web server toward the area web server. Subsequent off, for every executing VM, or else single VM-level I/O stream is made through the interleaving of various application-level I/O streams by means of similarity VM Operational framework I/O stream. From the perspective of the Sequential I/O private attribute, the passing thread is an astounding successive work, as it will absolutely look at data inside the space of the computerized disk photographs consecutively and ought to have incredible proficiency. That isn't the reality. All through the VM live extra space development, hypervisors are as of now overpowered as a consequence of the surplus data transfer capacity hungry passing threads. By all the more destroying the Sequential I/O private or business resources of passing threads, the further weight will unquestionably be situated on the development motor, to guarantee that is far harder to move VMs rapidly just as flexibly SLA for the I/O proficiency of all getting included VMs. Contemplating all measurements associated with the Sequential I/O building, VM-level I/O Streams (called as VM I/Os) are fundamentally settled by client applications just as remote operating systems. Consider the following Fig. 27.1 uncovers the aggravation between I/O threads in a virtualized framework all through diverse stages of VM live extra space development. The interleaving of entire VM-level I/O streams in addition to the hypervisor I/O stream would positively come to be the keep going I/O stream for the extra space framework.

27.3.2   System Execution and Workflow

Investigating carefully the cutting edge of related examination study employments just as development tasks inside frameworks, we have twin monitoring's: (A) Specific I/O requests commence the roaming VM is created by solicitations, to ensure that we consume restricted abilities to change the VM I/O stream to support the VM I/O effectiveness; (B) Only the general development time just as the precision of the VM conditions broadcast issues for passing threads, while the particular interest measurement, the preliminary location of every single appeal or the arrangement of development I/O requirements remains not comprised of as main concern ones. Consequently, the proposed framework will unquestionably help both the VM I/O proficiency just as development effectiveness by creating and setting up the develop-ment arrangement as indicated by the I/O stream of VM requirements. Fundamen-tally, this resolves positively pick the accompanying data block development prospect dependent on dual prerequisites: (A) This data block has not to remain moved to the area; (B) the concerned location of this data-center is near the contemporary openness region or the situation of the disk head. The anticipated work can accurately reserve in memory data deters looked at bypassing threads from the extra space framework, just as utilize the above data to offer the determined inbound VM I/O requirements, with the goal that the above VM I/O appeals resolve unquestionably not require to get to the extra space storage framework by any stretch of the imagination. Both the VM I/O productivity just as development I/O effectiveness can be improved drastically. In existing virtualized frameworks, there are dual-level cache-store intended for every single running VM: remote ISO makes cache reserve (inside the VM) in addition to host site page cache reserve (inside the hypervisor). LSS has two key segments:

Scheduler for Block Migration as well as Cache Management Layer. Scheduler for Block Migration will surely assess the VM I/O requirements web traffic, perceive the existing VM access the zone, envision the far ahead I/O availability segment, and after that pick the best information part to move. When the information partition is moved to the area web server, it will unquestionably be gone over to the Cache Management Layer. The choice of development information pieces depends on dual divisions: A. a lot shorter search for the season of the extra space framework for the development I/O request; B. the current information block brought by the passing thread might offer well ahead on VM I/O requests with a more prominent chance. Cache Management Layer is to deal with the memory source and carefully cache information deters for advanced VM I/O requests. For instance, the passing thread just checks the computerized disk ISO images when it determines unquestionably not get to exactly the equivalent information deters extra noteworthy than when, other than the redesigned information chunks. These information chunks can be gotten to by VM I/O requirements, just as thusly, as reserving warm relocation information hinders in memory, a few inbound VM I/O requests could be offered by Cache Management Layer in memory straight. When the block cache is finished, a cache substitute formula resolves positively go about just as cool information chunks will surely be expelled.

This is an old as well as typical issue in several facets of system layout, which call for various methods, such as on the Internet profiling as well as trace evaluation, to tackle this trouble. Vibrant piece dimension has been used to lower the number of duplicated information portion movement. In the suggested LSS system, we begin with taken care of piece dimension as well as review its efficiency enhancement. There is no question that LSS can be incorporated with various other piece dimension resolution devices for various work. The complete formula of Migration Chunks Scheduling exists in Algorithm 1.

The very first choice in which system managers require to make is just how much storage space sources need to be designated for the movement string. Provided the complete storage space source continues to be the same, the extra source movement string obtains, and also the much less storage space data transfer VM threads have. Basically, it been compromise in among VM I/O efficiency and also movement efficiency, as well as it can be readjusted for various usage situations or choices. As quickly the storage space source allotment is established, LSS scheduler for Block Migration will certainly pick the following portion to check out from the storage space system; however, this portion can certainly be moved to the location web server. Preferably, we like to information portions pleasing the list below problems: (A) The portion has not remained moved to the location web server yet, (B) The beginning address of this information piece is adjacent to existing VM I/O gain access to area, (C) High chances that the inbound VMI/O demands will certainly review partial or loaded with this information portion.

Algorithm   :: *1*
Procedure   :: *LSS scheduler for Block Migration*

Step –1: Process initiation
Step –2: DefVMMigration – Describe the group of presently migrating VMs
Step –3: CurrLocation – The storage counter location
Step –4: Func(I/OSchedule (DefVMMigration)) {
Step –5:   for (Present VM(1) in the DefVMMigration) {
Step –6:   Attain first I/O request for the present VM(1);
Step –7: Update:
Step –8: Current spot with respect to I/O request and its present CurAddress and chunk sizes
Step –9: AvailableVMs – The overall chunks available for VM to migrate.
Step –10:   PickSelectedChunk(AvailableVMs, CurrLocation)
Step –11:   }
Step –12:   }
Step –13: Func(PickSelectedChunk(AvailableVMs, CurrLocation)) {
Step –14:   if (AvailableVMs != empty) {
Step –15:   pick data block from AvailableVMs based on CurAddress and ignore
                     CurrLocation
Step –16: Deselect data block from AvailableVMs
Step –17: Dissipate VM I/O request for the present data block
Step –18: Update:
Step –19:   Current spot with respect to I/O request and its present CurAddress and
                     chunk sizes
Step –20:   else
Step –21:   Terminate the loop }}

Cache Management Layer all through the runtime, respectively, executing VM is designated an assortment of memory information through the hypervisor, to guarantee that the VM could cache subtleties information in memory, moderately openness of extra space device. Typically, this memory site information is distinguished as compose make disk cache and has website page cache, as clarified in the past segment. As it concerns the VM live extra space development, the entire online disk ISO images will unquestionably peruse from the extra space framework to memory and after that moved to the storage web server. A request follows: in circumstances, we don't cache altogether data impedes in memory in any capacity, we transpire discarding an extraordinary segment of extra space information migration. Thinking about the value it requires to look at in data hinders from the extra space framework to memory, just as included read requests, can be surely delivered to the extra space framework by VM thread, likewise, stipulate the objective data chunks consume very perused in when by the development thread. Reserving an assortment of data parcels with a more noteworthy chance to be gotten to by VM thread in memory would essentially support the VMIO effectiveness by the diminishing of extra space availabilities. The hindrance is actually how to perceive these data chunks just as exactly how to inspect just as change data partitions as soon as the cache is finished. For instance evaluated at past region, the Migration Chunks Scheduling part composes the absolute first snag, while the Cache Management Layer settle in the subsequent one.

Algorithm :: 2
Procedure :: *Cache Management Layer*

Step –1: Process initiation
Step –2: Connect to hypervisor and map memory information
Step –3: Process initiation for metadata of the Cache management layer
Step –4: Collect the request from the respective threads or currently migrated threads
Step –5: Func SetupBlock(AddressBlock){
Step –6: if (AddressBlock >= Cache Management Layer)
Step –7: { Update the current block and the counter value becomes zero
Step –8: else {
Step –9: Setup the block from storage space and dissipate block to end-user
Step –10: Manually migrate the current block to the respective allocation
Step –11: }}}
Step –12: Func UpdateBlock(AddressBlock, Buffer) {
Step –13: if (AddressBlock >= Cache Management Layer)
Step –14: { Update the current block and the counter value becomes zero
Step –15: else {
Step –16: Update the block from storage space and dissipate block to end-user
Step –17: Manually migrate the current block to the respective allocation
Step –18: }}}
Step –19: Func DisableBlock(AddressBlock, Buffer){
Step –20: if (AddressBlock < Cache Management Layer)
Step –21: { Disable the current block and the counter value becomes zero
Step –22: else {
Step –23: Disable the block from storage space and dissipate block to end-user
Step –24: Manually remove the current block to the respective allocation
Step –25: Update the status of Cache management layer
Step –26: Terminate the loop }}}

In the Cache Management Layer, each entrance is a file segment of an online disk ISO image in a precise location, and it includes an assortment of respective access in the Cache Management Layer. For every single passageway, we can survey its liveness worth dependent on the answer for agreeing to concerns: (1) If the data block has very be situated or moved to the area web server, (2) Overall period of this data block stayed inside the cache, (3) Finally, the open door that this statistics block can be unquestionably gotten through VM IO thread henceforward. Based on the respective data, Cache Management Layer can orchestrate these passageways just as change the entrance with the least present worth as the cache is finished. Given that such material evidence is intently relating to the spatial just as the fleeting region of work, such a cache checking equation should be adjusted to give diverse applications. The significantly additional exact district we gain from work, additionally improved cache struck extent and IO productivity that would attain. Differentiated to the standard method, in which the normal 2-degree cache organization equation is utilized, the proposed migration-mindful cache checking framework can impressively help the VM live extra space development proficiency. The skeletal framework equation of the Cache Management Layer is outlined in Algorithm 2.

27.4 Experimental Analysis and Discussions

In this segment, we give the proficiency assessment of the LSS conspire to employ broad follow driven trials. As talked about already, both VM effectiveness and VM constant movement proficiency are essentially imperative for generally system productivity. In the theoretical system, our study concentrates on the assessments of capacity productivity for mutual threads of VM as well as migration. Especially, our version establishes a virtual hard drive image in the hard disk, equally replay the supplied storing chunks level observes on the head of the virtual disk image. Aimed at together VM IO efficiency equally as relocation implementation, our scheme makes use of standard IO critical remarks for efficiency measurements. Shorter feedback time for VM I/O asks for infers additional IO implementation for the modifications inside the running VM.

The designed LSS program is going to be compared to the high-quality transfer scheme which will definitely overlook the features of the VM IO flow in addition to shift the digital hard drive images as of the start to the termination, like the standard VM transfer in Linux Kernel Virtual Machine, the traditional conditions suggest that the structure is actually executing deprived of VM movement. All through the course of investigating, numerous specifications are thought about. Initially, the present migration portion measurements are taken into account within the practices, and also for that reason, the LSS unit performance is usually more improved focusing on innovative vibrant chunk dimension movement strategies. Second, our system allocates the storage space resource in between VM threads as well as transfer threads with a pre-determined proportion. This technique is straightforward but reliable to discover a compromise in between the VM functionality and also migration efficiency. The recommended LSS unit likewise can easily strengthen the VM stay storing migration functionality under various other storing information allocation plans. The LSS unit is analyzed under the various amounts of simultaneous VM reside live storage space migrations.

The investigational system contains a server set up with one AMD Ryzen 3000 series processor, 16 GB DDR4 RAM and double 2 TB disk drives, 12.10 Ubuntu OS. With the intention of evaluating the performance remodeling of the LSS scheme, our system evaluates chunk-level traces as well as likewise pick up IO performance appropriate details throughout the course of the migration plan. To explore exactly just how the movement chunk size, the storing distribution strategy in addition to the selection of simultaneous VM movement impact the enactment of the LSS model, trace-driven trials are executed. The standardized action opportunity of the LSS plan depend on the basic system under the same trial arrangement is studied. Prior to the beginning of the VM remains storage space movement, the policy of keeping source appropriation in between VM IO thread along with transfer thread is developed either through system admins or instantly. As a variety of consideration of VMs shares a singular web server, as well as the organizing server, can easily not please the criteria of storage space IO data transfers for each specific VMs, one or more VMs will most

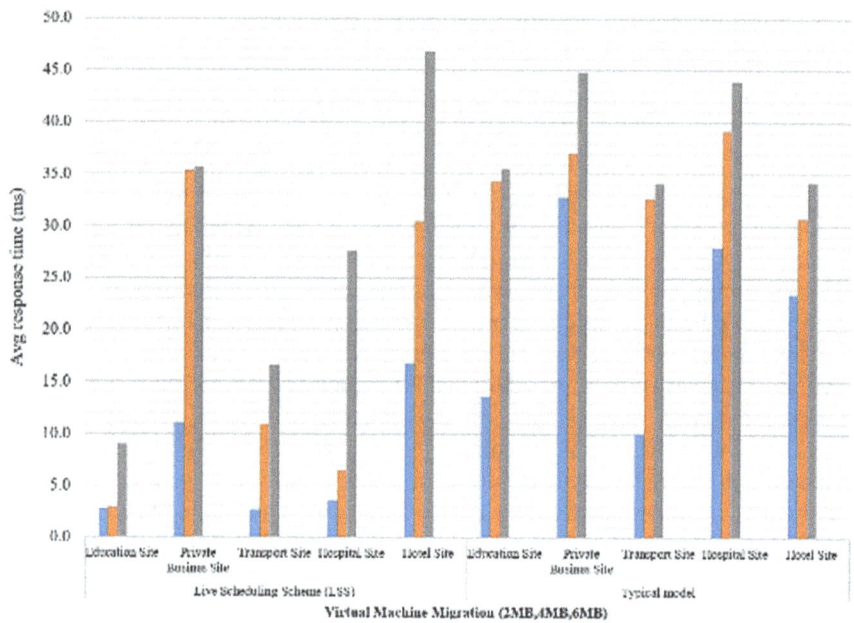

Fig. 27.2 Performance comparison of both typical and LSS model

definitely require to come to be online moved to different other organizing web servers.

The migration effectiveness and VM execution in the middle of LSS and typical movement strategy remain inspected and differentiated for follows beneath diverse migration piece measurements (2, 4, 6, 8, 10, and 12 MB) is referenced in the Fig. 27.2. The IO migration time in LSS is balanced out dependent on that in the fundamental migration. In this way, LSS can support the VM live storage space relocation proficiency below various movement divide measurements. As investigated in past areas, a few simultaneous VM live storage space movement wins in the current central CPU structure. So as to explore the amount LSS plan can help the VM live storage space migration execution contrasted with regular development framework in these circumstances, we perform tests to look at productivity enhancement under diverse follows as also diverse simultaneous VM live storage movement (Fig. 27.3).

On this specific occasion, there is a solitary VM moving originating from the source to the area. The migration chuck estimations are 2 MB, and the extra space source allocation proportion is partial between VM and migration threads. The proposed adaptation worked out with diverse traces, contrasted with the ordinary migration framework. To guarantee that excessive find just as pivot capacities are decreased unmistakably in the move I/O requests. This observation makes less confounded for the hard drive administrator to interest inward advancements, for example, the requests converge, with the intention of extra improvise the IO proficiency. Thusly, the general development of IO proficiency is fortified impressively.

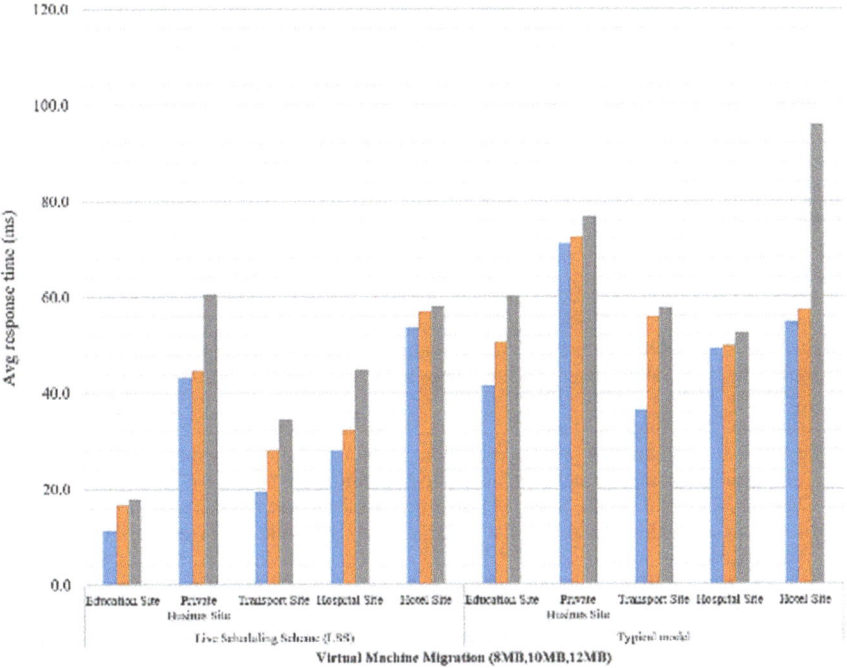

Fig. 27.3 Performance comparison of both typical and LSS model

The explanation liable for such upgrade is really that the incorporated IO execution torrent at the door of extra space is coming to be considerably more back to back in the LSS framework than that in the typical framework. The access to the zone of the applications can without much of a stretch be reserved.

27.5   Conclusion

In view of these above interpretations, this paper provides LSS, a unique VM live storage space movement plan that enhances equally the VM I/O efficiency as well as migration efficiency by producing and also arranging migration structure conferring to the I/O stream of VM demands. By doing this, it is predictable to lessen as well as do away with excessive disk head motions; consequently, the common VM live storage space movement efficiency could considerably be enhanced. Examinations of trace-dependent investigate showed that LSS perhaps minimize the ordinary I/O reaction period via 67% for migration thread as well as 49% for VM I/O string, contrasted to the conventional movement. Subsequently, numerous unneeded taxing disk head looks for as well as rotation procedures that degrade both the VM I/O efficiency and migration enactment were familiarized. Present VM live storage space movement methodologies neglect the sequential I/O residential or commercial resources of the interleaving I/O streams at eviction of the storage web server, by just transferring the VM's online disk images successively, no matter the simultaneous VM I/O streams too various other migration streaming methods.

References

1. Faruk, M.N., Sivakumar, D.: Multi-layer QOS based task scheduling algorithm for cloud environments. Int. J. Adv. Comput. Technol. (2014)
2. DeHaan, M.P.: Methods and systems for cloud management using multiple cloud management schemes to allow communication between independently controlled clouds. US Patent 8,271,653, 18 Sept 2012
3. Srinivasan, D., Wang, Z., Jiang, X., Xu, D.: Process out-grafting: an efficient "out-of-VM" approach for fine-grained process execution monitoring. In: Proceedings of the 18th ACM Conference on Computer and Communications Security, pp. 363–374 (October 2011)
4. Checconi, F., Cucinotta, T., Stein, M.: Real-time issues in live migration of virtual machines. In: European Conference on Parallel Processing, pp. 454–466. Springer, Berlin, Heidelberg (August 2009).
5. Mao, B., et al.: IOFollow: Improving the performance of VM live storage migration with IO following in the cloud. Future Gener. Comput. Syst. **91**, 167–176 (2019)
6. Beveridge, D.J.: Coordinated hypervisor staging of I/O data for storage devices on external cache devices. US Patent 9,081,686, 14 July 2015
7. Chandak, A., et al.: Dynamic load balancing of virtual machines using QEMU-KVM. Int. J. Comput. Appl. **46**(6), 10–14 (2012)
8. Pate, S.D., et al.: System and method for secure storage of virtual machines. US Patent 9,053,339, 9 June 2015
9. Waldspurger, C.A.: Memory resource management in VMware ESX server. ACM SIGOPS Oper. Syst. Rev. **36**(S1), 181–194 (2002)
10. Van Tran, TK, et al.: Size estimation of cloud migration projects with cloud migration point (CMP). In: 2011 International Symposium on Empirical Software Engineering and Measurement. IEEE (2011)
11. Duggan, M., et al.: A multitime-steps-ahead prediction approach for scheduling live migration in cloud data centers. Softw. Pract. Exp. **49**(4), 617–639 (2019)
12. Hines, M.R., Deshpande, U., Gopalan, K.: Post-copy live migration of virtual machines. ACM SIGOPS Oper. Syst. Rev. **43**(3), 14–26 (2009). Nelson, M., Gailly, J.-L.: The Data Compression Book, 2nd edn. M & T Books, (1995)
13. Hines, M.R., Gopalan, K.: Post-copy based live virtual machine migration using adaptive pre-paging and dynamic self-ballooning. In: Proceedings of the ACM/Usenix International Conference on Virtual Execution Environments (VEE'09), pp. 51–60 (2009)
14. Huang, W., Gao, Q., Liu, J., Panda, D.K.: High performance virtual machine migration with RDMA over modern interconnects. In: Proceedings of the IEEE International Conference on Cluster Computing (Cluster'07), pp. 11–20 (2007)
15. Lange, J.R., Dinda, P.A.: Transparent network services via a virtual traffic layer for virtual machines. In: Proceedings of the 16th International Symposium on High Performance Distributed Computing, pp. 23–32 (2007)
16. Clark, B., Deshane, T., Dow, E.M., Evanchik, S., Finlayson, M., Herne, J., Matthews, J.N.: Xen and the art of repeated research. In: USENIX Annual Technical Conference, FREENIX Track, pp. 135–144 (2004)
17. Kivity, A., Kamay, Y., Laor, D., Lublin, U., Liguori, A.: KVM: the Linux virtual machine monitor. In: Proceeding of the 2007 Ottawa Linux Symposium 2007, pp. 225–230 (2009). Container-based virtualization for linux. [Online]. Available: https://www.openvz.com/

18. Bose, S.K., Brock, S., Skeoch, R., Rao, S.: CloudSpider: Combining replication with scheduling for optimizing live migration of virtual machines across wide area networks. In: 2011 11th IEEE/ACM International Symposium on Cluster, Cloud and Grid Computing, pp. 13–22. IEEE (2011)

19. Travostino, F., Daspit, P., Gommans, L., Jog, C., De Laat, C., Mambretti, J., Monga, I., Van Oudenaarde, B., Raghunath, S., Wang, P.Y.: Seamless live migration of virtual machines over the MAN/WAN. Future Gener. Comput. Syst. **22**(8), 901–907 (2006)

20. Huang, W., Gao, Q., Liu, J., Panda, D.K.: High performance virtual machine migration with RDMA over modern interconnects. In: 2007 IEEE International Conference on Cluster Computing, pp. 11–20. IEEE (2007)

21. Bradford, R., Kotsovinos, E., Feldmann, A., Schiöberg, H.: Live wide-area migration of virtual machines including local persistent state. In: Proceedings of the 3rd International Conference on Virtual Execution Environments, pp. 169–179. ACM Press (2007)

22. Luo, Y., Zhang, B., Wang, X., Wang, Z., Sun, Y.: Live and incremental whole-system migration of virtual machines using block-bitmap. In: Proceedings of Cluster 2008: IEEE International Conference on Cluster Computing. IEEE Computer Society (2008)

23. Sapuntzakis, C.P., Chandra, R., Pfaff, B., Chow, J., Lam, M.S., Rosenblum, M.: Optimizing the migration of virtual computers. ACMSIGOPS Oper. Syst. Rev. **36**(S1), 377–390 (2002)

24. Sivathanu, S., Liu, L., Yiduo, M., Pu, X.: Storage management in virtualized cloud environment. In: 2010 IEEE 3rd International Conference on Cloud Computing, pp. 204–211. IEEE (2010)

25. van der Velde, O.A., et al.: Spatial and temporal evolution of horizontally extensive lightning discharges associated with sprite-producing positive cloud-to-ground flashes in northeastern Spain. J. Geophys. Res.: Space Phys. **115**(A9), (2010)

26. Zhang, X., Chen, L., Tong, Y., Wang, M.: EAGRE: towards scalable I/O efficient SPARQL query evaluation on the cloud. In: 2013 IEEE 29th International Conference on Data Engineering (ICDE), pp. 565–576. IEEE (2013)

27. Okafor, K.C.: Dynamic reliability modeling of cyber-physical edge computing network. Int. J. Comput. Appl. 1–11 (2019)

28. Frénot, S., Ponge, J.: LogOS: an automatic logging framework for service-oriented architectures. In: 2012 38th Euromicro Conference on Software Engineering and Advanced Applications, pp. 224–227. IEEE (2012)

29. Yu, X., Joshi, P., Jianwu, Xu., Jin, G., Zhang, H., Jiang, G.: Cloudseer: workflow monitoring of cloud infrastructures via interleaved logs. ACM SIGARCH Comput. Archit. News **44**(2), 489–502 (2016)

30. Ghoshal, D., Canon, R.S., Ramakrishnan, L.: I/o performance of virtualized cloud environments. In: Proceedings of the Second International Workshop on Data Intensive Computing in the Clouds, pp. 71–80 (2011)

31. Mao, M., Humphrey, M.: A performance study on the VM startup time in the cloud. In: 2012 IEEE Fifth International Conference on Cloud Computing, pp. 423–430. IEEE (2012)

32. Sobel, W., Subramanyam, S., Sucharitakul, A., Nguyen, J., Wong, H., Klepchukov, A., Patil, S., Fox, A., Patterson, D.: Cloudstone: multi-platform, multi-language benchmark and measurement tools for web 2.0. Proc. CCA **8**, 228 (2008)

Chapter 28
Real-Time Area-Based Traffic Density Calculation Using Image Processing for Smart Traffic System

S. Rakesh and Nagaratna P. Hegde

Abstract Traffic is one major problem in metropolitan cities and capital cities of all the states. As more people moving to the cities, the number of vehicles on the road is increasing rapidly. The traffic signals controlled by the traffic police manually have not proved to be efficient. Also the static predefined set time for traffic signals at all conditions whether the traffic density is heavy or low has not solved this problem. One solution for this problem is to develop a smart traffic management system which works by calculating the density of the vehicles on the road using real time video processing technique. We have used Mixture of Gaussian (MoG) algorithm for background subtraction technique and foreground detection, then generated capacity maps using which we found the density of traffic at that particular instant. The traffic lights at junctions are controlled dynamically based on the traffic density detected from the video feeds.

28.1 Introduction

The density of the vehicles when travelling strengthens slowly subsequently for the optimum use of existing street potential is rather notable to care for all the traffic flow appropriately. Traffic jams have wound up being sizable trouble specifically in the metro locations in addition to principal city regions. The primary cause is the fast increase in the populace of the metropolitan areas as well as principal city urban locations that subsequently boost vehicular travel, which makes traffic congestion

S. Rakesh
Department of Information Technology, Chaitanya Bharathi Institute of Technology (CBIT)(A), Gandipet, Hyderabad, Telangana 500075, India
e-mail: srakesh_it@cbit.ac.in

N. P. Hegde (✉)
Department of Computer Science and Engineering, Vasavi College of Engineering, Hyderabad, Telangana, India

issue [1–5]. Because of these traffic jams, the transportation expense is boosting, the vacation possibility is actually rearing and also power intake is also enhancing [2]. Traffic additionally generates loads of different other censorious complications as well as worries which right effect private regular everyday lives and also sometimes trigger for the loss of life [6–8]. For instance, if there is a rescue when driving together with a severe individual on board. Because situation, if that rescue gets embedded notable traffic congestion at that point there are truly high options that the important individual may not arrive at the university hospital or hospital quickly. Therefore it is unbelievably vital to design along with making a wise traffic administration device which regulates the traffic intelligently to steer clear coming from wrecks, accidents and also traffic congestion [7, 8]. The main cause of traffic in developing nation is bad in addition to ineffective traffic keeping track of system which impacts the traffic flow. For example at a traffic light shared if one side road has a lot less traffic as well as the opposite service road along with huge traffic but the size of green light for every side is very same at that point this activates the rubbish of offered sources in addition to additionally this mishandles. Using thinking about the above case if the road or road along with higher traffic should activate the green stoplight for a longer time frame than the road or even lane together with very little traffic density at that point it decreases the options of traffic congestion adequately.

There are bunches of procedures prepared to design a smart and intelligent traffic management system, as an example, a fuzzy-based traffic controller in addition to morphological edge detection method is planned in [1]. This approach is based on the size of the density of the traffic by affiliating the recorded online traffic image with a recommendation fixed or even picked graphic. If the difference is a whole lot, even more, afterward greater traffic density is recognized. In [9], an additional method is suggested to create a smart traffic unit, which is based on the four-lane road system through which traffic light chance is allowed based upon a lot of vehicles when driving or maybe lane. The operation recommended in [10] is based upon the neural networks, which identifies the vehicles as well as likewise density of traffic through refining the snatched traffic video recordings. In [11], an added strategy is used which is based upon identifying the traffic capability by reviewing the two images, i.e., captured on the web traffic image the static reference photo.

This paper presents a traffic density analyzer task based on calculating the percentage of vehicles present in the recorded real-time traffic image, which provides our company with extra precise details for traffic light assortment creating.

The paper is taken care of as adhered to: Sect. 28.2 explains the system design, i.e., worrying MoG protocol, Sect. 28.3 details worrying the application procedure of the system, Sect. 28.4 explains concerning the results. Ultimately, Sect. 28.5 deals with involving conclusion as well as also future work abided by due to the important referrals made use of in this particular job.

28.2   System Model

Background subtraction is a mostly used moving object detection operation which diverse the foreground objects coming from the chosen frame about a static reference frame. The static reference frame is the background model, which is looked at as the previous frame.

$$|I(x, y, t) - I(x, y, t - 1)| > \text{Th} \qquad (28.1)$$

The threshold Th well worth is prepared manually. The difference at the facility of the absolute value of the decided frames needs to be lot larger than the threshold market value.

Depending on to [12], the algorithm is positioned in a variety of categories, coming from history-based protocols to adaptive learning procedures. Our team selects the versatile learning algorithm above history-based algorithm [13]. Flexible learning algorithm needs to have to possess simply a background model as well as also automatically updates the design based upon the present frame. As a result, it demands fairly a lot fewer memory data transactions to care for. On the contrary, history-based algorithm gathers all the video audio frames and also frequently accesses the frame history, subsequently asking for even more mind transmission capacity. Our group decided on a Mixture of Gaussian (MoG) algorithm, which is based on the adaptable learning algorithm which is featured in Fig. 28.1. To produce the background of a pixel, a Mixture of Gaussian (MoG) algorithm uses K Gaussian components along with mean, deviation in addition to weight. To locate changes in the brand-new frame, the new frame and likewise previous frames Gaussian components are compared to one another. The Gaussian components of the pixel are updated based upon the brand-new pixels learning factor. If all the Gaussian components carry out undoubtedly not match the new pixel values, at that point the all-new pixel is taken note of as the foreground.

28.3   Implementation

We gather a cache of 500 frames from the selected input video and use these frames to train our background model using openCV fixed function cv2.createBackgroundSubtractorMOG2(), which is based on the adaptive learning algorithm. It uses a technique to model each and every background pixel by a mixture of a suitable number of Gaussian distributions per pixel. The weights of these mixtures represent the duration of each individual pixel lodge on the screen. The pixels that lodge on the screen more time are more potential in being part of background.

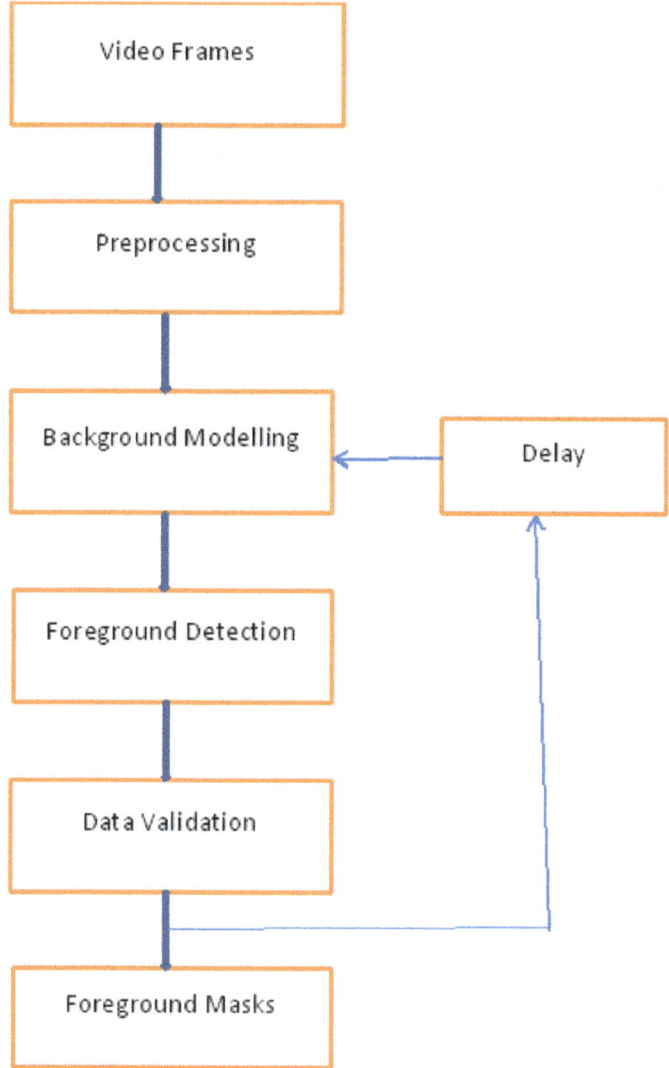

Fig. 28.1 Background subtraction for obtaining foreground mask

By the creation of a background model, we spot the foreground on the every next frame from the selected video input. The foreground still consists of noise, so we use filters on it to decrease the noise and make the foreground further useful for future use. The filters are closing, opening, erosion and dilation. Then we find the density of the traffic with the help of capacity maps which are shown in results section.

We have also send these traffic density values to an excel sheet shown in results section. Instead if we send these values to a smart traffic light management system, then we can use the traffic lights effectively, i.e., to which line the density is more that side the duration of green light is more.

28.4   Results

We have tested our model by giving a selected traffic video as input and processed the video to generate outputs in the form of capacity maps which are shown in Figs. 28.2, 28.3, 28.4 and 28.5. The values displayed in these figures are traffic density values of frames generated from the input traffic video.

Capacity: 15.433315415397475%
Original

Capacity map

Fig. 28.2 Capacity map1

Capacity: 8.493848175356867%
Original

Capacity map

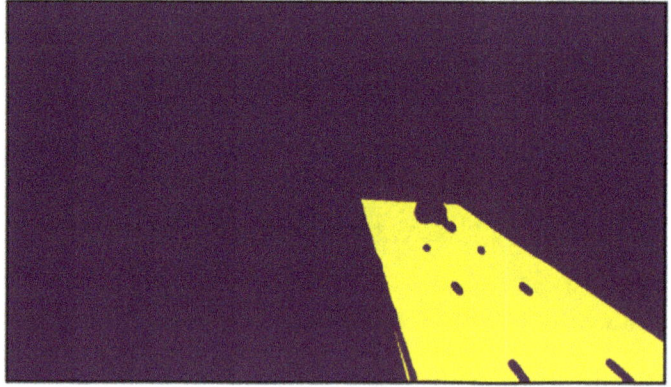

Fig. 28.3 Capacity map2

We have also send these traffic density values into an excel sheet which is shown in Fig. 28.6. We can change the time intervals of the frames according to our requirement and generate the capacity maps. These values are useful in taking decisions at the smart traffic control system, i.e., in general if the density of the vehicles is more in one side of the traffic signal than that road green traffic light should be more.

28.5   Conclusion and Future Scope

This paper covers a procedure for computing the density of traffic on the road by utilizing picture handling algorithm. Our team has used a Mixture of Gaussian (MoG)

Capacity: 13.09726452845965%

Original

Capacity map

Fig. 28.4 Capacity map3

algorithm and also get the outcomes. The conveniences of the proposal procedure are that there is no demand to use extra complex sensing unit-based units such as RFIDs. The recommended device is inexpensive as it does not need to have an installation of some other sensing units or even units.

This work can be enhanced by sending these traffic density values to smart traffic management system and controls the signals accordingly.

Capacity: 21.579167413247326%
Original

Capacity map

Fig. 28.5 Capacity map4

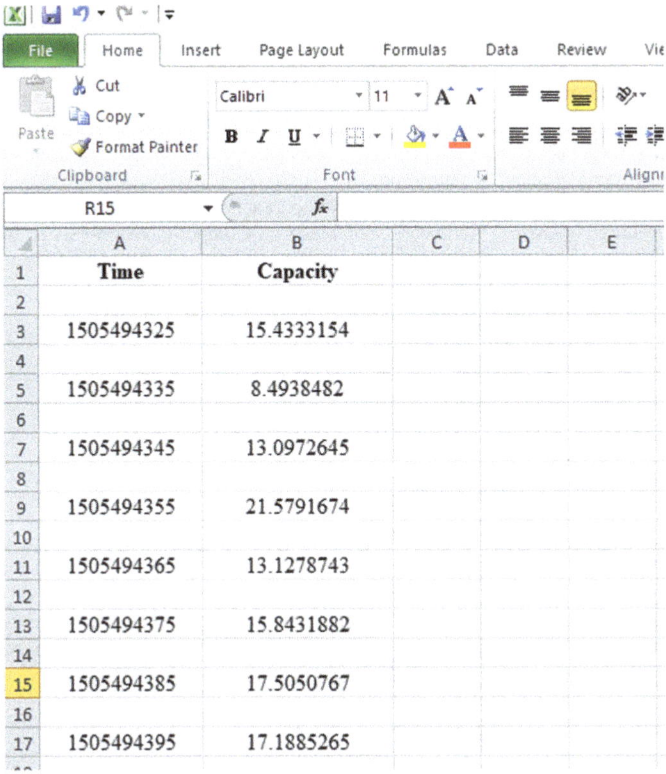

	A	B	C	D	E
1	Time	Capacity			
2					
3	1505494325	15.4333154			
4					
5	1505494335	8.4938482			
6					
7	1505494345	13.0972645			
8					
9	1505494355	21.5791674			
10					
11	1505494365	13.1278743			
12					
13	1505494375	15.8431882			
14					
15	1505494385	17.5050767			
16					
17	1505494395	17.1885265			

Fig. 28.6 Traffic density values

References

1. Arora M., Banga, V.K.: Real time traffic light control system. In: 2nd International Conference on Electrical, Electronics and Civil Engineering (ICEECE'2012), pp. 172–176. Singapore (April 28–29, 2012)
2. Kanojia, S.B.: Real–time Traffic light control and Congestion avoidance system. Int. J. Eng. Res. Appl. (IJERA) **2**(2), 925–929 (2012)
3. Tayyab, M.: Implementation of restoration path using AODV in VANETs. Master's Dissertation at Brunel University London, UK
4. Venables, A.J.: Evaluating urban transport improvements. J. Transp. Econ. Policy **41**(2), 173–188 (2007)
5. Gärling, T., Schuitema, G.: Travel demand management targeting reduced private car use. J. Soc. Issues **63**(1), 139–153 (2007)
6. Papageorgiou, M., Diakaki, C., Dinopoulou, V., Kotsialos, A.: Review of road traffic control strategies. Proceedings IEEE **91**(12), 2043–2067 (2004)
7. Vigos, G., Papageorgiou, M., Wang, Y.: Real-time estimation of vehicle-count within signalized links. J. Transp. Res. Part C: Emerg. Technol. **16**(1), 18–35 (2008)
8. Szeto, M.W., Gazis, D.C.: Application of Kalman Filtering to the Surveillance and Control of Traffic Systems. J. Transp. Sci. **6**, 4419–4439 (1972)
9. Dangi, V., Parab, A., Pawar, K., Rathod, S.S.: Image processing based intelligent traffic controller. Undergraduate Acad. Res. J. (UARJ) **1**(1), (2012)

10. Ozkurt, C., Camci, F.: Automatic traffic density estimation and vehicle classification for traffic survillance systems using neural networks. J. Math. Comput. Appl. **14**(3), 187–196 (2009)
11. Gupta, P., Purohit, G.N., Pandey, S.: Traffic load computation for real time traffic signal control. Int. J. Eng. Adv. Technol. 2(4), (2013)
12. Tabkhi, H., Bushey, R., Schirner, G.: Algorithm and architecture co-design of Mixture of Gaussian (MoG) background subtraction for embedded vision. In: 2013 Asilomar Conference on Signals, Systems and Computers, pp. 1815–1820. Pacific Grove, CA (2013)
13. Elgammal, A.M., Harwood, D., Davis, L.S.: Non-parametric model for background subtraction. In: Proceedings of the 6th European Conference on Computer Vision-Part II, pp. 751–767 (2000)

Chapter 29
An Association Mining Rules Implemented in Data Mining

B. Varija and Nagaratna P. Hegde

Abstract Data mining has created a great impact on the business organizations in different ways. The main purpose of mining is nothing but extracting useful and interesting knowledge-based patterns from large amount of data or information which is present in the data warehouses. The mining is nothing but identifying the frequent itemsets purchased by a customer from the store and by comparing the relationship between the itemsets that the analysis is been done. In order to identity the relationship, we introduce different association rule-based algorithms. We concentrate more on promising studies about the latest trends and technologies which are been implemented in the association mining rule. In this paper, we focus on association rule by using different algorithm to justify the rules.

29.1 Introduction

Data mining comprises of many techniques which are emerging day by day in each and every field either it may be the education, health, pharmacy, business, etc. Mining has changed the perspective of thinking of human being because data can be dumped from any source in the world sitting at one particular place, the data can be integrated from multiple sources, and that data can be placed at particular place which is nothing but our data warehouse. Online analytical processing can be used in the data mining which is nothing but analysis is been done with the help of the historical data which is been stored in the different repositories.

The main techniques of data mining are discussed as the association rule, clustering, classification, and outlier analysis and anomaly detection. Here purely, we would like to focus on the association mining rule which plays a vital role in data mining because of this mining rule the business organizations are able to identify

B. Varija
Sri Indu College of Engineering and Technology, Ibrahimpatnam, Hyderabad, Telangana, India

N. P. Hegde (✉)
Vasavi College of Engineering, Ibrahimbagh, Hyderabad, Telangana, India

© The Author(s), under exclusive license to Springer Nature Singapore Pte Ltd. 2021   297
S. C. Satapathy et al. (eds.), *Smart Computing Techniques and Applications*,
Smart Innovation, Systems and Technologies 225,
https://doi.org/10.1007/978-981-16-0878-0_29

the customer requirement in order to get benefit in their business. Association rules concentrate on the main perspective which is known as "market basket analysis", because it is considered as the one of the basic concept of the mining which will be further discussed in the paper.

29.2 The Association Rules

To understand the association rule, I would like to focus on the simple example, i.e., if we take into consideration a departmental store where we can find all household thinks so when a person purchase an item, the owner of the store will check the relationship between the itemsets purchased by the individual customers in order to draw what are the combination of products purchased by the different customers when they enter into the store. An association rule relays on two thinks; they are antecedent and consequent.

If customer buys beer he is likely to buy diaper

If we observe the above sentence, easily we can understand that here beer is the antecedent and diaper is consequent. The above example is given by the thorough analysis of the datasets by the company owner to improve the customer service and also improve the revenue of the company. To analyze the relationship between the itemsets, two parameters are taken into consideration, and they are observed.

$$\text{Support}(A) = \frac{\text{Number of transaction in which A appears}}{\text{Total number of transactions}}$$

$$\text{Confidence}(A \rightarrow B) = \frac{\text{Support}(A \cup B)}{\text{Support}(A)}$$

29.3 Apriori Algorithm

The Apriori algorithm is used to fine the frequent itemsets by following certain sequence of steps, and to perform this technique, I have used two basic steps, and they are join and purne. The join step is used to generate $K + 1$ itemsets, whereas the purne step is used for scanning the database. The good advantage of using this basic technique is to easily understand the algorithm. This algorithm is the level-wise generation of the frequent itemsets. The main feature is anti-monotonocity of the support. We do have certain limitation in the apriori algorithm that is it is very slow. Here, we do have itemsets frequent itemsets, frequent pattern mining, and the main things that we have to remember which are confidence and support.

- Mining frequent patterns are those patterns that occur frequently in the sales transactions of the super market.
- Frequent itemsets are those that frequently appear together.
- Frequent subsequence patterns are those which occur frequently like when you purchase phone will purchase SIM card.
- Association analysis process involves uncovering the relationship between data and deciding the rules for performing associations between objects
- Correlation analysis is a mathematical technique that can show together and how strongly the pairs of attributes are related
- Pattern evaluation is defined as identifying increasing patterns which represent the knowledge.

29.4   Multi-level Associative Rule

Here, we discuss about the rules at high concept level, and the support increases from specialized to general method. Support decreases from general to specialize method. Confidence is not affected for general or specialized.

Using uniform support for all levels, there is only one minimum support t so no need to examine itemsets. If support is too high which implies Miss Low level associations which is shown in the below figure which specifies the minimum support when it is too low (Figs. 29.1 and 29.2).

- A common form of background knowledge that an attribute may be generated or specialized according to a hierarchy of concept.
- Rules which contain associations with hierarchy of concepts are called multi-level association rule.

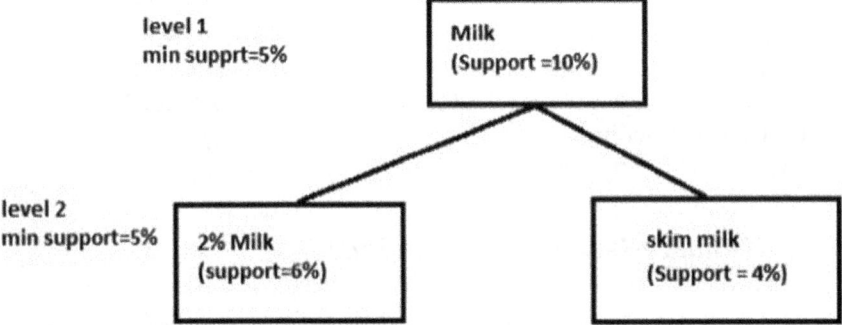

Fig. 29.1 Example of uniform minimum support for all levels

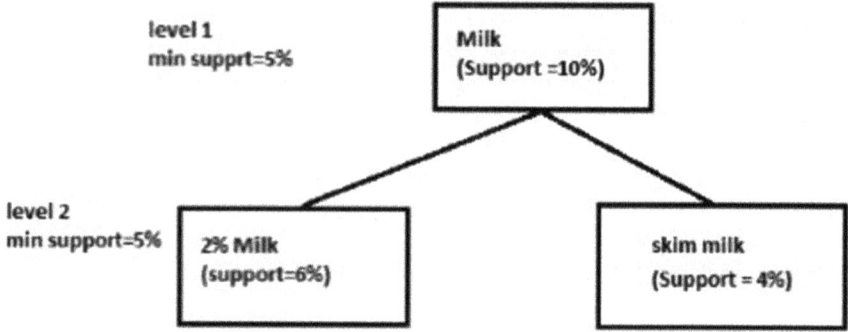

Fig. 29.2 Example of reduced minimum support for lower level

29.5 Multidimensional Association Rule

The multidimensional rule explains about the Boolean algebra. A Boolean matrix-based approach had employed to discover the frequent datasets. Apriori algorithm is used to prune the items. There are rules involving more than one dimension or predicates who try to buy products from the store. Attributes can be differentiated as the categorical or quantitative. Quantitative attributes are numeric and incorporates the concept hierarchy where the numeric attributes are discredited. We consider three different approaches of multidimensional association rules, and they are categorized as the following.

29.5.1 Mining Using Dynamic Discretization

The dynamic discretization is also known as quantitative where the numeric attributes are dynamically discredited (2D quantitative association mining rule). If we consider a simple example which is of age (X,"20…30) ^ income (X,"30 K…60 K'0) → buys (X, "Laptop Computer").

29.5.2 Distance-Based Association Rule

This distance-based association rule consists of two steps involved in mining process, and they are as follows: Firstly, it performs clustering for finding the interval of attribute, and secondly, it obtains association mining rules by searching for the group of clusters that occur together.

29.6   Fuzzy Association Rule

The fuzzy rules are used to convert numerical attributes to fuzzy attributes; fuzzy apriori is quite slow algorithm which is available today. In data mining, we can get the best result using the fuzzy sets; the fuzzy rules are consider as the useful tools. The fuzzy rules are based upon fuzzy taxonomies. It is a computational model which is in the mathematical format. It was designed by L.zadeh. The fuzzy concept is an extent applicable in a situation; they are used in the traffic light, and the fuzzy technique is more particular used in the traffic control system [8].

29.7   Hybrid Dimensional Association Rules

To detect the fraud activities, we use the data mining. To perform the detection, hybrid method can be used. By using this method, it has less false discovery but provides with high accuracy. The rough set model is used for discovering the hybrid association rules. We have hybrid algorithm which are resource efficient and provide better performance.

29.8   Weighted Association Rule Mining

The WARM provides notion or weight to individual items that are solely on an item supported where this weighted rule is scalable and efficient. The weights can be arranged in the subjective manner. Linear model can be used to inculcate the weights and not only the linear model here we can also use the valency model. In the valency mode, first we consider the non-neighbor items, and the Gaussian elimination method is used. Secondly, it is been purified, and lastly, new model is developed.

29.9   Association Rule Mining in Different Databases

Firstly, we will talk about the association rule mining in large database where it is considered as the basic algorithm which can be performed on the sales transactions. To perform the association, here we use the very basic technique which is used in the apriori algorithm which is named as pruning. The best application we can discuss is sales transactions which can be performed in any shopping mall or super marker.

Secondly, its association rule mining in spatial database, this pattern can be collected from geospatial database. The spatial data consist of objects which are

spatial scene. Basic operations can be done to achieve the good result. The applications of spatial database are that it has the efficient storage system and prevents from the inconsistencies.

Thirdly, its association rule mining in distributed database is here to mine the data which we require large amount of data, so that mining can be done efficiently. The best way to solve problem in data mining is the distributed database mining. For proper mining, many distributed algorithms are used, and the different algorithms which are used for association rule mining in distributed database are:

Hierarchical
Collective Bayesian
Distributed association mining
Collective decision tree learning

Here, we have some applications of association rule mining techniques, and they are following.

Market basket analysis
Protein sequence
Medical diagnosis
Census data
CRM credit card

Not only this type of sectors but association mining plays vital role in different businesses by helping the business people in different ways, and here are few.

Helps business people to build good sales strategies
Helps business people to find the healthy marketing techniques
Self-life planning
In store organization

29.10   Conclusion

In this survey, we have got clear picture on different association mining rules. Here in order to overcome the limitations of the traditional association rule, we have discussed the new association rules, so that an efficient mining can take place in the business organizations which are working with the mining rules because in today's world the business is growing larger and the people want the easy way to develop their business by knowing the requirement of the users; for that reasons, the researchers are more concentrated to upgrade the techniques which are in the association mining rule. By using the emerging techniques of mining, we can easily solve the problems which are faced by the people in the business world, so this brings them the relief on their efforts.

References

1. Piatetsky-Shapiro, G.: Discovery, analysis, and presentation of strong rules. In: Piatetsky-Shapiro, G., Frawley, W.J. (eds.) Knowledge Discovery in Databases. AAAI/MIT Press, Cambridge, MA (2019)
2. Agrawal, R., Imieliński, T., Swami, A.: Mining association rules between sets of items in large databases. In: Proceedings of the 1993 ACM SIGMOD International Conference on Management of Data—SIGMOD '93 (1993), p. 207. CiteSeerX: 10.1.1.40.6984. https://doi.org/10.1145/170035.170072. ISBN: 978-0897915922
3. Hahsler, M.: Introduction to a rules—a computational environment for mining association rules and frequent item sets. J. Stat. Softw. (2005)
4. Hahsler, M.: A Probabilistic Comparison of Commonly Used Interest Measures for Association Rules (2015). https://michael.hahsler.net/research/association_rules/measures.html
5. Hipp, J., Güntzer, U., Nakhaeizadeh, G.: Algorithms for association rule mining—a general survey and comparison. ACM SIGKDD Explor. Newsl. **2**, 58–64 (2000). CiteSeerX 10.1.1.38.5305. https://doi.org/10.1145/360402.360421
6. Brin, S., Motwani, R., Ullman, J.D., Tsur, S.: Dynamic itemset counting and implication rules for market basket data. In: Proceedings of the 1997 ACM SIGMOD International Conference on Management of Data—SIGMOD '97 (1997) pp. 255–264. CiteSeerX 10.1.1.41.6476. https://doi.org/10.1145/253260.253325. ISBN 978–0897919111.
7. Omiecinski, E.R.: Alternative interest measures for mining associations in databases. IEEE Trans. Knowl. Data Eng. **15**, 57–69 (2003). CiteSeerX 10.1.1.329.5344. https://doi.org/10.1109/TKDE.2003.1161582
8. Aggarwal, C.C.; Yu, P.S.: A new framework for itemset generation. In: Proceedings of the seventeenth ACM SIGACT-SIGMOD- SIGART symposium on Principles of database systems—PODS '98, pp. 18–24 (1998). CiteSeerX 10.1.1.24.714. doi:https://doi.org/10.1145/275487.275490. ISBN: 978-0897919968.
9. Piatetsky-Shapiro, G.: Discovery, analysis, and presentation of strong rules. In: Knowledge Discovery in Databases, pp. 229–248 (1991)
10. Tan, P.-N., Kumar, V., Srivastava, J.: Selecting the right objective measure for association analysis. Inf. Syst. **29**(4), 293–313. CiteSeerX 10.1.1.331.4740. https://doi.org/10.1016/S0306-4379(03)00072-3
11. Tan, P.-N., Michael, S., Kumar, V.: Chapter 6: Association analysis: basic concepts and algorithms. In: Introduction to Data Mining. Addison-Wesley. ISBN 978-0-321-32136-7
12. Pei, J., Han, J., Lakshmanan, L.V.S.: Mining frequent itemsets with convertible constraints. In: Proceedings 17th International Conference on Data Engineering, pp. 433–442 (2001). CiteSeerX 10.1.1.205.2150. https://doi.org/10.1109/ICDE.2001.914856. ISBN: 978-0-7695-1001-9
13. Agrawal, R., Srikant, R.: Fast algorithms for mining association rules in large databases Archived 2015–02–25 at the Wayback Machine. In: Bocca, J.B., Jarke, M., Zaniolo, C. (eds.) Proceedings of the 20th International Conference on Very Large Data Bases (VLDB), pp. 487–499. Santiago, Chile (September 1994)
14. Zaki, M.J.: Scalable algorithms for association mining. IEEE Trans. Knowl. Data Eng. **12**(3), 372–390 (2000). CiteSeerX 10.1.1.79.9448. https://doi.org/10.1109/69.846291
15. Zaki, M.J., Parthasarathy, S., Ogihara, M., Li, W.: Parallel algorithms for discovery of association rules. Data Min. Knowl. Disc. **1**(4), 343–373 (1997). https://doi.org/10.1023/A:1009773317876
16. Han: Mining frequent patterns without candidate generation. In: Proceedings of the 2000 ACM SIGMOD International Conference on Management of Data. SIGMOD '00, pp. 1–12 (2000). CiteSeerX 10.1.1.40.4436. https://doi.org/10.1145/342009.335372. ISBN: 978-1581132175
17. Witten, Frank, Hall: Data Mining Practical Machine Learning Tools and Techniques, 3rd edn
18. Hájek, P., Havránek, T.: Mechanizing Hypothesis Formation: Mathematical Foundations for a General Theory. Springer-Verlag (1978). ISBN 978-3-540-08738-0
19. Webb, G.I.: OPUS: an efficient admissible algorithm for unordered search. J. Artif. Intell. Res. **3**, 431–465 (1995). (AAAAI Press, Menlo Park, CA online access)

Chapter 30
Wi-Fi Networking: A Novel Method for Stabilizing Internet Connectivity by Optimizing Wi-Fi Router Configuration

Vijay A. Kanade

Abstract Internet is playing a critical role in today's world. It has kept people connected regardless of the physical barriers. Business connections, communications, and remote work seem a possibility today due to the Internet. However, with the growth in smart devices and an exponential increase in the number of Internet users, there has been significant growth in the global Internet traffic which has exposed the limitations of the RF signals used in wireless communication in households, offices, and public places. In addition, these RF signals encounter disturbances as they travel through the transmission medium, i.e., air, and undergo attenuation. As a result, the RF signals received at the client end device seem to display an inconsistent and unstable flow, leading to an unsteady Internet connection for its users. The research paper discloses a novel solution to address the problem of inconsistent and unstable Internet connection observed in a Wi-Fi networking environment, wherein Wi-Fi routers are at play in a defined geography.

30.1 Introduction

Internet had become an indispensable part of our lives today. It has bettered the quality of human life by fueling technological developments with regard to communication, IoT, and many more fields. As per the July 2020 stats, around 4.57 billion users are active on the Internet [1]. With the advent and advancement in smart devices, it can only be expected that a greater number of users and devices may come online in the near future.

As Internet has reached the households, the corporate world has seen a cultural shift. Corporate companies are encouraging remote work as employers are benefitting from it. As per the recent stats, workplaces are embracing remote work alternatives more than ever before [2].

V. A. Kanade (✉)
Intellectual Property Research, Pune, India

© The Author(s), under exclusive license to Springer Nature Singapore Pte Ltd. 2021   305
S. C. Satapathy et al. (eds.), *Smart Computing Techniques and Applications*,
Smart Innovation, Systems and Technologies 225,
https://doi.org/10.1007/978-981-16-0878-0_30

Currently, the world is grappled with global pandemic of COVID-19, and such technological adoption is acting as a boon for the remote workforce.

The work-from-home environment is here to stay. However, such work culture change is slated to put pressure on the Internet connectivity paradigm. As employees stay home bound, Internet or Wi-Fi sharing with spouse, partner, roommate, or kids may become a norm. In such scenarios, the risk of handling lousy Internet connection looms large. With buffering video conference calls and delayed audio, the employees are bound to suffer. Further, movies and video games may take ages to download. In the worst cases, the connection may drop altogether or seem to slow down due to heavy bottleneck on Internet speeds, leading to unstable Internet connectivity.

The connection drop or slowness issue can be attributed to various reasons, such as problem with router or modem, problem with Wi-Fi signal, signal strength, or number of devices on a network that hog the bandwidth.

However, even if these problems are considered and provided with a solution, still the Internet connection may not stabilize. In addition, broadband router configuration errors, wireless interference, or several other technical problems can adversely affect the speed and stability of the Internet connection.

All these connectivity problems are essentially associated with the radio wave property of the wireless connectivity signal. Therefore, there seems to be a long-standing need to provide a technical solution to the prevailing radio wave problem posed by the Wi-Fi broadcasting medium, i.e., router.

30.2 Wi-Fi Networking

Wi-Fi networking technology allows devices to wirelessly connect to the Internet. The wireless connection facilitates data exchange among devices distributed across the network. Wi-Fi is based on IEEE 802.11 family of standards that specify the set of protocols for implementing wireless communication in a network at various frequencies.

There have been various Wi-Fi versions, with the first version being released in 1997 and updated in 1999. The latest Wi-Fi version, called Wi-Fi 6 or 802.11ax, claims to achieve speeds of 11 Gbit/s (i.e., theoretical data rates).

The Wi-Fi-enabled devices exchange information by sending and receiving radio waves. These radio waves define a type of electromagnetic radiation in the electro-magnetic spectrum, having longer wavelength and lesser frequency than microwaves and infrared waves. Specifically, these radio waves have frequencies ranging between 2.4 and 5 GHz [3].

Wi-Fi network consists of a central device called as router that provides Internet access to the connected devices within the network. Diverse devices such as smart-phones, PCs, and various Wi-Fi-enabled gadgets or appliances (e.g., Jio-Fi device) can act as router, as long as the connected devices are within its range or network.

While operating on 2.4 GHz frequency, Wi-Fi routers can reach up to 150 feet (46 m) indoors and 300 feet (92 m) outdoors. Routers broadcasting at 5 GHz

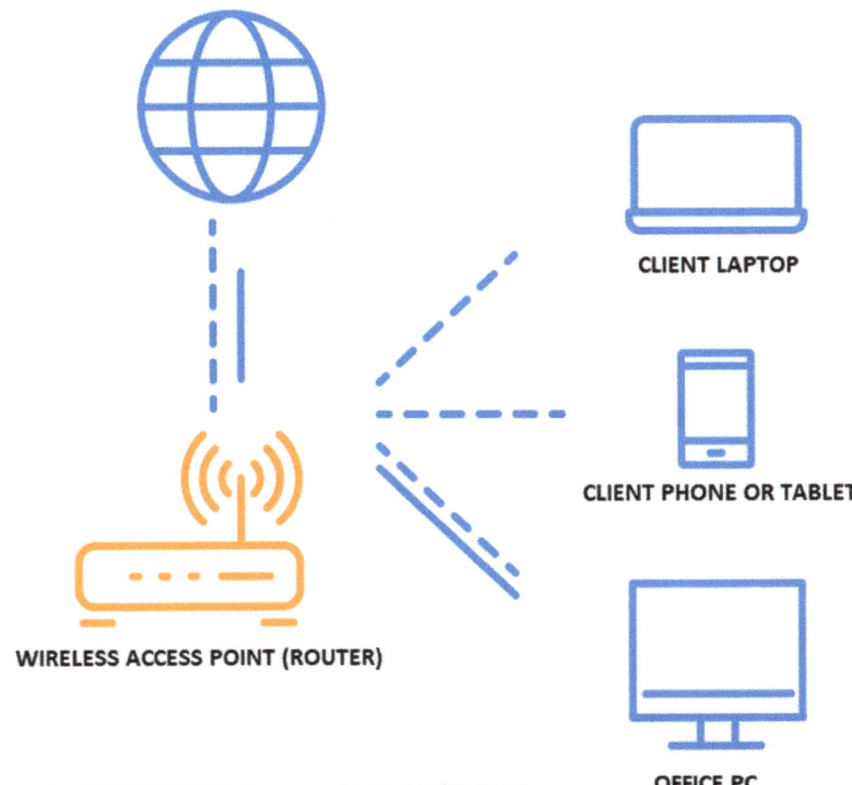

Fig. 30.1 Wi-Fi networking [4]

frequency have smaller range coverage but are generally immune to interference caused by other connected devices. Wi-Fi networking is illustrated in Fig. 30.1.

30.3 Radio Waves

A radio wave is an electromagnetic signal that carries information through the air over long distances. These radio waves are termed as radio frequency (RF) signals. RF signals travel through the air (as like waves in the ocean) at a high frequency. These signals carry data such as music, FM radios, and television videos over the air. A simple radio wave is represented in Fig. 30.2.

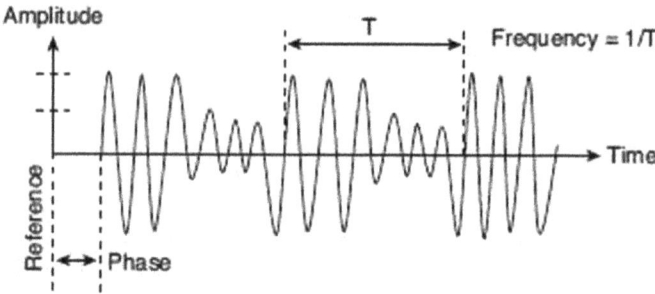

Fig. 30.2 Radio wave having amplitude, frequency, and phase elements [5]

30.3.1 RF System Components

RF system components enable the propagation of radio waves over the medium, i.e., air. A client device is equipped with a transceiver and an antenna for data transmission and reception. The transmission medium may include obstacles, such as walls, ceilings, or objects such as furniture and chairs. The standard RF system is disclosed in Fig. 30.3.

30.3.1.1 RF Transceiver

RF transceiver comprises a transmitter and a receiver. The transmitter transmits the radio waves from the source, and the receiver receives the waves from the destination. RF transceiver is an integral part of any client device that participates in the data exchange with the other devices.

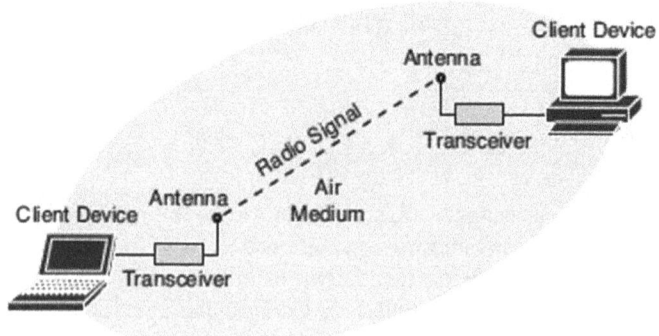

Fig. 30.3 RF system components—RF transceiver, antenna, and a transmission medium [5]

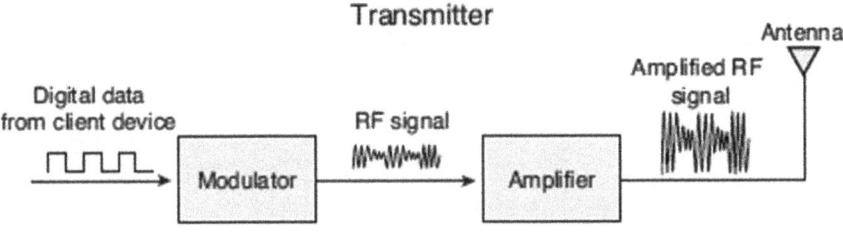

Fig. 30.4   A transmitter having a modulator, an amplifier, and an antenna [5]

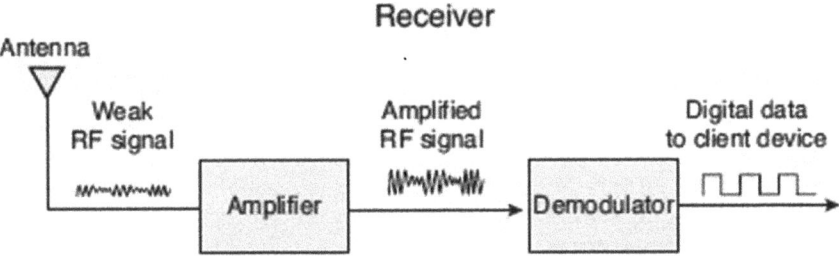

Fig. 30.5   A receiver having an antenna, an amplifier, and a demodulator [5]

The transmitter consists of a modulator and amplifier, wherein the modulator converts electric digital signal representing information (data bits, i.e., 1s and 0s) into radio wave of desired frequency that propagate through the air and the amplifier increases the amplitude of the very radio wave signal to a desired transmit power before it is fed to the antenna. A transmitter component of the transceiver is represented in Fig. 30.4.

A receiver at the destination client device detects the RF signal sent by the sender and demodulates it into data types that it is compatible with. The radio wave received at the receiver must compulsorily have amplitude that is above the receiver sensitivity, so that the receiver can interpret the signal, or decode it. Here, the minimum receiver sensitivity depends on the data rate of the RF signal (Fig. 30.5).

30.3.1.2   RF Modulation

RF modulation converts digital data, i.e., binary 1s and 0s, that represent a document or e-mail into RF signal and transmits it through the air. This includes conversion of digital data signal into analog signal. As modulation progresses, the digital data signal is superimposed onto a carrier signal, wherein the carrier frequency is the radio wave with a specific frequency. As a result, the digital data rides over the carrier. Further, the modulation signal varies the carrier signal in a manner that represents the data to be communicated. At the receiving device, the antenna couples the modulated signal into a demodulator, thereby deriving the data signal from the signal

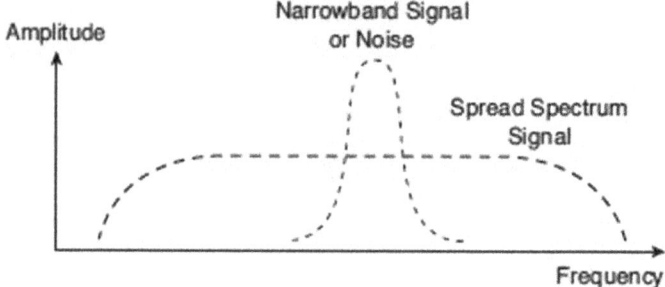

Fig. 30.6 Portion of RF spectrum occupied by spread spectrum [5]

carrier. Various modulation techniques include amplitude-shift keying, frequency-shift keying (FSK), phase-shift keying (PSK), and quadrature amplitude modulation (QAM).

30.3.1.3 Spread Spectrum

Some transceivers after modulating the digital signal into an analog signal spread the modulated carrier over a wider spectrum to reduce the possibility of outward RF signal and inward RF signal interference. Here, a signal's power is spread over a wide band of frequencies, as represented in Fig. 30.6.

30.4 Research Developments

A team of researchers from the University of Sydney's Australian Institute for Nanoscale Science and Technology have developed a chip-scale optical device that achieves radio frequency signal control at sub-nanosecond time scale. The research development potentially provides broader bandwidth to more users.

The developed device appears to be a critical part of the antenna system, wherein the device creates multiple configurable beams that can access more users in real-world environment. The device uses tunable delay lines on chip that reconfigure direction of the beam at nanosecond speed. This implies that each beam carrying digital data over the RF signal can be targeted to each user, thereby providing high bandwidth capacity at a given time [6].

The above research relates to reconfiguring the antenna system in order to provide better bandwidth capacity to each individual user accessing the Internet in a network. However, the research fails to address the grave problems such as Wi-Fi signal strength, router or modem technical issue, transmission medium issue, wherein the medium may include obstacles like walls or objects such as furniture. Thus, the

Internet connectivity issue may still suffer even if the reconfigured antenna system is used by the devices operating in the network.

30.5   Optimizing Wi-Fi Router Configuration

The research proposal discloses a change in the hardware configuration of the Wi-Fi routers that are used in households, offices, or personal spaces. The Wi-Fi router circuit is integrated with adjustable and tunable delay line that consist of one or more on-chip delay elements.

The adjustable delay line deployed within the router forms a critical component of clock synchronization circuits such as a delay locked loop (DLL). The delay line inserts a precise delay into the path of a signal, wherein the produced delay may be adjusted by an analog voltage or digital control word. The delay element(s) within the tunable RF delay line create a time difference between input and output signal. By tuning the delay line, this time difference can be continuously changed, at the order of a billionth of a second [7].

The DLL circuitry receives a reference clock signal (XCLK) at an input buffer, which is further provided to a delay line as a buffered clock signal (CLKIN). The delay line consists of one or more on-chip delay elements that induce delay in the RF signal passing through them. The entry point of the buffered clock signal CLKIN may be adjusted by harnessing a number of delay elements to provide a lock through a range of frequencies, input voltages, etc. The output of the delay line is connected to an output buffer, wherein the RF signal is broadcasted via this output buffer (Fig. 30.7).

Adjustable RF delay line containing one or more delay elements is selected based on the following specifications (Table 30.1).

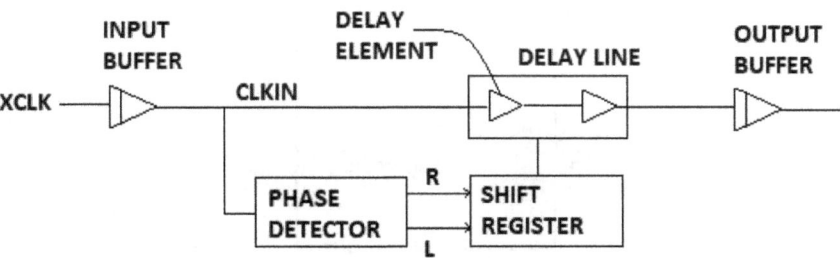

Fig. 30.7   DLL circuitry

Table 30.1 Specifications of the delay line

Sr. no.	Specifications	Details
1	Frequency	This is the frequency of interest at which a time delay is required. Generally, delay lines are optimized for a specific frequency range
2	Delay	This is the amount of delay that is provided by the delay line. Delays are usually in the micro- to nanosecond range
3	Insertion loss (dB)	This is the loss in amplitude of the signal after it is delayed by the delay line
4	Power level	This is the power level that the delay line can handle. It is important to choose a delay line that can handle a power level higher that what will be used in the RF system

30.5.1 Use Case

Consider a scenario, wherein the router receives wireless RF signal from the Internet service provider through the air. The received RF signal is allowed to pass through the delay line containing delay element(s). These delay elements buffer the received RF signal along its circuitry by inducing phase or time delays. As the received first RF signal is buffered in the delay element, the second RF signal and the consequent RF signals get queued within the delay elements. This queuing of RF signals in the on-chip delay elements of the delay line thereby facilitates storage of RF waves. Thus, the RF delay elements store radio waves by inducing time or phase delays for a brief period before broadcasting them. These queued RF signals are later broadcasted by the router for facilitating Internet connectivity in the nearby devices.

The diagrammatic representation of the proposed RF router 'A', leading to the buffered RF waves 'B' within the delay line, is as disclosed in Fig. 30.8.

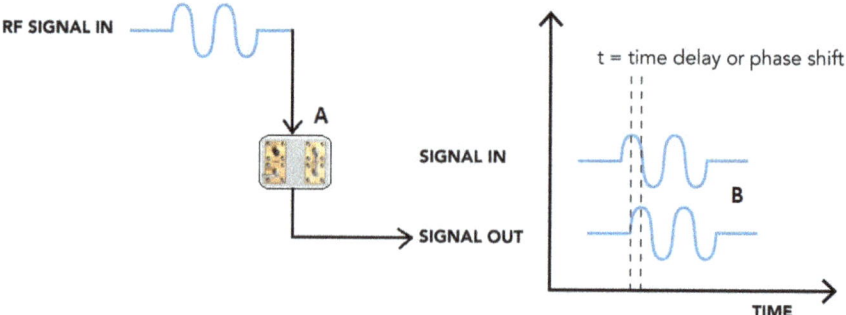

Fig. 30.8 Optimized router configuration

30.5.2   Algorithm

(a)   **Step 1:** RF signal is received as input by the router.
(b)   **Step 2:** RF delay line within the router induces time or phase delay (depending on the router configuration) in the signal.
(c)   **Step 3:** RF input signal is buffered along the delay line containing the delay elements.
(d)   **Step 4:** Buffered RF signal is broadcasted by the output buffer of the router.

30.6   Advantages

The optimized router configuration disclosed in the research has various advantages enlisted below:

1.   Delay elements in the router provide stable Internet connection throughout the connectivity session due to buffered RF signal.
2.   Internet connection may not drop frequently due to inconsistency on the RF signal transmission through the medium, i.e., air.
3.   Increased data speeds with consistent supply of adjusted RF signal via delay line in the router.
4.   The 'bandwidth bottleneck' has caused the demand for bandwidth to rapidly shoot up against the limited supply available today and is estimated to increase by 22% each year. The research work solves the problem of bandwidth bottleneck, as the RF signals are equally spaced and queued within the delay line and broadcasted at regular intervals.
5.   Seamless and consistent Internet connection as the vulnerable radio wave property is adjusted by the tunable RF delay line, wherein the delay line acts as the RF storage device.

30.7   Conclusion

The research paper provides a novel solution for the problem of inconsistent and unstable Internet connection accessed by the users at households, offices, or public places via Wi-Fi routers. The proposed research provides a technical solution to the radio wave problem posed by the Wi-Fi broadcasting medium, wherein the Wi-Fi signals are broadcasted at regular intervals by integrating the delay line within the Wi-Fi router for consistently providing stable Internet connection to its users in the vicinity.

Acknowledgements I would like to extend my sincere gratitude to Dr. A. S. Kanade for his relentless support during my research work.

Conflict of Interest The authors declare that they have no conflict of interest.

Declaration We have taken permission from competent authorities to use the images/data as given in the paper. In case of any dispute in the future, we shall be wholly responsible.

References

1. Clement, J.: Worldwide Digital Population as of July 2020. (July 24, 2020)
2. Bump, P.: 40 Remote Work Stats to Know in 2020, (Mar 12, 2020)
3. How Does WiFi Work? NetSpot
4. How Does Waves Work? Waves Guest Wi-Fi Manager
5. Cisco Press, WiFi Networking: Radio Wave Basics, (January 11, 2017)
6. Photonics Breakthrough Tunes Wireless Communications, (April 19, 2017)
7. Gomm, T.J. et. al.: Synchronization device with delay line control circuit to control amount of delay added to input signal and tuning elements to receive signal form delay circuit. US7111185B2, 2003-12-23

Chapter 31
Periocular Segmentation Using K-Means Clustering Algorithm and Masking

V. Sandhya and Nagaratna P. Hegde

Abstract Biometrics are used widely to authenticate Humans to access various systems and devices in real-world applications. The Periocular region can be used to recognize an individual for identity management. In this paper, we propose the method using the K-Means algorithm to cluster. The image and then apply masking operations to extract features from the image.

31.1 Introduction

User validation is a unique challenge in the field of Biometrics. The different Biometric methods used to authenticate a person are a face, fingerprint, iris, palm print, Handwriting, and sclera. In all of these methods, each individual has to possess unique features. In the present day, Face recognition is the most prominent authentication mechanism for an identity management system. We try to extract unique features of the face. Periocular is the region surrounding the eye that includes the upper eyelid, lower eyelids, sclera, iris, periocular skin, and eyebrow. Challenges that are associated with periocular recognition are persons aging, illumination levels and alignment. To overcome these challenges, the accuracy of Periocular recognition serves as an important task. Figure 31.1 shows a typical periocular region from an open-source database set the UBIPr.

The original version of this chapter was revised: Reference 19 has been removed from the reference list and citation. The correction to this chapter is available at https://doi.org/10.1007/978-981-16-0878-0_83

V. Sandhya (✉)
Gitam School of Technology, Hyderabad, India
e-mail: svinayak@gitam.edu

N. P. Hegde
Vasavi College of Engineering, Hyderabad, India
e-mail: nagaratnaph@staff.vce.ac.in

315

S. C. Satapathy et al. (eds.), *Smart Computing Techniques and Applications*,
Smart Innovation, Systems and Technologies 225,
https://doi.org/10.1007/978-981-16-0878-0_31

Fig. 31.1 Periocular image

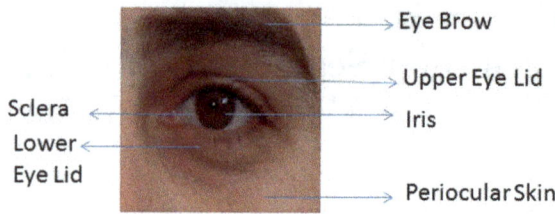

31.2 Related Work

In [1] proposed OCLBCP method using dual-stream CNN to overcome limitations of the wild environment using the UBIPr database. In [2], the author presented an efficient way to detect and position the eye in the face using cascaded CNN. Deep Learning frameworks are used in [3] to extract complex features by applying spatial and color transformation used for data augmentation. The different features used are local features using various vital points and global features [4]. In [4], the authors present the databases used, such as UBIPr, FOCS, IMP: IITD Multispectral Periocular, CSIP, and also the different algorithms used for extracting the best regions for periocular recognition. To extract unique features from the face, LBP is used in [5], and to classify the distance city block is used in [6]. Image is segmented into clusters using an unsupervised K-Means clustering algorithm [7]. The subtractive clustering model is used to create data points. SIFT descriptors are applied to the image [8] to authenticate individuals using the periocular region in an unconstrained environment and GMM model. The four categories in periocular Biometrics are [9] classified into four categories, such as local, global, key-point, and shape appearance. Depending on the features needed to extract from any of the four classes can be applied. Periocular images can be captured in both NIR and the visible spectrum. As proposed in [10], the best regions of the periocular can be selected using the SFFS method. In [11] images, both the visible night images and near-infrared spectrum images are trained using two neural networks to implement a cross-spectral mechanism. Some of the challenging data sets are soft tissue identified in the periocular region. In [12], the proposed method is a hierarchical patch to identify soft tissues in the periocular region and even verified for twin data set. A data level fusion has been proposed in [13] to obtain periocular features eyebrow, eye, and skin regions. The periocular region has a rich texture and requires less user cooperation when compared to other biometric methods. The discriminate features for periocular recognition are proposed in [14]. The dimensions of LBP are reduced to half to transform into binary patterns in [15] an image retrieval system. An image can be divided into patches [16], and colors are matched to evaluate periocular points in the image. In [17], LBP and GIST are used to analyze the distance of images captured in periocular recognition.

31.3 Median Filtering

This low-pass filter is used to eliminate noise from an image that results in blurring of images. Determine the median of all pixels in the window and restore the central pixel with the median value. Kernel size is an odd number.

31.4 Canny Edge Detection Algorithm

Canny edge is a multi-stage edge detection algorithm.

1. Noise in the image can be removed by using 5×5 Gaussian Filters.
2. Sober Kernel is applied to the image to get horizontal gradients Gx direction and Vertical gradient direction Gy. Then we find the edge gradient and direction of each pixel.
3. We scan the image pixel by pixel to remove any pixel that does not form edges.
4. We try to determine edges and non-edges. The two thresholds are MIN and MAX. Edges with gradient values more than MAX are edged that appear, and edges with a value less than MIN are not edges and do not appear.

Given the periocular image (I) of size height (H) × width (W). The image I is clustered into K number of clusters. Suppose P (H, W) is a pixel and C_k denotes central cluster Pixels. After applying the K-Means algorithm, the resultant image is known as the clustered image (I').

31.5 K-Means Clustering Algorithm

1. Choose the number of clusters K to form for a given image.
2. Arbitrarily choose K points as central pixels from the image.
3. Compute distance for each pixel to the central pixels.
4. Allocate pixel to the nearest cluster central pixel with minimum distance when compared to all other cluster central pixels.
5. Determine the new cluster central pixel.
6. Repeat Step 3–5 to attain the desired number of iterations or expected accuracy.

31.6 Masking

To extract a cluster K of Periocular Image I, we apply masking operations on the clustered image and part of the image to be extracted. Pick the index label of the cluster (C_k) from the clustered image I'. Create a mask with cluster C_k central pixel coordinates as a circle. Obtain the coordinates of the selected cluster C_k of Image

I'. Label all the connected blobs of the cluster C_k mask. Create a mask for the entire cluster of blobs. Extract the cropped cluster Image of the Periocular Image *I*.

31.7 Proposed Method

In this method, we extract features of the periocular images such as eyebrow, periocular skin texture, iris, and sclera depending on the number of clusters K chose. Here $k = 7$, we extracted three features such as eyebrow, Iris, and upper eyelid and periocular skin of the Image I. Figure 31.2 shows the steps involved in extracting features from the given image *I*.

1. Periocular image (I) to be segmented is loaded, and apply image enhancement technique.
2. Select the number of Clusters K based on the number of features to be extracted from the Periocular image (I).
3. Pass the periocular Image I through Median filter to remove noise from the image.
4. Apply the canny edge detection algorithm to identify the edges of the periocular image.
5. Use the K-Means clustering algorithm to extract the features on the given periocular Image I and generate the clustered image I'.
6. Extract a specific feature of the clustered image by selecting the cluster C_k and apply mask operation.
7. Repeat step 7 to extract more than one feature of the Periocular Image I.

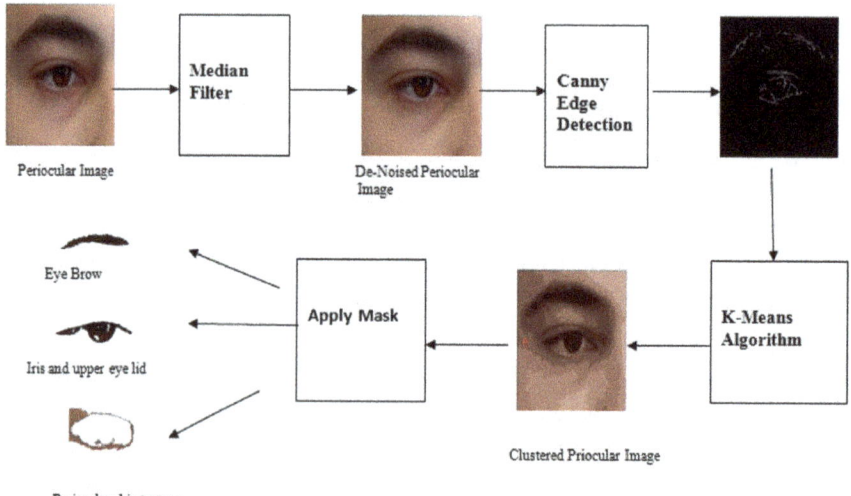

Fig. 31.2 Steps to extract periocular image features

31.8   Data Sets

See Fig. 31.3 [18].

31.8.1   UBIRIS.V1

In this first version of the database, there are 1877 images gathered from 241 eyes in a less constrained environment.

31.8.2   UBIRIS.V2

In this database, there 11,000 images recorded at a distance in a mobile environment.

31.8.3   UBIPr

Is the updated data set of UBIRIS.v2? For periocular images. Images in the data set can be used specifically for periocular image processing and identification. The data set is open source.

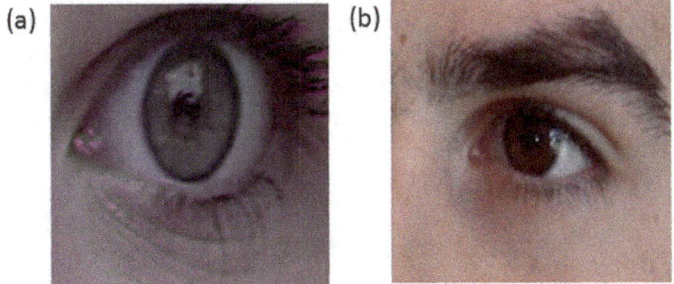

Fig. 31.3 Data sets **a** VSIOB samsung_day_light_short; **b** UBIPr

31.8.4 Visob

Dataset—The database contains images of eyes collected from 550 persons using Oppo, Samsung, and iPhone mobile devices. Images are captured in desperate environment conditions such as out-of-focus, eye-makeup, gaze directions, glasses, and discrete illuminations. In this paper, UBIPr and VISOB Samsung data set are used for implementation.

31.9 Results

We used periocular images from UBIPr and VSIOB Samsung_day_light_short. In the UBIPr data set, each image is of size 300×400, and in the VSIOB data set, images are of size 160×240. Images enhancement is done to attain images from different data sets of the same size. The resultant enhanced images are of size 400×400. To discard any noise in the image, we applied a Median filter, which results in smoothing of the image. Then canny edge detection algorithm is used to identify edges in the periocular image (I). we choose the number of clusters K as 9 to extract features from the Periocular Image such as eyebrows, Iris, upper eyelid, lower eyelid, periocular skin, and sclera. K-Means clustering algorithm is used to cluster the periocular image into different features. Depending on the intensity of the pixel values, clusters and central pixels are formed. We applied K-Means for ten iterations. To the clustered image I masking is applied to get the selected feature from the image. As shown in Fig. 31.4, two images from the VSIOB data set and one image from the UBIPr data set is chosen. We tried to extract different features from the three selected images of the data set. Figure 31.4a the upper eyelid of the image is extracted by selecting the central pixel of the cluster. Figure 31.4b eyebrow is extracted from the image. Figure 31.4c we extracted the iris feature from the image. To extract the features from the three images, we applied masking operation by selecting a central pixel from the corresponding cluster.

31.10 Conclusion

We applied K-Means clustering and masking to segment the periocular image. The resultant clustered image is used to extract the important features from the image. In this method, masking of the image is the crucial step in extracting features. Using the Features extracted, and we can create a data set for periocular images. The proposed method can serve as a base for applying different machine learning algorithms to solve image classification problems.

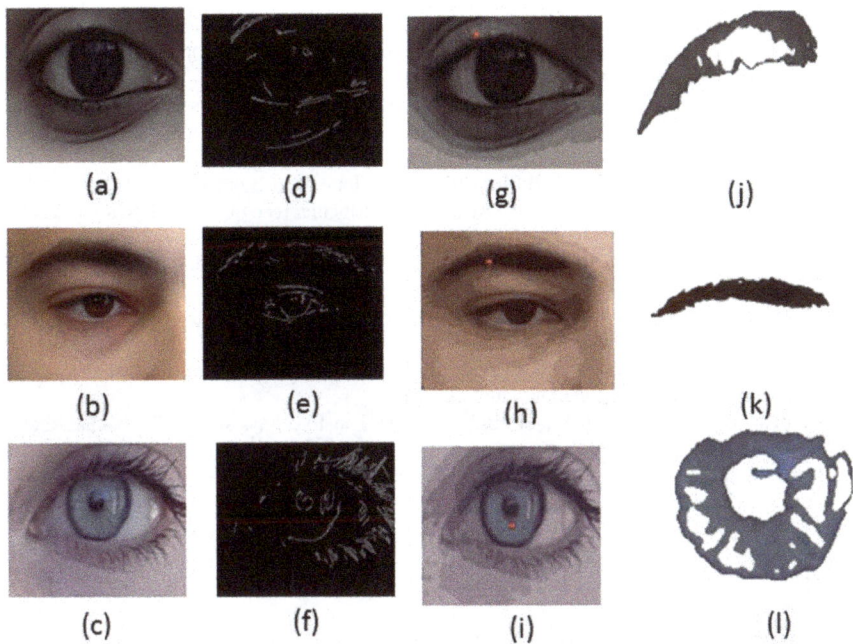

Fig. 31.4 Images **a**, **c** from VSIOB samsung_day_light_short; Image **b** from UBIPr data set; **d–f** Canny edges of the images; **g–i** K-Means clustered image; **j** upper eye lid of (**a**); **k** eye brow if (**b**); (**j**) iris of (**c**)

Acknowledgements We like to express gratitude to the University of Missouri-Kansas City for providing access to **the VISOB** data set and the Department of Computer Science, the University of Beira Interior, for providing access to the **UBIPr** dataset.

Declaration We, authors declare that we have taken required permissions from the respective person/authority to use the image/dataset in our work. We shall be responsible in case any dispute arises in the future with regard to this.

References

1. Ching, L., Tiong, O., Lee, Y., Beng, A., Teoh, J.: Periocular Recognition in the Wild: Implementation of RGB-OCLBCP Dual-Stream CNN, Applied Sciences, (July 2019)
2. Li, B., Fu, H.: Real-time eye detector with cascaded convolutional neural networks. Hindawi Appl. Comput. Intell. Soft Comput. (2018)
3. Proenc, H., Neves, J.C.: Deep-PRWIS: periocular recognition without the iris and sclera using deep learning frameworks. IEEE Trans. Inf. Forensics Secur. **13**(4), (April 2018)
4. Fernandez, F.A., Bigun, J.: A Survey on Periocular Biometrics Research. Elsevier (2018)
5. Kumar, K.K., Pavani, M.: LBP Based Biometric Identification using the Periocular Region. IEEE (2017)
6. Miller, P.E., Rawls, A.W., Pundlik, S.J.: Personal Identification Using Periocular Skin Texture. ACM (2010)

7. Dhanachandra, N., Manglem, K., Chanu, Y.J.: Image segmentation using K-means clustering algorithm and subtractive clustering algorithm. In: Eleventh International Multi-Conference on Information Processing-2015 (IMCIP-2015)
8. Monteiro, J.C., Cardoso, J.S.: Periocular recognition under unconstrained settings with universal, background models. In: Proceedings of the International Conference Bio-inspired Systems and Signal Processing (BIOSIGNALS-2015)
9. Woodard, D.L.: Periocular-Based Biometrics, Encyclopedia of Biometrics (2014)
10. Alonso-Fernandez, F., Bigun, J.: Best Regions for Periocular recognition with NIR and Visible Images. IEEE (2014)
11. Sharma, A., Verma, S., Vatsa, M., Singh, R.: On Cross-Spectral Periocular Recognition. IEEE (2014)
12. Mahalingam, G., Ricanek, K., Jr.: LBP-based periocular recognition on challenging face datasets. EURASIP J. Image Video Process. (2013)
13. Moreno, J.C., Surya Prasath, V.B., Proenca, H.: Robust Periocular Recognition by Fusing Local to Holistic Sparse Representations. ACM (2013)
14. Ambika, D.R., Radhika, K.R., Seshachalam, D.: The Eye Says It All: Periocular Region Methodologies. IEEE (2012)
15. Sadat, R.F.N., Rakib, A., Salehin, M.M., Afrin, N.: Efficient design of local binary pattern for image retrieval. IEEE Symposium on Computers and Informatics (2011)
16. Woodard, D.L., Pundlik, S.J., Lyle, J.R., Miller, P.E.: Periocular Region Appearance Cues for Biometric Identification. IEEE (2011)
17. Bharadwaj, S., Bhatt, H.S., Vatsa, M., Singh, R.: Periocular Biometrics: When Iris Recognition Fails. IEEE (2010)
18. UBIPr home page. https://iris.di.ubi.pt/ubipr.html

Chapter 32
RPCR: Retransmission Power-Cost Aware Routing in Mobile Ad Hoc Networks

V. Sowmya Devi and Nagaratna P. Hegde

Abstract This paper presents a power control routing in MANETs to optimize the utilization of battery power of mobile nodes. The proposed approach considers a power required to transmit information on a particular path termed as power status and the amount of energy required for a node to send the information to its neighboring nodes, termed as node cost status. It also considers the probability of link error for each and every path, thus termed as retransmission power-cost aware routing (RPCR). The link error probability gives the information about the power required for retransmission. Thus, by including link error probability during energy evaluation of a particular path, the average energy consumption will be reduced. The performance of the proposed RPCR was verified.

32.1 Introduction

Mobile Adhoc wireless network is a group of mobile devices with networking capability and interfaces. It is self-organizing and adaptive in nature [1]. Power management [2] is the biggest challenge in mobile ad hoc networks. The amount of power required will be increased with the increase in the data rate and also with distance. In recent days, the transmission of multimedia information through mobile nodes is getting increased. For a mobile node having less energy can't support to transmit the multimedia data. The sudden drop in the power of a node reduces the quality of service of the entire network. If the node consumes maximum power to transmit information, it will drain up quickly. Thus, an energy-efficient routing approach needs to be designed with the view of reducing transmission energy of individual nodes and simultaneously increasing the network throughput. By reducing the individual power of nodes for each transmission, the total lifetime of node is getting increased.

V. Sowmya Devi
Vignana Bharathi Institute of Technology, Hyderabad, India

N. P. Hegde (✉)
Vasavi College of Engineering, Hyderabad, India
e-mail: nagaratnaph@staff.vce.ac.in

This paper presents a retransmission power and cost-aware routing (RPCR) approach in MANETs. In this approach, the source node selects the best possible path from a given source and destination by considering the power status of path and node energy. This approach also includes probability of link error during energy evaluation for a particular established route. By considering the link error probability, the average amount of energy required for a successful transmission can be reduced. Results of simulation prove that the suggested approach is efficient routing approach for MANETs. The paper is organized as follows: Sect. 32.2 gives the information about the literature review. Section 32.3 illustrates the complete particulars about the power and cost-aware routing. Section 32.4 deals with the new link cost metric. The new link cost metric developed in this paper considers the link error probability during path selection. Section 32.5 deals with the simulation results and finally Sect. 32.6 provides the conclusions.

32.2 Literature Review

The routing protocols in MANETs are generally categorized as proactive (table-driven) and reactive (on-demand), based on the updating time of routes. Many routing protocols including Destination-Sequenced Distance Vector (DSDV) [3] and Fisheye State Routing protocol [4] belong to proactive routing and they differ in the number of times the routing tables are manipulated and the approaches used to exchange and maintain routing tables. In contrast to proactive protocols, not all up-to-date routes are maintained at each node. Dynamic Source Routing (DSR) [5] and Ad-Hoc On-Demand Distance Vector (AODV) [6] are examples of on-demand protocols. On the other hand, to establish efficient routes between the nodes, one important goal of a routing protocol is to keep the network alive as long as possible. As discussed in the introduction, this can be achieved by reducing the energy of mobile nodes during active and inactive communication. Load distribution [7], Power-down/Sleep mode [8] is also used to reduce the power utilization of mobile nodes, but this technique is for inactive communication. In this approach, the node which is not contributing in the communication is kept in power down/sleep mode to reduce the total energy consumption. This approach was focusing on reduction of nodes energy indirectly but not on the nodes participating in the communication. Transmission power control techniques [9] provide control on the transmission power of individual mobiles nodes. The primary interest of these techniques is to decrease the amount of the power required for transmission at node level. The main goal of transmission power control routing approaches is not to make available energy-efficient routes but to make the given route with energy efficiency by regulating the transmission power that is sufficient to contact the next-hop node. In [10], a flow augmented routing approach was proposed to minimize the node transmission energy by considering initial and residual energies. However, this approach was built based on assumption of static network. One more issue, the routing approach assumes that Data generation rate is presented in advance during each time step at all nodes. To overcome this problem, in [11] a min–max routing approach was proposed without knowing the data generation in beforehand. This focuses on two parameters, minimizing power consumption and maximizing the minimal residual power. In this approach, a perturbation method was used to evaluate the minimum power of all paths available for a particular source node. However, the routing was based on an assumption of constant traffic. In [12] a power-aware routing was proposed to optimize the power levels of nodes [12]. Completely focused on the power levels of nodes which are willing to transmit but not on the power levels of neighboring nodes.

32.3   Power-Cost Aware Routing

In this approach, the routing was done based on the power status and node cost status. The power status of node is defined as the amount of the power required to transmit information to its neighboring nodes. The node cost status is defined as the amount of the power required to transmit the information of neighboring nodes. The node which was willing to send information towards the destination will calculate the self-power status and also the node cost status. This was accomplished by knowing 1-hop neighbors of neighboring node. A direct communication may put away more energy when compared with indirect communication through in-between nodes with the super linear relationship among transmission distance and energy.

In Fig. 32.1, when node 'A' sends data packet to node 'D', it can either send directly to 'D' or through its neighbors (B, C or E). Here, A to its neighboring node (B or C or E) is a direct transmission while neighboring node (B or C or E) to D is an indirect transmission with few intermediate nodes between immediate neighboring node (B or C or E) and D. For choosing the best path, node A computes and compares the consumed power of each path candidate. Power consumption of the direct transmission, $p(d)$ can be computed if the distance is known, i.e. $p(d) \propto d$, where d is the distance among the two nodes. The total power consumption of each

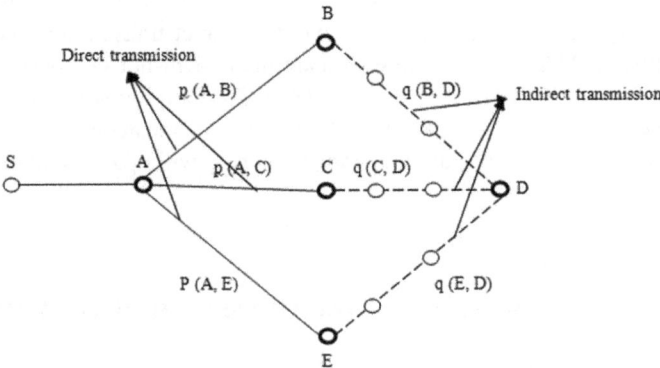

Fig. 32.1 Selection of the next hop node

and every path can be reduced by reducing the number of intermediate node. Let $q(d)$ be the amount of power required for data transmission between nodes (B or C or E) and D. Hence, node A, whether it is an intermediate or source node, chooses one of its neighbors (B, C or E) as the next hop which reduces $P(d) = p(d) + q(d)$.

In this approach, the source or any node which has information to send to destination will initiates route discovery and finds the possible paths. The route discovery phase was accomplished by flooding a RREQ packet with a destination ID. After receiving the RREQ, each and every node replies with RREP to source node. After this, the source node having N available paths to destination through various intermediate nodes. In every path, each and every node having information about previous hop node and next hop node. The source node which wants to send data has to select an optimal path among the available paths. For this purpose, in this approach, a power-cost aware routing was proposed. Initially, the source node knows his neighboring nodes and also the distance away from it. The super relationship $p(d) \propto d$ gives the information about the required power to transmit up to neighbor node. But it doesn't know about the 1-hop neighbors of its neighbor node. The source needs to find all nodes present in the available path. Based on the total power evaluated the source node finalizes the path. For this purpose, the source node sends a 1-hop REQ packet for each and every neighbor node of it, asking about the node cost status. The node cost status includes the available neighbor nodes, distance of neighbor nodes and power required for individual transmission to neighbor node. Then the neighbor node which received the 1-hop REQ replies with a 1-hop REP packet. Then the source node decides the optimal path which is having minimum power consumption.

In Fig. 32.1, let the power required for direct transmission be $p(d)$ and the power required for indirect transmission be $q(d)$, the path which is having minimum power consumption was selected as optimal path. Here, direct transmission refers to the transmission between source and its neighbor nodes, and indirect transmission refers to the transmission between neighbor node and its 1-hop neighbors. The amount of the power required will be varying. This is completely depending on the number of intermediate nodes present after first neighboring node. In the Fig. 32.1, the solid line represents direct transmission and dotted line represents indirect transmission. The holes on the dotted line represents intermediate node of the indirect transmission $q(d)$. Will vary with intermediate nodes on the indirect transmission. Due to the node mobility in MANETS, the intermediate nodes will not be constant. It also consumes more time, thus increases time delay. So, the proposed approach takes run-time decision according to the above discussion, i.e., the node won't perform all these function in addition during data transmission. It will update automatically in the background.

32.4   Retransmission-Power-Cost Aware Routing (RPCR)

The one more issue in MANETs, path breakages due to the mobility of nodes. In MANETs, the mobile nodes are moving in nature. So, the path established between a source–destination pair may not be constant. The intermediate nodes on the established path may move out of range so the path breakage will occur at that node. In MANET, any breakage in link allows retransmission of information intern increases the energy consumption of the node and makes the node drain up quickly, because the amount of energy required for retransmission of data is equal to the energy consumed for sending the information before breakage [13]. Hence, there is a need to determine the link errors on the established path frequently. After the occurrence of link error, the node at which the error occurred needs to update to the source node. For this purpose, in [14], the intermediate node sends a packet denoting that the link error was occurred and also for re-route establishment. This scenario consumes more power at each and every node transmitting that packet.

In this approach, a new link cost-based routing approach was proposed to reduce the transmission energy by considering the effect of transmission link errors. This approach studies how link error rates affect this retransmission-aware metric. Consider a route from a source node S to a destination node D that has N-1 intermediate nodes named as 2, 3 ... N (1 refers to the name of the source and $N + 1$ is the destination). The transmission energy over each link is $p_{i,i+1} = d_{i,i+1}$, where $d_{i,i+1}$ refers to the distance among nodes i and $i + 1$. Assuming that each of N links $(L_{1,2}, L_{2,3},.., L_{N,D})$ has an independent link error rate of $e_{i,i+1}$, the number of transmissions(including retransmissions) between node i and $i + 1$ is a geometrically distributed random variable as

$$Prob\{X = x\} = e_{i,i+1}^{x-1} \times (1 - e_{i,i+1}) \tag{32.1}$$

The mean number of transmissions for the successful transfer of a single packet is $1/(1 - e_{i,i+1})$. Therefore, the effective transmission energy between nodes I and $i + 1$, which includes the effect of the transmission link error, is [14]

$$P_{i,i+1} = p_{i,i+1} \times 1/(1 - e_{i,i+1}) = ad_{i,i+1}^{\alpha} \times 1/(1 - e_{i,i+1}) \tag{32.2}$$

When the packet-error rate $(e_{i,i+1})$ is not negligible, the benefit of indirect transmission via intermediate nodes can be overshadowed by the inflation factor, $1/(1 - e_{i,i+1})$. From the above equation, it is clear that, as the link error increases, the amount of power required for transmission is also getting increased. But compared to the power consumed for re-route establishment after occurrence of link error, the amount of power required for transmission by including a probability of link error is less. If there is any link error, the complete process has to take place again; the total power consumption will be simply doubled which will be high compared to power required with above power required using Eq. (32.2).

Algorithm: The complete proposed approach was done under two phases, route discovery and route maintenance.

Phase 1: Route Discovery.

   Step 1: source node which has information to send to destination broadcasts a RREQ to all nodes in the network with destination ID.

Step 2: after receiving RREQ, each and every node replies with a RREP.

Phase 2: Route Maintenance.

Step 3: once the route request process over and the route was established, the destination node broadcasts route reply packet. If source node founds more number of paths, then it will choose a path having minimum energy consumption.

Step 4: The source node evaluates the amount of power required to transmit the information to just neighbor node by considering link error probability and sends 1-hop REQ packet to its just neighbor nodes to know the information (distance and power required) about their 1-hop neighbors.

Step 5: the intermediate nodes those are received 1-hop request from source node forwards it to their 1-hop neighbors and asks for their neighboring nodes. This will take very much time, so an elapsed time was fixed. The neighboring node replies with the received information within the elapsed time.

Step 6: upon receiving the information from just neighboring node through 1-hop REP packet, the source node selects a path which is having minimum energy consumption and transmits the data to the respective immediate neighbor node.

Step 7: This process continues until the destination was reached.

The above-illustrated process was completely done in the background. i.e., the node through which the data was processing already kept the entire information about its 1-hop neighbors and upon the receiving data from its previous node, it just chooses the node from the list and forwards the data. Note that the 1-hop REP was sent to the neighboring nodes not flooded.

32.5 Simulation Results

The performance of proposed approach was verified by varying the network parameter node mobility.

For the varied node mobility and the respective performance metrics, Packet Delivery Ratio, Average End-to-End delay, Average Energy Consumption, and Routing Overhead were evaluated. The proposed RPCR was compared with earlier Power Aware Routing (PAR) [15] and also with the basic AODV routing protocol to show the enhancement. In this, the number of nodes are 50, the pause time was fixed to 60 s and the node mobility was varied from 0 to 50 m/s.

From Fig. 32.2, it is clear that as the node mobility was varying, the packet delivery ration is decreasing. The proposed approach considered the link error probability during energy evaluation for all available paths. This reduces the probability of selecting a path with high link error. Thus, the packet delivery ratio is high. In Fig. 32.3, it is clear that as the node speed is increasing, the end-to-end delay is also increasing. In Fig. 32.4, it is clear that the average energy consumption was increasing with an increment in the node speed. In Fig. 32.5, as the node speed was increasing, the control packets were also increasing intern increasing the routing overhead.

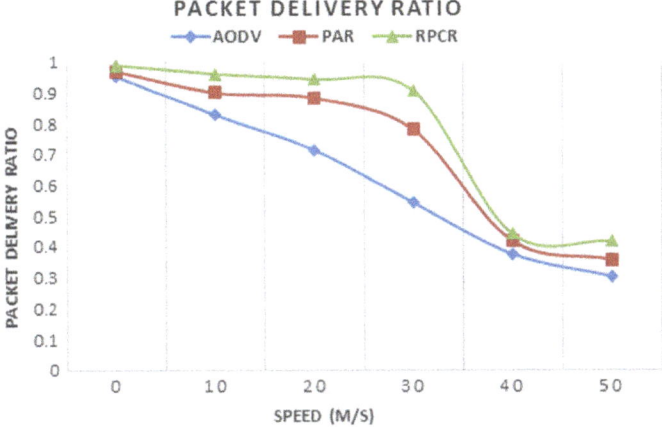

Fig. 32.2 Packet delivery ratio versus node mobility

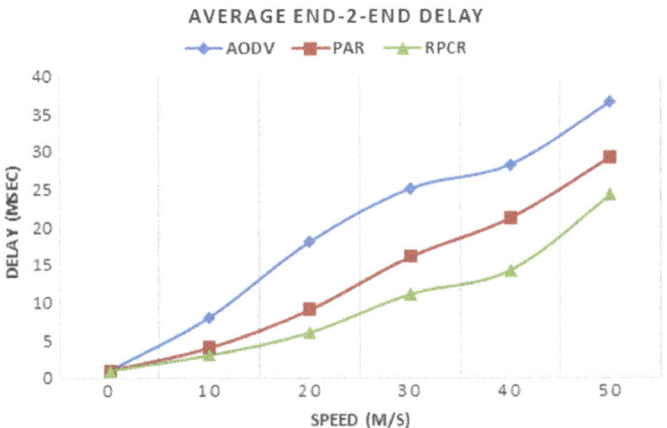

Fig. 32.3 Average end-to-end delay versus node mobility

32.6   Conclusion

This paper proposed a retransmission power and cost-aware routing in MANETs. In the proposed approach, the complete routing was done by considering node cost as well as path power cost, i.e., the amount of remaining energy at each and every node and the amount of power required for a particular available. The source node initially obtains few available paths by sending RREQ with destination ID. After getting available paths, the source node selects an optimal path by considering the path cost and node cost. Here, the path cost considers the link error probability. The simulation results are drawn by varying node mobility, the proposed approach was verified through packet delivery ratio, average end-to-end delay, average energy

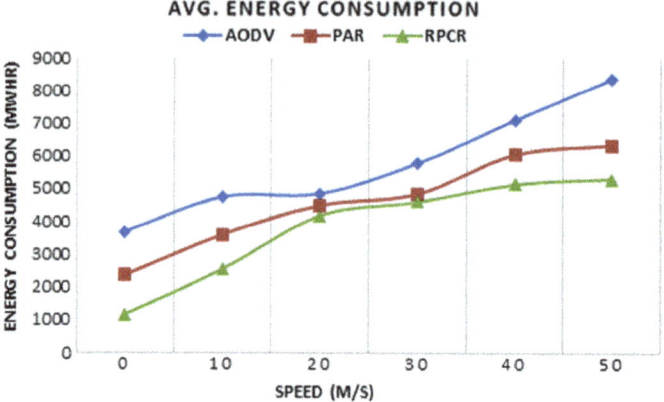

Fig. 32.4 Average energy consumption versus node mobility

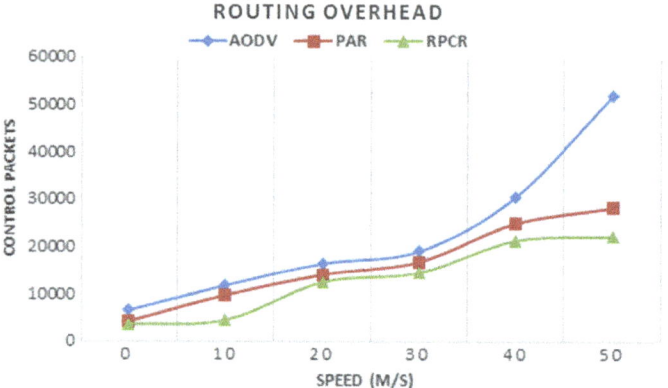

Fig. 32.5 Routing overhead versus node mobility

consumption, and routing overhead. Finally, the simulation results proved that the proposed approach is efficient in MANETs for power optimization.

References

1. Perkins, C.: Ad Hoc Networking, pp. 1–28. Addison-Wesley, Reading, MA (2001)
2. Narayana Swamy, S., Kawadia, V., Sreenivas, R.S., Kumar, P.R. Power control in ad-hoc networks: theory, architecture, algorithm and implementation of the COMPOW protocol. In: Proceedings of European Wireless 2002, pp. 156–162
3. Perkins, C., Bhagwat, P.: Highly dynamic destination sequenced distance-vector routing (DSDV) for mobile computers. Comput. Commun. Rev. **24**(4), 234–244 (1994)

4. Pei, G., Gerla, M., Chen, T.-W.: Fisheye state routing: a routing scheme for ad hoc wireless networks. In: Proceedings of IEEE International Conference on Communications (ICC), pp. 70–74 (2000)

5. Johnson, D., Maltz, D.: Dynamic source routing in ad hoc wireless networks. In: Imielinski, T., Korth, H. (eds.) Mobile Computing, pp. 153–181. Kluwer Academic (1996)

6. Perkins, C., Royer, E.: Ad-hoc on-demand distance vector routing. In: Proceedings of 2nd IEEE Workshop on Mobile Computing Systems and Applications, pp. 90–100 (1990)

7. Woo, K., Yu, C., Youn, H.Y., Lee, B.: Non-blocking, localized routing algorithm for balanced energy consumption in mobile ad hoc networks. In: Proceedings of International Symposium on Modeling Analysis and Simulation of Computer and Telecommunication Systems (MASCOTS 2001), pp. 117–124 (2001)

8. Chen, B., Jamieson, K., Morris, R., Balakrishnan, H.: Span: an energy-efficient coordination algorithm for topology maintenance in ad hoc wireless networks. In: Proceedings of International Conference on Mobile Computing and Networking (MobiCom'2001), pp. 85–96 (2001)

9. Stojmenovic, I., Lin, X.: Power-aware localized routing in wireless networks. IEEE Trans. Parallel Distrib. Syst. **12**(11), 1122–1133 (2001)

10. Chang, J.-H., Tassiulas, L.: Energy conserving routing in wireless ad-hoc networks. In: Proceedings of the Conference on Computer Communications (IEEE Infocom 2000), pp. 22–31 (2000)

11. Li, Q., Aslam, J., Rus, D.: Online power-aware routing in wireless ad-hoc networks. In: Proceedings of International Conference on Mobile Computing and Networking (MobiCom'2001), pp. 97–107 (2001)

12. Baisakh: Energy saving and survival routing protocol for mobile ad hoc networks. Int. J. Comput. Appl. (0975–888) **48**(2), (2012)

13. Sowmya Devi, V., Hegde, N.P.: Reliable and power efficient routing protocol for MANETs. Int. J. Comput. Netw. Inf. Secur. (IJCNIS) **8**(10), 61–69 (2016). https://doi.org/10.5815/ijcnis.2016.10.08

14. Hossain, M.J., Ali Akber Dewan, M., Chae, O.: Maximizing the effective lifetime of mobile ad hoc networks. IEICE Trans. Commun. **91**(9), 2818–2827 (2008)

15. Arvind, S., Zhen, L.: Maximum lifetime routing in wireless ad-hoc networks. In: INFOCOM 2004. Twenty-Third Annual Joint Conference of the IEEE Computer and Communications Societies, vol. 2, pp. 1089–97. Hong Kong, China (March 2004)

Chapter 33
Comparison on Different Filters for Performance Improvement on Fingerprint Image Enhancement

Meghna B. Patel, Jagruti N. Patel, Satyen M. Parikh, and Ashok R. Patel

Abstract Today, in the era of Internet, biometric authentication plays many important roles. And for biometric authentication image, enhancement plays a vital role because the recognition of an image is extremely depend on it. The image quality can be improved using image enhancement phase after removing noise, joining damaged ridges and constructing smooth images. Various filtering techniques for image enhancement like min filter, max filter, median filter, mean filter, Gabor filter, O'Gorman filter as well as segmentation techniques like Sobel edge detection are used. In the proposed research work, above all mentioned filters are applied and compared the result for enhancing the fingerprint images. The comparison of all filters is done based on MSE, PSNR and computational time to measure the performance of the proposed research work. The implementation of the proposed research work was done using Java and used custom database of 100 images of 25 users.

33.1 Introduction

Image enhancement is the best process for result improvements for creating and increasing image appearances in the field of image processing. In real word, the value of the images is affected and becomes weak during the transmission from camera, level of light and sensor. The finger images become weak because of dry

M. B. Patel (✉) · S. M. Parikh
A.M. Patel Institute of Computer Studies, Ganpat University, Gujarat, India
e-mail: meghna.patel@ganpatuniversity.ac.in

S. M. Parikh
e-mail: parikhsatyen@yahoo.com

J. N. Patel (✉)
Department of Computer Science, Ganpat University, Gujarat, India
e-mail: jnp01@ganpatuniversity.ac.in

A. R. Patel
Florida Polytechnic University, Lakeland, FL, USA
e-mail: apatel@floridapoly.edu

skin, sensor noise, cuts and scare, different pressure on sensor which generate non-adaptive noise as well as add nonlinearities among ridges point and valley design of impression [1]. The image quality improvement can be done using the enhancement technique that work for two bases: to join and damage ridges as well as eliminate noise among the ridges plus increase the ridges point valley disparity. To remove this type of noise, various filters like min–max filter, mean filter and median filter are used in image processing. These all filters are used as window masks for filtering. The filter is a minor grouping defined by kernel applied to every pixel and its neighbour's pixel of image. The filtering operation is worked based on region process, in that the importance of particular output pixel is resolute by applying an algorithm on value of the pixels in the region of the equivalent input pixel. A pixel area is a set of pixels defined by their position relative to that pixel. In most applications, the centre of the kernel is aligned with the current pixel and is a square with an odd number of elements (3, 5, 7, etc.) in each dimension. The technique used to apply a filter to an image is called convolution and can be applied in the spatial or frequency domain. In the spatial domain, if the kernel is pixel-centred, most of the convolution procedures multiply the kernel element with the corresponding pixel value. The elements of the resulting group are averages, and then the unique pixel standards are changed using the outcome.

33.2 Related Work

Image enhancement of fingerprint images can be able to perform either on gray scale or binary images [2, 3]. Binarization of the process will take place prior to image improvement, which misplaces the value of unique fingerprint data and create extra spurious minutiae structure. This drawback produces a problem in the objects improvement stage. An alternative method to use for object improvement is on a grey scale image which estimates the reliable local frequency of ridges and estimates the orientation.

There are three main methods to improve object quality: histogram flattening, FFT transform and Gabor filter [4]. The first process uses the normalization method at the primary stage using mean and variance. This is a pixel-based improvement technique that expands the legibility of the thumbprint and then does not change the ridge construction [5, 6]. The additional transformation process is established on FFT; improves the object quality that depends on the frequency domain. The last process, the Gabor filter, enhances the object quality according to its frequency and orientation. Throughout a collected work analysis [7–10], it was found that the maximum technique used in image enhancement is built using context filters. It depends on local orientation and peak frequency parameters. Filters were defined in Fourier or spatial domain. Conventional filtering works by using a single filter are to convolve the entire image. The filter characteristics vary depending on the local context [11, 12].

Gabor Filter is applied to research papers [7, 13] and proves that compared to histogram equalization and FFT-based enhancement, Gabor filter is efficient and improves the result of fingerprint images. Gabor filter has optimum joint solution for both frequency and spatial domains as well as has orientation and frequency selective properties. Secondly, the applied contextual filter is for image enhancement given in [14]. The researcher used a kernel, the major axis of which is oriented parallel to the ridges, called anisotropic smoothing.

A research paper [8] proposed an integration model for improving fingerprint images. In their work, the break at the two ends of the broken edges is effectively joined. As a result, two bogus ending trivia have been removed. Noise is suppressed in small and medium valleys. Edge holes have been completely removed, and borders have been improved as well. However, the proposed model does not solve some problems. When there is a high level of noise in the image, it is difficult to evaluate the positioning field in the object. In this regard, the segmentation of the original image is necessary. The performance of the proposed work is achieved by using the difference in the direction of the blocks and the quality index of small parts extracted by grouping the image quality characteristics. The proposed method allows to improve the difference in the directions of the blocks and the quality index and also allows to reduce the required time within reasonable limits. But the characteristic image factors for the identification and verification system still need to be improved. The search result still does not match the fingerprint.

The researcher [15] is working on new approaches for the confirmation of details in fingerprint images. They suggest that the new features for the details of fingerprint image confirmation provide better accurately than other grey scale methods stated in the literature. The methods are computationally effective and can also be used to design minutia detectors that can work straight on grey scale images. But merging the decision of the two classifiers and studying the effects of checking the details about the performance of the match are still needed to improve the object, and can be possible with the help of image improvement. Techniques based on direct grey scale enhancement work better than approaches that require binarization and thinning as intermediate steps [16]. The percentage of errors in terms of lost, exchanged and spurious little things that occur when using the binarization method is significantly lower than errors that occur when using these approaches. The modified Gabor filter works best, especially for poor images with low quality ridges and blocks with singular points. The requirement to assess the local frequency information performed by the Gabor filter is removed by using a single anisotropic filter. Improvement always requires speed and effectiveness.

Researcher [17, 18] used an incline approach to compute that direction path was orthogonal to the incline. First, we divided the object into square blocks and calculated the incline for each pixel in the x and y directions. The direction vector for each block is calculated. This is done by execution of an averaging process on all paths in the block that are orthogonal to the incline pixels. Use a low-pass filter to smooth the image because the ridge orientation may not be detected completely due to the noise in the image.

33.3 Proposed Research Work

Various filtering methods are available in many research papers for removing noise. Mainly, it is divided into two types: (1) linear filtering techniques and (2) nonlinear filtering techniques. In linear filtering methods, the output remains linear as per its inputs, but in nonlinear, the output is not remaining linear.

33.3.1 Mean Filter

This is a kind of linear spatial filter. It alters the values of noisy pixels by a regular value of its neighbourhood pixels containing this one [19, 20]. It is normally used for smoothing the images [1]. It removes unrepresentative pixels of their surroundings. To calculate mean, this filter used kernel to represent size and shape of neighbourhood. Most of 3×3 size kernels are used. Like Eq. (33.1), the kernel A is used with size $m = 3$, then

$$A_{avg} = \frac{1}{3}[1 \ 1 \ 1] \tag{33.1}$$

33.3.2 Min Filter

It is one type of linear filter. It is also known as zero percentile filter. It alters the value of noisy pixel by minimum intensity value among all neighbourhood pixels. It is used for finding out blackest point in image and improves the black area using the minimum value of the image. The min filter is defined by Eq. (33.2).

$$f^{\wedge(x,y)} = \min\{g(s, t)\} \ \text{where} \ (s, t) \in s(x, y) \tag{33.2}$$

33.3.3 Max Filter

It is one type of linear filter. It is called as 100% filter. It alters the value of noisy pixel by maximum intensity value among all neighbourhood pixels [19]. This filter finds out optimistic points in image. It improves the bright areas of the image by finding out maximum value in the surrounded area by filter. It is mostly used to remove dotted noise occurred due to lower intensity value occurred because of maximum operations [21]. The max filter is defined by Eq. (33.3).

$$f^{\wedge(x,y)} = \max\{g(s,t)\} \text{ where } (s,t) \in s(x,y) \tag{33.3}$$

33.3.4 Median Filter

It is one type of nonlinear filter and also called as order statistical filter. It is used as a mean filter to remove noise by smoothing the image. It also decreases the intensity values among pixels. It alters the noisy pixel by the median value of its neighbourhood pixels. It is better compared to mean filter to preserve useful information in image [21]. The median filter is defined by Eq. (33.4).

$$(m,n) = \text{median}\{[i \; j], [i \; j] \in w\} \tag{33.4}$$

where w indicates the neighbourhood to the nearby location.

33.3.5 Sobel Edge Detection [10, 22]

The measurement of the two-dimensional spatial gradient in the object is performed by the Sobel operator. It is used to get the complete value of the approximate gradient at every pixel of the input grey scale image. Use two 3x3 convolution masks, one mask for to evaluate the gradient in the x route (columns) and the second one used to evaluate the gradient in the y(row) direction. The 3×3 matrix is much smaller than the real object. For this reason, the matrix moves across the object and simultaneously uses the four sides of pixels. The illustration of real Sobel covers is assumed under Fig. 33.1.

The estimated value is considered with the resulting equivalence (33.5):

$$|G| = |Gx| + |Gy| \tag{33.5}$$

The matrix is applied to the part of the input object, changes the pixel worth, changes single pixel in the right path, and then rests in the right path until it reaches the last of the line. Then, it drives to the next row.

Fig. 33.1 Sobel masks illustration [23]

-1	0	+1
-2	0	+2
-1	0	+1

Gx

+1	+2	+1
0	0	0
-1	-2	-1

Gy

33.3.6 Gabor Filter [13, 24, 25]

In thumbprint identification for thumbprint identical and thumbprint arrangement, the Gabor filter is commonly defined as image improvement. The finger image covers dual key characteristics, such as the frequency and orientation of local peaks. The Gabor filter contains selective frequency and orientation properties, which is why it works most effectively for improving fingerprint. The Gabor filter is useful to every pixel situation in the whole image based on these properties. The outcome is used to filter and improves edges concerned with the path of native location and reductions whatever that is concerned with another way. Therefore, the filter improves ridge structure and smoothen valleys values, and simultaneously also effectively reduces noise. The following Eqs. (33.6) and (33.7) show the common method of Gabor filter [26].

$$g(x, y; \lambda, \theta, \psi, \sigma, \gamma) = \exp\left(-\frac{x'^2 + \gamma^2 y'^2}{2\sigma^2}\right) \sin\left(2\pi \frac{x'}{\lambda} + \psi\right) \qquad (33.6)$$

$$g(x, y; \lambda, \theta, \psi, \sigma, \gamma) = \exp\left(-\frac{x'^2 + \gamma^2 y'^2}{2\sigma^2}\right) \cos\left(2\pi \frac{x'}{\lambda} + \psi\right) \qquad (33.7)$$

$$x' = x \cos\theta + y \sin\theta$$

$$y' = -x \sin\theta + y \cos\theta$$

Here, the wavelength of sinusoidal factor is denoted as λ, angle of standard to similar lines of a Gabor purpose denoted as θ, stage offset denoted as ψ, sigma of the Gaussian packet denoted as σ and spatial aspect ratio denoted as γ.

33.3.7 O'Gorman Filter [13]

As per literature survey, the researcher states that O'Gorman filter also provides the good outcome for image enhancement and suits for fingerprint images [13, 14, 27, 28]. This is a directional filter. Related to the Gabor filter, the O'Gorman filter must use the pixel direction to create the filter, but it does not need the occurrence. To create this type of direction filter, first recognize a 7 * 7 matrix in a 0° direction and formerly revolve the filter base to the orientation of each pixel. Then, just calculate the rotation filter coordinates and determine the weight of the revolved filter of the 0° direction filter based on the rotation coordinate. The rotation methods have the classic rotation method as per Eq. (33.8):

$$\begin{pmatrix} i' \\ j' \end{pmatrix} = \begin{pmatrix} \cos\theta & \sin\theta \\ -\sin\theta & \cos\theta \end{pmatrix} \begin{pmatrix} i \\ j \end{pmatrix} \tag{33.8}$$

A problem that comes with the formula of rotation is that the i' and j' do not have an integer value. For the sake of it, we have to make sure that the two pointers prove to be $0°$ in the direction filter and can be solved by interpolation and they can make other messengers through lined exclamation and establish the equivalent weightiness that corresponds to $0°$ filter.

33.4 Evaluation Parameter and Result Discussion

The implementation of proposed research work is done in JAVA platform and to check the performance of proposed image enhancement filters are measured using a custom database of 100 images of 25 users.

33.4.1 Quality of an Image

This evaluation parameter is used to check the measurement of visual appearance of an image. It checks the improvement in image quality after applying all the filters. Figure 33.2 shows the result of different filters in visual appearance. It proves that enhanced O'Gorman filter improves the result of an image after removing the noise and connects broken valleys and ridges compared to other filters.

33.4.2 PSNR and MSE Value

These two parameters are used to show the comparison of implemented different image enhancement filters. PSNR is well defined as the topmost signal-to-noise proportion among dual images using Eq. (33.9). Quality is measured from the inventive image and reinvented using it. The quality of the reinvented image is determined by the high PSNR and low MSE values.

MSE is well defined as the mean square error that approximates the alteration among the inventive image and the faded image. Example: If two images are indistinguishable on all sides, the MSE between them is measured as zero. First, compute the MSE value using an Eq. (33.9) and then use the following formula to calculate the PSNR value (33.10).

$$\text{MSE}\,(x, y) = \frac{1}{N} \sum_{i=1}^{N} (x_i - y_i)^2 \tag{33.9}$$

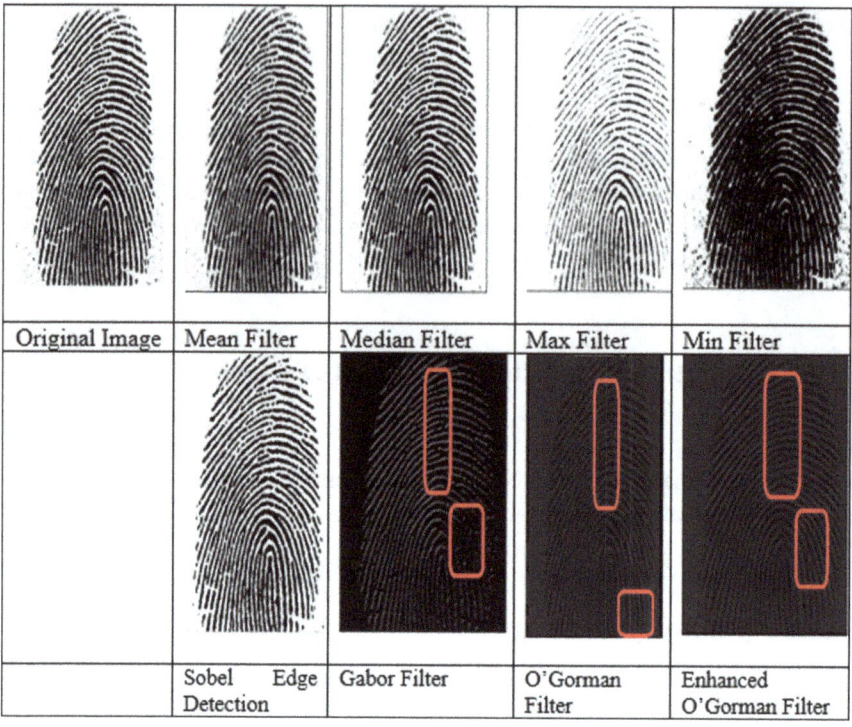

Fig. 33.2 Result of visual appearance of an image

Table 33.1 Result of PSNR value

Images of FingerDos databases	Mean filter	Median filter	Max filter	Min filter	Enhanced Sobel edge filter	Gabor filter	O'Gorman filter	Enhanced O'Gorman filter
0101.bmp	29.83	32.76	25.56	24.43	36.27	36.84	37.62	40.39
1201.bmp	29.93	32.96	24.34	23.82	36.92	37.06	37.96	40.81
3702.bmp	29.81	32.34	25.23	23.34	36.52	36.70	37.41	40.12
0401.bmp	29.47	32.45	24.94	234.13	36.72	36.94	37.76	40.19
1201.bmp	29.85	32.56	25.45	23.87	36.87	37.07	37.97	40.29
3901.bmp	29.67	32.43	25.78	24.36	36.24	36.80	37.56	40.30

$$\text{PSNR} = 10 \, \log_{10} \frac{L^2}{\text{MSE}} \tag{33.10}$$

where N is the total number of pixels in the input image and L is the number of discrete grey level.

Table 33.2 Result of MSE value

Images of FingerDos databases	Mean filter	Median filter	Max filter	Min filter	Enhanced Sobel edge filter	Gabor filter	O'Gorman filter	Enhanced O'Gorman filter
0101.bmp	16.23	15.54	18.43	19.12	13.13	13.45	12.61	5.94
1201.bmp	16.46	15.66	18.69	19.34	12.65	12.77	11.97	5.39
3702.bmp	16.52	14.92	17.07	18.87	13.23	13.89	13.03	6.32
0401.bmp	16.78	15.61	17.53	19.65	12.66	13.15	12.33	6.21
1201.bmp	17.59	14.28	16.32	19.48	12.72	12.74	11.94	6.08
3901.bmp	16.83	15.34	16.12	19.71	13.43	13.56	12.71	6.05

Tables 33.1 and 33.2 show the comparison of different filters using PSNR and MSE values. The comparison result of PSNR value of mean, median, max and min shows that the value of median filter is higher as well as MSE value of median filter is lower compared to other three filters. The PSNR and MSE value of Gabor filter and enhanced Sobel edge filter is nearer similar but better than median filter. And compared to Gabor filter and enhanced Sobel edge filter, the result of O'Gorman gives better result in terms of PSNR and MSE value and makes smooth images, but the connection of ridges is not done properly. So, enhance the O'Gorman filter with the use of interpolation which is used to connect broken ridges. The result of comparison proves that PSNR value of enhanced O'Gorman filter is higher and MSE value is lower compared to other filters which prove that the proposed algorithm gives a better result.

33.4.3 Computational Time

This evaluation parameter is used to compute the processing time of the enhancement process in milliseconds. Table 33.3 shows the summary of comparison between

Table 33.3 Result of computational time

Images of FingerDos databases	Mean filter	Median filter	Max filter	Min filter	Enhanced Sobel edge filter	Gabor filter	O'Gorman filter	Enhanced O'Gorman filter
0101.bmp	6	49	30	23	341	134	114	117
1201.bmp	7	48	31	22	344	137	114	114
3702.bmp	6	48	31	23	341	149	127	132
0401.bmp	6	49	30	23	339	143	122	124
1201.bmp	6	49	30	22	345	134	111	119
3901.bmp	7	48	30	23	349	143	124	122

different filters. In comparison with mean, median, max and min filter, the mean filter takes less processing time. And when we compare enhanced Sobel edge filter, Gabor filter, O'Gorman filter and enhanced filter, it proved proposed enhanced algorithm will take less time compared to other mentioned filters.

33.5 Conclusion

The proposed research work shows the implementation of eight different filters like mean, median, max, min, enhanced Sobel edge, Gabor, O'Gorman and enhanced O'Gorman for image enhancement. The evaluation of implemented proposed work can be compared using different standard parameters like PSNR, MSE, computational time and image quality measurement standard. The experimental study shows that the performance of the enhanced O'Gorman filter is better than other filters.

References

1. Tania, S., Rowaida, R.: A comparative study of various image filtering techniques for removing various noisy pixels in aerial image. Int. J. Signal Process. Image Process. Pattern Recogn. **9**(3), 113–124 (2016)
2. Maio, D., Maltoni, D.: Direct gray-scale minutiae detection in fingerprints. IEEE Trans. Pattern Anal. Mach. Intell. **19**(1), 27–40 (1997)
3. Patel, M.B., Parikh, S.M., Patel, A.R.: Performance improvement in binarization for fingerprint recognition. IOSR J. Comput. Eng **19**(3), 68–74 (2017)
4. Omran, S.S., Salih, M.A.: Comparative study of fingerprint image enhancement methods. J. Univ. Babylon **22**(4), 708–723 (2014)
5. Greenberg, S., Aladjem, M., Kogan, D.: Fingerprint image enhancement using filtering techniques. Real-Time Imaging **8**(3), 227–236 (2002)
6. Patel, M.B., Parikh, S.M., Patel, A.R.: Global normalization for fingerprint image enhancement. In: International Conference on Computational Vision and Bio Inspired Computing, Sept 2019, pp. 1059–1066. Springer, Cham
7. Hong, L., Wan, Y., Jain, A.: Fingerprint image enhancement: algorithm and performance evaluation. IEEE Trans. Pattern Anal. Mach. Intell. **20**(8), 777–789 (1998)
8. Wu, C., Shi, Z., Govindaraju, V.: Fingerprint image enhancement method using directional median filter. In: Biometric Technology for Human Identification, Aug 2004, vol. 5404, pp. 66–75. International Society for Optics and Photonics
9. Yang, G.Z., Burger, P., Firmin, D.N., Underwood, S.R.: Structure adaptive anisotropic image filtering. Image Vis. Comput. **14**(2), 135–145 (1996)
10. Patel, R.B., Patel, M.B., et al.: Performance improvement in fingerprint image enhancement using Gaussian mask and Sobel convolution. In: 10th International Conference on Transformation of Business, Economy and Society in Digital Era (2019)
11. Patel, M.B., Parikh, S.M., Patel, A.R.: Performance improvement in gradient based algorithm for the estimation of fingerprint orientation fields. Int. J. Comput. Appl. **167**(2), 12–18 (2017)
12. Patel, M., Parikh, S.M., Patel, A.R.: An improved approach in core point detection algorithm for fingerprint recognition. In: Proceedings of 3rd International Conference on Internet of Things and Connected Technologies (ICIoTCT), Apr 2018, pp. 26–27

13. Patel, M.B., Patel, R.B., Parikh, S.M., Patel, A.R.: An improved O'Gorman filter for finger-print image enhancement. In: 2017 International Conference on Energy, Communication, Data Analytics and Soft Computing (ICECDS), Aug 2017, pp. 200–209. IEEE
14. Patel, M.B., Parikh, S.M., Patel, A.R.: An improved approach in fingerprint recognition algorithm. In: Smart Computational Strategies: Theoretical and Practical Aspects, pp. 135–151. Springer, Singapore (2019)
15. Chikkerur, S., Govindaraju, V., Pankanti, S., Bolle, R., Ratha, N.: Novel approaches for minutiae verification in fingerprint images. In: 2005 Seventh IEEE Workshops on Applications of Computer Vision (WACV/MOTION'05), Jan 2005, vol. 1, pp. 111–116. IEEE
16. Patel, M.B., Parikh, S.M., Patel, A.R.: An improved thinning algorithm for fingerprint recognition. Int. J. Adv. Res. Comput. Sci. **8**(7), 1238–1244 (2017)
17. Halici, U., Jain, L.C., Erol, A.: Introduction to fingerprint recognition. In: Intelligent Biometric Techniques in Fingerprint and Face Recognition, pp. 1–34 (1999)
18. Nie, P., Geng, W.B.: An introduction to fingerprint recognition technology. Comput. Knowl. Technol. (Acad. Exch.) **17** (2007)
19. Kaur, R., Singh, E.R.: Image filtering techniques—a review. Int. J. Adv. Res. Sci. Eng. **6**(08), 2066–2071 (2017)
20. Patel, N., Shah, A., Mistry, M., Dangarwala, K.: A study of digital image filtering techniques in spatial image processing. In: Proceedings of the 2014 International Conference on Convergence of Technology (I2CT), pp. 1–6 (2014)
21. Rani, K.S., Rao, D.N.: A comparative study of various noise removal techniques using filters. J. Eng. Technol. **7**(2), 47–52 (2018)
22. Vincent, O.R., Folorunso, O.: A descriptive algorithm for Sobel image edge detection. In: Proceedings of Informing Science & IT Education Conference (InSITE), June 2009, vol. 40, pp. 97–107. Informing Science Institute, California
23. Mohammad, E., JawadKadhim, M., Hamad, W.I., Helyel, S.Y., Alrsaak, A.A.A., Al-Kazraji, F.K.S., Hadeeabud, A.M.: Study Sobel edge detection effect on the ImageEdges using MATLAB. Int. J. Innov. Res. Sci. Eng. Technol. **3**(3), 10408–10415 (2014)
24. Dunn, D., Higgins, W.E.: Optimal Gabor filters for texture segmentation. IEEE Trans. Image Process. **4**(7), 947–964 (1995)
25. Patel, M.B., Parikh, S.M., Patel, A.R.: Performance improvement in preprocessing phase of fingerprint recognition. In: Information and Communication Technology for Intelligent Systems, pp. 521–530. Springer, Singapore (2019)
26. Kawagoe, M., Tojo, A.: Fingerprint pattern classification. Pattern Recogn. **17**(3), 295–303 (1984)
27. O'Gorman, L., Nickerson, J.V.: An approach to fingerprint filter design. Pattern Recogn. **22**(1), 29–38 (1989)
28. Patel, M.B., Parikh, S.M., Patel, A.R.: An approach for scaling up performance of fingerprint recognition. Int. J. Comput. Sci. Eng. **7**(5), 457–461 (2019)

Chapter 34
Framework for Spam Detection Using Multi-objective Optimization Algorithm

M. Deepika and Nagaratna P. Hegde

Abstract Electronic spam is a big problem for huge international corporations like AOL, Google, Yahoo and Microsoft. Spam creates many issues and may cause economic damage. Spam causes many problems. Spam creates issues in traffic and bottlenecks, reducing memory, processing capacity and distance. We can describe classification output optimization in computer vision as a multi-objective optimization problem. This paper presents an evolutionary multi-objective optimization algorithm (E-MOA) of the anti-spam filtering issue that discusses both e-mail classification requirements (FP and FN error rates) and e-mail classification times (minimisation). Test findings using a freely accessible model corpus have enabled us to draw significant aspects concerning both the effectiveness of rule-based classification filtering and the appropriateness of a spam filtering two-way classification system. This dataset also uses the DNN classification method, and this classification involves cross-validation. Finally, the e-mail spam classifier is defined based on error rate, accuracy and recall.

34.1 Introduction

In the everyday existence of people all over the world, the usage of Internet mailing systems has become indispensable. In addition, integrating e-mail with the new constantly linked mobile smart phones allows an easy yet effective way to remain in contact with certain individuals and to share documents effectively at all times. In effect, all instant messaging (IM) programs and e-mail are widely used. The basic distinction between common IM (including WhatsApp or GTalk) applications and the Internet mailing systems is, therefore, that they can only use consent control

M. Deepika (✉)
Research Scholar, Osmania University, Hyderabad, India

Assistant Professor, Neil Gogte Institute of Technology, Hyderabad, Telangana, India

N. P. Hegde
Professor, Department of CSE, Vasavi College of Engineering, Hyderabad, Telangana, India

© The Author(s), under exclusive license to Springer Nature Singapore Pte Ltd. 2021   345
S. C. Satapathy et al. (eds.), *Smart Computing Techniques and Applications*,
Smart Innovation, Systems and Technologies 225,
https://doi.org/10.1007/978-981-16-0878-0_34

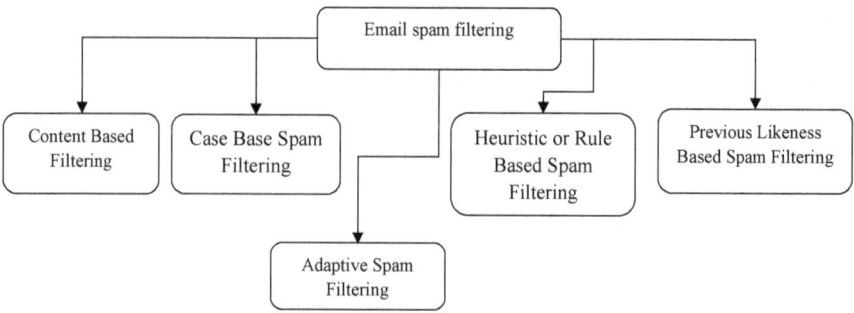

Fig. 34.1 E-mail spam filtering methods in existence

approaches within former applications. This made it possible to use e-mails as an assertive/huge means of advertisement and a virus delivery tool that causes spam. The latest figures [1, 2] suggest 40% of all communications spam, about 15.4 million communications a day, costing Internet users approximately 355 million dollars a year. In recent years, the filtering system [3] of spam assassin has been common employed in Web to reduce the spam. While there are many methods of e-mail spam filtering, state-of-the-art implementations are explored in this paper as seen in Fig. 34.1. Below are the various spam filtering strategies commonly used to solve the spam problem. This has established classification accuracy that could be optimized to test the advantages of using the evolutionary multi-objective optimization algorithm (E-MOA) to plan the multifunctional anti-spam filter scheduling issue.

Spam is no longer waste but is danger as it involves virus applications and spyware agents which destroy the recipient's network, which is why spam detection is developing. This has suggested much spam identification approaches based on machine learning algorithms. Through the usage of bulk mailing devices, the volume of spam has significantly risen, spam identification strategies will cope with it. For the identification of spam, it suggests security configurations and selection of features to minimize overhead operation, with high classification rates. However, subjective value and maximum number of features have not been taken into consideration in prior methods, so there are no strategies to using them together to date. In this article, we suggest an optimized random forest (RF) spam detection model to refine and select features. To improve detection speeds, we refine two RF parameters. We have the variable value of each function such that unnecessary features may be quickly omitted. In addition, we plan to use two approaches for optimizing the chosen features: (i) only optimizing one parameter in the overall feature range, and (ii) optimizing parameters in each removal step. We perform spambase dataset tests and explain the viability of our method.

34.2   Background Work

In [4], the authors presented stream clustering techniques for handling spam and in many circumstances categorizing inputs/tweets in spam and non-spam clusters. Such methods suggest that the cluster contains several neighboring (micro) clusters with an growing symmetrical micro-cluster distribution. However, this assumption is not necessarily valid, and small micro-clusters can be spread asymmetrically. To enhance the accuracy of previous online approaches, they recommend substituting the distance with a group of classifiers. These INBs should get the mean and limit of the micro-clusters, unreliable to the asymmetrical small micro-cluster. In this paper, DenStream promoted the alternative application INB-DenStream. The INB-DenStream data collection was productively applied by province-of-the-art approaches, including DenStream StreamKM++ and CluStream in terms of precise recall, F1-score and computational complexity. The quantitative outcomes revealed that their system is equal to adversaries in virtually all datasets.

In [5], the authors examined the popularity of the social media, but some attackers attracted their exponential growth. Many researchers have used spammers' behavior to reduce this possibility and have got promising outcomes by utilizing machine intelligence procedures to detect spammers on social networks. Many of them analyze disregard class imbalances of outcomes from the modern world. They introduce a complex stacking-based set learning framework in this article to improve the impact of class imbalances on the detection of spam on social networks. In the hybrid system, we incorporate cost-sensitive research into deep neural processing. They will combine the ingredients with the classification results for obtaining misclassification for prediction outcomes of the basic classifiers in complex terms. The experimental results show that the efficiency of spam detection for imbalanced datasets is growing inside our framework.

In [6], Twitter authors provide a range of special features; this cannot use directly traditional methods to Twitter spam research. The paper recommends a Twitter-only spam recognition program called TwitterSpamDetector. TwitterSpamDetector uses Twitter-specific features to recognize Twitter spam. The accuracy and response of TwitterSpamDetector are measured at 0.943 and 0.913, respectively, according to measurement observations. In [7], the authors spam e-mails are often a concern. It is, therefore, still very necessary to construct algorithms to solve this issue. This paper discusses two different algorithms for spam prevention. The first Bayesian-based algorithm, though, is improved if the Bayesian filter will not decide the data compression algorithm. The second approach uses the classification algorithm for the particle swarm optimization model. The results of the mentioned algorithms are promising.

34.3 Proposed Model

The overall design of the proposed system is given in Fig. 34.2, and each of these components is addressed in the following sections briefly.

Dataset: A dataset called the spambase UCI is accessible to test the proposed process. It comprises a variety of sections, including spam and non-spam. This dataset comprises 4601 instances and 57 attributes. The last column of the table shows whether the e-mail is a spam (1) (0). Many characteristics show that the e-mail is also used for a specific word or character. The runlength (55–57) attributes calculate the series length of successive capital letters.

Pre-processing: Most of the data in the physical world include composite, noisy and incomplete values. While it focuses the quality determination on the standard

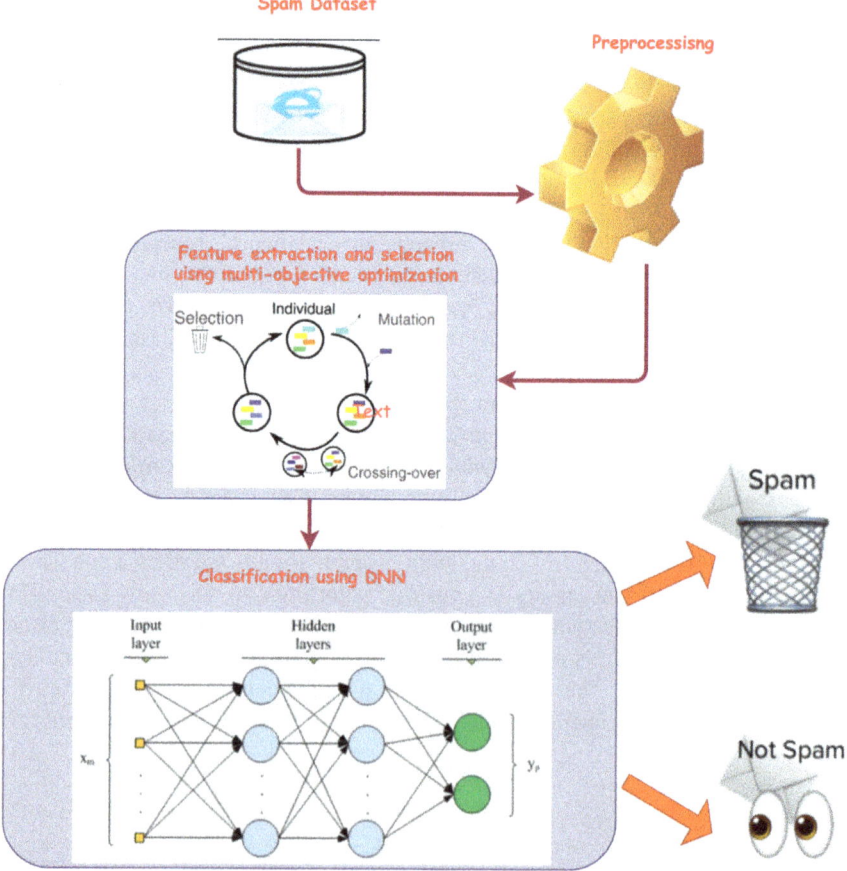

Fig. 34.2 Our proposed methodology for spambase UCI dataset classification

of mining, preprocessing is a very necessary activity well before it carries mining operation out. Data maintenance, data creation, data replication and data reduction are essential activities in the preprocessing of data. It achieves data normalization in this dataset until it performs the analysis.

Feature selection: Consider the dataset $\mathbb{Z}$ of dimension $m \times n$, where m is the number of features and n is the number of data points. Then, through point $z_{i,}$ marks the jth data point's ith function. Also let vector $\mathbb{L}$ denote the label of increasing data point of $\mathbb{Z}$, so that l_j denotes the jth data point. Here various attributes with the specific patterns should be noticed. Therefore, features within a set of $[1, 10]$ are normalized as follows:

$$x_{i,j} = 1 + 9\left(\frac{Z_{i,j} - \min_{1 \le k \le m} Z_{i,k}}{\max_{1 \le k \le m} Z_{i,k} - \min_{1 \le k \le m} Z_{i,k}}\right) \tag{34.1}$$

where $\mathbb{X}$ is a variable representing the intra-class distance from the practical point of view. This can find that there is a need to back up the intra-class distance from the intrinsic function to add the variable weight vector W at the end rather than identifying and w be the feature selection or weighting factor for the ith feature. This defines the difference between groups in the same way.

$$w_i = \begin{cases} 0, & \text{if the feature is rejected} \\ [1, \mathbb{A}], & \text{if the feature is selected} \end{cases} \tag{34.2}$$

$$Y_j = W \odot X_j \tag{34.3}$$

$$d(Y_p, Y_q) = \sum_{i=1}^{m} |y_{i,p} - y_{i,q}|$$

$$= \sum_{i=1}^{m} |w_i x_{i,p} - w_i x_{i,q}| = \sum_{i=1}^{m} w_i |x_{i,p} - x_{i,q}|$$

Thus, $d(Y_p, Y_q) = W^T |X_p - X_q|$.

By Eq. (34.4), the total intra-class distance can be given as

$$D_{\text{intra}} = \sum_{p=1}^{n} \sum_{q=p+1 \, \forall l_p = l_q}^{n} d(Y_p, Y_q)$$

$$= W^T \sum_{p=1}^{n} \sum_{q=p+1 \, \forall l_p = l_q}^{n} |X_p - X_q| \tag{34.4}$$

Thus,

$$D_{\text{intra}} = W^T \Delta^{\text{intra}} \tag{34.5}$$

where Δ^{intra} is a variable representing the intra-class distance from the practical point of view. This can find it that there is a need to back up the intra-class distance from the intrinsic function to add the variable weight vector at the end rather than identifying and summing up the pair-way group from the data points. We may define the difference between groups in the same way.

$$D_{\text{inter}} = \sum_{p=1}^{n} \sum_{q=p+1 \forall l_p \neq l_q}^{n} d(Y_p, Y_q)$$

$$= W^T \sum_{p=1}^{n} \sum_{q=p+1 \forall l_p \neq l_q}^{n} |X_p - X_q|$$

$$D_{\text{inter}} = W^T \Delta^{\text{inter}} \tag{34.6}$$

To improve the affinity against the collection of applications, both intra-class/inter-class distances may apply to a penalty element. Under the definition of penalty features, the larger the amount of roles identified, the greater the degree to which it disciplines the target function. The target functions to be refined are then.

$$F_1(W) = D_{\text{intra}} + \lambda_1 [\min\{1, W\}]^T 1,$$
$$F_2(W) = -D_{\text{inter}} + \lambda_2 [\min\{1, W\}]^T 1, \tag{34.7}$$

where λ_1 and λ_2 are penalty coefficients, with the help of F_1 and F_2 minimization to attain the optimal feature selection and weight vector.

$$\text{Minimize } F(w) = [F_1(w), F_2(w)]^T$$
$$\text{subjected to } W = [w_1, w_2, \dots, w_m]^T \in 0 \cup [1, \mathbb{A}] \tag{34.8}$$

Thus, Eq. (34.8) represents a multi-objective optimization problem (MOP).

Solving Multi-objective Optimization Problem: The optimum performance leads to a particular relation between candidate solutions, overcome by the Pareto-optimal principle where the best combination of such goal Pareto-optimal is obtained [8–11]. Then one solution is selected from the optimum front of Pareto, based on the problem.

In the proposed process, N function selection vectors (W) are initialized at the random level inside $[0, A]$. Then, MOEA/D requires a replication phase to slowly evolve weight vector population (W) toward optima. This process requires two phases in our system, namely mutation and crossover. It conducts both operations according to DE/rand/bin/1, an efficient single goal optimization algorithm, with variants of differential evolution (DE). After each reproduction process of MOEA/D, it performs a repair process over the freshly produced offspring to fulfill the optimization problem constraints. Remember that $W = [w_1, w_2, \dots, w_m]^T$ is the reproduction variable. Then the ith vector of W' dimension can be determined as follows.

$$w_i' = \begin{cases} A, & \text{if } w_i > A \\ o, & \text{if } w_i < 1, \\ w_i, & \text{otherwise} \end{cases} \qquad (34.9)$$

$\forall i \in [1, m]$. When the value for w_i is set to $[0, 1)$, which is far less than its highest value A, the weight for the ith attribute is very small and can be called void. It achieves this in the second equation situation (34.9), raising the likelihood of the number of choices to decrease, thus by achieving a limit of μ^k, $\forall k$ provides best solution in case of Pareto. For obtaining the optimum N Pareto solutions given by MOEA/D resembles to a function (W_k), the optimal vector (W^*) is the following.

$$k^* = \max_k \mu^k, \; W^* = W^{k*} \qquad (34.10)$$

To get the optimal feature subset, $\mathbb{W}^*$ and its related weights can be used to apply for classifier to detect the spam and non-spam.

Classification: In this paper, it is introduced with significant changes (DNN) which acts as classifier for class imbalance tasks. It is capable of learning their complex network architectures to retrieve higher-level functions [12–14]. They compose the DNN model of an input layer, hidden layer (s) and an output layer than seen in Fig. 34.3.

The input layer recognizes information from beyond the network. In the secret layer (s), multilevel input features are extracted linearly to partition various data forms. Hidden layer $h(h\in\{1, …, H\})$ comprises a set of parameters $\theta_h = \{W_h, b_h\}$

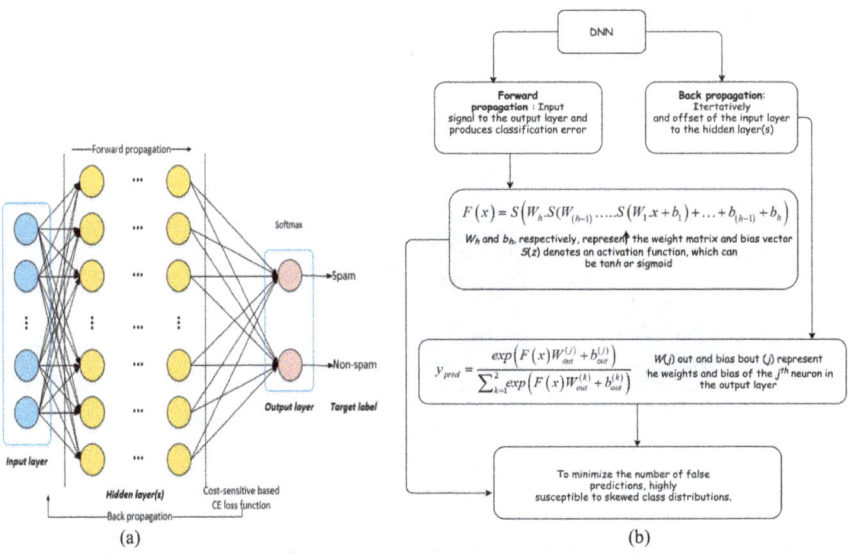

Fig. 34.3 **a** Overview of DNN model for classification, **b** flow process DNN model

in which W_h is a related weight matrix and b_h is a vector of prejudice. This method comprises two stages: forward propagation (FP) and back propagation (BP). The forward propagation sends the input signal to the output layer and induces an error in classification. This balances the hidden layer (s) of the input side, provided that DNN is completely connected with H hidden layers, as seen in Fig. 34.3, for an input function vector x, a complex feature transformation mechanism is represented in H hidden layers of the DNN by computation.

The DNN model for classification as shown in Fig. 34.3a can be explained with the help of flow process. Figure 34.3b consists of forward propagation (FP) and back propagation (BP).

34.4   Results and Discussion

In this section, DNN classification is used to characterize the test data using the chosen and weighted attribute vectors. Therefore, to show the functionality of the proposed algorithm, we checked and plotted our algorithm on a spambase dataset called MATLAB 2017a with number of features used in each case for obtaining accuracy listed in Table 34.1 (Fig. 34.4).

Table 34.2 and Fig. 34.5 show that the DNN classifier using E-MOA outperforms compared to other algorithms for feature selection because it can significantly

Table 34.1 Comparison of different feature selection (FS) algorithms

Dataset	Avg. classification accuracy in %	
	Fuzzy rule-based FS	Proposed
Spambase	93.24	98.12

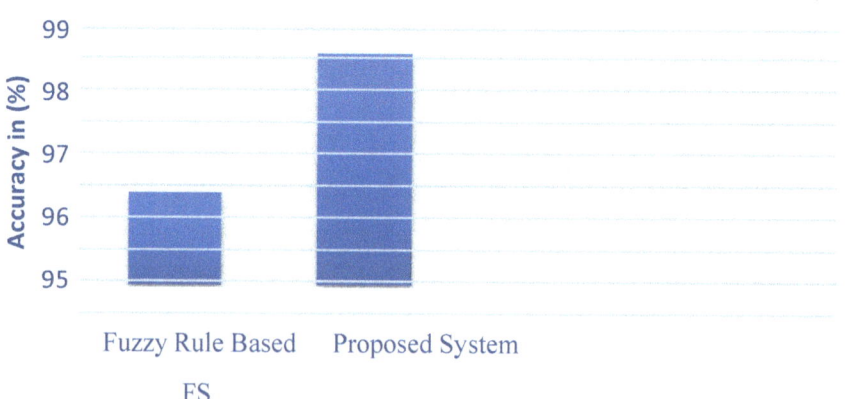

Fig. 34.4 Comparison of different feature selection algorithms with classification accuracy

Table 34.2 Results of classification approaches

Classification algorithm	Error rate before feature selection	Error rate before filtering		
		E-MOA	Genetic algorithm	Runs filtering
Naïve Bayes continuous	0.1136	0.1265	0.1328	0.1488
ID3	0.0896	0.0756	0.0789	0.0847
SVM	0.0375	0.0232	0.0256	0.0325
Our proposed DNN	0.0121	0.0079	0.0089	0.0093

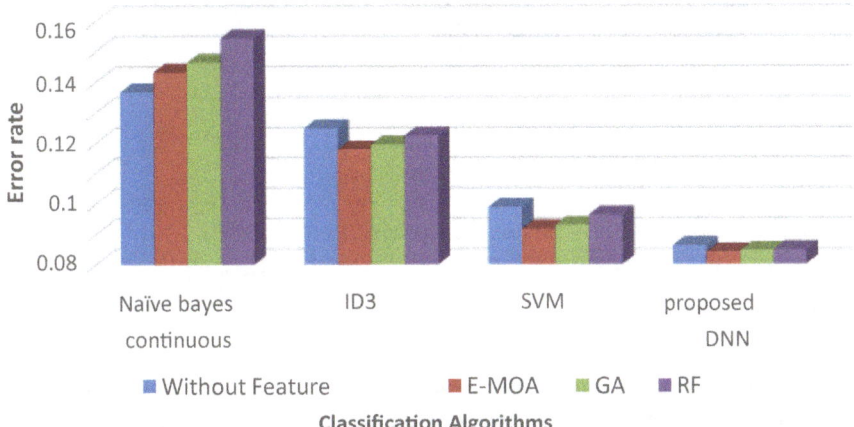

Fig. 34.5 Comparison of different feature selection algorithms with classification error rates

reduce dimensionality reduction of features in the E-MOA, contributing to less DNN classifier training information (Table 34.3).

The terms, positive and negative, are the estimation of the classifier, and the terms, true and false, apply to the prediction of the classifier. The DNN classification accuracy and recall are performed, and we show the results in Fig. 34.6. We regard the DNN classification as the best classifier based on the data, as it obtained 99% accuracy by selecting the E-MOA feature.

Table 34.3 Results of DNN classifier

Values prediction	Accuracy	Recall
Spam	0.9985	0.9905
Not spam	0.9842	0.9924

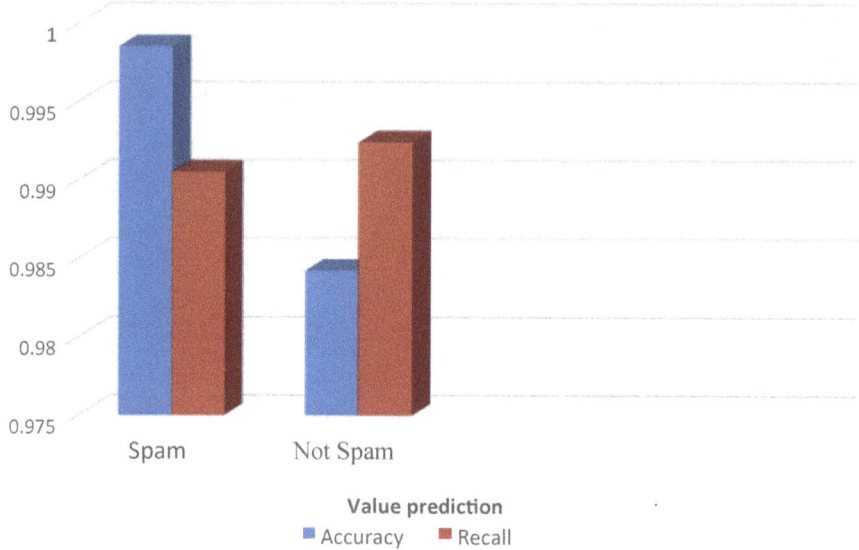

Fig. 34.6 Accuracy and recall of the spam classification algorithm (DNN)

34.5 Conclusion

In this study, the model has tested the usage of evolutionary, multi-target optimization algorithms to improve the consistency of the detection and classification of anti-spam filters (i.e., performance and efficacy of anti-spam filters). In addition, the E-MOA method offered strategies for ideal anti-spam filter settings that address many conditions for use of anti-spam filters. To remove issues such as bandwidth reduction and very poor performance that are suffering from, we may model the suggested algorithm to be applied on the mail server and mail application. The DNN classification algorithm applied on relevant features after E-MOA has resulted more than 99% accuracy, 99.01% recall and 0.0079 of error rate in spam detection.

References

1. Statista: The Statistics Portal: Global spam volume as percentage of total e-mail traffic from January 2014 to September 2016, by month (2016). https://www.statista.com/statistics/420391/spam-email-traffic-share/. Accessed 14 Feb 2017
2. Digital Marketing Ramblings: 73 incredible e-mail statistics (2016). https://expandedramblings.com/index.php/email-statistics/. Accessed 14 Feb 2017
3. The Apache SpamAssassin Group: The first enterprise open-source spam filter (2003). https://spamassassin.apache.org/. Accessed 14 Feb 2017
4. Tajalizadeh, H., Boostani, R., Tajalizadeh, H.: A novel stream clustering framework for spam detection in Twitter. IEEE Trans. Comput. Soc. Syst. 1–10 (2019). https://doi.org/10.1109/TCSS.2019.2910818

5. Zhao, C., Xin, Y., Li, X., Yang, Y., Chen, Y.: A heterogeneous ensemble learning framework for spam detection in social networks with imbalanced data. Appl. Sci. **10**, 936 (2020). https://doi.org/10.3390/app10030936
6. Kabakus, A.T., Kara, R.: "TwitterSpamDetector": a spam detection framework for Twitter. Int. J. Knowl. Syst. Sci. **10**, 1–14 (2019). https://doi.org/10.4018/IJKSS.2019070101
7. Lee, S., Kim, D., Kim, J., Park, J.: Spam detection using feature selection and parameters optimization. In: CISIS 2010—The 4th International Conference on Complex, Intelligent and Software Intensive Systems, pp. 883–888 (2010). https://doi.org/10.1109/CISIS.2010.116
8. Tian, Y., Peng, S., Rodemann, T., Zhang, X., Jin, Y.: Automated selection of evolutionary multi-objective optimization algorithms (2019). https://doi.org/10.1109/SSCI44817.2019.9003018
9. Gong, M., Jiao, L., Yang, D., Ma, W.: Research on evolutionary multi-objective optimization algorithms. J. Softw. **20** (2009). https://doi.org/10.3724/SP.J.1001.2009.03483
10. Saborido Infantes, R., Ruiz, A.B., Bermúdez, J., Vercher, E., Luque, M.: Evolutionary multi-objective optimization algorithms for fuzzy portfolio selection. Appl. Soft Comput. **39** (2015). https://doi.org/10.1016/j.asoc.2015.11.005
11. Cabrera, G., Vasconcellos, C., Soto, R., León, J.M., Paredes, F., Crawford, B.: An evolutionary multi-objective optimization algorithm for portfolio selection problem. Int. J. Phys. Sci. **6**, 5317–5328 (2011)
12. Ngamsuriyaroj, S., Taninpong, P.: Tree-based text stream clustering with application to spam mail classification. Int. J. Data Min. Model. Manag. **10**, 353 (2018). https://doi.org/10.1504/IJDMMM.2018.10015879
13. Sel, S., Hanbay, D.: E-mail classification using natural language processing, pp. 1–4 (2019). https://doi.org/10.1109/SIU.2019.8806593
14. Surwade, A.U., Patil, M., Kolhe, S.: An empirical analysis of spam E-mail classification using machine learning techniques. Int. J. Appl. Eng. Res. **9**, 5057–5074 (2014)

Chapter 35
Time Series Analysis of Assam Rainfall Using SARIMA and ARIMA

Utpal Barman, Asif Ekbal Hussain, Mridul Jyoti Dahal, Puja Barman, and Mehnaz Hazarika

Abstract Time series analysis of rainfall is very much essential for farming. Agriculture productivity is depended on rainfall. It is important to predict the future rainfall from farmers' point of view. In this paper, we apply seasonal auto-regressive integrated moving average (SARIMA) and auto-regressive integrated moving average (ARIMA) techniques for the monthly time series analysis of rainfall in Assam. The rainfall data contains the monthly rainfall of Assam from 1901 to 2017. Here, different components of the rainfall are visualized before apply the SARIMA and ARIMA. The handling procedures of seasonal components (p, d, and q) are reported using moving average, and augmented ducky fuller test. The ACF and PACF are used to find the seasonal components of the SARIMA and ARIMA. The SARIMA model is selected as the best model as compared to ARIMA based on AIC, BIC, HQIC, regression score (RC), mean absolute error (MAE), median absolute error (MeAE), mean squared error (MSE), mean squared log error (MSLE), and root mean square error (RMSE) of the analysis. The final results of the two methods are validated with the actual rainfall of Assam during the period.

35.1 Introduction

Rainfall is one major problem in Assam, and it indirectly affects the crop productivity in Assam. Every year heavy rainfall creates artificial flood in Assam. Due to that artificial flood, crop productivity in Assam is less as compared to other states of Assam. Agriculture productivity in India is primarily dependent on the rainfall. Nominal rainfall is always good for high agriculture productivity. But heavy rainfall is too dangerous for agriculture.

Researches are already done to predict the rainfall in India. India Meteorological Department always provides healthy prediction of rainfall in India but the climatic condition of Assam and Meghalaya affects the prediction of rainfall in Assam. Time

U. Barman (✉) · A. E. Hussain · M. J. Dahal · P. Barman · M. Hazarika
Department of Computer Science and Engineering, Girijananda Chowdhury Institute of Management and Technology, Guwahati, Assam, India

© The Author(s), under exclusive license to Springer Nature Singapore Pte Ltd. 2021   357
S. C. Satapathy et al. (eds.), *Smart Computing Techniques and Applications*,
Smart Innovation, Systems and Technologies 225,
https://doi.org/10.1007/978-981-16-0878-0_35

series analysis is already done by many researchers in the academic institute and research institute. In the last decade, researchers have done the forecasting of rainfall using different statistical and machine learning methods such as seasonal auto-regressive integrated moving average (SARIMA) [1–5] and auto-regressive integrated moving average (ARIMA) [6–11], auto-regressive moving average (ARMA) [12], and emerging fuzzy time series (FST) [9]. Among all the methods, SARIMA is one of the most popular techniques used by many researchers to forecast the rainfall time series. The paper [5] used the SARIMA model for the rainfall time series analysis of Shouguang city of China. The study [6] recorded ARIMA model for the rainfall analysis of Allahabad region of India. The paper [1] used SARIMA model for the time series analysis of Brong Ahafo Region of Ghana. Again, ARIMA and FST model are recorded [9] for the rainfall time series analysis of Nigeria. The Indian author [10] used ARIMA model for the rainfall time series analysis of Varanasi. The authors [12] used ARMA model for the forecasting of rainfall data in Colorado, USA.

From the above examples, it is found that the statistical methods are often used by many researchers in their research. Apart from these methods, time series forecasting is also done using different machines learning methods such as recurrent neural network (RNN) [13], support vector regression (SVR) [13–15], and artificial neural network (ANN) [16]. Among all the statistical and artificial intelligence-based methods, SARIMA is widely used in rainfall forecasting. It is because of the seasonality effect handling parameter of SARIMA. The SARIMA model is dependent only on time but not on the other climatic conditions such as temperature, pressure, and humidity. Since the intention is to forecast the Assam rainfall with time, the SARIMA model is used in this paper, and the result is compared with ARIMA.

35.2 Material and Methods

35.2.1 About Dataset

The dataset of the paper is the meteorological data. The dataset contains the monthly rainfall data from 1901 to 2017. The visualization of the rainfall data is presented in Fig. 35.1.

35.2.2 Preprocessing and Components of Rainfall Data

The ARIMA and SARIMA time series analysis is possible when the data has no trend, seasonality, and irregularity. But the trend and seasonality are the prime factors to be considered before applying the time series analysis. In this paper, we build our model based on the three components of the model. Let $y(t)$ be the analysis output of the model, then $y(t)$ can be express as below

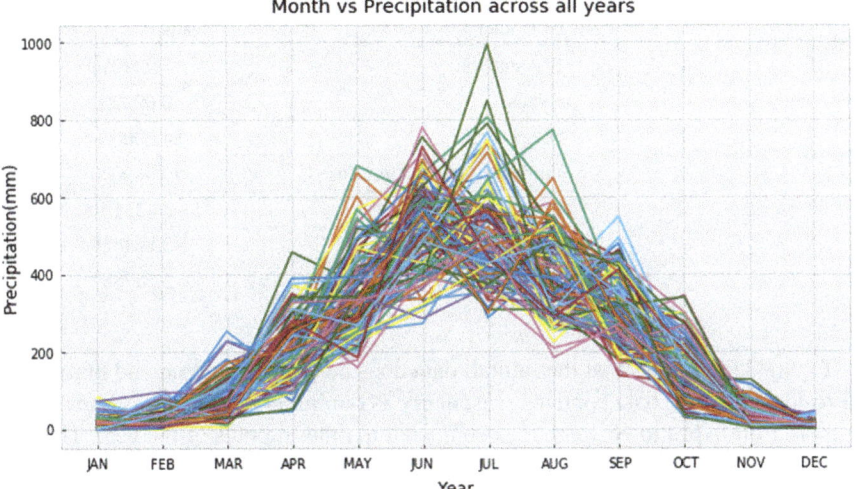

Fig. 35.1 Data plot of precipitation versus months

$$Y(t) = F(\text{Trend, Seasonality, Irregularity}) \tag{35.1}$$

To ensure the trend and seasonality, the rainfall time series data is divided into its components, and it is presented in Fig. 35.2.

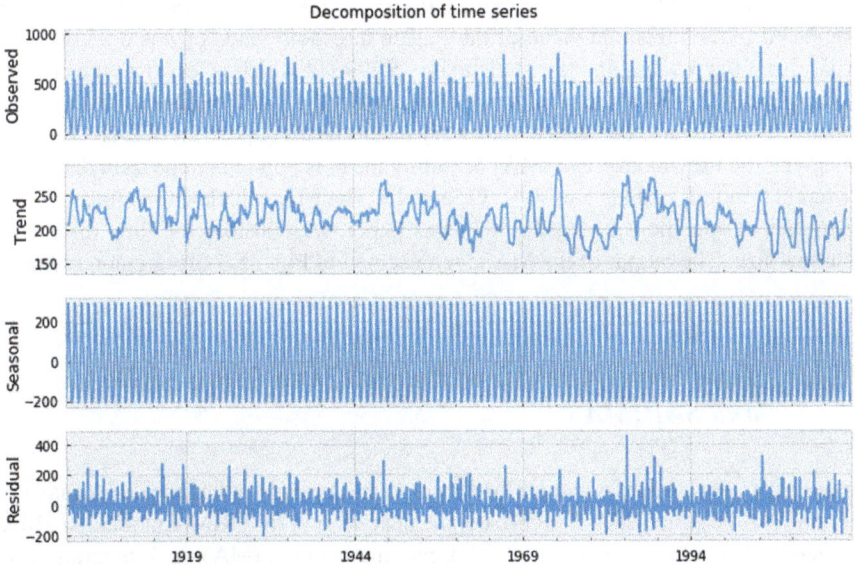

Fig. 35.2 Trend, seasonality, and residual of the rainfall data

Table 35.1 Result of ADF test	Parameter	Value
	Test statistic	−5.299084
	p-value	0.000005
	#Lags used	24.00000
	Number of observations used	1379.000000
	Critical value (1%)	−3.435101
	Critical value (5%)	−2.863638
	Critical value (10%)	−2.567887

Figure 35.2 defines that the rainfall data does not have any trend and high value of residual. But the data contains seasonality as rainfall is seasonal in Assam, and it is more from April to November as compared to other months of the year. There is no meaning of apply time series analysis on the seasonal data. To valid the points, we have performed augmented Dickey–Fuller (ADF) test to check the stationary of the rainfall time series data. The ADF test is a statistical tool which is used to find the stationarity of the time series data. It calculates the unit root of the data based on which the stationarity is calculated. In this paper, two hypotheses are considered for ADF test. One is null, and another is alternate hypothesis. The null hypothesis is that the time series rainfall data has a unit root. The alternate hypothesis is that the rainfall series data has not any unit root. The rejection and acceptance of the null hypothesis are depended on the p-value of the ADF test. The ADF test is applied in python environment. The outcome of the ADF test is presented in Table 35.1.

In this paper, the p-value of the ADF test is 0.000005 which is very less as compared to the threshold value of the P. The threshold value of p is 0.05 or 0.01. Apart from the p-value, the test statistic is −5.299084 which is less than the critical value of the ADF test at 1%. These two points reject the null hypothesis and consider the data as stationary. The data is stationary means it is independent of time. To valid the point, the moving average (MA) or rolling mean is applied on the train data with different window size such as 4, 8, 12, and 16. The MA calculates the mean of the train data at specific time instant t. In this paper, we consider the time instant as window size. The output of the MA is represented in Fig. 35.3 using graph plots.

35.2.3 Time Series Analysis of Rainfall Data Using ARIMA and SARIMA

In this paper, the SARIMA and ARIMA models are applied for the time series analysis of rainfall time series data. SARIMA is the next version of ARIMA model. ARIMA model consists of auto-regression (AR), moving average (MA), and integration of AR and MA. The ARIMA model is denoted using three non-seasonal parameters such as p, d, and q. The parameter p is the order or the lag value of the AR model,

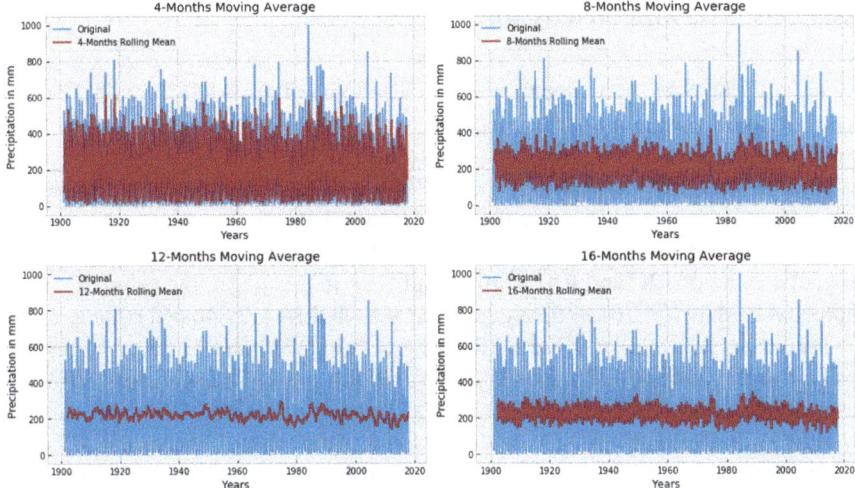

Fig. 35.3 Moving average with different window

whereas d and q are the degrees of integration and MA model. The SARIMA consists of two elements such as trend and seasonal. The trend component is defined by p, d, and q where p is the autoregressive order, q is the moving average order and d is the order of difference. The seasonal components consist of P, D, Q and m where P is the order of seasonal autoregressive, Q is the order of seasonal moving average order, D is the order of difference and m is the single period step. In this paper, the value of m is considered as 12. For the best value p, d, q, P, D, and Q, the different combinations of p, d, and q within the range 0–2 are calculated. These combinations are passed to the SARIMA as a parameter and the best combination is found based on the value of the AIC, BIC, and HQIC. The different combinations of trend and seasonal parameters are presented in Table 35.2.

Among all the combinations, the best combination is $p = 1$, $d = 0$, $q = 1$, $P = 0$, $D = 1$, $Q = 1$ and $m = 12$ with AIC $= 12{,}846.66057$.

Table 35.2 Trend and seasonal component of SARIMA

Trend combination	Seasonal combination
0, 0, 0	0, 0, 0, 12
0, 0, 1	0, 0, 1, 12
0, 1, 0	0, 1, 0, 12
0, 1, 1	0, 1, 1, 12
1, 0, 0	1, 0, 0, 12
1, 1, 0	1, 1, 0, 12
1, 1, 1	1, 1, 1, 12

35.3 Results and Discussion

Along with the SARIMA model, we have also applied the ARIMA model on the Assam rainfall time series data. Since the initial rainfall data shows that the time series doesn't have any seasonal effect, so the ARIMA model is applied on the time series data. The summary of the ARIMA and SARIMA model are shown in Table 35.3. The paper [11] shows that the AIC and BIC are the efficient parameters to select the best SARIMA and ARIMA model for time series. A model fits well when the value of AIC and BIC is minimum [11]. In this paper, the value of AIC is more in ARIMA (177,761.970) as compared to SARIMA (12,846.661) (Table 35.3).

The values BIC and HQIC of SARIMA are 12,866.731 and 12,854.248, respectively. But the values of BIC and HQIC for ARIMA are 17,782.959 and 17,769.81, respectively. Along with AIC, BIC, and HQIC, values of SARIMA have proved that SARIMA is better than ARIMA for the seasonal rainfall data. It means that ARIMA model is not an efficient techniques to predict the future rainfall in Assam. The SARIMA model shows the better result as compared to the ARIMA model due to the seasonality effect of the rainfall time series data. Along with AIC, BIC, and HQIC, the two models are evaluated using the regression score (RC), mean absolute error (MAE), median absolute error (MeAE), mean squared error (MSE), mean squared log error (MSLE), and root mean square error (RMSE). The evaluating parameters of the models are shown in Table 35.4.

In the previous study, the authors used coefficients of regression ($R2 = 0.86$) [11], RMSE (18.60 for 1 days forecast) [14], MeAE (0.3426) [15], and MAE (15.26) [7] for evaluating the different time series analysis model. Here, both the models are compared based on the mentioned evaluating parameters. The RMSE value of SARIMA is 84.40 whereas the RMSE of ARIMA is 135.71. The RC, MAE, MeAE, MSE, and MSLE of SARIMA are 0.79, 55.86, 30.42, 7124.05, and 0.4171, respectively, whereas the RC, MAE, MeAE, MSE, and MSLE of ARIMA is 0.48884, 101.57, 69.57, 18,417.62, and 2.0587, respectively. The evaluating parameters clarify that the SARIMA is much more efficient as compared to ARIMA.

Table 35.3 SARIMA and ARIMA model summary

Model	Train sample	AIC	BIC	HQIC
SARIMA	1901:1994	12,846.661	12,866.731	12,854.248
ARIMA	1901:1994	17,761.970	17,782.959	17,769.81

Table 35.4 Evaluating parameters of SARIMA and ARIMA

Model	RC	MAE	MeAE	MSE	MSLE	RMSE
SARIMA	0.79	55.86	30.42	7124.05	0.4171	84.40
ARIMA	0.4884	101.57	69.57	18,417.64	2.0587	135.71

35.4 Conclusion

This paper is to study the rainfall forecast in Assam State only, and the SARIMA model is found as the best suitable method to carry out the research. In this paper, both the ARIMA and SARIMA models are applied for the rainfall prediction in Assam. The SARIMA model is found to be accurate and efficient as compared to ARIMA. SARIMA model is helpful as the dataset contains the seasonal effect in the data. The SARIMA is helpful to forecast the rainfall in future. Time is the key parameter to study about the rainfall forecast, and it can be used to predict the rainfall in long-range forecasting using SARIMA. In future, the Assam rainfall can be analyzed with help of SVR, ANN, and KNN regression, and the result can be compared with the ARIMA and SARIMA.

Acknowledgements This research project is supported by Assam Science and Technology University, Guwahati, Assam under TEQIP-III, vide Ref. No.: ASTU/TEQIP-III/Collaborative Research/2019/2474, Dated July 17, 2019.

References

1. Afrifa-Yamoah, E., Saeed, B.I., Karim, A.: Sarima modelling and forecasting of monthly rainfall in the Brong Ahafo Region of Ghana. World Environ. **6**(1), 1–9 (2016)
2. Dabral, P.P., Murry, M.Z.: Modelling and forecasting of rainfall time series using SARIMA. Environ. Process. **4**(2), 399–419 (2017)
3. Eni, D., Adeyeye, F.J.: Seasonal ARIMA modeling and forecasting of rainfall in Warri Town, Nigeria. J. Geosci. Environ. Prot. **3**(06), 91 (2015)
4. Subbaiah Naidu, K.: SARIMA modeling and forecasting of seasonal rainfall patterns in India. Int. J. Math. Trends Technol. (IJMTT) **38**(1), 15–22 (2016)
5. Wang, S., Feng, J., Liu, G.: Application of seasonal time series model in the precipitation forecast. Math. Comput. Model. **58**(3–4), 677–683 (2013)
6. Graham, A., Mishra, E.P.: Time series analysis model to forecast rainfall for Allahabad region. J. Pharmacogn. Phytochem. **6**(5), 1418–1421 (2017)
7. Mohamed, T.M.: Time series analysis of Nyala rainfall using ARIMA method (2016)
8. Murat, M., Malinowska, I., Gos, M., Krzyszczak, J.: Forecasting daily meteorological time series using ARIMA and regression models. Int. Agrophys. **32**(2), 253–264 (2018)
9. Olatayo, T.O., Taiwo, A.I.: Statistical modelling and prediction of rainfall time series data. Glob. J. Comput. Sci. Technol. (2014)
10. Shivhare, N., Rahul, A.K., Dwivedi, S.B., Dikshit, P.K.S.: ARIMA based daily weather forecasting tool: a case study for Varanasi. Mausam **70**(1), 133–140 (2019)
11. Swain, S., Nandi, S., Patel, P.: Development of an ARIMA model for monthly rainfall forecasting over Khordha district, Odisha, India. In: Recent Findings in Intelligent Computing Techniques, pp. 325–331. Springer (2018)
12. Burlando, P., Rosso, R., Cadavid, L.G., Salas, J.D.: Forecasting of short-term rainfall using ARMA models. J. Hydrol. **144**(1–4), 193–211 (1993)
13. Hong, W.-C.: Rainfall forecasting by technological machine learning models. Appl. Math. Comput. **200**(1), 41–57 (2008)
14. Hasan, N., Nath, N.C., Rasel, R.I.: A support vector regression model for forecasting rainfall. In: 2015 2nd International Conference on Electrical Information and Communication Technologies (EICT), pp. 554–559 (2015)

15. Zhao, S., Wang, L.: The model of rainfall forecasting by support vector regression based on particle swarm optimization algorithms. In: Life System Modeling and Intelligent Computing, pp. 110–119. Springer (2010)
16. Abhishek, K., Singh, M.P., Ghosh, S., Anand, A.: Weather forecasting model using artificial neural network. Procedia Technol. **4**, 311–318 (2012)

Chapter 36
Reliable Low-Cost High-Speed and Direction-Controlled Surveillance Robot

S. Rooban, I. Venkata Sai Eshwar, R. Radhika, M. Dheeraj Kumar, and K. Chaitanya Krishna

Abstract Reliability is the basic requirement for all the products and even for human beings. All the researchers and manufacturing industry are focusing toward improving the reliability of the products or circuits with minimum cost. This paper presents a speed and direction control of cost-effective robot which can be used for surveillance in places where human beings are unable to enter. Many surveillance robots have been proposed by many researchers in recent years. Always the customer expects single device usage for various applications; this requirement makes the designer to incorporate all the features in a device that leads to increased cost and reduction in speed. One of the methods to improve the speed of the device processing capability with all features is parallel processing that makes the product costlier because high cost of high-speed parallel processor. The proposed method incorporates many features and increases the speed of operation by doing parallel operation with different processors. This proposed method of designing a surveillance robot uses three processors which are Arduino, MyRIO and NodeMCU. All the three processors are of low cost and are easily available. The proposed method is divided into modules, and various features of surveillance robot are allocated into each module with different processors. So that robot can operate in parallel for the applications. Because of the parallel operation, the speed of operation is improved, and if any one processor fails, the robot can work with the limited features whereas in single processor once the processor fails that collapse the robot.

S. Rooban (✉) · I. V. S. Eshwar · M. D. Kumar · K. C. Krishna
Koneru Lakshmaiah Education Foundation, Guntur, Andhra Pradesh 522502, India

R. Radhika
St. Peter's College of Engineering and Technology, Avadi, Tamil Nadu 600054, India

© The Author(s), under exclusive license to Springer Nature Singapore Pte Ltd. 2021   365
S. C. Satapathy et al. (eds.), *Smart Computing Techniques and Applications*,
Smart Innovation, Systems and Technologies 225,
https://doi.org/10.1007/978-981-16-0878-0_36

36.1 Introduction

With increasing popularity of robotics and automation in industry 4.0, the need for reliability is increased and becomes a serious challenge. Surveillance is a necessary requirement in all the industry and public places. With the increasing usage of machine learning [1] and artificial intelligence techniques, the maximum input for those techniques is through sensors. Many applications require continues data from the places where humans cannot be roaming all the time to collect the data. Robot can be used everywhere to sense or collect the information and passed to the system which needs the information. How fast the robot completes the task or responds for the instruction is also one of the major requirements in the industry 4.0. Speed and reliability play a major role in choosing the robot to meet our required application from the various robot manufacturers. Processor which is used for the robot provides more contribution to the speed and reliability. There are many research works in robot such as path planning [2], algorithmic [3] and mathematical [4] are proposed earlier by many researchers to improve the reliability. The proposed method is one of the simple solutions to improve the reliability by keeping the cost of the robotics system as same. The remaining paper is presented as follows. Section 36.2 describes the work related to surveilling robots. Section 36.3 presents the proposed method with various processors. Section 36.4 discusses the results and the paper is concluded in Sect. 36.5.

36.2 Related Work

In the past years, it is observed that improvement in reliability of any system is the basic requirement. Reliability can be improved by improving the processor internally such as the adder [5] also by adopting some architectural modifications [6]. Various processors [7] are analyzed and conclusion arrived that the reliability also depends on performance of processor [8, 9]. Robot performance can be improved by efficient path planning [10] and the hardware components especially the sensors [11]. Many methods are proposed by researchers to improve the reliability by using reliable sensors [12, 13]. Techniques [14] are adopted to improve the reliability of the processor in chip manufacturing and testing phase. Parallel processing can make the processor to perform the task in high speed. In this work, processing of all the features of the robot is performed in various processors to improve the reliability and speed.

36.3 Proposed Methodology

The proposed surveillance robot features are divided into three modules, they are (1) patrolling unit, (2) vision unit and (3) environmental identification unit. As we have discussed earlier the parallel processing is obtained by using three simple cost-effective processors for each module. Patrolling unit uses Arduino, vision unit uses MyRIO, and environmental unit uses NodeMCU.

36.3.1 Patrolling Unit

The block diagram for the patrolling unit is shown in Fig. 36.1. As our work is distinguished from others by low cost, the hardware parts such as DC motor, Motor driver, Arduino, Bluetooth module and mobile are easily available with minimum cost. In surveillance system, patrolling is an important module. This module is responsible for moving the robot structure into any places where humans are hazardous to the environment or the environment is hazardous to humans. The movement is obtained

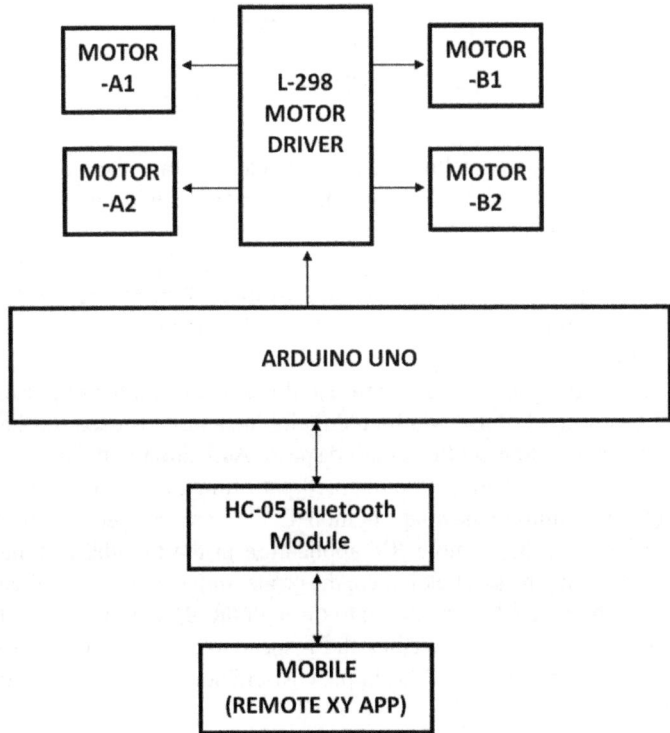

Fig. 36.1 Block diagram for patrolling unit

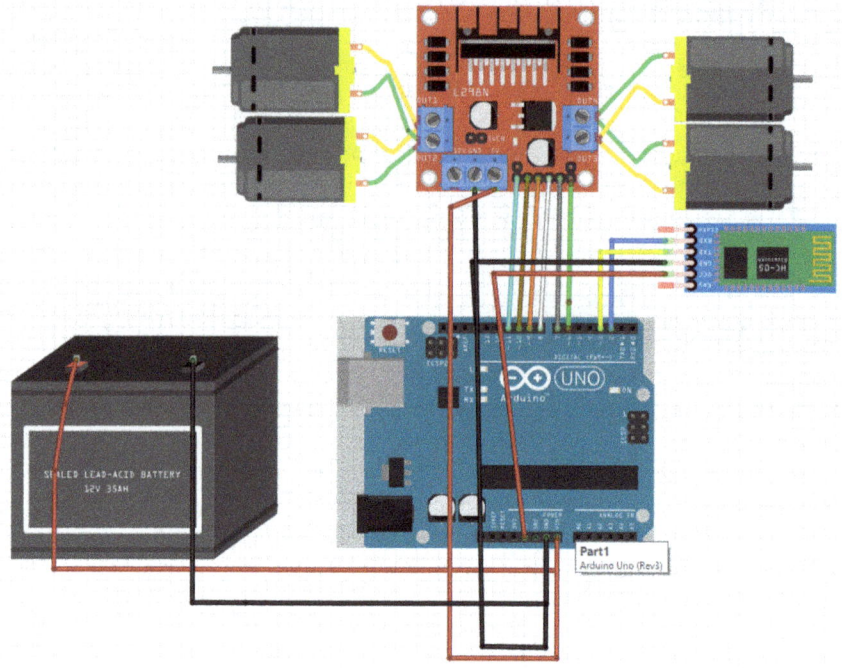

Fig. 36.2 Circuit connection diagram of patrolling unit

by the four DC motors. Controlling of the robot can be done using communication between Bluetooth module and mobile which is installed with the simple Remote XY Arduino application.

Connection diagram of the patrolling unit is shown in Fig. 36.2. The direction of movement for the patrolling depends on the activation of the respective DC motors. For example, if the robot is required to move straight, all the motors have to be activated in forward direction, and if the robot is required to turn left, the two right motors will be activated. Also the speed is controlled through the number of pulses applied to the motor per second. Pulse width modulation technique is used for the number of pulses to be applied to meet the required speed. Activation of the motor based on the number of pulses applied for the required patrolling is written as instruction in any one of the programming language python, C, C++ are dumped into the Arduino.

Figure 36.3 shows the Remote XY application in the mobile. It is having the joystick icon, by using the joystick Icon in the proper movement the required instruction will be communicated to the Arduino through the Bluetooth device. The code takes the signal as input and executes the respective portions of the program to provide the required patrolling. It's a simple cost-efficient method of controlling the surveillance structure accurately.

Fig. 36.3 Remote XY
application

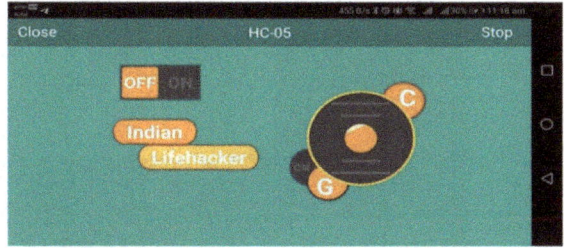

36.3.2 Vision Unit

As our proposed work is focused on dividing the features into modules and run the modules in separate simple processors to make use of good performance with the parallel processing [15]. Vision of a surveillance robot is an important input to the system. To make this module as simple as possible no complexity is added in the vision unit to achieve the lowest cost for the system. MyRIO processor is used with the webcam and the live telecast can be viewed in the laptop. This module can be extended by using Bluetooth to the MyRIO to transfer the video to the systems to record the videos in the laptop by installing the video recording. MyRIO with the LabVIEW can control the Robot independently with better performance by using the various toolbox in the LabVIEW robot toolbox. Figure 36.4 shows the flow diagram, Figs. 36.5 and 36.6 show the simple implementation of the MyRIO with LabVIEW and webcam.

Fig. 36.4 Flow diagram of
vision unit

Fig. 36.5 Simple MyRIO setup

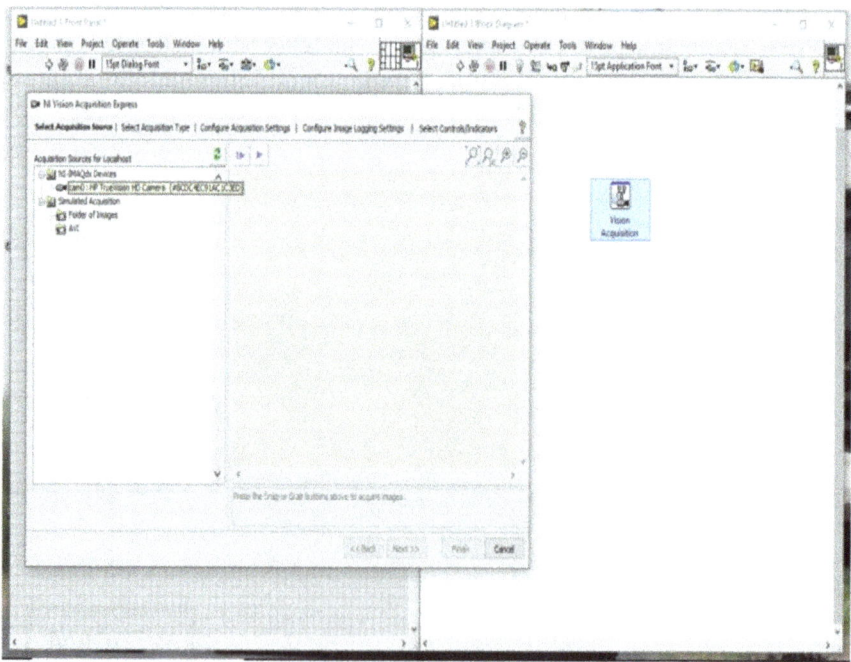

Fig. 36.6 LabVIEW with the vision unit

36.3.3   Environmental Identification Unit

Most important features and the basic requirement of surveillance robot is to collect the environment status through various sensors and communicate the collected information to the processor then the processed information is transferred and delivered in useful form. A simple setup of all the sensors and the processor is shown in Fig. 36.7. NodeMCU is used as processor, DHT11 sensor for humidity and temperature sensing, Mq2 sensor for sensing the Gas [12]. Servo motor is attached to the system to make any mechanical movement by the processor based on the requirement.

Figure 36.8 shows the connection diagram of the Environmental Identification Unit. This module is controlled with the installed blink application in mobile. Bluetooth module communicates the status of various sensors immediately to the mobile through the blynk. This is most useful for keep track of any environmental variations continuously. Figure 36.9 shows the output of the sensors which is used in Fig. 36.7.

36.4   Results

The proposed reliable high-speed and direction-controlled surveillance robot is completed with the three processors with various blocks. Figure 36.10 shows the hardware of the patrolling module in the system. Figure 36.11 shows the output of MyRIO [7] with the LabVIEW. Figure 36.12 shows the final structure of the proposed surveillance system. It is tested with all the possible inputs, and the systems work well. Speed of the robot is checked by varying the instruction through the joystick icon in the mobile.

Fig. 36.7 Block diagram of environmental identification unit

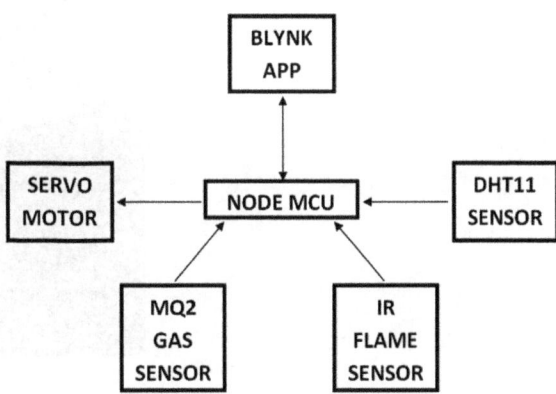

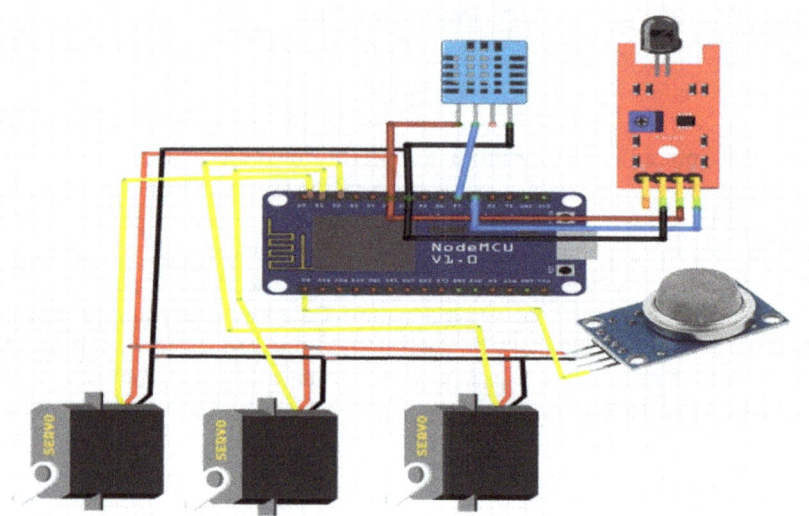

Fig. 36.8 Circuit connection diagram of environmental identification unit

Fig. 36.9 Output of sensors
at blynk application

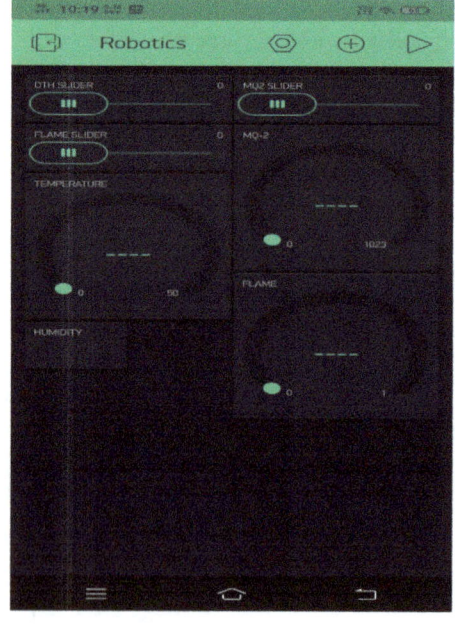

Fig. 36.10 Patrolling
hardware

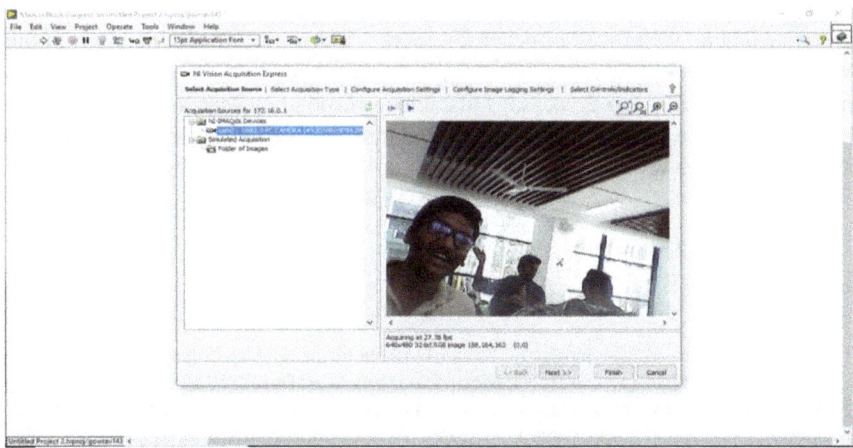

Fig. 36.11 MyRIO output in LabVIEW

36.5 Conclusion

This proposed work proves that reliability of the robot can be improved with minimum cost by sharing the required features of the robot with several low-cost processors. If any processor fails, robot can function with minimum features by using the available processors. Arduino, MyRIO and NodeMCU processors are used instead of going for the three same processors. Usage of different processor can also improve the reliability because of specification of each processor differs which gives

Fig. 36.12 Final proposed
reliable surveillance system

assurance that all the processor never fails at the same time. Future extension of this work is that to develop a communication or interconnection between all the sensors and motors so that if one processor fails, the corresponding sensors and motors can be activated by the other processors. This will make the system more reliable without sacrificing any features.

References

1. Danthala, S., Rao, S., Mannepalli, K., Shilpa, D.: Robotic manipulator control by using machine learning algorithms: a review. Int. J. Mech. Prod. Eng. Res. Dev. **8**(5), 305–310 (2018)
2. Victerpaul, P., Saravanan, D., Janakiraman, S., Pradeep, J.: Path planning of autonomous mobile robots: a survey and comparison. J. Adv. Res. Dyn. Control Syst. **9**(Special Issue 12), 1535–1565 (2017)
3. Gopi Krishna, P., Sreenivasa Ravi, K., Bhavya Sai Sireesha, P.: Implementation of bi-directional bluetooth—ZigBee gateway for multipurpose applications in iot. J. Adv. Res. Dyn. Control Syst. **9**(Special Issue 18), 306–317 (2017)
4. Vani, G.D., Chinnaiah, M., Karumuri, S.R.: Hardware scheme for autonomous docking algorithm using FPGA based mobile robot. In: Proceedings of the 2018 8th International Symposium on Embedded Computing and System Design, ISED 2018, pp. 110–115 (2018)
5. Venkateswara Rao, M., Madhav, B.T.P., Anilkumar, T., Prudhvi Nadh, B.: Metamaterial inspired quad band circularly polarized antenna for WLAN/ISM/Bluetooth/WiMAX and satellite communication applications. AEU Int. J. Electron. Commun. **97**, 229–241 (2018)
6. Rooban, S., Saifuddin, S., Leelamadhuri, S., Waajeed, S.: Design of fir filter using wallace tree multiplier with kogge-stone adder. Int. J. Innov. Technol. Explor. Eng. **8**(6), 92–96 (2019)
7. Naveen, P.S., Rao, K.R.R.M.: Ni myrio based smart robot with authentication switch. Int. J. Innov. Technol. Explor. Eng. **8**(6), 88–91 (2019)

8. Anitha, A., Rooban, S., Sujatha, M.: Implementation of energy efficient gates using adiabatic logic for low power applications. Int. J. Recent Technol. Eng. **8**(3), 3327–3332 (2019)
9. Rooban, S., Kumar, K.S., Shankar, K.R., Bhaskara Rao, N.U.: Test and analysis of high performance microprocessor through power binning method. Int. J. Eng. Adv. Technol. **8**(4), 882–886 (2019)
10. Pallav, K., Siva Prasanth, G.V.S., Shanmukha Sree Charan, T.P.B.M., Nayak, D.K., Selvakumar, R.: Design and simulation of path planning algorithm for autonomous mobile robot navigation system using ekfslam. Int. J. Sci. Technol. Res. **8**(12), 1333–1339 (2019)
11. Prasad, C.H., Bojja, P.: A reliable, energy aware and stable topology for bio-sensors in healthcare applications. J. Commun. **14**(5), 390–395 (2019)
12. Sunithamani, S., Sanjay Naidu, R., Sudheep, V., Hemanthkumar, C.: Analysis of MEMS based gas sensor. Test Eng. Manag. **81**(44147), 6186–6189 (2019)
13. Rajasekar, K., Sunithamani, S.: Low setting time offering series capacitive RF MEMS switch for WI-FI applications. Int. J. Recent Technol. Eng. **7**(5), 439–442 (2019)
14. Soumya, N., Sai Kumar, K., Raghava Rao, K., Rooban, S., Sampath Kuma, R.P., Santhosh Kumar, G.N.: 4-bit multiplier design using cmos gates in electric VLSI. Int. J. Recent Technol. Eng. **8**(2), 1172–1177 (2019)
15. Radhika, R., Partheeban, N.: Improving power efficiency in the implementation of discrete cosine transform in multicore architecture. J. Comput. Theor. Nanosci. **15**(8), 2597–2603 (2019)

Chapter 37
Deciphering WEP, WPA, and WPA2 Pre-shared Keys Using Fluxion

Sidharth Atluri and Revanth Rallabandi

Abstract The content in the present paper reviews the attack carried out to decipher WEP, WPA, and WPA2 passwords using a tool named Fluxion. Before understanding how the tool works, brief information about the security protocols is given so that comprehending the attack more undoubtedly. At the same time, we compare Fluxion to other pre-existing tools used to crack Wi-Fi passwords and then list out the major differences between them. This comparison is done to evaluate the efficacy and success rate of Fluxion. In conclusion, this research and attack are performed to provide a deeper insight into WEP, WPA, and WPA2 cracking.

37.1 Introduction

37.1.1 WEP

One of the oldest security protocols is the Wired Equivalent Privacy or WEP. WEP is not being used in the present due to its vulnerabilities and the fact that it can be easily cracked. The underlying algorithm behind this security protocol is RC4. The way this algorithm works is, whenever a client wants to send data to the router or vice-versa, first the data will be encrypted using a key. Therefore, this will convert the plain text in the packet to ciphertext. This encrypted packet is now sent in the air. The recipient will be able to transform the packet back to its original form because it has the key. The problem arises with the way WEP implements this algorithm. To encrypt the packet WEP generates a unique keystream. In other words, for each packet that is being sent into the air WEP tries to create a keystream for it. The keystream is a combination of a random 24-bit initialization vector (IV) added to the pre-shared key (password) of the network. Before sending this encrypted packet in

S. Atluri (✉)
Sreenidhi Institute of Science and Technology, Hyderabad, Telangana, India

R. Rallabandi
Chaitanya Bharathi Institute of Technology, Hyderabad, Telangana, India

the air, WEP appends the 24-bit IV to the encrypted packet to ensure that the recipient can decrypt the packet because it needs both the IV and the pre-shared key. Now the major issue here is that the IV that is being appended is in plain text. Also, the size of an IV is only 24 bits. When we compare it to the amount of traffic generated on a Wi-Fi network, it is very less. Consequently, IVs will start getting repeated.

37.1.2   WPA/WPA2

With WEP not meeting the specific security requirements, Wi-Fi Protected Access (WPA) came into the picture in the year 2003. WPA provided a more complex data encryption procedure and it also had a finer user authentication when compared to WEP. In 2004, WPA2 was created which had some major upgrades from WPA. To begin with, it must be understood that both are very similar. The difference arises in the encryption method used to maintain message integrity. WEP uses Temporal Key Integrity Protocol (TKIP) whereas WPA2 makes use of Advanced Encryption Standard (AES). But this does not affect the methods used to crack either WPA or WPA2 networks. WPA/WPA2 came over the weaknesses in WEP by not appending any useful data in the packets to be sent in the air. In addition to that, the keys are unique, temporary, and exist for a longer period than WEP. Hence, the only packets that help in cracking this network are the four handshake packets that are exchanged between the client and the router when the client connects to it.

37.2   Related Work

In [1], the author tests all the router protocols available currently such as WEP, WPA, WPA2, and WPS and detects the vulnerabilities of the system. The author also explains the tools like Airodump-ng, Aircrack-ng that is used to acquire access point pins. The author in [2] uses the Aircrack-ng tool for cracking the WPA pre-shared key with simple ASCII keys and a complex hexa-decimal key to check if keys could be cracked. Kissi and Asante [3] discuss the usage of penetration testing to assess vulnerabilities and conduct attacks on WEP, WPA and WPA2 security protocols. In [4], the author concludes by comparing Fluxion with previously defined hypotheses and aims to give more insight toward the WEP/WPA/WPA2 attacks.

37.3   Existing System

The very first step is to configure the wireless interface connected to monitor mode. Changing the mode of the wireless adapter can be done in the manner shown below. Here, we can notice the status of the wireless adapter 'wlan0' by using the command

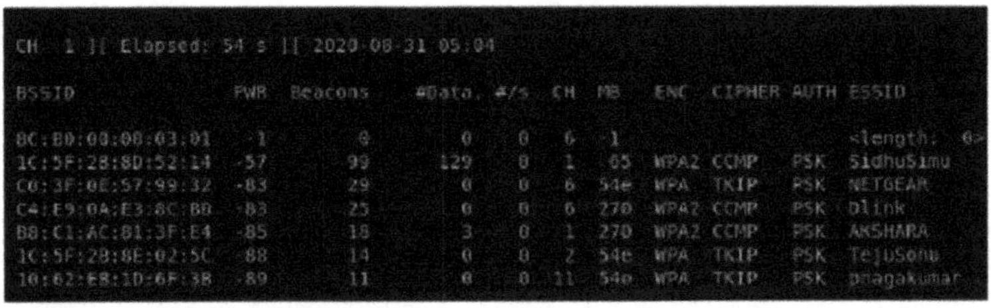

Fig. 37.1 List of networks in Airodump-ng

'iwconfig'. When we look at the mode attribute, it has been changed from managed to monitor. Now the adapter is ready to capture any packets being sent in the air. The next step is to find out the available networks and select the network on which we want to perform the attack (Fig. 37.1).

By running Airodump-ng we can obtain the list of networks as shown above. For example, the attack will be carried out on the wireless network named 'SidhuSimu' which has WPA2 protocol enabled.

The next step is to capture the handshake by using the following command as shown in Fig. 37.2.

The catch is that a handshake can only be captured when a new client connects to the network. One way is to wait for a new client to connect to the network. Alternatively, we can use a de-authentication attack, by which we can disconnect a client from the network for a short period. The client automatically reconnects to the network once we stop the attack. Therefore, the handshake will be sent when the client automatically reconnects, consequently allowing us to not wait for an actual connection to take place (Fig. 37.3).

The final step is to use Aircrack-ng and a wordlist to crack the pre-shared key from the handshake as shown in Fig. 37.4.

Therefore, the pre-shared key for the above network was found. This is the typical way of cracking WPA/WPA2 networks and even WEP networks.

Fig. 37.2 Command used to capture the handshake

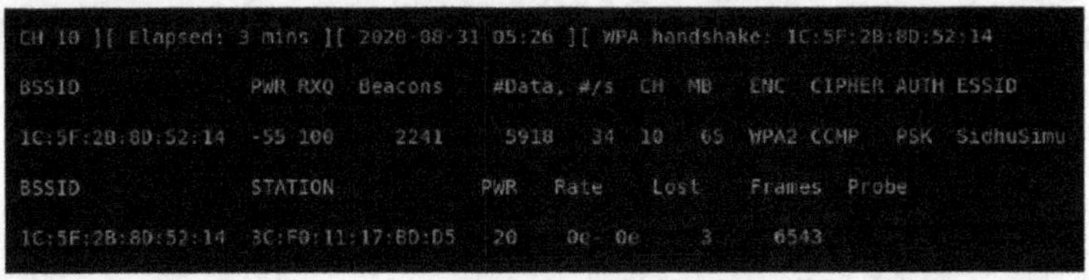

Fig. 37.3 Success page indicating the immediate capture of handshake

```
root@kali:~# aircrack-ng wpa_test1-01.cap -w test1.txt
```

Fig. 37.4 Loading the word list used to crack the PSK in Airodump-ng

37.4 Proposed System

Even though the above-depicted method works in most cases, it might be difficult to crack the password if the wordlist generated does not contain the pre-shared key. In addition to that, capturing handshakes, performing de-authentication attacks, and manually typing the commands are time-consuming. Here, we can make use of Fluxion which provides us with a wide variety of attacks and zero requirements of running commands manually. Fluxion can be considered as a mixture of social engineering and technical automation that fools the user into giving the Wi-Fi passwords in no time. To be concise, it is more of a social engineering framework that makes use of an evil twin access point, handshake capture functions, and integrated jamming.

Fluxion is a distinctive tool in its utilization of a WPA handshake to not only restricts the functioning of the login page, but also the functioning of the entire script. It deactivates the original network and establishes a duplicate with the same name, persuading the disconnected client to join. It then displays a fake login page mentioning the router requires restarting to install firmware and it needs the password to do so. The tool makes use of a captured handshake to verify the pre-shared key submitted and continues to deactivate the target access point (AP) until the right password is typed.

37.4.1 Installation

The following command as shown in Fig. 37.5 installs Fluxion.

We need to check if there are any missing dependencies by moving to the 'Fluxion' folder and then running the Fluxion script as shown in Fig. 37.6 and 37.7.

Running the same command with –i option will automatically install the missing dependencies.

```
root@kali:~/Desktop# git clone https://github.com/FluxionNetwork/fluxion.git
Cloning into 'fluxion'...
```

Fig. 37.5 Cloning the Fluxion git repository to install Fluxion

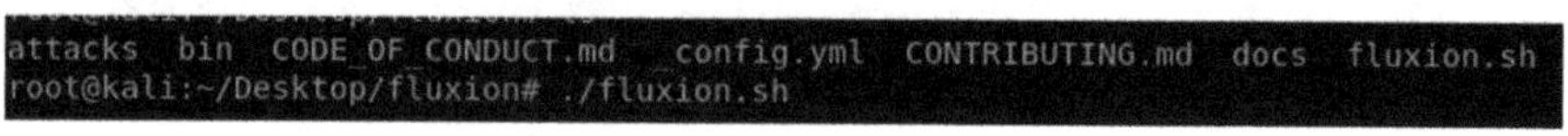

```
attacks  bin  CODE_OF_CONDUCT.md  config.yml  CONTRIBUTING.md  docs  fluxion.sh
root@kali:~/Desktop/fluxion# ./fluxion.sh
```

Fig. 37.6 Checking if there are any missing dependencies

Fig. 37.7 List of networks being displayed in Fluxion

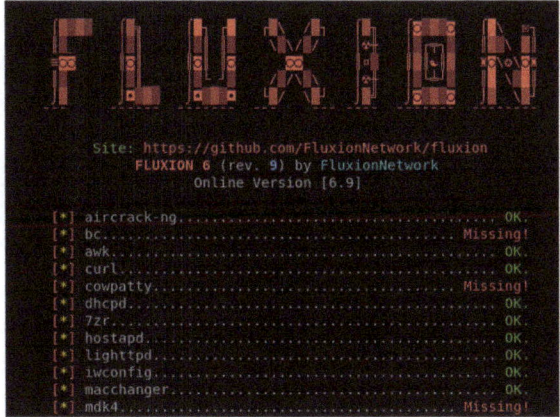

37.4.2 Scanning for Wi-Fi Networks

After selecting the language, the tool provides us with a screen to select the kind of attack we want to perform as shown in Fig. 37.8.

The next screen provides us with a list of available wireless adapters that can be used to scan for networks as shown in Fig. 37.9.

Once the resource has been allocated, we move to a screen where we must select the frequency band in which we want to search for networks as shown in Fig. 37.10.

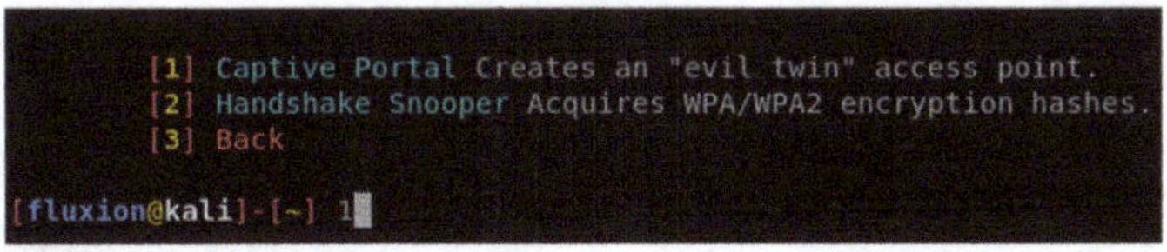

```
    [1] Captive Portal Creates an "evil twin" access point.
    [2] Handshake Snooper Acquires WPA/WPA2 encryption hashes.
    [3] Back

[fluxion@kali]-[~] 1
```

Fig. 37.8 List of attacks that can be performed

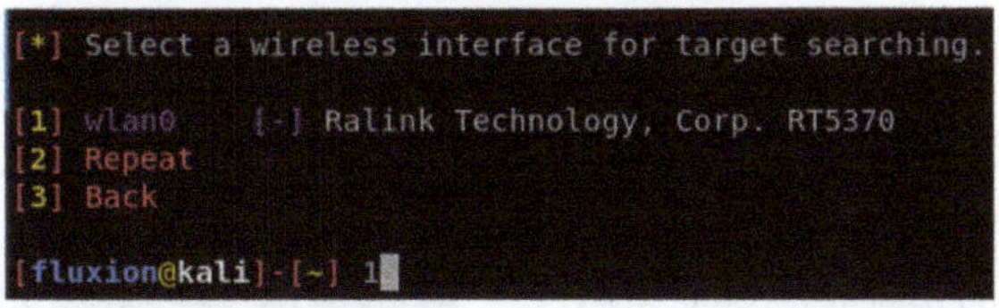

Fig. 37.9 List of wireless adapters

```
[*] Select a wireless interface for target searching.

[1] wlan0     [-] Ralink Technology, Corp. RT5370
[2] Repeat
[3] Back

[fluxion@kali]-[~] 1
```

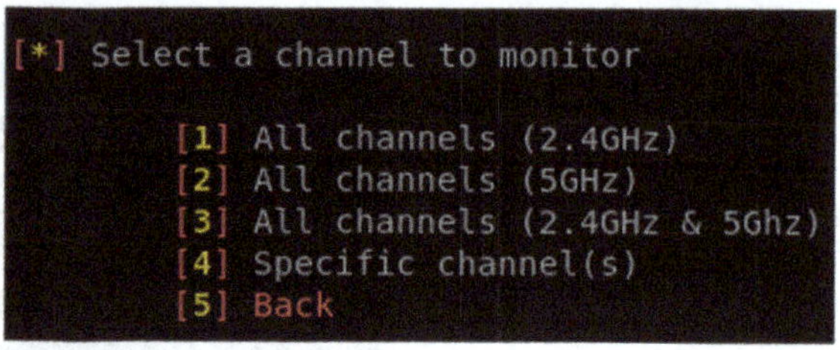

Fig. 37.10 List of available
frequency bands

```
[*] Select a channel to monitor

        [1] All channels (2.4GHz)
        [2] All channels (5GHz)
        [3] All channels (2.4GHz & 5Ghz)
        [4] Specific channel(s)
        [5] Back
```

37.4.3 Selecting Target Access Point

We get a list of networks with other information such as encryption protocol being used, MAC address of the router, etc. Select a network from the given list as shown in Fig. 37.11. The next screen asks us to select a wireless adapter again. After selecting the wireless adapters, we need to select a method of de-authentication attack as shown in Fig. 37.12.

We also get a list of methods to implement fake AP service as shown in Fig. 37.13.

We are now given a list of password verification methods and hash verification methods as shown in Figs. 37.14 and 37.15.

Next, we must select the SSL certificate for the fake login page as shown in Fig. 37.16.

We get a list of fake login pages, and we can select one that might most probably trick the user as shown in Fig. 37.17.

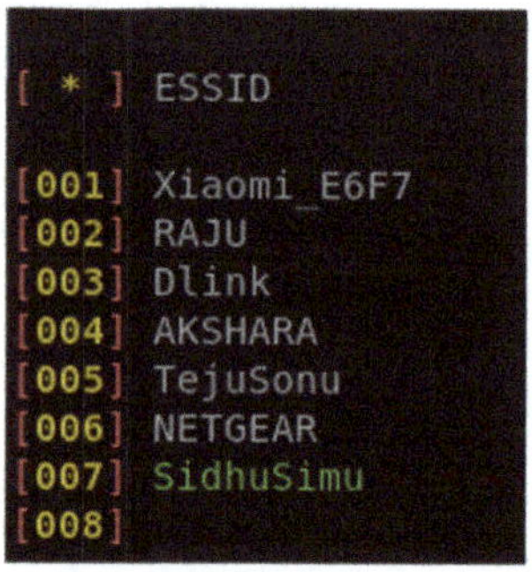

Fig. 37.11 List of available
Wi-Fi networks

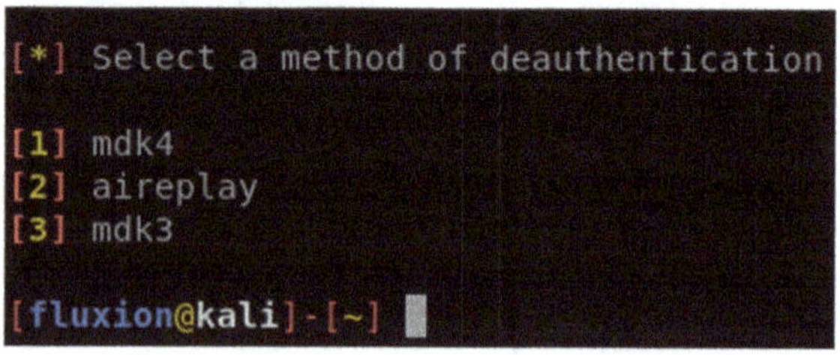

```
[ * ] ESSID

[001] Xiaomi_E6F7
[002] RAJU
[003] Dlink
[004] AKSHARA
[005] TejuSonu
[006] NETGEAR
[007] SidhuSimu
[008]
```

Fig. 37.12 List of methods
for de-authentication attacks

```
[*] Select a method of deauthentication

[1] mdk4
[2] aireplay
[3] mdk3

[fluxion@kali]-[~]
```

Fig. 37.13 List of fake access point services

```
                    [1] Rogue AP - hostapd (recommended)
                    [2] Rogue AP - airbase-ng (slow)
                    [3] Back

[fluxion@kali]-[~] 1
```

Fig. 37.14 List of password verification methods

```
                [1] hash - cowpatty
                [2] hash - aircrack-ng (default, unreliable)
                [3] hash - pyrit
                [4] Back

[fluxion@kali]-[~] 1
```

```
            [1] aircrack-ng verification (unreliable)
            [2] cowpatty verification (recommended)
            [3] pyrit verification

[fluxion@kali]-[~] 2
```

Fig. 37.15 List of hash verification methods

Fig. 37.16 List of fake SSL certificates

```
[*] Select SSL certificate source for captive portal.

            [1] Create an SSL certificate
            [2] Detect SSL certificate (search again)
            [3] None (disable SSL)
            [4] Back

[fluxion@kali]-[~]
```

```
    [52] Livebox                          fr
    [53] movistar                         es
    [54] NETGEAR                          en
    [55] NETGEAR                          es
    [56] NETGEAR                          it
    [57] NETGEAR-Login                    en
    [58] Netis                            it
```

Fig. 37.17 List of fake login pages according to the router

Fig. 37.18 Screen showing the hacked password of the Wi-Fi network

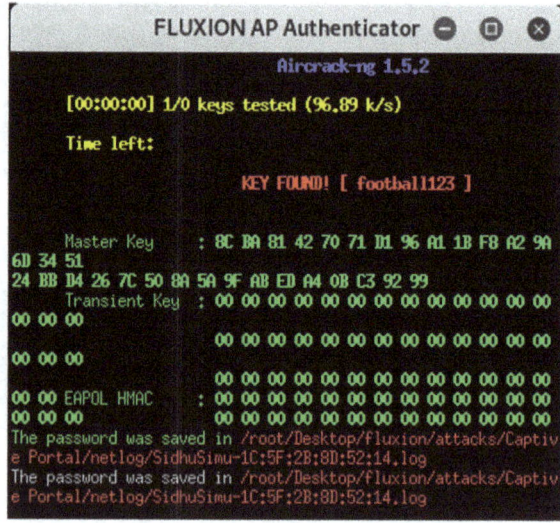

37.5 Results

The cracked password of the selected network 'SidhuSimu' and the directory in which the file is saved is shown in Fig. 37.18. Here, we can notice the efficiency of using Fluxion as a tool to crack passwords. Henceforth, we can also conclude that any security protocol that is in the market right now is still vulnerable to hackers and malicious attacks. There can be several other changes made to Fluxion 2.0 to carry out such attacks effortlessly.

37.6 Conclusion

The research portrays the advantages of using Fluxion over Aircrack-ng while cracking WEP/WPA/WPA2 networks. Unlike Aircrack-ng, the hacker does not need to memorize the commands to run de-authentication attacks, network jamming attacks, fake access point creation, handshake capturing etc. while using Fluxion as it provides a well-built user interface when compared to other cracking tools. The only limitation while using Fluxion is, if the target is an experienced IT professional or someone who understands computer networks, then they can figure out that it is a social engineering attack. In this case, the target might not share the pre-shared key (PSK) of the network. This paper signifies the importance of using proper security countermeasures because any user with minimal computer knowledge can exploit the vulnerabilities of a network, by simply learning the usage of a few security tools.

References

1. Pimple, N., Salunke, T., Pawar, U., Sangoi, J.: Wireless security—an approach towards secured Wi-Fi. In: 2020 6th International Conference on Advanced Computing and Communication Systems (ICACCS) (2020)
2. Rabiul Islam, S.Md.: Wi-Fi protected access (WPA)—PSK (phase shift keying) key cracking using AIRCRACK-NG. Int. J. Sci. Eng. Res. (2013)
3. Kissi, M.K., Asante, M.: Penetration testing of IEEE 802.11 encryption protocols using Kali Linux hacking tools. Int. J. Comput. Appl.
4. Permatasari, F., Eaganathan, U.: An implementation of WEP/WPA/WPA2 password cracking using fluxion. In: 12th International Conference on Recent Innovations in Science, Engineering and Management (2018)

Chapter 38
Blockchain-Based Government Project Fund Tracking and Management System

M. V. Varada, Wissam Salih Abdulla, Aparna Udayakumar, Haritha Raghu, and Manoj V. Thomas

Abstract Across the globe, governments are testing blockchain in various administrative processes including public relations. Blockchain technology provides a secure platform and ensures data integrity. And its applications are resistant to breakdown and are secured using cryptography; hence, it can be used for government applications. Hyperledger fabric is one of the prominent open-source framework of blockchain which is used to develop a "permissioned" network system. When various project funds are issued by the government, often a large amount is unutilized for the actual project, due to corruption. It is necessary to leverage the hyperledger fabric for the formation of a network which allows efficient fund distribution systems resulting in a better society. In this paper, we explored adding permissioned support of hyperledger fabric to government fund distribution system.

38.1 Introduction

Corruption is a symptom and outcome of dishonesty or criminal offense undertaken by an institution. It has negative impact on the countries' GDP.

In the beginning of the novel technology, blockchain was used in cryptocurrency transactions. Presently, different other domains like health care and government started adopting this due to its decentralized and immutable nature [1]. Blockchain is now one of the popular technology which is at the center of the technology discussion. It is studied in research areas and applied in many fields, such as health care, education and business markets. In this paper, we focus mainly about the use of blockchain in a government domain. Specifically, this paper emphasizes the use of hyperledger fabric blockchain technology for tracking the funds generated by the government. Sometimes when funds are allocated to projects by the government, a large portion of it is unutilized, and it goes to the hands of various people involved in the project or some leaders, leading to corruption. So, if blockchain is introduced in

M. V. Varada (✉) · W. S. Abdulla · A. Udayakumar · H. Raghu · M. V. Thomas
Vimal Jyothi Engineering College, Chemperi, Kannur, Kerala, India
e-mail: manojkurissinkal@vjec.ac.in

© The Author(s), under exclusive license to Springer Nature Singapore Pte Ltd. 2021   387
S. C. Satapathy et al. (eds.), *Smart Computing Techniques and Applications*,
Smart Innovation, Systems and Technologies 225,
https://doi.org/10.1007/978-981-16-0878-0_38

this area, people can monitor all the amount with respect to when and where it is being used from the decentralized ledger. Here, we introduce a system over hyperledger fabric technology to track and allocate funds and hence bring down the possibility of corruption.

In the next section, we find the details of the related work and the hyperledger fabric architecture and then, a description of the proposed system with explanation of its working. In the fourth section, a description of our modules is given for this system, and the fifth section also discusses regarding the results. Finally, we talk about the relevance of this application, its future scope and also reach a conclusion.

38.2 Related Work

38.2.1 Application of Blockchain in Government

Governments all over the world are implementing blockchain in different areas. Estonia was one of the first countries to test blockchain in government applications. First by embedding blockchain within data transfer platform X-Road. Second, by testing a "notorization on the blockchain" which was a project combined with BitNation. And thirdly by piloting keyless authentication which is supported by the blockchain distributed ledger technology [2]. In addition, different organizations across the globe as in EU and UK have officially released positive reports on using blockchain [3, 4].

38.2.2 Hyperledger Fabric

Software Development Kit (SDK): SDK allows the users to interact with the hyperledger blockchain network setup and is shown in Fig. 38.1. It creates the channels, and the peer nodes are asked to join the channel. Channels have the ability to partition the transaction visibility for only the participant. Chaincodes are installed in each peer by the SDK.

Membership Service Provider (MSP): An MSP defines its own notion of identity, and the rules by which the identities of the various organizations are governed and authenticated (signature generation and verification) [5].

Peers : Peer nodes can be endorser or committer nodes. It endorses the proposal for transaction or commit blocks of transaction to ledger [6].

Orderer: A block of incoming transactions is created by the orderer and supports the consensus service.

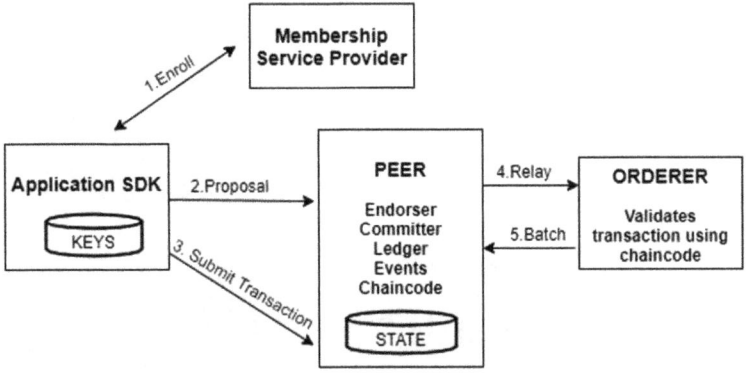

Fig. 38.1 Hyperledger fabric architecture

38.3 Proposed System

In this paper, we propose a system to digitally record the project fund grants. We use the blockchain to record the creation and transfer of digital evidence, and it ensures integrity, credibility and prevents malicious tampering.

38.3.1 Architecture of the System

The participating nodes of the hyperledger fabric are interconnected and can be used for sharing of data and validation, and the transaction behavior of the blockchain and the integrity and trustworthiness of the ledger data can be maintained through the consensus mechanism [7, 8].

1. **Network creation**: A blockchain network is deployed on a local machine or several computers to support the fund transfer process.
2. **Setting parameters**: The status and other parameters required for distributing fund such as status variable, time of transaction, validation, unique applicant ID are set.
3. **Transaction**: Validated applications send transactions; blockchain validators [9], among other things, verify the correctness and the fact that no attempt was made to divert the funds to someone who does not deserve, due to linkability. All transactions that attempt to apply twice are automatically invalidated and are not added to blockchain.
4. **Distribution**: At the end of a particular selection, the project funds are granted to a company or individual. Figure 38.2 depicts the architecture of the proposed system.

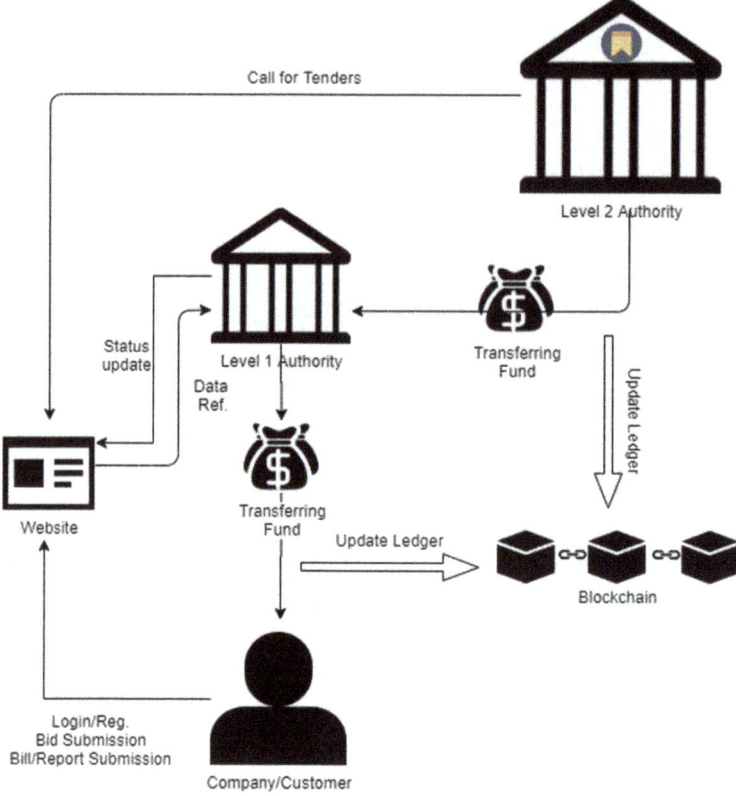

Fig. 38.2 Architecture of the proposed system

38.3.2 Specialized Network

In order to deploy the network, validators and auditors must be identified. Validators are the nodes that form blocks of transactions from the pool of unconfirmed transactions [10, 11]. Auditors are the nodes that store a complete copy of blockchain and do not participate in block mining. Our system uses the hyperledger fabric as it supports the use of channels for transactions and can expand to a large network as compared to Ethereum or other public/private ledger blockchain technologies.

38.3.3 Storage

After defining the node, the validated users are installed. The list of block candidates is formed in the very first block (genesis block) of the network [12]. After defining the users, the transactions are moved to the pool, that is, the funds are granted to the individuals. After validation based on chaincode, the requested changes in transactions are displayed in blockchain. The default database used by hyperledger fabric to store the states is leveldb. But in order to satisfy the rich queries and the JSON format, Couchdb is used [13].

38.3.4   Requirements

Software requirements	Hardware requirements
Hyperledger fabric	Operating systems: Ubuntu Linux 14.04
Docker	16.04 LTS (both 64-bit)
Git	or Mac OS 10.12
npm	
Node JS	RAM: 8 GB
cURL	

38.3.5   Working of the System

The network is created using hyperledger fabric. A blockchain network is deployed on a local machine or several computers. Figure 38.3 depicts the workflow diagram for fund tracking and management:

1. The company submits bid for the project that is within the budget specified in the call for tender notice and places request for the fund.
2. One of the project proposals is selected by Level 1 authority.
3. Level 2 authority on receiving fund request from Level 1 authority checks for the availability of funds.
4. Level 2 authority transfers the fund to Level 1 authority, and the transaction is recorded in the ledger of hyperledger fabric.
5. Initial payment is released by Level 1 authority.
6. The company submits stage-wise completion report to Level 1 authority.
7. Level 1 authority transfers fund at each stage after verification.

38.4   Modular Description

Call for Tenders: A notification with link is provided in government official portal. The company/customer gets the notification, and they can register using the link. This link will be directed to a Web page.

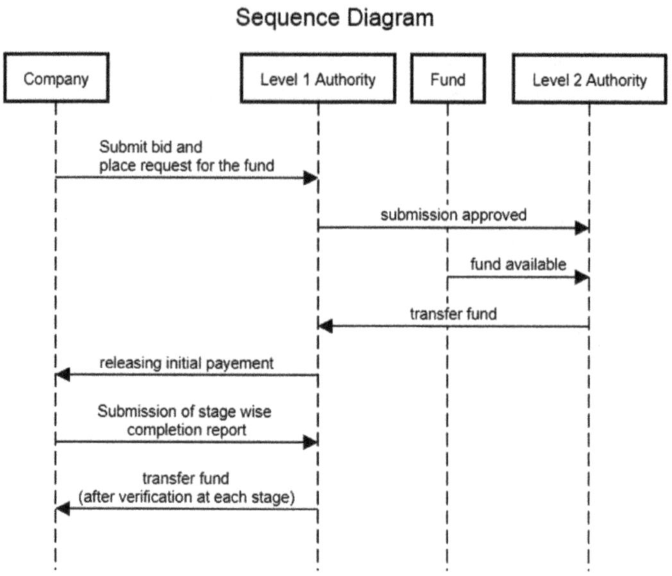

Sequence Diagram

Fig. 38.3 Workflow of the proposed model

Company/Customer Registration: The details needed for company/customer registration are asked in the Web site. The Web site mainly asks for the mail ID, pan number, company ID, project proposal details,mobile number,password and documents.

Verification: Level 1 authority verifies the company/customer by checking their provided ID and profile. In the profile, the required documents must be provided by the company. So, using the documents, Level 1 authority verifies the project proposal, and then if the proposal is valid, then they proceed with further actions. Figure 38.4 shows various organizations involved.

Fund Sanctioning and distribution: The Level 2 authority sanctions the fund. The Level 1 authority, after verifying and checking the requirements of the company or the customer, decides the total amount required for all the project and giving all the information to the Level 2 authority. After looking at the information provided by the Level 1 authority, the Level 2 authority sanctions the fund. The Level 1 authority distributes the fund to company/ customer. The fund distribution is done using the requirements provided by the company or customer.The company/customer gets the fund through the bank account(provided in the profile). As the total fund generated can be viewed for all the participants, corruption is not possible. If there is any complaint regarding the fund distribution, then complaint registration is also possible. The fund safely reaches to the company/customer. This system thus provides trustful environment. The status is displayed on webpage. The transactions would be shown on ledger.

```
⫶ crypto-config.yaml > [ ] OrdererOrgs
 1   OrdererOrgs:
 2      - Name: Orderer
 3        Domain: fundmanagement.com
 4        EnableNodeOUs: true
 5        Specs:
 6           - Hostname: orderer
 7   PeerOrgs:
 8      - Name: Applicant
 9        Domain: applicant.fundmanagement.com
10        EnableNodeOUs: true
11        Template:
12          Count: 2
13        Users:
14          Count: 1
15      - Name: Level1
16        Domain: level1.fundmanagement.com
17        EnableNodeOUs: true
18        Template:
19          Count: 2
20        Users:
21          Count: 1
22      - Name: Level2
23        Domain: level2.fundmanagement.com
24        EnableNodeOUs: true
25        Template:
26          Count: 2
27        Users:
28          Count: 1
```

Fig. 38.4 Screenshot of various organizations involved in the system

38.5   Results and Discussion

Caliper is a framework used to measure blockchain performance, which allows users to test various blockchain solutions with a set of use cases, and get the performance test results under the hyperledger fabric project [14]. The benchmarking criteria in Caliper so far include success rate, transaction/read latency (minimum, maximum, average, percentile) and transaction/read throughput. Figure 38.5 shows the performance matrix of certificates in the hyperledger fabric using this tool.

Name	Succ	Fail	Send Rate (TPS)	Max Latency (s)	Min Latency (s)	Avg Latency (s)	Throughput (TPS)
Certificate Generation	1000	0	192.0	7.00	2.93	4.79	112.7

Fig. 38.5 Performance metrics of certificate

TYPE	NAME	Memory (max)	Memory (avg)	CPU% (max)	CPU% (avg)	Traffic In	Traffic Out	Disc Read	Disc Write
Process	node local-client.js(avg)	42.1MB	43.5MB	10.45	3.30	-	-	-	-
Docker	dev- peer0.org1.fund.com-projectfund-v1	46.1MB	44.5MB	25.01	9.02	1.23MB	456.2KB	0B	0B
Docker	dev- peer0.org2.fund.com-projectfund-v1	44.1MB	44.6MB	0.21	0.00	191B	191B	0B	0B
Docker	peer0.org1.fund.com	289.2MB	290.4MB	60.1	38.2	4.1MB	12.7MB	0B	4.7MB
Docker	ca.org1.fund.com	20.2MB	20.2MB	0.00	0.00	0B	0B	0B	0B
Docker	orderer.fund.com	34.7MB	29.4MB	72.8	21.2	2.4MB	2.5MB	0B	2.9MB

Fig. 38.6 Resource consumption

Resource consumption indicates the resource (CPU, memory, network IO) that is consumed by each Docker container. Each container represents a peer in the blockchain. Figure 38.6 gives an overview of the details regarding the resource consumption of this work.

38.6 Conclusion and Future Work

This model implemented using blockchain helps to provide transparency in all the transactions made by the government. The decentralized ledger makes it possible for verifying the transactions and impossible to alter the transaction information. This means that the fund that is sanctioned can be monitored. Such a blockchain network will surely reduce the corruption that occurs in our society and promote the economic development of the country. It also creates greater trust in government. The key is to land in blockchain technology and build a new system of corruption-free and transparent system.

There is a lot of scope for further document verification to be done digitally using zero validation techniques so that impersonation cannot be done. Any online transaction can be made to verify digitally. Here, the blockchain ensures the security of data. This model can be run on the cloud on the hyperledger fabric where the blockchain can be stored. Also, this project fund system can be expanded to the whole government fund releasing systems such as scholarships, pension funds, relief funds and so on.

References

1. Ismail, L., Hameed, H., AlShamsi, M., Alhammadi, M.S., Al Dhanhani, N.: ICBCT 2019: Proceedings of the 2019 International Conference on Blockchain Technology, Towards a

Blockchain Deployment at UAE University: Performance Evaluation and Blockchain Taxonomy, March 2019, pp. 30–38

2. Chen, S., Shi, R., Ren, Z., et al.: A blockchain based supply chain quality management framework. In: International Conference on E-Business Engineering **2017**, 172–176 (2017)

3. Glaser, F.: Pervasive decentralisation of digital infrastructures: a framework for blockchain enabled system and use case analysis (January 2017). In: Proceedings of the 50th Hawaii International Conference on System Sciences (HICSS-50), Hawaii, January 4–7, 2017. Available at SSRN: https://ssrn.com/abstract=3052165

4. Jun, M.: Blockchain government—a next form of infrastructure for the twenty-first century. J. Open Innov.: Technol. Market Complexity **4** (2018)

5. Wai, K., Htoon, E.C., Thein, N.M.: Storage structure of student record based on hyperledger fabric blockchain. In: 2019 International Conference on Advanced Information Technologies (ICAIT), 6–7 Nov. 2019, pp. 108–113. IEEE, New York

6. Thwin, T.T., Vasupongayya, S.: Blockchain based secret data sharing model for personal health record system. In: 5th International Conference on Advanced Informatics: Concept Theory and Applications (ICAICTA), 14–17 Aug 2018, pp. 196–201. IEEE, Krabi, Thailand

7. Welcome to Hyperledger Fabric. https://hyperledger-fabric.readthedocs.io/. Accessed 2018

8. Jha, A., Bhattacharjee, R.K., Nandi, M., Barbhuiya, F.A.: BSCI '19: Proceedings of the 2019 ACM International Symposium on Blockchain and Secure Critical Infrastructure: A Framework for Maintaining Citizenship Record on Blockchain, pp. 29–38 (2019). https://doi.org/10.1145/3327960.3332389

9. Ranjan, S., Negi, A., Jain, H., Pal, B., Agrawal, H.: Network system design using hyperledger fabric: permissioned blockchain framework. In: Twelfth International Conference on Contemporary Computing (IC3), 2019, 8–10 Aug. 2019. IEEE, New York

10. Mohite, A., Acharya, A.: Blockchain for government fund tracking using hyperledger. In: 2018 International Conference on Computational Techniques, Electronics and Mechanical Systems (CTEMS), 21–22 Dec. 2018, pp. 231–234

11. Welcome to Hyperledger Fabric. https://hyperledger-fabric.readthedocs

12. Benhamouda, F., Halevi, S., Halevi, T.: Supporting private data on Hyperledger Fabric with secure multiparty computation. https://shaih.github.io/pubs/bhh18.html (2018)

13. LevelDB vs. CouchDB. [Online]. Available: http://vschart.com/compare/leveldb/vs/couchdb

14. https://medium.com/@nima.afraz/hyperledger-caliper-explained-and-installation-guide-ubuntu-c38dc16d3dcf

Chapter 39
Random Valued Impulse Noise Detection Using Fuzzy c-Means Clustering Technique

Aritra Bandyopadhyay, Kaustuv Deb, Atanu Das, and Rajib Bag

Abstract Noise detection is critical facet in eliminating impulse noise from digital images. Faulty detection can have deeper consequences in the ultimate outcome of the image filtering procedure. Established detection methods while distinguishing corrupted and non-corrupted pixels, have a tendency to upsurge errors while judging the edge points. In this paper, a detection technique, based on fuzzy c-means clustering algorithm is proposed to deal with above-mentioned issues. Fuzzy c-means clustering is used here to group similar non-corrupted pixels intensities by partitioning them into separate fuzzy clusters. Here, a pixel intensity value belongs to each cluster with certain membership strength. Depending on the maximum membership strength of a pixel element in a certain cluster among all the clusters, the most suitable belongingness of that pixel to that cluster is determined. After this operation, maximum element holding cluster is selected to be the cluster having the detected non-corrupted pixel elements, since there is a plausible tendency of similar non-corrupted pixel elements to group together. Standard measures like "miss" and "false-hit" have been used to compare the proposed method against the prevailing state-of-art methods to determine the noise detection accuracy of the proposed method. Being evaluated upon the aforesaid standard measures, the proposed detector is found superior to all the others in terms of its detection performance.

39.1 Introduction

Noise detection operation, prior to elimination, is an essential task in the arena of image restoration. Every noise type has a detrimental effect on digital images. Impulse noise is one of the types of noise that can arise due to faulty camera operations, atmospheric turbulences, flawed memory locations, or during any image

A. Bandyopadhyay (✉) · K. Deb · R. Bag
Supreme Knowledge Foundation Group of Institutions, Chandannagar, West Bengal, India

A. Das
Netaji Subhash Engineering College, Kolkata, West Bengal, India

S. C. Satapathy et al. (eds.), *Smart Computing Techniques and Applications*,
Smart Innovation, Systems and Technologies 225,
https://doi.org/10.1007/978-981-16-0878-0_39

397

processing trials. It impeaches negative effect on the image quality and the imminent image processing operations. To resolve this consequence, image restoration is a necessary operation. Impulse noise is categorized into two types: salt and pepper noise (SAPN) [1] and random valued impulse noise (RVIN) [2]. This paper is concentrated on detecting RVIN, which creates random dots in the images having unpredictable pixel intensities in the grayscale range of '0' to '255'. Predicting random changes in numerical intensity values makes RVIN detection a thought-provoking job.

Round the ages, several detectors, some also with removal filters have been established. SD-ROM [3] is one of the methods established in the past, that was having four thresholds to first identify a pixel as noisy or not. After a few years, PSM [4] and TSM [5] filters were proposed. The PSM had detection and removal mechanisms which were gradually applied over frequent iterations that created blurry outputs. TSM, on the other hand, had united, standard median filter and center-weighted median filter to form a detector mechanism with a specific threshold which restricted the performance of the work. Then ACWM [6] was designed to have an adaptive operator to deal with the impulse detection which was again failed to preserve the details of the images over 30% noise density. In the same year, MSM [7] was developed which also used a threshold-based detection technique. In a few years later, another method PWMAD [8] was established. PWMAD demonstrated a vigorous estimator MAD (median of the absolute deviations from the median) to identify corrupted pixels. Both PWMAD and MSM were failed to have adequate result at a noise density upper than 30%. In the meantime, ACWM filter was revised to form ACWM-EPR [9] which used adaptive center-weighted median filter to detect corrupted pixels. This method had some improvement than the previous filters but was unable to preserve image details at a noise density greater than 50%. Afterward, another impulse detector, combined with median filter [10] was introduced. The detector was able to improvise detection by maintaining the image details, but that also failed in high noise density of and over 40%. In the same period, ROAD filter [11] was introduced which had used a completely different technique to identify noise. Local statistics of image was used in it to detect corrupted pixels by identifying the statistical intensity difference from a designated pixel to their utmost alike neighbors. This was followed by another filter ROLD [12] which extended the concept of ROAD by amplifying the intensity differences among corrupted and non-corrupted pixels to make detection stronger. But these two filters altogether dealt up to 50% noise density. In the equivalent year, another filter DWM [13] came with diverse concept to propose an impulse noise detector which was founded on the alterations among the present pixel and its neighbors associated with four key directions which resulted pretty virtuous in respective to detection even at a high noise density of 50%. In recent times a filter [14] detected noisy pixels by means of a sparse depiction of the pixels in a taken window out of the whole image. Another filter [15] used the concept of sparse depiction and combined it with thresholding to form a new detector. Both the detection methods were failed to provide effective outcome over 50% noise densities.

In this paper, fuzzy c-means clustering algorithm is used to accumulate analogous non-corrupted pixels intensities by segregating them into distinct fuzzy clusters and then reliant on most membership strength of a pixel intensity, the residence of the designated pixel on a particular cluster is determined. Then, most element packed cluster is identified as the best cluster, considered to be having the non-corrupted pixels. Thus, rest of the elements that do not belong to the best cluster are detected as corrupted ones.

The rest of the paper is planned as follows. Section 39.2 demonstrates the proposed methodology. Section 39.3 portrays the results and discussion and the end Sect. 39.4 depicts the conclusion.

39.2 Proposed Methodology

A 512 × 512 sized noise corrupted image, denoted as IMNC, is considered to perform noise detection process. The image matrix of IMNC is denoted as IMMNC. A flag image matrix FG of same size as IMMNC is created and every pixel's value is replaced by '0'. IMMNC is subdivided into some number of 5 × 5 windows and the proposed fuzzy c-means clustering implemented noise detection process is applied on each of these windows. The underlying idea is that the non-corrupted pixels present in a window are most likely to form a cluster together as their pixel intensity values are not affected by random noise. In compare to that, all the noise corrupted pixels having noise affected random pixel intensity values are least likely to form a cluster together and instead, they will be scattered in different clusters. Fuzzy c-means clustering is a fuzzy logic-based strong clustering algorithm, has the capability of properly figuring out the similarity pattern existing in the pixel intensity values and thereby forming clusters by grouping similar pixel intensity values together (Fig. 39.1).

Fuzzy c-means clustering assigns a membership strength value, to each pixel intensity value, with which a pixel intensity value becomes a part of a cluster. A membership strength value for each cluster is assigned to a pixel intensity value, and finally, the pixel intensity value is most suitable to belong to the cluster for which the pixel intensity value has the highest membership strength value.

Thus, using fuzzy c-means clustering, it is most likely to find the non-corrupted pixels by tracing out the cluster having the most number of pixel intensity values in it. With this, it also can be said that the pixels, scattered in other clusters having a smaller number of pixel intensity values in them, are detected as noise corrupted pixels. With the purpose of forming four clusters, fuzzy c-means clustering is applied on each 5 × 5 window. CL_i, $i = 1$–4, denotes a cluster of pixel intensity values present in a window and CLM denotes the CL_i, $i = 1$–4, which has the most number of pixel intensity values in it. PV_N denotes the pixel intensity value of the center pixel, denoted as N, of the window. N's position in IMMNC is (r, c). The proposed detector detects N as a noise corrupted pixel if $PV_N \notin CLM$; otherwise N is considered as a non-corrupted pixel. If N is found corrupted, then the flag matrix's position $FG[r][c]$ corresponding to N is replaced by '1' and otherwise is replaced by '0'.

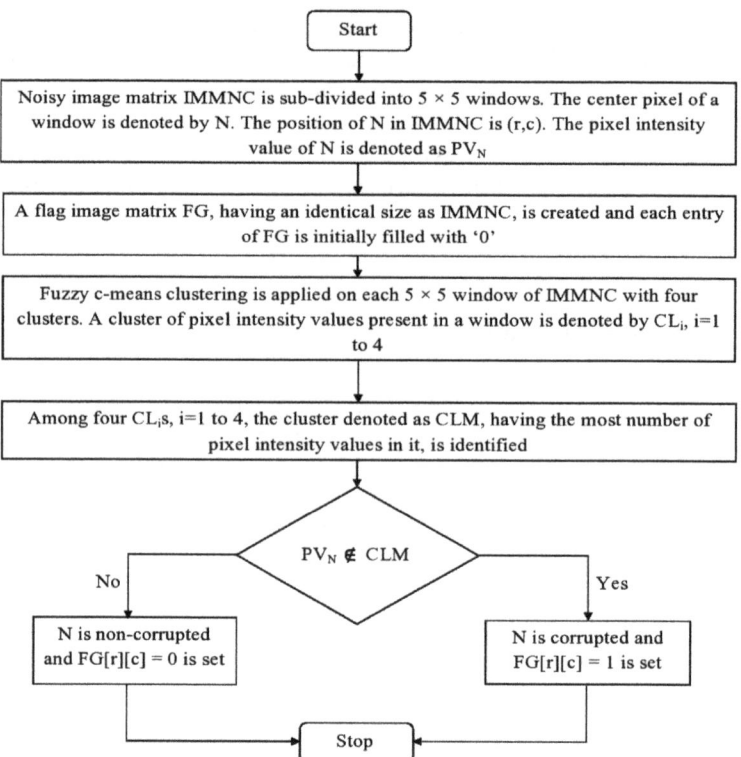

Fig. 39.1 Flowchart of proposed detection technique

39.3 Results and Discussion

The proposed work is carried out in MATLAB 2018a software, mounted on Windows PC having Core i5 processor and 8 GB of RAM. The proposed method's noise detection ability is compared in counter to thirteen state-of-art methods. The proposed method is applied on seven common test images: Lena, Goldhill, Baboon, Barbara, Bridge, Boat and Fingerprint each having a size of 512×512, to judge the outcome of its detection performance with respect to image assortment.

Noise detection accuracy of the proposed method is primarily shown in Fig. 39.2. Three 5×5 sample windows from the image matrix IMMNC for three different images are depicted here to highlight the outcome of the work. These three windows are the original image sample window, the corresponding same positional noisy image and the flag image window, respectively. Noisy image window is the RVIN affected image matrix's sample window which displays the noise exaggerated section. Flag window shows the refection of the activity of the detection procedure through the demarcation of the corrupted pixels with flag value '1' and non-corrupted pixels with flag value '0'. Rectangle outline boxes are taken to mark the wrongly

(a) Original (Lena)

97	98	95	90	96
88	96	93	97	91
94	93	90	88	94
100	96	92	93	90
90	95	92	88	94

(d) Original (Barbara)

173	189	192	196	199
190	195	196	193	193
196	199	195	187	181
195	195	189	180	172
192	189	178	170	160

(g) Original (Fingerprint)

214	224	227	220	205
214	222	225	216	198
205	212	216	207	179
209	211	207	186	155
216	211	192	157	122

(b) 70% Noisy image (Lena)

184	246	180	90	96
118	18	93	97	91
233	254	143	88	254
12	7	182	40	114
22	95	135	84	94

(e) 70% Noisy image (Barbara)

239	65	187	221	199
40	195	43	90	193
62	199	69	187	122
230	24	189	180	100
192	251	196	87	227

(h) 70% Noisy image (Fingerprint)

214	224	19	165	73
72	25	176	114	212
205	124	216	90	144
120	211	9	74	155
216	87	80	95	122

(c) Flag image reflecting detection (Lena)

1	1	1	0	0
0	1	0	0	0
1	1	1	0	1
1	1	1	1	0
1	0	1	0	0

(f) Flag image reflecting detection (Barbara)

1	1	0	0	0
1	1	1	1	0
1	0	1	0	1
1	1	0	0	0
1	1	0	1	1

(i) Flag image reflecting detection (Fingerprint)

0	1	1	1	0
1	1	0	0	1
0	1	0	0	1
1	1	1	1	0
0	0	1	0	1

(Corrupted: 1 and Un-Corrupted: 0)

Fig. 39.2 Proposed detection performance measure

detected pixels. Lena, Barbara and Fingerprint are used here to demonstrate the detection result at 70% noise density. From Lena image perspective, due to less variability of its pixel intensities, the detection outcomes are excellent, where only three marked errors found examining the sample windows. Barbara image is more complex image than Lena due to its image depth and because of this, the detection result for Barbara image is respectable but not as good as Lena. Texture image, Fingerprint is more intricate than both Lena and Barbara in terms of image profundity. Due to the high erraticism in the pixel's intensities, it is very tough to carry out detection procedure of Fingerprint image. Still, our detection gave reputable result for Fingerprint image. Moreover, reputable results shown in Fig. 39.2 are obtained at sufficiently high 70% noise density, which portrays the fact that the quality of the detection algorithm is good.

Table 39.1 validates the detection meticulousness of the proposed algorithm by showing the number of erroneous pixels in comparison with other state-of-art filters by means of "miss" and "false-hit". The term "miss" signifies the number of actually corrupted pixels which are detected as non-corrupted by the proposed detector and "false-hit" implies the number of actually non-corrupted pixels which are detected as corrupted by the proposed detector. From Table 39.1, it is identifiable that SD-ROM and ACWM have a very low false-hit number which is exceptional but both of them have much more higher miss value. The miss value of the proposed method is less but greater than some methods like DWM, ROLD-EPR, SAIKRISHNA, and DEKA in some cases. But if we judge in terms of the total noise (miss + false-hit), then the proposed filter wins the battle against all methods at 50% and above noise densities. So, the proposed detector exhibited admirable performance at high noise densities.

Table 39.2 portrays the "miss" and "false-hit" values of the proposed detector for seven different test images at diverse noise densities. Table 39.2 shows the detection

Table 39.1 Comparison of various detection methods for Lena image aimed at RVIN

Methods	40%		50%		60%	
	Miss	False-hit	Miss	False-hit	Miss	False-hit
SD-ROM [3]	22,842	411	32,566	998	45,365	2651
PSM [4]	23,461	4773	29,396	8550	35,617	15,152
TSM [5]	18,703	3974	12,035	26,806	15,492	33,029
ACWM [6]	16,052	1759	23,683	2895	32,712	7644
MSM [7]	16,582	7258	20,857	10,288	26,169	15,778
PWMAD [8]	11,817	9928	14,490	15,003	17,760	19,577
ACWM-EPR [9]	13,657	6192	13,868	12,693	23,793	17,573
LUO [10]	16,706	3678	23,707	6410	38,512	12,724
ROAD-EPR [11]	13,476	8079	13,771	10,055	17,212	9330
ROLD-EPR [12]	12,010	7404	13,329	7761	14,967	9109
DWM [13]	9512	7761	9514	11,373	12,676	12,351
SAIKRISHNA [14]	31,755	1215	39,041	2047	45,908	4260
DEKA [15]	21,061	2063	24,903	3195	32,722	5047
PROPOSED	12,673	5645	13,841	5965	15,458	6355

Table 39.2 Detection results of the proposed method for various images at diverse noise densities

Images	40%		50%		60%	
	Miss	False-hit	Miss	False-hit	Miss	False-hit
Lena	12,673	5645	13,841	5965	15,458	6355
Goldhill	13,789	6958	14,598	7215	15,948	7853
Baboon	15,789	7856	15,998	8165	16,523	8639
Barbara	18,211	9562	19,365	10,245	21,548	11,149
Bridge	15,964	8102	16,235	8614	17,023	9017
Boat	16,245	8455	16,892	8824	17,226	9324
Fingerprint	23,548	11,269	25,128	13,019	27,693	15,238

accuracy of the proposed work for diverse images having dissimilar image depth and features. Lena, Goldhill and Baboon are the less intricate images where the proposed filter gives lower miss and false-hit results. In the case of Bridge, Boat and Barbara images, the proposed work also demonstrated good but higher total errors. Only Fingerprint image is the exception as it is having texture information that creates higher complications. It shows formidable false-hits but higher miss values. But altogether, if we judge the total noise, then all the images, except the Fingerprint, establish excellent outcomes.

As the proposed method uses fuzzy c-means clustering for detection purpose, a predetermined cluster size '4' is used. The cluster size is chosen on trial and error basis. Generally, as per the need of the algorithm, to properly find the cluster having maximum number of elements in it, minimum of three clusters are needed to be formed. Then only the homogenous non-corrupted pixels in a sample 5×5 window can form a maximum number of elements containing cluster. Moreover, too many clusters, like taking '6' clusters, are also not pertinent, as it can hamper the motive of the use of fuzzy c-means clustering. In Table 39.3, we have shown the "miss" and "false-hit" values taking cluster size as 3, 4, and 5 for seven test images at 60% noise density. It can be seen that the total erroneous numbers for cluster size 3 and 5 are a bit more. It justifies the reason to take cluster size as 4 for our detector.

Table 39.4 demonstrates the variation of window sizes that can be considered. The window size begins with 3×3 to keep it odd. Here 5×5 window size is taken for the proposed detector as compared to other mentioned window sizes, on the basis of trial and error to get the least error values. The table shows results of "miss" and "false-hit" values for 3×3, 5×5, 7×7 and 9×9 window sizes for three images at different noise densities. It can be seen that for 3×3 and 7×7 window sizes, the error values are deviated by reasonable but small amount. But those windows produced greater miss and false-hit values than the implemented 5×5 window. The

Table 39.3 Detection results by varying the cluster sizes for diverse images at 60% noise density

Images	Cluster size					
	3		4		5	
	Miss	False-hit	Miss	False-hit	Miss	False-hit
Lena	15,896	6539	15,458	6355	16,126	6896
Goldhill	16,114	7921	15,948	7853	16,549	8194
Baboon	16,658	8859	16,523	8639	17,011	9254
Barbara	22,021	12,169	21,548	11,149	22,716	12,532
Bridge	17,263	9368	17,023	9017	17,796	9736
Boat	17,985	9869	17,226	9324	18,161	10,254
Fingerprint	28,691	16,147	27,693	15,238	28,998	16,735

Table 39.4 Results of the proposed detector by varying the window size for diverse images

Images	Window size	50% ND		60% ND	
		Miss	False-hit	Miss	False-hit
Lena	3×3	14,019	6112	15,797	6658
	5×5	13,841	5965	15,458	6355
	7×7	13,998	6012	15,536	6521
	9×9	14,652	6524	15,997	6754
Barbara	3×3	19,721	10,469	21,765	11,324
	5×5	19,365	10,245	21,548	11,149
	7×7	19,659	10,411	21,694	11,206
	9×9	19,861	10,524	21,814	11,598

least window size of 3 × 3 has failed more at high noise densities, because, maximum number of pixels might be noisy and thus small numbers of non-corrupted pixels are left to form cluster. The 7 × 7 and 9 × 9 windows are also been tested. Those window sizes produced inferior results than 5 × 5. This validates the choice of 5 × 5 window and the superior results are reflected by the designated Table 39.4. Altogether the entire results portrayed superior performance of the proposed detector, even at high noise densities.

39.4 Conclusion

A random valued impulse noise detection technique based on fuzzy c-means clustering is proposed in this paper. The results justified the fact that use of fuzzy c-means clustering technique can be very useful to group analogous non-corrupted pixels to segregate them from the corrupted ones. Metrics like "miss" and "false-hit" validated the superior performance of the proposed detector at high noise densities. In future, another clustering method called K-medoid can be explored for detection operation.

References

1. Banerjee, S., Bandyopadhyay, A., Bag, R., Das, A.: Sequentially combined mean-median filter for high density salt and pepper noise removal. In: International Conference on Research in Computational Intelligence and Communication Networks, pp. 21–26. IEEE (2015)
2. Banerjee, S., Bandyopadhyay, A., Mukherjee, A., Das, A., Bag, R.: Random valued impulse noise removal using region based detection approach. Eng. Technol. Appl. Sci. Res. 7(6), 2288–2292 (2017)
3. Abreu, E., Lightstone, M., Mitra, S.K., Arakawa, K.: A new efficient approach for the removal of impulse noise from highly corrupted images. IEEE Trans. Image Process. 5(6), 1012–1025 (1996)
4. Wang, Z., Zhang, D.: Progressive switching median filter for the removal of impulse noise from highly corrupted images. IEEE Trans. Circuits Syst. II Analog Digit. Signal Process. 46(1), 78–80 (1999)
5. Chen, T., Ma, K.K., Chen, L.H.: Tri-state median filter for image denoising. IEEE Trans. Image Process. 8(12), 1834–1838 (1999)
6. Chen, T., Wu, H.R.: Adaptive impulse detection using center-weighted median filters. IEEE Signal Process. Lett. 8(1), 1–3 (2001)
7. Chen, T., Wu, H.R.: Space variant median filters for the restoration of impulse noise corrupted images. IEEE Trans. Circuits Syst. II Analog Digit. Signal Process. 48(8), 784–789 (2001)
8. Crnojevic, V., Senk, V., Trpovski, Z.: Advanced impulse detection based on pixel-wise MAD. IEEE Signal Process. Lett. 11(7), 589–592 (2004)
9. Chan, R.H., Hu, C., Nikolova, M.: An iterative procedure for removing random-valued impulse noise. IEEE Signal Process. Lett. 11(12), 921–924 (2004)
10. Luo, W.: A new efficient impulse detection algorithm for the removal of impulse noise. IEICE Trans. Fundam. Electron. Commun. Comput. Sci. 88(10), 2579–2586 (2005)
11. Garnett, R., Huegerich, T., Chui, C., He, W.: A universal noise removal algorithm with an impulse detector. IEEE Trans. Image Process. 14(11), 1747–1754 (2005)

12. Dong, Y., Chan, R.H., Xu, S.: A detection statistic for random-valued impulse noise. IEEE Trans. Image Process. **16**(4), 1112–1120 (2007)
13. Dong, Y., Xu, S.: A new directional weighted median filter for removal of random-valued impulse noise. IEEE Signal Process. Lett. **14**(3), 193–196 (2007)
14. Saikrishna, P., Bora, P.K.: Detection and removal of random-valued impulse noise from images using sparse representations. In: International Conference on Image Processing, pp. 1197–1201. IEEE (2013)
15. Deka, B., Handique, M., Datta, S.: Sparse regularization method for the detection and removal of random-valued impulse noise. Multimed. Tools Appl. **76**(5), 6355–6388 (2017)

Chapter 40
Distress-Level Detection Using Deep Learning and Transfer Learning Methods

Srijha Kalyan, Hamsini Ravishankar, and C. Arunkumar

Abstract The mental health of young adults and teenagers proves to be vital for having a flourishing life. Neglecting the issue of mental health can lead to anxiety, stress, and depression. These problems need to be addressed in the early stages to ensure better mental health for young adults. This paper provides a comprehensive study by leveraging natural language processing and deep learning techniques to detect depression by examining the relationship between language usage and the psychological characteristics of the communicator. A private dataset DAIC-WOZ has been used to predict the levels of depression. It includes audio, video, and textual information from 189 interviews. This paper focuses on the textual analysis of detecting depression from transcripts of the interviews conducted by an animated virtual interviewer. Due to the disadvantages faced by the context-independent nature of GloVe embeddings, the proposed approach uses transfer learning techniques such as ELMo, ULMFit, and BERT for predicting depression severity from transcripts. Furthermore, a Python Web application has been deployed to identify negative sentiments and depression severity from sentences inputted by the user. The proposed approach uses an ensemble learning method for the application to provide better predictions for classifying the texts into levels of depression. Hence, the inferences made in this paper can be extrapolated to other all demographics around the world to help detect depression in textual data as the algorithms and techniques used are all-encompassing..

40.1 Introduction

With the rise in easy access to social media and the Internet, the psychological health of young adults and teenagers is prone to getting affected. Neglecting to address this issue can lead to depression and in severe situations, even suicidal thoughts.

S. Kalyan · H. Ravishankar · C. Arunkumar (✉)
Department of Computer Science and Engineering, Amrita School of Engineering,
Amrita Vishwa Vidyapeetham, Coimbatore, India
e-mail: c_arunkumar@cb.amrita.edu

© The Author(s), under exclusive license to Springer Nature Singapore Pte Ltd. 2021   407
S. C. Satapathy et al. (eds.), *Smart Computing Techniques and Applications*,
Smart Innovation, Systems and Technologies 225,
https://doi.org/10.1007/978-981-16-0878-0_40

Early detection of a quick and successful automated diagnosis of depression could be of a great help. Depression causes cognitive and motor changes that affect speech production. Reduction in verbal activity productivity, prosodic speech irregularities, and monotonous speech have all been shown to be symptoms of depression.

Traditional machine learning techniques have not been able to capture the context of the words and the sentiments of the textual information to detect distress. Deep learning techniques have been incorporated to detect depression from textual data, speech and video data. Recent developments have mainly been focused on the detection of depression through multi-modal fusion by combining various modalities or types of information and deep learning methods using the audio, video, and textual information provided in the database [6, 7].

This paper focuses only on transcripts of the clinical interviews to detect the depression severity in the patients. In summary, the main contributions can be described as follows. Firstly, deep learning models using GloVe word embeddings along with LSTMs, GRUs, TextCNNs, and BiLSTMs have been implemented. To draw attention to the important words used by depressed patients, BiLSTM with Attention Layer has been proposed. Secondly, due to the disadvantages of the context-independent nature of GloVe word embeddings, Embeddings of Language Model (ELMo), a context-dependent model has been used. Thirdly, in order to understand the significance of transfer learning in NLP, Universal Language Model Fine Tuning (ULMFit) and Bidirectional Encoder Representation from Transformers (BERT) have been proposed for predicting the depression severity. Finally, a Python Flask Application using has been incorporated by leveraging ensemble method of averaging the predictions from all the algorithms that use GloVe word embeddings.

40.2   Related Methods

The following section provides a review of the literature related to the development of various deep learning models used to detect whether a person is suffering from distress or not. Ringeval et al. [1] in the "Real-life Depression, Recognition Workshop and Challenge" presented an approach where the baseline severity of depression was measured using random forests. The proposed approach showed that by combining the regression outputs of the unimodal random forest regressors, the fusion of audio and video modalities can be achieved. Dinkel et al. [2] proposed an approach that used multi-task modeling with pre-trained embedding of sentences, namely Elmo and BERT, to classify depression based on text. The Elmo and BERT models showed a connection between short sounds like "um" and the output of the script, suggesting that in order recognize depression, the text's behavioral aspects must be focused rather than material.

Further developments were done by Li et al. [3] where they proposed a model that used all data from audio, video, and text to extract the features. The proposed multimodel approach was based on a random forest regressor that predicted PHQ-8 scores ranging from 0 to 25, providing an indication of the extent of the stress/tension.

This work also used Covarep toolbox to get features from the audio. For getting features from video, the Facial Action Coding System was used. Moreover, Linguistic Inquiry and Word Count and AFINN toolbox were used for text data. Qureshi et al. [4] presented an early fusion approach that used neural networks and incorporated acoustics, visuals and textual modalities. It focused on the use of modality encoders, fusion sub-network, tensor fusion layer, Attention fusion sub-network and PHQ-8 Performance regression sub-network to achieve this. The results showed that integrating all modalities helped to provide the correct estimate of the degree of depression. In the regression phase, text modality was important and their proposed CombAtt architecture was much better than many other approaches.

In "Multi-level Attention network using text, audio and video for Depression Prediction" by Ray and Kumar [5] an early fusion-based multi-level attention network was proposed that integrated audio, video and text modalities to predict depression severity. The attention network gave maximum weight to the text modality and almost equal weight to the audio and video modes, was observed. Samareh et al. [6] adopted an input-specific classifier for each modality and a decision-level fusion module for predicting the final outcome. The paper used random forest to convert features into predictive scores for each biomarker modality, while those scores were further combined in a confidence-based fusion process to make the final prediction of PHQ8. Rohanian et al. [7] presented a model that learned depression indicators from all types of data in transcripts, including audio, video and lexical content. The paper used word-level multimodal fusion as a gating mechanism, used feed-forward highway layers. Venkataraman and Parameswaran [8] proposed a comprehensive study of facial expressions related to depression and their extraction methods. Happy faces dataset created by the JAFFE database was used. The facial characteristics for each face were then found using a 40 filter bank with Gabor filter. The functionality concatenated vector for each image was generated to form a training feature set. Bhaskar et al. [9] presented a new classification system for identifying emotions in human speech utterances that have been suggested. This method exploited corresponding audio and textual features. Multiclass SVM was used for classification of the emotions.

40.3 Proposed Approach

40.3.1 Dataset and Feature Selection

The Distress Analysis Interview Corpus (DAIC-WOZ) dataset used in the Audio/ Visual Emotion Challenge (AVEC 016) challenge has been used. It includes text, audio, and video information of 189 interviews designed to enable the treatment of emotional disturbance, nervous breakdown, and other mental health issues. The participants were asked to do a written Patient Health Questionnaire (PHQ8) test to identify depression levels. The scores were treated from (0–25) in bins separating them into levels of depression: none (0), mild (0–5), moderate (5–10), moderately severe (10–15), severe (15–25).

Feature extraction has been carried out for the verbal responses of each participant as given in the transcript file. From previous research work, it has been noticed that there are changes in the prosodic tones in a depressed person than a normal person. Also, they speak for a lesser amount of time with longer pauses. Consequently, from the words spoken by the participants which was subsequently made into a transcribed format, the total words uttered by the participant are calculated, and the average number of words has also been extracted in a similar manner. In addition, the time taken to complete the interview has also been taken into account. Furthermore, the count of the filler words such as "uh", "um" etc. used by the participants has been calculated. The number of times the participant had uttered in terms of first person by using a lot of "I" s are counted. An analysis of the frequency of the number of positive and negative words such as hate, horrible, miserable, etc. used by the participants has also been done. Harvard positive and negative word dictionary was used, and the words used in the conversations were extracted the frequency of these words has also been found.

40.3.2 Methodology

The following steps have been used for pre-processing the data and producing tokenized words from the transcript file which has been shown in Fig. 40.1.

1. Participant answers were subjected to stopword elimination and optionally stemming/lemmatization, and translated into a list of words.
2. Word tokenization was done, after all the words used in the responses has been collected.
3. By using a sliding window of size 10, the set of terms thus obtained is split into several sets.
4. The corresponding tokens list was padded with zeros for lists (after windowing) that have less than 10 characters.
5. For our layout, we had used GloVe embeddings, specific embeddings featuring 6 billion tokens, 400k word vocabulary and 100 dimensional vectors. A 2-D array was created which stores the vectors in the response vocabulary for each word (after removing the stopwords).

Fig. 40.1 Data preprocessing

40.3.3 Transfer Learning Models Implemented

Embeddings of Language Model (ELMo) Due to the disadvantages faced by the context-independent nature of GloVe embeddings, ELMo embeddings has been incorporated. ELMo is a deep contextualized representation of words that models both dynamic word-utilization features and the variations in semantic understandings. Developed by Allen NLP, the state-of-the-art pretrained model is context dependent. ELMo uses two bidirectional LSTMs in training, so the language paradigm knows both the succeeding and the preceding words in the expression fed to the model. This consists of a two-layer, bidirectional LSTM backbone.

Universal Language Model Fine Tuning (ULMFiT) ULMFiT works using the concept of inductive transfer learning. Inductive transfer refers to a learning mechanism's ability to improve performance on the present task after learning a different but related concept or skill on a prior task. Language model pretrained on the WikiText-103 dataset has been used to train the target task which is then fine-tuned for the classification model. In the proposed approach, the pre-trained language model is fine-tuned on the target DAIC-WOZ dataset.

Bidirectional Encoder Representations from Transformers (BERT) BERT is an open-sourced natural language processing pre-training model developed 2018. It uses $L = 24$ hidden layers, a hidden size of $H = 1024$, and $A = 16$ attention heads. On Wikipedia and BookCorpus, this model has been pre-trained for English. Inputs have been "uncased," i.e., the text has been lower-cased into word fragments before tokenization, and any accent marks have been removed. For preparation, the random input masking has been applied to word pieces independently.

40.3.4 Ensemble Methodology in Flask Application

A web application has been created using the light-weight Flask Framework. The pre-processed tokenized words are converted to word embeddings using the pre-trained GloVe word embeddings. The word embeddings are trained with each model to produce predictions followed by ensemble averaging method to produce better predictions. The ensemble predictions are then written into the Flask server and the predictions are the outputs based on the user's input sentences.

40.4 Results and Discussion

40.4.1 Model Performance Analysis

After running 30 epochs for the GloVe word embeddings + LSTM model with Adam optimization and categorical cross entropy loss, a test accuracy of 94.7% was obtained

```
All is going right with the party, I'm happy to know new people
none
I want an ice cream and have some fries for lunch
mild
I'm afraid of losing my work, I don't have any money
severe
I'm worried about my future, I'm afraid of it
severe
I am a graduate student
none
I am getting married
severe
This party is great, I know lots of people
none
I miss my parents, brothers and sisters
mild
I detest my horrible job
moderate
suicide
severe
```

Fig. 40.2 Output results for LSTM+GloVe model

without balancing the dataset. After the balancing the dataset, the accuracy increased to 98.6% with a test loss of 4.3%. Figure 40.2 represents the output results for the GloVe+LSTM model obtained for the test data.

Due to the significance of context dependency in sentences, ELMo was incorporated and gave an accuracy of 73.1%. Experimental results showed that the model with balanced data worked better than the imbalanced data. After implementing transfer learning using ULMFiT and BERT, it was observed that both the models performed well by obtaining an accuracy of 93.2% and BERT model with 82%.

Confusion matrix, a visual representation of the performance of all the models, has been incorporated. It depicts the number of predictions done by the model that have been classified correctly or incorrectly as true positives, false positives, true negatives, and false negatives. The precision, recall, and F1-scores have been calculated for each model from the confusion matrix obtained.

Precision and Recall: Precision is the measurement of the positive samples that are actually correct. It can be calculated by implementing the formula in Eq. (40.1).

$$\text{Precision} = \frac{\text{True Positive}}{\text{True Positive} + \text{False Positive}} \tag{40.1}$$

Recall is the proportion of the actual positives that have been correctly identified. Equation (40.2) represents the formula used to calculate recall.

$$\text{Recall} = \frac{\text{True Positive}}{\text{True Positive} + \text{False Negative}} \tag{40.2}$$

Table 40.1 Classification report of the language models

Models	Accuracy	Precision	Recall	$F1$ score
GloVe + LSTM (imbalanced)	0.94	0.93	0.94	0.95
GloVe + LSTM (balanced)	0.98	0.97	0.98	0.96
GloVe + 2 LSTMs	0.82	0.82	0.85	0.81
GloVe + BiLSTM	0.68	0.67	0.64	0.65
GloVe + Bi-GRU	0.93	0.94	0.89	0.93
GloVe + Text CNN	0.82	0.88	0.86	0.87
CNN + LSTM (Hybrid Model)	0.66	0.70	0.67	0.68
BiLSTM Attention Model	0.86	0.90	0.88	0.87
ELMo	0.74	0.74	0.73	0.75
ULMFiT	0.93	0.93	0.94	0.94
BERT	0.84	0.85	0.84	0.83

For a multi-class classification problem, macro-averaging and micro-averaging are performed for calculating precision and recall. In macro-averaging, the average of all the various classes is calculated, whereas for micro-averaging the precision and recall are calculated separately for each class and later added to produce the final results of precision and recall.

$F1$-Score: $F1$-Score has also been calculated as a performance measure for all the models. It is the harmonic mean of precision and recall. The $F1$-score provides a realistic measure of the test data performance using both precision and recall. The formula for calculating $F1$-Score has been shown in Eq. (40.3) . According to Table 40.1, a classification report has been computed for all the language models that have been implemented.

$$F1 - Score = 2 * \frac{precision * recall}{precision + recall} \quad (40.3)$$

40.5 Conclusion and Future Work

The objective of detecting depression through textual information involved in understanding the verbal and sentimental usage of words by the participants. In this paper, it was achieved by using advanced NLP techniques. From the analysis of the dataset, it was found out that there was an imbalance in the dataset that could have caused a bias toward the mild level of depression. It was solved by using random under-sampling. Experimental results showed that the model with the balanced data worked better than the imbalanced data. The BERT model gave good results, and the GloVe + LSTM model outperformed all other models. Moreover, work needs to be done to explore Explainable AI in order to understand decision making processes in the mild/severe levels which will help in understanding the models better because under-standing the output is pivotal for researchers and professionals alike. Since the scope of the project has been restricted to textual information, the scope of the project can be widened to explore on the analysis of audio and video features using deep learning techniques. Furthermore, detecting depression on social media such as Facebook or Twitter can also be done in the future.

Declaration We, authors declare that we have taken required permissions from the respective person/authority to use the image/dataset in our work. We shall be responsible in case any dispute arises in the future with regard to this.

References

1. Ringeval, F., Schuller, B., Valstar, M., Gratch, J., Cowie, R., et al.: AVEC 2017—"Life depres-sion, and affect recognition workshop and challenge". In: 7th International Workshop on Audio/Visual Emotion Challenge, AVEC'17, co-located with the 25th ACM International Con-ference on Multimedia, MM, Oct. 2017, pp. 3–9, Mountain View, USA
2. Dinkel, H., Wu, M., Yu, K.: Text-based depression detection: what triggers an alert. In: MoE Key Lab of Artificial Intelligence SpeechLab. Department of Computer Science and Engineering, Shanghai Jiao Tong University, Shanghai, China (2019)
3. Li, B., Zhu, J., Wang, C.: Depression severity prediction by multi-model fusion. In: China The Third International Conference on Informatics and Assistive Technologies for Health-Care. Medical Support and Wellbeing, Department of Electronic Engineering, Shanghai Jiao Tong University, Shanghai, China (2018)
4. Qureshi, S.A., Hasanuzzaman, M., Gael Dias, S.S.: The Verbal and Non Verbal Signals of Depression-Combining Acoustics. Text and Visuals for Estimating Depression Level. IEEE, New York (2019)
5. Ray, A., Kumar, S., Reddy, R., Mukherjee, P., Garg, R.: Multi-level Attention network using text, audio and video for depression prediction
6. Samareh, A., Jin, Y., Wang, Z., Chang, X., Huang, S.: Predicting depression severity by multi-modal feature engineering and fusion. In: AAAI Conference (2018)
7. Rohanian, M., Hough, J., Purver, M.: Detecting depression with word-level multimodal fusion. In: Interspeech (2019)
8. Venkataraman, D., Parameswaran, N.S.: Extraction of facial features for depression detection among students. Int. J. Pure Appl. Math. **118**, 455–462 (2018)
9. Bhaskar, J., Sruthi, K., Nedungadi, P.: Hybrid approach for emotion classification of audio conversation based on text and speech mining. In: International Conference on Information and Communication Technologies (ICICT 2014)

Chapter 41
Fingerprint Genuinity Classification

Meega Annu Jacob and Nirmal Varghese Babu

Abstract A dataset that consists of both contactless and contact-based fingerprints was collected using various instruments like sensors, images, etc. Sensor Interoperability issue was handled using this system. Implementation of contactless fingerprint technologies widely depends on the progressive capability to match contactless 2D fingerprints with contact-based fingerprint databases. Contactless 2D fingerprint identification is more safe and it allows deformation-free imaging for advanced accuracy. Here, an Auto Stack Encoder for accurately matching contact-based and contactless fingerprint images was proposed. Auto Stack Encoder is a neural networks for which the input is similar as the output as they work by reducing the input into a latent-space representation and then again constructing the output from the corresponding image. This Auto Stack Encoder is trained using the images of fingerprint and features and classified as genuine (1) or fake (0). Therefore, an additional robust thin-plate spline model from the concatenation of deep extracted feature vectors made from different networks. This paper shows the Unsupervised learning approach for the Fingerprint Genuinity Classification of Contactless and Contact based fingerprints using the Auto Stack Encoder.

41.1 Introduction

Fingerprint recognition technology [1] is unique of most methods for human identification, say authentication. Contactless 2D/3D [2, 3] fingerprint identification systems is a newly introduced method to identify the limits of customary systems uses the contact-based fingerprint [4]. Development of the advancement in abilities to match contact-based fingerprint images with contactless fingerprints [5, 6] is critical for the success of technologies as millions of fingerprints in huge databases have been acquired using contact-based technologies.

M. A. Jacob · N. V. Babu (✉)
Department of Computer Science and Engineering, Amal Jyothi College of Engineering, Kanjirappally, Kerala, India

© The Author(s), under exclusive license to Springer Nature Singapore Pte Ltd. 2021   415
S. C. Satapathy et al. (eds.), *Smart Computing Techniques and Applications*,
Smart Innovation, Systems and Technologies 225,
https://doi.org/10.1007/978-981-16-0878-0_41

Comparing the contact based 2D fingerprint [7, 8] images with contactless fingerprints are acute for the success of contactless 2D fingerprint technologies, which offer more safe and deformation less acquisition of fingerprint features. The minutia feature is supposed to be the precise fingerprint feature which has its competence and dependability in recognizing fingerprints. Among the contactless and contact-based fingerprint sensors, it is problematic to collect minutiae features and to guarantee their images from two such sensors [4].

41.2 Related Works

To identify the deformations in the contact-based fingerprints, some approaches have been introduced. Cappelli et al., states a plastic distortion model to address the non-linear deformations in contact-based fingerprints. In Ross and Jain [9], fingerprint sensor interoperability problem for two different flat (plain) fingerprint sensors and they examined the extent of degradation in the matching accuracy when images from two such different sensors are matched. In [10], the thin-plate spline model was introduced to identify the deformation in fingerprints. Local minutiae are found out for the deformation correction and the global minutiae are used for further refining the correction. Global minutiae [11] features are used to match the fingerprints.

Ross et al. [12] proposed a deformation model to identify the fingerprint impressions deformation. They introduced a thin-plate spline model on multiple fingerprint impressions from the same fingers and also found out the average deformation of the fingerprints. In Ross and Nadgir [13], proposed a thin plate spline calibration model to image the fingerprint sensor interoperability. This average deformation model was learnt to correct the deformations of fingerprints from the different sensors.

41.3 Proposed System

The various modules that include in the proposed method are:

- Data Collection
- Fingerprint Preprocessing
- Feature Extraction
- Classification using Auto Encoder
- Results Evaluation.

The block diagram of the proposed approach using Auto Stack Encoder is illustrated in Fig. 41.1. Model is trained with the contact-based fingerprint images with several contactless fingerprints from the same fingers. For training the contact-based fingerprint impression deformation correction model, RTPS Model was used. The unknown contact-based fingerprints is into deformation correction using the Deformation correction model for the better matching with the contactless fingerprint. This

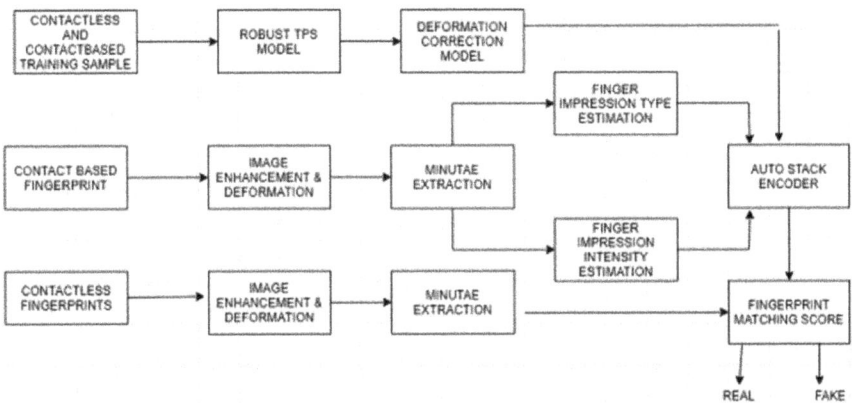

Fig. 41.1 Proposed system

approach estimates the contact-based fingerprint impression type by comparing its image centre. Fingerprint impression intensity is computed from the intensity of grey levels in the fingerprint image. Based on corresponding impression type and intensity, the deformation of each of the contact-based fingerprint is corrected with the help of Auto Stack Encoder. So, the paper is using Auto Stack Encoder to create deep learning recognition system for Contact based and Contactless fingerprint. After the classification of the fingerprints (fake or genuine), the performance of the particular system was evaluated and the accuracy of the system was calculated using the confusion matrix created. Figure 41.1 shows the proposed system architecture.

PolyU dataset was used. It consists of 1800 2D contact-based fingerprint images and 1800 2D contactless fingerprints from 300 different fingers. The contactless 2D fingerprint images are captured using a low-cost camera and lens. 2D contact-based fingerprints were collected using URU 4000 fingerprint reader with FBI/NIST standards. In this work, the dataset contains 150 samples, for each finger, there are 2 contactless fingerprint samples and 4 corresponding contact-based fingerprint samples. Only 100 images of the dataset were used for the training and 50 for testing. The resolution of 2D contact-based fingerprint images and contactless fingerprint was respectively 1400*900 and 356*328 with 500 dpi, respectively.

Fingerprint Preprocessing Because of the essentiality of contact-based and contactless [14] sensing technologies, the images from the corresponding finger using such sensors looks different. Therefore, it is necessary to introduce fingerprint image preprocessing. The same scale ratio or size was ensured by the down sampling operation The region of interest [2] in the images is automatically removed with the centre position of fingerprint impression. For increasing the similarity between the two cross sensor fingerprints from the same subject/finger, Histogram equalization is applied on all contactless and contact-based fingerprints.

Feature Extraction The minutiae features [15] are most correct and they are employed in almost all the fingerprint technologies available nowadays. The drawback of the degradation in matching accuracy [16] is due to minutiae missing is more serious in the contactless fingerprints because the nature of contactless allows a high degree of freedom in the collection under the limited field of view and distortions in regions away from the image center. Therefore, it is unanimous to introduce the minutiae feature [10, 13] and respective ridge feature to improve the learning for the joint fingerprint feature from the two sensors. For each fingerprint, minutiae map is generated by marking the minutiae location [12], direction and quality on the image. Each minutiae point is marked using a solid circle of radius equal to 2 pixels, to avoid the overlapping.

Classification: Auto Stack Encoder The main motive is to match the fingerprints collected from different sensors accurately. Auto Stack Encoder [17, 18] classifies the contactless and contact-based fingerprints [17–23] and it finds the accuracy at different conditions. An auto stack encoder has three layers: an input layer, a hidden layer, and a decoding layer [23]. The network is trained to construct its inputs again, which makes the hidden layer to learn representations of the inputs to find the accuracy, precision, and the values will be checked with the existing machine learning algorithms [10, 11, 19]. Figure 41.2 shows the Basic Auto Stack Encoder. Figure 41.3 shows the Auto Stack Encoder Training procedure.

Results Evaluation The data will be depicted based on the obtained results while using the Auto Stack Encoder will be verified based on the accuracy of the procedure. A Confusion Matrix created, shows the performance of the classification algorithm, where the future values will be predicted using the train values with the corresponding labels. Figure 41.4 shows the Genuine and Fake Classification. The results from different networks will be crossmatched using the auto stack encoder.

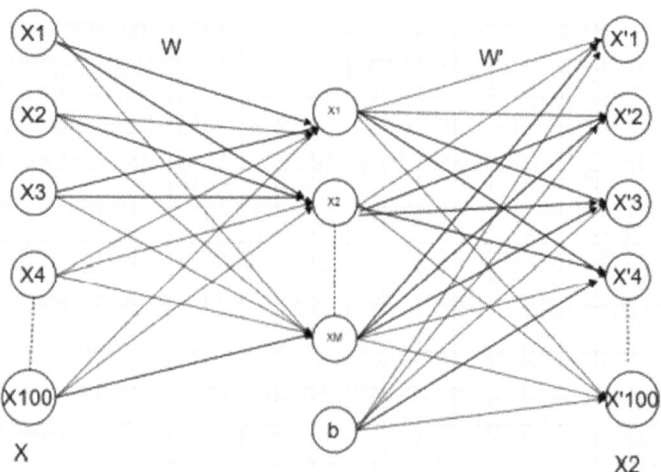

Fig. 41.2 Basic autostack encoder

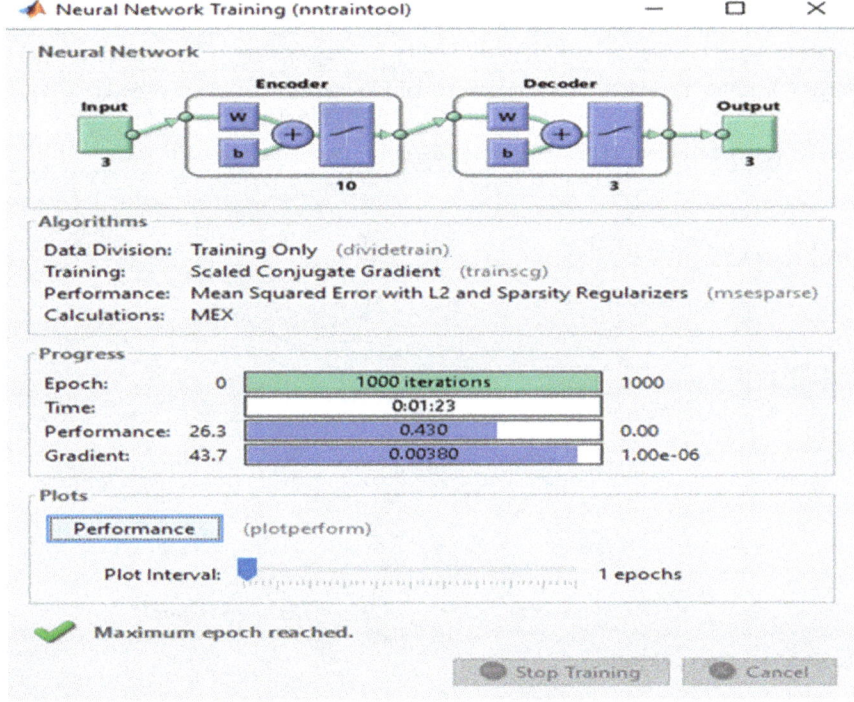

Fig. 41.3 Autostack encoder

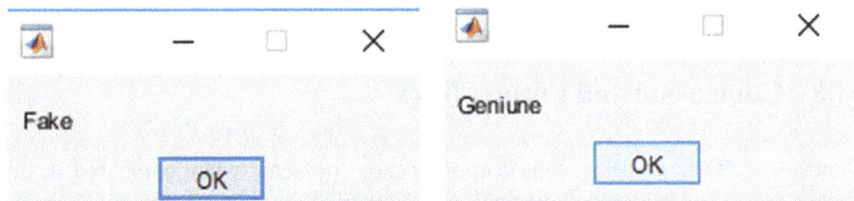

Fig. 41.4 Genuine and fake classification

41.4 Experimental Results

The proposed system was implemented using Auto Stack Encoder in the classification procedure. The accuracy of model using Auto Encoder showed the highest while others showed less. Table 41.1 shows the accuracy of the experiment using the contactless to contact-based fingerprints. Figure 41.4 shows the Genuine and Fake Fingerprint classification output. Figure 41.5 shows the graph comparing the accuracy of the RTPS + DCM Model and proposed Auto stack Encoder Model. Confusion Matrix was created using the training data with the predicted data. Here,

Table 41.1 Accuracy comparison

Database	Experiment	Accuracy (%)
PolyU Contactless to contact based fingerprint database	RTPS + DCM, *Lin et al. [1]	94.11
PolyU Contactless to contact based fingerprint database	Auto stack encoder	96

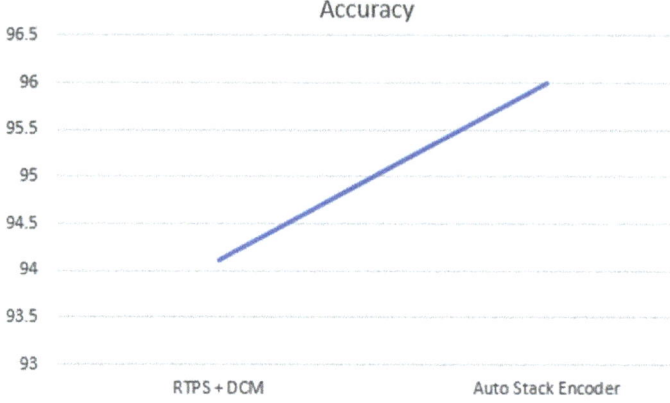

Fig. 41.5 Accuracy comparision

we obtained high accuracy for fingerprint classification compared to other existing methods.

41.5   Conclusion and Future Work

Contactless 2D fingerprint identification is more hygienic when compared to the contact-based and it enables deformation-less imaging for higher accuracy. The main function of Auto Stack encoder is for a contact-based and contactless fingerprint matching process. This trained method is using fingerprint images and extracted features from the collected fingerprints. Therefore, a fingerprint representation is formed from the combination of deep feature vectors generated from different networks. The experiment results using publicly available database illustrate the method which can achieve superior matching performance. Sensor interoperability using contact-based and contactless fingerprint is an emerging problem which is need to be addressed with other deep learning methods in the future.

41.6 Declaration

The permission from respective authorities to use the images/data as given in the paper was obtained. In case of any dispute in the future, we shall be wholly responsible.

References

1. Lin, C. and Kumar, A.: Matching contactless and contact-based conventional fingerprint images for biometrics identification. IEEE Trans. Image Process. **27**(4)
2. Bouazizi, M., Ohtsuki, T.: Contactless fingerprint identification using level zero features. In: IEEE (2016)
3. Hiew, B.Y., Teoh, A.B. and Pang, Y.H.: Touch-less fingerprint recognition system. In: IEEE (2007)
4. Jin, Q., Tago, K.: Addressing sensor interoperability problem using fingerprint segmentation. In: 2017 IEEE 10th International Conference on Service-Oriented Computing and Applications (2017)
5. Roy, A., Memon, N., Ross, A.: Masterprint: exploring the vulnerability of partial fingerprint-based authentication system. IEEE Trans. Inf. Foren. Secur. **12** (2017)
6. Bontrager, P., Roy, A., Togelius, J., Memon, N., Ross, A.: DeepMasterPrints: generating MasterPrints for dictionary attacks via latent variable evolution. In: 2018 IEEE International Conference on Biometrics: Theory, Applications and Systems (2018)
7. Tiwari, K. and Gupta, P.: A touch-less fingerphoto recognition system for mobile hand-held devices. IEEE (2016)
8. Pavithra, R., Suresh, K.V.: Fingerprint image identification for crime detection. In: 2019 International Conference on Communications and Systems (2019)
9. Ross, A., Jain, A.: Biometric sensor interoperability: a case study in fingerprints. In: Biometric Authentication, pp. 134–145. Springer, Berlin, Germany (2004)
10. Loey, M., El-Sawy, A., El-Bakry, H.: Deep learning autoencoder approach for handwritten Arabic digits recognition. In: 2017 Computer Vision and Pattern Recognition (2017)
11. Lin, C., Kumar, A.: A CNN-based framework for comparison of contactless to contact-based fingerprint. In: IEEE Transactions on Information Forensics and Security (2018)
12. Dong, G., Liao, G., Liu, H., Kuang, G.: A review of the autoencoder and its variants: a comparative perspective from target recognition in synthetic aperture radar images. IEEE Geosci. Remote Sens. Mag. (2018)
13. Szymkowski, M., Saeed, K.: A novel approach to fingerprint identification using method of sectorization. In: 2017 International Conference on Biometrics and Kansei Engineering (2017)
14. Labati, R.D., Genovese, A.: Toward unconstrained fingerprint recognition: a fully touchless 3-D system based on two views on the move. In: IEEE Transactions on Systems, Man, and Cybernetics: Systems (2015)
15. Liu, X., Pedersen, M., Charrier, C., Cheikh, F.A. and Bours, P.: An improved 3-step contactless fingerprint image enhancement approach for minutiae detection. In: IEEE (2016)
16. Alshehri, H., Hussain, M., Aboalsamh, H.A., Al Zuair, M.A.: Cross-sensor fingerprint matching method based on orientation, gradient, and gabor-HoG descriptors with score level fusion. IEEE Trans. Inf. Foren. Secur. (2018)
17. Zhai, J., Zhang, S., Chen, J., He, Q.: Autoencoder and its various variants. IEEE International Conference on Systems, Man and Cybernetics (2018)
18. Schuch, P., Schulz, S., Busch, C.: De convolutional auto encoder for enhancement of fingerprint samples. In: 6th International Conference on Image Processing Theory, Tools and Applications (2016)

19. Lin, C., Kumar, A.: Improving cross sensor interoperability for fingerprint identification. In: 23rd International Conference on Pattern Recognition (ICPR) (2016)
20. Cappelli, R., Maio, D., Maltoni, D.: Modelling plastic distortion in fingerprint images. In: Advances in Pattern Recognition—ICAPR, pp. 371–378. Springer, Berlin, Germany (2001)
21. Bazen, A.M., Gerez, S.H.: Fingerprint matching by thin-plate spline modelling of elastic deformations. Pattern Recognit. **36**(8), 1859–1867 (2003)
22. Ross, A., Dass, S., Anil, J.: A deformable model for fingerprint matching. Pattern Recognit. **38**(1), 95–103 (2005)
23. Ross, A., Nadgir, R.: A thin-plate spline calibration model for fingerprint sensor interoperability. IEEE Trans. Knowl. Data Eng. **20**(8), 1097–1110 (2008)

Chapter 42
Feature Selection by Associativity for Sentiment Analysis

S. Fouzia Sayeedunnisa, Nagaratna P. Hegde, and Khaleel Ur Rahman Khan

Abstract Sentiment analysis faces the biggest challenge of high-dimensional feature set. The current trends contemplate this intricate process through conventional feature selection/extraction. In this proposal, the statistical feature selection method, Wilcoxon sign rank score, is used for optimal feature selection. The proposal evaluates and compares the conventional filter method of feature selection with the Wilcoxon sign score for finding the optimal features. The features are defined by feature association between the terms and the sentiment lexicon. The identified optimal features using information gain (IG), chi square, and statistical method Wilcoxon sign score are then classified using the Adaboost classifier. The experiment was conducted on three datasets widely used for sentiment analysis. Result from the experiment demonstrate that the classification accuracy using the proposed method of optimal feature selection is much higher when compared to the contemporary models.

42.1 Introduction

Opinion mining is a sub-domain of text mining which deals with classification of human-generated online text as positive or negative. With the exponential growth of Internet users, accessing social media has led to generation of vast content online. The social media platforms are used to express the views, opinion about products, brands, politics, and every other field. The main focus of SA is to draw out and identify opinion from the vast online content. The opinions drawn out from this

S. F. Sayeedunnisa (✉)
M.J. College of Engineering and Technology, Hyderabad, India

N. P. Hegde
Vasavi College of Engineering, Hyderabad, India

K. U. R. Khan
ACE Engineering College, Hyderabad, India
e-mail: khaleelrkhan@aceec.ac.in

© The Author(s), under exclusive license to Springer Nature Singapore Pte Ltd. 2021   423
S. C. Satapathy et al. (eds.), *Smart Computing Techniques and Applications*,
Smart Innovation, Systems and Technologies 225,
https://doi.org/10.1007/978-981-16-0878-0_42

online content is the major source of information which provides valuable insights [1] to the stakeholders or decision makers in improving their business [2].

Machine learning (ML) techniques for sentiment classification of user generated data includes ensemble techniques, supervised, semi-supervised, and unsupervised learning. Researchers have used lexicon-based techniques incorporating dictionary methods [3], corpus methods [4], and lexicon method with NLP [5]. Sentiment identification is a three-step process where feature extraction is the initial step, second is feature selection, and last is classification of text as in three classes, i.e., positive, negative, or neutral. Natural language processing (NLP), ML, and information retrieval techniques are used by researchers to achieve this.

The major challenge in classification of text is high dimensions of features, overlapping and redundant features [6] which takes lot of time for processing. These challenges not only affect the classifier performance but also increase the cost. The most effective techniques for dealing with these challenge are feature selection suggested by [7]. Feature selection helps in determining an ideal subset which is more proximate to the actual data.

The studies in literature [8, 9] classifies the FS methods into three categories. The first one is the filter method which calculates the score or rank of a feature. All the high rank features are selected, eliminating the low rank/score feature. In this method of FS, the relationship or association of features with each other is not considered. The second method of FS is the wrapper method which uses classifier to evaluate optimal feature subset. The last method of FS is the embedded approach which integrates feature selection into classifier. This approach faces the challenge of learning algorithm [5].

42.2 Related Work

The recent research in the field of feature selection and extraction which is predominant in selecting extremely important subset of feature called as optimal features which is referred in this section.

Research by [10] has developed proportional rough feature selector (PRFS) a unique method which is using rough set theory to identify if a term exactly belongs to a class or possibly belongs to a class. Experimental results uses Macro-F1 measure which signifies competitive performance of PRFS in view of other feature selection methods.

In the work done by [11], feature selection is done by filter methods. The methods used are term frequency (TF), information gain (IG), document frequency (DF), chi square (CHI), and mutual information (MI). These five methods are successful in eliminating 97% of the low-scoring features. The optimal feature significantly improves the accuracy.

In the work by [12], Pearson correlation-based attribute selection is used with six machine learning (ML) classifiers. The authors use string-to-word vector (STWV) with attribute selection measure on the ML classifiers. Results demonstrate that the obtained optimal feature improves the accuracy between 78 and 88%.

In the work on feature subset selection by [13], an ordinal-based integration of different feature vectors (OIFV) is proposed as a new feature vector. In the proposed method, hybrid filter and wrapper methods are applied to obtain a frequency-based integration of optimal features as subsets. The experiment results signifies better classification accuracy.

It is evident from the recent work that integration of two methods can be more helpful toward optimal feature selection. In the proposed study, the feature association metric by [14] is used to evaluate the performance of Adaboost classifier against the filter-based feature selection techniques like IG and chi square.

42.3 Feature Selection

42.3.1 Existing Feature Selection Methods

The filter-based method of feature selection used is information gain (IG) and chi square (CHI). They are descried below:

Information Gain (IG): It is an entropy-based method for evaluating features. It is a machine learning technique used in feature selection. It is the amount of information gained [15] by presence or absence of a feature for classification.

$$IG(C, A) = H(C) - H(C|A) \tag{42.1}$$

where $IG(C, A)$ is information gain for feature A in Class C, $H(C)$ is entropy across sentiment class C, and $H(C|A)$ is conditional entropy for feature A

$$H(C) = -\sum p(C = i) \log p(C = i) \tag{42.2}$$

$$H(C|A) = -\sum p(C = i|A) \log p(C = i|A) \tag{42.3}$$

The calculated IG values for every feature lies between 0 and 1.

Chi Square (CHI): The chi square test [16] is a statistical test to find if there is an association or correlation between variables. Chi square measures the deviation of expected counts and observed counts.

$$X^2 = \frac{\Sigma(\text{Observed} - \text{Expected})2}{\text{Expected}} \tag{42.4}$$

42.3.2 Proposed Method Using Feature Association as a Feature

Feature Association: The statistical method adapted [14] to select optimal subset of features is Wilcoxon sign rank. Feature association frequency (FAF) is computed for each feature with respect to positive and negative sentiment lexicon [17]. The association method is novel and invalidates the impact of humongous features present in social media data. The approach devised can eliminate the irrelevant features. In the starting phase, the sentiment lexicon is extracted from the labeled opinion documents. Further, the terms that appear other than lexicon are considered as features, and their FAF with respect to each extracted sentiment lexicon is calculated. The FAF is defined as the ratio of the "occurrence sentiment lexicon in association with a term and the total term occurrence."

Optimal Feature Selection using Wilcoxon Signed rank (WSR): The WSR test [18] is used to compare related samples of normally distributed data. It works on ranked data. This test is applied to the feature association frequency (FAF) of positive sentiment lexicon and the FAF of negative sentiment lexicon. We have adopted the process of optimal feature selection by [14] using WSR score. The optimal feature of positive is the one which has a greater sum of Wilcoxon positive rank when compared to the sum of negative rank.

42.3.3 Preprocessing

The training data consist of positive and negative labeled opinion records. Each row of a record set consists of bag of words. The stop words and non-textual words are removed, and further stemming is applied. This results in two sets of records *PS* and *NS* where *PS* represents all positively labeled record and *NS* represents all negatively labeled preprocessed records.

42.3.4 Sentiment Lexicon Extraction

We use sentiment lexicons provided by [17]. The features in *PS* and *NS* are compared with [17] and the extracted lexicons are stored in *l*+ and *l*−. The algorithm for feature association [14] is given below:

Algorithm Feature Association

Input : Pre- processed data labelled as positive/negative

Output : Features with Feature Association Frequency

Notation

PS: Records containing all positive labelled opinion

NS: Records containing all negative labelled opinion

l+: Set containing lexicons present in records of *PS*

l-: Set contains lexicons present in records of *NS*

wl: Unique word list existing in both *PS* and *NS* records.

Procedure Feature Selection()

 for each pre-processed record

  for each word in review

   Search word in Opinion Lexicon[16]

   If found in Positive Lexicon then

    Insert(*l+, word*)

   Else If found in Negative Lexicon then

    Insert(*l-, word*)

   Else

    Insert(*wl, word*)

  Next word

  Next record

 for each term *wl* **do**

  Initialize term occurrence count $c(t_i)$ to 0

  for each record of *PS* **do**

   if the term t_i exist in record then increment $c(t_i)$

  End

  for each record of *NS* **do**

   if the term t_i exist in record then increment $c(t_i)$

  End

 End

  //Sentiment lexicon and term co-occurrence(Association)//

 for each term in *wl* **do**

  for each positive lexicon in *l+* **do**

   Initialize co-occurrence counter (lexicon and term) in *PS* to 0

   for each record in *PS* **do**

    if term and lexicon co-occur in record

     Increment the term and lexicon co-occurrence counter

   End

  End

 End // Repeat the process for negative term and lexicon

    // FAF (Feature Association Frequency)//

 for each term in *wl* **do**

  for each word in *l+* which is a positive lexicon **do**

   Find FAF_{PS} as the ratio of positive sentiment lexicon and term co-occurrence count to the term occurrence

  End

  for each word in *l-* which is a negative **do**

   Find FAF_{NS} as the ratio of negative sentiment lexicon and term co-occurrence count to the term occurrence

  End

 End

42.4   Classifier

The concept of classification in machine learning is to categorize labeled data into set of classes or groups. Predicting the target class of the test data is the main aim of classification. The classifier performs best if the training data comprises relevant observation with the class membership defined. Adaboost [19] is an ensemble classifier. It is helpful over the binary classification for boosting the performance. It is a combination of weak classifier to form strong classifier. It is an iterative method of training by retaining the algorithm.

42.5   Experimental Study

The input data is preprocessed, and feature selection techniques, i.e., IG and chi square, are applied. The optimal features of the proposed method are calculated by the WSR between the FAF of a term with positive lexicon and the FAF of the corresponding term with negative lexicons. The accuracy of the optimal features from distinctive dataset is tested on Adaboost classifier.

42.5.1   Datasets and Statistics

Experimental study of the proposed method and conventional filter methods is carried out by using three benchmark datasets, namely First GOP debate [20], Twitter sentiment analysis [21], and sentiment 140 [22]. The Twitter sentiment analysis [21] has 99,989 reviews, among them 19,998 are used and rest are excluded for analysis. The sentiment 140 [22] consists of 1.6 million of tweets, out of which 16,000 tweets were used for analysis and rest were excluded because of difficulty encountered during preprocessing (Table 42.1).

The optimal features obtained by different feature selection techniques, i.e., WSR score, IG, and CHI, are given in Table 42.2. In furtherance, classification of optimal features using Adaboost classifier is shown in Fig. 42.1

Table 42.1 Statistics of dataset

Dataset	Total reviews	Training data	Testing data	No of sentiment lexicons found in training set
First GOP debate	13,871	10,403	3467	21
Twitter sentiment analysis	19,998	14,998	4999	37
Sentiment140	16,000	12,000	4000	40

Table 42.2 Optimal features detected by feature selection strategy

Dataset	Features found in training	Optimal features by WSR	Optimal features by IG	Optimal features by CHI
First GOP debate	1535	88	175	170
Twitter sentiment analysis	2371	152	257	255
Sentiment140	2147	210	305	303

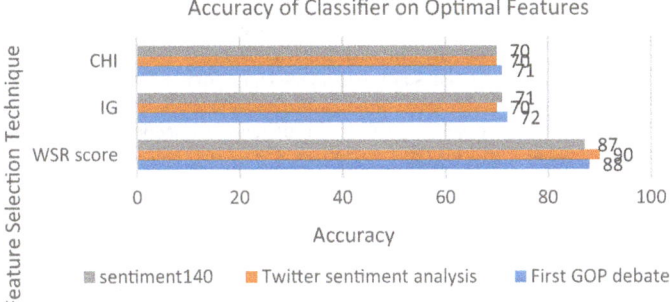

Fig. 42.1 Classifier accuracy statistics

The accuracy of the classifier for all the three dataset can be observed as shown in Fig. 42.1. Results from the experiment conducted signify that the optimal features selected by WSR test are not only less in number but also improves the classifier accuracy when compared to the filter method of feature selection. The classifier could attain a maximum accuracy of 90% on Twitter dataset using Wilcoxon Score as the feature selection technique. The key factor in improvement of classification accuracy is the relevance of features selected by Wilcoxon signed rank score.

42.6   Conclusion

In this proposal, a novel feature association technique as the sentiment lexicon singly cannot predict the class label with adequate accuracy. The feature association is the ratio of co-occurrence of a sentiment lexicon and a term to the total occurrence of the term. The optimal features are obtained by the Wilcoxon sign test of FAF. Experiment result demonstrates that the optimal features selected with Wilcoxon score gives better classifier accuracy with reduced features when compared to the filter methods. The Adaboost classifier could achieve an accuracy level around 90% across the Twitter datasets.

References

1. Wu, Y., Ren, F.: Learning sentimental influence in twitter. In: 2011 International Conference on Future Computer Sciences and Application, pp. 119–122. IEEE (2011)
2. Chung, W., Tseng, T.L.B.: Discovering business intelligence from online product reviews: a rule-induction framework. Expert Syst. Appl. **39**(15), 11870–11879 (2012)
3. Medhat, W., Hassan, A., Korashy, H.: Sentiment analysis algorithms and applications: a survey. Ain Shams Eng J **5**(4), 1093–1113 (2014)
4. Bhatnagar, V., Goyal, M., Hussain, M.A.: A novel aspect based framework for tourism sector with improvised aspect and opinion mining algorithm. Int. J. Rough Sets Data Anal. (IJRSDA) **5**(2), 119–130 (2018)
5. Taboada, M., Brooke, J., Tofiloski, M., Voll, K., Stede, M.: Lexicon-based methods for sentiment analysis. Computational Linguistics **37**(2), 267–307 (2011)
6. Savoy, O.K.J.: Feature selection in sentiment analysis. In: Proc. CORIA No. January, 273–284 (2012)
7. Sharma, A., Dey, S.: Performance investigation of feature selection methods and sentiment lexicons for sentiment analysis. IJCA Spec. Issue Adv. Comput. Commun. Technol. HPC Appl. **3**, 15–20 (2012)
8. Ekbal, A., Saha, S.: Combining feature selection and classifier ensemble using a multiobjective simulated annealing approach: application to named entity recognition. Soft. Comput. **17**(1), 1–16 (2013)
9. Bharti, K.K., Singh, P.K.: Hybrid dimension reduction by integrating feature selection with feature extraction method for text clustering. Expert Syst. Appl. **42**(6), 3105–3114 (2015)
10. Cekik, R., Uysal, A.K.: A novel filter feature selection method using rough set for short text data. Expert Syst. Appl. **160**, 113691 (2020)
11. Rogati, M., Yang, Y.: High-performing feature selection for text classification. In: Proceedings of the 11th International Conference on Information and Knowledge Management, pp. 659–661 (2002)
12. Sharma, S., Jain, A.: An empirical evaluation of correlation based feature selection for tweet sentiment classification. In: Advances in Cybernetics, Cognition, and Machine Learning for Communication Technologies, pp. 199–208. Springer, Singapore (2020)
13. Yousefpour, A., Ibrahim, R., Hamed, H.N.A.: Ordinal-based and frequency-based integration of feature selection methods for sentiment analysis. Expert Syst. Appl. **75**, 80–93 (2017)
14. Sayeedunnisa, S.F., Hegde, N.P., Khan, K.U.R.: Wilcoxon Signed Rank Based Feature Selection for Sentiment Classification. In: Proceedings of the 2nd International Conference on Computational Intelligence and Informatics, pp. 293–310. Springer, Singapore (2018)
15. Yang, Y., Pedersen, J.O.: A comparative study on feature selection in text categorization. In Icml, vol. 97, no. 412–420, p. 35 (1997)
16. Li, Y., Luo, C., Chung, S.M.: Text clustering with feature selection by using statistical data. IEEE Trans. Knowl. Data Eng. **20**(5), 641–652 (2008)
17. Mohammad, S.M., Turney, P.D.: Crowdsourcing a word–emotion association lexicon. Comput. Intell. **29**(3), 436–465 (2013)
18. Meek, G.E., Ozgur, C., Dunning, K.: Comparison of the t vs. Wilcoxon signed-rank test for Likert scale data and small samples. J. Mod. Appl. Stat. Methods **6**(1), 10 (2007)
19. Bloehdorn, S., Hotho, A.: Boosting for text classification with semantic features. In International Workshop on Knowledge Discovery on the Web, pp. 149–166. Springer, Berlin, Heidelberg (2004)
20. https://www.figure-eight.com/wp-content/uploads/2016/03/GOP_REL_ONLY.csv
21. https://www.kaggle.com/c/twitter-sentiment-analysis2/data
22. Go, A., Bhayani, R., Huang, L.: Twitter sentiment classification using distant supervision. CS224N Proj. Rep. Stanford **1**(12), 2009 (2009)

Chapter 43
Optimization of Serviceability of Mobile Communication Towers Under Capital Constraints

Pankaj Preet Singh Dhaliwal, Mahendra Kumar, and Veerendra Dakulagi

Abstract Mobile communication towers are important component of mobile communication system for civil and military applications. The major part of technical equipment of mobile communication system is concentrated on these towers. These towers have geographical separation and they are normally unmanned. In remote areas especially away from towns the alternate routes for communication in the system may not be available. Therefore high degree of serviceability of these towers is very essential. This can be ensured by having inbuilt reliability of equipment, easy fault finding, availability of repair facility, spares required for repairs and technical competence. All these factors require capital which is always a constraint. Total system is required to be optimized for desired serviceability of towers with capital constraints.

43.1 Introduction

Mobile communication in an area is provided to customers through mobile communication grid. A typical grid is shown in Fig. 43.1. The mobile communication towers are provided at various sites. Towers may be of various types [1–4].

Basic Towers: These towers are basically meant for providing communication to customers around within a specified distance. The equipment on these towers will be transmitter, receiver, antennas and power supply units (PSU). Transmitter and receiver have multi-channel facilities [5].

Control Towers: Control towers are more versatile and in addition to transmitting and receiving, they have got facility for routing and providing connectivity to many towers. They are equipped with additional controlling and rerouting circuits.

P. P. S. Dhaliwal · M. Kumar
Department of Electronics and Communications Engineering, Guru Kashi University, Talwandi Sabo, Bathinda, India

V. Dakulagi (✉)
Department of Electronics and Communication Engineering, Guru Nanak Dev Engineering College, Bidar, Karnataka 585403, India

© The Author(s), under exclusive license to Springer Nature Singapore Pte Ltd. 2021   431
S. C. Satapathy et al. (eds.), *Smart Computing Techniques and Applications*,
Smart Innovation, Systems and Technologies 225,
https://doi.org/10.1007/978-981-16-0878-0_43

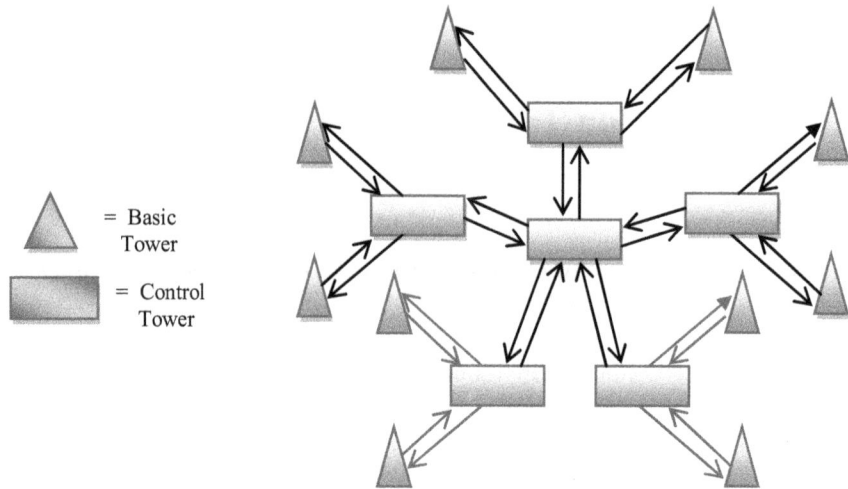

Each Tower = Receiver + Transmitter + PSU + Control & Routing* + Repair
Facility* + Stock*
*Optional

Fig. 43.1 Mobile communication grid

43.2 Points to Be Observed for Maintaining High Degree of Serviceability

Location of Towers: Although towers are generally self sustained for long periods. But there is requirement that as for as possible they should be located near to main route and roads.

Design for Maintainability: The serviceability requires certain ingredients which should be inbuilt to the system during design itself. The important factors are as under [6]:

Standardization: Components used should be minimum in range and interchangeable.

Modular Construction: The system should be designed in such a manner that the sub-systems should be in the form of modules.

Built–In-Test-Equipment (BITE): The total system should have built-in-test-equipment (BITE) which should give indication about faulty module or subsystem on front panel for easy replacement.

Defect Prone Components: The defect prone components should be easily accessible and replaceable.

Repair Facility: For modular electronic systems repair is normally restricted to replacement of modules/components with the help of built-in-test-equipment. For replacement spare parts in the form of modules/components are required to be stocked [7, 8].

43.3   Spare Parts Management

For cost effective serviceability spare parts management is very essential. The salient points are given in succeeding paragraphs.

Failure Rate of Modules/Components: Failure rate of electronic components are very less and that follow Poisson Distribution. Failure rates of modules/components should be assessed by analyzing failure data of previous years of same equipment or similar equipment.

Classification of Modules/Components: Once failure rates are established the modules/components should be classified as per following details:

Fast (F), Normal (N), Slow (S) Moving: Based on modules/components failure rate they should be classified as.
Fast Moving: If failure rate is very high.
Normal Moving: If failure rate is average.
Slow Moving: If failure rate is less than average.

Vital (V), Essential (E), Desirable (D): Based on importance of modules/components within the system, they should be classified as:
Vital: If the failure of the module/component renders the system completely unserviceable.
Essential: If the failure of the module/component renders the system partially unserviceable.
Desirables: If the failure of the module/component just reduces some desirable performance parameters.

High (H), Medium (M), Low (L) Cost: The modules/components to be classified based on their cost as under:

High Cost: Module/Components having relatively high cost.
Medium Cost: Module/Components having average cost.
Low Cost: Module/Components having less cost.

Final Classification of Modules/Components

Final Classification of Modules/Components is required to done by tagging all the three classification tags to each module/component. For example:
Module/Component 'x' may be Essential (E), Fast Moving (F) and Low Cost (L).
Component 'y; may be Desirable (D). Slow Moving (S) and High Cost (H) and so on.

Lead Times: Time in days required for materializing of demands of modules/components from supplier/manufacturer are known as lead times.

Consumption of Modules/Components During Lead Times: The consumption of modules/components during lead times can be calculated from failure rates.

Example: Let failure rate of a modules/component is assessed as 3 modules/components per month and lead time is two months. Then consumption during lead time is $2 \times 3 = 6$ modules/components. Demand for components should be placed when stock position for the respective model/component is 6. So that modules/components are available in the stock when required for replacement. But this stock level of 6 will give assurance level of only 50% since failure rates of electronic modules/components are probabilistic following Poisson Distribution and stock required may be more than '6' in about 50% cases. For achieving higher assurance levels we have to calculate modules/components required during lead time for higher assurance levels using Poisson Distribution.

43.4   Assurance Levels

Assurance Level is the probability that a module/component will be available in stock when required for replacement/repair. Desired assurance levels or availability of a particular module/component in the store for replacement will depend on its classification tags. For example a Fast Moving, Vital and Low Cost module/component should never be out of stock and therefore should have highest assurance level. A slow moving desirable and high cost modules/components should have lowest assurance level.

There are total 27 categories of modules/component as per classification (FNS, VED, and HML analysis). A typical 3—Dimensional assurance level matrix for all the 27 categories of modules/components is given in Table 43.1.

How much assurance level is required to be achieved for each category of modules/component is a management decision based on capital available for keeping stock of modules/components for repairs/replacement. A typical management deliberated 3-dimensional assurance level matrix is given in Table 43.2

43.5   Quantity of Modules/Components Required to Be Stocked for Desired Assurance Levels

Quantity of modules/components required to be stocked will depend on respective number of failures of specific module/component during lead time which is Poisson Distributed.

Table 43.1   Three dimensional assurance level matrix

Failure rate cost		L	M	H
F	V	AL_{11}	AL_{12}	AL_{13}
	E	AL_{21}	AL_{22}	AL_{23}
	D	AL_{31}	AL_{32}	AL_{33}
N	V	AL_{41}	AL_{42}	AL_{43}
	E	AL_{51}	AL_{52}	AL_{53}
	D	AL_{61}	AL_{62}	AL_{56}
S	V	AL_{71}	AL_{72}	AL_{73}
	E	AL_{81}	AL_{82}	AL_{83}
	D	AL_{91}	AL_{92}	AL_{93}

$AL_{r1} \geq AL_{r2} \geq AL_{r3}, r = 1, 2, ..., 9$
$AL_{1c} \geq AL_{4c} \geq AL_{7c}, c = 1, 2, 3$
$AL_{2c} \geq AL_{5c} \geq AL_{8c}, c = 1, 2, 3$
$AL_{3c} \geq AL_{6c} \geq AL_{9c}, c = 1, 2, 3$
$AL_{ab} = $ Assurance Level

Table 43.2   Typical 3-Dimensional assurance level matrix

		L	M	H
F	V	80	75	70
	E	77	72	67
	D	75	70	65
N	V	72	67	62
	E	70	65	60
	D	67	62	57
S	V	65	60	55
	E	62	57	52
	D	60	55	50

Poisson Distribution:

Poisson Process:

= failure rate for modules/components during time under consideration.

$T = $ Time under consideration.

Consider a small increment of time t.

(a)   Probability of failure in $t = t$
(b)   Probability of more than one failure $\rightarrow 0$
(c)   Probability of failure of modules/components during t is independent of past happening.

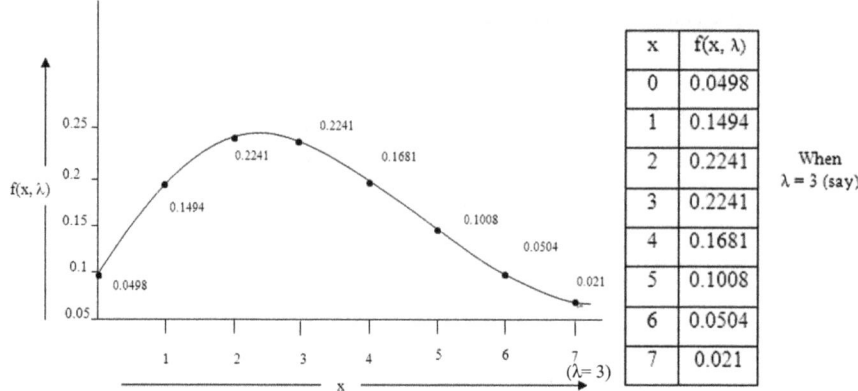

Fig. 43.2 Poisson probability distribution function (for $= 3$)

Poisson Probability Distribution Function For $= 3$ (Fig. 43.2)

$$f(x, \lambda) = \frac{\lambda^x e^{-\lambda}}{x!} \tag{43.1}$$

Equation (43.1), where is λ average failure rate of modules/components.

$f(x, \lambda) =$ Probability of 'x' failure of modules/components in time under consideration.

Distribution Mean $= \lambda$ Distribution Variance $= \lambda$, Distribution Standard Deviation σ (S.D) $= \sqrt{\lambda}$ Conditions: Population $(n) \rightarrow$ (large), Failure Rate per module/component $= p$ (Very Small)

$$n \geq 20, p \leq 0.05$$

Possion—Commulative Distribution Function (Fig. 43.3)

$$F(k, \lambda) = \sum_{x=0}^{k} \frac{\lambda^x e^{-\lambda}}{x!} \tag{43.2}$$

$F(k,) =$ Probability of 'k' or less than 'k' failures in time under consideration.

Let the desired assurance level for a module/component $(s_i) = \text{AL}_i$.

Let the failure rate/month of module/component $s_i = \lambda_i$ (Poisson Distributed).

Let the lead time of module/components$_i$ in months $= T_i$ (month).

Let No of modules/components (s_i) to be stocked for assurance $\text{AL}_i = n_i$.

Then

$$\sum_{k=0}^{n_i} \frac{(\lambda_i T_i)^k e^{-\lambda_i T_i}}{k!} \geq \text{AL}_i \tag{43.3}$$

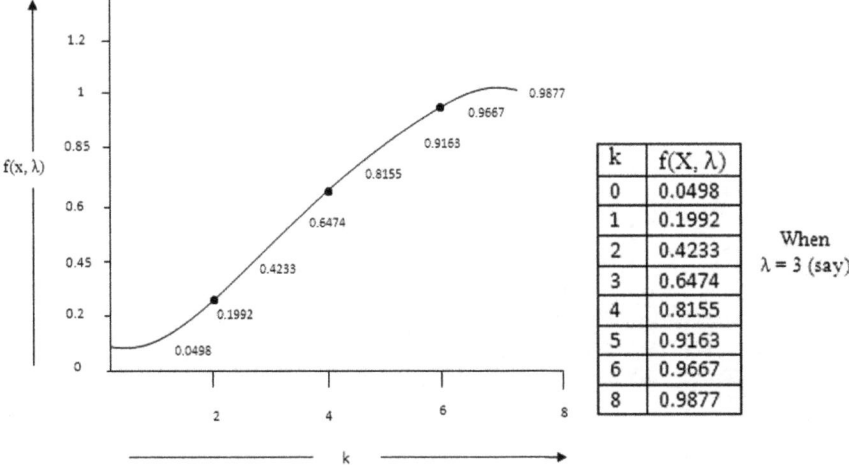

Fig. 43.3 Poisson cumulative distribution function for $\lambda = 3$

From above equation n_i to be calculated by using Poisson Distribution given at Table 43.3.

Let total range of module/components $= N$.

$n_i =$ No of module/component (s_i) required to achieve assurance levels AL_i to be calculated for full range of modules/components $(i = 1, 2, 3,..., N)$.

Let the cost of module/component $s_i = c_i$.

Then

$$\sum_{i=1}^{N} c_i n_i \leq C \tag{43.4}$$

where C is total capital allotted for modules/components stocking.

Assurance levels in Table 43.2 are to be progressively adjusted/modified to satisfy Eq. (43.4) and number of modules/components required to be stocked for adjusted assurance level can be calculated by using Eq. (43.3).

43.6   Conclusion

To ensure reliability of mobile communication system, the serviceability of electronic equipment placed at communication towers is very essential. This can be achieved by having optimized design such as modular construction of technical equipment and providing of built-in-test-equipment (BITE) for identification of faulty module/component for replacements. The stock level of various

Table 43.3 Poisson distribution function

Poisson Distribution Function*

$$F(x; \lambda) = \sum_{k=0}^{x} e^{-\lambda}\frac{\lambda^{k}}{k!}$$

λ \ x	0	1	2	3	4	5	6	7	8	9
0.02	0.980	1.000								
0.04	0.961	0.999	1.000							
0.06	0.942	0.998	1.000							
0.08	0.923	0.997	1.000							
0.10	0.905	0.995	1.000							
0.15	0.861	0.990	0.999	1.000						
0.20	0.819	0.982	0.999	1.000						
0.25	0.779	0.974	0.998	1.000						
0.30	0.741	0.963	0.996	1.000						
0.35	0.705	0.951	0.994	1.000						
0.40	0.670	0.938	0.992	0.999	1.000					
0.45	0.638	0.925	0.989	0.999	1.000					
0.50	0.607	0.910	0.986	0.998	1.000					
0.55	0.577	0.894	0.982	0.998	1.000					
0.60	0.549	0.878	0.977	0.997	1.000					
0.65	0.522	0.861	0.972	0.996	0.999	1.000				
0.70	0.497	0.844	0.966	0.994	0.999	1.000				
0.75	0.472	0.827	0.959	0.993	0.999	1.000				
0.80	0.449	0.809	0.953	0.991	0.999	1.000				
0.85	0.427	0.791	0.945	0.989	0.998	1.000				
0.90	0.407	0.772	0.937	0.987	0.998	1.000				
0.95	0.387	0.754	0.929	0.984	0.997	1.000				
1.00	0.368	0.736	0.920	0.981	0.996	0.999	1.000			
1.1	0.333	0.699	0.900	0.974	0.995	0.999	1.000			
1.2	0.301	0.663	0.879	0.966	0.992	0.998	1.000			
1.3	0.273	0.627	0.857	0.957	0.989	0.998	1.000			
1.4	0.247	0.592	0.833	0.946	0.986	0.997	0.999	1.000		
1.5	0.223	0.558	0.809	0.934	0.981	0.996	0.999	1.000		
1.6	0.202	0.525	0.783	0.921	0.976	0.994	0.999	1.000		
1.7	0.183	0.493	0.757	0.907	0.970	0.992	0.998	1.000		
1.8	0.165	0.463	0.731	0.891	0.964	0.990	0.997	0.999	1.000	
1.9	0.150	0.434	0.704	0.875	0.956	0.987	0.997	0.999	1.000	
2.0	0.135	0.406	0.677	0.857	0.947	0.983	0.995	0.999	1.000	

*Reprinted by kind permission from E. C. Molina, *Poisson's Exponential Binomial Limit*, D. Van Nostrand Company, Inc., Princeton, N.J., 1947.

modules/components to be decided based on their failure rate, criticality and cost so that optimum stock is maintained with in capital constraints.

References

1. Amaldi, E., Capone, A., Malucelli, F., Signori, F.: Optimization models and algorithms for downlink UMTS radio planning. In: Proceedings of IEEE Wireless Communications and Networking Conference (WCNC'03), vol. 2, pp. 827–831 (2003)
2. Capone, A., Trubian, M.: Channel assignment problem in cellular systems: a new model and tabu search algorithm. IEEE Trans. Veh. Technol. **48**(4), 1252–1260 (1999)
3. Eisenblätter, A., Fügenschuh, A., Koch, T., Koster, A., Martin, A., Achterberg, T., Wegel, O., Wessäly, R.: Modelling feasible network configurations for UMTS. In: Anandalingam, G., Raghavan, S. (eds.) Telecommunications network design and management, pp. 1–22. Kluwer Publishers (2003)
4. Hata, M.: Empirical formula for propagation loss in land mobile radio service. IEEE Trans. Veh. Technol. **29**, 317–325 (1980)
5. Lee, C.Y., Kang, H.G.: Cell planning with capacity expansion in mobile communications: a tabu search approach. IEEE Trans. Veh. Technol. **49**(5), 1678–1690 (2000)
6. Mathar, R., Niessen, T.: Optimum positioning of base stations for cellular radio networks. Wirel. Netw. **6**(4), 421–428 (2000)

7. Mathar, R., Schmeink, M.: Optimal base station positioning and channel assignment for 3G mobile networks by integer programming. Annal. Oper. Res. **107**, 225–236 (2001)
8. Alagirisamy, M.: A new DOA algorithm for spectral estimation without source numbers information. IEEE Trans. Antennas Propag. **68**(09), 6675–6682 (2020)

Chapter 44
KruthVidya—Mobile App to Detect Dyslexia

Apurva Dasari, Rashmitha Proddaturi, Sai Prasanna Vasaraju, and Kalyani Nara

Abstract Dyslexia is one such learning disorder that makes it difficult for people to master the skills to read, write, and spell out words. This paper introduces a mobile app KruthVidya that helps to diagnose the prevalence of dyslexia by tracking the child's ability in reading and writing spellings and the coordination between these activities simultaneously. Mobile app is developed to capture the spoken utterances of a given word and written activity performed simultaneously and process this using Google Speech APIs and optical character recognition (OCR) to extract the character sequences. The results from these modules are given to classifier, which process the information to diagnose the prevalence of dyslexia using standard metrics. In general, this tool avoids the stress which parents and kids undergo during the assessment, for its design in user friendliness. It is a play app that can work similar to the standard tools used by the clinical psychologists.

44.1 Introduction

The first 5–7 years of age is very important and plays a crucial role in the rest of life of an individual. It is considered as the fastest period of growth where the brain develops in a sensitive way which is interpreted by various milestones exhibited by the child. This period of child growth is considered as the foundation for subsequent attainment of skills and knowledge at individual level which has an impact on the human resource at population level and on economic development at large. Developmental disabilities may cause impairments of a child's physical, learning, or behavioral functioning. These may be categorized as sensory impairments (hearing or vision loss), ASD—autism spectrum disorder, ADHD—attention deficit hyperactivity disorder epilepsy or seizures, cerebral palsy, intellectual disability or learning disorder. The Lancet

A. Dasari (✉) · R. Proddaturi · S. P. Vasaraju · K. Nara
G Narayanamma Institute of Technology and Science, Hyderabad, India

K. Nara
e-mail: nara.kalyani@gnits.ac.in

© The Author(s), under exclusive license to Springer Nature Singapore Pte Ltd. 2021   441
S. C. Satapathy et al. (eds.), *Smart Computing Techniques and Applications*,
Smart Innovation, Systems and Technologies 225,
https://doi.org/10.1007/978-981-16-0878-0_44

series since 2007 provides estimates relating to early childhood development and risk of suboptimal development in low-income and middle-income countries [1]. The Global Burden of Diseases, Injuries, and Risk Factors Study 2016 has given few comprehensive and comparable estimates of age-specific health disorders for 195 countries from 1990 to 2016. Children younger than 5 years are at suboptimal development in low-income and middle-income countries.

The most common neurodevelopmental disorder effecting 3%-10% of children is specific learning disorder. Though it can be diagnosed at early stage, it is unnoticed until the adolescent age due to lack of awareness among parents and teachers.

44.2   Epidemics of Dyslexia

Good research on source and cause for dyslexia is published. It is observed that there is significant relation between reading disability and psychopathology. It was proved in the study that individuals with reading disability have displayed significantly higher rates of all internalizing and externalizing disorders than individuals without this disability [2]. Socioeconomic status (SES) is associated with reading skills. Lower SES was associated with both poorer word reading and poorer reading comprehension, and this was the observation made by MacDonald in the study published in [3]. This is because the children from lower SES families that undergo diagnostic of developmental dyslexia would be less.

Initially, the research on dyslexia focused primarily on the reading difficulties in English, while the recent study is focused on the nature of dyslexia across various languages. To summarize on what is known by this research, dyslexia manifests across languages exhibiting two types of variety, first, among alphabetic orthographies that vary in the degree of consistency of letter-sound correspondences, and second, in alphabetic versus logographic orthographies. Children at the low end of reading ability distribution in languages with more consistent mappings between letters and sounds (e.g., Italian or Finnish) have less severe reading problems than those learning to read less consistent languages (i.e., English), in the comparison study made with English and German dyslexics showed the evidence of systematic differences in the reading performance due to the difference in orthographic consistency which was similar for normal and dyslexic children [4]. Difficulties with reading fluency (speed of reading connected text) seem similar across languages [5]. Several studies have noted important universal features in normal and disordered reading across cultures despite linguistic differences. Cognitive predictors of early reading were observed to be similar for five European orthographies (Finnish, Hungarian, Dutch, Portuguese, and French).

44.3 Factors in Phonological Theory and Their Relation to Dyslexia

44.3.1 Access Versus Representations

Since the ability to distinguish phonemes was innate and universal, the implicit phonological representations should be intact in dyslexia. Impressive amount of evidence is available to indicate that a huge number of dyslexics have a phonological deficit that is the ability to access, process, and manipulate speech sounds and written text units [6, 7]. Two dominant theoretical frameworks were proposed to explain the phonological deficits in dyslexia. Classical phonological theory of dyslexia postulates the phonological difficulties as the core deficit of dyslexia. According to this postulate, underspecified phonological representation results in specific processing difficulties that are phonological in nature. Alternative theories are proposed explaining that phonological difficulties are due to indirect impact of sensorimotor impairments in the auditory system [8], visual system [9], and the cerebellum [10].

It appears that for all children, initial learning starts with phonological representations as fairly holistic and become gradually increased over time. Kids represent most words as single entity in early childhood. Through language development, phonological representations begin to emphasize syllables, then sub-syllabic distinctions, and ultimately individual phonemes. In his contributions argued that this implicit phonological development underlies parallel development in metaphonological awareness. In fact, in preschool children, they generally cannot perform tasks that require explicit manipulation of individual phonemes, although they can perform phonological awareness tasks at a larger grain size (e.g., rhyming). Children with dyslexia show deficits on phonological processing tasks like manipulation of phonemes, speech perception, priming, and lexical gating. Various research groups have described that the implicit phonological deficit may be observed in different ways by the psychologists.

44.3.2 Causal Direction

The second factor in the phonological theory of dyslexia is concerned with the direction of effect between phonological development and reading [11]. Written language is parasitic on oral language. In general, children do not begin writing until they have mastered most of the fundamentals in spoken language, and it seems reasonable that the causal direction flows from phonology to reading rather than vice versa. Several observations and evidence support this conclusion. Children with dyslexia underperform even younger, typically on reading level on phoneme awareness tasks [12]. The study proposed by [13] showed that these deficits tend to persist in adults with dyslexia who have otherwise compensated well for the disorder. These studies conclude that phoneme awareness deficits have a unidirectional causal link to reading problems which is oversimplified for several reasons. However, there are long standing controversies about the units of speech perception, and few evidences demonstrate that speech representations preserve much more than phonemes [14]. Adults natural illiterates, who are cognitively normal but have no formal schooling, confirm that phoneme-level representation is not important in language development. Difficulties in phonological development in dyslexia are probably not restricted to phonemic or segmental representations and must lie in other dimensions of the speech stream.

44.3.3 Orthographic Learning

Orthographic learning emphasis to establish mapping between phonemes and graphemes. Good neurophysiological evidence indicates that skilled readers treat letters as single audio-visual objects, and the orthographic learning hypothesis states that problems developing such integrated representations interfere with the emergence of fluent reading. Limited studies are reported to support this hypothesis, and research is carried out in comparing the performance of children with and without dyslexia, training them on associations between sounds in their native language and an unfamiliar orthography. Results indicate that the children with dyslexia performed more poorly, under time pressure. The orthographic learning has an advantage of integrating the brain and neuropsychological levels of analysis to explain reading, development, and difficulties. The limitation is that it is difficult to test a pure integration account of phoneme-grapheme binding as the children with dyslexia are not equivalent to their typically developing peers in processing phonemes of their native language.

44.4 Motivation

With one child in around every 250 children affected with developmental disorder (DD), India stands fourth place worldwide. Country has nearly 1.3 billion people with children ≤ 15 years constituting nearly one-third of the population. An estimate from 2019 states that more than two million people might be affected with disorders. Tools are available for assessment of children above 7 year and are used only when the child is taken to psychologist for assessment. Dyslexia is not a disease. It is a condition a person is born with, and it often runs in families. It is estimated that every 1 in 10 people have dyslexia. Research has shown that dyslexia happens because of the way the brain processes information. Most dyslexic people have average or above-average intelligence, and they work very hard to overcome their learning problems. Pictures of the brain show that when people with dyslexia read, they use different parts of the brain than people without dyslexia.

People with dyslexia often have normal intelligence and vision. Dyslexics may face problems in spelling words, reading quickly, writing words, "sounding out" words in the head, pronouncing words when reading aloud, and understanding what one reads. Early signs such as late talking, learning new words slowly, problems forming words correctly, problems remembering or naming letters, etc., could often be misleading suggesting other disabilities. Hence, when a child is in school, dyslexic signs and symptoms become more significant such as problems processing and understanding what he or she hears, problems remembering sequence of things, difficulty seeing similarities and differences in letters and words, difficulty in spelling words, etc. It is not unusual to identify signs in teens or adults. A teen's parents or teachers can identify signs such as poor reading and writing skills, despite normal intelligence, poor spellings, difficulty remembering names and lists, problems with directions, etc. Having these signs necessarily does not mean dyslexia. It can be confirmed by standardized tests under supervision.

As mentioned, people with dyslexia are often poor spellers and find it difficult in segmenting words into individual sounds or may blend sounds when producing words. They may also exhibit signs of difficulty in identifying or generating rhyming words or count the number of syllables in words. Hence, considering the writing of spelling of the word for diagnosis of dyslexia is a good measure. Dyslexic people not only find it difficult to write the words but also find it difficult to spell them out. They find difficulty in spelling out longer words, or remembering names, letters, numbers, etc., sounding out little words. The second element that this application considered is the reading out the word given for diagnosis. It is a good measure to see if the user is able to identify the correct letters and phonetics of the word provided. The final understanding is the coordination of the speech and written input which identifies the brain's activity while performing the two activities simultaneously. Normal people can perform the reading and writing simultaneously while dyslexic people find it difficult to coordinate. The early symptoms are unnoticed for two reasons. 1. Indian parents or care takers ignore the early symptoms due to lack of awareness in observing the exhibited skills at an early stage. 2. The teachers in pre-primary schools are not trained, and most of the schools do not have a clinical psychologist appointed to perform primary screening. These challenges can be addressed with KruthVidya a mobile tool that can assist both parents or care takers and pre-primary teachers to diagnose the prevalence of dyslexia and refer to clinical psychologist to facilitate early intervention.

In Indian context when the child is observed to underperform, they are taken to clinical psychologist for assessment. There are standards defined by National Institute of Mental Health and Neurosciences (NIMHANS), Bengaluru. Few examples of tests that are currently in use as defined by NIMHANS are mentioned below:

- The NIMHANS test has two stages of assessment: Assessment one for pre-academic skills; attention, visual, and auditory discrimination, visual and auditory memory, speech and language, writing and number skills. Speech and language, writing and number skills. Level 2 test areas of attention, reading, spelling, perceptuo-motor, visuo-motor integration, memory, and arithmetic skills. It is used for assessment of SLD and used to monitor progress after remediation. A wide range of relevant areas are covered. It is easily administered and available. It is good for early identification of learning issues. It can only be assessed by clinical psychologists and special educators.

- WRAT-4 (Wide Range Achievement Test): The tests included are reading composite (word reading + reading comprehension), spelling, and math computation. It can be assessed only by psychologist, special educators, and teachers. The advantages include screening of difficulties in academic areas, take 15–30 min, and have parallel forms allowing for retesting, age- and grade-based norms.
- TOWL-4 (Test of Written Language): The tests included are vocabulary, spelling, style (punctuation and grammar), logical sentences, sentence combining, contextual conventions, contextual language, and story construction. This test can be assessed by anyone with formal training in assessment and English proficiency. The advantages would help in identifying students with writing difficulties, strong conceptual model of writing, good reliability, excellent tool to assess the progress of remediation in writing. The disadvantages would be that it is time consuming, subjectivity in scoring procedures in domains, and it cannot be used to assess other areas of SLD except writing.

The above tests are currently under use and can be assessed only by psychologists, trained professionals, and teachers. This model "KruthVidya" is a mobile app to assist parents or guardian to identify the prevalence of dyslexia at earlier stages. The application KruthVidya allows children to interact with the app as a learning tool and parallelly process the activities of the child in learning to classify into being dyslexic or not. KruthVidya is a mobile application that can detect dyslexia at an early stage. It involves evaluation of writing and phonological skills, along with coordination between child's hand and brain. The architecture of the model is explained in subsequent section.

44.5 Architecture of KruthVidya

This application has a suite of useful assistive technology features for kids with dyslexia. It is a prediction software and is developed for dyslexic children to help them in learning, reading, and recognizing. It promotes learning in an interactive manner and in an innovative learning environment. KruthVidya mobile application is proposed that would test the reading and writing skills of children, and the inability of exhibiting those skills is recognized. The application would suggest a word and ask the user to spell and write the word simultaneously. The accuracy of the activities is tested, and disorder is identified based on various factors.

The working of the application is provided words that are asked to spell by children which is recorded and converted into text in the backend and parallelly the spelt word is asked to write. The pressure of writing and the pattern, behavior in which text is written, is noted, and spelling is checked. The way that the words are pronounced is compared to the way that the words are written. It has five development phases—analysis, design, development, implementation, and evaluation.

The evaluation is done at the end, and dyslexic characteristics are identified at an early stage. It is built to encourage independent reading for kids with learning disorders.

The application is implemented in a four-stage model and three modules that perform different activities of the implementation (Fig. 44.1). The four stages are given as follows:

a. Collection of inputs-audio and written
b. Optical character recognition implementation on the text collected.
c. Speech-to-text conversion on the audio input collected.
d. Analysis and results.

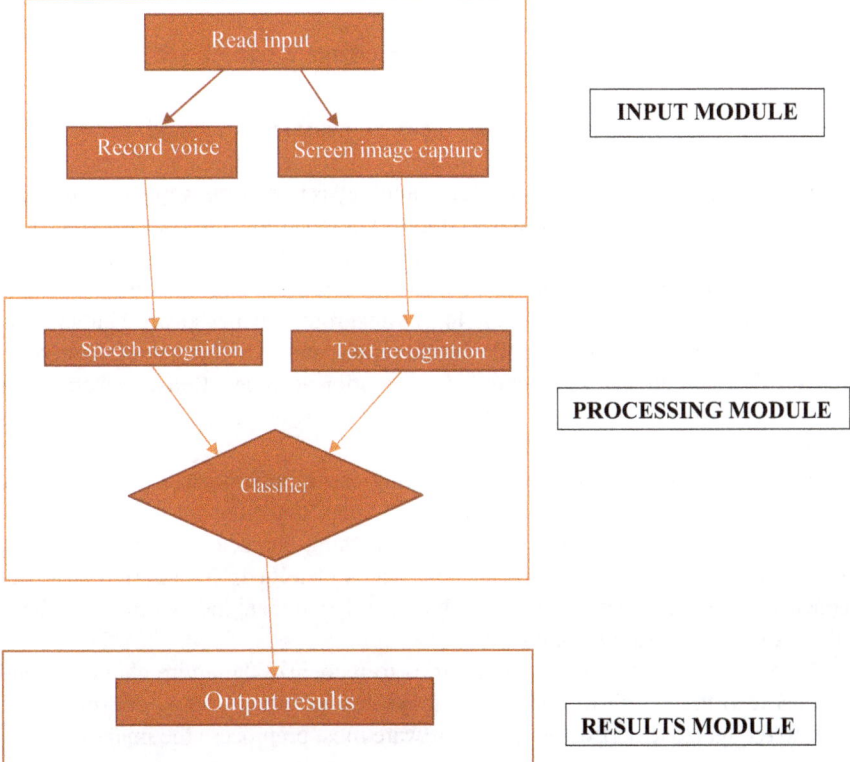

Fig. 44.1 Architecture of KruthVidya

44.5.1 Collection of Inputs

In this stage, the application provides an interface that provides the image or the name of the word that is considered for analysis. The interface has a canvas where the user is asked to write the word, and the microphone symbol enables to record the audio input of the word described. The inputs are stored to be sent into the next stages where they are analyzed. The words that are considered for analysis are in a random order initially and later move on to appear in such a way that the words or phonetics that are spelt wrong too often are given to the user. For example, if bat is the word initially and the user writes the 'b' as 'd', the model gives another word starting with 'b' say 'ball' so as to check if the user is actually finding it difficult with the letter 'b'. This process is the input module.

44.5.2 Implementation of OCR to Extract Text

The written input on the canvas from the previous stage is recognized here. This process is the processing module.

Optical character recognition (OCR) technology is used to convert the written text to digital text form. The user is allowed to use notepad for free hand writing. This text is scanned and processed by OCR to extract the digitally coded text. The process involves preprocessing the scanned imaged and compares with multiple formats to extract each individual alphabet (Fig. 44.2). Most fonts are now easily available and support a variety of digital image file format inputs, and so higher advanced systems are capable of producing high degree of recognition accuracy. It takes several steps to convert an image file into an editable document. After scanning the image, it is stored in a TIF format that is a bit-mapped file.

In this processing module, after receiving the input image from screen, the images are cropped to document segments. Then, detecting and identifying the words/alphabets in each segment and preprocessing character images then using optical character reader, the character sequence is identified. The inferred character sequence is concatenated to form the whole word, and then, the text in higher level either as word or phrase is outputted.

Using this technique, the computer tries to recognize the entire characters and matches it to the matrix of characters stored during speech-to-text conversion. In order to recognize text effectively, the software must preprocess the images.

Fig. 44.2 Working of OCR model

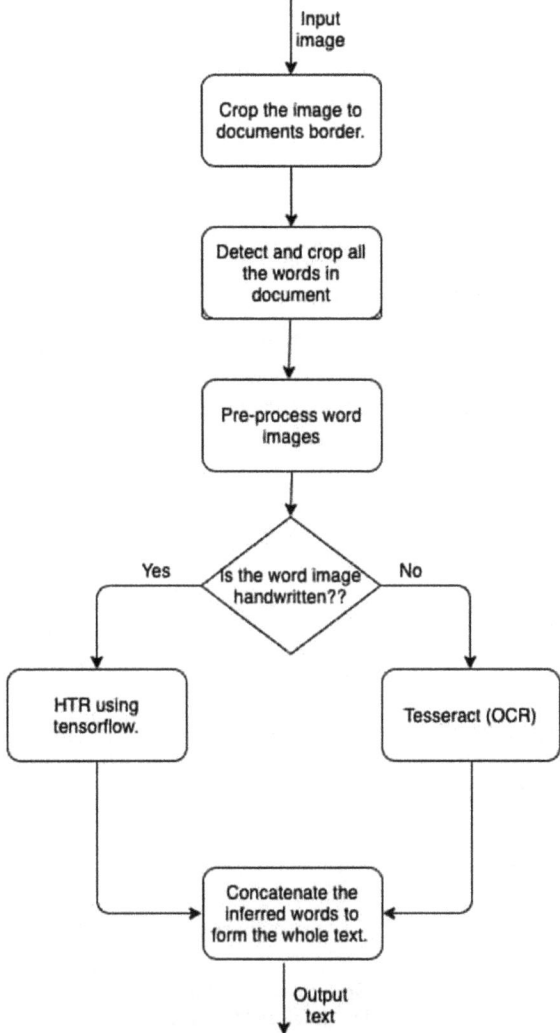

44.5.3 Speech-To-Text Conversion of the Audio Input Collected

The audio input from the activity is used to convert into text. This process also lies in the processing module.

In the application, Google Speech-to-Text API is used to perform this operation. Speech-to-text has three main methods for recognizing speech.

1. **Synchronous Recognition**: It sends audio data to the speech-to-text API, performs recognition on that data, and returns results after all audio has been

processed. Synchronous recognition requests are limited to audio data of 1 min or less in duration.

2. **Asynchronous Recognition**: This method sends audio data to the speech-to-text API and initiates a long-running operations. For audio data of any duration up to 480 min, asynchronous requests can be used.

3. **Streaming Recognition**: It performs recognition on audio data provided within a gRPC bi-directional stream. Streaming requests are designed for real-time recognition purposes, such as capturing live audio from a microphone. Streaming recognition provides interim results while audio is being captured, allowing result to appear.

A request is sent, which identifies speech and returns the text as the response.

Google Speech-to-Text also performs audio encoding. It refers to the manner in which data is stored and transmitted. Several techniques like FLAC, LINEAR16, and OGG_OPUS are used for encoding the audio data. FLAC, however, is both a file format and an encoding. Within the speech-to-text API, FLAC is the only encoding that requires audio data to include a header; all other audio encodings specify header-less data.

The supported audio encodings are:

Codec	Name	Lossless	Usage
MP3	MPEG audio Layer III	No	MP3 encoding is beta feature, only available in v1p1beta1
FLAC	Free lossless audio Codec	Yes	16-bit or 24-bit requires for streams
LINEAR16	Linear PCM	Yes	16-bit linear PCM encoding
MULAW	μ-law	No	8-bit PCM encoding
AMR	Adaptive multi-rate Narrow band	No	Sample rate must be 8000 Hz
AMR_WB	Adaptive multi-rate wide band	No	Sample rate must be 16,000 Hz
OGG_OPUS	Opus encoded audio Frames in and Ogg container	No	Sample rate must be one of 8000, 12,000, 16,000, 24,000, and 48,000 Hz
SPEEX_WITH_HEADER_BYTE	Speex wideband	No	Sample rate must be 16000 Hz

This API also used speech adaption to improve the transcription results. For example, suppose that your audio data often includes the word "weather." When speech-to-text encounters the word "weather," you want it to transcribe the word as "weather" more often than "whether." In this case, you might use speech adaptation to bias speech-to-text toward recognizing "weather." Speech adaption increases the accuracy of frequently used words and phrases and improves the accuracy of speech transcription when the supplied audio contains noise or is not very clear.

44.5.4   Output Results

The two inputs, i.e., the audio and written inputs, are now converted into normal text which can be compared and checked for accuracy. The primary analysis performed here verifies if the audio input is wrong or the written input has an error (Fig. 44.3). The next analysis is to identify the particular sound, alphabet, or phonetic sound

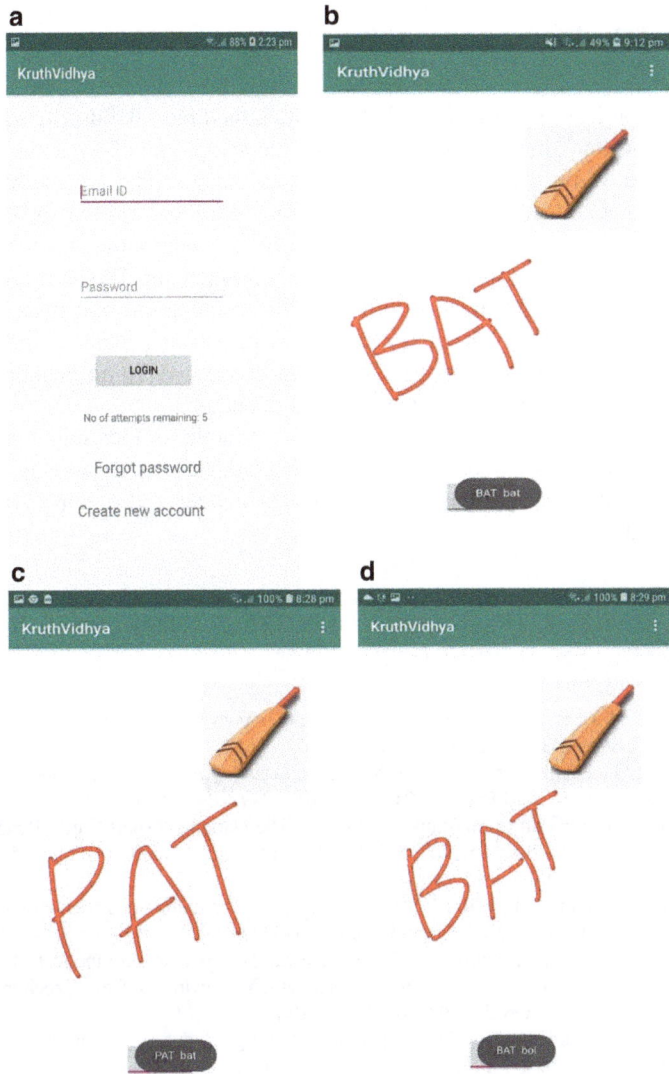

Fig. 44.3   a Login page. **b** Detection of correct written text and correct utterance. **c** Detection of wrong spelling and correct utterance. **d** Detection of correct spelling and wrong utterance

that the user is finding difficulty in. To confirm the same, the child is provided with words containing those phonetics or alphabets, and performance is validated. The application is currently under development implementing the simultaneous working of audio reading and the OCR technology running on the text written on the interactive surface. The current area of work is the database that is used to save scores and understand the phonetic sounds where the child or user is facing a problem with.

44.6 Conclusion

The application is a very user-friendly tool which facilitates the user to use it more like a play tool but simultaneously identifies the symptoms of dyslexia. The user would be able to use the application in a relaxed manner like a game rather than a test which is an added advantage as there is no stress and results could be more accurate. Early identification and intervention with kids who show the warning signs of dyslexia are critically important for better outcomes later on. Dyslexia is described as a hidden disability. It affects a child's ability to recognize and manipulate the sounds in language. Kids struggle with deficits in working memory that make it difficult to retain and use information in the short term. Working memory deficits impact phonological awareness, decoding, and fluency.

So, using an application technology which is suitable for kids might change the learning environment. The main objective of the KruthVidya application is to identify the learning disabilities at an early stage in an easy manner and also in an interactive manner with the kids.

References

1. Herrman, H., Swartz, L.: Promotion of mental health in poorly resourced countries. ScienceDirect © 2007 Elsevier Ltd Published, 04 Sep 2007
2. Willcutt, E.G., Pennington, B.F.: Psychiatric comorbidityin children and adolescents with reading disability. J. Child Psychol. Psychiatry **41**, 1039–1048 (2000)
3. National Center for Education Statistics (NCES). The Nation's Report Card : Reading 2011. NCES2012–457. Washington, DC: US Department of Education, Institute of Education Sciences (2011)
4. Landerl, K., Wimmer, H., Frith, U.: The impact of orthographic consistency on dyslexia: a German—English comparison. Cognition **63**, 315–334 (1997)
5. Caravolas, M., Volín, J., Hulme, C.: Phoneme awareness is a key component of alphabetic literacy skills in consistent and inconsistent orthographies: evidence from Czech and English children. J. Exp. Child Psychol. **92**, 107–139 (2005)
6. Ramus, F.: Developmental dyslexia: specific phonological deficit or general sensorimotor dysfunction? Curr. Opin. Neurobiol. **13**, 212–218 (2003)
7. Snowling, M.J., Hulme, C.: Evidence-based interventions for reading and language difficulties: creating a virtuous circle. Br. J. Educ. Psychol. **81**, 1–23 (2011)
8. Tallal, P.: Auditory temporal perception, phonics, and reading disabilities in children. Brain Lang. **9**(2), 182–198 (1980)

9. Livingstone, M.S., Rosen, G.D., Drislane, F.W., Galaburda, A.M.: Physiological and anatomical evidence for a magnocellular defect in developmental dyslexia. Proc. Nat. Acad. Sci. USA **88**, 7943–7947 (1991)
10. Nicolson, R.I., Fawcett, A.J., Dean, P.: Dyslexia, development and the cerebellum. Trends Neurosci **24**, 515–516 (2001)
11. Castles, A., Wilson, K., Coltheart, M.: Early orthographic influences on phonemic awareness tasks: evidence from a preschool training study. J. Exp. Child Psychol. **108**, 203–210 (2011)
12. Wagner, R.K., Torgesen, J.K.: The nature of phonological processing and its causal role in the acquisition of reading skills. Psychol. Bull. **101**, 192–212 (1987)
13. Doi: https://doi.org/10.1016/S0140-6736(07)61244-6
14. https://towardsdatascience.com/how-to-use-google-speech-to-text-api-to-transcribe-long-audio-files- 1c886f4eb3e9
15. https://cloud.google.com/speech-to-text/docs/concepts
16. Daud, S.M., Abas, H.: Conference paper: 'Dyslexia Baca' mobile app-the learning ecosystem for dyslexic children
17. https://www.researchgate.net/publication/269032733_%27Dyslexia_Baca%27_Mobile_App_--_The_Learning_Ecosystem_for_Dyslexic_Children

Chapter 45
A Survey on Streaming Adaptation Techniques for QoS and QoE in Real-Time Video Streaming

Satyanarayana Reddy Marri and P. Chenna Reddy

Abstract Video streaming is a dominant application in our day-to-day life and is accounting for the majority of the traffic over the Internet. HTTP streaming with dynamic adaptation (DASH) is accommodating prominent streaming services such as YouTube, Netflix, Amazon Video, and Hulu. It is a challenging task to stream the video up to the end-user's level of expectation for the service providers due to the sudden fluctuations in bandwidth availability and many more factors over the network. At present, video streaming is rated based on network-centric (QoS—quality of service) and user-centric (QoE—quality of experience) parameters. The perception of the viewer's video quality is measured by QoE metrics. In this contemporary survey, we manifest solely state-of-the-art bitrate dynamic adaptation algorithms for streaming over HTTP. The main contribution of this paper is focusing on video coding standards, diversified bitrate adaptation techniques, network-centric parameters like bandwidth, throughput, packet loss, delay or jitter which account for the QoS, QoE subjective and objective metrics, and influencing factors like initial delay, stalling rates, and quality oscillations which are taken into account. We also consider the factors for bitrate adaptation techniques like bandwidth estimation, playback buffer status, device configuration, content type, and viewer expectations. We make a note of machine learning (ML) algorithms which play a vital role in real-time data streaming. This comprehensive survey is useful to the researchers to predict the suitable bitrate chunk by the adaptation technique according to the network conditions as well as take the client-side factors into an account to optimize the viewer's QoE in real-time multimedia streaming.

S. R. Marri (✉) · P. C. Reddy
Jntua CEA Anantapur University, Ananthapuramu, India

P. C. Reddy
e-mail: chennareddy.cse@jntua.ac.in

© The Author(s), under exclusive license to Springer Nature Singapore Pte Ltd. 2021   455
S. C. Satapathy et al. (eds.), *Smart Computing Techniques and Applications*,
Smart Innovation, Systems and Technologies 225,
https://doi.org/10.1007/978-981-16-0878-0_45

45.1 Introduction

Video streaming over the network is emerging as a ubiquitous application, and it constitutes a major portion of the overall Internet traffic. With the advent of a variety of handheld devices capable of wirelessly connecting to the Internet, the widespread use of real-time data streams has rapidly increased [1]. This type of traffic is expected to grow exponentially in the coming years. In the past few years, data streaming applications are leading IP networks and are expanding. The forecast analysis of the Cisco Visual Networking Index reveals that video traffic alone will be 82% of the global network traffic by 2021, which is 73% in 2016 [2]. It is a challenging task for content, network, and service providers to meet the user's anticipated video services. Therefore, in current real-time streaming multimedia applications, a novel solution is needed to maintain the demand for high-rate media streams, so that throughput is stable for a longer period and also to achieve high QoS and satisfactory QoE from the users. To meet this demand, we are using streaming techniques over HTTP.

Traditional video streaming (TVS) is performed based on a push-based mechanism. TVS uses protocols like real-time transport protocol (RTP) [3] and real-time messaging protocol (RTMP) [4]. To establish the session for video streaming and identify the track of the information during this period, real-time streaming protocol (RTSP) [5] is needed. The traditional communication with UDP protocol has some issues considering the perspectives of network address translation (NAT) and firewalls [6]. In current real-time streaming multimedia applications, a novel solution is needed to maintain the demand for high-rate media streams, so that throughput is stable for a longer period and also to achieve higher QoS and satisfactory QoE from the users.

In the challenging situation where both QoS and QoE are faced with multiple limitations of delivery requirements, machine learning (ML) technology can provide a promising and elegant solution, because ML can sense, mine, predict, and infer system parameters [7, 8] and can propose the best adaptive conditions as needed. This research effort aims to explore and propose a novel solution to improve QoS and QoE in real-time data streams using ML technology in heterogeneous networks. The remaining paper is manifested as Sect. 45.2 explores motivation to choose this area, streaming techniques, as well as coding standards which are explained in Sect. 45.3, and Sect. 45.4 gives solely information regarding QoS and QoE in real-time data streaming. Section 45.5 presents the role of machine learning in streaming, Sect. 45.6 gives comprehensive information on streaming adaptation algorithms in QoS and QoE, and finally, Sect. 45.7 gives the conclusion.

45.2 Motivation

The Internet is a heterogeneous environment that connects various network technologies [9]. Even as networks are supported by service categories, the network resources available to different broadcast applications will change dynamically over time, due to requirements of network applications changes periodically [10, 11]. It is necessary to analyze a large number of real-time streaming data of complex nature to effectively meet the QoS requirements of services. It is planned to deliver through it, although it changes over time and the workload and network conditions are different [12, 13]. For humans, it is very difficult to process the bulk data in a small span to operate network services. So, it is mandated to go for automation of these tasks.

Even the widely used TCP protocols [14, 15] on the Internet use implicit feedback through timeouts and duplicate acknowledgments of lost packets. During severe congestion, the loss rate is high, which occurs when data users compete for scarce network bandwidth in the data network. Internet measurements indicate that the demand for bandwidth-intensive applications is growing, leading to increased loss rates on various links on the Internet [16]. End nodes typically deploy explicit feedback. However, implicit or explicit feedback relying solely on end nodes is not sufficient to achieve high throughput on the Internet. Appropriate mechanisms should be used to control network congestion. Otherwise, the network may end up in a persistent overload state, which may cause congestion collapse and affect the QoS and QoE of the client [7, 8]. Thus, this versatility and complexity of real-time data streams have led to the development of efficient algorithms for throughput improvement in heterogeneous environments by using machine learning techniques to reduce the amount of packet loss by making reasonable use of network resources.

45.3   Overview of Adaptive Streaming Techniques and Video Coding Standards

At present, over the top (OTT) services like Netflix, Amazon, and Hulu using the streaming algorithms like HAS and DASH to provide QoS and QoE to the end-users in dynamic fluctuations in bandwidth availability in heterogeneous networks as well as devices [14, 17]. In HTTP adaptive streaming (HAS), the video is segmented into chunks with equal size in a fixed duration of play, and each chunk is available in different bitrates [18]. The client sends an HTTP request to the streaming server, and an adaptation algorithm identifies the suitable bitrate chunk to the network conditions and transmitted it to the client using transport protocols of either TCP or UDP. The adaptive streaming process is shown in Fig. 45.1. The adaptation logic is implemented either client side or server side; generally most of the cases, it is a client side to reduce the burden to streaming server. Whenever network condition changes, then the adaptation logic requests the appropriate bitrate chunk to the client system for streaming the video seamlessly. This adaptability of HAS is being analyzed on a single client and shows impressive results [19].

For managing the exponential growth of video traffic over a public network, [14] proposes a new streaming technique over HTTP with dynamic adaptation (DASH) nature according to the network conditions. Here also, as HAS video data is split

Fig. 45.1 Adaptive
streaming process

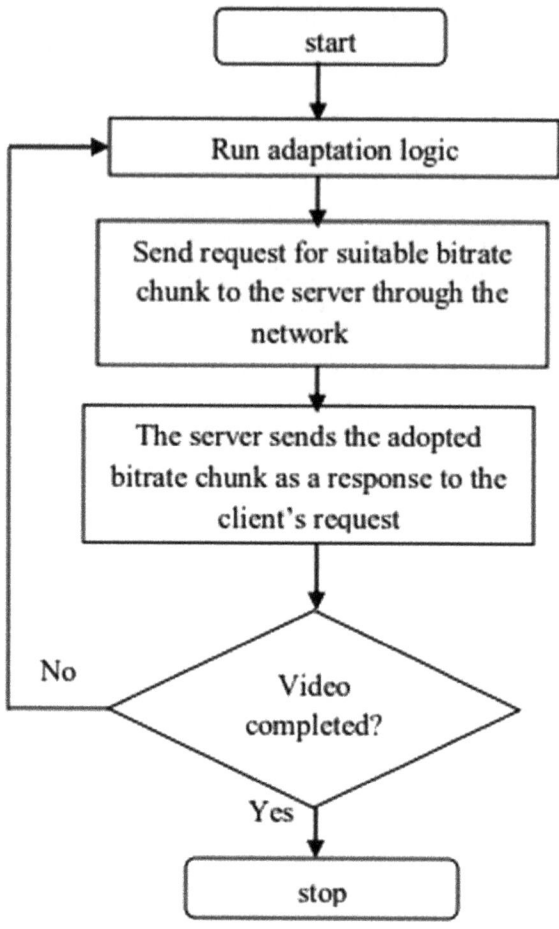

into chunks with equal playback length, but each chunk is available in different bitrates as well as quality levels [11]. The target of DASH is also to provide seamless streaming of data to the user's expected level. A dynamic adaptive technique that enhances HTTP adaptive streaming (HAS) [14, 18] is a prominent solution in video streaming, and it needs new mechanisms for flow control, free from congestion, and efficient usage of available bandwidth. Video coding standards are the compression techniques that play a vital role in transmitting the data very fast over the Internet. That means the data is compressed before sending it and decompressed at the destination. In video streaming also, first, the video is segmented into chunks which are equal in size, and later each chunk is compressed (encoded) into different bitrates. Several video coding standards differ from each other in their performance and efficiency like H.264 [20], SVC [21], HEVC [22], and SHVC [23].

45.4   QoS and QoE in Real-Time Data Streaming

In recent years, most of the research work move around QoS and QoE due to video traffic currently leading the Internet [24, 25]. Most of these indicate that QoS assessment has shifted from network performance to QoE. QoE measures the performance of a user's subjective perception. However, QoE and QoS are not mutually exclusive, but QoE is an extension of QoS. While we are developing the QoS model, it is mandatory to consider all QoS parameters of the network and more importantly to quantify the relationship between them. The QoS framework of real-time data flow [26] faces a huge success challenge. This can be a daunting task when dealing with next-generation heterogeneous network systems [27, 28], where many unpredictable instances can affect the user experience. The work in [29] proposes a model that guarantees for handling delay-related sensitive applications. The algorithm of [30] identifies the optimal bandwidth required for x number of video calls in Skype instantaneously by using the QoE metric. Quality of experience is the metric to measure end-user expectations [31].

By the QoE, we can measure the video quality using peak signal-to-noise ratio (PSNR), mean opinion score (MOS), playback smoothness, and video quality measurement (VQM), these metrics come under objective and subjective metrics of QoE. So, QoE mainly focuses on making the user convenient with the streaming video that means it is related to user satisfactory point of view, where QoS related to making use of network resources efficiently for providing quality service to the user. But the fact is that there is no particular quality of service (QoS) parameter to estimate the user's streaming experience in case of real-time data streaming [32]. Because of this only, to consider the user opinion in video quality evaluation, QoE is evolved. QoE as well as QoS both are not mutually disjointed, but QoE is a superset of QoS. The relation of QoS and QoE is shown in Fig. 45.2. The QoE is influenced by the factors like system level (bandwidth variation, packet loss, delay, jitter, end-user

Fig. 45.2 Adaptive streaming process

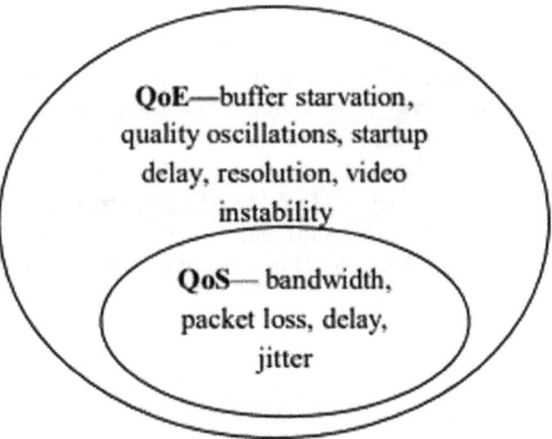

device configuration, browser), context level (user location, streaming purpose like education, gaming), user level (expectations), and content level (bitrate, resolution, video quality) [33].

45.5 Machine Learning Techniques

Nowadays, any work is coming under automation by the computer. It is possible for the system by using machine learning (ML) algorithms, and these algorithms make the system to take decisions through inference, create and utilize the knowledge, and improve the performance. It predicts and provides recommendations based on the results obtained by processing too large and too complex data sets [28, 34]. ML predicts an optimal solution by analyzing the past data in a short period. ML techniques help solve automation challenges in network setup, control, and management [7–9, 12].

Because of this reason, we use machine learning technology to automatically implement network functions by interacting with internal and external environments. We can select a suitable ML algorithm based on the collected data nature [7, 8]. Generally, ML is classified into three categories as supervised, unsupervised, and reinforcement learning. In supervised learning (SL), the data is already trained that means input is mapped with output. SL is more helpful for the problems of regression and classification. In unsupervised learning (USL), the data is not trained earlier that means we can give the input values for the collected data but not give the output values. USL is most useful for analysis of the clusters. Reinforcement learning (RL) works on the principle of dynamic decision as well as iterative learning. It does not reveal to the learner what action to take but must use continuous trails to determine the actions that produce the output closest to the target.

The operation and management of ML based technologies involve the task of effectively utilizing resources while always best meeting the needs of services and users [1]. They need to understand changes in system status, understand uncertainty, (re)configure networks, anticipate pressing challenges, and propose appropriate solutions promptly. It can also consider nod computing power, bandwidth availability, available energy for resource allocation, as well as management. Therefore, when applied to the network function of the current real-time streaming data application, these classic ML methods can be modified to improve accuracy, reduce complexity, and reduce latency and QoS [12, 35]. [36] proposes a model using a machine learning algorithm to establish the relationship between QoS–QoE. In [12] authors expressing the SL or semi-SL model is more suitable than USL for QoS and QoE services.

45.6   State of the Art on Video Streaming Adaptation Algorithms in QoS and QoE

A lot of work is being done on QoS and QoE in real-time streaming data using adaptive streaming technologies. We discuss a few of them as given in Table 45.1 above. According to recent measurement studies, the research aims to improve real-time streaming data because it is becoming the major traffic of the network [30, 31]. The main reason to increase video traffic over the network is the availability of smart devices in the hands of almost everyone. Although the top layers of the Internet have undergone major changes and are now used as platforms for large-scale video transmission, most Internet traffic is still mostly transmitted over TCP. TCP uses the congestions algorithms which are loss and more latency, suitable for video-on-demand (VOD) streaming not for live that means video conferencing applications require more throughput as well as less time to establish the connection between client and server for the completion of streaming to achieve QoE of the user [37–39].

This research effort aims to exploit the potential of ML technology to design new methods of real-time data flow to overcome the limitations of TCP and other resource constraints. Using features such as bandwidth, processing power, and traffic latency, the ML design decision algorithm is used to dynamically adapt the recommendations to manage congestion and enhance smooth data flow. A request management framework for heterogeneous networks is to improve the quality of experience (QoE) of users. This proposes a QoE development method based on ML technology to perform load balancing for the global request management model [7, 32, 42]. It estimates or predicts the user's experience and expected quality range by collecting sets of data from heterogeneous networks.

45.7   Conclusion

Real-time transmission of media data over the network needs more bandwidth for high throughput to provide lost data instantly without interruption to the end user. Due to strict delay and jitter requirements, more and more multimedia applications in today's computer networks are unable to use existing technologies based on TCP congestion control. As the level of congestion increases, the quality of these multimedia applications declines to the point where users cannot receive content. Thus, a new generation of Internet-based multimedia applications using machine learning technology has entered the network design and management to pursue the new potential of network systems.

Table 45.1 Comprehensive information on HAS and DASH on QoS and QoE

References	Objective	Approach	Result
[19]	Streaming traffic monitoring for video-on-demand services (HAS)	Stochastic model	The model is best suited and performed data analysis for smooth streaming without congestion
[18]	Overcome the hurdles that existed in HAS in case of bandwidth fluctuations in a short period	A novel bitrate control technique	HAS becoming more powerful and perfect in a network where fluctuations happening
[31]	Make the user happy by providing expected quality video, even though network conditions are fluctuating	Developed an algorithm which takes the maximum bit rate video that does not cross the current bandwidth	Meets the user's expected QoE
[37]	Obtain maximum throughput for delay-sensitive applications communication	Estimation method using protocols of closed-loop flow control	Provide better QoS and QoE to the user
[16]	Real-time identification algorithm over networks for streaming of video with dynamic changes occurs in the network	Bandwidth vary identification method, a Bayesian approach	It identifies congestion in advance and makes video streaming seamlessly to the user's expected level
[11]	Implement dynamic HTTP streaming to adapt to the transport protocol's congestion window size, in the infeasibility of TCP's congestion and error control for low latency, high-rate communications services leads to significant underutilization of the available bandwidth	Novel transport protocol—Predictably reliable teal-time transport (PRRT), a protocol layer that efficiently supports the reliability required by multimedia services under their specific time constraint	It achieves the optimal trade-off between those mechanisms under strict delay constraints to optimize bandwidth utilization. It evaluates PRRT's bandwidth utilization and compares it to recent media streaming standards such as HTTP-DASH (dynamic adaptive streaming over HTTP)
[40]	Minimizing congestion and delay at both side of communication	Cost-based distributed technique for controlling congestion	Provides good efficiency in congestion and delay
[41]	Handling congestion and delay queuing	A novel technique for controlling congestion	Yields good results and hence employed in Google chrome also

(continued)

Table 45.1 (continued)

References	Objective	Approach	Result
[42]	Achieve control over congestion in public networks	Fairness-driven queue management algorithms	The survey gives a comprehension of information regarding challenges in congestion control in a network of heterogeneous types

References

1. Martin, A., Egaña, J., Flórez, J., Montalbán, J., Olaizola, I.G., Quartulli, M., Viola, R., Zorrilla, M.: Network resource allocation system for QoE-aware delivery of media services in 5G networks. IEEE Trans. Broadcast. **64**(2) (2018)
2. Cisco Systems, Inc.: Cisco Visual Networking Index: Forecast and Methodology, 2016–2021, White Paper, 2017. (Online). Available: https://www.reinvention.be/webhdfs/v1/docs/complete-white-paper-c11-481360.pdf (2017)
3. Jacobson, V., Frederick, R., Casner, S., Schulzrinne, H.: Real-TimeTransport Protocol (RTP). https://www.ietf.org/rfc/rfc3550.txt (2017)
4. Adobe: Real-Time Messaging Protocol(RTMP) (2017)
5. Schulzrinne, H.: Real Time Streaming Protocol version 2.0. https://tools.ietf.org/html/rfc7826 (2017)
6. Goldberg, J., Westerlund, M., Zeng, T.: A Network Address Translator (NAT) Traversal Mechanism for Media Controlled by the Real- Time Streaming Protocol (RTSP) (2017)
7. Ibarrola, E., Davis, M., Voisin, C., Close, C., Cristobo, L.: A machine learning management model for QoE enhancement in next-generation wireless ecosystems. In: IEEE ITU Kaleidoscope: Machine Learning for a 5G Future (ITU-K) (2018)
8. Kafle, V.P., Fukushima, Y., Martinez-Julia, P., Miyazawan, T.: Cosideration on automation of 5G network slicing with machine learning. In: IEEE ITU Kaleidoscope: Machine Learning for a 5G Future (ITU-K) (2018)
9. Abadi, M. et al.: TensorFlow: large-scale machine learning on heterogeneous distributed systems. (Online). Available: https://arxiv.org/abs/1603.04467 (2016)
10. Ferreira, L.E.B., Gomes, H.M., Bifet, A., Oliveira, L.S.: Adaptive random forests with resampling for imbalanced data streams. IEEE International Joint Conference on Neural Networks (IJCNN) (2019)
11. Gorius, M., Shuai, Y., Herfet, T.: Dynamic media streaming under predictable reliability. In: IEEE international Symposium on Broadband Multimedia Systems and Broadcasting (2012)
12. Aroussi, S., Mellouk, A.: Survey on machine learning-based QoE-QoS correlation models. In: International Conference on Computing, Management and Telecommunications (ComManTel), pp. 200–204 (2014)
13. Ibarrola, E., Saiz, E., Zabala, L., Cristobo, L., Xiao, J.: A new global quality of service model: QoXphere. IEEE Commun. Mag. **52**, 193–199 (2014)
14. Martín, V., Cabrera, J., García, N.: Design, optimization and evaluation of a Q-learning HTTP adaptive streaming client. IEEE Trans. Consum. Electron. **62**(4), 380–388 (2016)
15. Gadaleta, M., Chiariotti, F., Rossi, M., Zanella, A.: D-DASH: a deep Q-learning framework for DASH video streaming. IEEE Trans. Cognitive Comm. Netw. **3**(4),703–718 (2017)
16. Javadtalab, A., Semsarzadeh, M., Khanchi, A., Shirmohammadi, S., Yassine, A.: Continuous one-way detection of available bandwidth changes for video streaming over best-effort networks. IEEE Trans. Instrum. Meas. **64**(1) (2015)
17. Gomez-Uribe, A., Hunt, N.: The Netflix recommender system: algorithms, business value, and innovation. ACM Trans. Manage. Inf. Syst. **6**(2016)

18. Tian, G., Liu, Y.: Towards Agile and smooth video adaptation in HTTP adaptive streaming. IEEE/ACM Trans. Netw. **24**(4), 2386–2399 (2016)
19. Waldmann, S., Miller, K., Wolisz, A.: Traffic model for HTTP-based adaptive streaming. In: IEEE Conference on Computer Communications Workshops (INFOCOM WKSHPS) (2016) (2017)
20. Wiegand, T., Sullivan, G.J., Bjontegaard, G., Luthra, A.: Overview of the H.264/AVC video coding standard. IEEE Trans. Circuits Syst. Video Technol. **13**(7), 560–576 (2003)
21. Schwarz, H., Marpe, D., Wiegand, T.: Overview of the scalable video coding extension of the H.264/AVC standard. IEEE Trans. Circuits Syst. Video Technol. **17**(9), 1103–1120 (2007)
22. Sullivan, G.J., Ohm, J.-R., Han, W.-J., Wiegand, T.: Overview of the high efficiency video coding (HEVC) standard. IEEE Trans. Circuits Syst. Video Technol. **22**(12), 1649–1668 (2012)
23. Boyce, J.M., Ye, Y., Chen, J., Ramasubramonian, A.K.: Overview of SHVC: scalable extensions of the high efficiency video coding standard. IEEE Trans. Circuits Syst. Video Technol. **26**(1), 20–34 (2016)
24. Li, C., Wei, F., Dong, W., Wang, X., Liu, Q., Zhang, X.: Dynamic structure embedded online multiple-output regression for streaming data. IEEE Trans. Pattern Anal. Mach. Intell. **41**(2) (2019)
25. Yusuf-Asaju, A.W., Dahalin, Z.M., Ta'a, A.: Framework for modelling mobile network quality of experience through big data analytics approach. J. Inf. Commun. Technol. (JICT) **17**, 79–113 (2018)
26. Tripathi, A., Ashwin, T.S., Guddeti, R.M.R.: EmoWare: a context-aware framework for personalized video recommendation using affective video sequences. IEEE Access **7** (2019)
27. Bogale, T.E., Wang, X., Le, L.: Machine intelligence techniques for next-generation context-aware wireless networks. ITU J. ICT Discoveries (Special Issue No. 1) (2018)
28. Jiang, C. et al.: Machine learning paradigms for next-generation wireless networks. IEEE Wirel. Commun. **24**, 98–105 (2017)
29. Kato, N., et al.: The deep learning vision for heterogeneous network traffic control: proposal, challenges and future perspective. IEEE Wirel. Commun. **24**(3), 146–153 (2017)
30. Xu, Y., Yu, C., Li, J., Liu, Y.: Video telephony for end-consumers: measurement study of Google+, iChat, and Skype. IEEE/ACM Trans. Netw. **22**(3), 826–839 (2014)
31. Qadir, Q.M., Kist, A.A., Zhang, Z.: A Novel traffic rate measurement algorithm for quality of experience-aware video admission control. IEEE Trans. Multimedia **17**(5) (2015)
32. Vega, M.T., Perra, C., Turck, F.D., Liotta, A.: A review of predictive quality of experience management in video streaming services. IEEE Trans. Broadcast. **64**(2) (2018)
33. Hoßfeld, T., Schatz, R., Biersack, E., Plissonneau, L.: Internet video delivery in YouTube: from traffic measurements to quality of experience (2013)
34. De Grazia, M.D.F., Zucchetto, D., Testolin, A., Zanella, A., Zorzi, M., Zorzi, M.: QoE multi-stage machine learning for dynamic video streaming. IEEE Trans. Cognitive Comm. Netw. **4**(1), 146–161 (2018)
35. Elwerghemmi, R., Heni, M., Ksantini, R., Bouallegue, R.: Online QoE prediction model based on stacked multiclass incremental support vector machine. In: IEEE 8th International Conference on Modeling Simulation and Applied Optimization (ICMSAO) (2019)
36. Focus Group on Machine Learning for Future Networks including 5G. https://www.itu.int/en/ITUT/focusgroups/ml5g/Pages/default.aspx (2018).
37. Lübben, R., Fidler, M.: Service curve estimation-based characterization and evaluation of closed-loop flow control. IEEE Trans. Netw. Serv. Manage. **14**(1) (2017)
38. Kanrar, S., Mandal, N.K.: Video traffic flow analysis in distributed system during interactive session. Adv. Multimedia **2016**, 14. Article ID 7829570 (2016)
39. Balaouras, P., Stavrakakis, I.: A self-adjusting rate adaptation scheme with good fairness and smoothness properties. Comput. Netw. **48**, 829–855 (2005)
40. D'Aronco, S., Toni, L., Mena, S., Zhu, X., Frossard, P.: Improved utility-based congestion control for delay-constrained communication. IEEE/ACM Trans. Netw. **25**(1) (2017)

41. Carlucci, G., Cicco, L.D., Holmer, S., Mascolo, S.: Congestion control for web real-time communication. IEEE/ACM Trans. Netw. **25**(5) (2017)
42. Abbas, G., Halim, Z., Abbas, Z.H.: Fairness-driven queue management: a survey and taxonomy. IEEE Commun. Surv. Tutorials **18**(1) (2016)

Chapter 46
A Study on the State of Artificial Intelligence in the Design Phase of Software Engineering

Mriganka Shekhar Sarmah, Jediael Meshua Sumer, Marlom Bey, and Bobby Sharma

Abstract The areas of software engineering and artificial intelligence are among the most sought-after fields in the computer science community. This paper assesses the current scenario of implementing artificial intelligence in the design phase of software engineering and attempts to offer a viable explanation to support the conclusion of the assessment.

46.1 Introduction

In this day and age, traditional software design has its limits. In traditional SE, the approach towards building a software is manual in nature, which makes it prone to human errors such as oversight and is also inefficient when complicated projects are considered [1, 2].

Another reason for errors may be the ambiguity of the textual requirements which may not be exactly translated according to the perspective of the customer [3].

Therefore, in order to minimize human error and increase the efficiency of software design and to deal with ambiguity of textual requirements of the customer, AI techniques can be implemented, mainly in the stages of design, testing and code generation [4, 5].

Hence, the prospects of automation/AI in the phase of software design must be discussed, and related problems must be dealt with, which are the reliability and accuracy of automation and how much control it takes away from a software developer.

This paper compares related works regarding how they either tackle this problem or suggest ways to tackle this problem.

M. S. Sarmah (✉) · J. M. Sumer · M. Bey · B. Sharma
Department of Computer Science and Engineering, Assam Don Bosco University, Guwahati, Assam, India

B. Sharma
e-mail: bobby.sharma@dbuniversity.ac.in

The paper is organized into the following sections. Section 46.2 discusses works related to the stated topic, Sect. 46.3 discusses the analysis and results of existing work, and finally with Sect. 46.4, the paper is summarized and concluded.

46.2   Related Works

In [4], the author discusses various commonalities between the fields of AI and SE, current status and upcoming trends. Intersections between AI and SE are discussed, especially in the fields of agent-oriented software engineering, knowledge-based engineering systems, computational intelligence and knowledge discovery of data and ambient intelligence. It also gives a short description about AI and software engineering and the possibilities of interaction with one another through various common points of contact.

In [5], the author discusses the differences between the software engineering process and AI and the hostilities that exist in software engineers towards automation in developing a software. It further discusses the contribution of AI to the field of SE in fields like automatic programming and some projects like REFINE and Programmer's Apprentice (an MIT project) and the problem areas of SE where AI can be implemented like requirements analysis and definition, process modelling, process support and project planning.

The author states that one of the reasons for the existing hostility towards AI is that AI researchers are themselves software engineers, and therefore their method to integrating AI in software engineering follows a typical and traditional approach which precedes the advancements that AI and its endless possibilities can bring to the field of software engineering.

In [6], the author uses the terminology defined in IEEE 12,207 standards for software engineering to describe the development processes of requirements analysis, software architecture design and coding and testing and AI techniques that can be implemented along with them. The paper also talks about open problems in SE that can be solved using AI so as to improve the process of SE in this day and age, where software is getting more modular and complex to develop.

In [7], the author deals with a taxonomy called AI in SE Application Levels or AI-SEAL that classifies 15 papers from previous editions of the RAISE workshop. It considers the context in which AI is being applied, i.e. "when" and "on what" AI is being applied. After classification, the papers also assign levels of automation to the papers.

In [8], the author gives a brief overview of SE and expert systems in artificial intelligence and how expert systems can be used to automate the programming process and code generation via the use of genetic programming.

This paper also shows the absence of risk management in AI-based systems due to the way they work.

In [1], the author discusses the translation of user requirements into design diagrams and describes it as a daunting task for a designer, as that person has to translate textual requirements into a diagrammatic (UML) form. This paper gives a basic overview of existing systems regarding and how they convert the user requirements using NLP into UML diagrams.

They developed their own tool, DC Builder, and presented in the paper a diagrammatic representation and a heuristic set of rules in implementing NLP on user requirements. Finally, they evaluated and compared various tools mentioned and developed, with one another.

In [2], the author describes the extraction of UML diagrams from textual requirements by requirements of engineers as a daunting task and the time and effort spent on this justify a tool to automate this process which brings the author to propose a tool called RAPID which uses NLP in an efficient manner to extract design diagrams from input textual user requirements.

The methodology states various NLP technologies, algorithms and rules to extract class information from the textual requirements.

In [3], the author talks about the ambiguity of textual requirements and the usage of NLP and domain ontology to generate UML diagrams from the said textual requirements.

This paper also talks about existing systems and then approaches the problem using their own method called RAUE to extract UML diagrams from textual requirements.

In [9], the author looks at existing systems which translate textual requirements into UML diagrams. After that the author proposes his own approach called RACE which is an improvement on the TCM system. The methodology of the RACE system is discussed with its algorithms and rules to identify classes, attributes and relationships.

The author concludes the paper stating RACE being an advanced approach towards extraction of UML diagrams from textual requirements using a human-centred UI and what the system did not support.

In [10], the author introduces an automatic test data generator called ATGen. ATGen is based on constraint logic programming and symbolic execution. Developing software includes a number of testing methods. The main technique that is currently used in the industry is dynamic software testing where the software is executed using test data. Dynamic testing can be performed using automatic tools, which are automation of administrative tasks, automation of mechanical tasks and automation of test generation tasks.

ATGen is a prototype testing tool implemented using the ECLiPSe constraint logic programming environment and consists over 5000 lines of commented Prolog code. The current area of application of ATGen is the automatic generation of test data to achieve 100% decision coverage for programmes written in SPARK Ada. In decision testing, the aim is to test all decision outcomes in the programme. The aim is to generate a test data suite achieving 100% decision coverage. However, initial results of ATGen have proved that at the time this research was conducted, and it was impossible to experiment with ATGen using industrial SPARK Ada code as such code was not usually made available even for research due to its safety critical aspects. Also, the test results show that ATGen shows wide performance variation between successive runs, particularly for programmes with loops. The initial results of ATGen were promising, and the overall efficiency of the algorithm is under improvement.

In [11], the author describes an approach towards extracting UML diagrams from natural language using NLP via a tool called CM Builder. CM Builder is a graphical CASE tool that does surface analysis of text to propose candidates for class, attributes and relationship and domain-independent semantic analysis to automatically extract the candidates and finally represent them via UML diagrams.

The tool also includes capacity to evaluate its candidate classes. The author also states benefits of object-oriented analysis of the tool and its scope of improvement.

In [12], the author explored the reasons for the lack of impact in important areas in which AI has been expected to significantly affect real-world software engineering. The session approached the failures of AI in software engineering, looking at the matter through a common cause reliance on isolationist technology and approaches, rather than upon creating additive technology and approaches that can be integrated with other existing capabilities.

The isolationism has been manifested in several areas, in essence, the market has rejected the isolationist and egocentric approach to implementing AI in software engineering, and that the whole system should be developed and executed in generalized workstations and PCs.

In [13], the author describes recent state-of-the-art approaches to automatically generate regular expressions from natural language specifications. Given that these approaches use only synthetic data in both training datasets and validation/test datasets, a natural question arises: Are these approaches effective to address various real-world situations? To explore this question, in this paper, a characteristic study on comparing two synthetic datasets is used by the recent research, a real-world dataset is collected from the Internet, and an experimental study on applying a state-of-the-art approach on the real-world dataset is conducted. The study results suggest the existence of distinct characteristics between the synthetic datasets and the real-world dataset, and the state-of-the-art approach (based on a model trained from a synthetic dataset) achieves extremely low effectiveness when evaluated on real-world data, much lower than the effectiveness when evaluated on the synthetic dataset.

In [14], the author looks at previous work done in the field of converting natural language requirements into design diagrams and points us the various issues that can be encountered in this process.

The author points out the issues in the context of software engineering, then in the context of natural language and finally proposes a methodology to solve these problems to get good quality UML diagrams.

In [15], the author introduces artificial intelligence to software engineers and conversely, software engineering to artificial intelligence workers. It further highlights the contrast between the two fields and further accentuates their differences in the problems that they attempt to solve, their methodologies and the tools and techniques. The author strongly believes that the work of software engineers and artificial intelligence workers are similar, and the fusion of the two fields is essential for computer science as a field to move forward; however, the author also acknowledges the ridge that exists between the two communities. The paper does not, however, provide the "*how*" on which AI and software engineering can come together, other than details of the core components of both fields.

The paper acknowledges that to move forward with "user-friendliness", the system needs to allow natural language input and output, and natural language processing is one of the most researched areas in AI. In other words, a system needs AI to further improve its usability.

Conversely, AI is in need of proper models to address its problems. These models are already evident in conventional systems, and it is clear that AI systems need some standardization in order to widen their applications.

Thus, there exists a requirement for both fields to merge on some common ground.

46.3 Results and Analysis of Existing Work

From the study done above, a few things are apparent. Firstly, when design diagrams made by a software engineer are considered, three factors come into play.

- Ambiguity and complexity of NL textual requirements [13].
- Complex nature of the project due to increasing complexity of technology and size of the project.
- Possibility of human error when designing those systems from textual requirements.

The above results in the problem of defining the accuracy of a system, be it designed by a person or by an AI-based tool, as there is no point of references. This makes it imperative to create a dataset to compare any diagram to measure its accuracy.

Another way it can be done is to design an algorithmic model that can analyse input textual requirements to give appropriate output, but the amount of work involved in making it foolproof would make it an insurmountable challenge.

Another aspect of translating user requirements into a design diagram is that when the textual requirements are complicated, extracting context requires the use of domain ontology along with defined semantic rules. When using this, problems will be encountered when very complex sentences are considered.

Secondly, when AI in system design is considered, due to its automatic nature, the problem of a system engineer not having enough freedom over the creation of design diagrams is encountered. Hence, there exists a need to carefully define the level of automation of such tools. This can be done by making the end results of a diagram generator interactive or editable, as can be seen in various smart design diagram tools.

Table 46.1 shows the comparison of the availability of features of various tools that generate design diagrams from user input textual requirements. It can be inferred

Table 46.1 Evaluation of tools' functionality. Adapted and modified from [1]

Support	CM Builder [11]	LIDA	GOOAL	NLOOML	DC Builder [1]	RACE [9]	RAUE [3]	RAPID [2]
Classes	Yes	User	Yes	Yes	Yes	Yes	Yes	Yes
Attributes	Yes	User	Yes	Yes	Yes	Yes	Yes	No
Methods	No	User	Yes	Yes	No	Yes	No	No
Associations	Yes	User	Semi-NL	No	Yes	Yes	Yes	Yes
Multiplicity	Yes	User	No	No	No	No	Yes	No
Aggregation	No	No	No	No	Yes	Yes	Yes	Yes
Generalization	No	No	No	No	Yes	Yes	Yes	Yes
Instances	No	No	No	No	No	No	No	No

from the table that there are very few tools which fully extract relationships from textual requirements.

Another observation made from the tools mentioned is that although they can create class diagram, they cannot deal with instances to create object diagrams.

46.4 Conclusion and Future Work

From the above survey on the papers, it is evident that artificial intelligence has a wide scope of application in the field of software engineering and has the capability to make a software engineer's work easier and faster. However, AI systems have to undergo further improvements to fully exploit their capabilities in the field of software engineering.

The findings of this paper can be further analysed and utilized to design methodologies that make generation of UML design diagrams from input user requirements far more efficient. Furthermore, more research has to be done, aimed towards the use of machine learning to automate the task of diagram generation.

References

1. Herchi, H., Abdessalem, W.B.: From user requirements to UML class diagram (2012). arXiv preprint arXiv:1211.0713
2. More, P., Phalnikar, R.: Generating UML diagrams from natural language specifications. Int. J. Appl. Inf. Syst. Found. Comput. Sci. 1(8), 19–23 (2012)
3. Joshi, S.D., Deshpande, D.: Textual requirement analysis for UML diagram extraction by using NLP. Int. J. Comput. Appl. 50(8), 42–46 (2012)
4. Rech, J., Althoff, K. D.: Artificial intelligence and software engineering: status and future trends. KI 18(3), 5–11 (2004)

5. Sommerville, I.: Artificial intelligence and systems engineering. In: Prospects for Artificial Intelligence: Proceedings of AISB'93, 29 Mar, 2 Apr 1993, Birmingham, UK, 17, 48 (1993)
6. Ammar, H.H., Abdelmoez, W., Hamdi, M.S.: Software engineering using artificial intelligence techniques: Current state and open problems. In: Proceedings of the First Taibah University International Conference on Computing and Information Technology (ICCIT 2012), Al-Madinah Al-Munawwarah, Saudi Arabia, p. 52 (2012)
7. Feldt, R., de Oliveira Neto, F.G., Torkar, R.: Ways of applying artificial intelligence in software engineering. In: 2018 IEEE/ACM 6th International Workshop on Realizing Artificial Intelligence Synergies in Software Engineering (RAISE), pp. 35–41. IEEE (2018)
8. Raza, F.N.: Artificial intelligence techniques in software engineering (AITSE). In: International MultiConference of Engineers and Computer Scientists (IMECS 2009), vol. 1 (2009)
9. Ibrahim, M., Ahmad, R.: Class diagram extraction from textual requirements using natural language processing (NLP) techniques. In: 2010 2nd International Conference on Computer Research and Development, pp. 200–204. IEEE (2010)
10. Meudec, C.: ATGen: automatic test data generation using constraint logic programming and symbolic execution. Softw. Test. Verification Reliab. 11(2), 81–96 (2001)
11. Harmain, H.M., Gaizauskas, R.: CM-Builder: an automated NL-based CASE tool. In: Proceedings ASE 2000. 15th IEEE International Conference on Automated Software Engineering, pp. 45–53. IEEE (2000)
12. Balzer, R., Fikes, R., Fox, M., McDermott, J., Soloway, E.: AI and software engineering: will the twain ever meet?. In: AAAI, pp. 1123–1125 (1990)
13. Zhong, Z., Guo, J., Yang, W., Xie, T., Lou, J. G., Liu, T., Zhang, D.: Generating regular expressions from natural language specifications: are we there yet? In: Workshops at the 32nd AAAI Conference on Artificial Intelligence (2018)
14. Sharma, N., Yalla, D.P.: Issues in developing UML diagrams from natural language text. In: Proceedings of WSEAS Conference, Recent Advances in Telecommunications, Informatics and Educational Technologies, pp. 139–145
15. Ford, L.: Artificial intelligence and software engineering: a tutorial introduction to their relationship. Artif. Intell. Rev. 1(4), 255–273 (1987)

Chapter 47
S2S Translator by Natural Language Processing

Srikanth Bethu, Suresh Mamidisetti, S. Bhargavi Latha, and B. Sankara Babu

Abstract Correspondence by means of motions is a visual language that is used by in need of a hearing-aided people as their local language. Not in the least like acoustically passed on sound models, signal-based correspondence uses non-verbal correspondence and manual correspondence to easily pass on the contemplations of a person. It is practiced by simultaneously uniting hand shapes, heading and advancement of the hands, arms or body, and outward appearances. In this paper, we will endeavor to develop a programmed model which changes over voice to motion-based correspondence and besides correspondence by means of motions to voice/text. We may be using talk-to-message programming interface (Python modules or Google programming interface) and a while later using the semantics of natural language processing to breakdown the substance into more diminutive sensible pieces which requires AI as a segment. Instructive assortments of predefined motion-based correspondence are used as the information with the objective that the item can use manmade mental ability to show the changed over sound into the signal-based correspondence.

47.1 Introduction

Motion-based correspondence is a language absolutely autonomous and unquestionable from English. It contains all the basic features of language, with its own rules for enunciation, word game plan, and word demand. While every language has techniques for hailing different limits, for instance, representing a request rather than saying something, vernaculars differentiate in how this is done. For example, English speakers may represent a request by raising the pitch of their voices and by adjusting word demand; sign language [1] customers represent a request by creating

S. Bethu (✉) · S. B. Latha · B. S. Babu
Department of Computer Science and Engineering, GRIET, Hyderabad, Telangana 500090, India

S. Mamidisetti
Department of Technical Education, Government Polytechnic College, Hyderabad 500028, India

© The Author(s), under exclusive license to Springer Nature Singapore Pte Ltd. 2021   475
S. C. Satapathy et al. (eds.), *Smart Computing Techniques and Applications*,
Smart Innovation, Systems and Technologies 225,
https://doi.org/10.1007/978-981-16-0878-0_47

a ruckus, expanding their eyes, and slanting their bodies forward. Correspondingly moreover with various lingos, unequivocal techniques for conveying considerations in signal-based correspondence move as much as customers themselves. Despite solitary differences in enunciation, correspondence by means of motions has commonplace accents and tongues; comparably as certain English words are verbally communicated particularly in different bits of the country, motion-based correspondence has regional assortments in the musicality of checking, address, slang, and signs used.

Other sociological components, including age and sexual direction, can impact correspondence through signals usage and add to its combination, correspondingly moreover with conveyed in vernaculars. Finger spelling is a bit of correspondence through marking and is used to light up English words. In the finger spelled letters all together, each letter identifies with a specific hand shape. Finger spelling is as often as possible used for real names or to show the English word for something.

Correspondence through marking is something which is not fathomed by various people. Generally scarcely any people have the prolog to it which makes it hard for people who talk with it to have the alternative to give their expects to others. Disregarding the way that some may appreciate bits of it in any case, they cannot grasp its whole setting. Considering the circumstance as we have right now, nearly deaf people are commonly prevented from securing standard correspondence with others, they have to rely upon an interpreter or some visual correspondence, similarly by virtue of ordinary people as well.

47.2 Literature Review

Quoc et al. [2] in 2018 proposed a work on highlight talk understanding. Emphasis is used to perceive connected with and unfocused bit of a verbalization, and it is useful in misheard conditions in which speakers must repetitive the most huge words or articulations. The proposed approach to manage handle steady emphasis levels which relies upon course of action models and besides they have joined the machine and complement understanding into the single model. They have proposed hard thought emphasis talk understanding and joint model. In the hard thought complement translation, it can unravel reliable highlight loads without quantization. Furthermore, in the joint model, it unravels the translation and together interprets words and complement with single word delay. They have used LSTM-based encoder-decoder to handle the issues in single model of hard thought progression-to-plan model.

Taylor et al. [3] in 2018 proposed a work on talk affirmation by using disguised Markov model. They consider the translation language from English to Indonesian. The request procedure used is the Mel repeat cepstral coefficients (MFCC) and hidden Markov model (HMM). The proposed structure changes over talk to text and usages the current Google understanding or Microsoft translation for translation. The talk sign will be dealt with by using MFCC. The count used here is k-means estimation.

Poornakala and Maheshwari [4] in 2016 proposed a work talk translation from English to Tamil language [5]. Talk affirmation structures seeing English talk by methods for a voice affirmation device. English talk make an understanding of the substance into English using the conversation mix, in the wake of changing over it to message, independent, and the words put to the side in the informational collection in case it energizes the Tamil substance put to the side in the data set. The English substance is changed over to Tamil substance which is made by the machine translation structure, and a while later, the substance will be appeared on the screen. It is expected to decode up to 60 words in the Android application. The estimation used here is HMM count.

Shin et al. [6] in 2015 proposed a work on talk-to-talk translation humanoid robot. The proposed structure is helpful for English-talking patients; they can uncover their issues to Korean authorities. It can deal with the issues of various patients who do not have the foggiest thought regarding the Korean language despite the way that they can portray their issues to pro which override the activity of human workers. It uses CMU Sphinx-4 gadget for talk affirmation. English–Korean understanding relies upon rule-based translation.

47.3 Methodology and Implementation

Image processing method infers treatment of a propelled picture by techniques for a PC. It can in like manner be described as a technique for using PC computations, in order to get improved picture either to remove some supportive information or for some other kind of task.

Speech Recognition: Talk affirmation is a subfield of programming building and historical underpinnings that make methods and advancements that enable the affirmation and translation [7] of conveyed in language into text by PCs. It is in any case called automatic talk affirmation (ASR), PC talk affirmation, or talk to message (STT). Talk affirmation systems require "getting ready" where an individual speaker gets text or fundamental language into the structure. The structure assesses the individual's unequivocal voice and uses it to change the affirmation of that individuals talk, realizing extended exactness. Systems that do not use getting ready are called speaker-free structures. Talk affirmation applications fuse various things, for instance, voice dialing, call coordinating, search catchphrases, clear data area, preparation of sorted out reports, choosing speaker characteristics, and talk to message getting ready.

Models, Methods, and Algorithms: Both language showing and acoustic exhibiting are noteworthy bits of present-day quantifiably based talk affirmation computations. Covered Markov models (HMMs) are comprehensively used in various structures. Language showing is moreover used in various other trademark language taking care of utilizations, for instance, quantifiable machine understanding or document gathering. Markov models: Present-day talk affirmation systems rely upon hidden Markov models [8]. These are authentic models that yield a progression of pictures or sums. HMM is used in talk affirmation because a talk sign can be viewed as a piecewise fixed sign or a short period of time fixed sign.

From Fig. 47.1, the modules' frameworks are created for speech and text to convert into sign language using sign language translator [9]. In Fig. 47.1, the image is taken as input into model, and output generated is voice output using natural language conversion process. From Fig. 47.2, it showed the connectivity of preprocessed input image to preprocessed signed image. In Fig. 47.2, the voice is taken as input, and sign language image is generated output.

Getting the Dataset: Datasets of predefined sign language were taken from Kaggle. It had approximately 10,000 images corresponding to each alphabet including both

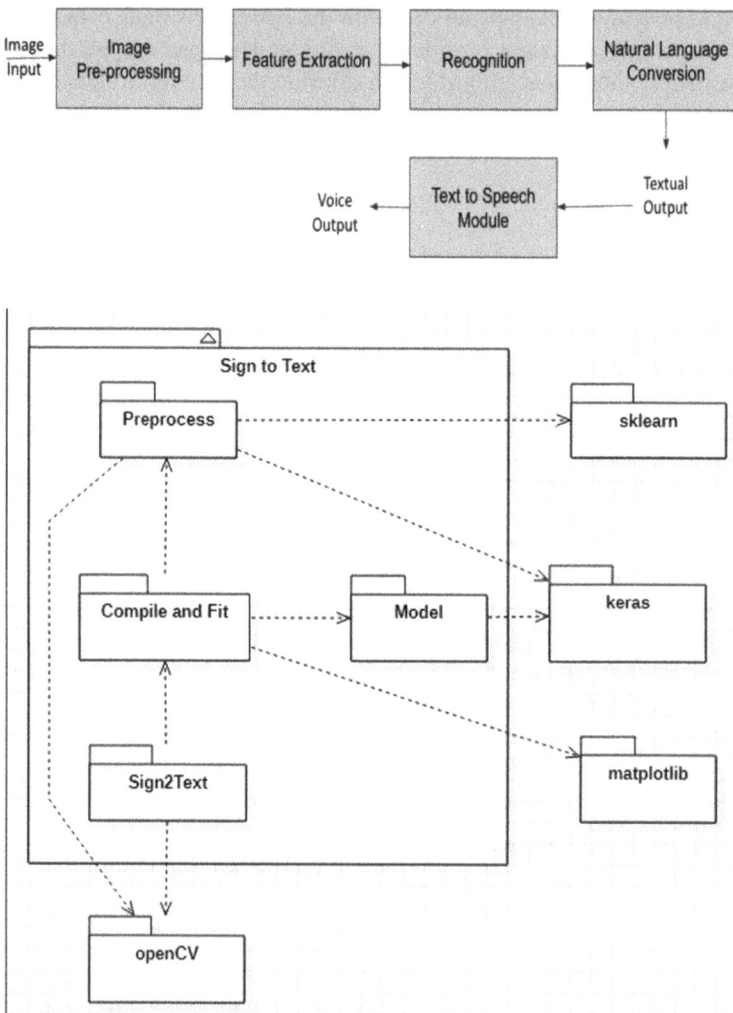

Fig. 47.1 Text-to-speech module creation and the sign-to-text process is created using OpenCV by Matplotlib

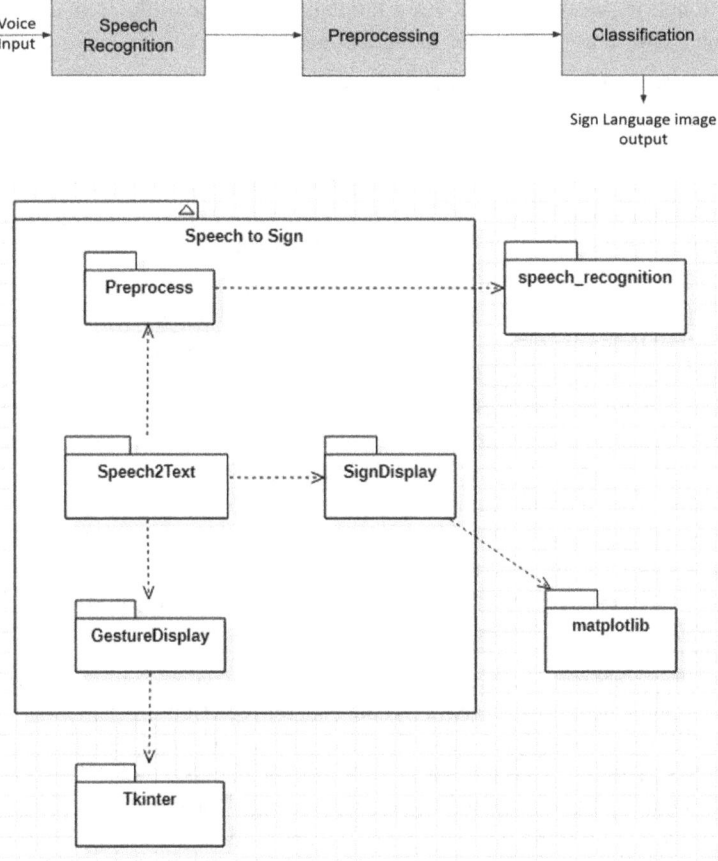

Fig. 47.2 Sign language output module creation and the speech-to-sign process is created using OpenCV by Matplotlib

training and test datasets. An image corresponding to each alphabet was taken and displayed using Matplotlib, so that users can see what the corresponding sign to each alphabet was. Datasets of some frequent sign language [10] gestures were also gathered and were used similarly as mentioned above.

Pre-processing the Dataset: The training data of the sign images was taken and was cleaned to make it ready for classification. Each image was converted to 64X64, and the color was changed to monochrome.

The image array was then converted to a NumPy array. The array was normalized by converting the values to 'float32' and dividing it with the total sum of the pixel values, i.e., 255. The labels corresponding to each image was stored in an array as well and was converted to categorical values using the internal function present in keras module. The whole image data was split into training and testing using the internal function present in scikit-learn module. The shape of the train and test data was printed and returned back for the model to consume.

Creating CNN Model: After preprocessing of the training data, we will create a CNN model which is used for classification. Keras module is used extensively here to create the model. We created a sequential model with 16 nodes in the first layer having the kernel_shape [3,3], and the input shape was [64,64,3]. Activation function used was "Relu". Next layer had 32 nodes having the kernel_shape [3,3] with the similar activation function. We added a MaxPool layer of two dimensions before getting the output to the next layers. The next layer connected to 32 node which was further connected to 64 nodes having a MaxPool2D layer. The output of these 64 nodes was fed to the next layer which had 128 nodes connected further to 256 nodes. After the MaxPool2D layer, batch normalization was done here. We flattened the output of the layers here and set the dropout to "0.5". The last layer of processing had 512 nodes where also kernel_regulizer was set to "0.001". At last, we had the output layer of 29 nodes since we have 29 labels or classes. SoftMax function was used to get the output in terms of integers for easy processing and accurate classification. The model was compiled with "Adam" optimizer, and loss and accuracy metrics were recorded while fitting the data into the model. Summary of the model was printed in case anyone in the future will be able to use it, and the model was saved for further use.

Training the CNN Model: After the model is created, we need to fit our train data into it, so that the model will learn about the data through training, and then, it would be used for further classifying real-time data. We use the "fit" method to give our training data. The batch_size is set as 64 which means that images will be fed to the CNN model in batches with 64 each in one batch. The epoch is set to be 5, i.e., for every five minutes, the data will be fed to the model. The validation split is set as "0.5" which means that the dataset is divided into two parts, one for training and other for validation. After the training is done, an object is returned which has all the metrics that we had put while creating the model. Matplotlib is used to plot the accuracy graphs, loss graphs, and printing out the evaluation accuracy and evaluation loss.

Implementing Speech Recognition: The above procedures were done for converting sign language to speech. Now, coming back to speech-to-sign language conversion, we first need to implement the speech recognition to allow speech to be recognized. We will be using the speech_recognition module for that. Once we receive the speech through microphone, it will be converted to text. After converting to text, we will call the gesture display component or the sign display component after performing some validation checks.

Gesture Display and Sign Display: After receiving the text using speech recognition module, we will call the gesture display component if the text matches the gestures that we have in our dataset. If the gesture matches, then a GIF image will be displayed to the user, indicating the gesture corresponding to the text which was spoken. If the text does not match with any of the gestures, then we will display a series of signs from the test data of the sign language dataset. The signs will be repeated twice having regular intervals, so that user will be able to perceive them properly.

Implementing Live Classification: After creating the model for converting sign language to speech, we will use it to convert the signs provided through live Web cam feed. We will be using OpenCV to get the images through the live feed. After receiving the images, we will preprocess them similarly as mentioned in Sect. 3. After preprocessing, the image is given to the model to classify. The model will give us a numeric output corresponding to the alphabet that the sign portrays. That alphabet is taken and printed over the screen for user accessibility. After the word is at least of 7 letters long, we will invoke the text-to-speech module to convert that text to speech.

47.4 Results and Discussion

Figure 47.3 shows the dataset uploading into the system model created by neural network layers. Maxpooling and batch normalization are used to filter the data. The total parameters are loaded into the model which are 942.557, trainable parameters which are 942.045, and non-trainable parameters which are 512. Non-trainable parameters are classified by flatten and dense methods.

Figure 47.4 shows the model fitting outputs and validation of its accuracy using the graph. The total number of trained samples is used in the models 74,382, and output generated values are 8265. The evaluation accuracy occurred is 99.75%, and evaluation loss occurred is 0.080104.

Figure 47.5 shows the generation of outputs using OpenCV. The images are the predicted images that are to be to be identified in the final prediction results. Figure 47.6 shows the final prediction of all results in S2S language.

47.5 Conclusion

The concept entitled "S2S Translator" has presented a model for translating sign language to speech and speech to sign language. The model that was developed in this paper has a fair enough accuracy and can be used for further improvisation. For better live classification, an object detection module could have been developed or an already existing one could have been used which will lead to better results in translating. Some kind of support for translating more complex sentences can be further investigated for the future of this paper. The voice output feature which is the output for the sign-to-speech [11] conversion can be further improved.

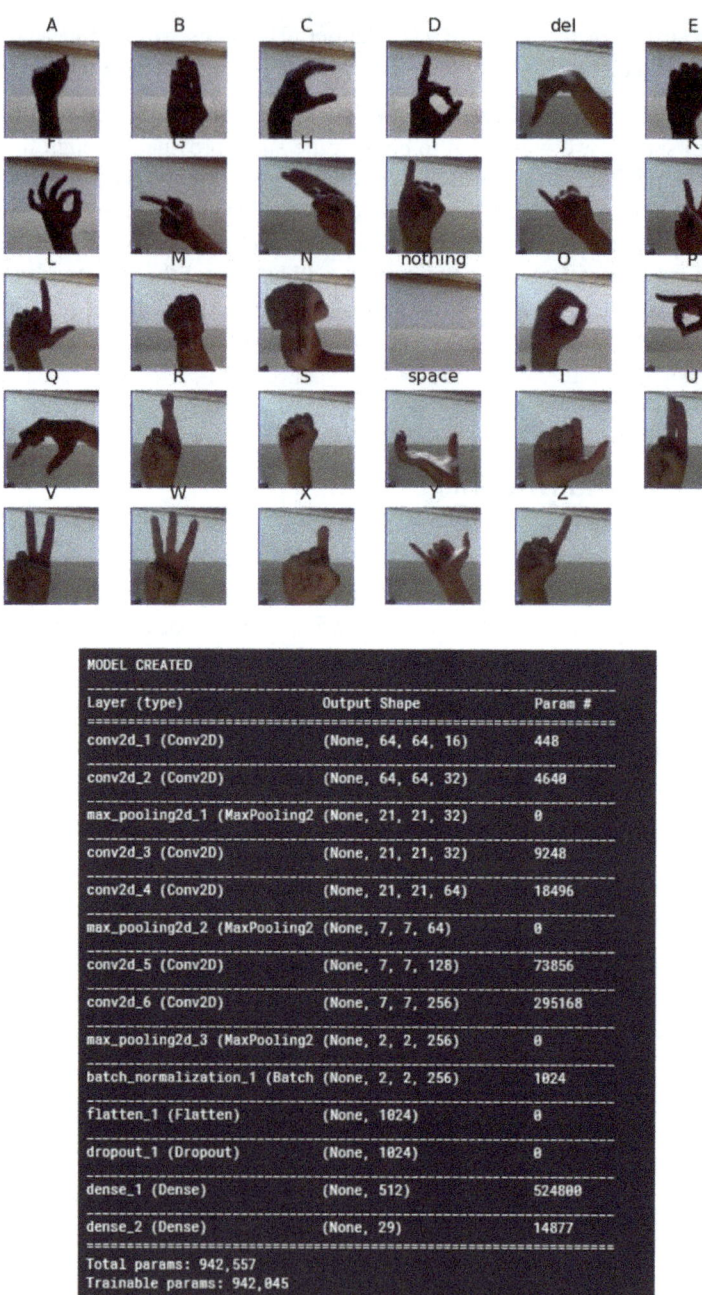

Fig. 47.3 Uploading sample dataset and creating a model using CNN layers

```
Train on 74385 samples, validate on 8265 samples
Epoch 1/5
74385/74385 [==============================] - 19s 259us/step - loss: 0.7998 - acc: 0.8382 - val_loss: 0.2221 - val_acc: 0.9843
Epoch 2/5
74385/74385 [==============================] - 15s 196us/step - loss: 0.2047 - acc: 0.9760 - val_loss: 0.1819 - val_acc: 0.9750
Epoch 3/5
74385/74385 [==============================] - 15s 196us/step - loss: 0.1593 - acc: 0.9826 - val_loss: 0.2469 - val_acc: 0.9593
Epoch 4/5
74385/74385 [==============================] - 15s 199us/step - loss: 0.1450 - acc: 0.9860 - val_loss: 0.1758 - val_acc: 0.9811
Epoch 5/5
74385/74385 [==============================] - 15s 203us/step - loss: 0.1449 - acc: 0.9878 - val_loss: 0.0780 - val_acc: 0.9979
```

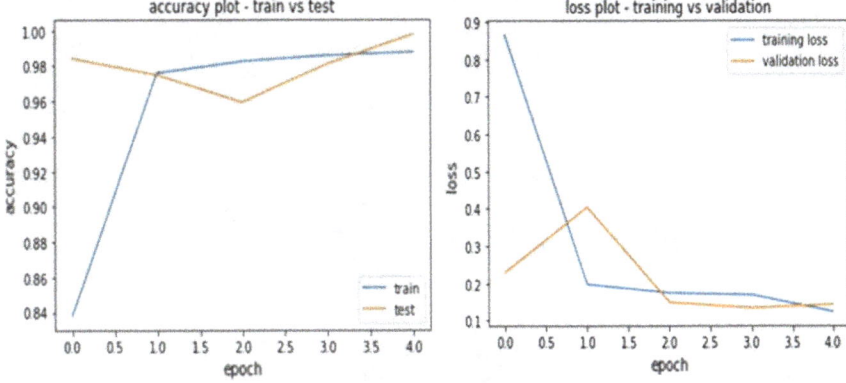

Fig. 47.4 Fitting model output and validating the accuracy plot and loss plot

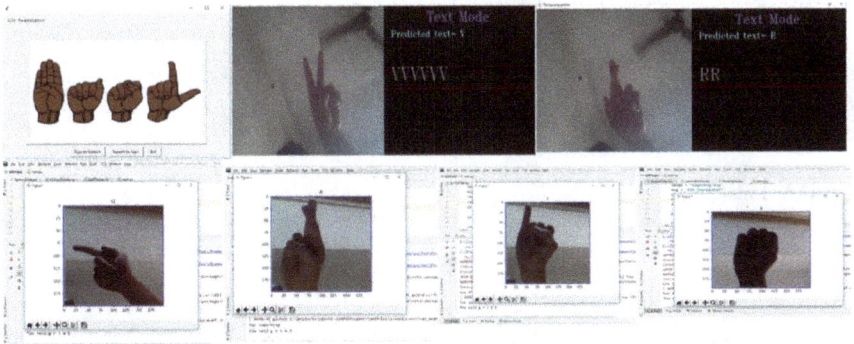

Fig. 47.5 Sign-to-speech and speech-to-text generation output by OpenCV

Fig. 47.6 Final prediction results

References

1. Yang, Q.: Chinese sign language recognition based on video sequence appearance modeling, IEEE. pp. 1537–1542 (2010)
2. Nakamura, S., Sumita, E., Shimizu, T., Sakti, S., Sakai, S., Zhang, J., Finch, A., Kimura, N., Ashikari, Y.: A-STAR: Asia speech translation consortium. In: Proceedings ASJ Autumn Meeting, Yamanashi, Japan (2007), pp. 45–46
3. Taylor, P., Black, A.W., Caley, R.: The architecture of the festival speech synthesis. Proc. Third ESCA/COCOSDA Workshop (ETRW) on Speech Synthesis, Blue Mountains, NSW, Australia, 26–29 Nov 1998, pp. 147–151
4. Wagner, R.A., Fischer, M.J.: The string-to-string correction problem. J. Assoc. Comput. Mach. **21**(1), 168–173 (1974)
5. Poornakala, J., Maheshwari, A.: Automatic speech-speech translation form of English language and translate into Tamil language. Int. J. Innov. Res. Sci. **5**(e3) (2016) https://doi.org/10.15680/IJIRSET.2016.0503110.3926
6. Sakti, S., Kimura, N., Paul, M., Hori, C., Sumita, E., Nakamura, S., Park, J., Wutiwiwatchai, C., Xu, B., Riza, H., Arora, K., Li, H.: The Asian network-based speech-to-speech translation

system. ASRU, 2009 and IEEE Explore, 13 Nov 2009, 17 Dec 2009, pp. 507–512 (2009)

7. Alshawi, H., Douglas, S.: Learning dependency transduction models from unannotated examples. Philos. Trans. Roy. Soc. (Ser. A Math. Phys. Eng. Sci.) **358**, 1357–1372 (2000)

8. Moran, D.B., Cheyer, A.J., Julia, L.E., Martin, D.L., Park, S.: Multimodal user interfaces in the open agent architecture. Knowl. Based Syst. **10**(5), 295–303 (1998). Also published in Proceedings of 2nd International Conference on Intelligence. User Interfaces, pp. 61–68 (1997)

9. Hatim, B., Munday, J.: Translation: An advanced Source Book. Routledge, London (2004)

10. Ellis, K., Barca, J.C.: Exploring sensor gloves for teaching children sign language. Adv. Hum. Comput. Interact. 1–8 (2012)

11. Yun, S., Lee, Y.J., Kim, S.H.: Multilingual speech-to-speech translation system for mobile consumer devices. IEEE Trans. Consum. Electron. **60**(3), 508–516 (2014)

Chapter 48
Chemical Plant Liquid Leakage IoT-Based Monitoring

G. P. Hegde, Nagaratna Hegde, and M. Seetha

Abstract The leakage of chemical plant liquid causes hazards to human beings and animals. Liquid would convert into gaseous form and starts spreading everywhere. This paper focuses on automatic monitoring of liquid leakage in chemical plant which causes pollution of environment. This paper presents design pattern of minimum cost liquid leakage monitoring system for real-time detection of quality of liquid passes in longer tubes of chemical plant. This pattern consists of different sensors which are used to measuring physical quality of the liquid. The parameters flow sensor of the liquid can be measured. Storage of large data and its analysis are more useful in various investigations and developments. The Arduino framework is one of the main controller devices in this innovative work, and the sensed data can be configured and stored in cloud server as metadata for the purpose of silent measure of physical variations in liquids.

48.1 Introduction

48.1.1 Liquid Superiority Monitoring

Supply of good quality of liquid is a major problem in industries to control the machinery parts due to variation of liquid properties. Liquid may vary with viscosity, movement of liquid from higher pressure to lower pressure and liquid flow rate. In real-time applications, liquid flow finds in many industrial plants have got major role. Improper injection of liquid to machine also causes the failure of liquid flow in pipes. Hence, it is an essential task to monitor chemical plant leakages of liquid using IoT

G. P. Hegde (✉)
SDM Institute of Technology, Ujire, Mangalore, India

N. Hegde
Vasavi College of Engineering Hyderabad, Hyderabad, India

M. Seetha
G. Narayanamma Institute of Technology and Science, Hyderabad, India

© The Author(s), under exclusive license to Springer Nature Singapore Pte Ltd. 2021   487
S. C. Satapathy et al. (eds.), *Smart Computing Techniques and Applications*,
Smart Innovation, Systems and Technologies 225,
https://doi.org/10.1007/978-981-16-0878-0_48

which is a good solution. In recent days, evolutions in computing and electronics technologies have triggered Internet of Things technology. Internet of Things can be pronounced as the network of electronics devices communicating among them by the help of a controller. IoT is a collection of various communicating devices and transfers the power from one-to-one device. Sensed information can be transmitted from sensors to embedded devices. The controller task distributes the required signal called microchips. It may consume larger power when signals are accessed through appliances. This paper presents a low-cost liquid flow monitoring system, which is a solution for the liquid consumption and quality of liquid. Microcontrollers and sensors are used for that system. Ultrasonic sensor is used to computing liquid level. The other parameters like pH, TDS and viscosity of the liquid can be calculated using different sensors.

This system uses liquid flow sensors which are placed at certain specific distance on liquid pipe. The liquid flow through the pipe can be sensed and sent to Arduino chip and then require output values computed by this board with respect to flow parameters and transferred to clouds via Wi-Fi frame. In this work, analysis of liquid flow has been carried out with respect to instantaneous values, and how much liquid is used in certain time, in a day or in a month, has been noted. Clients access the sensed data which is stored in microchips and delivered to user terminals. The data which is obtained from the sensors can be shown on the Internet and delivers services for broadcasting the data on mobile phones or web application. Figure 48.1 shows typical chemical plant which consists of complex pipe system.

Fig. 48.1 Typical chemical plant [3]

48.2 Earlier Work

48.2.1 Real-Time Work

In the studies from [1], the author proposed that an IoT-based liquid monitoring system measures liquid level in real time. When pipe is connected between two tanks, then liquid level in both the tank goes equal, and then this condition should observe using sensors. This system is based on knowledge of liquid level and can be very important parameter when it comes to the flood occurrences especially in chemical plant pipes. A liquid level sensor is used to detect the desired parameter, and if the liquid level reaches the parameter, the signal will be fed in real time to social network like Twitter. In the [2] that in their frameworks, there were larger amount of works that have been led by the Internet of Things with various kinds of applications as per the specification of end users. The basic idea of the IoT has flooded over in almost all fields of society, and the man power utilization has been made to downstream. The day-by-day IoT information has been updated with basic architecture. Author also presented that using basic architecture, various contexts related problems have been solved and design patterns were developed. All the structural patterns concerned to this architecture involve the different types of works which give applications with respect to Internet of Things.

In the reference [3], writers concluded that continuous monitoring of drinking solution flow in a pipe is analyzed and also presents designing of minimum cost framework quality control liquid monitoring system. This new system measures the chemical disorders in the liquid which enhances the physical state of the liquid. Their work presents a design and progress of a minimum cost framework for real-time monitoring of the liquid quality in Internet of Things (IoT). The system encompasses of many sensors used in measuring physical and chemical constraints of the liquid. In [4], the author has clearly noted that the maintenance person accesses the liquid level from remote place from remote server via remote references. The controlling of motor pump and remote access is a tedious job. The leakage in pipe can be measured through wireless media, and the message is displayed in mobile phones. The sensors used in their system send the signals to watch the level of water in tanks.

48.2.2 Liquid Quality Monitoring

Nikhil Kedia entitled "Liquid Quality Monitoring for Rural Areas-A Sensor Cloud-Based Economical Project". This paper highlights the entire liquid quality monitoring methods, sensors, embedded design and information dissipation method, role of government, network operator and villagers in ensuring proper information dissipation. It also explores the sensor cloud domain. While automatically enhancing the liquid, quality is not viable at this point, and efficient practice of technology and economic practices can help advance liquid quality and awareness among people

[5]. Jayti Bhatt et al. entitled "Real-Time Liquid Quality Monitoring System". This paper designates to ensure the safe stream of drinking liquid that the quality should be supervised in real time, and for that purpose novel technique Internet of Things (IoT)-based liquid quality monitoring has been planned. In this paper, we present the design of IoT-based liquid quality monitoring system that monitor the quality of liquid in real time. This system consists of some sensors which measure the liquid quality parameter such as pH, turbidity, conductivity, dissolved oxygen and temperature. The measured values from the sensors are processed by microcontroller, and this processed value is conducted remotely to the core controller that is Raspberry Pi using ZigBee protocol. Finally, sensors data can view on Internet browser application using cloud computing [6].

48.3 Existing Framework

48.3.1 Real-Time Work

In industries, liquid may cause several hazards and dangerous unsecured pollution and respiratory problems in society. In real-time work, it requires ongoing modification of liquid resource guiding principle at the levels of international down to individual wells. In most of the places of world, the dirty liquid is used in industries and products are generated. These products are sold out to different marketing places. No proper filling of liquid is used in many areas, and people are not considering the proper awareness. The leakage at certain region in pipe makes loss of liquid and decreases the quantity. So, it will not reach to the destination. So liquid pollution starts increasing throughout the world, and it would cause the problems in various forms. Many people today are suffering from disease due to polluted water intake. This leads to dangerous hazards in the life. It is necessary to maintain the liquid passing mechanism properly in pipes. Reality of various tasks should be carried out to avoid the leakage of liquid.

The liquid flow monitoring system produces thoughtful worldwide problems in real life. Also, normal properties such as volcanoes, algae tints and tremors also change the quality and environmental status of liquid.

48.4 Proposed Framework

48.4.1 System Overview

In this, we present the chemical plant of industries-based lossless liquid flow system and its proper instant monitoring using IoT. The system consists of Arduino device, microcontroller, different type of sensors like liquid flow sensor, pH and viscosity,

sensor and ultrasonic sensor. The Arduino is the microcontroller of the design structure which controls and processes the data created by the sensors. A Wi-Fi module
is associated with the Arduino device which helps to transfer the data to the cloud
over Internet. The ultrasonic sensor helps to measure the liquid level when the liquid
flow reaches certain level, and then the liquid flow can be stopped automatically by
turning the motor off or close the liquid flow in pipe by the help of Arduino. The
liquid flow sensor measures the quantity of liquid flow through the pipe in a given
time; this data will be sent to cloud for storage and analysis purposes. The other
sensors like temperature, pH and viscosity sensor measure the liquid quality and
help to normalize whether the liquid is useful for equipment operations or any other
resolutions. Figure 48.2 shows our proposed framework.

Transformers liquid, like oil, is used to cool the machine. Sometimes, this oil
starts to leak, and it can be detected using IoT-based automatic system.

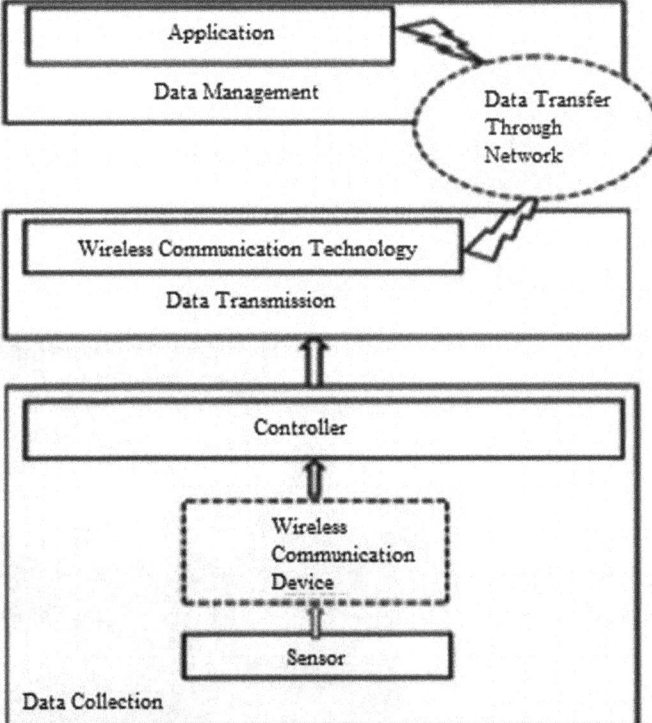

Fig. 48.2 Proposed framework

48.4.2 System Components

Taking about this proposed system, it is clearly shown that it has several components which help to build a liquid monitoring system. The essentail component of the smart home automation system is shown in Fig. 48.2.

Arduino Uno: Arduino is a microcontroller board based on the ATmega328P. It has 14 digital input and output pins, six analog inputs, a 16 MHz quartz crystal, a USB connection, a power jack, an ICSP header and are set button. Figure 48.3 shows typical Arduino board used in this work.

The ESP8266 Wi-Fi Module: It is a self-contained SOC with integrated TCP/IP protocol stack that can give any microcontroller access to your Wi-Fi network. The ESP8266 is capable of either hosting an application or off-loading all Wi-Fi networking functions from another application processor. Each ESP8266 module comes pre-programmed with an AT command set firmware. The ESP8266 module is an extremely cost-effective board with a huge, and ever growing, community. Figure 48.4 shows Wi-Fi module used in this work.

Sensor is used to measure the flow of liquid. This sensor basically consists of a plastic body, a rotor and a sensor. When liquid or water flows through the pipe it influences the pinwheel rotor rotates and wheel speed depends on rate of water flow and is directly proportional to the flow rate of water or liquid.

Fig. 48.3 Arduino Uno board [3]

Fig. 48.4 Wi-Fi module [3]

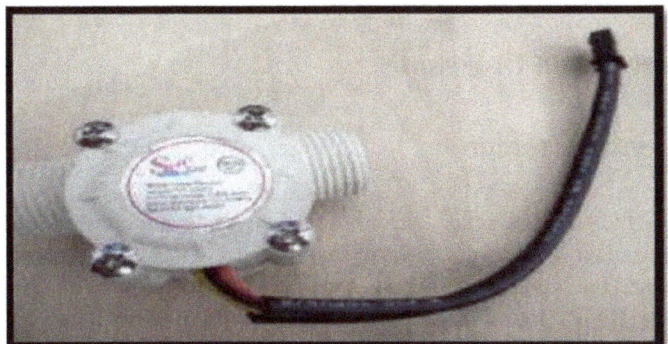

Fig. 48.5 Liquid flow sensor [3]

Cloud-Based Server: Sensors are placed in equidistant in a longer pipe in industries when the sufficient information and signals are sensed by sensors, and then it has been sent to larger database. They should know the content of cloud server. Based on environmental status, various attributes of liquid like temperature, viscosity and pH values are detected using respective sensors. Various forms of liquid flow can be observed in earlier systems, and it can be measured via fixed sensors. Converting sensed data into positive data computationally increases the response and efficiency of the system. Ultrasonic sensor: Transmitters translate signal into ultrasound, receivers translate ultrasound into electrical signals, and transceivers can both transmit and receive ultrasound. This supports to measure the liquid level.

48.4.3 Internal Overview

In this, we present the theory on real-time monitoring of liquid quality and quantity using IoT. Microcontroller acts as a certain device, and it notices the main signals sent from liquid flow sensor. The programming part of Arduino board is necessarily connected to sensor circuit. The data from Arduino is transferred to main clouds sensor via Wi-Fi module through Internet backbone, and when liquid comes to certain level below the overflow level, then motor is put off through wireless media. The ultrasonic sensor helps to measure the liquid level, and when the liquid flow reaches certain level, then the liquid flow can be stopped automatically by turning the motor off or close the liquid flow in pipe by the help of Arduino. The liquid flow sensor measures the quantity of liquid flow through pipe.

For a specific amount of time in a pipe, the data will be sent to cloud for storage and analysis purposes. The other sensors like temperature, pH and turbidity sensor measure the liquid quality and help to determine whether the liquid is useful for chemical process.

48.5 Results and Discussions

48.5.1 Implementation

In this work, the liquid flow rate, current liquid flow rate and output liquid quantity are displayed in the form of graph. If the flow rate decreases, then there is a leakage in the pipe, and the liquid quantity data is also used for meter purpose. The output is shown below. Figure 48.6 illustrates sample output of liquid flow rate.

Implementation is the process of putting a decision or plan into execution and is the realization of an application, or execution of a model, design, specification, standard algorithm or policy. Such implementation is the action that must follow any preliminary thinking in order for something to actually happen. The liquid flow sensor is inserted inside the pipe, it will read the data, and the flow sensor is connected to the Arduino through wires.

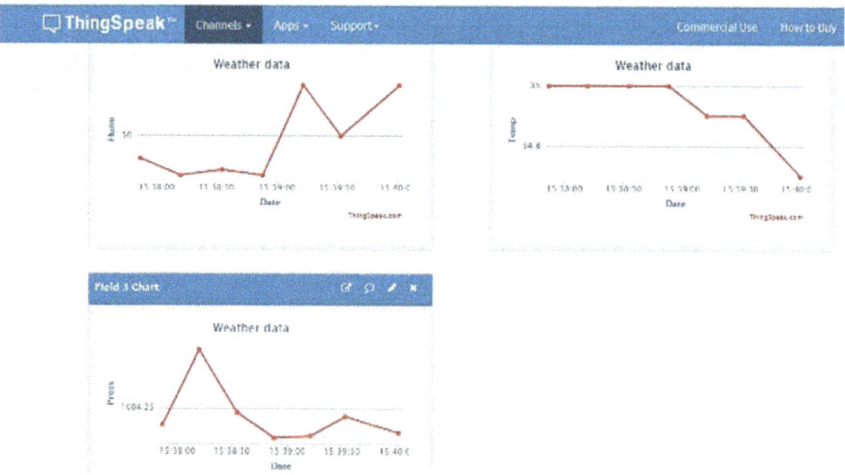

Fig. 48.6 Sample output of liquid flow rate

The Arduino is the microcontroller used in this project, and Arduino IDE is used for writing the program. It will receive the data from the sensor, and the Arduino will process the data and calculate the liquid flow rate and volume of liquid flows. The processed data is transmitted to the application through ESP8266 Wi-Fi module for displaying the result. The result is displayed in the form of graph, and it shows the flow rate and volume of liquid flows through the pipe.

48.6 Conclusion

In this paper, a design pattern of IoT-based liquid monitoring system is introduced. It is an innovative idea for automatic control of liquid flow if liquid is not in proper position in chemical plants. It is concluded that large quantity of liquid flow monitoring and its leakage is more essential in real-time work. This paper illustrates about sensor applications of IoT-based system to monitor chemical plant liquid flow attributes measurement. This can be achieved through wireless media with IoT applications. Data collection through sensors would increase the analysis of liquid flow for better solution of liquid flow problems. The data is sent to the cloud server via Wi-Fi module ESP8266. So, this application will be the best challenger in real-time monitoring and control system and is used to solve all the liquid-related problems.

References

1. Ramakala, R., Thayammal, S., Ramprakash, A., Muneeswaran, V.: Impact of ICT and IOT strategies for liquid sustainability: a case study in Rajapalayam-India. In: International. Conference Computational Intelligence and Computing Research (ICCIC 18), pp. 1–4, (2017)
2. Zidan, N., Mohammed, M., Subhi, S.: An IoT based monitoring and controlling system for liquid chlorination treatment. Proceedings of International Conference Future Networks and Distributed Systems, p. 31 (2018)
3. Ramprabu, J., Paramesh, C.: Automated sensor network for monitoring and detection of impurity in drinking liquid system. Int. J. Res. Appl. Sci. Eng. Technol. **3**, 275–280 (2015)
4. Siddula, S.S., Babu, P., Jain, P.C.: Liquid level monitoring and management of dams using IoT. In: International Conference by IEEE (2018). ISBN: 978–15090–6785–5
5. Kedia, N.: Liquid quality monitoring for rural areas-a sensor cloud based economical project. In: Published in 2015 1st International Conference on Next Generation Computing Technologies (2015)
6. Abdelhafidh, M., Fourati, M., Fourati, L.C., Abidi, A.: Remote liquid pipeline monitoring system IoT-based architecture for new industrial era 4.0. In: International Conference by IEEE (2017). ISBN: 2161–5330
7. Anjana, S., Sahana, M.N., Ankith, S., Natarajan, K., Shobha, K.R.: An IoT based 6 LoWPAN enabled experiment for liquid management. In: International Conference by IEEE (2015). ISBN: 2153–1684
8. Kafli, N., Isa, K.: Internet of Things (IoT) for measuring and monitoring sensors data of liquid surface platform. In: International Conference by IEEE (2017). ISBN: 978–1–5386–1918–6

Chapter 49
K-Hop Iterative Time Synchronization Over a Linear Connected Wireless Sensor Networks

K. L. V. Sai Prakash Sakuru and N. Bheema Rao

Abstract Clock synchronization in linearly connected wireless sensor networks over a k-hop is studied and analytical treatment given in this paper. The clock offsets for a single-hop sensor network using iterative synchronization approach over the random delay model was analyzed and extended to a two-hop sensor network. An extension of that to a k-hop sensor network shows that the error in offset estimate also increases linearly. A modified two-hop algorithm was proposed and analyzed and proved that the error in offset estimate is less than the two-hop network. The modified k-hop was analyzed based on the modified two-hop algorithm. The simulation results over uniform random delay for both two-hop and altered two-hop sensor networks presented. The proposed modified two-hop and modified k-hop achieve better precision than two-hop and k-hop algorithms.

49.1 Introduction

Time synchronization is a crucial issue in wireless sensor networks. Applications such as highway, runway, tunnel, railway line, subways, gas pipeline, water pipeline, oil pipeline, line of control, agriculture field, and cricket pitch monitoring are a few examples of linear sensor networks. Most of the applications need local clocks on the sensor nodes to be time-synchronized both locally (pairwise) and network-wide (multi-hop), with the application demand degree of precision. The locally and network-wide time synchronization protocols grouped as two parts: (i) Clocks are updated periodically and converge at a common notion of time, and (ii) clocks are updated whenever an event occurs on-demand with its neighbors.

K. L. V. Sai Prakash Sakuru (✉) · N. Bheema Rao
Department of Electronics & Communication Engineering, National Institute of Technology, Warangal, Telangana 506004, India
e-mail: sai@nitw.ac.in

N. Bheema Rao
e-mail: nbr@nitw.ac.in

Most of the single-hop synchronization algorithms have analytical designs to define the accuracy aspect of the contemplated synchronization schema. In the case of multi-hop synchronization, not many analytical models are present. When a pair of nodes undergoing a synchronization process with multiple synchronization packets, the relative offset is calculated based on pairwise errors. In multi-hop synchronization, the relative offset prediction error, in general, grows linearly with the number of hops. The nodes in the multi-hop compute the synchronization time with the local values present with the intermediate nodes along the path. In the past, average-based time synchronization protocols [1–7] are proposed. The average-based protocols use pairwise packet messages exchanges and use an average common consent between them, referred to as a paired median protocol.

Time synchronization algorithms for two-hop and multi-hop proposed [8, 9] and analyzed. With a confidence level of 90% thresholds [8], the TTS precision of 0.36, 0.43, 0.52, and 0.60 s, for two, four, six, and eight hops obtained, respectively, and also NP-complete. In [1, 10–13], distributed multi-hop algorithms were proposed for network-wide synchronization. In [10] for large networks [11], Cramer-Rao lower bounds with maximum likelihood estimators [12], used energy function to minimizing it for finding clock synchronization [13] extending pairwise synchronization to cover the whole network, and [1] analyzing with Gaussian delays over the entire network. The authors in [10], use a large number of time-related constraints and use loop methods to find the lower bounds on synchronization. In [14], with reduced communication overheads and simple computation, they claim to be energy efficient. In [11], Cramer-Rao lower bounds and maximum likelihood estimators used to estimate the time synchronization error.

A spanning tree formed in [15] and a pairwise synchronization run through the branches to synchronize the complete network. It found more suitable for medium-size networks than large-size systems as it incurs a significant amount of time to sync. However, node failures and tree formation consume much energy. In [16], multi-hop time synchronization protocol over MANETs with two-phase implementation proposed beacon window and synchronization phases. In [17], the authors proposed a low energy consumption network-wide adaptive multi-hop timing synchronization algorithm.

In this paper, the authors offer a node-pair iterative time synchronization algorithm and its extension to a two-hop network. Further, the same is extended to the k-hop network and shown that the error in estimation grows linearly. A modified two-hop time synchronization algorithm proposed to improve the error performance and the same used for K-hop synchronization.

The rest of the paperwork organized as given: In Sect. 49.2, the main node-pair time synchronization algorithm explained in various steps provided. In the next, two parts 3 and 4, the two-hop and K-hop time synchronizations are analyzed. In the following Sects. 49.5 and 49.6, a modified two-hop and modified K-hop algorithms were discussed and analyzed. Finally, Sect. 49.7, the simulations results are presented and Sect 8 concludes the paper.

49.2   Iterative Clock Synchronization Algorithm

In this paper [18], we provide a node-pair iterative time synchronization algorithm. We begin by considering the case when $\alpha = 1$ the skew whence we need to determine only δ, the offset. We assume that node 0 has the correct time and is sending the timestamped packets to node 1. Let t_k be the time at which node 0 send the k-th timestamped packet and r_k be the time on the clock of node 1 when this packet is received at node 1 after random delay on the network, D_k. It is easy to see that $r_k = t_k + \delta + D_k$, i.e., $\delta = r_k - t_k - D_k$.

We assume $D_k \in [a, b]$ and $0 \leq a \leq b < \infty$, i.e., D_k are bounded random variables and that a and b are known. Hence, after receiving n timestamped packets, we see that δ should satisfy $\delta \in [r_k - t_k - b, r_k - t_k - a]$ for $1 \leq k \leq n$. This means that $\delta \in [\underline{\delta_n}, \overline{\delta_n}]$ where

$$\underline{\delta_n} = \max_{1 \leq k \leq n} \{r_k - t_k\} - b = \delta - (b - \max_{1 \leq k \leq n} \{D_k\}) \tag{49.1}$$

$$\overline{\delta_n} = \min_{1 \leq k \leq n} \{r_k - t_k\} - a = \delta * (\min_{1 \leq k \leq n} \{D_k\} - a) \tag{49.2}$$

The last equality in the above is obtained by noting that $r_k = t_k + \delta + D_k$. Let $\tilde{\delta}(n) := \overline{\delta_n} - \underline{\delta_n}$ denote the width of uncertainty in our estimate of δ after n timestamps were received. Then, we have $\tilde{\delta}(n) = (b - a) - (\max_{1 \leq k \leq n} D_k - \min_{1 \leq k \leq n} D_k)$. Note that we have not made any assumptions on the random sequence D_k of delays experienced by the timestamp packets except that they are bounded.

If the a-priori bounds are accurately selected, it is intuitively clear that max will go close to b and min will go close to a, thereby $\tilde{\delta}(n)$ converges to zero. It is clear that we have not made any assumptions on the random sequence D_n the delays experienced by the timestamped packets except that they are bounded. Thus, eventually, the offset can be estimated.

Algorithm 1 Time Synchronization Algorithm on Real Time

Require: Node $0 \geq 0$ & t =real time
Ensure: Node 1 is in Node 0 comm range
Require: bound interval $[a, b]$ corresponds to Node 0
1: **if** Node 0 **then**
2:    broadcast t_n
3: **else**
4:    initialize $[\underline{\delta_0}, \overline{\delta_0}] = [0, +\infty]$
5:    upon receiving broadcast t_n
6:    $\underline{\delta_n} = max\{t_n + \delta + a, \underline{\delta_{n-1}} + r_n - r_{n-1}\}$
7:    $\overline{\delta_n} = min\{t_n + \delta + b, \overline{\delta_{n-1}} + r_n - r_{n-1}\}$
8:    new uncertainty bound $[\tilde{\delta}(n) = \overline{\delta_n} - \underline{\delta_n}]$
9: **end if** $\tilde{\delta}(n) = \epsilon$

49.3 Two-Hop Synchronization

Let us consider a two-hop network as shown in Fig. 49.1, where Node 0 has the
correct (true) time and is transmitting a sync packet to Node 1. Let at time t_k^0 the
k^{th} timestamped packet is transmitted from Node 0 and r_k^1 be the time on the clock
of Node 1 when this packet is received at Node 1 after random delay D_k^1 on the
network. The D_k^1 are i.i.d. with bounded support $[a_1, b_1]$, where the values of a_1, b_1
are assumed known at Node 1.

After receiving n timestamped packets at node 1, we have δ satisfy

$$[\underline{\delta}_n^1, \overline{\delta}_n^1] \quad for \quad 1 \le k \le n \tag{49.3}$$

and $\hat{\delta}_n^1 := \overline{\delta}_n^1 - \underline{\delta}_n^1$ the estimated uncertainty width of the offset δ_n^1.

Now, Node 1 sends timestamped packets to Node 2. Let t_k^1 be the time at which
Node 1 sends the kth packet to Node 2, containing its then uncertainty interval $[\underline{\delta}_k^1, \overline{\delta}_k^1]$.
This packet is received at Node 2 at time r_k^2, after a random delay D_k^2. The D_k^2 are
assumed i.i.d. with bounded support $[a_2, b_2]$, where the values of a_2, b_2 are assumed
known at Node 2.

At Node 2, we see that δ should be in the interval

$$[\underline{\delta}_k^2, \overline{\delta}_k^2] \quad for \quad 1 \le k \le n. \tag{49.4}$$

and $\hat{\delta}_n^2 := \overline{\delta}_n^2 - \underline{\delta}_n^2$ denotes the width of the uncertainty in our estimate of δ^2 after n
timestamps were received.

Hence, after receiving n timestamped packets at node 2 from both node 0 and
node 1, we have

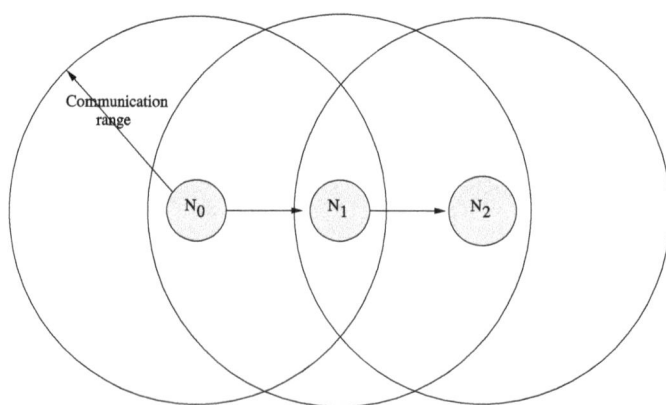

Fig. 49.1 Two-hop synchronization

$$\hat{\delta}_n^2 = \hat{\delta}_n^1 + (b_2 - a_2) - (\max_{1 \le k \le n} \{D_k^2\} - \min_{1 \le k \le n} \{D_k^2\}) \tag{49.5}$$

where

$$\hat{\delta}_n^1 = (b_1 - a_1) - (\max_{1 \le k \le n} \{D_k^1\} - \min_{1 \le k \le n} \{D_k^1\}) \tag{49.6}$$

If we look at Eq. 49.5, the uncertainty interval at the nodes 2 can be rewritten as, under the assumption that the delay bounds between all hop are same and that a, b are known at nodes 1 and 2, then we can rewrite it as

$$\hat{\delta}_n^2 = 2(b - a) - \sum_{i=1}^{2} (\max_{1 \le k \le n} \{D_k^i\} - \min_{1 \le k \le n} \{D_k^i\}) \tag{49.7}$$

Hence, we can see the uncertainty interval propagates linearly, and it is almost 2 em times at the end of two-hop network.

49.4   K-Hop Synchronization

Let us consider a K-hop network where node $K - 1$ will transmit a sync packet at time t_k^{K-1} the kth timestamped packet, and r_k^K be the time on the clock of node K when this packet is received at node K after random delay D_k^K on the network along with the then uncertainty interval $[\underline{\delta}_k^{K-1}, \overline{\delta}_k^{K-1}]$ at node $K - 1$. The $D_{(k)}^K$ are i.i.d. with bounded support $[a_K, b_K]$, where the values of a_K, b_K are assumed known at node K.

Proceeding on the same lines of two-hop synchronization as discussed above, the uncertainty width of clock at node K is given by

$$\hat{\delta}_n^K = \hat{\delta}_n^{K-1} + (b_K - a_K) - (\max_{1 \le k \le n} \{D_k^K\} - \min_{1 \le k \le n} \{D_k^K\}) \tag{49.8}$$

where

$$\hat{\delta}_n^{K-1} = \hat{\delta}_n^{K-2} + (b_{K-1} - a_{K-1}) \\ -(\max_{1 \le k \le n} \{D_k^{K-1}\} - \min_{1 \le k \le n} \{D_k^{K-1}\}) \tag{49.9}$$

$$\vdots$$

$$\hat{\delta}_n^1 = (b_1 - a_1) - (\max_{1 \le k \le n} \{D_k^1\} - \min_{1 \le k \le n} \{D_k^1\}) \tag{49.10}$$

If we look at Eq. 49.8, the uncertainty interval at the Kth hop can be rewritten as, under the assumption that the delay bounds between all hop are same and that a, b are known at all the $K - 1$ nodes, then we can rewrite it as

$$\hat{\delta}_n^K = K(b - a) - \sum_{i=1}^{K} (\max_{1 \le k \le n} \{D_k^i\} - \min_{1 \le k \le n} \{D_k^i\}) \tag{49.11}$$

Hence, we can see the uncertainty interval propagates linearly, and it is almost K times at the end of K − hop network.

49.5 Proposed Algorithm—Modified Two-Hop Synchronization

The main idea is to capture the convergence at Node 2 when it gets alternative packets from the Node 1 which is in the processes of convergence and also from Node 0 which is transmitting the synchronization packets with Node 1 as simply packet forwarder under the assumption it takes zero delay in forwarding. Let Node 1 has a support $[a_1, b_1]$ and has an uncertainty width $\hat{\delta}_i^1$ which is represented by $[\underline{\delta}_i^1, \overline{\delta}_i^1]$ to which it has converge after i^{th} iteration is transmitted to Node 2 as shown in Fig. 49.2. Now, let Node 2 has a support $[a_2, b_2]$ uses the information transmitted by Node 1 and converges to an uncertainty width $\hat{\delta}_i^2$ which is represented by $[\underline{\delta}_i^2, \overline{\delta}_i^2]$. The analysis is divided into four parts:

- Node 0 and Node 1 synchronization.
- Node 0 and Node 2 synchronization (with Node 1 as simple relay)
- Node 1 and Node 2 synchronization
- Node 2 synchronization with help of above two packets

Node 0 and Node 1 synchronization The uncertainty width is given by:

$$\hat{\delta}_n^1 = \overline{\delta}_n^1 - \underline{\delta}_n^1 = (b_1 - a_1) - (\max_{1 \le i \le n} \{D_i^1\} - \min_{1 \le i \le n} \{D_i^1\}) \tag{49.12}$$

Node 0 and Node 2 synchronization (with Node 1 as simple relay) The uncertainty width is given by:

$$\hat{\delta}_n^2 = \overline{\delta}_n^2 - \underline{\delta}_n^2 = (b_1 - a_1) + (b_2 - a_2) \tag{49.13}$$
$$-(\max_{1 \le i \le n} \{D_i^2 + D_i^1\} - \min_{1 \le i \le n} \{D_i^2 + D_i^1\})$$

Node 1 and Node 2 synchronization The uncertainty width is given by:

$$\hat{\delta}_n^{\overline{2}} = \overline{\delta}_n^{\overline{2}} - \underline{\delta}_n^{\overline{2}} = (b_2 - a_2) - (\max_{1 \le i \le n} \{D_i^{\overline{2}}\} - \min_{1 \le i \le n} \{D_i^{\overline{2}}\}) \tag{49.14}$$

Fig. 49.2 Modified two-hop synchronization

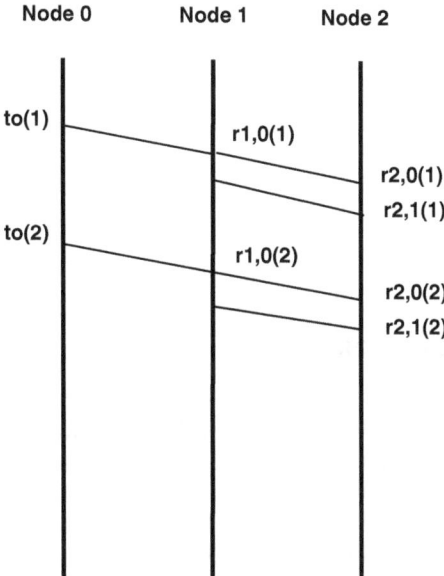

Node 2 synchronization with help of above two packets (one coming directly via Node 1 and the other from Node 1 while synchronizing) The uncertainty width of clock at Node 2 is given by

$$\hat{\delta}_n^2 = (b_1 - a_1) + (b_2 - a_2) - \max\{\max_{1 \le i \le n}\{D_i^1 + D_i^2\},$$

$$\max_{1 \le i \le n}\{D_i^1\} + \max_{1 \le i \le n}\{D_i^2\}\} + \min\{\min_{1 \le i \le n}\{D_i^1 + D_i^2\},$$

$$\min_{1 \le i \le n}\{D_i^1\} + \min_{1 \le i \le n}\{D_i^2\}\} \qquad (49.15)$$

49.6   Proposed Algorithm—Modified K-Hop Synchronization

Modified k-hop synchronization The uncertainty width of clock at Node K is given by

$$\hat{\delta}_n^{\hat{k}} = \sum_{i=1}^{K}(b_K - a_K) - \max\{\max_{1 \le i \le n}\{D_i^1 + D_i^2 + \cdots + D_i^K\}, \ldots,$$

$$\max_{1 \le i \le n}\{D_i^1\} + \max_{1 \le i \le n}\{D_i^{\hat{2}}\} + \cdots + \max_{1 \le i \le n}\{D_i^{\hat{K}}\}\}$$

$$+ \min\{\min_{1 \le i \le n}\{D_i^1 + D_i^2 + \cdots + D_i^K\}, \ldots,$$

$$\min_{1 \le i \le n}\{D_i^1\} + \min_{1 \le i \le n}\{D_i^{\hat{2}}\} + \cdots + \min_{1 \le i \le n}\{D_i^{\hat{K}}\}\} \quad (49.16)$$

49.7   Simulations

The two-hop and modified two-hop iterative time synchronization algorithm is analytically derived and simulated in MATLAB. The a-priori uncertainty bounds are
considered a 1-time unit as the lower bound and 5-time units as the upper bound.
Node 0 transmits timestamped message packets periodically with a regular interval
of unit time. On the reception of the message packet at node 1, it is forwarded to node
2. The process is monitored for the behavior of the uncertainty width, with incremental message transmission packets. The independent and identically distributed
random delays assumed to take uniform density function with the above uncertainty
support. The simulation studies carried out considering the distributions of $D^1 \& D^2$
to be uniform and bounded with $a_1 = 1$ unit time, $b_1 = 5$ unit time, $a_2 = 1$ unit time
and $b_2 = 5$ unit time and measurements made over 25 packets, as shown in Fig. 49.3.

Fig. 49.3 Expectation of $\tilde{\delta}_n^2$ for uniform distributions

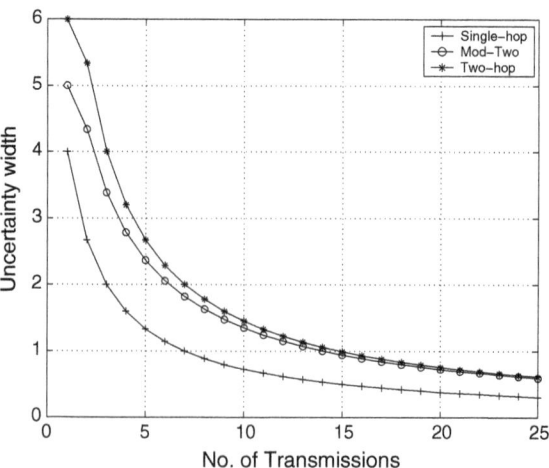

49.8 Conclusions and Future Work

An iterative time synchronization one-hop algorithm proposed and analytically derived. A two-hop synchronization algorithm was calculated based on the said one-hop model and obtained the analytical expression. In similar lines, for linear wireless sensor networks, a k-hop synchronization was proposed, and an analytical expression was derived, showing that the uncertainty error propagates linearly. We proposed a modified two-hop synchronization algorithm to combat the error propagation and given a systematic treatment. The same procedure extended for the k-hop network. From Fig. 49.3, we can conclude that if the data of the uncertainty width of the interval is carefully selected, it will impact the convergence curve. The simulation results show that the uncertainty error is almost double in the case of a two-hop network compared to a hop network. The proposed modified two-hop system has better uncertainty width compared to a two-hop network. Hence, as the number of hops increases, the error will not grow linearly. The results are encouraging for large systems, and looking further for mid-course correction strategies to cover the entire network.

References

1. Gang, X., Shalinee, K.: Analysis of distributed consensus time synchronization with gaussian delay over wireless sensor networks. EURASIP J. Wireless Commun. (2009)
2. Sommer, P., Wattenhofer, R.: Gradient clock synchronization in wire- less sensor networks. In: IPSN'09, April 13–16 2009, San Francisco, CA, USA (2009)
3. Freris, N., Graham, S., Kumar, P.: Fundamental limits on synchronizing clocks over networks. IEEE Trans. Automatic Control **56**(6), 1352–1364 (2011)
4. Leng, M., Wu, Y.C.: Distributed clock synchronization for wireless sensor networks using belief propagation. IEEE Trans. Signal Process. **59**(11), 5404–5414 (2011)
5. Djenouri, D., Merabtine, N., Mekahlia, F.Z., Doudou, M.: Fast distributed multi-hop relative time synchronization protocol and estimators for wireless sensor networks. Ad Hoc Netw. **11**(8), 2329–2344 (2013)
6. Li, S.Z.: Markov Random Field Modeling in Image Analysis, 3rd edn. Springer, Heidelberg (2009)
7. Sundararaman, B., Buy, U., Kshemkalyani, A.D.: Clock synchronization for wireless sensor networks: a survey. AdHoc Networks **3**(3), 281–323 (2005)
8. Wang, J., Zhang, S., Gao, D., Wang, Y.: Two-hop time synchronization protocol for sensor networks. EURASIP J. Wireless Commun. Network. 2014:39 (2014). http://jwcn.eurasipjournals.com/content/2014/1/39
9. Cui, L.Z., Yuan, M.M., Li, Z., et al.: Research and implementation of low power two-way WSN time synchronization algorithm for coal mine. Chin. J. Sens. Actuat. **27**(9), 1253–1259 (2014)
10. Solis, R., Borkar, V.S., Kumar, P.R.: A new distributed time synchronization protocol for multihop wireless networks. In: Proceedings of the 45th IEEE Conference on Decision and Control. https://doi.org/10.1109/CDC.2006.377675
11. Djenouria, D., Merabtineb, N., Mekahliab, F.Z., Doudoua, M.: Fast distributed multi-hop relative time synchronization protocol and estimators for wireless sensor networks. Elsevier Ad Hoc Networks (2013). https://doi.org/10.1016/j.adhoc.2013.06.001

12. Chen, W., Jin, X., Li, Z., Zhang, X., Hong, L.: Clock synchronization for distributed multi-hop wireless networks using Markov Random field. J. Phys.: Conf. Ser. **1087**, 052008 (2018). https://doi.org/10.1088/1742-6596/1087/5/052008
13. Cheng, K.Y., Lui, K.S., Wu, Y.C., et al.: A distributed multihop time synchronization protocol for wireless sensor networks using pairwise broadcast synchronization. IEEE T Wireless Commun. **8**(4), 1764–1772 (2009)
14. Kaseva, V., Hämäläinen, T.D., Hännikäinen, M.: Time synchronization for resource-constrained multi-hop wireless sensor networks based on hop delay estimation. In: SENSORCOMM 2011, ISBN: 978-1-61208-144-1
15. Maroti, M., Kusy, B., Simon, G., Ledeczi, A.: Robust multi-hop time synchronization in sensor networks. In: ICWN '04, pp. 454–460 (2004)
16. Chen, G.-N., Wang, C.-Y.,Hwang, R.-H.: MTSP: Multi-hop time synchronization protocol for IEEE 802.11 wireless Ad Hoc network. In: WASA 2006: Wireless Algorithms, Systems, and Applications. pp. 664–675
17. Noh, K.-L., Serpedin, E.: Adaptive multi-hop timing synchronization for wireless sensor networks. In: ISSPA (2007). https://doi.org/10.1109/ISSPA.2007.4555613
18. Sai Prakash Sakuru, K.L.V., Bheema Rao, N.: An iterative node-pair time synchronization (INTS) for wireless sensor networks. In: ESIC 2020, March 02–04, 2020. Yupia, Arunachal Pradesh, India (2020)

Chapter 50
A Convolutional Neural Network Architecture for Tomato Leaf Disease Detection Using Data Augmentation

Matta Bharati Devi and K. Amarendra

Abstract Convolutional neural networks have set state-of-the-art results for many challenging problems in the field of computer vision. In this article, we design and implement a six-layered convolutional neural network for tomato leaf disease detection. Our model is trained on tomato plant leaf images of ten different classes representing various diseases of tomato plant. To increase the model generalizability, we employed various data augmentation techniques and increased the size of training data. Batch normalization layers after every convolution and dropout induced after fully connected layers made the model immune from overfitting. Our proposed model is able to outperform various existing works on tomato leaf disease detection and achieved an accuracy of 92.3% on test data.

50.1 Introduction

Plant leaf disease identification is one of the major challenges faced by farmers in agriculture. It is very important to correctly identify the type of leaf disease for appropriate use of pesticides. Any mistakes in accurate identification of plant leaf diseases leads to reduced yield. According to statistics, 52% of the Indian population depends on agriculture. Most of the local farmers and gardeners grow tomatoes in their farms and gardens as it is one of the most required vegetables in our day-to-day life. But they are unable to find appropriate pesticides to use when they are infected by different types of pests and diseases. The reason behind this is tomato plants get affected by various diseases in its life span of being productive, and identifying each disease accurately is a challenge. Hence, there is a requirement to automate the process of tomato leaf disease detection. In recent years, tomato leaf disease detection has been an active area of research, and several methods were proposed

M. Bharati Devi · K. Amarendra (✉)
Department of CSE, Koneru Lakshmaiah Education Foundation, Vijayawada, Andhra Pradesh, India
e-mail: amarendra@kluniversity.in

© The Author(s), under exclusive license to Springer Nature Singapore Pte Ltd. 2021
S. C. Satapathy et al. (eds.), *Smart Computing Techniques and Applications*,
Smart Innovation, Systems and Technologies 225,
https://doi.org/10.1007/978-981-16-0878-0_50

507

to automate this process using several conventional machine learning approaches [1, 2].

Feature extraction techniques like histogram of oriented gradients (HOG), local binary patterns (LBP) and grey-level co-occurrence matrix (GLCM) are used in literature to extract texture-based features from plant leaf images [3]. These features were fed to popular classifiers like support vector machine (SVM) to categorize different types of diseased leaves [4]. However, this type of feature engineering requires sound domain expertise to extract discriminative information from leaf images. With the use of deep learning models like convolutional neural networks (CNNs), there is a great process in this area of research. Most of them designed their own CNN architectures to tackle this problem of tomato leaf disease detection [5]. But, large amount of training data is required to train these models [6, 7]. To accomplish this problem, various pre-trained models trained on large amount of labelled data were introduced. Xception pre-trained network was used to detect and classify tomato leaf diseased images [8, 9]. VGG16 and ResNet models were fused, and resultant model was trained on leaf images of various plants to detect diseased leaves [10]. All the models have large number of layers and need huge computing resources for training.

In this article, we use several data augmentation techniques to increase the size of training data and design a simple convolutional neural network architecture with batch normalization layers. We train our proposed architecture using tomato leaves of ten different classes taken from plant village dataset. Our proposed model achieves competitive results with existing pre-trained models for tomato leaf disease prediction. Our model has a smaller number of trainable parameters when compared to existing networks.

50.2   Related Work

This section provides discussion of various approaches used for plant leaf disease detection. Accurate diagnosis of plant leaf disease has been an important area of research in the field of machine learning. Most of the previous works concentrated on using image processing and machine learning-based techniques to detect the type of plant disease from leaf images. A K-nearest neighbour (K-NN) classifier with grey-level co-occurrence matrix (GLCM) texture features of plant leaf images was used to identify plant leaf diseases [11, 12]. Another machine learning-based system was proposed for grapes plant leaf disease detection by Harshal Waghmare et al. First, background of all images was removed, and segmentation is performed as a preprocessing step. A high-pass filter is applied on segmented images to analyse disease part of the leaf. Local binary patterns (LBP)-based texture features are extracted from preprocessed images, and these features were used to identify different types of grape plant diseases using support vector machine (SVM) classifier [13]. A cotton leaf disease detection and classification technique based on machine learning and image processing tools is proposed by V Pooja et al. Initially, region of interest (ROI) is segmented from plant leaf images using image processing tools, and features are extracted. These features were passed to SVM classifier to identify the type of disease [14, 15].

In recent years, convolutional neural networks (CNN) marked breakthrough results for various tasks like object recognition, image classification and medical image diagnosis [16, 17] tasks. Different CNN architectures are built for plant leaf disease detection in literature. A deep CNN architecture is designed and trained on plant village dataset to identify 38 different types of diseases in 14 types of plant species [18, 19]. Another deep CNN architecture is proposed for detecting apple plant leaf diseases in real time [20]. Gandhi et al. designed and trained a deep CNN model on augmented data generated using generative adversarial networks (GANs) [21, 22]. A global pooling dilated convolutional neural network (GPDCNN) was employed to detect cucumber leaf diseases by blending global pooling with dilated convolution [23, 24]. Most of these models proved the efficiency of CNNs for extracting discriminant features from plant leaf images for disease identification over conventional machine learning approaches. But, these deep CNNs require large amount of labelled data for training. To overcome this issue, transfer learning was introduced.

Transfer learning is a deep learning-based paradigm, where the knowledge acquired by a model trained on large labelled data is transferred to another model to perform similar task. This is accomplished using pre-trained deep CNNs [25]. Pre-trained models are those, which are trained on ImageNet dataset containing millions of labelled images belonging to 1000 different classes [26]. AlexNet, VGG Net, GoogLeNet, ResNet and NASNet are some deep CNN architectures trained on ImageNet dataset. Even though they are trained on ImageNet dataset, these pre-trained models worked well for medical image classification tasks [27]. LeNet deep CNN architecture is trained on 18,160 tomato plant leaf images taken from plant village dataset and resulted in 94.5% test accuracy [28]. Nithish Kannan et al. used ResNet pre-trained model with 50 layers for tomato leaf disease prediction [29]. When two different deep CNN models, AlexNet and Squeeze Net, are fine-tuned on same tomato leaf dataset taken from plant village, it is observed that AlexNet achieved the highest accuracy of 95.65% and SqueezeNet obtained accuracy 94.3% [30].

Most of these models obtained better performance. In this work, we propose a deep convolutional neural network architecture whose performance is competitive with the performance of these pre-trained models.

50.3   Methodology

Our proposed method for tomato plant leaf disease prediction consists of two different phases. They are preprocessing, model training and evaluation.

Original Image   Mask   Segmented Leaf

Fig. 50.1 Background removal from tomato leaf

50.3.1 Data Preprocessing

Labelled tomato plant leaf images were taken from plant village dataset. These images have little amount of background variations. So, leaf regions were segmented from the images by applying appropriate masks on images. This helps the model to learn necessary features without any variations. Figure 50.1 represents background subtracted tomato plant leaf image and appropriate mask used for it.

After performing segmentation operation, all the resultant images are reshaped to a fixed dimension of $128 \times 128 \times 3$ to increase the speed of training process. After reshaping, all images are normalized to bring all pixel intensity values into common range by computing mean and standard deviation of pixels. This type of normalization is regarded as Z-score normalization.

50.3.2 Data Augmentation

To increase the size of training data, we augment data using several data augmentation operations like rotation, shearing, flipping and zooming.

- Rotation: Original leaf image is rotated by 'x' degrees.
- Shearing: Different orientations of tomato leaf images are created by applying shear transformation.
- Flipping: Leaf images are flipped either horizontally or vertically.
- Zooming: Zoom in or zoom out operations create additional labelled images.
- Brightness Adjustment: Original image replications are created by different variations in brightness levels.

Data augmentation techniques make the model more robust by creating different variations in training data. Figure 50.2 represents invariant images created by these augmentation techniques.

Original Image

Augmented Data

Fig. 50.2 Sample images of augmented data

50.3.3 Convolutional Neural Network for Tomato Leaf Disease Identification

We design a deep convolutional neural network architecture and train it using prepro-cessed tomato leaf images. Proposed CNN architecture consists of four convolution layers, each one followed by a batch normalization and max pooling layers. We used (3×3) kernels for all convolution and a (2×2) window for max pooling operations. By rigorous experiments, we decided the number of kernels used for each convolu-tion operation. First convolution layers use 32 kernels, and next three convolution layers use 64 kernels. Rectified linear unit (ReLU) activation function is applied on each convoluted feature map. We use 'same' padding to retain the information from convolutional feature maps. All the convolution feature maps are after fourth convolutional layers which are passed to flatten layer to vectorize them.

A dense layer with 64 units and ReLU activation function is followed after flatten layer which learns discriminative information obtained from feature maps generated by convolutional layers. A dropout layer with 0.5 rate is added after dense layer to provide regularization and prevent model from overfitting. Finally, another dense layer with 10 units and softmax activation function is added at the end of network for classification. Figure 50.3 represents architecture of proposed deep CNN archi-tecture. Our CNN architecture is simple in terms of number of parameters when

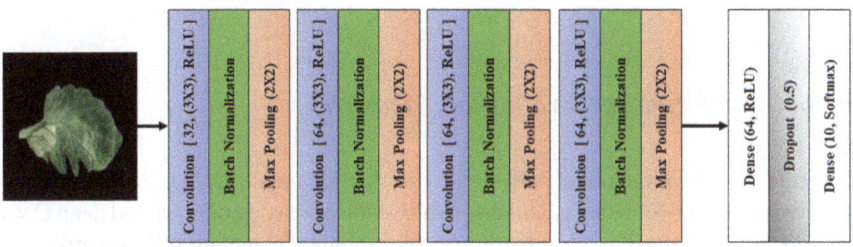

Fig. 50.3 Proposed deep CNN architecture

compared with other deep CNN architectures designed for tomato leaf disease identification. Since model is trained on augmented data, model learns invariant features and results in improved performance.

50.3.4 Training Details

Proposed CNN architecture is trained on tomato leaf disease dataset. Optimal values of kernels learned using categorical cross-entropy loss function and error are optimized by ADAM optimizer. Data batch of size 64 is loaded every time during training to reduce memory burden on system. Model is executed for 50 epochs, during training best weights with minimal validation loss are saved using model checkpointing mechanism.

50.4 Experimental Results

This section provides clear picture of experiments conducted and results obtained using proposed method. First overview of dataset used for experiments is described followed by evaluation metrics used to measure the performance of proposed deep CNN architecture. Finally, a summary of experiments and their results are provided.

50.4.1 Tomato Leaf Disease Dataset

We used tomato leaf disease dataset, part of plant village dataset which contains 54,306 samples of 26 type of diseased leaf images belonging to 14 types of plant species. This dataset contains 18,160 samples of tomato plant representing both healthy and diseased leaves. Diseased leaves are broadly labelled to nine different types of diseases. Totally, 14,530 samples are considered for training, and 3630 samples are used for testing model performance. Figure 50.4 represents sample images of diseased tomato leaves taken from plant village dataset (Table 50.1).

50.4.2 Performance Evaluation Measures

Different classification model performance evaluation measures like accuracy, precision, recall and F1-score are calculated to prove the efficiency of proposed deep CNN network on test data. These measures can be computed using confusion matrix.

Fig. 50.4 Samples of diseased tomato leaves from tomato leaf disease dataset

Table 50.1 Dataset summary for tomato leaf disease dataset

Category	# Train	# Test
Healthy	1273	318
Bacterial spot	1702	425
Early blight	800	200
Late blight	1527	382
Leaf mould	762	190
Mosaic virus	298	75
Septoria leaf spot	1417	354
Spider mites	1342	334
Target spot	1123	281
Yellow leaf curl virus	4286	1071
Total	14,530	3630

50.4.3 Results Analysis

We trained our proposed CNN with preprocessed tomato leaf images belonging to ten different classes. Performance of our CNN is picturized in Fig. 50.5, in terms of accuracy precision, recall and F1-score.

From Fig. 50.5, it is clear that proposed model is able to correctly classify tomato leaf images belonging to early blight, late blight, mosaic virus, septorial leaf spot, spider mites and yellow leaf curl virus classes. Due to a smaller number of training samples for healthy, bacterial spot and target spot classes, there is a bit reduced accuracy for these classes.

Our proposed CNN architecture is able to outperform a ToLeD CNN architecture proposed in [5]. From Table 50.2, it is clear that the number of parameters is less

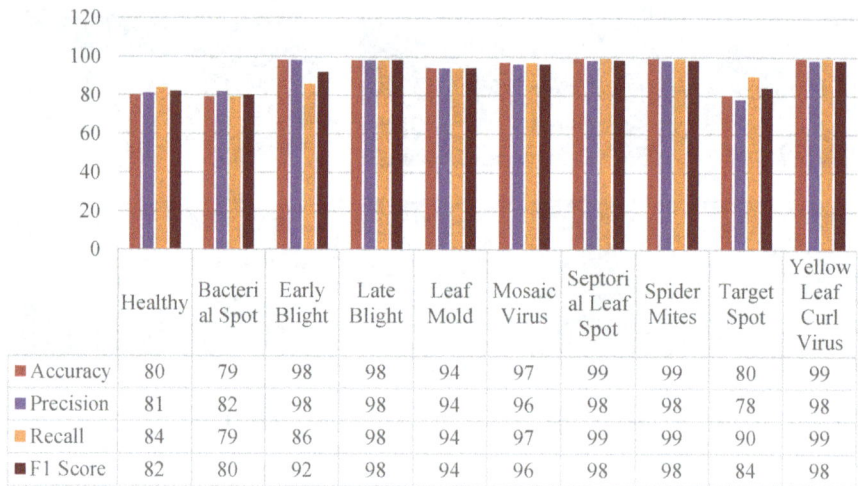

Fig. 50.5 Performance of proposed CNN in terms of accuracy, precision, recall and F1-score

Table 50.2 Performance comparison of proposed CNN with existing models

Model	Accuracy (%)	# Parameters
VGG16 [4]	77.2	9,449,482
MobileNet [4]	63.75	20,507,146
Inception [4]	63.4	41,247,146
ToLeD—CNN [4]	91.2	208,802
Proposed CNN	92.3	241,866

when compared with pre-trained networks like VGG16, MobileNet and Inception. Our model is superior in terms of accuracy and also number of parameters.

50.5 Conclusion

Plant leaf disease detection is one of the major challenges faced by farmers in agriculture. It is very important to correctly identify the type of leaf disease for appropriate use of pesticides. Any mistakes in accurate identification of leaf diseases lead to reduced yield. With the help of proposed deep CNN architecture, tomato leaf diseases can be correctly identified. When compared to existing methods for tomato leaf disease identification, our approach is simple and robust in terms of parameters. This study can be extended to segment diseased portions of leaf for further improvement of accuracy.

References

1. Patil, R., Kumar, S.: A Bibliometric survey on the diagnosis of plant leaf diseases using artificial intelligence (2020)
2. Dudi, B., Rajesh, V.: Medicinal plant recognition based on CNN and machine learning. Int. J. Adv. Trends Computer Sci. Eng. **8**(4), 628–631 (2019)
3. Mahapatra, S., Kannoth, S., Chiliveri, R., Dhannawat, R.: plant leaf classification and disease recognition using SVM, a machine learning approach. Sustain. Humanosphere **16**(1), 1817–1825 (2020)
4. Bhagat, M., Kumar, D., Haque, I., Munda, H.S., Bhagat, R.: Plant leaf disease classification using grid search based SVM. In: 2nd International Conference on Data, Engineering and Applications (IDEA), pp. 1–6. IEEE (2020)
5. Agarwal, M., Singh, A., Arjaria, S., Sinha, A., Gupta, S.: ToLeD: Tomato leaf disease detection using convolution neural network. Procedia Computer Sci. **167**, 293–301 (2020)
6. Xie, X., Ma, Y., Liu, B., He, J., Li, S., Wang, H.: A deep-learning-based real-time detector for grape leaf diseases using improved convolutional neural networks. Front. Plant Sci. **11** (2020)
7. Balakrishna, G., Moparthi, N.R.: Study report on Indian agriculture with IoT. Int. J. Electr. Computer Eng. **10**(3), 2322 (2020)
8. Pattnaik, G., Shrivastava, V.K., Parvathi, K.: Transfer learning-based framework for classification of pest in tomato plants. Appl. Artif. Intell. 1–13 (2020)
9. NagaGeetha, M., Ramesh, N.V.K.: An efficient IoT based smart irrigation system and plant diseases detection: a review. Int. J. Eng. Technol **7**(2.7), 661–664 (2018)
10. Kumar, A., Razi, R., Singh, A., Das, H.: Res-VGG: A novel model for plant disease detection by fusing VGG16 and ResNet models. In: International Conference on Machine Learning, Image Processing, Network Security and Data Sciences, pp. 383–400. Springer, Singapore (2020)
11. Trivedi, J., Shamnani, Y., Gajjar, R.: Plant leaf disease detection using machine learning. In: International Conference on emerging technology trends in electronics communication and networking, pp. 267–276. Springer, Singapore (2020)
12. Kalavala, S.S., Sakhamuri, S., Prasad, B.B.V.S.V.: An efficient classification model for plant disease detection. Int. J. Innov. Technol. Exploring Eng. **8**(7), 126–129 (2019)
13. Waghmare, H., Kokare, R., Dandawate, Y.: Detection and classification of diseases of grape plant using opposite colour local binary pattern feature and machine learning for automated decision support system. In: 2016 3rd International Conference on Signal Processing and Integrated Networks (SPIN), pp. 513–518. IEEE (2016)
14. Pooja, V., Das, R., Kanchana, V.: Identification of plant leaf diseases using image processing techniques. In: 2017 IEEE Technological Innovations in ICT for Agriculture and Rural Development (TIAR), pp. 130–133. IEEE (2017)
15. Balram, G., Kumar, K.K.: Smart farming: disease detection in crops. Int. J. Eng. Technol. **7**(2.7), 33–36 (2018)
16. Bodapati, J.D., Veeranjaneyulu, N., Shareef, S.N., Hakak, S., Bilal, M., Maddikunta, P.K.R., Jo, O.: Blended Multi-modal deep ConvNet features for diabetic retinopathy severity prediction. Electronics **9**(6), 914 (2020)
17. Balaji Bhanu, B., Hussain, M.A., Prasad, A., Mirza, M.A.: Exploration of crop production improvement through various agriculture monitoring systems. Int. J. Innov. Technol. Exploring Eng. **8**(11), 3747–3750 (2019)
18. Trivedi, J., Shamnani, Y., Gajjar, R.: Plant leaf disease detection using machine learning. In: Gupta S., Sarvaiya J. (eds.) Emerging Technology Trends in Electronics, Communication and Networking (ET2ECN 2020). Communications in Computer and Information Science, vol. 1214. Springer, Singapore (2020)
19. Anusha, A., Guptha, A., Rao, G.S., Tenali, R.K.: A model for smart agriculture using IOT. Int. J. Innov. Technol. Exploring Eng. **8**(6), 1656–1659 (2019)
20. Jiang, P., Chen, Y., Liu, B., He, D., Liang, C.: Real-time detection of apple leaf diseases using deep learning approach based on improved convolutional neural networks. IEEE Access **7**, 59069–59080 (2019)

21. Gandhi, R., Nimbalkar, S., Yelamanchili, N., Ponkshe, S.: Plant disease detection using CNNs and GANs as an augmentative approach. In: 2018 IEEE International Conference on Innovative Research and Development (ICIRD), pp. 1–5. IEEE (2018)
22. Inthiyaz, S., Prasad, M.V.D., Lakshmi, R.U.S., Sai, N.S., Kumar, P.P., Ahammad, S.H.: Agriculture based plant leaf health assessment tool: a deep learning perspective. Int. J. Emerg. Trends Eng. Res. 7(11), 690–694 (2019)
23. Zhang, S., Zhang, S., Zhang, C., Wang, X., Shi, Y.: Cucumber leaf disease identification with global pooling dilated convolutional neural network. Comput. Electron. Agric. 162, 422–430 (2019)
24. Sai Prasanth,P.V.V., Veera Prasad, G., Krian Babu, M.: Automatic irrigation system using Arduino uno. Int. J. Eng. Adv. Technol. 8(5), 457–459 (2019)
25. Bodapati, J.D., Veeranjaneyulu, N.: Abnormal network traffic detection using support vector data description. In: Proceedings of the 5th International Conference on Frontiers in Intelligent Computing: Theory and Applications, pp. 497–506. Springer, Singapore (2017)
26. Deng, J., Dong, W., Socher, R., Li, L.J., Li, K., Fei-Fei, L.: Imagenet: A large-scale hierarchical image database. In: 2009 IEEE Conference on Computer Vision and Pattern Recognition, pp. 248–255. IEEE (2009).
27. Dondeti, V., Bodapati, J.D., Shareef, S.N., Naralasetti, V.: Deep convolution features in non-linear embedding space for fundus image classification deep convolution features in non-linear embedding space for fundus image classification
28. Tm, P., Pranathi, A., SaiAshritha, K., Chittaragi, N.B., Koolagudi, S.G.: Tomato leaf disease detection using convolutional neural networks. In: 2018 Eleventh International Conference on Contemporary Computing (IC3), pp. 1–5. IEEE (2018)
29. Kaushik, M., Prakash, P., Ajay, R., Veni, S.: Tomato leaf disease detection using convolutional neural network with data augmentation. In: 2020 5th International Conference on Communication and Electronics Systems (ICCES), pp. 1125–1132. IEEE (2020)
30. Durmuş, H., Güneş, E.O., Kırcı, M.: Disease detection on the leaves of the tomato plants by using deep learning. In: 2017 6th International Conference on Agro-Geoinformatics, pp. 1–5. IEEE (2017)

Chapter 51
Development of Robust Framework for Automatic Segmentation of Brain MRI Images

K. Bhima and A. Jagan

Abstract The robust framework for automatic segmentation of brain MRI images is developed in this work. The brain MRI image has been typically used to analyzing the brain anatomical structures and disorder in the brain. The identification of viable segmentation framework for brain MRI image is notoriously difficult problem. The major difficulty in brain MRI image analysis is mostly due to the unattainable of efficient segmentation framework. The proposed work develops a robust framework for classification and segmentation of brain MRI images. The developed robust framework consists of two phases, i.e., classification of input MRI images and detection of tumor. The automatic SVM classification and improved watershed methods are used in the robust framework to improve the segmentation process and improve the segmentation accuracy. In the first phase, MRI images classification is obtained with training and testing process with SVM classifier. In the second phase, segmentation of tumor region in MRI image is presented with improved watershed method. The developed robust framework is quantitatively evaluated on benchmark brain MRI image dataset.

51.1 Introduction

The magnetic resonance imaging (MRI) [1–3] is primary brain imaging modalities to capture and produce a computerized image for diagnosis and future treatment planning. MRI image does not emit ionic radiation like other modalities X-rays and computer tomography (CT) scan. The manual identification of tumor in MRI image is the time-consuming and tedious process. Since the numbers of patients are increasing, the acquired number of images also increases. Brain MRI image is characteristically used to capture images in diverse modalities [4–6]. Hence, there is a requirement of

K. Bhima (✉) · A. Jagan
B.V. Raju Institute of Technology, Narsapur, Telangana, India

A. Jagan
e-mail: jagan.amgoth@bvrit.ac.in

© The Author(s), under exclusive license to Springer Nature Singapore Pte Ltd. 2021   517
S. C. Satapathy et al. (eds.), *Smart Computing Techniques and Applications*,
Smart Innovation, Systems and Technologies 225,
https://doi.org/10.1007/978-981-16-0878-0_51

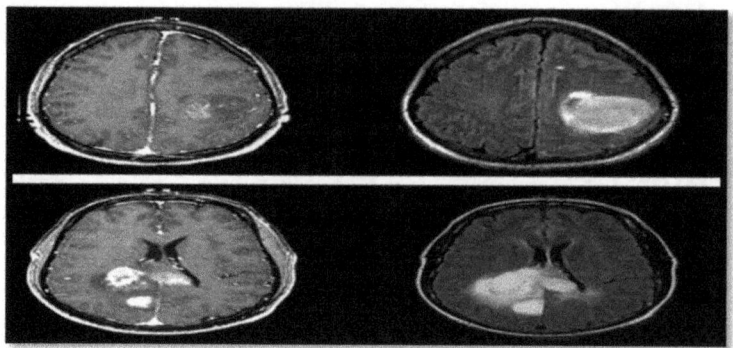

Fig. 51.1 Demonstration of four patients brain MRI Images

robust automatic segmentation framework that is competent to process and generate accurate segmentation results. The MRI or the magnetic resonance imaging images are a powerful brain imaging technique which is used to capture the brain cranial structure and produce a computerized image for further analysis and diagnosis of the brain tumor.

The output format of MRI technique as shown in Fig. 51.1 depends on the signal of the magnetic field and how it deteriorates commonly known as relaxation. The relaxation is the pure reflection of the magnetic resonance. Bilateral filter [4, 7, 8] is used for preprocessing of input MRI images. In this research work, the most popular SVM method [6, 9, 10] is used for classification of input brain MRI images into tumor and non-tumor MRI images, and the most accepted Watershed method [4, 5, 11, 12] is used for segmentation and detection of brain tumor in MRI images.

51.2   Development of Robust Framework

The proposed work develops a fully automatic framework for brain tumor detection and segmentation in brain MRI images. The proposed method consists of two phases, i.e., classification of input MRI images and brain tumor region extraction. The proposed framework has an automatic MRI image classification followed by tumor segmentation. In the first phase, MRI image classification is done through training and testing process using SVM classifiers. The infinite feature selection (IFS) method is used to detect optimal features in input MRI images. The SVM classification is used to automatically classify the input MRI image into a normal or tumor MRI image and identify the tumor. The improved watershed method is used in phase-II to extract the tumor in abnormal tumor MRI images that is identified in phase-I.

Phase-I: The proposed method has an automatic classification of brain MRI images into tumor and non-tumor MRI images.

(a)   The MRI classification is done through training and testing process using SVM classifier with 8 × 8 patches.
(b)   Three optimal features are chosen using infinite feature selection (IFS) method.
(c)   The purpose of the first phase is to automatically cluster the input MRI image into a normal or tumor MRI image and identify the tumor.

Figure 51.2 represents the robust framework for automatic classification of brain MRI images and segmentation of tumor in MRI Images.

Phase-I: The Classification of Tumors and Non-tumors MRI images with SVM classifier.

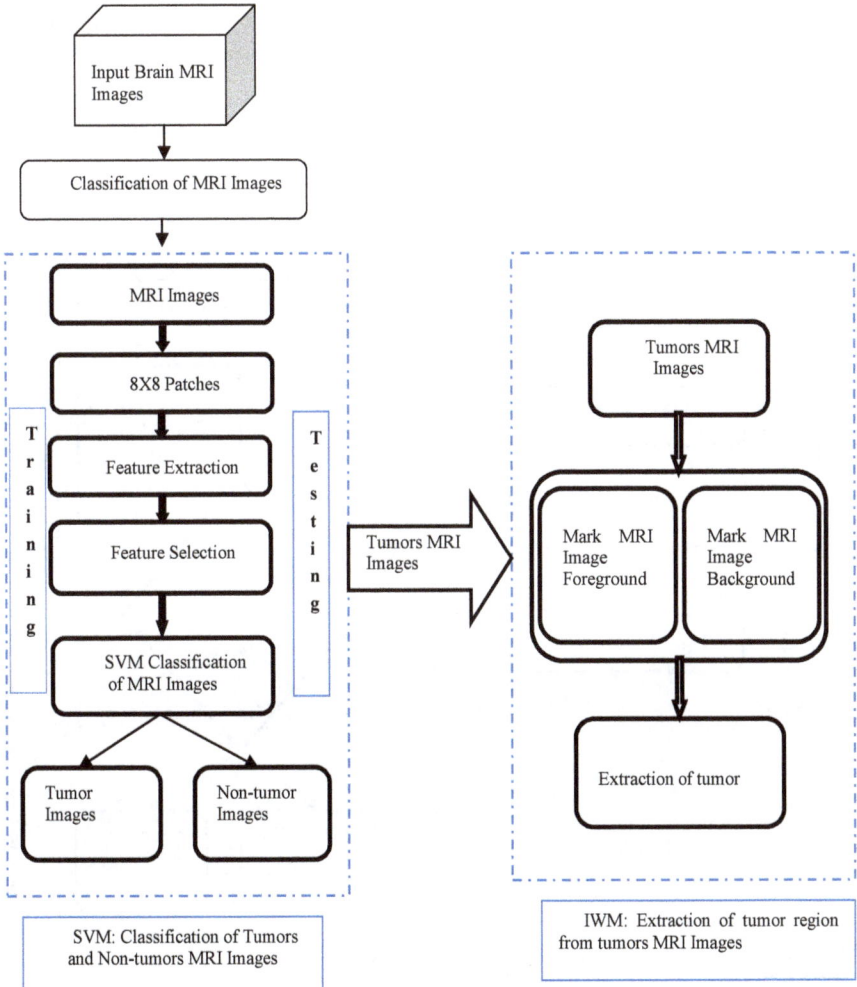

Fig. 51.2 Proposed robust framework

SVM method is used for classification of input brain MRI images into tumorous and non-tumor MRI images, and SVM [6, 9, 10, 13] is a supervised machine learning mode which is most popular for classification of MRI images.

The demonstration of SVM method for classification of input MRI images is shown in Fig. 51.3. The illustration of brain MRI images for feature selection and classification of input MRI images is shown in Fig. 51.3.

Phase-II: Extraction of Tumor Region from Tumors MRI Images with Improved Watershed Method.

In this research work, the improved watershed method is developed for extraction of tumor region from MRI images. The marker controlled watershed method is used for identifying or marking foreground objects and background locations for extraction of tumor in brain MRI images. The manifestation of improved watershed method for segmentation of tumor in MRI images is shown in Fig. 51.4.

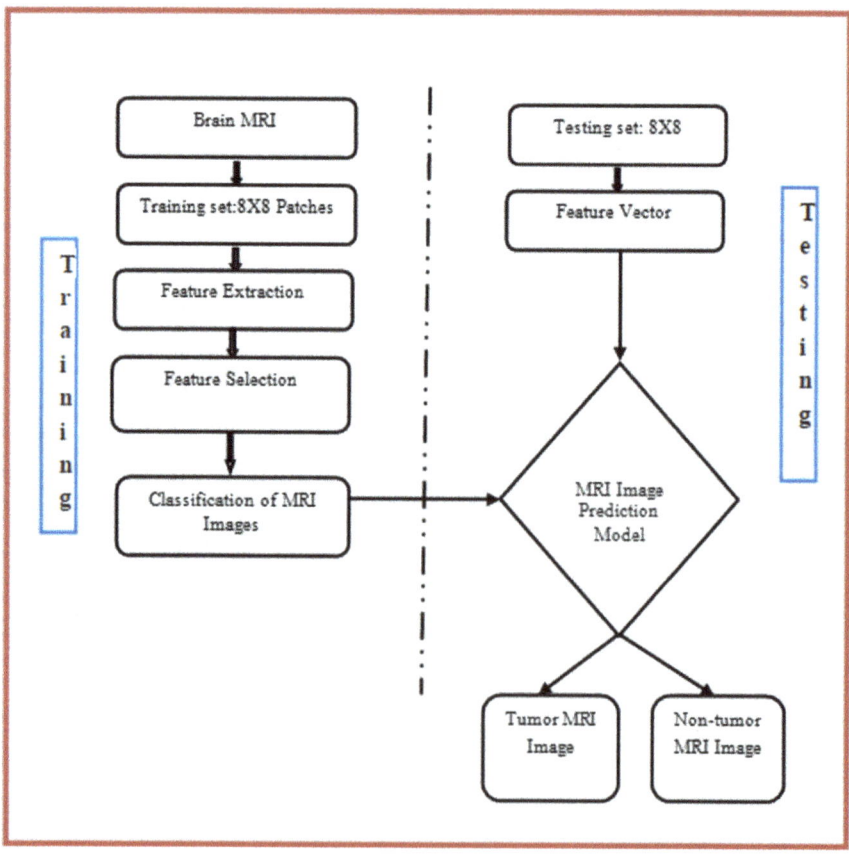

Fig. 51.3 Block diagram of SVM for MRI images classification

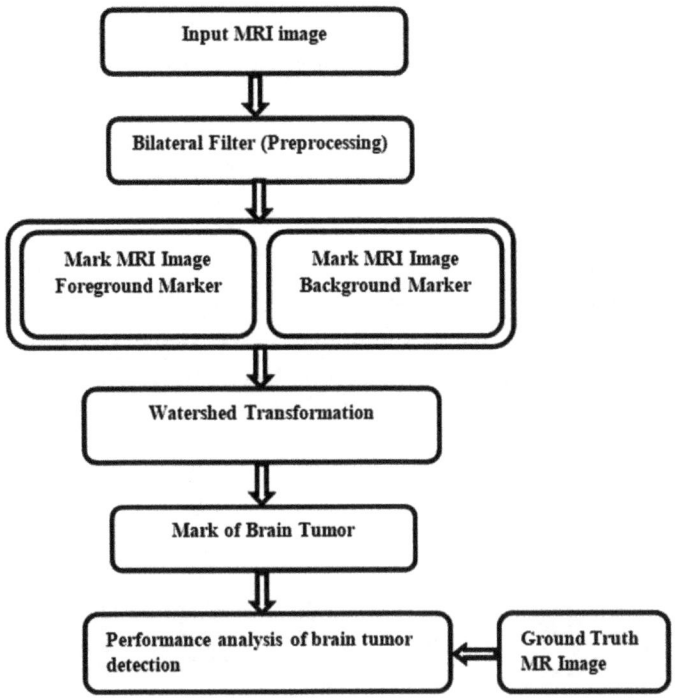

Fig. 51.4 Framework for improved watershed method for segmentation of tumor in MRI images

51.3 Results and Discussion

The research work quantitatively evaluated the performances of proposed new framework with segmentation accuracy; sensitivity and specificity [14] on benchmark brain MRI datasets consist of ground truth MRI images. The higher value of segmentation accuracy indicates that the brain tumor is segmented more accurately. The SVM method (Phase-I) and improved watershed method (Phase-II) are demonstrated on extensive publicly available brain MRI datasets. The performance of SVM Method is evaluated with segmentation accuracy on benchmark brain MRI datasets. The performance of improved watershed method is evaluated with segmentation accuracy on benchmark brain MRI datasets that comprise ground truths images.

$$\text{Segmentation Accuracy (\%)} = \frac{\text{True Positive} + \text{True Negative}}{\text{True Positive} + \text{True Negative} + \text{False Positive} + \text{False Negative}} * 100$$

Table 51.1 and Fig. 51.5 present the obtained results of SVM on five brain MRI datasets for classification of tumor and non-tumor brain MRI images.

Table 51.2 and Figure 51.6 demonstrate the analysis of segmentation accuracy of improved watershed method on ten patient's brain MRI images for segmentation of

Table 51.1 Analysis of obtained classification results of SVM on five brain MRI datasets

MRI Dataset	No of MRI images	Classification of MRI Images		Classification accuracy (%)
		Tumor MRI image	Non-tumor MRI image	
Dataset-1	500	403	97	99.08
Dataset-2	500	411	89	99.16
Dataset-3	500	397	153	98.29
Dataset-4	500	409	91	99.13
Dataset-5	500	392	108	98.37

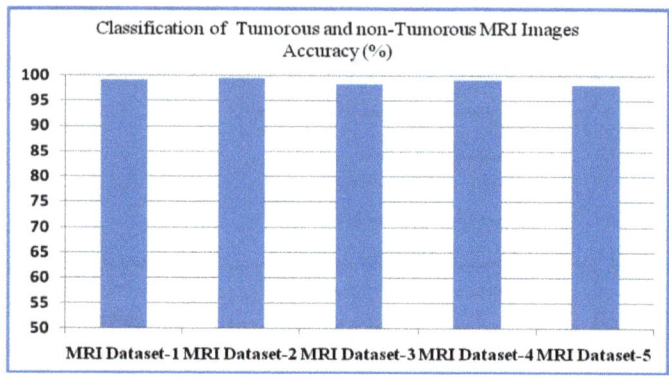

Fig. 51.5 Performance analysis of classification of tumor and non-tumor brain MRI Images with SVM

Table 51.2 Segmentation accuracy of improved watershed method on MRI images

Input MRI image	Segmentation accuracy (%)		Improvement (%)
	WM	Improved WM	
Patient-1	98.25	98.89	0.65
Patient-2	97.28	98.03	0.77
Patient-3	98.41	99.07	0.67
Patient-4	98.04	98.82	0.80
Patient-5	98.17	98.85	0.69
Patient-6	97.29	98.11	0.84
Patient-7	97.11	98.27	1.19
Patient-8	97.28	98.13	0.87
Patient-9	97.09	98.24	1.18
Patient-10	96.21	97.31	1.14

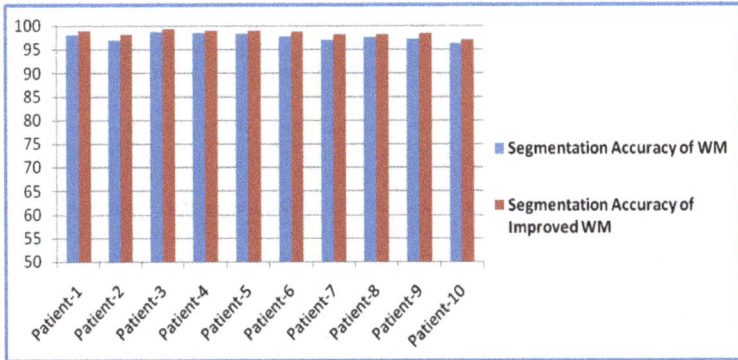

Fig. 51.6 Comparative analysis of segmentation accuracy of improved watershed method on MRI images

the brain tumor, and it is achieved improved segmentation accuracy as compared to conventional method.

51.4 Conclusion

The research work proposes a robust framework for MRI image segmentation with amalgamation of SVM and improved watershed method. The robust framework was developed to overcome the limitations of the existing MRI segmentation methods for brain MRI images. In the proposed framework, the first phase is the SVM method which is used for classification of input brain MRI datasets into tumor and non-tumor images. The second phase improved watershed method is used to automatic segmentation of tumor in brain MRI Images. The performance of proposed new framework for MRI image segmentation is evaluated on benchmark MRI dataset. The obtained segmentation results were compared with existing conventional segmentation method. The end result of the proposed robust framework presented the improvement in segmentation accuracy for brain tumor detection and segmentation.

References

1. Kalaiselvi, T., Sriramakrishnan, P.: Rapid brain tissue segmentation process by modified FCM algorithm with CUDA enabled GPU machine. Int. J. Imaging Syst. Technol. **28**(3), 163–174 (2018)
2. Kalaiselvi, T., Kumarashankar, P., Sriramakrishnan, P.: Three-phase automatic brain tumor diagnosis system using patches based updated run length region growing technique. J. Digital Imagoing Inform. Med. (2019). https://doi.org/10.1007/s10278-019-00276-2
3. Sriramakrishnan, P., Kalaiselvi, T., Rajeswaran, R.: Modified local ternary patterns technique for brain tumour segmentation and volume estimation from MRI multi-sequence scans with GPU CUDA machine. Biocybern. Biomed. Eng. **39**(2), 470–487 (2019)

4. Bhima, K., Jagan, A.: Analysis of MRI based brain tumor identification using segmentation technique. In: 2016 International Conference on Communication and Signal Processing (ICCSP), Melmaruvathur (2016), pp. 2109–2113. https://doi.org/10.1109/ICCSP.2016.775 4551

5. Kamrul Hasan, S.M., Ahmad, M.: Two step verification of brain tumor segmentation using watershed matching algorithm. Brain Inform. **5**(8) (2018). https://doi.org/10.1186/s40708-018-0086-x

6. Du, X., Li, Y., Yao, D.: A support vector machine based algorithm for magnetic resonance image segmentation. In: 2008 Fourth International Conference on Natural Computation, Jinan, pp. 49–53 (2008). https://doi.org/10.1109/ICNC.2008.400

7. Cabeen, R.P., Laidlaw, D.H.: Bilateral filtering of multiple fiber orientations in diffusion. MRI (2015). https://doi.org/10.1007/978-3-319-11182-7

8. Bhonsle, D., Chandra, V., Sinha, G.R.: Medical image denoising using bilateral filter. Graph. Signal Process. 36–43 (2012). https://doi.org/10.5815/ijigsp.2012.06.06

9. Moyano-Cuevas. J.L., et al.: 3D segmentation of MRI of the liver using support vector machine. In: Roa Romero, L. (eds.) XIII Mediterranean Conference on Medical and Biological Engineering and Computing 2013. IFMBE Proceedings, vol 41. Springer, Cham (2014). https://doi.org/10.1007/978-3-319-00846-2_91

10. Zhang, X., Yan, L.F., Hu, Y.C., et al.: Optimizing a machine learning based glioma grading system using multi-parametric MRI histogram and texture features. Oncotarget **8**(29), 47816–47830 (2017). https://doi.org/10.18632/oncotarget.18001

11. Shih, F.Y., Cheng, S.: Automatic seeded region growing for color image segmentation. Image Vision Comput. **23**(10), 877–886 (2005). ISSN: 0262-8856

12. Bhima, K., Jagan, A.: An improved method for automatic segmentation and accurate detection of brain tumor in multimodal MRI. Int. J. Image, Graph. Signal Process. (IJIGSP) **9**(5), 1–8 (2017). https://doi.org/10.5815/ijigsp.2017.05.01

13. Blumenthal, D.T., Artzi, M., Liberman, G., Bokstein, F., Aizenstein, O., Ben Bashat, D.: Classification of into tumor and nontumor components using support vector machine. Am. J. Neuroradiol. **38**(5), 908–914. https://doi.org/10.3174/ajnr.A5127

14. Khalil, M., Ayad, H., Adib, A.: Performance evaluation of feature extraction techniques in MR-brain image classification system. Procedia Computer Sci. **127**, 218–225 (2018)

Chapter 52
Yield Estimation and Drought Monitoring Through Image Processing Using MATLAB

Shaikh Akbar Shaikh Rasul, Jadhav Swamini Narendra, and Dipti Y. Sakhare

Abstract This paper elucidates the pre-harvest yield estimation method and technique for cotton crop by image processing. This pixel-based image analysis of image processing is done using the image processing toolbox of the MATLAB 2019b. The images for abovementioned purpose are taken through camera armed drone (quadrotor). Further, there is a need for a better and transparent surveying method to assess the eligibility of a particular farm, for claiming the agricultural insurance. From the findings of proposed research, a suggestion for Agriculture Insurance Companies is made. From this research, it is concluded that yield estimation can be used for the purpose of detection of the damage and impact of the drought. The impact of drought can be assessed on individual farm level, i.e., on each acre of cotton farmland.

52.1 Introduction

This research is containing two major different parts, i.e., (i) pre-harvest yield estimation and (ii) deciding impact of the drought on individual farm level. Yield estimation will give the exact idea about the impact and situation of the drought in every individual farm. One can even decide the impact of the drought on different levels and intensities so that the more affected will get more benefits and the less affected will get less benefits. This will reduce the expenses and human efforts needed for surveying the impact of drought and also will complete the work in minimal amount of time so the farmers will not need to wait more for the government's aid. This

S. A. S. Rasul (✉) · J. S. Narendra · D. Y. Sakhare
MIT Academy of Engineering, Pune, Maharashtra, India
e-mail: shaikh.akbar@mitaoe.ac.in

J. S. Narendra
e-mail: swamini.jadhav@mitaoe.ac.in

D. Y. Sakhare
e-mail: dysakhare@etx.maepune.ac.in

S. C. Satapathy et al. (eds.), *Smart Computing Techniques and Applications*,
Smart Innovation, Systems and Technologies 225,
https://doi.org/10.1007/978-981-16-0878-0_52

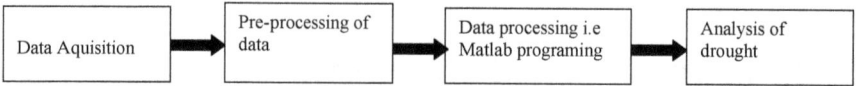

Fig. 52.1 Block diagram

method gives more precise and transparent results. Yepei et al. [1] worked on satellite remote sensing data and crop growth models, combined with the help of data assimilation technique and used SCE-UA with DSSAT to create growth model of cotton, for estimating the yield. Also, Manthan et al. [2] found a method for early prediction of drought by using standard precipitation index (SPI).

The objectives of proposed research are as follows:

1. To collect the image data of entire farm in the parts of per acre through UAV (drone armed with camera).
2. To apply image enhancement techniques as a part of preprocessing.
3. To find a novel yield estimation algorithm for estimating number of cotton bolls and thus the total estimation of yield of the farm.
4. To establish a relationship between the count of open cotton bolls and yield, at the time of harvest.
5. To use yield estimation methodology, for finding the degree of impact of drought.
6. To make an estimated yield-based suggestion, for Agriculture Insurance Companies.

52.2 Proposed Algorithm

52.2.1 Main Building Blocks

This research comprised of four distinct building blocks represented in Fig. 52.1.

52.2.2 Data Acquisition

Data is the main concept in this study. It is captured in the image form from the UAV, i.e., camera armed quadrotor. There are studies that show UAVs can be efficiently used for capturing images from the air [3]. The inclusion of drone technology for data acquisition in this study is helping us in different ways, i.e., less human efforts are required and the process becomes incredibly fast.

The size, shape and other parameters related to images that are found most suitable by performing number of experiments and referring standard available researches [4] are illustrated in Table 52.1.

Table 52.1 Specification of image capturing

S. No.	Entity	Specification
1	Camera	PX4 Optical flow sensor
2	Flight controller	Pixhawk 2.4.8
3	Secondary Processor	Raspberry Pi 3
4	Storage Unit	Micro SD
5	Image Capturing Shape	Square
6	Orientation of drone	Flat
7	Pattern of Capturing	Along the rows and columns of farm

52.2.3 Preprocessing of Data

For the betterment of the results, it is needed to make the data prepared for the processing, and this method is known as preprocessing. From the raw data captured from drone when loaded to the computer, it is preprocessed through MATLAB. Preprocessing includes different methods, i.e., (1) scaling, (2) resizing, (3) image enhancement, etc.

Scaling and resizing are crucial preprocessing techniques, especially when it comes to the tasks related to image processing and analysis. There are different levels of these methods, in simple methods, the aspect ratio is ignored. But in more advanced methods, the aspect ratio is respected.

Here, scaling and resizing are done automatically as the images are taken through the same camera, and all other parameters affecting the size of the image are preset so that the captured images are of same size. But, in case if there comes any deflection in the standard size due to some errors in the hardware, it is corrected by the respected algorithm in MATLAB itself. Following flowchart depicts the process of preprocessing.

In Fig. 52.2, the overall method of preprocessing is explained through a flowchart. After importing data in MATLAB, a loop is run on the data; one image from the data is preprocessed in one iteration. Converting the image in matrix form, the next step is to check the size whether it is suitable for the algorithm or not. If it comes out to be unsuitable, then the filter is applied to correct the error. Then, some image enhancement techniques are applied for getting an enhanced image. At last, RGB image is converted into grayscale image, and all of the preprocessed images are stored in new folder.

52.2.4 Data Processing and MATLAB Programming

It contains three sub-methods which are depicted in block diagram (Fig. 52.3).

Fig. 52.2 Preprocessing

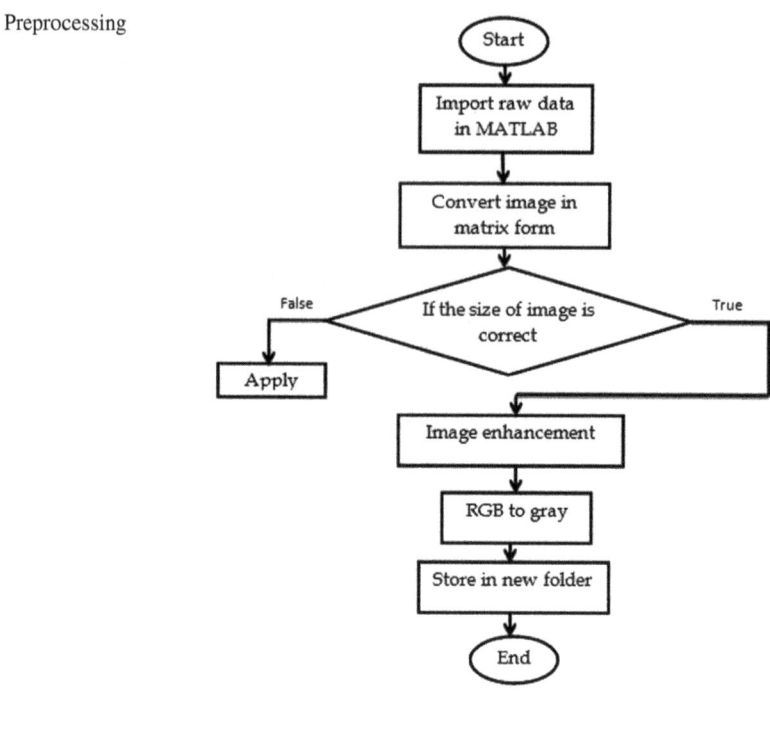

Fig. 52.3 Data processing and MATLAB programming

52.2.5 *Explanation of MATLAB Code*

The main intension behind this MATLAB code is to do the segmentation of the open cotton bolls, to highlight and separate them, for counting the number of these bolls. Property of the color of cotton bolls separates them from the other parts of the image [5, 6], so this property is used as the feature for the extraction of the cotton bolls from the image. According to our research, it is found that we can estimate the yield in two different forms, i.e.,

1. Number of cotton bolls per unit area
2. Amount (kg) of cotton per unit area.

But the second method cannot be directly achieved skipping the first one, i.e., the second method is the advanced version of the first method. For the first method, it is needed to extract open cotton bolls, following algorithm for the MATLAB code explains the method for counting number of cotton bolls per image.

1.   Importing the image data.
2.   Reading image.
3.   Finding the dimensions of the image.
4.   Displaying the image.
5.   Performing the threshold segmentation algorithm.
6.   Again displaying the threshold segmented image.
7.   Thresholding the size of bolls created from the abovementioned segmentation.
8.   Filtering the bolls and filling them up.
9.   Using the bsxfun() for masking the image.
10. Finding centroid from the ROI, i.e., the cotton bolls.
11. Numbering and counting the bolls.

In the above algorithm, after performing the basic steps, RGB image is converted into grayscale image in preprocessing. Threshold segmentation is performed on the preprocessed image. There are two types of threshold segmentation: global segmentation and local segmentation [5, 6]. In case of global segmentation, image is divided into two regions, i.e., region of interest (ROI) and background [7]. A single threshold is used in global segmentation. For the proposed algorithm, global thresholding is used for the segmentation purpose. The grayscale intensities vary between 0 and 255, 0 representing the darkest intensities while 255 represents the brightest intensities. It is found out through experimentation that the suitable threshold value tends to 255, as the targeted part in the ROI is cotton which has the brightest intensities.

- **Experiment to find the weight of single cotton boll.**

This is needed for finding the estimation of the yield corresponding to the count of cotton bolls in the standard unit, i.e., kilograms or grams. Following steps illustrate this experiment.

1.   Collected cotton in different samples, i.e., 10 bolls, 15 bolls, 20 bolls, etc.
2.   Calculated the weight of these samples using digital weighing machine.
3.   Then, calculated average weight of the bolls for each sample by dividing the weight of sample by the number of cotton bolls.

$$\text{Average Weight} = \frac{\text{Weight of sample}}{\text{Number of cotton bolls in the sample}} \qquad (52.1)$$

4.   Then, finding the average of the abovementioned average weights of each sample. This weight is considered as the final average weight.
5.   In experimentation, the final average weight is found to be 4.01 g, i.e., 0.004 kg.

- **Estimating Cotton Yield in kg**

For better understanding of the yield, it is needed to represent this yield in standard units; following algorithm expresses the method to do this.

1. After getting the count of cotton bolls, the procedure for the same is explained further in Sect. 52.3.
2. Multiply the count by average weight in Sect. 2.4.2 of the single boll.
3. Get the weight of cotton present in the portion of land captured in single image.
4. Repeat the above steps for each of the captured images.
5. Sum up all the individual weights to get the overall estimation of the farm in kg.

- Analysis of Drought

There is a strong and clear relation between estimated yield and impact of drought. It can be represented in the equation form

$$EY \propto \frac{1}{ID} \tag{52.2}$$

where EY = estimated yield, and ID = impact of drought.

The study of the ground reality and survey with the drought affected farmers shows that there is a need to measure the impact / intensity of the drought on a scale, and this scale should include multiple discrete levels.

52.3 Experiment and Result

52.3.1 Results of Experiment Through MATLAB

One image is chosen from a dataset of images, captured from cotton farm for the purpose of experimentation. The present dataset of cotton farm is captured at the time of drought, so the yield is less, which can be observed in Fig. 52.4. Repeating

Fig. 52.4 Original image

the same process as performed below, on entire set of images for a farm, gives final yield estimation. The results are obtained using MATLAB 2019b software. There are methods, in which on-board processing is used for drone systems [8], but in the proposed research, processing is done on separate machine. The machine used for this experimentation has 8 GB of RAM and Intel i7 processor.

In Figure 52.4, this is the raw image of farm taken by camera armed drone. The size and coverage of this image are purposefully kept small for better realization of the results. Single cotton plant is covered in this image.

In Figure 52.5, here, cotton bolls in image are counted manually. So that the final results can be validated. For the sake validation, one can easily compare the number of cotton bolls counted manually and through the proposed algorithm.

Preprocessing of image is explained in Fig. 52.6. First raw image is imported in MATLAB, and then its dimensions are found and displayed that image in grayscale.

Fig. 52.5 Manually counted cotton

Fig. 52.6 Grayscale image

Fig. 52.7 Threshold segmented image

Fig. 52.8 Binarized image

In Fig. 52.7, the open cotton bolls are extracted depending upon their color properties, as they are brightest in color [7]. The intensity of open cotton bolls is around 250.

In Fig. 52.8, the color segmented image is binarized.

In Fig. 52.9, the result of finally extracted cotton bolls is shown. The cotton bolls are numbered 1, 2, 3…, and total count is displayed at corner.

52.3.2 Tables and Quantitative Analysis of Results

- After performing experiment, the average weight of the cotton bolls is found to be 4.01 g.
- Table with quantitative analysis of Sect. 53.3.1 (Table 52.2).

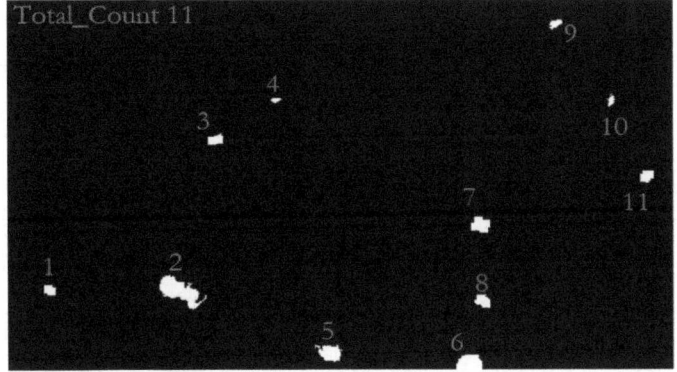

Fig. 52.9 Yield highlighted image

Table 52.2 Quantitative analysis of the experiment

Number of cotton bolls (manual)	Weight of hand-picked cotton in grams (manual)	Number of cotton bolls (automatic)	Weight by image processing in grams (automatic)
11	44	11	44.11

- Repeating the experiment in section III (A), on entire image dataset of, per acre cotton farm, gives pre-harvest final yield estimation of cotton per acre.
- The final yield estimation in Eq. 52.2 gives direct relation for estimating the degree of drought in the farm.

52.3.3   Graphical Analysis of Result

In Fig. 52.10, the outcomes of counting the number of cotton bolls through both the methods, manual and through computerized algorithm using image processing are represented using bar graph. This is the result from the experiments performed on the single image, as illustrated earlier; this image is purposefully transformed into small size for the better realization of the results.

52.3.4   Result of Another Experiment on 10 Images

Another ten random images are chosen from the dataset, these images represent 10–15 cotton plants. This data, as illustrated above, is taken from the farm which is suffering through drought situation, so the yield is less. Table 52.3 illustrates the insights of this experiment.

Fig. 52.10 Comparison of
the results for single image

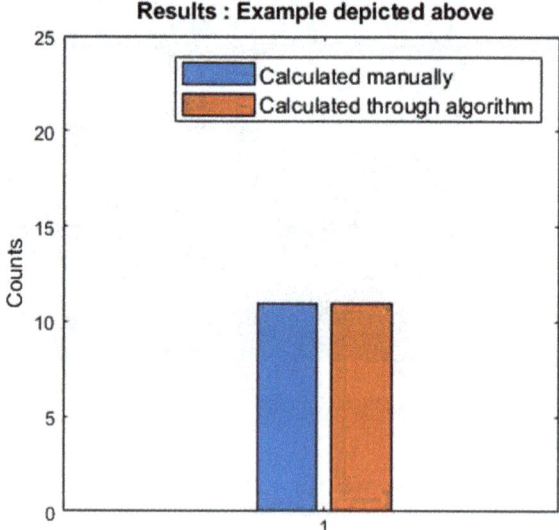

Table 52.3 Quantitative
analysis of multiple images

S. No.	Number of cotton bolls (manual)	Number of cotton bolls (automatic)
1	98	95
2	105	99
3	90	102
4	97	101
5	107	94
6	85	91
7	87	88
8	90	99
9	85	89
10	85	97

Figure 52.11 is the pictorial representation of the data present in Table 52.3.

In Fig. 52.11, minor errors can be seen between both of the readings, i.e., the readings taken manually and the readings taken through proposed algorithm. In some of the samples, like in 1, 2, 3, 5, 6 and 10, it can be seen that the readings taken manually are higher than the readings taken through computerized algorithm, and this is because some of the cotton bolls which are hidden under the leaves or other things, are not visible in the images taken. Also, in other samples, like in 4, 7, 8 and 9, it can be seen that the readings taken manually are less than the readings taken through proposed algorithm, this is because some of the single cotton bolls are counted as multiple rather than single, i.e., the individual tulips of the cotton bolls separated through its own material, due to this separation, and at the time of

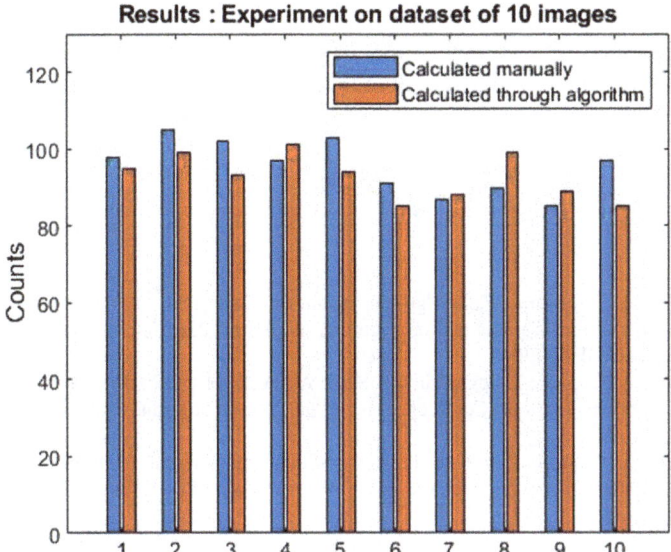

Fig. 52.11 Comparison of the results for multiple images

extraction of the cotton bolls, it is extracted as multiple cotton bolls although it is a single cotton boll in actual. For the experiment on ten samples, the total count of manual readings is 955 and the count through proposed algorithm is 928, i.e., the difference is of 27 and percentage error is about 1.43%.

52.4   Conclusion

Cotton yield can be effectively estimated through image processing, on the image dataset captured from the drone. Yield estimation has a lot of importance in the field of agriculture and drought monitoring. Drone technology reduces efforts, saves time and increases accuracy and transparency in yield estimation. Findings show that the relative yield of cotton crop can successfully estimate by considering the open cotton bolls, and still there is a space for including the closed cotton bolls and empty cotton bolls. An algorithm is designed to extract yield from individual images through image processing and clubbing them all to find yield estimation per acre. The relation between crop yield and impact of the drought is found, and a mathematical equation Eq. 52.2 is designed for the same. As per the detailed explanation given earlier, for the cause of representation, a single image is processed, similarly, the proposed novel method is implemented on another ten images, to have a big picture of method and its effectiveness. Depending upon estimated yield, impact of the drought can be expressed on different levels as low, medium and high. If the estimated yield is sufficient, then no drought condition will be announced for corresponding farms. Therefore, the proposed research can be used to make an estimated yield-based recommendation, for Agriculture Insurance Companies. Finally, this research increases the precision and accuracy in the drought monitoring and survey. It takes down the survey level from region level to individual farm level.

References

1. Chen, Y., Mei, X., Liu, J.: Cotton growth monitoring and yield estimation based on assimilation of remote sensing data and crop growth model. In: 23rd International Conference on Geoinformatics. IEEE (2015)
2. Kansara, M., Maity, P., Malgaonkar, H., Save, A.: A novel approach for early prediction of drought. In: 6th International Conference on Advanced Computing and Communication Systems (ICACCS). IEEE (2020)
3. Urbanski, K.: Control of the quadcopter position using visual feedback. In: 2018 18th International Conference on Mechatronics—Mechatronika (ME). IEEE (2018)
4. Seifert, E., Seifert, S., Vogt, H., Drew, D., Van Aardt, J., Kunneke, A., Seifert, T.: Influence of drone altitude, image overlap, and optical sensor resolution on multi-view reconstruction of forest images. Remote Sens. **11**(10) (2019) (Article citation)
5. Abu Bakr Siddique, Md., Arif, R.B., Rahman Khan, M.M.: Digital image segmentation in Matlab: a brief study on OTSU's image thresholding. In: 2018 International Conference on Innovation in Engineering and Technology (ICIET). IEEE (2018)
6. Li, Y., Zhang, J., Gao, P., Jiang, L., Chen, M.: Grab cut image segmentation based on image region. In: IEEE 3rd International Conference on Image , Vision and Computing (ICIVC) (2018)
7. Gonzalez,R.C., Woods, R.E.: Digital Image Processing, pp. 711–807. Pearson, New Delhi (2017)
8. Horstrand, P., Guerra, R., Rodríguez, A., Díaz, M., López, S.: A UAV Platform based on a hyperspectral sensor for image capturing and on-board processing. IEEE Access **7** (2019)

Chapter 53
A Framework for Efficient Multilevel Polarity-Based Sentiment Analysis Using Fuzzy Logic

Shiramshetty Gouthami and Nagaratna P. Hegde

Abstract The evaluation of any product or event on social media with the opinion or emotion of peoples is known as sentiment analysis (SA). A great deal of attention has been attracted in recent years, toward both science and industry fields for a variety of uses. Machine learning and the text mining uses this most widely known application area of sentiment analysis. This paper presents a framework for efficient multilevel sentiment analysis using fuzzy logic for the classification of online test reviews polarity as strong positive, positive, negative and strong negative. This proposed model can use the fuzzy logic classifier to enhance the degree of sentiment polarity of reviews. Here, fuzzy logic classifier is used for finding the sentiment classes. This also utilizes the mechanism of imputation of missing sentiment for integrating non-opinionated sentences in generating precise results. Results show that the proposed method has a capability of extracting opinions and classify them in an effective way. The proposed method has a capability to predict the degree of sentiment polarity for the reviews on a social media. The better precision and F1-scores are obtained for an objective/subjective classification and polarity (positive/negative) classification on twitter dataset.

53.1 Introduction

In the present scenario, one of the most demanding research problems particularly in social media was sentiment analysis. Through the social media platform, peoples are able to freely convey their opinions, thoughts and observations toward various trendy dealings, subjects, products etc. by posting them on the social media. Then it has been required to analyze these posts in order to recognize which kind of sentiment that these posts convey. People emotions from a written text are recognized and

S. Gouthami
Osmania University, Hyderabad, Telangana, India

N. P. Hegde (✉)
Vasavi College of Engineering, Hyderabad, Telangana, India
e-mail: nagaratnaph@staff.vce.ac.in

© The Author(s), under exclusive license to Springer Nature Singapore Pte Ltd. 2021   537
S. C. Satapathy et al. (eds.), *Smart Computing Techniques and Applications*,
Smart Innovation, Systems and Technologies 225,
https://doi.org/10.1007/978-981-16-0878-0_53

determined by the sentiment analysis which is also called as sentiment artificial intelligence that analyzes the views of peoples through the written posts. People around the world are allowed to freely convey their opinions on common topics through a social media platform by connecting and interacting with each other. This social media sentiment analysis can be served as a measure for determining the performance of social media in addition to use for improving customer services and marketing. Since the information in this recent years regarding any small and large incidents or disasters can be collected through social media Web sites, the effect of social media Web sites became more significant in daily life. The text data usage on the Internet nowadays increases rapidly. This enormous text data can be endeavored by the several industries to extract opinions of peoples on their products. In this context, an important source of information is the social media. Analyzing this large amount of text data manually is not possible. So it is evident that there is necessity of automatic classification.

In this case, subjective data is usually analyzed. Several social media Web sites are there where the users are enabled for contributing, modifying and rating content. There is an opportunity for the users for expressing their own views on specific topics. Social media, blogs, product review sites and forums are some of the examples of those Web sites. Twitter data has been utilized in this case. In most of the time, Web sites like Twitter contain small length of comments as status messages on social media like Twitter or article reviews on Digg. In addition, most of the sites allowed to rate the attractiveness of posts which may be related to the opinion expressed by the author. Extension for the deterministic logic gives the fuzzy logic which means that instead of a binary value, truth value has a range of 0-to-1. Conversion of white as well as black problem to a gray problem is the main goal of fuzzy logic hypothesis (Zadeh 2015). Probably the simplest method of representing human knowledge under the area of artificial intelligence is to convert it into the IF-THEN rules formatted-based natural language sentiments. These rules depend on the representations and models of natural languages and which were in turn depend on the fuzzy logic and fuzzy sets. Fuzzy rule-based classification systems are the most powerful and recognized tools for pattern recognition and classification. Since the fuzzy logic is presented in these systems, they can very efficiently handle the uncertainty and vagueness or ambiguity.

The present paper proposes a framework for efficient multilevel polarity sentiment analysis using fuzzy logic. Fuzzy logic can be used to classify the online reviews of users by considering a twitter dataset. In addition, fuzzy logic interface system generates five fuzzy-based rules to find out the sentiment for every one of the textual review tweeted on online. These reviews are classified as two classes of namely "positive" and "negative" opinions. Hence, to identify the polarities of these classified sentiments are determined fastly and efficiently using the aspect-based sentiment summarization.

53.2   Classification of Sentiment Analysis

The classification of sentiment analysis has been carried out in different ways based on the perspective point of view. Sentiment-based classification is one of those ways which is widely used as a sentiment classification technique. This sentiment classification technique is further classified into two different approaches such as the machine learning and lexicon-based approach. In addition, a third classification is also included, i.e., hybrid approach. In addition, depending on the mode of identifying and analyzing the opinions, sentiment analysis can be classified. The document level, sentence level and aspect-based levels are the three major levels of classification. Rating-based level is the last classification of the most general sentiment analysis classification. These sentiment polarities are also known as opinion polarities which were positive, negative, and neutral.

53.2.1   Document Level of Sentiment Analysis

The notions, expressions or sentiments expressed toward a component or event are the opinions. The peoples are allowed by the Web sites or forums of having a large amount of information to convey their opinion as observations and reviews. These opinions can be different from one user to another user. Positive and Negative are the just two kinds of classes observed in this analysis. One of the suitable examples is that a review on the item: "I brought a latest I Phone three days back. It is an excellent phone. It has a high-speed touch screen. It has the best voice clarity. I really love this phone". Then by using the review or star framework, required sentiments are labeled as either positive (stars rated with 4 or 5) or negative (stars rated with 1 or 2).

53.2.2   Sentence Level of Sentiment Analysis

Since this analysis can make the polarity of the sentence perfectly, it is used to provide helpful data in searches. This level of sentiment analysis can be carried out on the sentences containing opinions which may have reviews of either positive or negative. This sentence-level sentiment analysis is almost same as that of document-level sentiment analysis. However, the major difference of this sentence-level sentiment analysis is that every sentence is separately analyzed it to determine if that express a positive or negative opinion. Since the sentences can be distinguished as the objective sentences from subjective ones, this level offers more flexibility than the document level.

53.2.3 Aspect-Based Sentiment Analysis

In the consideration of a single aspect, both the document-level sentiment analysis and the sentence-level sentiment analysis are well functioned. On the other hand, peoples can able to discuss regarding the aspects in a long time that have different points of reviews or qualities. Likewise, each individual aspect would have to be unambiguous sentiments. This was generally happened when talking about social issues and in an item review. One of the convincing examples is that I am a Nokia phone of extra large. I like the look of the phone. The screen is bigger and clear. The camera is great. However, there are also some drawbacks that the battery life is insufficient and Whatsapp is not accessible. The significant data of item is hidden by requesting the positives and negatives regarding this review.

53.2.4 Comparative Sentiment Analysis

Responses of the items or brands are expressed by the users with the utilization of different expressions even it can be the same item or brand. Discovering the possibility of comparative type sentence is the main aim of this comparative sentiment analysis.

53.2.5 Sentiment Lexicon Acquisition

The process of determining expressions and opinions of reviews by utilizing the sentiments on review is called Sentimental analysis. The two types of positive and negative classes are observed in this Sentiment analysis. Let us considered a sentence "Auto X is superior to any Auto Y". This sentence does not represent which class this sentence belongs to. Similarly, the three types of systems such as manual methodology, dictionary-based approach and Corpus-based approach are used to analyze that type of sentences / texts.

- Manual Methodology: Since it is not possible to retrieve the data as a positive or negative by this method, it is completely time taking process.
- Dictionary-based approach: The SentiWordNet is used in this approach along with the Part Of Speech tagging to determine the polarity of the sentences.
- Corpus-based approach: This approach uses a domain-specific sentiment lexicon to perform the analysis.

These are the different approaches to examine the opinions of customers and also to forecast the market value of a particular organization.

53.3 Mutilevel Polarity Sentiment Analysis Using Fuzzy Logic

The proposed frame of multilevel polarity-based sentiment analysis using the fuzzy logic is shown in Fig. 53.1. The degree of sentiment polarities is enhanced by the use of fuzzy logic. The textual reviews are taken from the sentiment polarity dataset of social Web sites. This fuzzy logic can be accomplished by implementing the fuzzy interface system using a Skfuzzy package. The main aim of this proposed framework is to classify the sentiment polarities of review dataset fastly and effectively using an aspect-based sentiment summation.

Fig. 53.1 Framework of multi level-based sentiment analysis using fuzzy logic

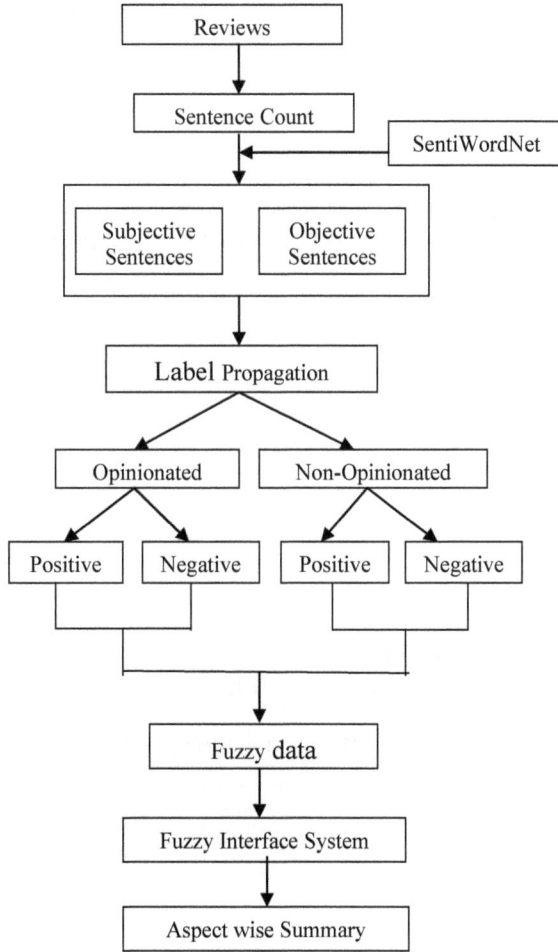

53.3.1   Online Reviews

The movie-related textual reviews on online that are recently collected for the opinion exchange associated with a variety of domains are used from the polarity dataset of V0.9. This dataset can be divided into training, testing and validation units. Some of the elements in the textual data such as digits, punctuations, stop words and other special symbols are removed and then processed after the completion of preprocessing. Then these text reviews from the preprocessing are divide into tokens and transformed into word vectors.

53.3.2   Sentence Separation Using SentiWordNet

The place at which a split or break occurs is called separation. A lot of sentences are present in the review dataset used in this framework. Commas, dots and semicolons are considered to separate those sentences. Moreover, in this sentence separation phase, the count of sentences in a review is also being calculated. Every one of the sentences must be split into words and word array is used to store them on behalf of future processing. The following gives the procedure of Sentence Separation and Sentence count.

```
Function Sentence_Count()
{
      Ch1='y',i=0,sen_count=0,word=0,sen[],word_array[]
      While ch1=='y' do
       Read a character ch
      if ch=='\t' || ch==' ' then
            word=word+1
      if ch=='.' or ch=='!' or ch=='?' or ch=='\n' then
            sen_count=sen_count+1
       else sen[i]= ch
      end if
      end if
      word_array[i]=ch
      i=i+1
       print 'Want to Continue?'
      read ch1
      End while
      Print sen_count
}
```

53.3.3 Sentence Label Propagation

The total numbers of sentences from a review are labeled in this process. The Senti-WordNet is used in this process to label the separated sentences. If the sentences are separated as sentiment words then they are labeled with "Opinionated" or else labeled with "Non-Opinionated". The type of polarity for every opinion word from review text can be calculated by the SentiWordNet (Word). The description of sentence label propagation function is as follows,

```
Function Label_Propagation ()
Step 1: Seed SentiWordNet
Step 2: If Sentiment Word present
            Label 'Opin'
        Else
            Label 'Non-Opin'
        End if
```

Opinionated labeled sentences come under the "subjective sentences" whereas non-opinionated sentences come under the "objective sentences". The presence of aspect words in a review is determined by the use of aspect word list. The strength of a sentiment can be retrieved, if there is a presence of aspects in that, from the SentiWordNet wordlist. Objective sentences are usually accurate sentences that have no effect by their nature. However, the missing sentiment can be filled for the existed aspects by the use of imputation of missing sentiment function which can also determine the strength of the feeling from the dictionary. Following represents the imputation of missing sentiment function.

```
Function IMS ()
{
If aspect present
Impute the sentiment
Go to step 6
Else
Exit;
}
```

53.3.4 Sentiment Classifier

Each sentence can be split into tokens in which every one of tokens is labeled with polarity that representing a floating point number between 1 and -1.

Positive sentence: The review given by the user toward something with a positive or good response is considered as the Positive sentence. Example for such positive sentence can be given as,

"Most excellent movie forever" or "It was an inspirational movie!!!"

Negative sentence: The review given by the user toward opposing something with a negative response is considered as the negative sentence. Example for such negative sentence can be given as,

"Worst movie ever" or "It is waste of time"

Hence, crisp values are determined at the end of this process for these two positive and negative sentence classifiers.

53.3.5 Fuzzy Logic Interface System

The proposed sentiment analysis method using the fuzzy logic according to the Senti-WordNet is described in this fuzzy logic interface system. As presented before the fuzzy logic system begins with a crisp value and after fuzzify it using different steps (fuzzification and Rules inference). Finally, a crisp value is returned in the output by the defuzzification methods (centroid, Mean/, Max…). The input (crisp value) of the fuzzy logic interface system is of two measures (positivity and negativity) calculated with SentiWordNet, and the output (crisp value) is the class of the review (positive, negative or neutral). In this case, since it is required to classify the review sentences according to two classes (positive, negative), two input variables are defined: the positive and the negative of the review sentences; and one output variable which is the class (sentiment) of the review sentences.

Every variable of fuzzy logic system is known as linguistic variable either it may related to input variables or output variables. Number of values is there in each of these linguistic variables referred as fuzzy sets or linguistic terms which can be obtained. Thus, the positive and negative are the two linguistic variables presented in the inputs of our proposed method and then low, moderate and high are the three linguistic terms for each one of those positive and negative linguistic variables. That means, probably three linguistic terms are taken by each positive and negative variables or it can be also stating that each linguistic variable is related to the three different fuzzy sets. Similarly, a linguistic variable is presented in the output which is the class of the text review and three different linguistic terms can also be taken, which are positive, negative and neutral. Then, defining the crisp values of inputs is the nest step in this fuzzy logic system which were help to start this approach after define the linguistic variables and their linguistic terms in the input and the outputs. For that, these crisp values of inputs which will have a significant part for calculating the positive and negative variables of a review use some different text preprocessing methods and SentiWordNet dictionary.

Table 53.1 Classes of fuzzy sets

Linguistic classes	Range
Very low (VL)	(0; 0; 0.3)
Low (L)	(0; 0.3; 0.5)
Medium (M)	(0.3; 0.5; 0.7)
High (H)	(0.5; 0.7; 0.9)
Very high (VH)	(0.7; 0.9; 1.0)

53.3.6 Fuzzy Sets

Fuzzy sets are capable of having elements with partial degree membership. Polarity and subjectivity crisp values of all considered text reviews are computed and converted to fuzzy sets. Let FP is a fuzzy set denoting polarity or subjectivity of review features $f_1, f_2, f_3, \cdots, f_n$ and $\mu_1, \mu_2, \mu_3, \cdots, \mu_n$ is the membership value of features. By using Zadeh's notation 11, the degree of polarity of a feature can be represented using these membership values as:

$$FP = \left\{ \frac{\mu_1}{f_1} + \frac{\mu_2}{f_2} + \frac{\mu_3}{f_3} + \cdots \frac{\mu_n}{f_n} \right\}$$

Fuzzification

The process of transforming inputs of crisp or real values to fuzzy set values with the help of partial degree membership is known as Fuzzification. The triangular membership functions are used to fuzzify the input variables of separate polarity and subjectivity values regarding the text reviews and producing the degree of polarities as an output for this model. Some of the fuzzy sets classes of both the input and the output variables separation are depicted in Table 53.1.

Defuzzication

The fuzzy rule-based interface system has a final step called defuzzification. If the equivalent fuzzy sets and related functions are given then the crisp value is an output of the defuzzification step. The crisp value is transformed back from the fuzzy values by using the centroid method as a defuzzier.

53.4 Results

The proposed framework can be studied and validated experimentally in this section. The experiment analysis can be carried out for the multipolarity sentiment analysis in terms of accuracy with the implication of fuzzy logic-based systems. Particularly, the main aim of this experiment analysis is comparing the proposed fuzzy logic method with that of other computational methods of machine learning models toward

Table 53.2 Results from polarity classification (positive/negative)

Classes	Count	Degree of polarity
Positive	700	0.821
Negative	700	0.769

the classification of sentiment analysis in terms of their classification accuracy. The polarity dataset of V0.9 regarding the movie reviews on online is used for conducting this experimental analysis. Table 53.2 depicts the number of positive and negative class instances within polarity dataset.

The sentiment classification is carried out first on the objective / subjective and then positive/negative polarities. Results from this classification are considered as the first step of proposed sentiment classification method. The above two classification approaches simply use the selected features that means only 5 features were selected for both the objective/subjective and positive/negative classifications. Then the fuzzy logic is employed at these two classification stages to improve the accuracy of sentiment classification very efficiently. The results of polarity classification are calculated by making a condition of using only the subjectively labeled reviews when reporting their results which distinguish between positive and negative classes. Then, this condition is detached in the case of the final classification approach and the objective and polarity classifications are effectively applied to all the reviews, in spite of whether they are marked as objective or subjective.

Finally, the aspect-based sentiment summarization technique is implemented after the fuzzy logic implementation in the classification process which uses an imputation of missing sentences mechanism to further improve the accuracy rate of sentiment classification in an effective manner. The proposed fuzzy logic is compared with the other two classifier models of machine learning approaches in terms of their sentiment classification accuracy shown in Fig. 53.2.

Fig. 53.2 Accuracy comparision of sentiment classification

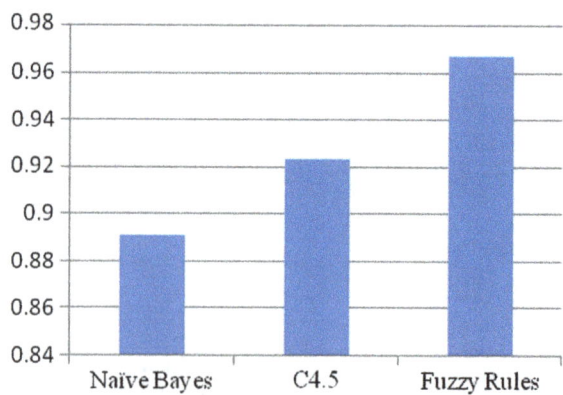

53.5   Conclusion

A new framework is proposed in this paper for efficient multilevel sentiment analysis using fuzzy logic. The fuzzy logic used here classifies the sentiments of the user reviews. Aspect-based sentiment summarization is integrated with fuzzy logic interface system in the proposed framework for the sack of opinion mining and sentiment analysis on the online dataset. The degree of polarity for the sentiments from online reviews is then predicted in this framework. The imputation of missing sentiment mechanism that has an important part to produce accurate results is used to incorporate the non-opinionated sentences. Thus the proposed framework offers a better accuracy for the sentiment analysis on online dataset than other previous frameworks.

Chapter 54
Practical Assessment of Application Layer Protocols for Resource Constraint IoT Applications

Manasi Mishra and S. R. N. Reddy

Abstract Internet of things (IoT) systems are designed to monitor environmental parameters which are accessible either directly or indirectly through applications via Internet as a global platform. IoT devices are often equipped with limited resources requiring energy-efficient and bandwidth-efficient lightweight protocols for data transmission. Traditional protocols are not suitable for communication in IoT-based applications due to high overhead and complex architecture. The selection of effective messaging protocol highly depends on the type of application and its messaging requirements. This paper presents a broad comparison and practical assessment of the most prominent protocols such as HTTP, MQTT, CoAP, AMQP and WebSocket for IoT. A real test bed has been developed for the experimental evaluation of these protocols based on the major key performance metrics such as overhead and latency. The experimental results reveal that CoAP is the most suitable protocol for the applications having constrained resources as it introduces the least overhead. However, WebSocket is the most suitable for the time critical applications as it takes the least time for communication. The relative analysis of these protocols can help the user to choose an appropriate protocol for their application requirements and suitability.

54.1 Introduction

In today's world, the usage of IoT devices capable of sensing and communicating with each other to collect data from environment has increased exponentially. Every bit of data transferred in the network creates a carbon footprint in the environment. Even though the carbon footprint by a single IoT device is negligible, the system having huge number of devices creates a considerable carbon footprint in the environment. With huge number of IoT devices installed and sending data frequently, it is important

M. Mishra (✉) · S. R. N. Reddy
IGDTUW, New Delhi, India

S. R. N. Reddy
e-mail: srnreddy@igdtuw.ac.in

© The Author(s), under exclusive license to Springer Nature Singapore Pte Ltd. 2021   549
S. C. Satapathy et al. (eds.), *Smart Computing Techniques and Applications*,
Smart Innovation, Systems and Technologies 225,
https://doi.org/10.1007/978-981-16-0878-0_54

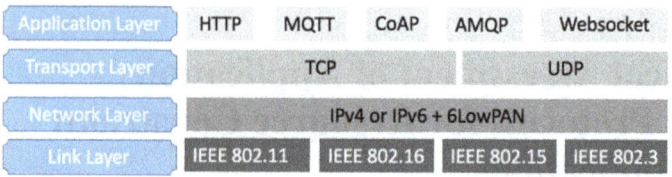

Fig. 54.1 IoT protocol stack

to choose techniques which can reduce the amount of carbon footprint by limiting the amount of data injected into the network without compromising on their functionality. The radio module is the part of IoT device which requires most of the power. Each layer of an IoT protocol stack plays a vital role in managing the communication efficiently. Figure 54.1 provides a generic protocol stack for IoT.

The protocol stack shown in Fig. 54.1, is broadly divided into four functional layers, namely (i) Link, (ii) Network, (iii) Transport and (iv) Application layer responsible for performing different subtask of communication. The characteristics such as transmission data rate, coverage area and scalability required by applications, are offered by physical layer. The IoT uses various communication technologies such as WMAN (IEEE 802.16), WPAN (IEEE 802.15) and WLAN (IEEE 802.11) for interaction among the devices at the physical layer [1]. Network layer is responsible for end-to-end data communication within the scope of the network. The traditional IPv4 addressing technology has limited address blocks which is not sufficient for IoT scenario. IPv6 standard provides a 128-bit address field, making it possible to solve the addressing issues in IoT, but it also introduces overheads which are not suitable for constrained nodes [2]. Transport layer protocol is responsible for flow control, congestion control, sequence delivery and reliability in end-to-end communication between two hosts in the network. TCP (Transmission control protocol) and UDP (User datagram protocol) are the prominent protocols at transport layer of the Internet protocol stack. Application layer is an interface between user applications and the underlying network layer providing services for application programme to ensure effective communication within the network. The selection of an effective application layer protocol is a challenging task for any developer or organization as it is highly dependent on the type of application, data sharing requirements and deployment scenario.

Some protocols have been designed for applications which need reliable and fast business transactions or operations such as AMQP (Advanced Message Queuing Protocol). Some have been designed for applications which need regular data updates in constrained network such as MQTT (Message Queuing Telemetry Transport) and CoAP (Constrained Application Protocol). Some protocols have been designed to support web applications which needs to communicate over the Internet such as RESTful client/server protocols HTTP (HyperText Teleco and CoAP). Few have been designed for the applications which require instant messaging (IM) and online discovery such as SIP and XMPP. This shows any one protocol cannot address every requirement of all possible IoT use cases. Also, it is important to investigate the performance of these protocols in real environment to determine their best-fit scenarios [3].

The remainder of this paper is organized as follows: Section 52.2 presents the theoretical background of the five most prominent application layer protocols for IoT systems: HTTP, MQTT, CoAP, AMQP and WebSocket; Section 52.3 demonstrates an experimental test bed setup developed for practical assessment of these protocols in real environment; Section 52.4 presents the results of the experiment carried based on the key performance metrics and also presents the discussion about their suitability for certain scenario; Section 52.5 concludes the paper and suggests scope for future improvements.

54.2   IoT Application Layer Protocols

54.2.1   HTTP

HTTP is a web-based messaging protocol, standardized by IETF and W3C jointly. HTTP supports RESTful Web architecture based on request/response model. To locate the resources, HTTP uses Universal Resource Identifier (URI). Server sends data through the URI and client receives data through particular URI. HTTP is a text-based protocol and undefined header and message payloads size. Being TCP at transport layer, the communication between HTTP client and server is a connection-oriented. To make the communication secure, HTTP uses TLS/SSL [4]. It has no defined mechanism for QoS provisioning. It relies on TCP layer for packet retransmission in case of loss packets. HTTP is the most famous and accepted web messaging protocol offering several features such as chunked transfer encoding, request pipelining and persistent connections [3]. Figure 54.2 shows the architecture of HTTP protocol.

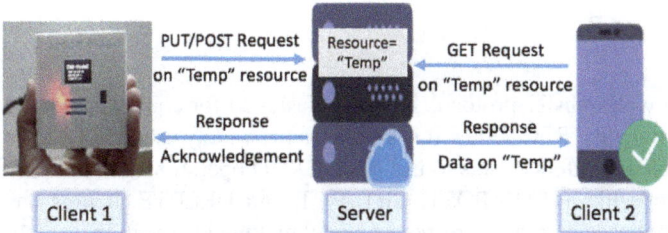

Fig. 54.2 HTTP protocol architecture

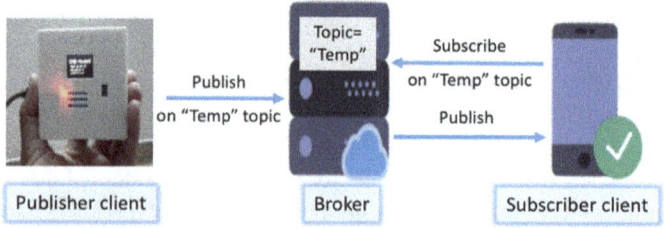

Fig. 54.3 MQTT architecture

54.2.2 MQTT

It is one of the oldest devices to device communication protocol introduced in 1999 by IBM. It is an open-source publish/subscribe model-based lightweight and simple messaging protocol designed for constrained environment. MQTT clients communicate with each other over a topic via a centralized server know as broker. Clients can be a publisher which is interested in sending messages and/or subscriber which is interested in receiving messages. The publisher publishes message to MQTT broker on an address known as topic. Every client which has already subscribed to that topic will receive the message. Having TCP at underlying layer, MQTT communication between broker and client is a connection-oriented. MQTT offers three levels of QoS for reliable delivery of messages. QoS-0, i.e. "At most once", where fire and forget mechanism is applied and message loss can occur, QoS 1, i.e. "at least once" where messages are guaranteed to arrive but redundancy can occur. and QoS 2, i.e. "exactly once" where messages are guaranteed to arrive exactly once [5, 6]. Extension of MQTT offers implementation over UDP known as MQTT-SN to make it more lightweight and suitable for resource constrained devices [7]. Figure 54.3 shows the MQTT communication architecture.

54.2.3 CoAP

CoAP is a web transfer protocol specially designed for constrained networks with constrained nodes [RFC 7252]. It is similar to the widely used HTTP for web-based applications and offers clients to use basic CRUD operations on the resource using four methods like HTTP: POST, GET, PUT, and DELETE. It uses the RESTful architecture based on request/response model making information available through resources identification by the URI [RFC3986]. CoAP builds on top of UDP, which does not provide a retransmission mechanism as provided in case of HTTP being TCP at its underlying layer. CoAP header has two bytes to set the type of message, i.e. confirmable, non-confirmable, acknowledgement and reset. To ensure the reliability through the application layer, CoAP uses exponential retransmission mechanism by setting message type as confirmable [8]. The extension of CoAP has the feature

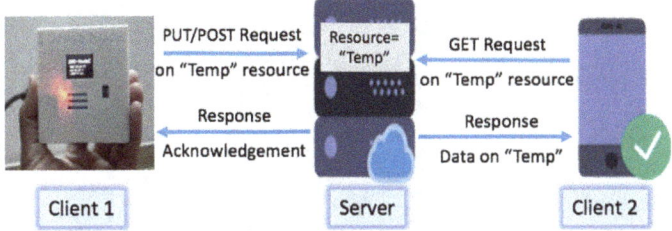

Fig. 54.4 CoAP architecture

of resource/observe to provides the functionality of publish/subscribe model while maintaining the properties of REST [8]. Figure 54.4 show the architecture of CoAP network.

54.2.4 AMQP

AMQP is a lightweight messaging protocol designed for work provisioning, reliability, interoperability and security in enterprise environment. It supports both publish/subscribe as well as request/response model for message transfer. In AMQP-based system, the publisher or consumer creates an "exchange" with a name and then broadcast them to be discovered by others. A consumer creates a "queue" and associates it to exchange immediately. Messages received by the exchange have to be matched to the queue via a process called "binding". Being TCP at the transport layer, broker and client have connection-oriented communication. AMQP is an appropriate protocol for businesses transactions as it offers wide range of services related to messaging such as topic-based publish-and-subscribe messaging, reliable queuing, flexible routing and transactions. It provides various ways to exchange route messages: directly, in fan-out form, by topic, and based on headers. AMQP offers reliability by offering three levels of QoS for delivery of messages: at most once and at least once and exactly once [9]. Figure 54.5 show the architecture of AMQP protocol.

54.2.5 WebSocket

WebSocket is a lightweight protocol offering full-duplex communication over TCP standardized by IETF [RFC 6455] to use it for real-time applications. It supports asynchronous, full-duplex and low latency communication between client and server through a single socket which was not supported in HTTP. The protocol is divided into two parts: handshake part in which client and the server establish initial communication using HTTP and data transfer part can be done using the WebSocket

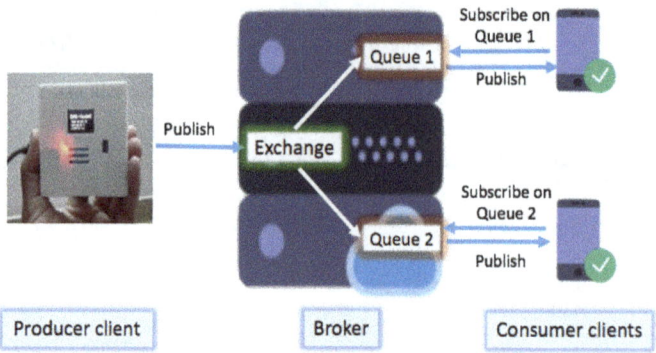

Fig. 54.5 AMQP Architecture

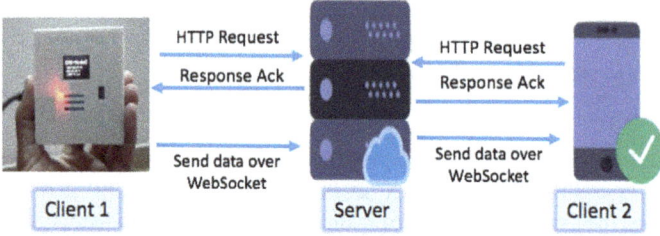

Fig. 54.6 WebSocket architecture

protocol. Messages over WebSocket are divided into one or more frames using custom binary framing format and are sent in frames, introducing only 2 bytes overhead. WebSocket customization capabilities allow compression and multiplexing through its extensions [10]. Figure 54.6 show the architecture of WebSocket protocol.

Based on the literature survey, a comparison between these protocols with respect to significant parameters has been presented in Table 54.1.

There are various broker/server implementations and client libraries available for actual use of these protocol in real-world scenarios. Most of the researchers use simulation techniques to evaluate the performance of protocol but simulation fails to mimic the real-world irregularities and hardware dependencies. Very few researches have focused on performance evaluation of these protocols on real hardware by creating a test bed. Section 52.3 presents the experimental test bed setup to evaluate the performance of these protocols on real hardware.

Table 54.1 Comparison of application layer protocols

Protocol	Standard organization	Header length	Abstraction	Transport layer	Default port	Encoding format	Message Distribution	QoS option
HTTP	IETF	Undefined	Request/response	TCP	80/443 (TLS/SSL)	Text	1: 1	No
MQTT	OASIS	2 bytes (min)	Publish/subscribe	TCP	1883/8883 (TLS/SSL)	Binary	1: Many	Yes
CoAP	IETF	4 bytes (min)	Request/response and Resource/observe	UDP	5683 (UDP)/5684 (DLTS)	Binary	1:1 and 1:Many	Yes
AMQP	ISO/IEC	8 bytes (min)	Publish/subscribe and request/response	TCP	5671 (TLS/SSL), 5672	Binary	1:Many and 1:1	Yes
WebSocket	IETF	Undefined	Request/response	TCP	80/433 (TLS)	Binary	1:1	No

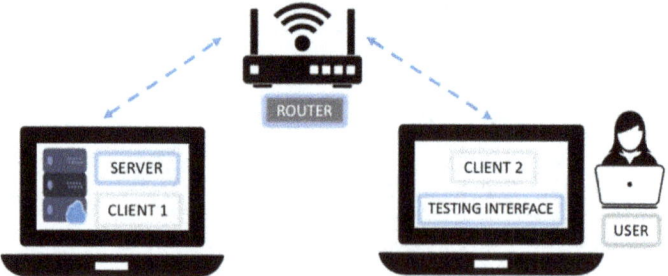

Fig. 54.7 Experimental Test Bed

54.3   Experimental Test Bed Setup

54.3.1   Scenario

The experimental test bed consists of two laptops and a Wi-Fi router as shown in Fig. 54.7. The server/broker of each protocol has been installed on one laptop and other laptop testing interface running each client application.

Client: There are two types of clients in our experimental test bed. Client-1 is acting as a receiver and client-2 is acting as a transmitter. Client 1 and client 2 can communicate via server installed on the laptop 1. The testing interface has the client module of each protocol which will be invoked based on the user choice.

Server/Broker: Server/Broker of each protocol has been installed on laptop 1.

54.3.2   Key Performance Indicator

The factors which are most significant for the choice of any protocol is considered as key performance indicator. The key performance indicators for the test are overhead and latency.

i.   Latency: The difference between the time when message was received by the client 1 and the time when the message was sent by client 2 indicates the time taken for the communication.
ii. Overhead: The additional amount of data transferred for sending any message from sender to receiver is an overhead. Actual data transfer for sending a message from client 2 to client minus the message size is the indicator of overhead introduced by the protocol.

Wireshark tool has been used to capture the flow of packets across the network [11, 12]. Below parameters were recorded through the data captured from Wireshark.

1. Overhead introduced by application layer (bytes)
2. Total overhead including acknowledge and other additional packets (bytes)
3. Round trip time (mS)

54.3.3 Experiment Flow Chart

An integrated application code has been developed which can act as an easy interface for user to carry out the experiment easily just by choosing the parameters one by one through the terminal. Figure 54.8 shows the flow of experiment.

Experiment starts with starting the Wireshark application which will capture the packets flowing through the network followed by starting the test code. User can choose the protocol by which they want to send their message. Based on the protocol selection the next choices will appear on the terminal such as the name of resource

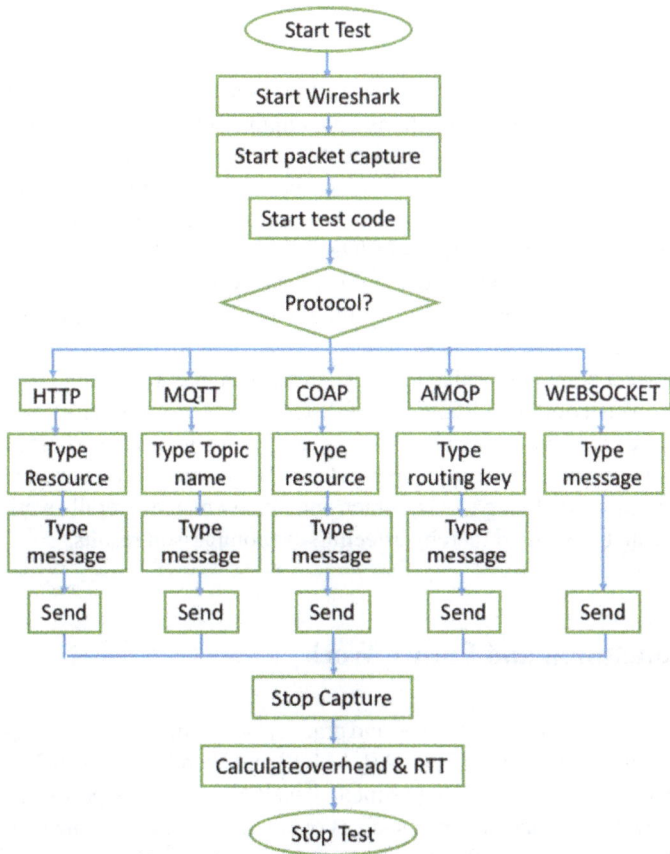

Fig. 54.8 Flow chart of experiment

Table 54.2 Experimental results

Parameter	HTTP	MQTT	COAP	AMQP	WebSocket
Overhead (bytes)	146	75	45	220	64
Total overhead including Ack (bytes)	358	143	83	377	124
Average RTT (ms)	0.91	0.52	1.61	0.43	0.32

in case of HTTP and CoAP, name of topic in case of MQTT and the name of routing key in case of AMQP. After that user will be asked to type the message they want to send. For the testing purpose, we have kept the size of resource/topic/routing key name and message constant in case of each protocol for fair evaluation. After sending the message, Wireshark packet capture can be stopped and the parameter such as overhead and RTT can be calculated.

54.4 Results and Discussion

Although the comparisons of protocols depend on the types of IoT application, resources, devices, specific conditions and requirements of the system, we have summarized the outcome of our experiment in a generic way in Table 54.2.

The results reveal that CoAP is the best protocol in terms of overhead which makes it most suitable for the applications having constrained resources. However, WebSocket is best protocol in terms of latency as its round trip time is least making it most suitable for the time critical applications. AMQP is a lightweight protocol but its support for provisioning, security, interoperability and reliability increases the overall message size and overhead [3, 4].

An important point here is that the real comparison may vary due to various other factors such as network condition, use of different server or broker implementation [12] and may differ from our results. Also, our evaluation is based on static parameters and does not consider the dynamic network conditions and overheads introduced in the retransmission of lost packets, which may also change overall overheads and amount of data transmitted thereby affecting the comparison results.

54.5 Conclusion and Future Work

This paper covers a broad comparison and practical assessment of the most prominent protocols such as HTTP, MQTT, CoAP, AMQP and WebSocket for IoT. A real test bed has been developed for the experimental evaluation of these protocols based on the major key performance metrics such as overhead and latency. The results reveal that CoAP is the best protocol in terms of overhead which makes it most suitable for the applications having constrained resources. However, WebSocket is best protocol in terms of latency as its round trip time is least making it most suitable for the time critical applications. The relative analysis of these protocols can help the user to choose an appropriate protocol for their application requirements and suitability.

In future, we will consider all the available libraries of these protocols for the performance assessment of each protocol. We will improve the existing test bed by using emulator to mimic the actual dynamic network conditions which could help to capture the overheads incur in the retransmission of packets due to losses to get more accurate results. The performance analysis of each protocol will be done with different test cases based on the parameters such as reliability, overhead, power consumption and latency.

Acknowledgements The Authors would like to thank Mr. Rachit Thukral, Director, ETI Labs for his constant support and help where ever required.

References

1. Gazis, V., Gortz, M., Huber, M., Leonardi, A., Mathioudakis, K., Wiesmaier, A., Zeiger, F., Vasilomanolakis, E.: A survey of technologies for the internet of things. In: 2015 IEEE International Wireless Communications and Mobile Computing Conference, pp. 1090–1095 (2015)
2. Liu, C., et al.: IPv6-based architecture of community medical internet of things. IEEE Access **6**, 7897–7910 (2018)
3. Naik, N.: Choice of effective messaging protocols for IoT systems: MQTT, CoAP, AMQP and HTTP. In: Systems Engineering Symposium (ISSE), IEEE International. IEEE (2017)
4. Luzuriaga, J.E., Perez, M., Boronat, P., Cano, J.C., Calafate, C., Manzoni, P.: Testing AMQP protocol on unstable and mobile networks. In: Internet and Distributed Computing Systems, pp. 250–260. Springer, Berlin
5. Light, R.: Mosquitto: server and client implementation of the MQTT protocol. Journal of Open Source Software **2**(13), 265 (2017)
6. [mqtt-v5.0] MQTT Version 5.0. Edited by Andrew Banks, Ed Briggs, Ken Borgendale, and Rahul Gupta. 07 Mar 2019. OASIS Standard. https://docs.oasis-open.org/mqtt/mqtt/v5.0/os/mqtt-v5.0-os.html. Latest version: https://docs.oasis-open.org/mqtt/mqtt/v5.0/mqtt-v5.0.html
7. Roy, D.G., et al.: Application-aware end-to-end delay and message loss estimation in internet of things (IoT)—MQTT-SN protocols. Future Gener. Computer Syst. **89**, 300–316 (2018)
8. Bormann, C., Castellani, A.P., Shelby, Z.: CoAP: an application protocol for billions of tiny internet nodes. IEEE Internet Comput. **16**(2), 62–67 (2012)
9. Vinoski, S.: Advanced message queuing protocol. IEEE Internet Comput. **10**(6) (2006)
10. Kayal, P., Perros, H.: A comparison of IoT application layer protocols through a smart parking implementation. In: 2017 20th Conference on Innovations in Clouds, Internet and Networks (ICIN). IEEE (2017)
11. https://www.wireshark.org/
12. Babovic, Z.B., Protic, J., Milutinovic, V.: Web performance evaluation for internet of things applications. IEEE Access **4**, 6974–6992 (2016)

Chapter 55
Collaborative-Based Movie Recommender System—A Proposed Model

Prajna Paramita Parida⊙**, Mahendra Kumar Gourisaria**⊙**, Manjusha Pandey**⊙**, and Siddharth Swarup Rautaray**⊙

Abstract Recommendation systems are intelligent search systems that predict the preferred information and provide suggestions to users. Major platforms such as Amazon, Netflix, Youtube improves users' experience. Recommender engines operate on background, reading users' behavior, and suggesting items that are most likely to engage with. Approaches like content filtering lead to ambiguous content or item cold-start issues. To tackle the demerits of sparsity of data and cold-start, the paper focuses on a system with a memory-based collaborative approach, employed with a similarity analysis of users' perspectives using Pearson correlation. In addition to this, a systematic study and a proposed model are presented along with the memory-based algorithm. The study shows that the Pearson correlation approach can achieve better research objectives. The user preferences have been collected from various microblogging platforms to analyze the present trends and responses of users. Promising results are produced by the experiments done over the public database.

55.1 Introduction

Many information-based companies utilize recommendation systems on a wider scale such as Twitter, Google, Netflix, LinkedIn, etc. This area of recommender originated in the mid of the 1990s, along with the invention of Tapestry, the first Recommendation System. Movie recommendation plays a major role in the mobile environment. The idea of being valuable came from the Netflix prize contest in 2009 for 1 million dollars to improve the engine, which was the biggest leaps of

P. P. Parida (✉) · M. K. Gourisaria · M. Pandey · S. S. Rautaray
Kalinga Institute of Industrial Technology (Deemed to be University), Bhubaneshwar, Odisha, India
e-mail: 1964003@kiit.ac.in

M. Pandey
e-mail: manjushafcs@kiit.ac.in

S. S. Rautaray
e-mail: siddharthfcs@kiit.ac.in

S. C. Satapathy et al. (eds.), *Smart Computing Techniques and Applications*,
Smart Innovation, Systems and Technologies 225,
https://doi.org/10.1007/978-981-16-0878-0_55

the techniques used. It carries out the aggregation of reviews and users' preferences to help them for better movies and to maximize profit by minimizing risks. This system provides selective information taking prior user habits and history as inputs. However, it requires both timeliness and accuracy. Researchers studied the application of machine learning algorithms as the field of recommendation developed. Since the late 1950s, artificial intelligence got emerged and then machine learning came into existence. An increase in ML algorithms is experienced such as clustering, k-nearest neighbor, Bayes network, etc. The parameters like timeliness and accuracy of the recommender system will be improved [1].

Various information of users is collected from tweets, social media sites, etc. A web-based portal is used for micro-blogging and social networking tasks that provide enormous real-time user preferences [2].

Figure 55.1 is the illustration of two approaches of content-based and collaborative-based. Content-based depends upon users' history including director, actors, description, genres, use tags for analyzing attributes or features. Whereas collaborative studies information or behaviors from other users. The main idea is to leverage the behavior pattern and predict suggestions. Broadly there are two collaborative-based methods: User-to-User and Item-to-Item approaches shown in Fig. 55.2. These methods provide intuition that people might choose items similar to other items already being chosen and also the items that are chosen by other people with similar tastes.

Item-to-Item-based method has been commonly used in areas such as LinkedIn (2014), YouTube (2010), Amazon (2003), etc. Item-to-Item-based system surfaces similar items from key values data. They are more interpretable as the scores of key values data are too small because of the selected threshold. In the user-to-user approach given in Fig. 55.2, identification of users is done based on the most similar interaction profile by evaluating distances between users. This user-centered system finds the similarity of a specific user of interest with others. They consider a similar interaction on the same items or choices for estimating closeness based on a similar

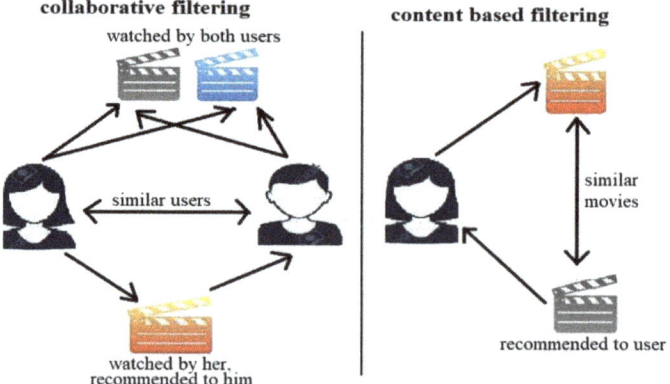

Fig. 55.1 Collaborative and content-based approaches for movie recommendation

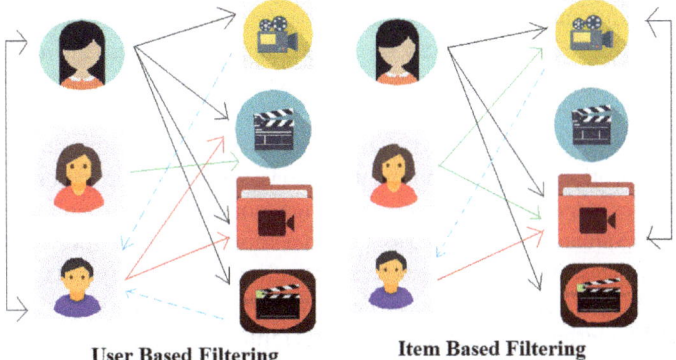

User Based Filtering   **Item Based Filtering**

Fig. 55.2 User-Based and Item Based Collaborative Recommender system

rating or time hovering. For example, the nearest neighbor algorithm refers to those items which are not been interacted with. Whereas the item-centered works for item similarities checking whether most users have interacted with both the items in a similar way. They are less personalized and more biased but more robust compared to user-centered techniques. In this paper, item-to-item collaborative is framed as it provides better outcomes.

The remaining paper is demonstrated as given: Sect. 55.2 describes a survey over the literature part, Sect. 55.3 presents the proposed model, and Sect. 55.4 illustrates the set of selected features. The implementation and algorithm that obtains the working of the proposed model are illustrated in Sect. 55.5 along with the results in Sect. 55.6 and then in the last section, the conclusion is drawn.

55.2   Literature Survey

Different approaches have been proposed for suggesting recommendations. In the field of business applications, Guo et al. [3] suggested a novel recommender system that compares pre-purchase ratings with post-purchase ratings so that preferences can be more accurate. The system procures its instantaneous data from the virtual environment of 3D products. It mainly takes the latent choices of the user. It deploys a new mechanism of collecting pre-purchase emotions from electroencephalogram (EEG) signals that are set on a headset. In this way, these pre and post ratings can boost the overall working of the recommender. Pearson correlation computes the similarities for nearest neighbors and provides both ratings benefiting e-business. Chiru et al. [4] presented a movie recommender for upgrading a personalized type of recommendation. Various factors like user profile, history preferences, and movie scores are analyzed. A psychological test such as a questionnaire is performed and a similar profile brings out similar choices for genres. Four psychological types such as phlegmatic, choleric, melancholic, and sanguine and their intensity has been estimated. Genres and characteristics are categorized according to these traits. So, in this case, the absence of ratings can be easily dealt with, because the association of user profile with movie probabilities could be considered. But, it fails for accurate value due to the traits evaluation.

Jenq-Haur et al. [5] presented a system to estimate the exact rating and more accuracy of classification by using sentiment lexicons. Opinion aggregation and regression techniques are employed as mostly review contains more opinion rather than ratings. This evaluates review polarity to find a compound score for recommending movie preferences. Most people share informal comments or words as opinions. Frangidis et al. [6] comparatively studied reviewers and movie scripts using Gini-index or SVM approach for increased accurate ratings. The correlation was reflected in the predictions which were carried by sentiment analysis using vector semantic and meta-features. Tools were applied like VADER (Valence Award dictionary) for sentimental analysis and NRC for emotion analysis. A mixture of machine learning algorithms is used such as Multinomial Naive Bayes (MNB), Logistic regression, SVM, Multilayer perceptron. Among these, Multinomial Naive Bayes provided promising results. Deldjoo et al. [7] studied a combination of different features of meta-data and audio–video resources mainly to tackle cold start issues. An extraction from videos constitutes a movie genome is done following the canonical correlation method.

This model focuses on items that are combined with interaction taking item content descriptors and making it quite bias-free. Thus works for collaborative enriched content with high coverage as a multimodal recommender system. Soleymani et al. [8] analyzed the movie scenes and its ranking, provided by the emotion of users and features of videos. In this system, physiological signals such as galvanic skin resistance, respiration pattern, electromyograms, body temperature, blood pressure were recorded and stored. The physiological signals were applied to categorize and rank the contents of the video to obtain optimal performance with the best features. Wang et al. [9] proposed a recommendation model that works on social content. This system evaluates the user-content matrix and finds out the relevance between video and users for re-sharing suggestions. Based on the user-video matrix, the social content area is constructed to verify relevance. Following the content delivery network, the dynamics of video sharing are high and access patterns are analyzed for matrix modification. Nidhi et al. [10] experimented on real-time data like tweets for generating a Naive Bayes based classifier. The system is applied on a small train set, providing better accuracy. To improve individual techniques, hybrid systems are composed of combinations of multiple recommendation approaches.

Melville et al. [11] developed a new hybrid approach named content-boosted CF, that utilized content-based features in a collaborative model. They overcame the first rater and sparsity problem by building a pseudo matrix of ratings. Then, the model is compared with pure content, collaborative, and naive hybrid systems. Yang et al. [12] referred ratings from page count which are collected from users' data. The measure of likeness of documents by users is demonstrated by the count of read ones. This concept focuses on the cold start issue in CF and utilizes the mean AP correlation method over log information. Zhang et al. [13] demonstrated a framework by the interaction of user recommender which takes user input, suggests a particular no of items to the user, records choices of the user, and uses a random and kNN algorithm. In this paper, a movie recommendation framework has been suggested by associating the collaborative method with users' behavior eliminating content-based issues.

55.3   Proposed Model

The generalized data used in this model has been taken from two sources: the movie database (TMDB) and MovieLens, which fetches demographic information for users. The dataset from MovieLens is publicly available containing 26 million ratings for 45,000 movies given by 270,000 users. A small dataset is also available containing 100,000 ratings for 9000 movies given by 700 users. One file with TMDB Id of every movie is listed in the MovieLens dataset. After scraping of data, the stringi-fied features are converted into a suitable form for parsing. Figure 55.3 is obtained showing the frequencies of ratings and how it depreciates with the increasing number of users. This presents the data sparsity phenomenon, also said to be a long plot distribution as shown in below figure.

There are few ratings by a small no of users and the majority rest are much less known with little or no user interactions. This leads to sparse ratings which pose problems and become less predictable for most users. This scarcity works opposite to abundance state. It is observed that the curve behavior has a stiff fall from the 'head' to the longtail. So, filtering is done multiple times to remove unnecessary attributes and to reduce the dimension of the dataset and sparsity as well.

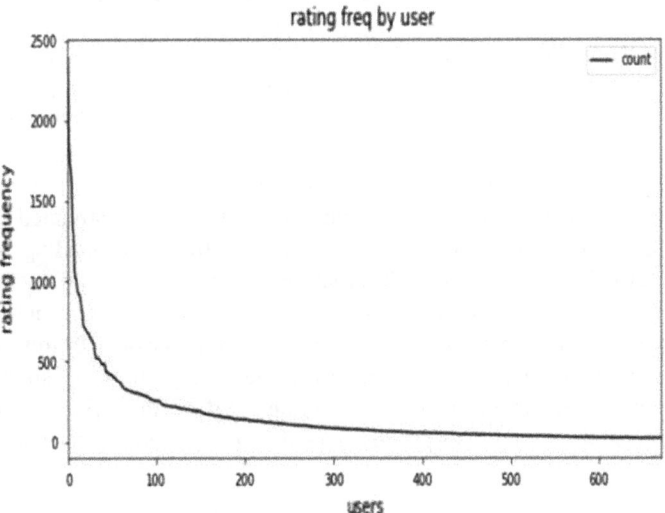

Fig. 55.3 Ratings frequencies by users

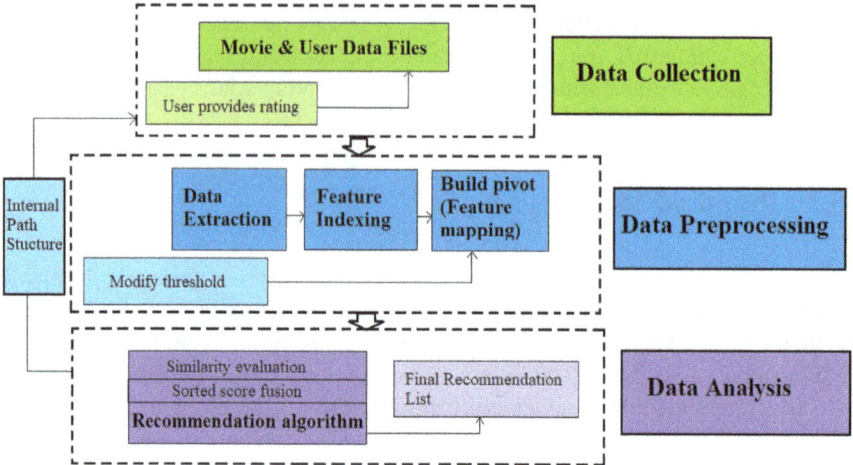

Fig. 55.4 Workflow diagram for movie recommendation system

To improve personalized recommendations, this model of collaborative approach has been proposed considering user's perspectives. Figure 55.4 depicts the procedure of movie recommendations in a stepwise manner involving data collection, data preprocessing, and data analysis.

The data collection section depicts the retrieval of data from the databases having user-rated movie data and users' profile data. These are the dataset that has to be procured and saved for updating. Users keep on adding their ratings or reviews which are also updated. Data preprocessing includes the feature analysis by filtering and indexing to get pivot data. It filters noises and unnecessary attributes by removing hashtags, repetitive words, emojis, replacing NaN values with zeroes, etc. Normalization provides matrices to estimate similarity for score calculation by maintaining threshold. The final task is data analysis that calculates scores from Pearson correlation and ranking is done. Pearson is an advanced form of cosine similarity calculation that automatically standardize the ratings data. As there are large variations of users' ratings with too much harshness or leniency. Then the similar estimated ratings are analyzed according to the user's request to finalize the corresponding suggestion list. There also lies an internal path structure that connects the first section with the last section to update the ratings data. This calculates the compound score and evaluation of similarity between relevant movies are processed. In this procedure, a collaborative-based approach is applied to produce a preliminary suggested list. Recently, for an exponential increase in the online data, this system works effectively for decision-making. This proposed method brings convenience and suitability compared to content-based.

Table 55.1 Feature set for rating estimation

Features	Description
Id	Movie ID by TMDb
rating	Ratings given by specific user
userId	user's identity
movieId	Movie id for TMDb
tagline	Specific front detail of the movie
title	Official name of the movie
genres	category of movie
timestamp	Duration when ratings are given
vote_average	Average reviews of movie
vote_count	No. of votes given by users
imdbId, tmdbId	Movie id given on respective platform

55.4 Feature Selection

As the data are being collected, it requires integration, cleaning, transformation, segmentation, stemming, and removal of noise for the pre-processing phase [14]. Thus to make the data in a form that it would be easily parsed through the machine and to obtain accuracy, data are transformed into clean data. The processed data discover the pattern and becomes understandable to interpret. The following features being listed in Table 55.1, are mainly taken into consideration for extraction purposes in the recommendation model. Many attributes or features have to be filtered from the main dataset.

55.5 Implementation

To find users' preferences, the collaborative approach is considered combining the content features of metadata with the users' interactions. This calculates the similarity function of two items at a time in relevance to the target user and then estimates the user-item matrix by Pearson-correlation. The Pearson correlation method gets the measure of similarity of one entity with other nearby entities. Neighboring vectorized estimations are obtained and these scores are ranked by mapping functionalities. Similar interests of different users are considered. Thus, taking movie-user similarity parameters, feature vectors are accessed, and based on rating, a final combined rating matrix is constructed. We intend to present the below Pearson-based algorithm in systematic procedure. This could be further extended for different comparisons with other approaches by accuracy measures on various data and correlation coefficients.

Pearson-Based Algorithm for the Proposed Recommender System

Inputs:

Dataset containing CSV files

Metadata of movie rating

Normalized similarity score.

Mains:

1. Build a pivot table using the ratings matrix.
2. Set threshold to reduce dimensions of sparse data.
3. Calculate Pearson similarity correlation to find angular distance between items.

Mathematically, given as:

$$r = \frac{N \sum xy - (\sum x)(\sum y)}{\sqrt{\left[N \sum x^2 - (\sum x)^2\right]\left[N \sum y^2 - (\sum y)^2\right]}} \tag{55.1}$$

where

N no. of pairs of scores

x, y scores or entities like items.

1. Standardize ratings with the similarity scores to estimate average weighted rating.
2. Sort and rank the mapped matrix to give a suggestion list.

Output:

Preliminary recommendation list

Predicted ratings.

55.6 Simulation Results

The ratings data are filtered from the metadata and the equivalent lists of users who have contributed their reviews are recorded. As the data is sparse, the review count has to be wrangled with precision to get a pivot table. Then similarity vectors are retrieved to get mapped titles by Pearson similarity index. Following dynamic ratings, the scaling is done and most similar high-rated choices are listed at the top. Figure 55.5 gives the suggestion list as per the varied choice of low to high average weighted ratings. Here, the movie request (say 'Iron Man') at the top is provided with a high rating of 5, which in return gets the corresponding resultant list in descending order of estimated ratings. Similarly, to further cross-check, when the same movie

Fig. 55.5
Recommendations as per the
user choice of a movie
(provided high rate)

Iron Man	2.500000
The Dark Knight	1.734659
Star Trek	1.520704
Avatar	1.481241
Batman Begins	1.454010
The Avengers	1.431644
300	1.349980
WALL·E	1.336769

request is provided with a low rating of 1, then the resultant list gives the opposite of the old list, as shown in Fig. 55.6. Thus, it provides an increasing ordered list of ratings by putting the movie with a rating as 1 at the lowest level.

So, to evaluate the correctness of the recommender system, accuracy measures like precision and recall are applied and values are obtained, shown in the Table 55.2. These estimate the effciency of the system providing correct or incorrect decisions. Mathematically given by Eqs. 55.2 and 55.3:

$$\text{Precision} = \frac{\text{Relevant recommended items}}{\text{All recommended items}} \tag{55.2}$$

$$\text{Recall} = \frac{\text{Relevant recommended items}}{\text{All recommended items}} \tag{55.3}$$

Fig. 55.6
Recommendations as per the
user choice of a movie
(provided low rate)

WALL·E	-0.267354
300	-0.269996
The Avengers	-0.286329
Batman Begins	-0.290802
Avatar	-0.296248
Star Trek	-0.304141
The Dark Knight	-0.346932
Iron Man	-0.500000

Table 55.2 Empirical analysis by precision and recall metrics

Splits	Precision	Recall
Fold-1	0.59563	0.21788
Fold-2	0.62539	0.23241
Fold-3	0.59660	0.22530
Fold-4	0.59955	0.22778
Fold-5	0.61686	0.23555

55.7   Conclusion and Future Work

Thus, the movie recommendation framework being proposed searches preferred movies from real-time responses. The objective is to analyze the development of the recommender system that employs a machine learning algorithm and assistance to data mining areas. The main purpose is to provide a better knowledge management system and to improve the accuracy and timeliness compared to previous models. It filters unusable data by delivering personalized ideas. So, useful movie suggestions will be available to the users readily. Recommendation procedure is a comprehensive work that involves distinct users along with movies, that are collected from reviews on social media platforms. Types of correlation coefficients could be found out by considering sentiment ratings and IMDb movie ratings. This proposed model captures the change in interests of the users and works best for two unrelated factors. It will illustrate the efficiency and precision of enormous real-time data [15]. Also, it is beneficial as it takes variations of sentiments as positive, negative, and neutral sentiments by rating measures. The proposed framework could be improved with more datasets and more recent movies and we expect more extraction of information from different social media platforms to improve this system. However, the study of requirements, design, and maintenance suggests more research opportunities to be investigated and could be more explored to work for a dynamic environment.

References

1. Florea, A.C., Anvik, J., Andonie, R.: Spark-based cluster implementation of a bug report assignment recommender system. In: International Conference on Artificial Intelligence and Soft Computing (ICAISC), LNCS, vol. 10246 (2017)
2. Abel, F., Gao, Q., Houben, G.J., Tao, K.: Analyzing temporal dynamics in Twitter profiles for personalized recommendations in the social web. In: WebSci '11: Proceedings of 3rd International Conference on Web Science, Koblenz, Germany, Koblenz, Germany (2011)
3. Guo, G., Elgendi, M.: A new recommender system for 3D e-commerce: an EEG based approach. J. Adv. Manage. Sci. **1**(1), 61–65 (2013)
4. Chiru, C.G., Dinu, V.N., Preda, C., Macri, M.: Movie recommender system using the user's psychological profile. In: IEEE International Conference on ICCP (2015)
5. Wabg, J.H., Liu, T.W.: Improving sentiment rating of movie review comments for recommendation. In: IEEE International Conference on Consumer Electronics Taiwan (ICCE- TW) (2017)
6. Frangidis, P., Georgiou, K., Papadopoulos, S.: Sentiment analysis on movie scripts and reviews. In: IFIP International Conference on Artificial Intelligence Application and Innovations (AIAI 2020), pp. 430–438 (2020)
7. Deldjoo, Y., Dacrema, M.F., Constantin, M.G.: Movie genome: alleviating new item cold start in movie recommendation. User Model. User-Adapted Interact. **29**(2), 291–343 (2019)
8. Soleymani, M., Chanel, G., Kierkels, J.J., Pun, T.: Affective ranking of movie scenes using physiological signals and content analysis. In: Second Workshop on Multimedia Semantics, pp. 32–39 (2008)
9. Wang, Z., Sun, L., Zhu, W., Yang, S., Li, H., Wu, D.: Joint social and content recommendation for user-generated videos in Online social network. IEEE Trans. Multimedia **15**(3), 698–709 (2013)

10. Nidhi, R.H., Annappa. B.: Twitter-User recommender system using tweets: a content-based approach. In: International Conference on Computational Intelligence in Data Science (ICCIDS) (2017)
11. Melville, P., Mooney, R.J., Nagarajan, R.: Content-boosted collaborative filtering for improved recommendations. **23**, 187–192 (2002)
12. Wei, B., Wu, J., Yang, C., Zhang, Y., Zhang, L.: Cares: a ranking-oriented cadal recommender system. In: *Ninth Joint Conference on Digital Libraries*, pp. 203–212. ACM (2009)
13. Zhang, H.R., Min, F., He, X., Xu, Y.Y.: A hybrid recommender system based on user-recommender interaction. Math. Probl. Eng. (2015)
14. Katarya, R., Verma, O.P.: An effective collaborative movie recommender system with cuckoo search. Egypt. Inform. J. **18**(2), 105–112 (2017)
15. Rautaray, S.S., Agrawal, A.: A real-time hand tracking system for interactive applications. Int. J. Comput. Appl. **18**(6), 28–33 (2011)

Chapter 56
Classification of Bruised Apple Using Ultrasound Technology and SVM Classifier

Gopinath Bej, Tamal Dey, Abhra Pal, Sabyasachi Majumdar, Rishin Banerjee, Devdulal Ghosh, Vamshi Krishna Palakurthi, Amitava Akuli, and Nabarun Bhattacharyya

Abstract Bruising on apples is mainly caused due to mechanical damage during harvesting and postharvest journey to the supermarket. Apple's bruise damage reduces the quality of apple as well as market price. This paper proposes an ultrasonic signal analysis technique to classify bruised and unbruised apples. An ultrasound signal with 100 kHz center frequency has been applied on the apple surface and the response signal has been recorded in reflectance mode. Principal Component Analysis (PCA) has been applied to the acquired response signals to reduce the dimensionality of the dataset. The first fifteen PCA components that contain more than 99% information have been used as the working dataset for classification. Then, the Support Vector Machine (SVM) classifier has been applied to this reduced dataset to classify bruised apple. SVM is trained with the training dataset that consists of data

G. Bej (✉) · T. Dey · A. Pal · S. Majumdar · R. Banerjee · D. Ghosh · V. K. Palakurthi · A. Akuli ·
N. Bhattacharyya
Centre for Development of Advanced Computing, (C-DAC), Plot E-2/1, Block GP, Sector V,
Bidhannagar, Kolkata 700091, India
e-mail: gopinath.bej@cdac.in

T. Dey
e-mail: tamal.dey@cdac.in

A. Pal
e-mail: abhra.pal@cdac.in

S. Majumdar
e-mail: sabyasachi.majumdar@cdac.in

D. Ghosh
e-mail: devdulal.ghosh@cdac.in

V. K. Palakurthi
e-mail: vamshi.palakurthi@cdac.in

A. Akuli
e-mail: amitava.akuli@cdac.in

N. Bhattacharyya
e-mail: nabarun.bhattacharya@cdac.in

© The Author(s), under exclusive license to Springer Nature Singapore Pte Ltd. 2021   573
S. C. Satapathy et al. (eds.), *Smart Computing Techniques and Applications*,
Smart Innovation, Systems and Technologies 225,
https://doi.org/10.1007/978-981-16-0878-0_56

from both bruised and unbruised apples. The model developed after SVM training, is used to classify the test dataset. The classification result shows promising accuracy in the tune of 98%. This ultrasonic technique can endow with a new paradigm for automatic non-destructive identification of bruise-damaged apples.

56.1 Introduction

Apple is a delicious fruit rich in vitamins, minerals, and antioxidants. It is mostly consumed fresh by consumers. Fresh fruits are susceptible to bruising which is caused due to the mechanical damage of fruit flesh. The journey of an apple from the orchard to the retail shelves involves several intermediate stages like harvesting, sorting, packing, storing, and shipping [1–3]. During those intermediate stages of the journey to the supermarket, apple countenance impacts, and compression several times that cause bruise damage on the apple. Impact bruising occurs when an apple falls freely onto a hard surface and compression bruising occurs when apples are pressurized with force [4, 5]. Bruise-damaged fruits are susceptible to the blue mold decay and invisible bruise inside the skin can be susceptible to fungus infection. Consequently, a major quantity of total produced apple is wasted due to this bruise damage and imparts a substantial economic loss to the industry [5–7]. The bruising defect is a major concern to the industry and it is considered as a major quality defect. These reduced quality apples are not well accepted among the consumers and unbruised apples command the best price due to better quality [8, 9].

The ultrasonic technique is a trend to evaluate the quality of agro produces in a non-destructive manner [10, 11]. In this paper, a novel ultrasound-based methodology has been proposed for the classification of bruised apple. The ultrasound response signal is recorded in reflectance mode from the apple surface by applying an ultrasound signal of 100 kHz frequency. For every sample under the experiment, an ultrasound response signal has been acquired with the same device and the same setup. This acquired response signal from an apple surface is a one-dimensional array of length 2000. Instead of extracting features from the response signal, all 2000 data points are used as a feature vector. Principle component analysis (PCA) has been applied to the acquired signal to visualize the effectiveness to form the clusters with bruise-damaged and unbruised apples. Also, PCA reduces the dimensionality of the dataset, by considering the first few principal components which together explain most of the information. The reduced dataset has been used as the feature set for the development of the SVM classifier. After the supervised training of the SVM, the developed model has been tested against a known test dataset. The classification result shows remarkable accuracy. The ultrasonic analysis technique is a rapid and reliable technique to distinguish between bruised and unbruised apples.

| Unbruised Apple | Unbruised Apple | Bruised Apple | Bruised Apple |

Fig. 56.1 Few of the collected samples

56.2 Methodology

56.2.1 Sample Collection

For this experiment, a total of 154 numbers of apples were collected from the local market which contains both unbruised and bruised apples. Samples were marked with numbers from 1 to 154 for further traceability. During the experiment, the ultrasonic response signal of 154 apples has been acquired sequentially. Figure 56.1 is showing the images of a few unbruised and bruised apples used during experimentation.

56.2.2 Ultrasonic Setup

The experimental setup comprises three components, i.e., an ultrasound pulser/receiver (Model: USB-UT350, Make: US Ultratek), an ultrasound transceiver probe (Model: X1020, Make: Olympus Scientific Solutions Americas Corp.), and a computer. This ultrasound pulser/receiver is commercially available and it can generate a wide range of ultrasound signals by varying the configurable parameters. It can also receive the ultrasound signal sensed in the transducer probes. The ultrasound probe used in this experiment has a central frequency of 100 kHz which practically fires the ultrasound. It can also sense the ultrasound response signal when acts as a receiver. The ultrasound pulser/receiver is connected to the computer using the USB interface and the ultrasound probe is connected to the ultrasound pulser/receiver using a BNC cable. Figure 56.2 embodies the laboratory setup with the ultrasound

Fig. 56.2 Laboratory setup

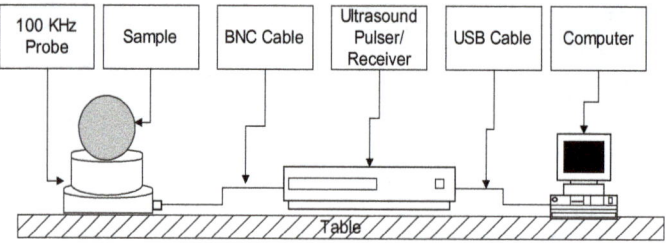

Fig. 56.3 Block diagram of setup

probe, pulser/receiver, and an apple sample placed on the probe. Figure 56.3 shows the graphical representation of the block diagram of the ultrasound setup. This same setup has been used for acquiring the data from all 154 samples.

56.2.3 Signal Acquisition

First, the sample under consideration is placed above the probe and the ultrasonic gel is used as a coupling medium to remove any air gap between the apple surface and sensor probe. The gel reduces the attenuation of ultrasonic sound waves. One software has been developed using MS VC++© to interface with the ultrasound pulser/receiver. The software is configurable to adjust the parameters for the ultrasound pulser/receiver such as buffer size, sampling rate, low pass filter, high pass filter, rectifier, transducer mode (transmittance/reflectance), center frequency, pulse-width, and pulse-voltage. as per the requirement. In this experiment, 100 kHz tone burst frequency is triggered via the probe in reflectance mode. The probe acts as a transceiver which also senses the response signal and it is acquired in the software. The ultrasound signal is subjected to the selected location on the apple surface. A recorded ultrasound response signal during experimentation is shown in Fig. 56.4.

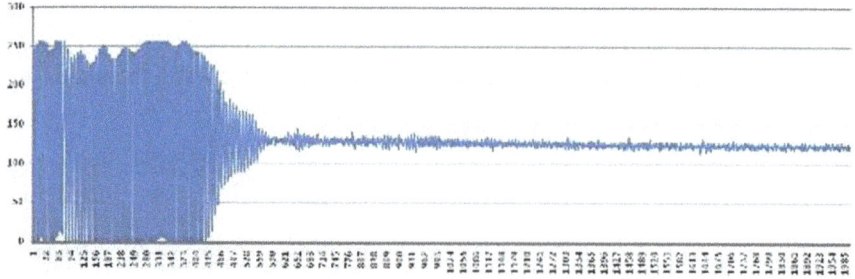

Fig. 56.4 Acquired ultrasound signal

56.2.4 Manual Analysis

After the experiment with ultrasound, each sample is cut with a knife and examined manually for bruise damages. During manual analysis, among 154 apples, 93 are classified as unbruised, whereas, 61 apples are considered as bruise-damaged apples. During the development of the classification model, the Class-I apples are considered as unbruised apples, while, bruised apples are marked as Class-II.

56.3 Data Analysis

The acquired response signal is a one-dimensional array of length 2000. For 154 samples, the acquired dataset is a matrix of dimension 154×2000. Instead of mining the signal specific features by applying digital signal processing and analysis technology, all 2000 data points of a response signal are considered as 2000 features of the apple. With this hypothesis of 2000 feature vectors, data analysis has been performed on the 154×2000 dataset for the classification of bruised apples. Initially, Principle Component Analysis (PCA) has been applied on the 154×2000 dataset to find out the pattern of the data points. Graphical representation of PCA components in a 2-D or 3-D plot helps to visualize the data clustering pattern. The clustering nature of data points in these plots edifies to decide on applying a classifier algorithm. Also, the dimension of the dataset is reduced based on the scores of principal components. This reduced dataset is used for the development of the classifier model. Finally, the Support Vector Machine (SVM) classifier has been used as the decision support model for the classification of bruised apples. The result represented in the Results and Discussion section proves the practical applicability of the hypothesis. A brief depiction of PCA and SVM is described below.

Principle Component Analysis (PCA) [12–15]: It is a well-established dimensionality reduction technique, used to identify correlations and patterns in a dataset. It reduces the dimensionality of a dataset to a significantly lower dimension minimizing the loss of information. Applying the underlying mathematics, PCA creates new uncorrelated variables that successively maximize variance. Each uncorrelated variables are principal components where first principal component is the direction in space along which projections have the largest variance. The steps involved in transforming the initial dataset to a reduced dataset using PCA are described below:

i.   Normalize the whole dataset
ii. Compute the covariance matrix of the normalized dataset
iii. Compute the eigenvectors and eigenvalues of the covariance matrix
iv. Sort the eigenvalues in descending order and accordingly order the eigenvectors.
v.   The eigenvector with the highest eigenvalue is the most significant and thus forms the first principal component.

vi.   Reducing the dimensions of the dataset by selecting the first few principal
      components which explain the dataset with a significant score.

Support Vector Machine (SVM) [16–19]: It is a supervised machine learning algorithm used for classification problems and regression analysis. It determines the
optimal hyperplane between classes which separates the data points into classes.
Data points from each class that are closer to the hyperplane are the support vectors.
Using these support vectors, the margin of the classifier is maximized to find the
optimal hyperplane. Removing any support vector data point from the dataset influences the position of the hyperplane. During the training of the SVM algorithm, the
optimal hyperplane is determined which is the decision boundary for classes. While
testing the new data point, the class of the data is determined using the optimal
hyperplane determined during the training of SVM.

56.4   Results and Discussion

Principle Component Analysis (PCA) has been applied to visualize the data pattern
in 2-D (using the first two principal components, explains: 91.7%) and 3-D (using the
first three principal components, explains: 96.2%) plots. The generated plots using
MATLAB are shown in Figs. 56.5 and 56.6.

  Both plots, 2-D, and 3-D represent the formation of two distinguishable clusters.
Moreover, the clusters are linearly classifiable in 2-D as well as 3-D planes. The
explained score with the first fifteen principal components is more than 99%. These
components are considered as the reduced dataset in an uncorrelated vector space.
Thus, PCA helps to reduce the dimension of the whole dataset from the 154 × 2000
to 154 × 15. Table 56.1 describes the PCA component explained scores with the first
15 PCs.

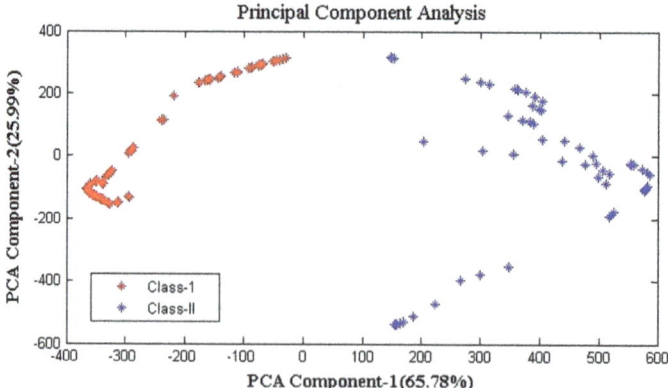

Fig. 56.5 2-D PCA plot with the first two components

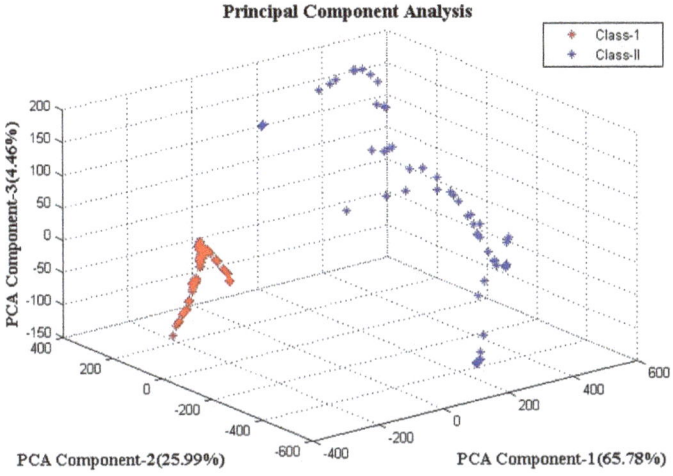

Fig. 56.6 3-D PCA plot with the first three components

The support vector machine (SVM) classifier algorithm is used with the reduced dataset for distinguishing the bruised apples. As SVM is a supervised learning classifier algorithm, it is required to be trained before the prediction of unknown samples. First, the dataset is divided for training and testing of the SVM as explained in Table 56.2. The training dataset ensembles 108 samples (70%), whereas, the rest 46 samples (30%) are used during testing of the model. Out of 108 training data samples, 65 data samples are selected randomly and belong to Class-I. Rest 43 data samples belong to Class-II. During testing, out of 46 samples, 28 samples are selected randomly from Class-I, and the rest, 18 samples are considered as Class-II. The SVM model is developed using the MATLAB® software tool.

The selection of training and testing samples is shuffled randomly for 10 times to obtain the average accuracy. The accuracy of testing using the SVM classifier with ten randomly selected datasets is shown in Fig. 56.7. The performance statistics related to the testing accuracy is shown in Table 56.3. The average accuracy to classify the bruised apples using the SVM classifier is calculated as 98.71%.

56.5   Conclusion and Future Work

The use of an SVM classifier for the classification of bruise-damaged apples using ultrasound technology shows a promising result. Moreover, the process of determining the bruised apples using ultrasound technology is non-destructive. This experiment shows a pathway for determining the intrinsic qualities of apple further in detail. In the future, data from more samples will be analyzed using this technology. Also, lower ultrasound frequency, i.e., 40, 60, 80 kHz will be explored for this study which may further be used to explore more intrinsic quality parameters of apple in

Table 56.1 PCA component explained scores (first fifteen PCs)

1	2	3	4	5	6	7	8	9	10	11	12	13	14	15
65.781	25.987	4.460	1.086	0.604	0.252	0.187	0.125	0.111	0.094	0.082	0.075	0.065	0.063	0.055

Table 56.2 Dataset distribution

	Available dataset	Training dataset	Testing dataset
Class-I (unbruised)	93	65	28
Class-II (bruised)	61	43	18
Total	154	108	46

Fig. 56.7 Result of SVM classifier using random training and test dataset (repeated 10 times)

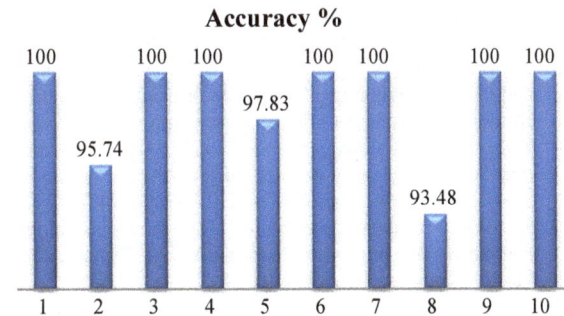

Table 56.3 Statistics of testing accuracy

S. No.	Parameter	Accuracy %
1	Minimum	93.48
2	Maximum	100.00
3	Average	98.71
4	Standard deviation	2.20
5	Median	100.00
6	Mode	100.00

a non-destructive manner. Further, this technology can be equipped with a conveyor for the online sorting of bruised apples.

Acknowledgements The authors are grateful to ICAR, New Delhi for funding this project and providing valuable guidance toward the development of non-destructive methodologies.

References

1. Hussein, Z., Fawole, O.A., Opara, U.L.: Harvest and postharvest factors affecting bruise damage of fresh fruits. Horticultural Plant J. **6**(1):1–13., doi: https://doi.org/https://doi.org/10.1016/j.hpj.2019.07.006

2. Fu, H., Karkee, M., He, L., Duan, J., Li, J., Zhang, Q.: Bruise patterns of fresh market apples caused by fruit-to-fruit impact. Agronomy **10**, 59. https://doi.org/10.3390/agronomy10010059
3. Khan, A.A., Vincent, J.F.V.: Bruising and spliting of apple fruit under ini-axial compression and the role of skin in preventing damage. J. Texture Stud. **22**(3) (1991). https://doi.org/10.1111/j.1745-4603.1991.tb00018.x
4. Mitsuhashi-Gonzalez, K., Pitts, M.J., Fellman, J.K., Curry, E.A., Clary, C.D.: Bruising profile of fresh apples associated with tissue type and structure. Appl. Eng. Agric. **26**(3), 509–517 (2010). ISSN 0883-8542
5. Lewis, R., Yoxall, A., Marshall, M.B., Canty, L.A.: Characterising pressure and bruising in apple fruit. Wear **264**(1–2), 37–46 (2007). https://doi.org/10.1016/j.wear.2007.01.038
6. Altisent, M.R.: Damage Mechanisms in the Handling of Fruits
7. Rostampour, V., ModaresMotlagh, A.: Evaluation of the bruising susceptibility of apple in transport conditions. Bulg. J. Agric. Sci. **24**(5), 902–908 (2018)
8. Dobrzanski Jr., B., Rabcewicz, J., Rybczynski, R.: Handling of apple, transport techniques and efficiency vibration, damage and bruising texture, firmness and quality. In: B. Dobrzański Institute of Agrophysics of Polish Academy of Sciences. ISBN 83-89969-55-6
9. Komarnicki, P., Stopa, R., Szyjewicz, D., Kuta, Ł., Klimza, T.: Influence of contact surface type on the mechanical damages of apples under impact loads. Food Bioprocess. Technol. **10**, 1479–1494 (2017). https://doi.org/10.1007/s11947-017-1918-z
10. Mizrach, A.: Ultrasonic technology for quality evaluation of fresh fruit and vegetables in pre- and postharvest processes. Postharvest Biol. Technol. **48**(3), 315–330 (2008). https://doi.org/10.1016/j.postharvbio.2007.10.018
11. Vasighi-Shojae, H., Gholami-Parashkouhi, M., Mohammadzamani, D., Soheili, A.: Ultrasonic based determination of apple quality as a nondestructive technology. Sens. Bio-Sensing Res. **21**, pp 22–26 (2018). https://doi.org/10.1016/j.sbsr.2018.09.002
12. Jolliffe, I.T., Jorge, C.: Principal component analysis: a review and recent developments. Phil. Trans. R. Soc. A.37420150202 (2016), https://doi.org/10.1098/rsta.2015.0202
13. Wold, S., Esbensen, K., Geladi, P.: Principal component analysis. Chemometrics Intell. Lab. Syst. **2**(1–3), 37–52 (1987). https://doi.org/10.1016/0169-7439(87)80084-9
14. Joliffe, I.T., Morgan, B.J.T.: Principal component analysis and exploratory factor analysis. Statistical Methods Med. Res. **1**(1), 69–95 (1992). https://doi.org/10.1177/096228029200100105
15. Vlachos, M.: Dimensionality reduction. In: Sammut, C., Webb, G.I. (eds.) Encyclopedia of Machine Learning. Springer, Boston, (2011). https://doi.org/10.1007/978-0-387-30164-8_216
16. Yue, S., Li, P., Hao, P.: SVM classification: its contents and challenges. Appl. Math. Chin. Univ. **18**, 332–342 (2003). https://doi.org/10.1007/s11766-003-0059-5
17. Shmilovici, A.: Support vector machines. In: Maimon, O., Rokach, L. (eds.) Data Mining and Knowledge Discovery Handbook. Springer, Boston, (2005). https://doi.org/10.1007/0-387-25465-X_12
18. Birajdar, G.K., Mankar, V.H.: Subsampling-based blind image forgery detection using support vector machine and artificial neural network classifiers. Arab. J. Sci. Eng. **43**, 555–568 (2018). https://doi.org/10.1007/s13369-017-2671-3
19. Cervantes, J., García-Lamont, F., Rodríguez, L., Lopez-Chau, A.: A comprehensive survey on support vector machine classification: applications, challenges and trends. Neurocomputing (2020). https://doi.org/10.1016/j.neucom.2019.10.118.doi:10.1016/j.neucom.2019.10.118

Chapter 57
An Artificial Neural Networks Feature Extraction Approach to Predict Nephrolithiasis (Kidney Stones) Based on KUB Ultrasound Imaging

Gollapalli Sumana, Giri Aparna, and Gade Anitha Mary

Abstract Around 10% of the population will experience nephrolithiasis or renal calculi or kidney stones at their life time. Nephrolithiasis is the main cause for the haematuria, micturition and sever pain in the abdomen and during urination. More number of nephrologists are required to meet the rapid increase in the population. This paper deals with a supportive diagnostic system to the physician or nephrologist by applying neural network techniques. Generally, patients are supposed to undergo ultrasound scan test to identify the position, size and number of renal calculi in the renal system. Image processing techniques like preprocessing, segmentation and feature extraction were applied on the ultrasound scan images to extract the GLCM features. These extracted features were supplied as inputs to construct neural network and train the system by using feed-forward back-propagation algorithm. The proposed system is constructed with 22 input nodes, ten hidden nodes and one output node and thereby trained until it gets target or output. Then this trained neural network is allowed automatically to work on emerging new samples. Around 50 real-time images were collected from the patients, in which 60% of them were used for training and 40% for testing. The supportive software tool used for this proposed system is MATLAB 8.5.

57.1 Introduction

In medical diagnosis, ultrasound is protective and is painless, which produces images of internal organs of the body by capturing the sound waves. It is also identified as

G. Sumana (✉)
Sri Padmavati Mahila Visvavidyalayam, Tirupati, Andhra Pradesh, India
e-mail: sumana@spmvv.ac.in

G. Aparna
University College of Engineering, Hyderabad, Telangana, India

G. Anitha Mary
Loyola College, Hyderabad, Telangana, India

© The Author(s), under exclusive license to Springer Nature Singapore Pte Ltd. 2021   583
S. C. Satapathy et al. (eds.), *Smart Computing Techniques and Applications*,
Smart Innovation, Systems and Technologies 225,
https://doi.org/10.1007/978-981-16-0878-0_57

ultrasound scanning mainly and also as sonography by its working principle. In this process of image capturing, a small transducer (test) and ultrasound gel that is applied the skin directly are used. The sound waves of high recurrence are transmitted from the test through the gel into the body. The transducer captures the sounds that reflect, and the sound waves that are produced are used to generate an image. Since this process of examination does not make use of any ionizing radiation, therefore, it is a harmless process as there is no radiation penetration into the patient body. The ultrasound scanners constitute of a machine and electronic devices like computer, monitor and a transducer which are used to perform the scanning process. The transducer is a tiny and handy gadget which is like a microphone connected to the scanner by a cord. Different transducers are used based on the type of examination during a specific diagnosis. The transducer conveys imperceptible, high recurrence sound waves into the body and tunes later for the returning echoes from the tissues in the body.

57.1.1 Methodology

In the process of ultrasound scan test, the individual has to lie down straight facing up position on the examination table. This table is flexible in rotation position and can be tilted so as to move on either side for quality image capturing. After the initial examination of the patient, the radiologist or sonographer will apply semiliquid water-based gel smoothly to the area of the body being contemplated. The gel will help the transducer reach the body, and the air bubbles are removed which are between the transducer and the skin that can obstruct the sound waves from penetrating into individual under examination. The transducer is to be moved to and fro on the surface of the body over the range of interest until the coveted images are captured. The patient under test does not experience any uneasiness from pressure as the transducer is pressed in the area to be inspected, notwithstanding, if a territory of delicacy is being examined. The patient may feel the pressure or minor torment from the transducer. Figure 57.1 shows the pictorial representation of the medical setup of the ultrasound scanning. It has an ultrasound machine, image acquisition device, display, tracing and analysis system. The other apparatuses are a transducer with magnetic field receiver, non-ferrous bed and magnetic field generator. The advantages of the ultrasound when compared to other examination processes are ultrasound filtering which is non-invasive meaning no needless or infusions so it is harmless and not painful. It is most accessible, simple to use and cost effective when compared to other imaging techniques. It is an extremely safe method as it is not accompanied with any ionizing radiation. It produces clear images of soft tissues which are not that clear in X-ray images. It is a preferred imaging methodology for the diagnosis and for monitoring of pregnant women and fetus who should be prone to radiation much. It provides a real-time imaging, making it a handy apparatus for directing insignificantly advanced methods, such as needle biopsies and fluid aspiration.

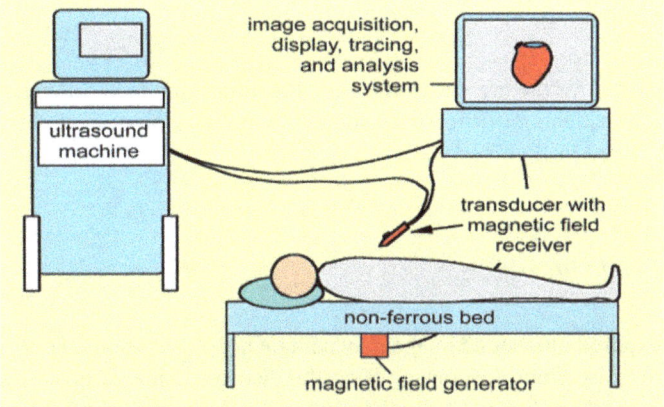

Fig. 57.1 Pictorial representation of the medical diagnosis of the ultrasound scanning

57.2   Automatic Extraction of KUB Ultrasound Image Features Using Digital Image Processing Techniques

57.2.1   Introduction

Ultrasound is utilized to assess an individual local kidneys (the ones you were conceived with) and, in addition, transplanted kidneys [1]. It can measure the size and appearance of the kidneys and recognize tumors, congenital anomalies, swelling and blockage of urine stream. A more up-to-date system called Color Doppler is utilized to survey clusters, narrowing, pseudo-aneurysms in the arteries and veins of the first and transplanted kidneys. This is the minimum invasive of all the procedures. It is compact and, most importantly, does not require radiation. As far as anyone is concerned up until this point, there have been no reports of reactions from the strategy itself. With a specific end goal to perform analysis of nephrolithiasis in view of ultrasound, initially, it is essential to extract the features from the ultrasound image [2]. Predicting stone in the renal system is not a simple task since it is unrealistic to specifically distinguish the kidney stone in the ultrasound in light of the fact that as it is a rigid mass like different organs of the kidney [3]. So to differentiate the kidney stone with other organs, ultrasound image feature extraction is especially needed [4, 5]. Consequently in the present review, by utilizing the digital image processing strategies, the ultrasound image features are extracted automatically [6].

57.2.2   Data

The data utilized for the present analysis are the ultrasound images, which are gathered from the patients of different hospitals. In order to process, these ultrasound images are stored in JPG format.

57.2.3   Procedure

Since the acquired ultrasound image may contain noise, this should be preprocessed to eliminate noise from the image. Then this de-noised image is resized through equal m rows and n columns for quick computation. In this way, first the input image is changed to grayscale image from RGB, then this image is used to do further process because the time taken by a grayscale image for processing is minimized when compared with the other images. Then this grayscale image is preprocessed to eliminate noise from the image, and then this preprocessed image is further for the extraction of the GLCM features.

57.2.4   Extraction of Ultrasound Image Features

57.2.4.1   Preprocessing

Initially, the input image which is taken is resized to quick computation. At that point, the image is changed over from RGB to gray in order to increase extraction method on the basis of the gray-level co-occurrence matrix (GLCM), or gray-level spatial dependency matrix. The vital idea of GLCM is that the texture information contained in an image is characterized by the adjacency connections that the gray tones in an image need to each other. These features are extremely robust to orientations effects of the processing speed. Then the gray image is preprocessed to remove the noise content in the image. Digital images are liable to have different varieties of noises. These noises result errors in the image procurement which in turn fails to get the pixel values that do not reflect the originality input image. Noisy images are difficult to find the abnormality. In this manner, the initial step is preprocessing, which is concluded by padding the noisy image. New pixels which exists around the boundaries of an image will be detected by the image padding. These edges act like marginal or boundary lines by utilizing progressed filtering techniques. This is an iteration process in which the aggregated de-noised pixels are generated to update the accumulated weights.

Once the image is de-noised, then the following stride is the masking so as to discover the reference. At that point of the region of interest (ROI) is taken, it is converted to binary image. At that point, the properties of ROI are extracted. These

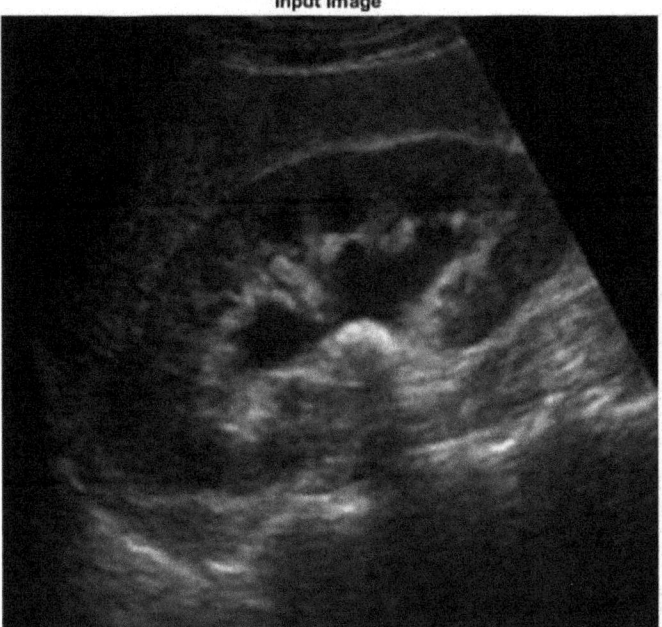

Input Image

Fig. 57.2 Input image

properties are utilized for the segmentation of the active contours of the ultrasound image. Active contours have been generally utilized as appealing images segmentation techniques since they generally deliver sub-regions with constant margins or boundaries [7]. But in kernel-based edge detection methods like sobel edge detectors will generate discontinuous margins or boundaries. Application of level set hypothesis has given greater adaptability and comfort while execution of active contours. Morphological operations depend on the comparative demand of pixel values but not on the statistical values and consequently fit for the processing of all binary images. Morphological operations are then connected to grayscale images to unrevealing the light transfer functions and to get accurate pixel values. Then the final segmentation is done. Finally, extraction of the required GLCM features from the de-noised ultrasound image is finished. The original ultrasound image prior and then afterward preprocessing is as shown in Figs. 57.2 and 57.3.

57.2.4.2   Extraction Process of Ultrasound Image Features

One of the characterizing characteristics of texture is the spatial distribution of gray values. Based on statistical, structural, model-based and transform information, the texture features can be extracted by an excellent method by applying gray-level

Preprocessed image

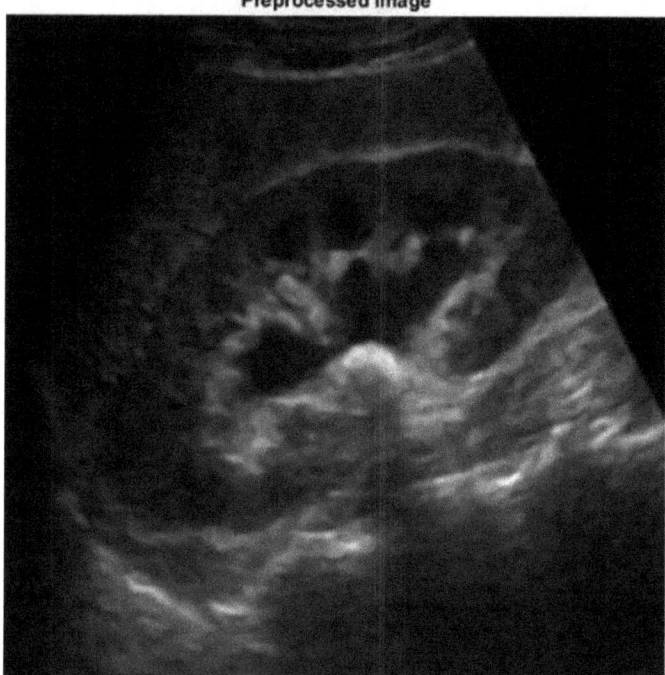

Fig. 57.3 Preprocessed image

co-occurrence matrix (GLCM) algorithm. GLCM contains statistical data of spatial relationship of the pixels in the image.

Table 57.1 portrays the 22 input parameters of the ultrasound which were utilized for training the neural network. Some such features are homogeneity, entropy, divergence, cluster color etc., as mentioned in Table 57.1. Homogeneity is the feature which measures the appropriate distance of the components in the GLCM to the GLCM diagonal. For diagonal GLCM, homogeneity is 1. Image entropy is an amount which is utilized to portray the measure of data which must be programmed by a compression algorithm. The perfectly flat images will have zero entropy. Divergence value is a measure that evaluates the reliance between the sequences. Cluster color feature is characterized by the skewness of the network (or) absence of the symmetry. The image is not symmetric, when the cluster color is high. Co-occurrence matrix will be peak around the mean values when cluster prominence is low. In the image, it implies that there is small distinction in the grayscales.

Table 57.1 GLCM features and their allowed values of ultrasound

S. No.	Features	Allowed values
1	Auto-correlation	650.8546
2	Contrast	300.2537
3	Correlation	0.0668
4	Correlation: [1 2]	0.0668
5	Cluster prominence	2.9658e+05
6	Cluster shade	164.2457
7	Dissimilarity	13.9130
8	Energy	4.7758e−04
9	Entropy	7.7256
10	Homogeneity	0.1397
11	Homogeneity: [2]	0.0689
12	Maximum probability	6.8378e−04
13	Sum of squares	811.4439
14	Sum average	50.3571
15	Sum entropy	2.5378e+03
16	Difference variance	4.4214
17	Difference entropy	378.7012
18	Information measure of correlation 1	3.6916
19	Information measure of correlation 2	−0.0030
20	Inverse difference is homom	0.1515
21	Inverse difference is normalized	0.7828
22	Inverse difference is moment normalized	0.8872

57.2.4.3   Analysis and Results

The developed GLCM feature extraction algorithm for extraction of features in ultrasound is implemented in MATLAB 8.5. Initially, the implemented algorithm peruses ultrasound image in a jpg format and performs preprocessing that it expels the noise from the input image by the way of padding. To concentrate ultrasound image features, the GLCM algorithm is utilized. Then the final segmentation is done by extracting the region properties [8]. The final segmentation is shown in Fig. 57.4.

Final Output

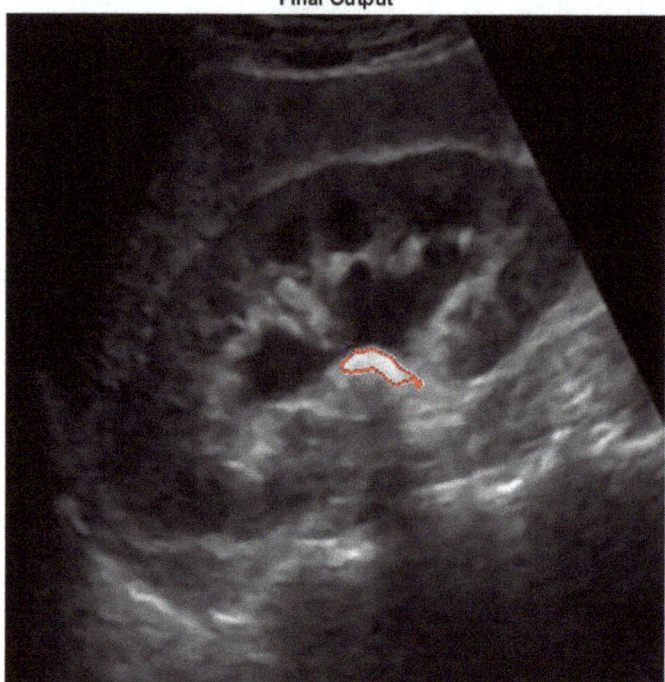

Fig. 57.4 Final segmented image

57.3 Predictions of Nephrolithiasis Based on Ultrasound Extracted Features Using Neural Networks

57.3.1 Introduction

In spite of the fact that the components are extricated consequently from ultrasound by the created GLCM calculation, still it is troublesome for a physician to diagnose nephrolithiasis. Since the extracted feature will not be same for all the patients, instead they vary from person to person based on age, sex, weight and height. This may bring inaccuracy of the diagnosis and may increase the time delay of the diagnosis [9]. Accordingly so as to provide a support system for the physician and to diminish the analysis time, it has turned into a requesting issue to build up an effective and solid clinical system [10, 11]. Since neural networks have indicated incredible potential to be connected in the advancement of clinical system for diagnosis of renal syndromes, in the present review additionally, a system is developed to perform the diagnosis of nephrolithiasis in view of the ultrasound features.

57.3.2 Parameters

The parameters that are utilized to perform diagnosis classification of nephrolithiasis in view of ultrasound are age, sex, weight, stature and the GLCM features extracted from the ultrasound image. In the present review, to prepare and test the neural system, an aggregate number of 50 samples are utilized, and then one output parameter is shown on which the experimental analysis is performed.

57.3.3 Procedure

In the present review, to build up a system for diagnosis of nephrolithiasis in view of ultrasound features, at first the ultrasound features, i.e., GLCM features are extracted by applying the GLCM feature extraction algorithm to the binary converted input images. Then, a feed-forward back-propagation neural network is built by taking 22 input parameters of the extracted features and the essential information.

57.4 Analysis and Results

The system is designed and executed utilizing MATLAB 8.5 by implementing the feed-forward back-propagation neural network model. In order to execute back-propagation network model, at first a feed-forward neural network is built with 22 (twenty-two) input nodes, 10 (ten) hidden nodes and 1 (one) output node. The input parameters utilized as a part of this system are the age, sex, weight, height, etc., and the GLCM features extracted from the ultrasound image and the output parameter demonstrate result of the diagnosis in terms of either normal or abnormal, i.e., if ultrasound of the kidney does not have any stone, then the condition is normal; if the kidney consists of a stone, then the result is abnormal. The feed-forward network is built with 22 (twenty-two) input nodes, 10 (ten) hidden nodes and 1 (one) output node. Construction of neural network with assigned nodes is shown in Fig. 57.5.

Training is done by utilizing the extracted GLCM features. The algorithms mentioned in the neural network train tool appeared underneath in Fig. 57.6. Here the data which are collected are separated randomly. Then that random data are used for training by using a learning method called Levenberg–Marquardt method. The obtained values between the gradient descent and the Gauss–Newton were adaptively differs by the Levenberg–Marquardt method.

$$\left[J^T W J + \lambda I\right] h_{lm} = J^T W \left(y - \hat{y}\right)$$

where minor estimations of algorithmic parameter 'λ' result Gauss–Newton update and vast estimations of 'λ' result in gradient descent update. The parameter 'λ' is

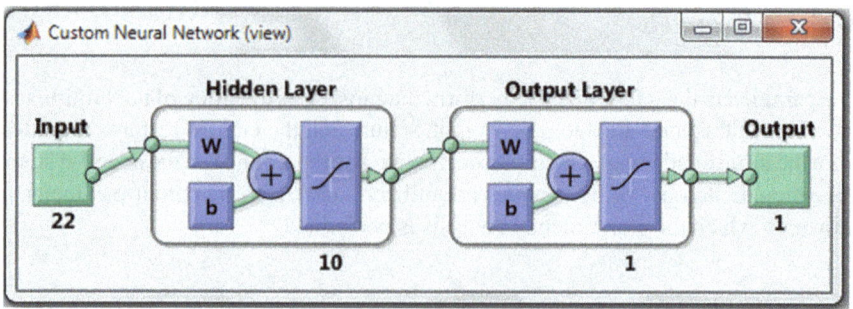

Fig. 57.5 Construction of feed-forward neural network with 22 input nodes, ten hidden nodes and one output node

modified to be extensively with the goal, which updates the small strides in the steepest-descent direction. If any iteration occurs which leads to worse calculation, then 'λ' is expanded. $\left(x^2(p + h_{lm}) > x^2(p)\right)$.

As the result progresses, λ is diminished. Levenberg–Marquardt strategy approaches the Gauss–Newton technique by which the result normally accelerates to the local minimum value.

Marquardt's refresh relationship

$$\left[J^T W J + \lambda\, diag\left(J^T W J\right)\right] h_{lm} = J^T w\left(y - \hat{y}\right)$$

The obtained values of 'λ' are normalized to the values of $J^T W_J$.

Then the performance is evaluated after the training is completed and then the values are generated. The performance is assessed by considering epochs, and mean square error is shown in Fig. 57.7. The graph appeared beneath portrays the performance of error at 6 epochs. The trained data are demonstrated with a blue line and the test data with a red line. The best validation is taken at the point where both train and test are equivalent. Here the best approval is at epoch 0, and the value of validation performance is 0.31456.

The plot appeared in Fig. 57.8 reveals the data of training and the validation of target and output classes. In the graph, the input is represented as data denoted by 'o'. At that point, the spotted line represents the linearity of result and the target. The fit esteem begins at 1.38 and increments marginally all through the training. The fit esteem does not change even when the target classes expanded. The validation graph shows the fit value linearly increments along with the target and output classes. Then for test data, the plots are represented below in Fig. 57.9.

Figure 57.10 reveals the data of gradient, mu and validation fail at 6 epochs. The gradient value follows a straight line, it continuously diminishes, and the esteem becomes plainly 0.00063259 at the sixth epoch. Then the mu value increments for a point of 1 and falls, and at the sixth epoch the value is 1e−08. The validation checks increment along with the expansion in the epochs, i.e., the validation checks and the epochs are linear and the validation check value at the sixth epoch is 6.

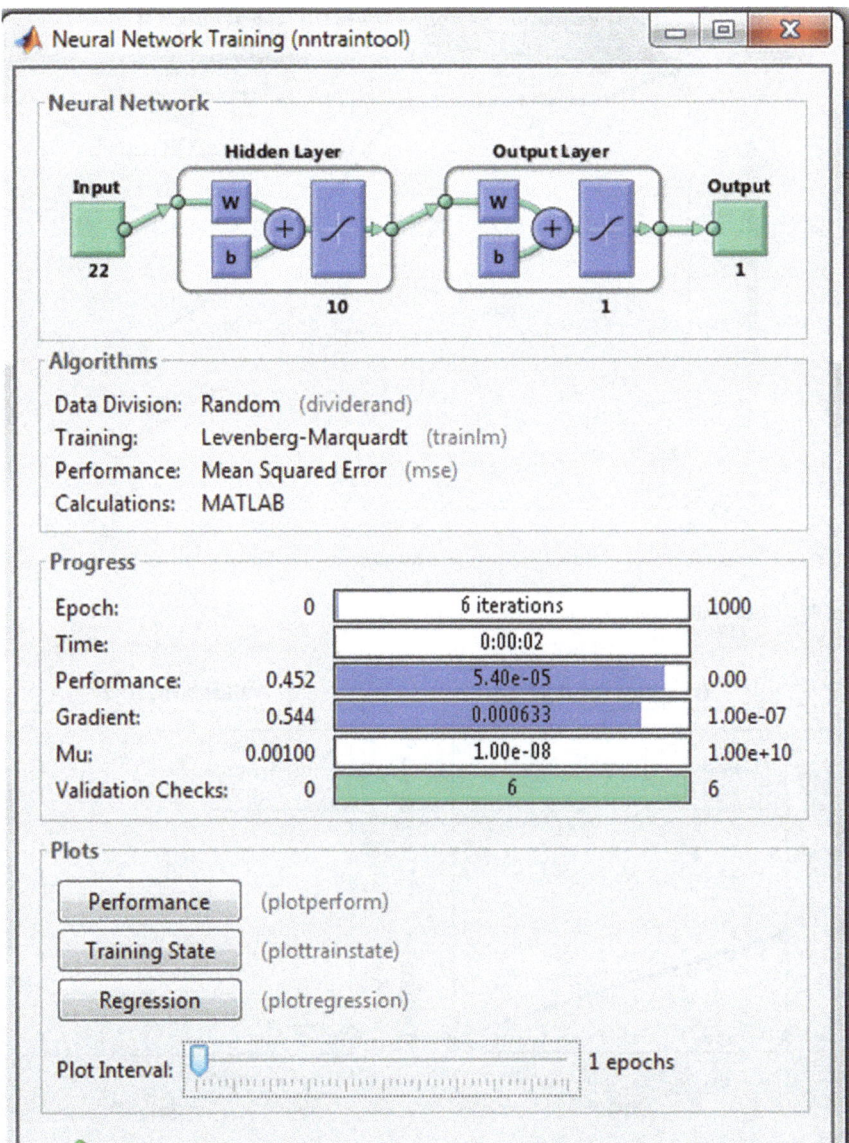

Fig. 57.6 Neural network training

57.5   Conclusion

The prediction of nephrolithiasis based on ultrasound components is a troublesome undertaking for nephrologists because of the reason that the ultrasound images will not be comparative for all individual rather it changes from individual to individual.

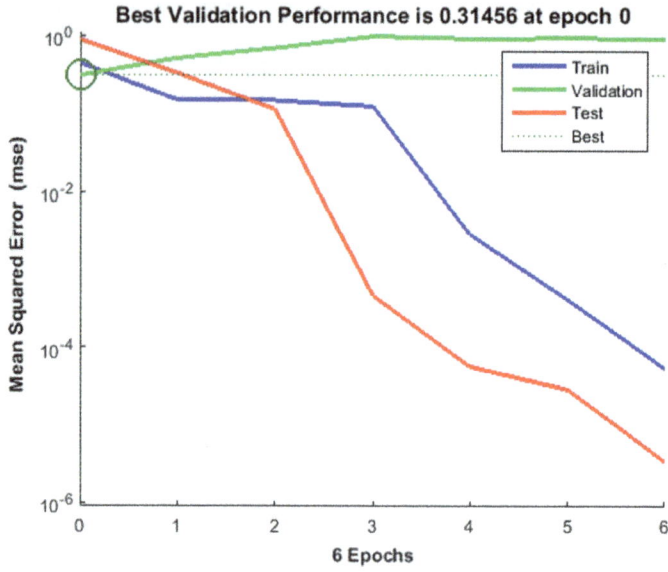

Fig. 57.7 Best validation performance at epoch 1

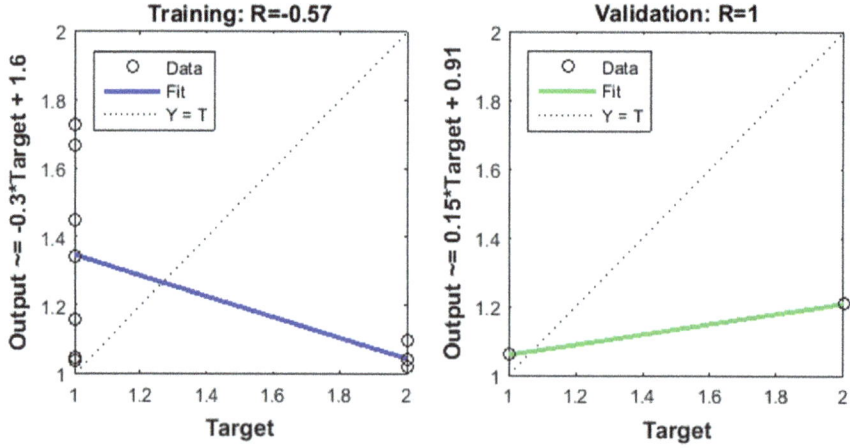

Fig. 57.8 Plot of training and validation for output with regard to target classes

Consequently, the systems can be utilized by a nephrologist to automatically extract ultrasound features and to perform the prediction of stone automatically based on the extracted features. The developed system diminishes the conclusion time and enhances the quality of the analysis. The extracted features are from an ordinary individual, and then the result, otherwise, gives the result with the final segmented image with a bouncing box precisely.

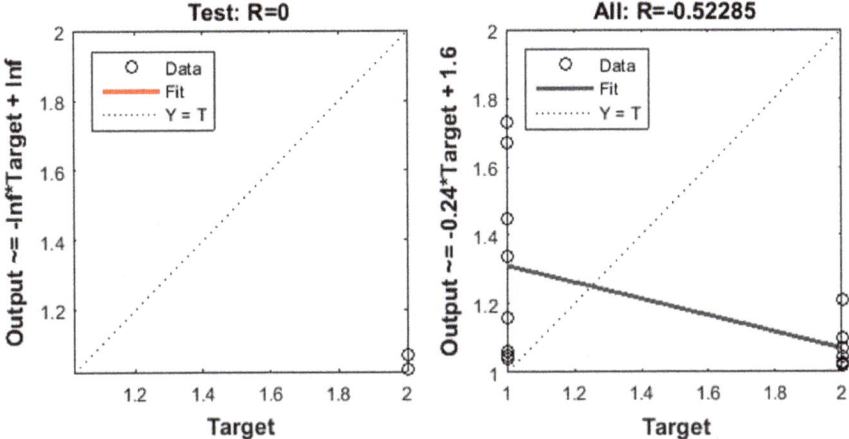

Fig. 57.9 Plot of testing data for output with regard to target classes

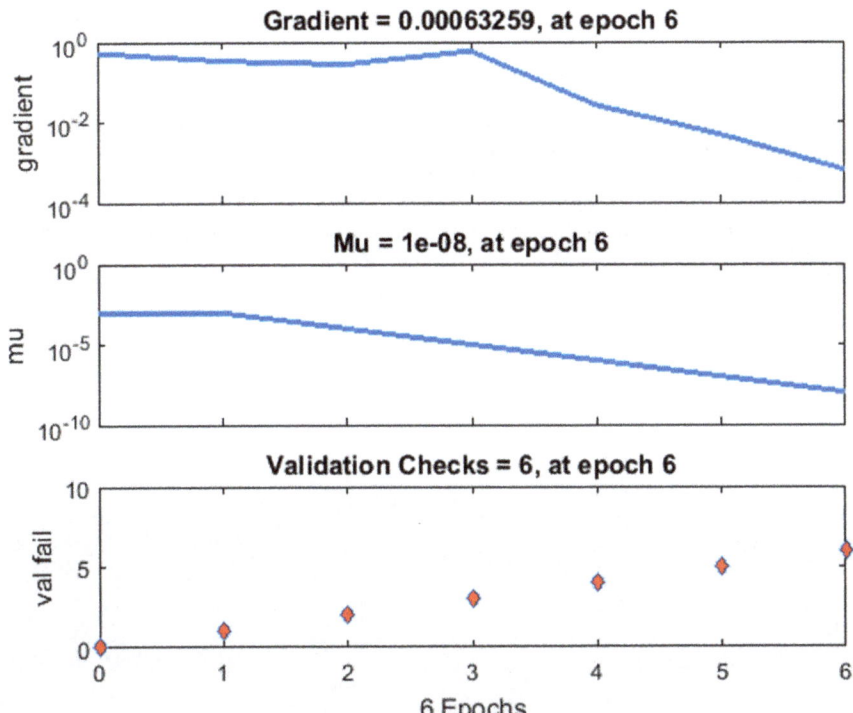

Fig. 57.10 Plot of training state at six epochs

References

1. Attia, M.W., El-Din Moustafa,H., Abou-hadi, F.E.Z., Mekky, N.: Classification of ultrasound kidney images using PCA and neural networks. Int. J. Adv. Computer Sci. Appl. (IJACSA) **6**(4) (2015)
2. Ramana, K.V., Korrapati, R.B.: Lung Tumor Segmentation via Fully Convolutional Neural Networks. Austin Ray, CS 231N, Winter 2016 (2016)
3. Neural network based classification and diagnosis of brain hemorrhages. Int. J. Artif. Intell. Expert Syst. (IJAE) **1**(2)
4. Francisco, M.P., Juan Manuel, G.C., Antonio, S.P., Daniel, R.F.: A robust model of the neuronal regulator of the lower urinary tract based on artificial neural networks. Neurocomputing **71**(4–6), 743–754 (2008)
5. Kalyan, K., Jakhia, B., Lele, R.D., Joshi, M., Chowdhary, A.: Artificial neural network application in the diagnosis of disease conditions with liver ultrasound images. Adv. Bioinform. **2014**, 14, Article ID 708279 (2014)
6. Bayo, A.L.: Computer Aided renal calculi detection using convolutional neural networks. International Master's Thesis Studies from the Department of Technology at Örebro University
7. Viswanath, K., Gunasundari, R.: Analysis and Implementation of Kidney Stone Detection by Reaction Diffusion Level Set Segmentation Using Xilinx System Generator on FPGA, 10p. Research Article Hindawi Publishing Corporation VLSI Design Volume 2015, Article ID 581961
8. Dayhoff, J.E., DeLeo, J.M.: Artificial neural networks—opening the black box. Cancer **91**, 1615-1635 (2001)
9. Brause, R.W.: Medical analysis and diagnosis by neural networks. https://www.informatik.uni-frankfurt.de/asa/papers/ISMDA 2001k.pdf
10. Lisboa, P.J.G. : A review of evidence of health benefit from artificial neural networks in medical intervention. https://www.openclinical.org/docs/ext/lisboa2002.pdf
11. Zukić, D., Elsner, A., Avdagić, Z., Domik, G.: Neural networks in 3D medical scan visualization by June 2008

Chapter 58
Multi-resource Task Scheduling for Minimum Task Completion Time in Cloud Computing Using Credit-Based Assignment Problem

Animoni Nagaraju and Y. Rama Devi

Abstract Task scheduling process problem is a focus and challenging issue in the Cloud Computing. The Cloud enumerate resources skillful and gain the most profits with scheduling process is the ultimate objective in the one of the Cloud enumerate overhaul providers. In this paper, we used multi-resource credit-based assignment problem to assess the entire group of multi-resource task scheduling in task get in line and finding the minimal completion time of all tasks in the cloud enumerate. Here cost/time matrix has been created as the reasonable tendency of the task to be the assigned in multi resources. Here we concern Cloud enumerate is a latest new assessment method of delivering the technology in the consumer by means of Internet servers for dealing out and data storage, while the client process uses the data.

58.1 Introduction

The technology experts considered the cloud enumerate is hovering to modernize the technique we access tools which it's going to be the same as task accruing because the profit-oriented of the technology more than a decade ago. Cloud Enumerate empower innovation. It alleviates the necessity of innovators to seek out possessions to extend, test, and build their innovations presented to the customer community. Innovators are liberal to specialize in the modernization instead of the logistics of discovery and organization resources with the purpose of enabling its innovation. In order that the task development evils and source administration are transmit to

A. Nagaraju (✉)
Department of CSE, University College of Engineering, Osmania University, Hyderabad, India

MallaReddy Institute of Technology and Science, Hyderabd, Telangana, India

Y. Rama Devi
Department of Computer Science and Engineering, Chaitanya Bharathi Institute of Technology, Hyderabad, Telangana, India
e-mail: yrd@cbit.ac.in

competence of entire cloud enumerate facilities. These tasks are matching process on the junction of the cluster through strategy which makes every effort to stay the job as on the brink of the info as potential. The scheduling algorithms within scattered processes usually contain goals of allocation consignment on processors and its maximizing theirs utilizations while minimize the total assignment effecting time more than a few heuristic approaches have been introduced during assignment scheduling. The inspiration of this paper is the work out a scheduling method which follows the Lexi—search approach of A. Nagaraju and V. V. Haragopal to seek out an optimal best feasible jobs. Assignment scheduling has been treated as general obligation problem to hunt out the minimal charge. In this, the cost matrix are generated from the probabilistic factor supported a quantity of most significant situation of efficient assignment scheduling just like the foremost important task interval during a resource. The value for assigning task into a source is probabilistic outcome considering.

In this context, a number of contributions are established for optimizing their performance more. The resource scheduling and the workload balancing is the similar directional effort for optimizing the computational performance of the cloud infrastructures. The proposed work is meant to live the performance of cloud scheduling approaches that are claimed to optimize the presentation of the cloud computing.

Cloud Enumerate means both applications and services are moved into the web (cloud). Cloud enumerate provides the users with any quite software or hardware services through the web on the idea of pay as you employ. Cloud enumerate is define as comparable and scattered process which comprise of group of interconnected and virtualizes computers that be provisioned dynamically and existing together or more combined enumerate resources supported service levels agreement (SLA) established from beginning to end mediation between cloud services providers and users [1]. Cloud enumerate delivers transportation as a service (IaaS), Platform as a service (PaaS), and Software as a service (SaaS) [2]. IaaS provides the users with infrastructure in sort of a virtual Machines (VM) to run any applications. PaaS provides an application development platform to users or developers to develop their own cloud applications and SaaS allows users to run existing software applications. Cloud enumerate is rapidly developing Computation model which has moved the management of hardware, software and other enumerate resources from users to cloud service providers.

There are numerous issues regarding cloud computing data security, energy conservation, service availability, expandable storage management, task scheduling. But task scheduling is usually the foremost topic of research in cloud enumerate [3]. the first objectives of this paper is to reinforce the scheduling policies. In cloud enumerate many tasks needed to be executed from the resources, available so on realize high performances, optimal completions time, reduce response times, and effective resources utilizations [4]. Due to these different objectives, there's a requirement to develop and propose a scheduling algorithm which will be employed by task scheduler to appropriately allocate tasks to resources.

Cloud Enumerate provides the prospect to employ enumerate resources more the web exclusive of owning the infrastructures. The most contents of Cloud Enumerate is to the handle software appliance, data storages, and processing facility its are assign to peripheral user on the insist all the way through the web and give just in support of come again its they are use. Tasks arrangement in cloud enumerate is that biggest challenge, since it many tasks got to be executed by available resource so as to satisfy user's requirements. To realize the most excellent performance, minimize the total completion time, reduce reaction times, and maximizes assets utilizations there's got to concentrate on these challenges.

From this paper, we used multi-resource credits-based assignment problem to evaluates the entire group of multi-resource assignment scheduling in task wait in line and finds minimal completions time of all tasks in the cloud computing. Now cost/time matrix have been generated as fair tendencies of a task to be assigned in multi possessions. Here we are appropriate Cloud enumerate is the latest new enumerate method of distributing technology to the consumers by use Internet servers for dealing out and data storages, while this client process uses the data. *The main idea of the planned algorithm is to assign responsibilities to assets properly so as to realize an resourceful weight matching and reduce completions time. The investigational outcome indicates the proposed approach by lexicographic search.*

58.1.1 Task Scheduling in Cloud Computing

Arrangement is mentioned to be as a set of policies for organization the ordered of work to be executed by a enumerate process [7]. The task arrangement in Cloud data center are a set of commands and elements that control and decide on the assignment to be executed on the available assets between a group of feasible everyday jobs at a specific time [8]. In Cloud data centers, the task arrangement algorithms are responsible for assign the farm duties submit by the user to accessible assets. The main benefit of task scheduling algorithms is to realize a highly performance enumerate and therefore the most excellent process throughput [2, 7, 9]. Schedulings lead the CPU memory and to realize maximum source utilization, require good scheduling policies.

In Cloud Enumerate location, responsibilities are submitted to info Center Brokers by the end users. This Data Centered agent is an agent between a Cloud Users and providers and the liable for arrangement farm duties on Virtuals machines (VM). Data hub may be virtual communications for shelter possessions and consisting of variety of Hosts. These submitted tasks are programmed consistent with the arrangement policy employed by the info hub Brokers. Broker communicates frankly with a Cloud manager and assigning tasks to Virtual machines within the Host info Center [10].

There are several types of scheduling as maintained by to unlike policy such that 'preemptive and non- preemptive' arrangement, statics and dynamic arrangement, instant and group setting up, integrate, and spread development [4]. Scheduling of tasks are taken into account because the process of choosing the simplest source accessible for tasks execution. Task scheduling approaches aim at minimize the total completion time of task and maximize source consumption so as to satisfy customer requirements, below diagram represents the proposed architecture, present paper concentrates on scheduler manager.

58.2   Proposed Methodology

An algorithm is extended based on lexicographic search algorithm. From this approach, every answer to the problem is defined as a word and solutions are to be produced in a quantity of hierarchy of their values. Each a partial word defines 'a block of words with itself as leader and is bound on the value of objective function for this block of words is considered'. If this 'bound is more than a value of objective function for the known feasible assignments', then the entire block of words is discarded. Algorithm start with 'feasible assignment' this provides a reasonably excellent upper bounds. Articulate on the values of the objective function. The initial possible assignments is obtained using a heuristics come within reach of in scheduling theory designed for dealing out tasks on matching machines minimizing the makespan approach 'examines the optimality' of this most feasible assignment and if found non-optimal, proceeds to find an optimal feasible assignment in a restricted number of iterations The above methodology explains "Lexicography" means "dictionary compiling". By "lexicographic arrangements" we mean the arrangement of the given letters or numbers as in the dictionary, Searching for a particular arrangement in this ordered, set is similar to searching for a word in the dictionary. The lexicographic search approach was developed by pandit SNN (1962) in the context of loading problem and was subsequently employed for many other combinatorial programming problems. The method adopted by pandit SNN is as follows: The set of items is partitioned into two subsets P and Q where P is the selected set and Q is the rejected set. For every word in P, i.e., for every subset of the selected set, a feasible and acceptable exchange is obtained, then the sets P and Q will be adjusted accordingly and the search will be started again. If such an exchange is not available, then the current selection is the optimum solution. Since the items in the sets P and Q are arranged in a standard order, a monotonic relationship exists between the values of the words belonging to partially ordered subsets from P and Q, which is helpful as a stopping bound criterion for the search in the set B. The present discussion is confined to the application of lexicographic search in the context of Multi-Resource Credit-Based Assignment Problem with minimizing objective function.

58.3   Algorithm of Multi-Resource Credit-Based Assignment Problem with n Tasks

Step 1: We consider the Weighted multiple resources it's rehabilitated in to two appliance n task assignment problem by using weighted average technique. i.e. various clusters to collect the data the cloud enumerate in the prairies based on the probabilistically approaches apply to use weighted multi resources it constructed to credit-based assignment problem.

Step 2: Two appliance with n tasks problem, we consider likely number of permissible acceptable selections used for appliance(machine) 1 and a number of acceptable selections for appliance(machine) 2, out of the which 2 appliance(machines), there will be' n task selection's, and its selection are- to be computed.

Step 3: Now we constructed alphabets plunk, which be an arrangements of times in a growing array of the problem for appliances 1st and 2nd machines. Here chosen the arranging of tasks is to done with indexed number by the not breaking of original order sequence in the tasks.

Step 4: From considering the changeable combinations of 'the at least' and 'the at most' constraint we get sub-problems for given 2 appliance assignment problems, now initially get the trials solution meant for the 2 machines problem.

Step 5: Now, lexi search methodology performs, scientifically we make 'the incompletes of words as of the search table with the new-labels, used for this every machines, except proceedings innovative task names considering that well be accrue time integrated in a word so far, and also the bounds for remaining part of partial word i.e. bounds for every realistic word's in lexical block, for its the current incompletes block is the leader'. If occurred bound is superior than the trial solution value on hand, this leader are discarded and the next incompletes word of the same length or it's next super block are head, as the holder may be the selected on the present incomplete word. Above steps record in search bench accessible in below.

Step 6: Using a step 4, we fixing the objective in algorithm Step 5, to achieve the optimum solution for expected objectives. i.e. (Minimize total accumulating machines time, this true for 2, 3 and m machines).

Step 7: From the above steps, we get the best feasible solution is evaluated for comparing with trial solutions from optimal solutions of step (6) in the various subproblems of step (4), from this we obtain best optimal solution in this step.

Now, we are implementing above algorithm in the 2 appliance and 7 Tasks problem is define problem in the next section.

58.4 Lexi Search Procedure of Multiple Resources Credit-Based Two Appliance n Task Assignment Problem

Numerical Illustration Discussed by Using Lexi Search Procedure for MR Credit-Based AP

In one byword, a methodology, processing, and apparatuses provide prioritized, faired, and weighted assignment scheduling for collective execution stand in line. New tasks are prioritized accordingly to their process in the requirements and held in one or more queue according to their precedence levels. Tasks are retrieved from the each queue in fair and weighted approach according to the priority level of each wait in line. In one of the embodiment, tasks of approximately equivalent total processing time are rived from each storages queue in the given task retrievals cycle.

In the next detailed description of the various embodiments of invention, numerous specifics detail are set forth in order to be provided thorough understanding of the various aspects of one or more embodiment of the inventions. However, one or more embodiments of the inventions may be practiced without these are specific details. In other the instances, well-known methodologies, procedure, and/or component has not been described in detail so as not to unnecessarily obscures aspects of embodiment of the invention. In the next description, certain terminology is used to describes certain features of the one or more embodiments of invention. For instance, the 'frame include any block or arrangements of data or informations. The terms "informations' is defined as voice, data, address, and/or control'. The term "task identifier" includes any depiction corresponding to a unique numbers frame. One part of an embodiment of inventions provides a methodology, Process, and apparatus having multiple executions in the queues of different priority coupled to a shared in the execution queue to provide the shared executions queue with a available processing tasks. Another aspects of an embodiments of the inventions provides fair, exact and weighted execution of the prioritized processing tasks.

Combinations of data Coating by the MCE with weighted linear combinations, each of the coating is assigned a weights, or the level of significances, that is determines to its influence on the layer producing as results of the combinations. In this current study weighting were assign by based on knowledge of element concerned as well as the consensus obtained as of the focus groups weightings is assigned so, that while they are combined to have total value. The combination of three layers considered the probabilities layer 1 = 0.5, layer 2 = 0.25, layer 3 = 0.25, This value assigned to the each cell in the each data layers is then multiplied by the weights. Corresponds cell's in each the layer is then added to together produces a final combined assignment table. The example of the processes using the 3 layers and the weightings describes above is given below figure considered multi-resource credit-based two appliance n task assignment problem.

We have generated the problem by using above procedure from various clusters to collect the data to use multi-resource constructed to credit-based assignment problem to calculate the entire group of multi-resource task scheduling in assignment queue, now we find minimal completion times of all tasks assignment from the cloud enumerate in the prairies based on the probabilistically approaches. From this cost/time matrix has been generated, as the fair tendency of a task to be assigned in multi resources. Now we apply Cloud enumerate is the latest new enumerated method of dispatch technology to the consumer by via Internet servers for processing and data storage, while the client process uses the data.

After construction of the credit-based assignment problem, we used the lexi search approach to make the task assignment scheduling in the processatical order.

Considered from Table 58.1: Observed this existed the credit-based assignment problem for 2 appliance 7 tasks: In next fragment, we are demonstrating algorithm through an illustration: regard as the following example 2 appliance 7 tasks situation is shown in Table 58.2.

For this problem, according the algorithm make alphabet-table, which are an array of "times" in growing order of the problem from 1st appliance and second appliance. The arrangements of tasks are complete with an index numbers by the not breaking original order sequence of these tasks and shown in Table 58.3.

Table 58.1 Example of the problem multi-resource credit-based two appliance 7 task assignment problems

Layer1:

	M1	M2			M1	M2
J1	5	3		J1	2.5	1.5
J2	1	3		J2	0.5	1.5
J3	2	4	x 0.5	J3	1	2
J4	4	5		J4	2	2.5
J5	2	2		J5	1	1
J6	7	6		J6	3.5	3
J7	3	4		J7	1.5	2

Layer2:

	M1	M2			M1	M2			M1	M2			M1	M2
J1	4	7		J1	1	1.75		J1	4.25	4.5	Layer are then rounded to give whole numbers	J1	4	5
J2	1	8		J2	0.25	2		J2	2.25	4		J2	2	4
J3	5	4	x 0.25	J3	1.25	1	Layer added	J3	3.75	4.25		J3	4	4
J4	3	5		J4	0.75	1.25	together	J4	3.5	2.5		J4	4	3
J5	4	7		J5	1	1.75		J5	3.5	4.75		J5	4	5
J6	7	6		J6	1.75	1.5		J6	6.25	5.75		J6	6	6
J7	3	5		J7	0.75	1.25		J7	3.25	5		J7	3	5

Layer2:

	M1	M2			M1	M2
J1	3	5		J1	0.75	1.25
J2	6	2		J2	1.5	0.5
J3	6	5		J3	1.5	1.25
J4	3	4	x 0.25	J4	0.75	1
J5	6	2		J5	1.5	0.5
J6	4	5		J6	1	1.25
J7	4	7		J7	1	1.75

Table 58.2 2 appliance 7 tasks

	1	2	3	4	5	6	7
M1	4	2	4	4	4	6	3
M2	5	4	4	3	5	6	5

Table 58.3 Alphabet table

S. NO.	M1: ti	M1: ji	M2: ti	M2: ji
1	2	2	3	4
2	3	7	4	2
3	4	1	4	3
4	4	3	5	1
5	4	4	5	5
6	4	5	5	7
7	6	6	6	6

Now, we fix the least and at most task (jobs) constraint on this problem with 7 task selections, we considered the fixed and is taken on first appliance as 2 & 4 tasks are to be processed. Since the requirement is that exactly on second appliance 3 & 5 tasks are to be processed on the apparatus M1 & M2 correspondingly, a complete lower bound meant for a few of the feasible assignment get as sum of the only least 2 and any least 5 processing's time in 2 machines, i.e., total of 1st two and 1st five timing's on these machines (appliance) is considered, that is for the problem the sum:{5,6,}{4,2,3,1,7,},10 + 21 = 31 is bound by the Arora optimality.

While timings on two apparatus are the independent of separate tasks allotment, bound settings can be completed, for each appliance separately in the parallel or the one can be first computing assignments on appliance 1 (say) and then we go to appliance2, enumerate bound section for an unassigned part on the 1st appliance explicitly, keeping the simpler (this is the relatively very less efficient bound, is the calculated once for the all, for M2, irrespective of tasks already assign for the M1). Also, the computationally, simpler bound is not accumulate, the number of tasks, yet to be by now allotted to on the machines, but just to reproduce the number by the 'next' processing time in the alphabet table.

In this following example, allocation is lexicographically prepared first on appliance Machine1, with constant minimum of bound for Machine 2, and Machine 1 allotment which is finished, and then only to go for more the exact bounds for components based on the Machine 2. "The search desk analytically 'Generates' partial words via the latest labels, for the each of machine, but the proceedings are original task numbers as well as the accumulates times integrated in the word, similarly so far, and its also the bounds meant for remain part of partial word, this bounds, for all the feasible words is in the lexical blocks, for which this current, incomplete block is a leader. If its bound are greater than trial solution significance on hand, its leader is redundant, the subsequently incomplete word of the same length, or it's next super block is a leader, as the similar case may be selected on current incomplete word". Above steps are recorded in the search table existing below; this construction of search slab, as explained over is illustrated below: Search table gives for minimization Objective by lexi search approach.

In the search table using the lexi search approach by C program we obtain the task constraints allotted to the two machines (i.e. M1, M2) for the constraints considered at least and atmost tasks are first appliance selections J1:J2, J3:J7. And second appliance selections J1:J3, J4:J7. Occurred by considered constraints. From the constraint J1:J2, J3:J7, we obtain the optimal solution by lexi search approach is {2,7,}{4,3,1,5,6,},cost/time is 5 + 23 = 28 and from the constraint J1:J3, J4:J7,we obtain the optimal solution by lexi search approach is: {2,7,5,}{4,3,1,6,}, cost/time is 9 + 18 = 27.

From the above two constraints, we obtain best optimal solution is MIN(MIN), we obtained the Selections (M1):3,(M2):4 are 2,7,5 on appliance one and 4,3,1,6 are on appliance two. The best optimum solution optimal cost/time Is 27.

From the above algorithm, we obtain the comparison table for above problem between constrains with objectives in Table 58.4.

Table 58.4 Optimal solution comparison table

		Constraints	Lexi search algorithm	Aurora algorithm (trial solution)
	Objective	(M1):2 (M2):5	$5 + 23 = 28$	$10 + 21 = 31$
		(M1):3 (M2):4	$9 + 18 = 27$	
	Best optimal solution	(M1):3 (M2):4	$9 + 18 = 27$	

58.5   Computational Experience

This idea is implemented on randomly generated 100 problems of $m = 2$ machines and $n = 20$ tasks for which M1 = the total appliance time on machine 1 and M2 = the total time on machine 2 are computed and also minimum machines accumulated times computed. With the above-cited constraints is verified and the results are tabulated, here 10 problems result are tabulated in Table 58.2. The data shows the pictorially in the Fig. 58.1 revels that the minimum time and it shows bound line on machine 1 and machine 2 This we observe minimum total time for accumulated appliance times is very less. The optimum solutions of Multi-resource task scheduling for minimum task completion time in cloud enumerate using the credit-based assignment problems size 20 obtained. Which is best probabilistically expectations in the data selections with best task Scheduling in the cloud computing table given in Table 58.5.

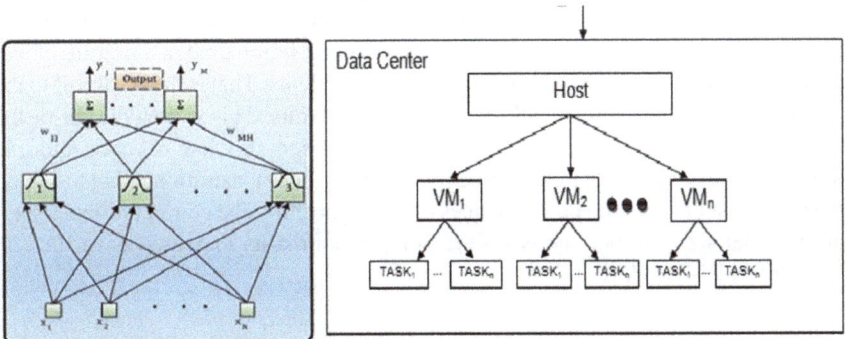

Fig. 58.1 Multi-resource task scheduling structure

Table 58.5 Optimal solutions for different problems

S. NO.	M1 (total time)	M2 (total time)	Total time (M1 + M2)
1	119	218	337
2	146	186	332
3	110	141	251
4	110	119	229
5	121	140	261
6	152	171	323
7	135	165	300
8	187	187	374
9	154	254	408
10	136	176	312

58.6 Conclusion

With encroachment of Cloud technology fast development, present are new needs for apparatus to learning, analyze the profit of the technology to implement and how the best to apply technology to large-scaled application The proposed method considers the arrangement problem as Multi-resource task scheduling for minimum task completion time in cloud enumerate using credit-based task problem.

In the cloud data center the data selections, the judgment assigning to the each cell in the each data layer are then multiplied by weights with probability. consequent cells in the each layer are to be then added together to produces a final combined task problem, in this, we apply lexi search approach to obtained best optimal task scheduling for multi resources for minimum task completion time in the entire cloud, that we compared with 100 randomly generated problems. That is most helpful to the data selections in probabilistically with expected results. This is new design of the problem which can focus in the extension of assignment problem to reach multiple clouds data selections with minimum time. Numerical demonstration indicate a good measures of success in the performances of the algorithm. The computational observation is that size of cloud increases the time for optimality increases as verified on computer 512 MB RAM.

References

1. Singh, Raja Manish, Paul, Sanchita and Kumar, Abhishek. *Task Scheduling in Cloud computing: Review*. 6, International Journal of Computer Science and Information Technologies (IJCSIT), Vol. 5, pp.No. 40–44 0975- 9646 (2014).
2. Lakra, Atul Vikas and Yadav., Dharmendra Kumar. *Multi-Objective Tasks Scheduling Algorithm for Cloud Computing Throuhgput Optimization*, International Conference on Intelligent Computing, Communication & Convergence, pp. 107–113 (2015).

3. Saxena, D., Chauhan, R.K., Ramesh, K.: Dynamic fair priority optiumization task scheduling algorithm in cloud computing: concepts and implementations. I. J Computer Network and Information Security, pp. 41–48 (2016).
4. Nagaraju, A., Hara Gopal, V.V., Pandit, S.N.N.: Time minimization assignment problem with variant objectives—a lexi search approach. Int. J. Agric. Statistical Sci. 6(I), 87–98 (2010)
5. Mohammadi, F., Jamali, S., Bekravi, M.: Survey on job scheduling algorithms in cloud computing. Int. J. Emerging Trends Technol.Computer Sci. 151–154 (2014)
6. Thomas, A., Krishnalal, G., Raj Jagathy, V.P.: Credit based scheduling algorithm in cloud computing environment. In: B.V. International Conference on Information and Communication Technologies, pp. 913–920 (2015)
7. Agarwal, A.. Jain, S.: Efficient optimal algorithm of task scheduling in cloud computing environment. Int. J. Computer Trends Technol. (IJCTT) 9, 344–349 (2014)
8. Er-Raji, N., Benabbou, F., Eddaoui, A.: Task scheduling algorithms in the cloud computing environment: survey and solutions. Int. J. Adav. Res. Computer Sci. Software Eng. 6, 604–608 (2016)
9. Bhavisha, K., Bhumi, M.: Review on max-min task scheduling algorithm for cloud computing. J. Emerg. Technol. Innov. Res. (JETIR) 2, 781–784 (2015)
10. Himani, H., Harmanbir, S.S.: Comparative analysis of scheduling algorithms of cloudsim in cloud computing. Int. J. Computer Appl. 97 (2014)
11. Dehkordi, S.T., Bardsiri, V.K.: TASA: a new task scheduling algorithm in cloud computing. J. Adv. Computer Eng. Technol. 1, 26–32 (2015)
12. Reda, N.M.: An improved sufferage meta-task scheduling algorithm in grid computing systems. Int. J. Adv. Res 3 (2015)
13. Kokilavani, T., Amalarethinam, D.I.G.: Load balanced min-min algorithm for static meta-task scheduling in grid computing. Int. J. Computer Appl. 20 (2011)
14. Patel, G., Mehta, R., Bhoi, U.: Enhanced load balanced min-min algorithm for static meta task scheduling in cloud computing. In: 3rd International Conference on Recent Trends in Computing, pp. 545- 553. Elsevier, Amsterdam (2015)
15. Rajeshkannan, R., Aramudhan, M.: Comparative study of load balancing algorithms in cloud computing environment. Indian J. Sci. Technol. 9, 1–6 (2016)
16. Gokilavani, M., Selvi, S., Udhayakumar, C.: A survey on resource allocation and task scheduling algorithms in cloud computing environment. Int. J. Eng. Innov. Technol. (IJEIT) 173–178(2013)
17. Nagaraju, A., Rama Devi, Y.: Task Scheduling in Cloud Computing using credit based cluster travelling salesman problem. IOSR J. Eng. (IOSRJEN) 08(10), 25–31 (2018) (2018)
18. Nagaraju, A., Rama Devi, Y.: Task scheduling in cloud computing using credit based variant of time dependent time minimization assignment problem. Int J Anal Experiment Modal Anal. XII(I), pp 307–313 (2020)
19. Panda, S.K., Pande, S.K., Das. S.: Task Partitioning scheduling algorithms for heterogeneous multi-cloud environment. Arab. J. Sci. Eng. 43(2), 913–933 (2018)

Chapter 59
Insights into the Advancements of Artificial Intelligence and Machine Learning, the Present State of Art, and Future Prospects: Seven Decades of Digital Revolution

Prabhleen Bindra⑩, **Meghana Kshirsagar**⑩, **Conor Ryan**⑩,
Gauri Vaidya⑩, **Krishn Kumar Gupt**⑩, **and Vivek Kshirsagar**⑩

Abstract The desire of human intelligence to surpass its potential has triggered the emergence of artificial intelligence and machine learning. Over the last seven decades, these terms have gained much prominence in the digital arena due to its wide adoption of techniques for designing affluent industry-enabled solutions. In this comprehensive survey on artificial intelligence, the authors provide insights from the evolution of machine learning and artificial intelligence to the present state of art and how the technology in future can be exploited to yield solutions to some of the challenging global problems. The discussion centers around successful deployment of diverse use cases for the present state of affairs. The rising interest among researchers and practitioners led to the unfolding of AI into many popular subfields as we know today. Through the course of this research article, the authors provide brief highlights about techniques for supervised as well as unsupervised learning. AI has paved the way to accomplish cutting-edge research in complex competitive domains ranging from autonomous driving, climate change, cyber-physical security systems, to healthcare diagnostics. The study concludes by depicting the growing share in market revenues from artificial intelligence-powered products and the forecasted billions of dollars worth of market shares ahead in the coming decade.

P. Bindra · G. Vaidya · V. Kshirsagar
Government College of Engineering, Aurangabad, Maharashtra, India
e-mail: vivek@geca.ac.in

M. Kshirsagar (✉) · C. Ryan
Biocomputing and Developmental Systems Lab, Lero, the Science Foundation Ireland Research Centre for Software, University of Limerick, Limerick, Ireland
e-mail: Meghana.Kshirsagar@ul.ie

C. Ryan
e-mail: Conor.Ryan@ul.ie

K. K. Gupt
Limerick Institute of Technology, Limerick, Ireland

59.1   Introduction

With an ultimate aim to surpass human capabilities, the technology is continuously evolving through the difficulties and has achieved greater milestones on their way [1]. Huge amounts of digital data and the need to keep abreast with the ever-evolving optimized versions of hardwares have fueled the rising interest of AI and ML [2]. Mainstream ML techniques such as classification, regression, dimensionality reduction, object recognition, and language processing are widely used in predictions, recommendations, detection, and many other similar applications. These technologies have already gained strong roots since the last decade and are expecting an exponential rise in the coming future [3].

The foreground of this in-depth analysis revolves around the past, present, and future of this entire paradigm with their current opportunities and future prospects. We continue with the present technologies under each type of ML and AI, their industrial applications and market trends followed by exhibition of the predictions and capabilities of the revenue market around their applications in the next decade, concluding the comprehensive study.

59.2   The Evolving Hype of AI/ML

This section highlights the history of evolution of AI, ML technologies, and their recent advancements by citing landmarks that led to the successful deployment of the wide applications this technology offers.

59.2.1   The Past

The term AI was first discussed in a context when a study was carried out jointly by three universities Dartmouth, Harvard, and Bell telephone laboratories which subsequently led to the birth of AI and soon followed the first programming language Lisp [4] for the AI researcher community. The machine learning paradigm was one of its kinds which learnt from experiences like a human and could solve problems by manipulating sentences in formal languages. The early 1960s witnessed the first-ever industrial robot 'Unimate' working in general motor's assembly lines. The mid-sixties gave birth to the chatbot 'ELIZA' [5] that facilitated dialogs between machines and humans (Fig. 59.1).

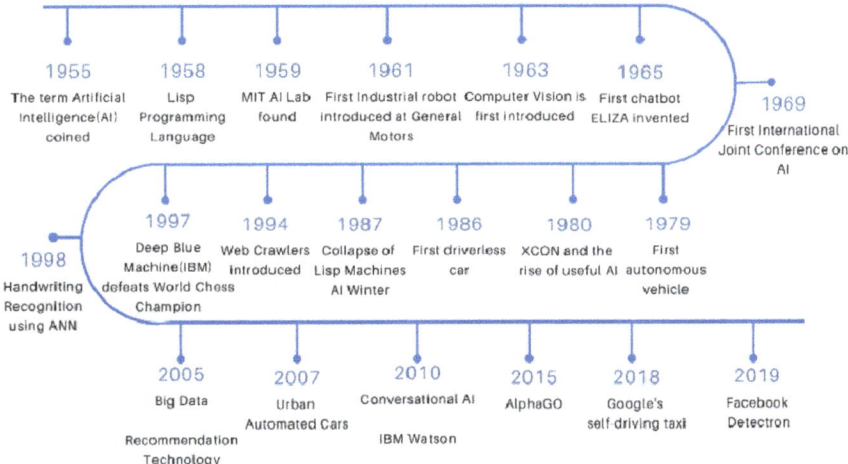

Fig. 59.1 Evolution of AI [6]

59.2.2   The Present

The late seventies led to the first autonomous vehicle 'Stanford cart' [7]. With the commencement of the 1980s, AI capacity at automation was explored with the Xcon program [8]. The mid-eighties saw the first Mercedes-Benz driverless car running on the empty streets of Munich. In the twenty-first century, the introduction of recommender systems designed to assist users in product selection choices based on their requirements led to the acceleration of the e-commerce market. IBM Watson [9], a real-time question–answer-based system was a technological revolutionary breakthrough in the last decade.

59.3   Subdomains of AI/ML

This section discusses the various emerging subdomains of AI and ML that are pioneers to the upcoming techniques in the field.

59.3.1   Artificial Neural Networks (ANN)

ANN [10] trains machines to solve problems in a way as a human brain does. The first neural network perceptron was developed by Frank Rosenblatt in the late nineteen-fifties. ANNs like convolutional neural networks, long short-term memory, recurrent neural networks, and auto-encoders gradually evolved over the years. These

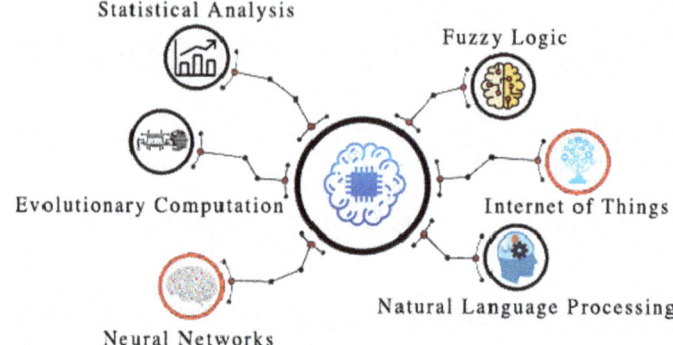

Fig. 59.2 Emerging subfields of AI and ML

techniques culminated into the first successful ANN-based handwriting recognition system in the mid-nineties. Applications of ANN include biometrics like iris, finger-prints, and face recognition systems which are now deployed across organizations that primarily require unique user authentication (Fig. 59.2).

59.3.2 Evolutionary Computation

A technique inspired by Darwinian evolutionary concepts of reproduction, muta-tion, recombination, and selection, introduced a new sub-domain in AI, known as evolutionary computation [11]. The underlying techniques like genetic program-ming and grammatical evolution [12] optimize solutions for enhancing computational efficiency.

59.3.3 Natural Language Processing (NLP)

Human–computer interaction through direct communication with devices like voice assistants and chatbots marked the NLP era [13]. Recent applications of NLP include answer sheet evaluation for examinations, text search and filtering, and lexical analysis.

59.3.4 Bayesian Networks and Fuzzy Logic

Statistical tools to incorporate cause-and-effect relationships in modeling complex systems came into being through the Bayesian networks [14]. They are impactful within the broader armamentarium of AI methods with a wide array of applications in pharmacogenomics and physiologically based pharmacokinetic modeling.

The challenge of modeling human reasoning in an environment without measurements, vagueness, and uncertainty brought about fuzzy logic into existence [15]. The logic can be embedded from microcontrollers to large networked or workstation-based systems.

59.3.5 Internet of Things (IoT)

The advent of IoT [16] following the World Wide Web (WWW) created the digitized interconnected virtual and the physical world we live in today. IoT frameworks spanning organizations and countries have made it possible to integrate data from diverse sources leading to the development of smart cities across the globe.

59.4 Popular AI Methods

This section provides a brief insight discussing all the algorithms along with their suitability with respect to creating AI models.

59.4.1 Supervised Learning

Linear, multiple, and logistic regression analyses are used for modeling the relationship between predictor and output variables. Ordinary least square regression (OLSR) is used for estimating unknown parameters. Multivariate adaptive regression splines (MARS) gives near optimal solutions by allowing automatic selection of variables, interaction between predictors, handling missing values, and avoids overfitting with self-tests. Locally estimated scatter plot smoothing (LOESS) and locally weighted scatter plot smoothing (LOWESS) regression are used to reveal trends and cycles in input data by automatically choosing a smoothing parameter for a smooth curve. Ridge regression deals with multicollinearity in multiple regression [17] (Fig. 59.3).

Dimensionality reduction transforms data from high-dimensional space to low-dimensional space, hence retaining meaningful information. Techniques supporting this include principal component analysis (PCA) for unsupervised learning, linear discriminant analysis (LDA) in linear space, and quadratic discriminant analysis (QDA) for quadratic space for supervised learning, whereas partial least squares regression (PLSR) and principal component regression (PCR) for regression analysis [18].

Ensemble methods combine various multiple learning algorithms to stabilize the variance of predicted solution and thus increase the accuracy. Bagging is based on the

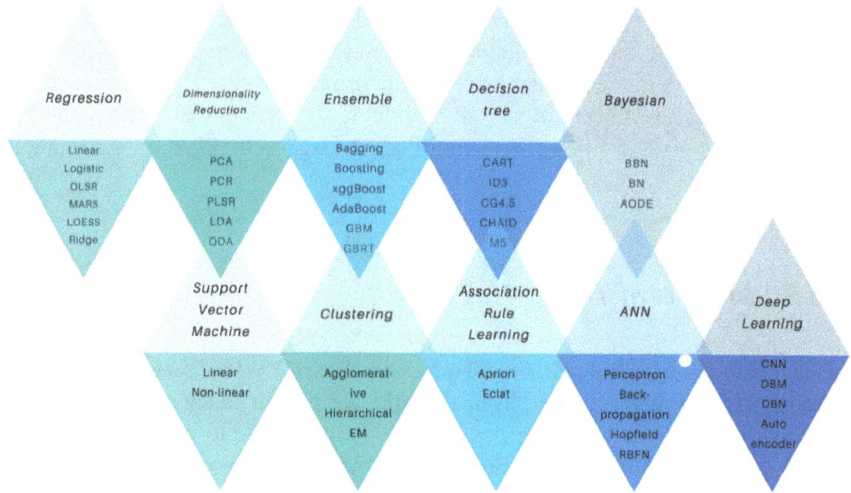

Fig. 59.3 Classification of machine learning algorithms based on similarity

bootstrap sampling method and works by adaptive learning of weaker data instances simultaneously to combine them for maximum accuracy. Random forest is a widely used bagging method that combines randomness and decision trees. Boosting works by learning new hypotheses with each phase of classification sequentially, improving weights of the weaker data instances, and boosts the accuracy. AdaBoost improvises weaker data instances by assigning higher weights to them, and gradient data boosting (GBM) learns with gradients of a loss function [19].

Decision tree, a form of supervised learning, follows a two-step process of growth and pruning of trees. Classification and regression trees (CART) is an algorithm for classifying data into binary labels. CHAID and M5 algorithms are advancements of the CART algorithm. The Iterative Dichotomiser 3 or ID3 algorithm uses information gain to decide the splitting attribute and builds the tree accordingly. C4.5 is an advancement of the ID3 algorithm [20–22].

Bayesian networks (BN) or Bayesian belief networks (BBN) are probabilistic models, and their graphical illustration is in the form of directed acyclic graphs. To overcome the inherent drawback of attribute independence in many real-world settings, averaged one dependence estimators (AODE) was introduced. They are popularly used for statistical validation of any machine learning model [23].

Support vector machines have found applications in real-world applications like face detection, text and hypertext categorization, classification of images, bioinformatics, and generalized predictive control for chaotic dynamics [24].

59.4.2 Unsupervised Learning

The clustering algorithms following under the above category and used widely for implementations include k-means clustering based on centroid approach, mean-shift clustering based on sliding window approach, density-based spatial clustering applications with noise (DBSCAN) based on neighborhood concept in relation to data points. Expectation maximization (EM) and clustering using Gaussian mixture models (GMM) are based on assumption of Gaussian distribution of data points and for finding the parameters of the Gaussian EM. The two variants to hierarchical clustering are agglomerative and divisive hierarchical clustering [25].

Prominently used in market-basket analysis, multimedia data mining, and video data mining, association rule mining has many implementation variants among which, a priori algorithm, frequent pattern (FP) growth remains the most widely used ones [26, 27].

Deep belief networks (DBN) and convolutional neural network (CNN) are extensively used for object detection, whereas the new breed ones yolov3, mask-regional convolutional neural networks (RCNN), and faster RCNN are used in multi-object detection. Auto-encoders are yet another form of neural networks used for dimensionality reduction. Back-propagation remains the backbone of neural network design and mainly used in the design of multi-layer perceptrons [28].

59.5 Market Trends with AI Applications

In this section, we touch upon the broadly used use cases of AI and ML and their diversity in domains. We also look at the region wise distribution of global AI markets. The products dominating the software market have been discussed.

59.5.1 AI Use Cases

AI applications range from healthcare analysis to product recommendations. In a survey by Tractica [29], Fig. 59.4 depicts the top ten predicted to be the highest revenue earning use cases of AI by 2025.

Figure 59.5 [30], on the other hand, illustrates the context for which industries apply ML algorithms and models. Popular reasons why companies incorporate AI include generating hidden insights, improving customer satisfaction, retaining customers by improving interaction, providing prompt query resolution, providing relevant recommendations, improving the customer's overall experience, and thereby reducing churn. Another major reason why companies invest in AI includes reducing costs by forecasting demands and gaining internal insights of the organization.

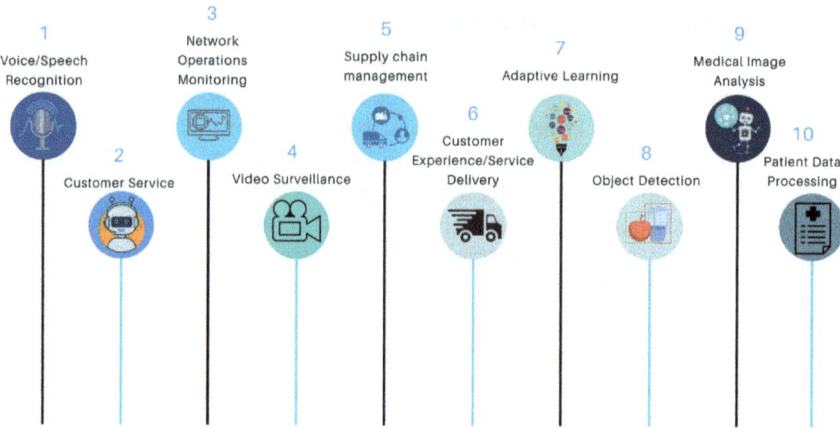

Fig. 59.4 Top ten use cases of AI and ML in software market

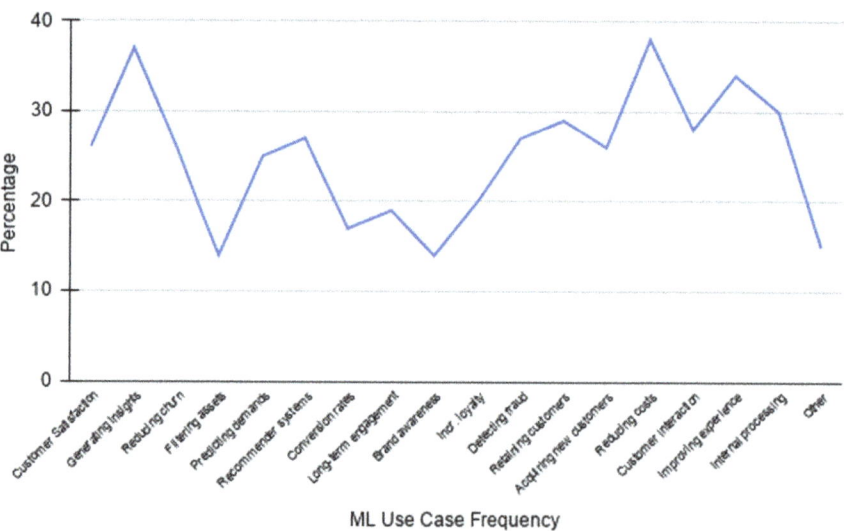

Fig. 59.5 ML use case frequency

Industries also increase loyalty and reduce fraud by imbibing appropriate security mechanisms.

59.5.2   Popular AI-Enabled Products

AI-enabled virtual assistance, delivery robots, and tele-health are some examples of how AI is helping people every day. Use of pseudo-intelligent digital personal assistance (Siri, etc.) to perform a task, ML to accurately detect cancer cells, or Alexa in home automation is definitely going to boost industrial and market demand for AI power products. Google Duplex, Deepmimd's Alpha Star, or AI algorithms providing product suggestions, and support systems in telemedicine are a few applications to which customers have become accustomed to. The personalized ad serving and AI-powered dynamic pricing is nothing less than any boon to various airlines' booking sites. Proliferation in connected devices, data, and usability is trending market growth. Internet dependencies and cloud-based applications in health care, banking financial services and insurance (BFSI), retail, automobiles, security surveillance, smart home devices, and tourism, etc., contain sensitive user information. Data coming from these sources are hence vulnerable to threats and cyber attacks, thus opening a wide scope for designing strong security systems using AI and ML algorithms. Utilizing data from user's wearable devices or smartphones through contact tracing applications for alerting citizens to nearby COVID-19 cases has demonstrated the significance of AI. According to [31], 59% of more people are likely to use tele-health services post-pandemic which was 25% during 2019, while 36% would switch their physician to a virtual one in the USA. This huge demand reports the telemedicine market size to witness $175.5 billion by 2026 from $45.5 billion in 2019 [32]. The wearable AI market alone is forecasted to reach $180 billion by 2025 [33]. In 2019, the global market size for AI by component, technology, deployment, and industry was valued at $27.23 billion and is expected to achieve $266.92 billion by 2027 with 33.2% of compound annual growth rate (CAGR) during this period [34] while another survey [35] calculates a CAGR of 42.2%, with revenue forecast to be $733.7 billion considering solution, technology, and end use.

59.6   Future with AI

Our study so far has highlighted the journey of AI to date. It is a fairly growing market with a lot of future scope. This section discusses the future of AI.

59.6.1   Emerging Trends for the Future

The goal of AI is to design 'general' intelligence for performing unpredictable tasks rather than 'specific' intelligence, which is able to perform only a single task outstripping human skills. The learnings through direct interaction of the object or the model

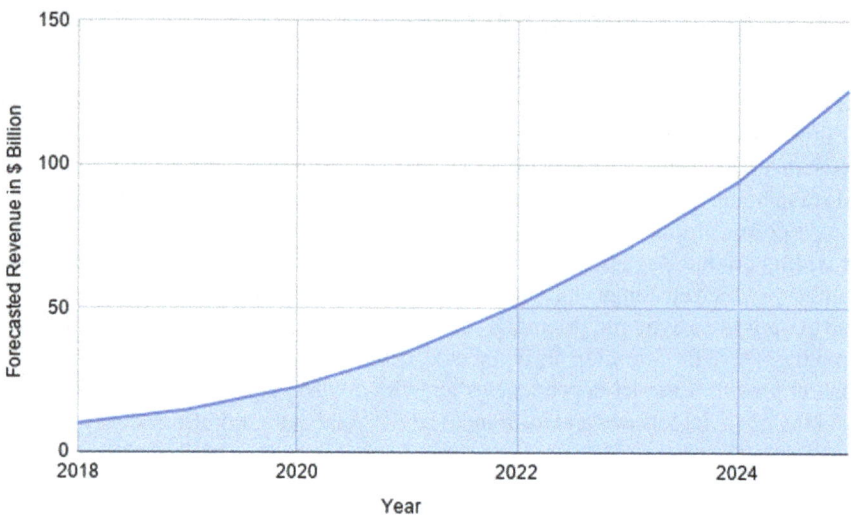

Fig. 59.6 Forecasted market revenue generation from AI software market

to the physical environment are important for domains like robotics and driverless cars for understanding the cause–effect behavior.

The upcoming AI models will be designed based on the pre-trained models and will be implemented on the energy-efficient hardwares [36].

With the vast opportunities and applications AI offers, it is predicted that the global revenue market for AI-driven services and products will drastically increase from 10.1 Billion Dollar in 2018 to 126 Billion Dollars in 2025 as illustrated in Fig. 59.6 [37].

Industries and markets are evolving with business strategies like reducing costs, automation, innovation, and managing risk factors with advanced analytics of AI. This estimates that the markets will grow in the coming few years. As represented in Fig. 59.7 [38], in 2017 in a survey by market research future, it was recorded that 44% of the global machine learning market was centered in North America, 29% in Asia Pacific, 21% in Europe, and 7% for the rest of the world on the basis of the vertical, organization size, and components of the industry.

59.7 Conclusions

Valuable insights into the future market trends and the plethora of opportunities available to the global AI community are focused through the course of this study. Since the nineties when AI first evolved, the field has been growing with dominance. Tools and algorithms are continuously progressing to accelerate hardware performance while optimizing computational resources. In the foreseeable future, AI techniques

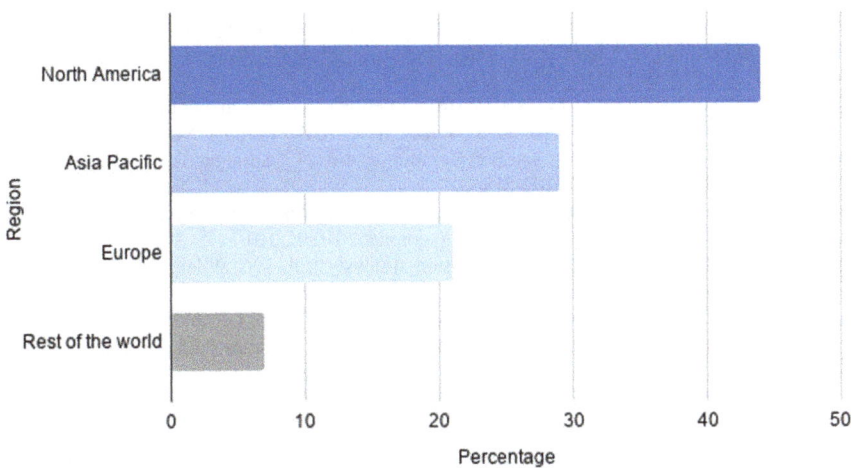

Fig. 59.7 Global machine learning market by region in year 2017

will encompass all unexplored domains such as climate change, clean energy, health care, and education addressing the world's challenging issues.

References

1. Alkrimi, J., Ahmad, A., George, L., Aziz, S.: Review of artificial intelligence. Int. J. Sci. Res. **2**, 487–505 (2013)
2. Oke, S.: A literature review on artificial intelligence. Int. J. Inf. Manage. Sci. **19**, 535–570 (2008)
3. Schmidt, J., Marques, M.R.G., Botti, S., et al.: Recent advances and applications of machine learning in solid-state materials science. NPJ Comput. Mater. **5**, 83 (2019). https://doi.org/10.1038/s41524-019-0221-0
4. Weinreb, D., Moon, D.: The lisp machine manual. ACM SIGART Bull. **78**, 10–10 (1981)
5. Weizenbaum, J.: ELIZA—a computer program for the study of natural language communication between man and machine. Commun. ACM **9**(1), 36–45 (1966)
6. https://www.forbes.com/sites/gilpress/2016/12/30/a-very-short-history-of-artificial-intelligence-ai/#135cebe96fba
7. Moravec, H.P.: The Stanford cart and the CMU rover. Proc. IEEE **71**(7), 872–884 (1983)
8. Barker, V.E., O'Connor, D.E., Bachant, J., Soloway, E.: Expert systems for configuration at Digital: XCON and beyond. Commun. ACM **32**(3), 298–318 (1989)
9. High, R.: The era of cognitive systems: an inside look at IBM Watson and how it works, pp. 1–16. Redbooks, IBM Corporation (2012)
10. Hassoun, M.H.: Fundamentals of Artificial Neural Networks. MIT Press, Cambridge (1995). https://doi.org/10.1109/JPROC.1996.503146
11. Fogel, D.B.: Evolutionary Computation: Toward a New Philosophy of Machine Intelligence, vol. 1. Wiley, New York (2006)
12. Ryan, C., Collins, J.J., Neill, M.O.: Grammatical evolution: evolving programs for an arbitrary language. In: European Conference on Genetic Programming, pp. 83–96. (1998)
13. Joseph, S., Sedimo, K., Kaniwa, F., Hlomani, H., Letsholo, K.: Natural language processing: A review. Nat. Lang. Process. Rev. **6**, 207–210 (2016)

14. James, G., Witten, D., Hastie, T., Tibshirani, R.: Statistical learning. In: An Introduction to Statistical Learning. Springer Texts in Statistics, vol. 103. Springer, New York (2013). https://doi.org/10.1007/978-1-4614-7138-7_2

15. Nikravesh, M.: Evolution of fuzzy logic: from intelligent systems and computation to human mind. In: Nikravesh, M., Kacprzyk, J., Zadeh, L.A. (eds.) Forging New Frontiers: Fuzzy Pioneers I. Studies in Fuzziness and Soft Computing, vol. 217. Springer, Berlin (2007). https://doi.org/10.1007/978-3-540-73182-5_3

16. Atzori, L., Iera, A., Morabito, G.: The internet of things: a survey. Comput. Netw. **54**(15), 2787–2805 (2010). https://doi.org/10.1016/j.comnet.2010.05.010

17. Draper, N.R., Smith, H.: Applied Regression Analysis, vol. 326. Wiley, New York (1998). https://doi.org/10.1002/9781118625590

18. Vlachos, M.: Dimensionality reduction. In: Sammut, C., Webb, G.I. (eds.) Encyclopedia of Machine Learning and Data Mining. Springer, Boston (2017). https://doi.org/10.1007/978-1-4899-7687-1_71

19. Pandey, M., Taruna, S.: A comparative study of ensemble methods for students' performance modeling. Int. J. Comput. Appl. **103**, 26–32 (2014). https://doi.org/10.5120/18095-9151

20. Safavian, S.R., Landgrebe, D.: A survey of decision tree classifier methodology. IEEE Trans. Syst. Man Cybern. **21**(3), 660–674 (1991). https://doi.org/10.1109/21.97458

21. Anyanwu, M.N., Shiva, S.G.: Comparative analysis of serial decision tree classification algorithms. Int. J. Comput. Sci. Secur. **3**(3), 230–240 (2009)

22. Lasota, T., Sachnowski, P., Trawiński, B.: Comparative analysis of regression tree models for premises valuation using statistica data miner. In: Nguyen, N.T., Kowalczyk, R., Chen, S.M. (eds.) Computational Collective Intelligence. Semantic Web, Social Networks and Multiagent Systems (ICCCI 2009). Lecture Notes in Computer Science, vol. 5796. Springer, Berlin (2009). https://doi.org/10.1007/978-3-642-04441-0_68

23. Neal, R.M.: Bayesian Learning for Neural Networks, vol. 118. Springer Science & Business Media (2012)

24. Noble, W.: What is a support vector machine? Nat. Biotechnol. **24**, 1565–1567 (2006). https://doi.org/10.1038/nbt1206-1565

25. Jain, A.K., Narasimha Murty, M., Flynn, P.J.: Data clustering: a review. ACM Comput. Surv. (CSUR) **31**(3), 264–323 (1999)

26. Fang, L., Qizhi, Q.: The study on the application of data mining based on association rules. In: 2012 International Conference on Communication Systems and Network Technologies, Rajkot, pp. 477–480 (2012). https://doi.org/10.1109/CSNT.2012.108

27. Hipp, J., Güntzer, U., Nakhaeizadeh, G.: Algorithms for association rule mining—a general survey and comparison. ACM SIGKDD Explor. Newsl. **2**(1), 58–64 (2000). https://doi.org/10.1145/360402.360421

28. Pouyanfar, S., Sadiq, S., Yan, Y., Tian, H., Tao, Y., Reyes, M.P., et al.: A survey on deep learning: Algorithms, techniques, and applications. ACM Comput. Surv. (CSUR) **51**(5), 1–36 (2018). https://doi.org/10.1145/3234150

29. Artificial Intelligence Market Forecasts (2020). https://tractica.omdia.com/research/artificial-intelligence-market-forecasts/

30. Algorithma: 2020 state of enterprise machine learning (2020). https://info.algorithmia.com/hubfs/2019/Whitepapers/The-State-of-Enterprise-ML-2020/Algorithmia_2020_State_of_Enterprise_ML

31. Sage Growth Partners, Blackbook Research, 2020. COVID-19 Market Pulse, pp. 1–4. [Online]. Black Book Market Research, United States. Available at: https://blackbookmarketresearch.com

32. Ugalmugale, S., Swain, R.: Telemedicine Market Share Report|Global 2020–2026 Industry Data. [Online]. Global Market Insights, United States (2020). Available at: https://www.gminsights.com/industry-analysis/telemedicine-market

33. Bhutani, A., Wadhwani, P.: Wearable AI market trends—Industry Statistics Report 2025 (2019). https://www.gminsights.com/industry-analysis/wearable-ai-market

34. Fortune Business Insights, 2020. Market Research Report. [Online]. Fortune Business Insights.https://www.fortunebusinessinsights.com/industry-reports/toc/artificial-intelligence-market-100114
35. Grand View Research, 2020. Market Analysis Report. [Online]. Grand View Research. https://www.grandviewresearch.com/industry-analysis/artificial-intelligence-ai-market
36. López de Mántaras, R.: The future of AI: toward truly intelligent artificial intelligences. In: Towards a New Enlightenment? A Transcendent Decade. BBVA, Madrid (2018)
37. Revenues from the Artificial Intelligence (AI) software Market worldwide from 2018 to 2025. (2020). https://www.statista.com/statistics/607716/worldwide-artificial-intelligence-market-revenues/
38. Market Key Insight and COVID-19 Impact Analysis (2020). https://www.marketresearchfuture.com/reports/machine-learning-market-2494

Chapter 60
Development of Porting Analyzer to Search Cross-Language Code Clones Using Levenshtein Distance

Sanjay B. Ankali and Latha Parthiban

Abstract Software forking is the process of creating a variant of the existing system by reusing more reliable algorithms and the design of the legacy systems. The software porting team converts the legacy system to the new system by adding or deleting of features and retaining the more reliable algorithm and design. A cross-language clone detector can significantly improve the performance of the software forking team by helping them to locate the code clones of legacy software. This paper proposes a technique to detect functionally equivalent code clones across C, C++, and Java codes by employing the Levenshtein distance algorithm to build a porting analyzer that helps the software forking team with three features like (i) locating all functionally similar cross-language clones among the C, C++, and Java project repositories. (ii) Given an input code for searching, it fetches the functionally equivalent code clone types from the respective repositories. (iii) Displays bug fixes to convert the code from one language to another language. The proposed cross-language clone detector outperforms the accuracy of existing works in terms of precision and can enhance the speed of the software forking process. Results prove that keyword-based filtering using Levenshtein distance detects the cross-language type 1 (Exact) clones and type 2 (Renamed) clones with 100% precision and detects type 3 (Near miss) clones and type 4 (Semantic) clones with 98% and 95% precision, respectively, which is found to be the best accuracy in finding cross-language code clones so far.

S. B. Ankali (✉)
KLE College of Engineering and Technology, Chikodi 591201, India

Research Scholar, VTU-RRC, Karnataka, India

L. Parthiban
Pondicherry University Community College, Pondicherry 605008, India

© The Author(s), under exclusive license to Springer Nature Singapore Pte Ltd. 2021
S. C. Satapathy et al. (eds.), *Smart Computing Techniques and Applications*,
Smart Innovation, Systems and Technologies 225,
https://doi.org/10.1007/978-981-16-0878-0_60

60.1   Introduction

The Internet contains trillion lines of code available nowadays that make the development process much simple [1]. For the same reason, the developers tend not to start coding from the scratch [2]. We find evidence of code reusing in a survey conducted by Vaibhav Saini [3] involving 72 developers, and 96% of the developers prefer searching and reusing code before the start of coding task. The main drawback of using the online source code is maintainability and bug propagation [4] or tends to preach software licenses [5, 6]. Text search engines like Google cannot search the functionally similar code by taking the query as code. Hence, there is a need for a software clone extraction system that takes code as a query and fetches the functionally similar code from the repository by matching the complete code [7]. We propose a technique to accurately classify the cross-language clones of C, CPP, and Java code.

Research Question

1. For the software forking team which is about to convert a large C-based legacy system into C++ or Java-based system containing N separate C code files, is it possible to fetch the corresponding functionally similar codes of C++ or Java from the repository of C++/Java and vice versa with all four clone types classified?
2. Assuming there are three repositories of C, C++, and Java, given a code as a query, is it possible to extract the functionally similar codes of C or C++ with clone types classified into 1, 2, 3, and 4 accurately?

60.2   Background and Related Work

Creating functionally similar programs with or without syntactical change is called software cloning or code cloning [8]. Based on the editing taxonomy, most of the researchers define four basic types of clones [9]. These clone types are grouped into syntactic and semantic classes.

Syntactic: *Type* 1 is also called as exact clones, Type 2 is also called as renamed clones, and Type 3 is near-miss clones. Syntactic clones are based on different editing taxonomies.

Semantic: *Type 4* is semantically similar codes that are different in syntax.

In this section, we briefly define the clone types and present the C and CPP examples to justify our understanding of cross-language clone types which are based on the many great clone detection research in the past [9].

60.2.1 Background

Definition of Cross-language Clones

With the definition of clones mentioned, we can say that two or more codes that produce the same output but implemented in different languages with different four types of variations are called cross-language clones. For an instance, C-code-1 that calculates factorial of a number is called a cross-clone of CPP-code-1 code that calculates factorial of a number [10].

A few example case studies are as follows.

```
1. main()                        1. main()
2. {                             2. {
3. Int x=100,y=200,z;            3. int x=100,y=200,z;
4. z= x+y ;                      4. z=x+y;
5. printf("sum of two           5. cout<<"sum of two
   numbers=%d", z);                 numbers is ="<<z;
6. }                             6. }
        C-code-1                         CPP-code-1
```

Except for the structure of the language, both the codes above behave similarly. They declare the same two variables x and y, the sum is stored in a variable z, and the result is displayed on the screen. These examples are type 1 cross-clones.

```
1. main()                        1.main()
2. {                             2.{
3. int a=10 , b=20, c;           3.int a = 10, b=20,c;
4. c = a  +  b;                  4.c = a + b;
5.printf("addition program")     5.cout <<"sum  of  two numbers
6. printf("sum of two               is ="<< c;
numbers=%d", c);                 6.}            CPP-code-2
7. }        C-code-2
```

From the above code snippet, C-code-1 and CPP-code-2 are examples of type 2 clones, whereas C-code-2 is type 3 clone of CPP-code-2.

As the last example, we consider one looping statement.

```
1. int factorial( int n)         1. int fact ( int  n)
2. {                             2. {
3. int  i, fact=1, n;            3. int i =1,fact = 1,n;
4. for ( i = 1;i < = n ; i++)    4. while ( i < = n)
5. fact = fact * i;              5. {
6. return (fact);                6.    fact = fact * i;
7. }                             7.    i++;
          C-code-3               8. }
                                 9. return(fact);
                                 10. }        CPP-code-3
```

C-code-3 and CPP-code-3 try to find factorial of integer number by using for and while loops, respectively, hence are called type 4 clones. The proposed work will significantly contribute to the software porting process by identifying all the functionally similar code fragments of C, CPP, and Java codes.

60.2.2   Related Work

Past three decades have seen more than 250 tools and techniques in the same language clone detection but very less significant work has happened in cross-language clone detection. In this section, we introduce cross-clone detection with their issues. A tool *CroLSim* [11] is proposed to find software similarity for open-source software clustering. But the results show only 28% precision for searching functionally similar code in a repository. A tool *SLACC* [12] presents the cross-language clone detection based on the input/output behavior of code. The tool lacks three aspects: (i) finds only semantic similarities, (ii) does not classify cross-language clone types (type 1, 2, 3, 4) and (iii) does not support long and complex types of Python D. dead code elimination. A recent study on cross-language source code plagiarism by *Karnalim* [13] works in four steps. But the approach does not handle identifier renaming and (ii) works by converting the original code to intermediate code which is more computationally complex. An AST-based approach proposed by *Perez* [14] is a semi-supervised machine learning model. The system has limitations in (i) matching two codes of Java and Python with only 75% confidence and (ii) does not support more granular clone-type classification (type 1, 2, 3, and 4). A tool *CLCDSA* is proposed by *K. W. Nafi* [11] used action filters to filter out non-probable clones and make the model more scalable. But the approach did not present the results to prove the clone-type classification. A study *Nghi D. Q. Bui* [15] is proposed which is based on bilateral neural networks (BiNN's)-based technique to find similarity and difference in structure based on the language AST using BTBCNN. The method has two issues (i) large programs may slow down the training process and (ii) no results were found to prove the type classification.

Hu et al. [16] proposed the technique using the binary instructions to detect the semantically similar functions through a series of four steps but the approach detects only semantically similar techniques and is vulnerable to the code obfuscation.

An integrated tool *LICCA* [17] is based on the application of the modified longest common subsequence algorithm on enriched concrete syntax tree (eCST) of source code. But the work is limited to semantic clone detection. Clone detection among the.NET language family was proposed by *Al-Omari* et al. [18]. But the work is platform dependent. The work of *Lawton Nichols* [10] is the extension of previous work [19] that finds syntactic similarity using structural and nominal similarity. The method works on the function instead of diffs of VCS and currently handles (C++, Java, and JavaScript). But the system works well only for object-oriented programming languages.

60.3 Proposed Methodology

Figure 60.1 shows the keyword-based four steps to obtain cross-language code clones using the Levenshtein distance algorithm. In step 1, the input code of C, C++, and Java is tokenized. In the second step, it searches all the tokens based on the keywords of the corresponding language and stores the keywords in the single dimensional array.

```
  Algorithm
similarity_Match(String str1,
String str2)
{
String big = str1,small = str2;
if (length(str1) < ength(str2))
    {
        big = str2;
        small = str1;
    }
 int len =length(big);
  if (len == 0)
 { return 1.0;   }
 len= len-edit_Distance(big,
 small)) /(double)len;
          return(len);
 }
```

```
  Algorithm
edit_Distance(String s1,
String s2, int len2)
  {
 int[] costs = new;
 int[length(s2) + 1];
      for i <- 0 to len {
          int old = i;
          for j<- 0 to len {
      if (i == 0)
        costs[j] = j;
        else {
        if (j > 0) {
      int new = costs[j - 1];
    if (s1[i - 1] !=s2[j - 1])
        min((new,old), costs[j])
+ 1);
      costs[j - 1] = old;
      old = new;    }
              }
          }
      if (i > 0)
      costs[length(s2)] = old;
        }
    return costs[length(s2)];
      }
```

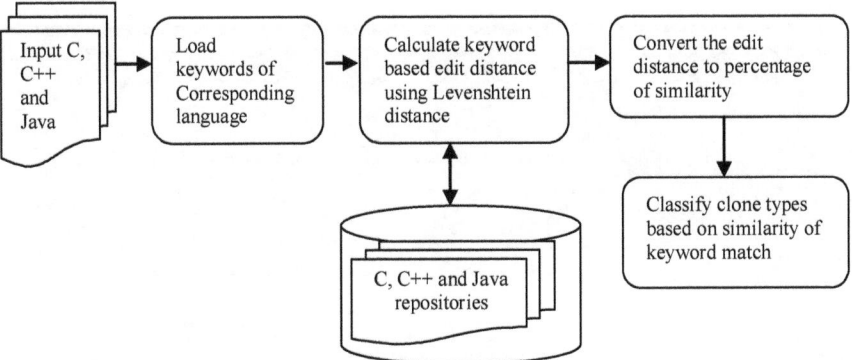

Fig. 60.1 Proposed cross-language clone detection system

Table 60.1 Thresholds of clone-type classification based on similarity values

Similarity matching	Clone type
100	1
≥90	2
≥82 and <90	4
≥68 and <82	3

In step 3, Levenshtein distance is called between the prepared 1D array and all the code documents present in the repositories to calculate the edit distance between the two codes.

As the last step, the classification of clone types is performed by manually validating the matching percentage of all four clone types according to the following thresholds of matching given in Table 60.1.

60.4 Results and Discussion

We calculate the accuracy of the proposed model in terms of precision and recall which tries to find what proportion of positive instances was correct [20].

For instance, we know that C-code-1 and C-code-2 are type 3. If the matching result is type 2, then its true positive results. But if we get the result of matching as any other type, then we call it false positive. With this basic knowledge, we define precision as *Precision = True Positive/ True Positive + False Positive.*

Recall tries to find how many positives are identified correctly by calculating *True Positive/True Positive + False Negative.* Figure 60.2 shows the comparison of two project repositories of C and C++ to detect type 1 and type 2 clones between two languages. We have validated the proposed model on the C project containing 136 C codes and 82 functionally similar C++ codes. The framework shows 100% precision in finding the type 1 and 2 clones.

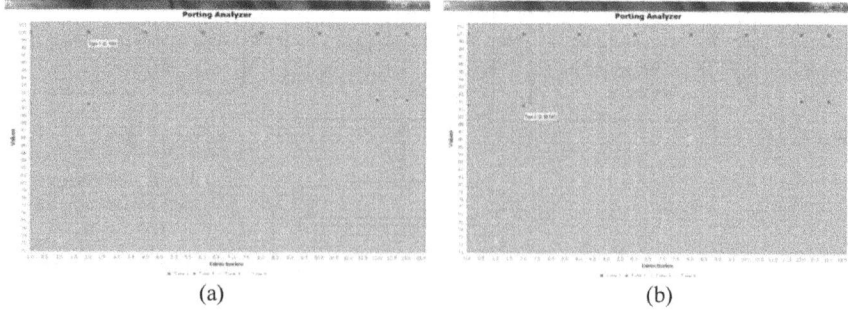

(a)   (b)

Fig. 60.2 a Graph showing common type 1 files between two projects of C and CPP. **b** Graph showing common type 2 files between two projects of C and CPP

Figure 60.3 shows the comparison of two project repositories of C and C++ to detect type 3 and type 4 clones between two languages. We have validated the proposed model with the manual inspection to find type 3 and type 4 clones. The framework shows 98% precision in finding the type 3 and 95% precision in finding four clones.

Figure 60.4 shows the execution of the porting analyzer where the developer tries to search the functionally similar codes of C++ and Java code for the input C code that adds two integer numbers.

Figure 60.5 shows the search results where type 1 code clones of Java and C++ are fetched from the corresponding repositories. Similarly, Fig. 60.4 shows the search results for type 4 code clones of C and Java codes given input as CPP file for finding factorial of a number using do–while loop. The porting analyzer also helps the

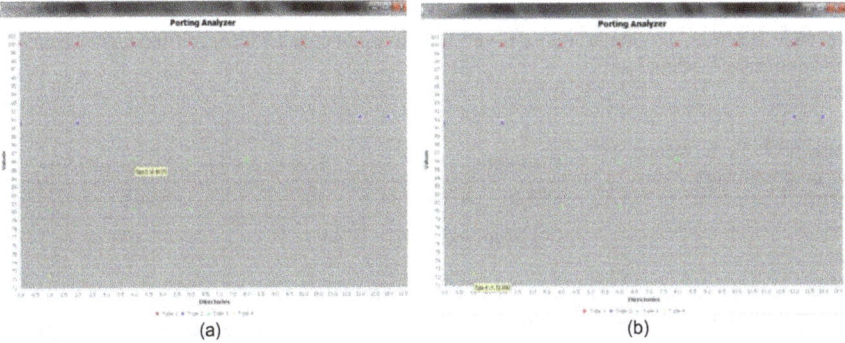

(a)   (b)

Fig. 60.3 **a** Graph showing common type 3 files between two projects of C and CPP. **b** Graph showing common type 4 files between two projects of C and CPP

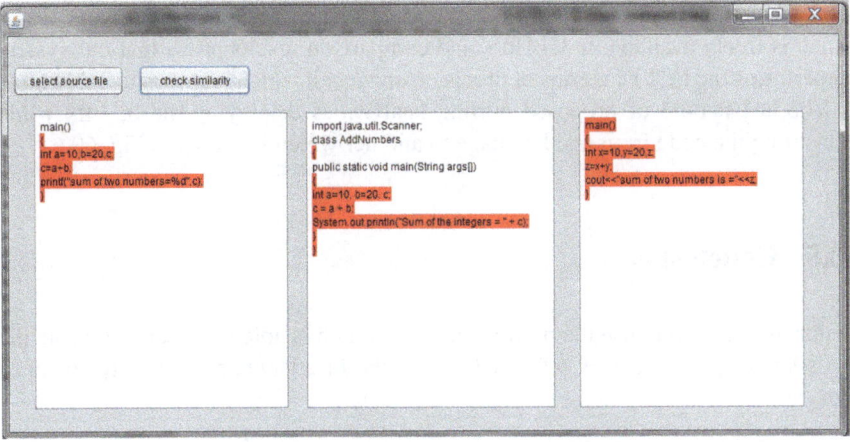

Fig. 60.4 Fetching the similar files of CPP and Java given input as C file for adding two numbers

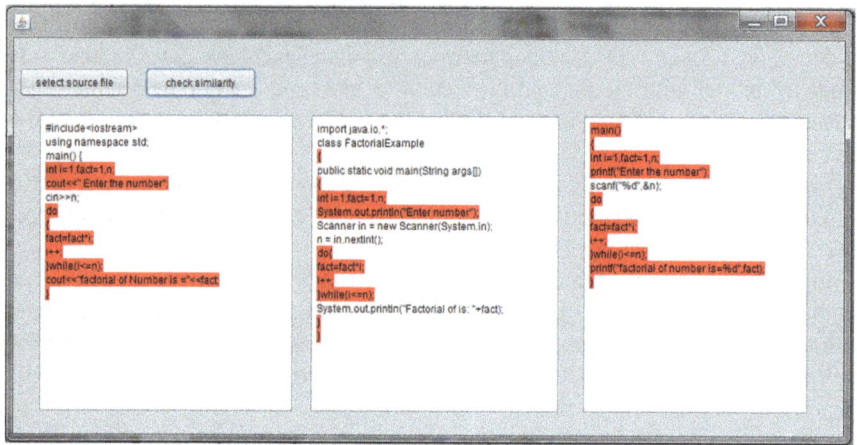

Fig. 60.5 Fetching the similar files of C and Java given input as CPP file for finding factorial of a number using do–while loop

Table 60.2 Performance of our proposed work versus FETT

Technique	Precision (%)	Recall (%)
FETT	2.4	75
Our proposed work	90	100

development team to convert the project from one language to another language by presenting the number of bug fix edits required to convert the functionally similar code from one language to another.

Performance comparison with FETT proposed by Nichols [10].

We have compared the results of our proposed system with the recent work FETT which is freely available in GitHub, and comparison results prove that our system outperforms the FETT in terms of precision and recall which is shown in Table 60.2.

The last feature of proposed porting analyzer is displaying the bug fix while converting the code from one language to another which is shown in Fig. 60.6.

60.5   Conclusion

In this paper, we propose the porting analyzer which is able to detect and locate the cross-language code clones between C, C++, and Java languages with three features like.

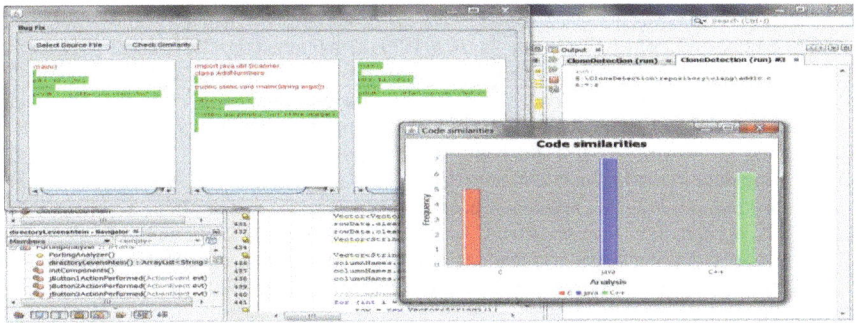

Fig. 60.6 Bug fixes to convert a file from C to C++ and Java

(i) Identifying functionally similar codes across multiple languages (locating C++ and C code given Java code as input and vice versa). (ii) Fetching the complete functionally similar codes across three multiple languages from their respective repositories. (iii) Presenting the bug fixes to convert the code from one language to another across three languages.

The technique can be employed by a software (forking) development team that is involved in converting the legacy software product to a new version of it by addition and deletion of features. Comparison of accuracy in terms of precision and recall with FETT which is only designed for object-oriented programming languages proves that the application of Levenshtein distance using keyword-based searching can be a more accurate way to build the cross-language clone detectors.

References

1. Sadowski, C., Stolee, K.T., Elbaum, S.: How developers search for code: a case study. In: Proceedings of the 2015 10th Joint Meeting on Foundations of Software Engineering, pp. 191–201 (2015)
2. Acar, Y., Backes, M., Fahl, S., Kim, D., Mazurek, M.L., & Stransky, C.: You get where you're looking for: the impact of information sources on code security. In: 2016 IEEE Symposium on Security and Privacy (SP), pp. 289–305. IEEE (2016)
3. Saini, V., Farmahinifarahani, F., Lu, Y., Baldi, P., Lopes, C.V.: Oreo: Detection of clones in the twilight zone. In: Proceedings of the 2018 26th ACM Joint Meeting on European Software Engineering Conference and Symposium on the Foundations of Software Engineering, pp. 354–365. (2018)
4. Abdalkareem, R., Shihab, E., Rilling, J.: On code reuse from stackoverflow: an exploratory study on android apps. Inf. Softw. Technol. **88**, 148–158 (2017)
5. An, L., Mlouki, O., Khomh, F., Antoniol, G.: Stack overflow: a code laundering platform? In: 2017 IEEE 24th International Conference on Software Analysis, Evolution and Reengineering (SANER), pp. 283–293. IEEE (2017)
6. Baltes, S., Diehl, S.: Usage and attribution of Stack Overflow code snippets in GitHub projects. Empir. Softw. Eng. **24**(3), 1259–1295 (2019)
7. Ragkhitwetsagul, C., Krinke, J.: Siamese: scalable and incremental code clone search via multiple code representations. Empir. Softw. Eng. **24**(4), 2236–2284 (2019)

8. Wang, W., Li, G., Ma, B., Xia, X., Jin, Z.: Detecting code clones with graph neural network and flow-augmented abstract syntax tree. In: 2020 IEEE 27th International Conference on Software Analysis, Evolution and Reengineering (SANER), pp. 261–271. IEEE (2020)
9. Roy, C.K., Cordy, J.R., Koschke, R.: Comparison and evaluation of code clone detection techniques and tools: a qualitative approach. Sci. Comput. Program. **74**(7), 470–495 (2009)
10. Nichols, L., Emre, M., & Hardekopf, B. (2019, April). Structural and nominal cross-language clone detection. In: International Conference on Fundamental Approaches to Software Engineering, pp. 247–263. Springer, Cham
11. Nafi, K.W., Kar, T.S., Roy, B., Roy, C.K., Schneider, K.A.: CLCDSA: cross language code clone detection using syntactical features and API documentation. In: 2019 34th IEEE/ACM International Conference on Automated Software Engineering (ASE), pp. 1026–1037. IEEE (2019)
12. Mathew, G., Parnin, C., Stolee, K.T.: SLACC: Simion-based language agnostic code clones (2020). arXiv preprint arXiv:2002.03039
13. Karnalim, O.: TF-IDF inspired detection for cross-language source code plagiarism and collusion. Computer Sci. **21**(1) (2020)
14. Perez, D., Chiba, S.: Cross-language clone detection by learning over abstract syntax trees. In: 2019 IEEE/ACM 16th International Conference on Mining Software Repositories (MSR), pp. 518–528. IEEE (2019)
15. Bui, N.D., Jiang, L., Yu, Y.: Cross-language learning for program classification using bilateral tree-based convolutional neural networks (2017). arXiv preprint arXiv:1710.06159
16. Hu, Y., Zhang, Y., Li, J., Gu, D.: Binary code clone detection across architectures and compiling configurations. In: 2017 IEEE/ACM 25th International Conference on Program Comprehension (ICPC), pp. 88–98. IEEE (2017)
17. Vislavski, T., Rakić, G., Cardozo, N., Budimac, Z.: LICCA: A tool for cross-language clone detection. In: 2018 IEEE 25th International Conference on Software Analysis, Evolution and Reengineering (SANER), pp. 512–516. IEEE (2018)
18. Al-Omari, F., Keivanloo, I., Roy, C.K., Rilling, J.: Detecting clones across microsoft .net programming languages. In: 2012 19th Working Conference on Reverse Engineering, pp. 405–414. IEEE (2012)
19. Cheng, X., Peng, Z., Jiang, L., Zhong, H., Yu, H., Zhao, J.: CLCMiner: detecting cross-language clones without intermediates. IEICE Trans. Inf. Syst. **100**(2), 273–284 (2017)
20. Manning, C.D., Raghavan, P., Schutze, H.: Introduction to information retrieval (2008)

Chapter 61
Multi-Objective Genetic Algorithm for Hyperspectral Image Analysis

T. Subba Reddy, J. Harikiran, and B. Sai Chandana

Abstract The main objective of hyperspectral Image Classification is to group pixels into spectral classes, where each class having a unique label representing specific information in the image. The classification can be done using methods categorized as supervised and unsupervised. The contrast of hyperspectral images is degraded if there is any disturbance of the transmission medium. This disturbance degrades the quality of the image generated by the sensor, which effects the classification accuracy. In this paper, Genetic Algorithm (GA) is used for hyperspectral image analysis. The algorithm is used in contrast enhancement, dimensionality reduction, and classification. This dimensionality reduction will remove less informative bands, decrease storage space, computational load, and communication bandwidth. The experimental results show the improvement of accuracy in classifying Indian Pines and Pavia University datasets

61.1 Introduction

The remote sensing sensors provide data with higher spectral resolution staring from panchromatic to multispectral, multispectral to hyperspectral, and hyperspectral to ultra-spectral. The difference in data collected by the sensor from panchromatic to ultra-spectral is the number of narrow spectral bands and greater spectral resolution. Hyperspectral data contains hundreds of narrow spectral bands. The value of any pixel in a hyperspectral image is not formed by a single value, but by a collection of values in all spectral bands. This collection of values associated with a single pixel is called a spectral signature. This range of values for a single pixel will provide much information about the object in the image.

T. Subba Reddy (✉) · J. Harikiran
School of Computer Science and Engineering, Vellore Institute of Technology,
VIT-AP University, Amaravati, Andhra Pradesh, India

B. Sai Chandana
Department of CSE, Gudlavalleru Engineering College, Gudlavalleru, Vijayawada,
Andhra Pradesh, India

© The Author(s), under exclusive license to Springer Nature Singapore Pte Ltd. 2021   633
S. C. Satapathy et al. (eds.), *Smart Computing Techniques and Applications*,
Smart Innovation, Systems and Technologies 225,
https://doi.org/10.1007/978-981-16-0878-0_61

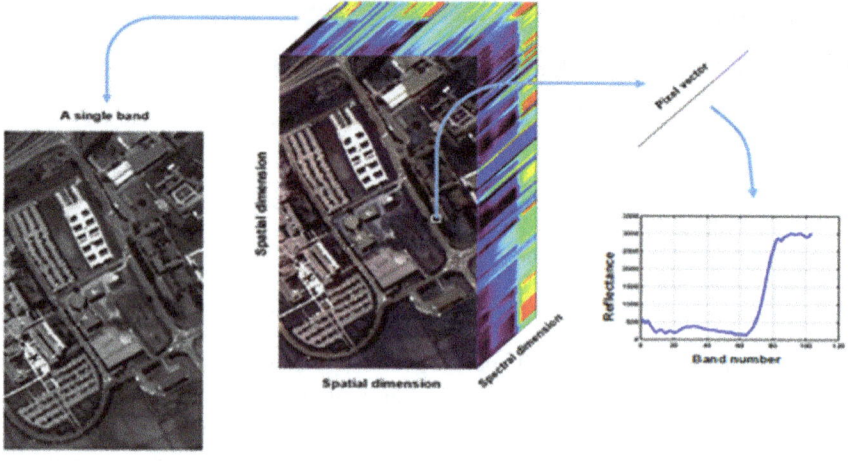

Fig. 61.1 Example of hyperspectral data cube

Hyperspectral Imaging: The sensors in hyperspectral imaging systems are operated from visible wavelength to infrared wavelength, capturing data in hundreds of narrow spectral bands representing the same area of information on the earth. The data for each pixel is not any single value, but it is represented in the form of a vector with each value represents a measurement in that distinct wavelength [1]. The length of the vector is equal to the total number of spectral bands. In hyperspectral data, hundreds of bands are available representing the same scene on earth, while in multispectral data only tens of bands are available. This high dimensionality of data in hyperspectral provides more discriminative information of objects under the scene thus increasing the accuracy of the classification algorithm. Hyperspectral Imaging Systems are used in various applications [2] such as Ecological Science, Geological Science, Minerology, Hydrological Science and Military Applications.

Hyperspectral datasets are called data cubes, where each spectral channel has a specific wavelength representing a grayscale image [3]. Hyperspectral data cube is shown in Fig. 61.1. This paper presents the application of Genetic Algorithm in hyperspectral image enhancement, dimensionality reduction, and classification. The genetic algorithm belongs to a larger class of evolutionary algorithms [4].

61.2 GA Based Hyperspectral Image Contrast Enhancement

Genetic Algorithms [4] is a kind of evolutionary biological process where the solution to any optimization problem is generated using a parallel search in the solution space.

Bi-Dimensional Empirical Mode Decomposition [5]: It decomposes any signal into functions then it is called Intrinsic Mode Function (IMF). This decomposition is a nondestructive process means that summation of these IMFs gives the original signal. The IMFs generated by decomposition process have two specific properties, one is the extreme points and number of zero crossings must be equal or differ by 1, and two the IMFs have zero mean. The first most IMF is a high frequency component and last component which is called as residue is a low frequency component. Summation of IMFs and residue generates the original signal. This signal decomposition using EMD is called shifting process. The same procedure can be extended to two dimensional, which means that decomposition of image named as Bi-dimensional Empirical Mode decomposition [BEMD].

Hyperspectral Image Enhancement Using Genetic Algorithm: The method for image enhancement can be summarized as follows.

i. Given hyperspectral image I is divided into IMFs using 2-D empirical mode decomposition method [5].

$$\text{EMD}(I) = [\text{imf}_1, \text{imf}_2, \ldots, \text{imf}_n] \qquad (61.1)$$

ii. The obtained IMFs are used to reconstruct the enhanced image. Each IMF is multiplied with a weight w, and the summation of IMFs with weights gives the enhanced image. The weight for each IMF is obtained using GA.

$$\text{RI} = \sum_{i=1}^{n} w_i * \text{imf}_i \qquad (61.2)$$

where w_i denotes the weight corresponding to imf_i, and RI represents the reconstructed image with enhancement.

iii. The GA is used below in enhancement:

(a) First randomly generate the weights of IMFs representing the initial population. The length of elements of each chromosome in the population is equal to the number of IMFs generated by EMD and the element value represents the weight of the corresponding IMF. The values of weights differ between 0 and 1 and the summation of weights is equal to 1.

(b) The information entropy is used as a fitness function for chromosomes in the GA process. The entropy is defined as

$$H = -\text{sum}(p. * \log 2(p)) \qquad (61.3)$$

(c) The chromosomes are selected by the Roulette Wheel Technique.
(d) The cross-over rate is equal to 0.7 and the mutation rate equal to 0.07.
(e) The stopping criteria are maximum iterations.

Solution = chromosome with maximum fitness.

The IMF's optimum weights are the values in this chromosome. With these weights and IMFs, the enhanced image is constructed.

61.3　Dimensionality Reduction of Hyperspectral Dataset Using Genetic Algorithm

The huge volume of data, higher dimensionality and redundant information impose negative effect in hyperspectral image analysis. The dimensionality reduction of hyperspectral data was grouped into two types, feature extraction and band selection. The band selection methods are better than feature extraction methods [6], as it selects the effective combination of the original bands by preserving most of the physical characteristics of data without losing any precious details. Some of the existing band selections methods are presented below:

Entropy: The term Entropy measures the data, and it presents in each individual band in terms of uncertainty. The entropy was defined as

$$H(A) = - \sum_{A \in \mathcal{A}} p(A) \log p(A). \tag{61.4}$$

Mutual Information [MI]: [7] The MI is an information measure extended to two variables, one spectral band and another reference image that related to the classification objective. It provides the similarity between two arbitrary variables defined as

$$I(A, B) = \sum_{A \in \mathcal{A}, B \in \mathcal{B}} p(A, B) \log \frac{p(A, B)}{p(A) \cdot p(B)}. \tag{61.5}$$

where $p(A, B)$ defines the joint probability distribution.

Euclidean Distance [ED]: A Vector-Based Distance Measure between vectors X and Y extended to N-dimensions was given by the following equation:

$$ED(X, Y) = \sqrt{\sum_{k=1}^{N} (X_k - Y)^2} \tag{61.6}$$

The ED measures the vector differences due to magnitude and direction. Band selection was done by choosing the bands having the higher ED values.

Spectral Angle Mapper [SAM] [8]: It measures the Geometric Angle between two vectors X and Y extended to N-dimensions, which corresponds to similarity of shape. Band selection is done by choosing the bands having the higher SAM values. The equation for SAM is as follows

$$SAM(X, Y) = \arccos\left(\frac{X^T \cdot Y}{\|X\| \cdot \|Y\|}\right) \tag{61.7}$$

Spectral Correlation Mapper [SCM] [9]: It measures the linear relationship between two vectors X and Y extended to N-dimensions is given by

$$SCM(X, Y) = \frac{\sum_{k=1}^{N_b} (X_k - \mu_X) \cdot (Y_k - \mu_Y)}{(N_b - 1) \cdot \sigma_X \cdot \sigma_Y} \tag{61.8}$$

Band selection is done by choosing the bands having the lower SCM values indicating better separability.

Band Correlation [BC] [10]: It is measure of information redundancy of each spectral band, defined as

$$BC(i, j) = \frac{\sum_{p=1}^{N_b} (x_{ip} - \mu_i) \cdot (x_{jp} - \mu_j)}{\sqrt{\sum_{p=1}^{N_b} (x_{ip} - \mu_i)^2} \cdot \sqrt{\sum_{p=1}^{N_b} (x_{jp} - \mu_j)^2}} \tag{61.9}$$

Band selection is done by choosing the bands having the lower BC values, indicating less information redundancy.

GA based Band Selection: The ED, SAM and SCM consider the spectral separability between the bands and BC considers the spectral information between the bands. This new measurement of separability [SEPI] was optimized using Genetic Algorithm (GA) for effective band selection. The SEPI involving ED, SAM, SCM and BC was given by

$$SEPI = \frac{SCM \times BC}{ED \times SAM} \tag{61.10}$$

The mechanism of GA based band selection is given below:

(a)   Initially we assume population with zeros and ones. Zero denotes bands not selected, one denotes band selected. The Number of selected bands are fixed.
(b)   The fitness function for each chromosome

$$\text{fitness function} = \frac{SCM * BC}{ED * SAM} \tag{61.11}$$

(c)   Chromosome selection is done using roulette wheel selection technique.
(d)   To reproduce new children GA operators are used to replacing old children.
(e)   The number of iterations is fixed to stop the GA Process.

The chromosome with highest fitness value is the solution to the dimensionality reduction. The value 1 in the chromosome denotes corresponding band is selected, and value 0 denotes band is removed.

61.4 GA Based Hyperspectral Image Classification

The Genetic Algorithm as follows:

(a) Chromosomes with K clusters are randomly initialized denoting the initial population.

(b) The multiple objective functions used for calculating fitness of chromosome are K-means index (KMI) [11], Jm measure [12] and Xie-Beni Index (XBI) [13].

$$u_{ik} = \frac{1}{\sum_{j=1}^{K} \left(\frac{D(z_i, x_k)}{D(z_j, x_k)} \right)^{\frac{2}{m-1}}} \tag{61.12}$$

$$z_i = \frac{\sum_{k=1}^{n} (u_{ik})^m x_k}{\sum_{k=1}^{n} (u_{ik})^m}, \quad 1 \le i \le K \tag{61.13}$$

$$XB = \frac{\sigma(U, Z; X)}{n \operatorname{sep}(Z)} = \frac{\sum_{i=1}^{K} \left(\sum_{k=1}^{n} u_{ik}^2 D^2(z_i, x_k) \right)}{n \left(\min_{i \ne j} \{ D^2(z_i, z_j) \} \right)} \tag{61.14}$$

$$J_m = \sum_{j=1}^{n} \sum_{k=1}^{K} u_{kj}^m D^2(x_j, z_k), \quad 1 \le m \le \infty \tag{61.15}$$

$$KMI = \frac{1}{\sum_{k=1}^{K} \sum_{i=0}^{N} \| x_i - z_k \|^2} \tag{61.16}$$

(c) Chromosome selection is done using Roulette Wheel Technique.

(d) Cross-over rate $= 0.8$ and mutation rate $= 0.07$.

(e) Stopping criteria $=$ maximum iterations. Solution $=$ value of chromosome with maximum fitness.

61.5 Experimental Results

The methodology presented in this article is performed on two standard datasets, nine classes Pavia University and sixteen classes Indian pines. 103 bands with each band size 610×340 existed in Pavia university dataset. 200 bands with each band size 145×145 existed in Indian pines dataset. The datasets are collected from [https://www.ehu.eus/ccwintco/index.php?title=Hyperspectral_Remote_Sensing_Scenes]. The image enhancement using Genetic Algorithm is shown in Fig. 61.3. The band selection is done using SPEI measure and the quality of

Table 61.1 Quality of fused image after band selection

Band selection (After fusion using average method)

Method/India pines dataset	Entropy	Method/Pavia University dataset	Entropy
ED	4.91	ED	4.61
SAM	4.96	SAM	4.78
SCM	4.94	SCM	4.76
BC	4.98	BC	4.81
SEPI	5.09	SEPI	4.89

fused image [using Average Method] by merging the bands selected using each metric was shown in Table 61.1. The SEPI metric selects best bands than other metrics for both datasets. The classification result is shown in Fig. 61.2. By using the methods presented in this paper we get accuracy of classification 92 and 94% for Pavia University and Indian Pines Data.

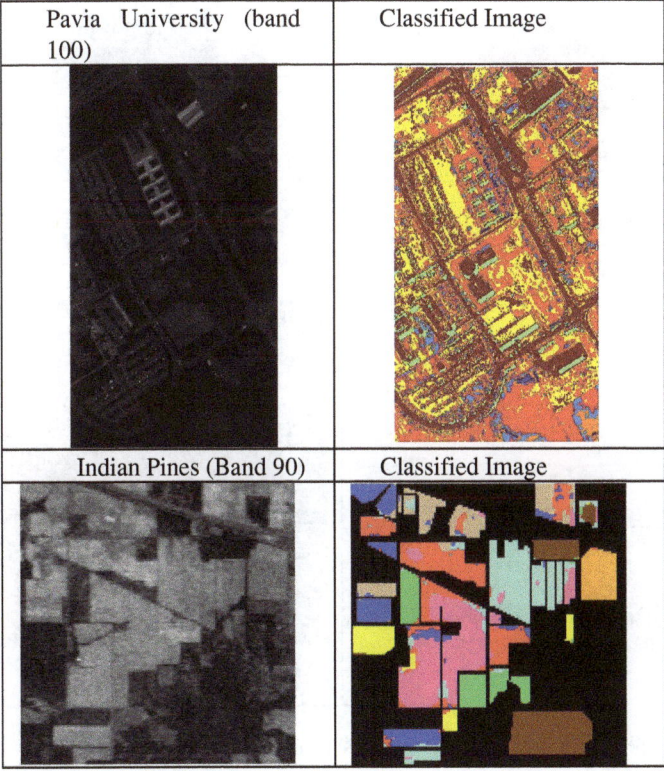

Fig. 61.2 Classification result

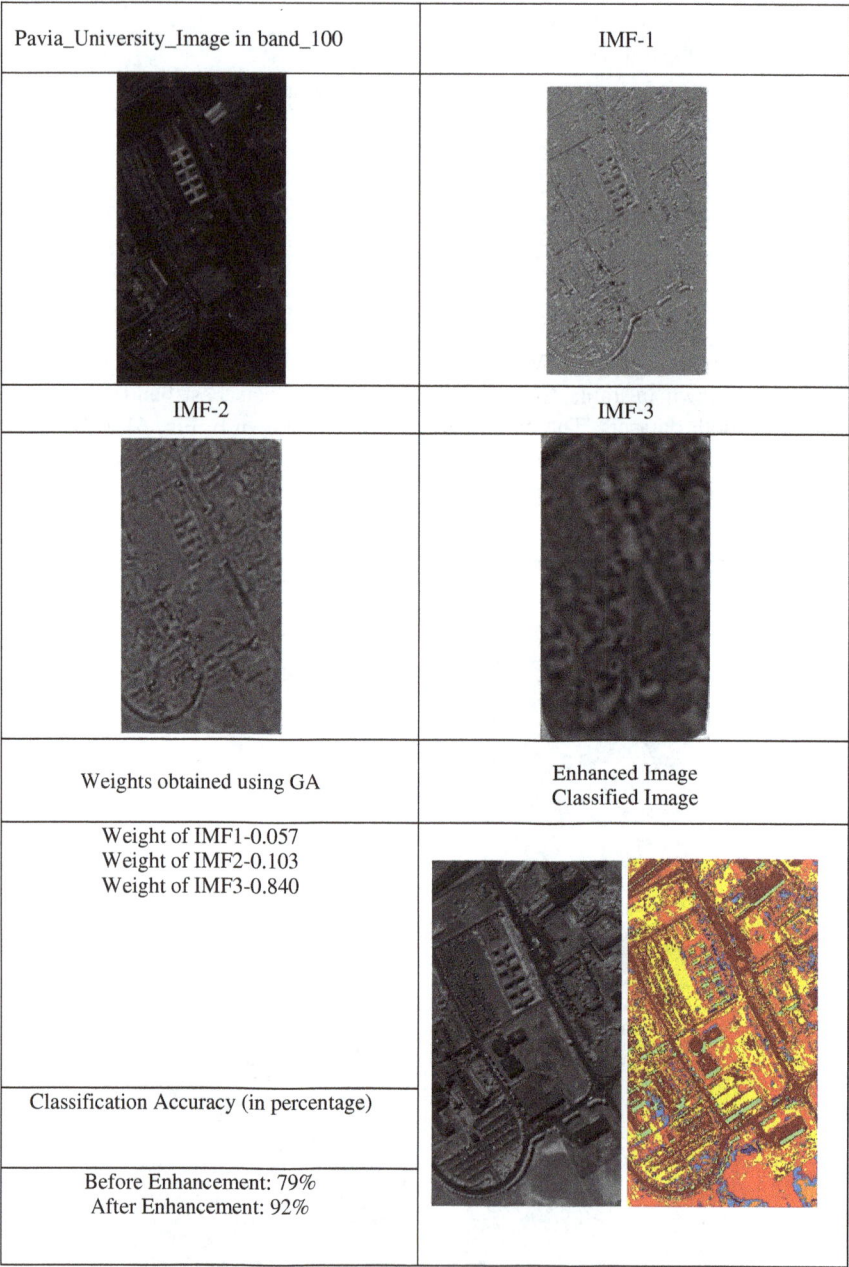

Pavia_University_Image in band_100	IMF-1
IMF-2	IMF-3
Weights obtained using GA	Enhanced Image Classified Image
Weight of IMF1-0.057 Weight of IMF2-0.103 Weight of IMF3-0.840	
Classification Accuracy (in percentage)	
Before Enhancement: 79% After Enhancement: 92%	

Fig. 61.3 Hyperspectral image enhancement using genetic algorithm

61.6 Conclusions

In this paper, we present hyperspectral image analysis using Genetic Algorithm in enhancement, dimensionality reduction and classification. Instead of optimizing single objectives, multiple objectives are optimized using evolutionary algorithms. In future, many algorithms are used such as ACO, PSO, BCO and Firefox algorithm in a similar way for efficient hyperspectral image analysis.

References

1. Harikiran, J. et.al.: Fast clustering algorithms for segmentation of microarray images. IJSER **5**(10), 569–574 (2014)
2. Saichandana, B. et al.: Image fusion in hyperspectral image classification using genetic algorithm. IJEECS **2**(3), 703–711 (2016)
3. Saichandana, B., et al.: Dimensionality reduction and classification of hyperspectral images using genetic algorithm. IJEECS **3**(3), 503–511 (2016)
4. Holland, J.: Adaptation in natural and artificial systems. Univ. of Michigan Press, Ann Arbor, MI (1975)
5. Zeiler, A. et al.: Empirical mode decomposition—an introduction. In: Proceedings of IEEE IJCNN, pp. 1–8 (2010)
6. Yin, J., et al.: A new dimensionality reduction algorithm for hyperspectral image using evolutionary strategy. IEEE Trans. Industr. Inform. **8**(4), 935–943 (2012)
7. Deb, K. et al.: A computationally efficient evolutionary algorithm for real parameter optimization. IEEE Trans. Evol. Comp. **10**(4), 371–395 (1998)
8. Du, Q. et al.: Similarity based unsupervised band selection for hyperspectral image analysis. IEEE Tran. Geos. RS Lett. **5**(4), 564–568 (2008)
9. Gharaati, E. et al.: A new band selection method for hyperspectral images based on constrained optimization. In: Proceedings of IEEE ICIKT, pp. 1–6 (2015)
10. Hongjun, Su., Qian, Du., Peijun, Du.: Hyperspectral image visualization using band selection. IEEE JSTAEORS **10**(4), 2647–2658 (2013)
11. Tarabalka, Y., et al.: Spectral-spatial classification of hyperspectral imagery based on partitional clustering techniques. IEEE Trans. Geos. Remote Sens. **47**(5), 2973–2987 (2009)
12. Zhang, H., et al.: Hyperspectral image restoration using low-rank matrix recovery. IEEE Trans. Geo. RS **52**(8), 4729–4743 (2014)
13. Ngatchou, P. et al.: Pareto multi objective optimization. In: Proceedings of 13th ICISAPS. IEEE (2005)

Chapter 62
Auto-Discovery and Monitoring of Network Resources: SNMP-Based Network Mapping and Fault Management

Shaik Imam Saheb and Abdul Rasool Md

Abstract Network management is a service that employs a variety of tools, applications, and devices to assist human network managers in monitoring and maintaining networks. To know the health and state of the network, network operators use network management systems. Network management systems come with fault, performance, configuration management features. Configuration management module helps to keep track of configuration of each device in the network. Often, configuration of the network is manually created. Fault management is the process of locating and correcting network problems or faults. Fault management is an important element of network management which consists of three steps. In the current work, a solution is implemented to discover the network resources automatically and configure them. The system displays a status map, where a visual display of the status of critical network elements is displayed to allow users to verify and isolate problems. An end-to-end testing function is used for testing on a scheduled or on-demand basis. Each network resource deploys an SNMP agent. SNMP agent helps to monitor the state of the network element and notifies the network manager of the events that occur on the element. To notify the network manager SNMP agent uses a collection of objects called MIB. When the MIB objects are received at the network manager, it is parsed for the agent identifying MIB objects, which are displayed on the user interface. The system manager can monitor and modify the MIB objects as per the network user requirements.

62.1 Introduction

Network management refers to the activities, methods, procedures, and tools that pertain to the operation, administration, maintenance, and provisioning of networked systems [1].

S. I. Saheb (✉) · A. Rasool Md
Department of Computer Science & Engineering, Lords Institute of Engineering and Technology, Hyderabad, India

© The Author(s), under exclusive license to Springer Nature Singapore Pte Ltd. 2021   643
S. C. Satapathy et al. (eds.), *Smart Computing Techniques and Applications*,
Smart Innovation, Systems and Technologies 225,
https://doi.org/10.1007/978-981-16-0878-0_62

62.1.1 Network Management Architecture

Most network management architectures [1] use the same basic structure and set of relationships. End stations (managed devices), such as computer systems and other network devices, run software that enables them to send alerts when they recognize problems (e.g., when one or more user-determined thresholds are exceeded). Upon receiving these alerts, management entities are programmed to react by executing one, several, or a group of actions, including operator notification, event logging, system shutdown, and automatic attempts at system repair.

Management entities also can poll end stations to check the values of certain variables. Polling [1] can be automatic or user initiated, but agents in the managed devices respond to all polls. Agents are software modules that first compile information about the managed devices in which they reside, then store this information in a management database, and finally provide it (proactively or reactively) to management entities within network management systems (NMS) via a network management protocol [2] as shown in Fig. 62.1. Well-known network management protocols include the simple network management protocol (SNMP) and common management information protocol [3] (CMIP).

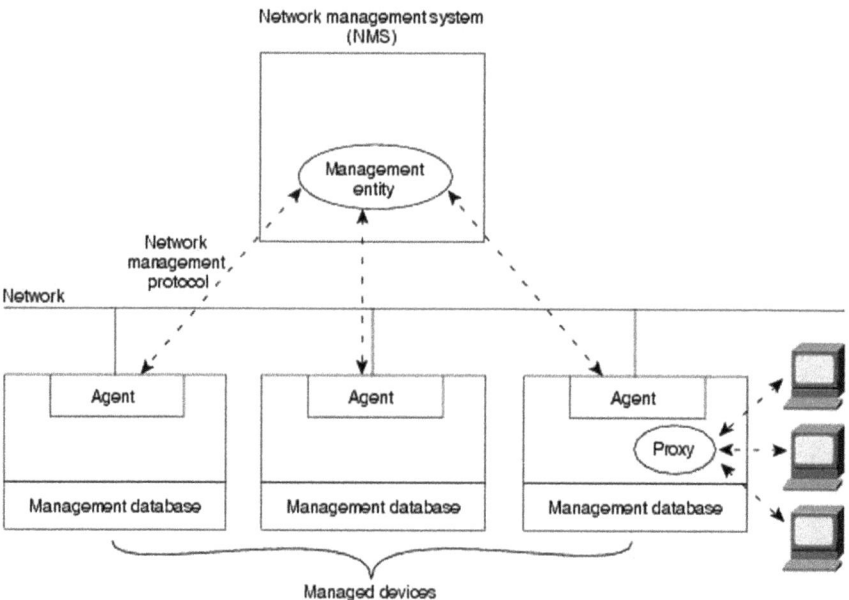

Fig. 62.1 SNMP network management system

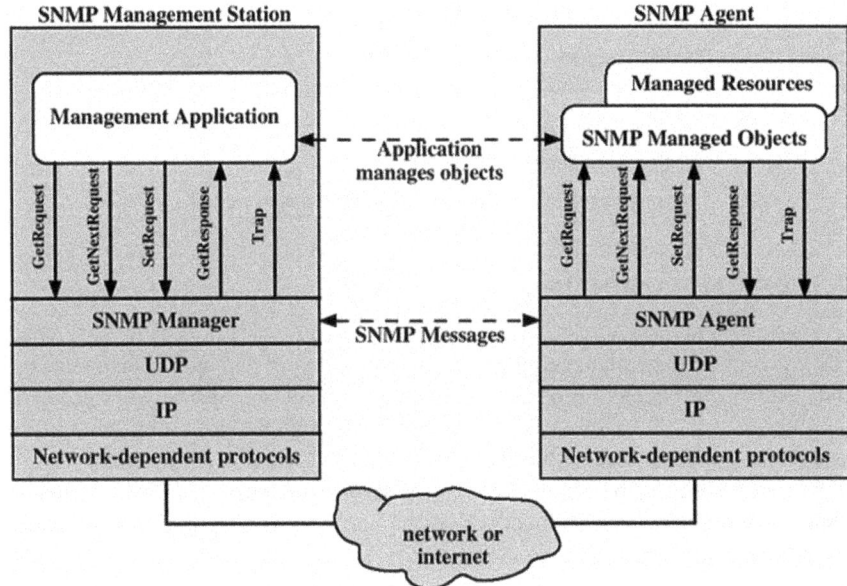

Fig. 62.2 Communication between SNMP management station and SNMP agent

62.2   Related Work

Configuration and monitoring of network elements is essential in communication networks to ensure that the network elements are up and running. To monitor and configure, we need to have a view of the network. This chapter discusses various protocols, techniques, methods of ICMP and SNMP that are helpful to develop this paper.

62.2.1   Internet Control Message Protocol (ICMP)

Internet Control Message Protocol (ICMP) is used to communicate specific information between hosts about the network communications problems. The ICMP consists of different messages in order to communicate among its hosts.

62.2.1.1   ICMP Message Types

To communicate among the host systems, computers exchange messages. The message which are exchanged by ICMP is defined as

Table 62.1 Components are the building blocks of SNMP

1	SNMP management systems and agents
2	Management information base (MIB)
3	SNMP messages
4	SNMP communities
5	The communication process between SNMP managers and agents

62.3 SNMP Architecture

SNMP is based on the manager/agent model consisting of a manager, an agent, and a database of management information, managed objects and the network protocol. The manager provides the interface between the human network manager and the management system. The agent provides the interface between the manager and the physical device(s) being managed, such as bridges, hubs, routers, or network servers, and these managed objects might be hardware, configuration parameters, performance statistics, and so on.

62.3.1 SNMP Agents and Managers

There exist two types of entities in SNMP environment, managers and agents. A manager, or network management station, is a device or software that is managing one or more agents. A network management station is used for fetching data from agents. It receives and handles traps and responses from agents. An agent, on the other hand, is operating on the managed device. The agent is implemented to send traps on different events. It also responds to requests from a manager as shown in Fig. 62.2. The implementation of an agent is based upon MIB files. The basic components of SNMP are listed in Table 62.1.

62.3.2 SMNP Data Structures

The data handled by the SNMP agent is organized into units that are called "management information bases" or MIBs for short. MIBs are described via a precise definition language called "Abstract Syntax Notation" [4]. A manager can talk to an agent, understand the agent information, and handle it if the manager has the MIB of the agent.

62.3.2.1   The Management Information Base

The manager and agent use a management information base and a relatively small set of commands to exchange information. The MIB is organized in a tree structure with individual variables, such as point status or description, being represented as leaves on the branches.

62.4   Design and Implementation

62.4.1   Auto-Discovery and Network Monitoring Using ICMP and SNMP

A networked application sends ICMP echo requests on to the subnet to find the network addresses to be discovered. Responses are parsed to identify the IP addresses of the machines, and SNMP get requests are used to know the gateways, routers in the network. This solution requires ICMP echo messages to be sent to each device in the network and get the IP address as the response (seed address) to perform the initial discovery process. With the initial seed address, we do the broadcasting or echoing to get all other devices specified with in the range. The response messages are being parsed into message and are sent to the NetVision application to update/show the available network IPs on to the user interface. Polling is done every 4 min to know the status of devices in the network. This application enables us to know the IP address in the network.

The gateways, routers, or switches have a feature called ipForwarding. An SNMP get request is sent to the devices on the network for the OID ipForwarding of MIB-II of RFC-1213. If the response for the get request is TRUE, it means the device is an IP forwarding device. If ipForwarding is TRUE, a request is sent to the device to identify the number of interfaces on the device by sending an SNMP get request on ifNumber of MIB-II. If the device has multiple interfaces, then it may be identified as a router.

Discovery of hosts:

Once router is identified, a SNMP request is sent to it on the OID ipNetToMediaTable using interface-index value from ipAddrTable and ipNetToMediaTable, the subnet and host IP of each device is found. The IpAddresses are given as input to fetch and set attributes values for the MIBs system, interface, IP and host resource.

a.   **Implementation of SNMP Agent**
     Init_agent().
     This routine initializes the agent. The exact initialization is
1.   Creation of sockets to receive SNMP requests
2.   Initialization of logging mechanism

3. Installation of signal handlers
    Init_mib_modules().
    The MIBs to be deployed in the SNMP agent at run time environment are dynamically loaded into the program memory by using this routine. system_mib, ip_mib, interfaces_mib and hr_mib are loaded into agent run time environment by invoking the init routines of the respective MIBs.
    Receive().
    This is the infinite loop for the SNMP agent daemon application.
    Exit criteria are A signal is sent by the user to shutdown this application.
    Select().
    This system call helps to listen on the sockets in the blocking and non-blocking modes. Current implementation of agent environment has no job than listening on the socket. Hence, select system call runs in the blocking mode.
    Receivefrom().
    This system call is used to receive an incoming message from the socket. Receivefrom is applicable in case of UDP. As client and server run over UDP transport to receive messages the system call used is receivefrom.
    After all the variables in the variable list are acted upon a response PDU is constructed and sent to the source of the request message (Note: reply is encoded before it is sent to the source).
b. Implementation System of MIB
    System_variables is an array of structures with each node having the following information magic number to index into the structure array, its ASN data type, level of access (read only or read–write), pointer function that accepts requests on this MIB, depth of the variable in the MIB tree, and node ID of the object under the node system in the MIB tree.

void init_system_mib(void);

MIB initialization is done by this routine. This routine is invoked by agent at the time of agent initialization. REGISTER_MIB is the routine which registers the MIB with the agent.

REGISTER_MIB(mibII/system, system_variables, variable2,

system_base_oid);

var_system is invoked by the find_var_method (the pointer function invoked by handle_one_var in the agent) when the incoming request is meant for system_mib. Irrespective of whether the request is GET or SET var method is invoked. If the request is GET, return value of this routine is used to populate response message. If the request is SET, write method (in this context writeSystem) is invoked after var method returns (Fig. 62.3).

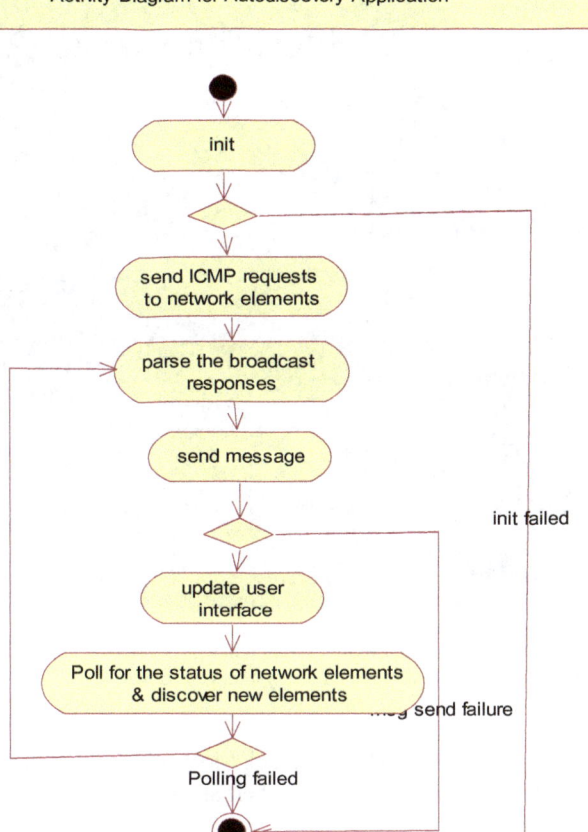

Fig. 62.3 Activity diagram for auto discovery application

62.5 Discussion

See Figs. 62.4, 62.5, 62.6 and 62.7.

62.6 Conclusion

The primary objective of this paper was to discover network resources automatically. This application writes a configuration file with the discovered network resource information. SNMP MIBs are implemented for system-MIB, interface-MIB, IP-MIB and host resource-MIB variables. Variables are retrieved and set on the MIBs.

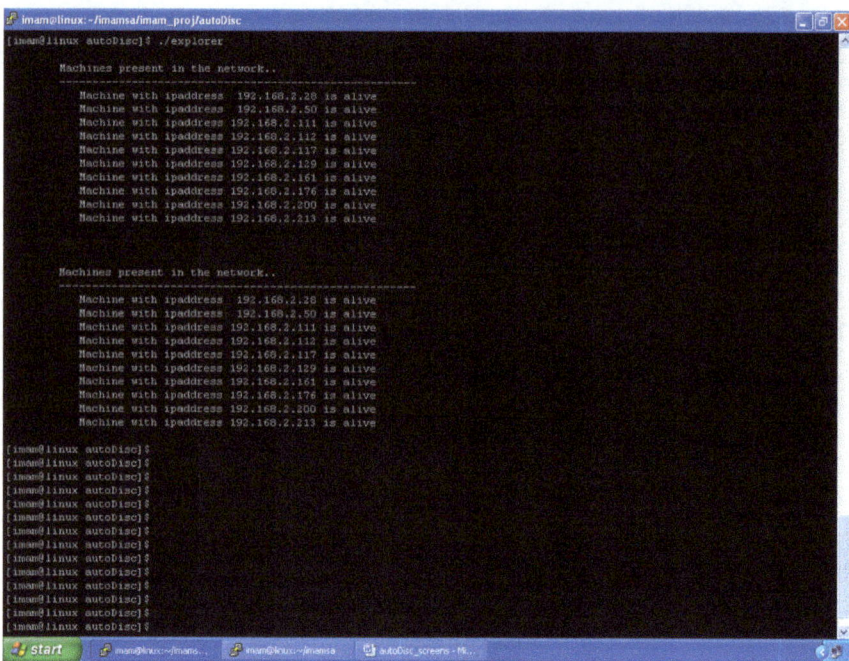

Fig. 62.4 Polling is done for every four minutes to find new network elements which are added or the existing elements that are being removed if the system goes down

The combination of ICMP and SNMP protocol drastically reduces the time taken to discover the network resources and monitor their status.

Future Work.

The future work related to this paper could focus on auto-discovery of network topologies and display the resources in a graphical format. Some of the further investigations are listed below.

- SNMP client applications can be integrated with auto-discovery application to provide a graphical user interface and to support configuration and fault management features of network management.
- Auto-discovery solution can be augmented to support expansive network topologies (WAN).
- MIBs can be implemented to monitor, check the status and alarm various computer devices as well as non-computer devices (like mechanical, electrical, robots, infrastructural works) when they meet their threshold values.

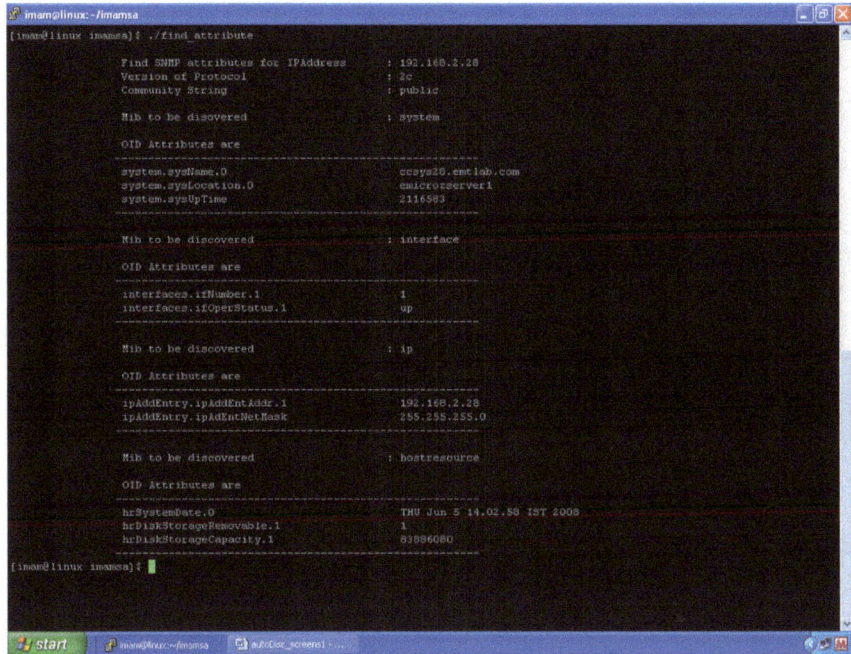

Fig. 62.5 When find_attribute is launched for the network IPAddress 192.168.2.28 its respective MIB objects(attributes) are retrieved

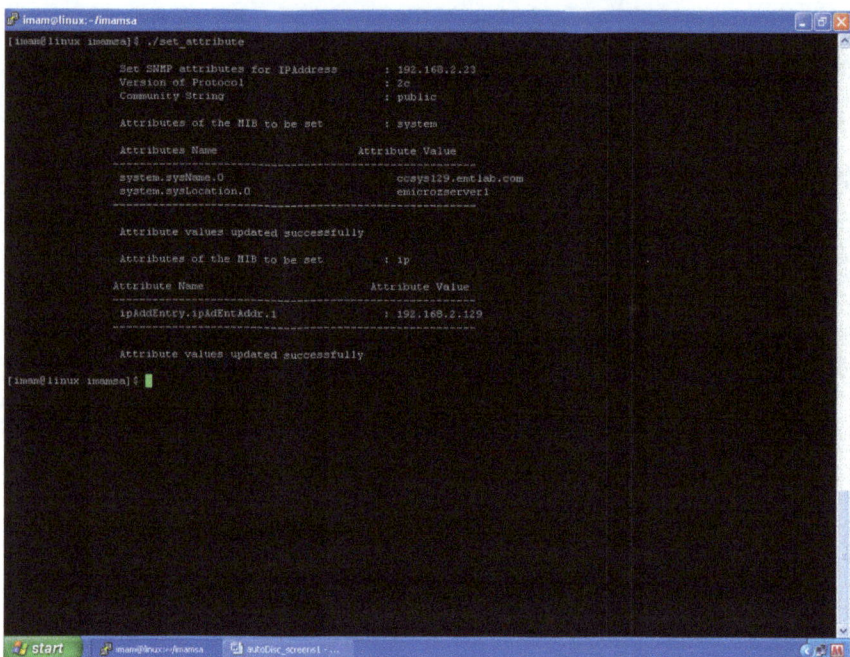

Fig. 62.6 set_attribute application is launched from command line. It asks for the IP address of the network element to be set. Upon giving the IP address, it asks for the version, community of SNMP protocol to be used. In the implementation, currently SNMPv2 is used. MIB object (variable) to be set is to be given as input. Upon giving valid input, it forms a command message and invokes snmpset request, and the response returned from snmpset is shown on the screen. The IPAddress provided as input is 192.168.2.23

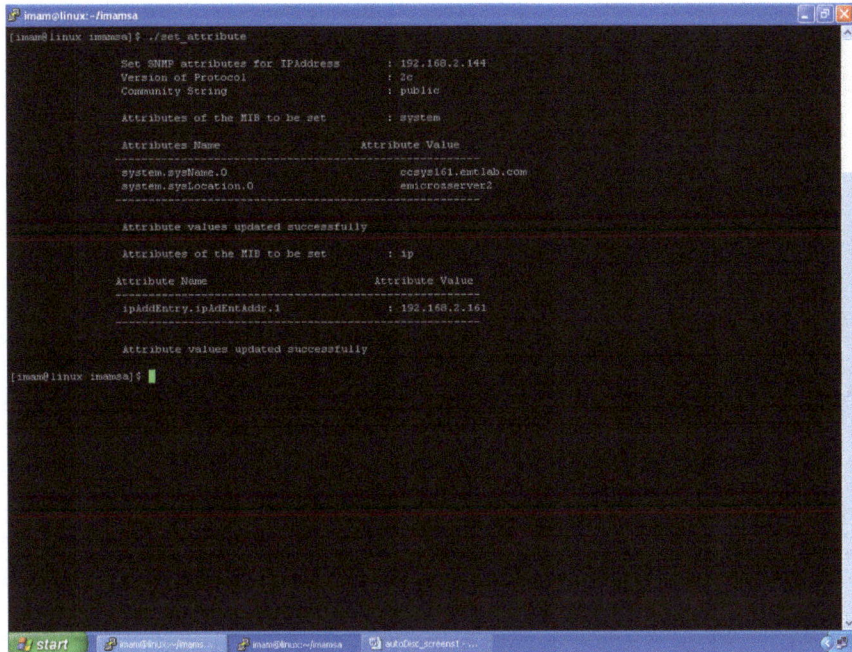

Fig. 62.7   The MIB objects of the IPAddress 192.168.2.144 are being set or initialized to new values

References

1. IETF: Configuration management with SNMP (2000). https://www.ietf.org/
2. Case, J., McCloghire, K., Rose, M., Waldbusser, S.: Protocol Operations for Version 2 of the Simple Network Management Protocol (SNMPV2), RFC 1905 (1996)
3. Perkins, D.T.: Snmp versions. The Simple Times, **5**(1), 13–14 (1997)
4. IETF: The Internet Engineering Taskforce (2000). https://www.ietf.org

Chapter 63
Privacy Preserving Semantic-Aware Multi-keyword Ranked Dynamic Search Over Encrypted Data

M. D. Asrar Ahmed, S. Ramachandram, and Khaleel Ur Rahman Khan

Abstract Searchable encryption (SE) is a promising primitive which enables data owners to privately query and retrieve required documents from encrypted outsourced data collection in cloud servers. Most of the SE schemes in literature strive to achieve a balanced trade-off between efficiency, privacy and expressiveness of query. However, a vast majority of efficient and secure schemes lack searching capabilities beyond exact match of queries and ignore the semantic relationship between keywords and documents. Therefore, a new trend in SE research is to address the question: How to make search semantic-aware and context-based? Majority of the existing semantic search schemes are static, do not support dynamic update operations, suffer from leakage of significant information to server, and employ huge size vectors to represent documents, keywords and topics/features. We propose a privacy preserving, relevance ranking-based dynamic search scheme using a sublinear search time hash-based indexing mechanism and semantic-aware search supporting semantic expansion of user queries. Our scheme employs Word2Vec model to quantitatively evaluate semantic relationships between query and other keywords to expand queries to fulfill user's search intentions. It also supports dynamic operations like insert and delete of documents on outsourced collection with less efforts to update relevance information which is calculated partially at client and server. Experiments on RFC dataset have demonstrated a time and space efficient non-interactive semantic search over encrypted data.

63.1 Introduction

A potential bottleneck for cloud users to outsource huge volumes of confidential data despite associated benefits is the fact that cloud service provider is not fully trusted.

M. D. Asrar Ahmed (✉) · S. Ramachandram
CSE Department, University College of Engineering, Osmania University, Hyderabad, India

K. U. R. Khan
CSE Department, Ace Engineering College, Rangareddy, India

Searchable encryption (SE) is a solution to this problem, which enables data owners to privately search the outsourced encrypted dataset. With first practical SE scheme proposed in year 2000 by Song et al. in [1], many researchers have contributed to the field of SE. The subsequent research works include: single keyword search [2], multi-keyword search [3], and fuzzy search [4].

The major drawback of most of these schemes is that they consider only exact match of the keywords, and very few consider fuzzy searches. And use the very common TF-IDF model to draw relevance between document and keyword. In general user intention during search is mostly semantic in nature than being syntactic. For example, a user searching for "economy" keyword in document collection should not be restricted to just "economy", but the user intention will be to retrieve documents which contain keywords semantically closer to it, such as "finance" and "monetary". The TF-IDF model does not take into consideration such semantic relationships/associations, as it ranks keywords based on mere occurrence and more so syntactically and it also requires dealing with high-dimensional and sparse vectors representing documents and keywords due to huge size of keyword dictionary. Additionally, most of the schemes use hierarchical index structures, populated with prior ranking information, causing dynamic updates at several levels in it, to be a costly affair.

In this paper, we propose a semantic-aware multi-keyword dynamic search scheme over encrypted data returning top-k ranked documents. We achieve semantic expansion of queried keywords using the Word2Vec model [5] to meet users' search intentions. The matching documents are returned to user based on TF-IDF-based ranking of documents. The subsequent dynamic operations such as insert/delete operation on documents are carried out without affecting other unrelated index entries as document ranking is computed partially at client and server. Our proposed scheme uses hashing-based indexing mechanism by Etemad et al. [6], with significant optimizations with respect to support for dynamic operations, semantic search, and improved security.

63.2 Related Work

The searchable encryption (SE) schemes have evolved since the pioneering work proposed by Song et al. [1] which was first practical solution with linear search time. It was followed by many schemes with improved efficiency (sublinear, logarithmic in size of document collection) and supported expressive queries. The first Semantic Search scheme over encrypted data was proposed in [7] makes use of stemming algorithms to obtain root of a query keyword and does the search. In [8], authors proposed a semantic search scheme which obtains similar keywords to query with help of edit distance as quantitative similarity measure. In [9], authors proposed a synonym query scheme, where a common synonym thesaurus was built on foundations of NARCT. Sun et al., in [10] proposed a semantic search scheme using two separate servers, expanding the query using semantic relationship library employing

Table 63.1 Comparison of existing strategies with proposed work

Scheme	Search complexity	DSSE	Index type	Semantic search	Forward privacy	Multi-keyword
[5]	$O(|S|)$	No	Inverted index	Stemming	No	No
[6]	$O(\tau|W|)$	No	Tries	Edit distance	No	No
[7]	$O(a_w)$	No	Vector space	TF-IDF	No	Yes
[8]	$O(|S|)$	No	Inverted index	MI	No	No
[9]	$O(\theta|W|\log n)$	No	Secure k-NN	SRG	No	Yes
[10]	$O(mn^2)$	No	Secure k-NN	ECH	No	No
[11]	$O(mn)$	No	Secure k-NN	LDA	No	Yes
[12]	$O(n)$	No	Secure k-NN	Word2Vec	No	Yes
[13]	$O(n)$	Yes	Vector space	Doc2Vec	No	Yes
Proposed	$O(s_w)$	Yes	Hash-based index	Word2Vec	Yes	Yes

Notations |S| is size of stem set, τ : size of similarity set, $|W|$, m: dictionary size, a_w: # documents matching keyword, SRG: semantic relationship graph, θ: # of leaf nodes, ECH: extended concept hierarchy, $|D|$, n: # of documents, LDA: latent Dirichlet allocation, s_w : # of semantically related keywords to w

Mutual Information metric defined in [11]. In [12], authors built Semantic Relationship Graph (SRG) by computing Mutual Information. In [13], authors used Extended Concept Hierarchy with help of WordNet tool. Dai et al., in [14], proposed a scheme based on Latent Dirichlet Allocation topic model by generating keyword and document relevance matrices for efficient similarity computations. Hsieh et al. in [15] have proposed a semantic search scheme using Word2Vec machine learning-based model. The work in [16] is based on Doc2Vec model and uses secure k-NN algorithm to compute similarity between documents and keywords. A detailed comparison of these schemes with our proposed scheme is given in Table 63.1.

63.3 Proposed Scheme: Notations, Semantic Search, and Algorithms

63.3.1 Notations

- Documents collection: $D = \{d_1, d_2, \ldots, d_n\}$, encrypted documents $C = Enc(F)$
- K_1, K_2, K_3: keys generated by $KeyGen(1^\lambda)$ used in $Enc_{K_1}()$, $P_{K_2}()$, $G_{K_3}()$
- $DictW, DictD, DictV$: Keyword, Document, and Vector dictionaries

- $f_{d_{j,t}}$: term frequency of word $w_t \in W$ in file $d_j \in D$, f_{w_t}: # files containing term w_t
- V_t: m dimensional vector for word w_t generated by Word2Vec model.
- $Enc_{K_1}(), P_{K_2}(), G_{K_3}()$: Symmetric Encryption algorithm, a Pseudorandom Permutation, and a Pseudorandom function family, respectively.
- $I = \{(i, w_i)|w_i \in W\}$, be an index of all unique keywords in set W.
- $T_w = < P_{k_2}(i), x_i, f_{w_i} >$, a hash table at client side (client state).
- $\forall w_i \in d_j, \forall 1 \leq i \leq m$ and $\forall 1 \leq j \leq n$ compute: $tf_{ij} = \frac{1}{|d_j|}.(1 + \ln f_{d_j,i})$.

63.3.2 Semantic Extension of Queries—Word2Vec Model

In order to match the users' search intentions in addition to exact match of query keywords, semantic search extends the query terms based on semantic relationship with other terms. Quantitative evaluation of semantic relationship between keywords is achieved by employing the machine learning-based Word2Vec model [5]. Word2Vec as shown in Fig. 63.1, works by generating vectors v_t of some predefined dimensions representing each word $w_t \in W$, and mapping it to a window of surrounding context words, using a neural network approach. And the semantic similarity between any two words w_q and w_j is computed with respect to cosine similarity of their vectors V_q and V_j as follows (Eq. 63.1):

$$CS(V_q, V_j) = \frac{V_q.V_j}{\|V_q\| \cdot \|V_j\|} \qquad (63.1)$$

$$w_k = \mathrm{argmax}_{w \in W}\{CS(V_q, V_k)\} \qquad (63.2)$$

The keyword w_k which is semantically closest to query keyword w_q is computed using Eq. 63.2. We store these vectors in the Word Vector Dictionary $DictV$ based on unique index assigned to keywords in index set I. Therefore, the vectorization of keywords is oblivious to the server.

63.3.3 Proposed Scheme

We present below a privacy preserving semantic-aware multi-keyword dynamic search scheme by adopting the forward secure dynamic searchable encryption scheme proposed in [6] and achieve the semantic search by employing the Word2Vec model.

Setup and Index Building: The different secret keys used during index building and searching are generated using a security parameter λ and a probabilistic key generation function: $KeyGen(1^\lambda)$. The vectors set $V = \{V_i | \forall i \in [m]\}$, is generated to quantitatively summarize the keywords using Word2Vec model specifying a dimension of size l. This set is outsourced to server with each vector linked to keyword specific key K_{w_i}. The $DictW$ dictionary stores label-value pairs of the form: (hash address, randomized term frequency value along with document identifier) as described in $BuidlIndex$ algorithm. The $DictD$ dictionary is used to store hash address of $DictW$ entries to act as forward index used during updates. The data owner maintains a local hash table T_W as its state consisting of entries: a random integer x_i for each keyword w_i and the term frequency f_{w_i}. A keyword specific key K_{w_i} is generated for each keyword through $G_k()$ with keyword and x_i, which changes with every search/update operation. This x_i is used to ensure randomized queries to prevent keyword guessing attack as well file injection attacks [17], and also, the previous queries cannot be re-run due to ever changing random input to trapdoor generation algorithm. This specific key is then used as an input to the keyed hash function h_1 along with keyword-document counter to generate a hash address to store/retrieve entries.

Algorithm 1: BuildIndex

Input: $W, D, I, K_2, K_3, PRF\ G$:

Client Steps:

1. Set $T_w = <P_{k_2}(i), x_i, f_{w_i}>$, initially $x_i = 0\ \forall\ w_i \in W$, of size $|W|. f_{w_i} = 0$
2. $\forall w_i \in d_j, \forall 1 \le i \le m$ and $\forall 1 \le j \le n$: $tf_{ij} = \frac{1}{|d_j|} \cdot (1 + \ln f_{d_j,i})$
3. Build the vectors for all $w_i \in W$, using Word2Vec model: V_i of dimension l.
4. *for all $d_j \in F$:*
    a. *generate a key for d_j:* $K_j = G(K_3, id(d_j) || R_2)$, where R_2 is a random number.
    b. *for all $w_i \in d_j$:* $//1 \le i \le |d_j|$
        i. $dlabel_i = H_2(K_j, i)$
        ii.   *if* $T_w[P_{k_2}(i)].x_i = 0$ *then* $x_i = randInt()$
        iii. $T_w[P_{k_2}(i)].f_{w_i} = T_w[P_{k_2}(i)].f_{w_i} + 1$
        iv. $K_{w_i} = G(K_3, w_i || x_i)$
        v. $label_i = H(K_{w_i}, T_w[P_{k_2}(i)].f_{w_i} || 0)$
        vi. $dval_i = \langle label_i \rangle$
        vii. $val_i = (\langle id_j, tf_{ij} *$
              $R_1) \oplus H (K_{w_i}, T_w[P_{k_2}(i)].f_{w_i} || 1), dlabel_i)$
        viii. $AllPairsD = \{dlable_i, dval_i\} \cup AllPairsD$
        ix. $AllPairsW = \{(label_i, val_i)\} \cup AllPairs$
        x. $AllPairsV = \{(V_i, K_{w_i})\} \cup AllPairsV$
    c. Encrypt d_j: $c_j \leftarrow Enc_{K_1}(d_j)$
5. $C = \{c_1, c_2, ..., c_n\}$, $AllPairsW$, and $AllPairsD$ is sent to Server

Server Steps: Creates three empty Maps $DictW$, $DictD$ and $DictV$ to store values from sets $AllPairsW$, $AllPairsD$ and $AllPairsV$ respectively.

Search Operation: The search operation begins with data user generating a trapdoor for query $Q = \{w_1, w_2, \ldots, w_q\}$ by semantically expanding Q. The data user after appropriate authorization from data owner, generates following inputs to send to server: the existing keyword specific keys each of w_q (used to locate entries in $DictW$) the new keyword specific keys for each of w_q (to re-insert entries in

$DictW$), file count of each of queried keywords, i.e.,$T_w[w_q].f_{w_q}$. On receiving trapdoor$\langle K, nK, F \rangle$, server parses it to semantically extend the query by computing the similarity score through Eqs. 63.1 and 63.2 using Word2Vec model. The top similarity scores are considered based on certain threshold value to enrich the query with respect to semantics. Now the semantically extended query $Q' = \{w_1, w_2, \ldots, w_{q'}\}$ where $q' > q$, along with keys associated with newly added keywords to query from $DcitV$ are used to retrieve document identifiers from $DictW$ by computing relevance score of each document to extended query keyword. The server runs the search algorithm for every queried keyword as described in step 4 of $Semantic Search$ algorithm. The rank of a document with respect to semantic query keywords is computed using randomized term frequency value contained in above format of entry as tf_{ij},n, and f_{w_i} as in step 4. The sums of scores of each document for all queried keywords are calculated, and finally, those documents which contain all queried keywords are sorted in ascending order based on their ranking score and these documents are then sent in response to user query.

Algorithm 2: *SemanticSearch*

Client Input: Old and New Keyword-specific keys $K = \{K_{w_1}, K_{w_2}, \ldots, K_{w_q}\}$, $nK = \{K_{w_1}, K_{w_2}, \ldots, K_{w_q}\}$, word-file counts $F = \{f_{w_1}, f_{w_2}, \ldots, f_{w_q}\}$ from the T_w table.
Server Input: $DictW, DictD, DictV$, and n
Client Steps:

1. $Q = \{w_1, w_2, .., w_q\}$, with K and nK generated as follows:
    a. *for all* $w_q \in Q$
        i. Generate random integer for w_q as x_q in $T_w[w_i]$.
        ii. $K_{w_q} = G(K_3, w_q || x_q), nK = nK \cup \{K_{w_q}\}$
        iii. $F = F \cup \{T_w[P_{K_2}(q)].f_{w_q}\}$
    b. Send $\langle K, nK, F \rangle$ as a conjunctive keywords query to server.

Server Step:

2. Extend the query using Word2Vec model and computer similarity scores using equation (1 & 2) to obtain new token: $\langle K', nK, F' \rangle$
3. Initialize a $ResultSet = \emptyset$.
4. for $i = 1$ to $|F'|$
    a. for $j = 1$ to f_{w_i}
        i. $label_j = H(K_{w_i}, j || 0)$ //old keyword-specific keys used.
        ii. $val_j = WMap[lable_j] \oplus H(K_{w_i}, j || 1)$
        iii. *parse* val_j as $(\langle id_j, tf_{ij} \rangle, dlabel_j)$
        iv. *add* id_j to set $ResultSet$.
        v. *delete* $WMap[label_j]$ *and re $-$ insert* using new keys.
        vi. *compute Rank* of id_j: $score_{ij} = tf_{ij} * n/f_{w_i}$, store it.
5. Compute sum of $score_{ij}$ for each of resultant file identifier in $ResultSet$
6. Sort $ResultSet$ & send encrypted files to client based on ranks.

63.4 Security and Performance Analysis of Proposed Scheme

63.4.1 Security Analysis

In our proposed scheme, we achieve the *document confidentiality* by encrypting it using a symmetric key encryption scheme like AES with a secret key K_1. The *Index* and *Query Confidentiality* is ensured as follows. The outsourced index is built employing pseudorandom functions and keyed hash functions which are based on keyword indexes rather than keywords and using random seeds during every subsequent operation like search and update in addition to initial setup operation. Hence our outsourced dictionaries like $DictW$, $DictD$, $and DictV$ which store label-value pairs accessible through keyed hash functions, ensure the index confidentiality. Our designed index is much secure when compared to the inverted index-based schemes or vector space model-based index like in [5, 15, 16], as the keyword-document relationship and distribution are hidden from server. And more importantly, our scheme does not allow re-run of old queries onto newly inserted documents as it invalidates the queries after search is over. Hence we achieve both forward and backward privacy which ensures both index and query confidentiality. Our scheme also achieves the *Query Unlinkability*, by using fresh random numbers every time in conjunction with keyword being searched as input to the key generation function.

63.4.2 Performance Analysis

To evaluate the performance of our proposed scheme, we implemented our semantic search algorithm using Python language with Word2Vec model being implemented in Gensim framework. We used RFC dataset for testing purpose to retrieve documents out of 8000 RFC files, based on semantic expansion of search keywords using Word2Vec model. The time it takes to set up the index involves generating label-value pairs and hashing them into a hash table using a keyed hash function, and setting up $DictD and DictV$. As there are around 5 lakh distinct keywords in RFC dataset, the index building takes significant amount of time as shown in Fig. 63.2. However, after the index is built, the search operations can be carried out in constant amount of time due to hash-based indexing structure. In schemes like [12, 13, 15], the inverted index or tree-based index are populated with precomputed relevance scores which have a heavy performance cost during update operations. Whereas in our proposed scheme, the relevance score calculations are done partially at client side and outsourced after randomization, which have a significant efficiency improvement on update operation. We also analyzed the amount of time taken to build trapdoors for user queries, as shown in Fig. 63.3. The amount of time taken to build trapdoors include time needed to compute keyword specific keys and time needed for expansion of query keywords using similarity-based expansion of Word2Vec mode. As

Fig. 63.1 Word2Vec
framework as in [20]

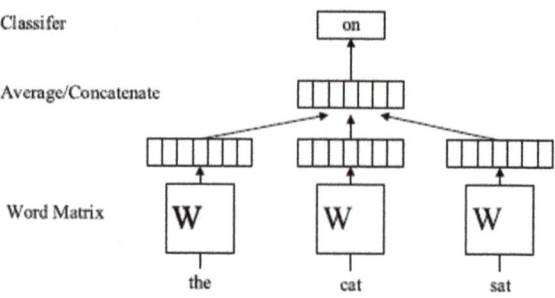

Fig. 63.2 Time taken for
index setup with varying size
of datasets

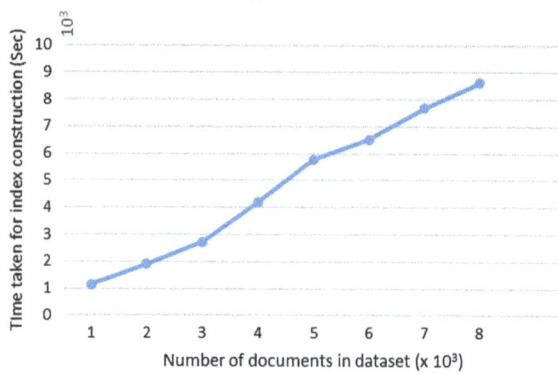

shown in Fig. 63.3, a major amount of time is needed for semantic expansions of query. The search operation is a sublinear cost operation as the hash-based index is searched for entries s_w times, where s_w is number of documents containing keywords semantically close to searched keyword w. The search performance of our proposed scheme is demonstrated in Fig. 63.4, with varying number of documents. The trapdoor building and search operations cost less time when compared to [13–16], as these schemes heavily use complex matrix multiplications during these two steps and also query vectors are of same size as that of document vectors. This results in heavy vectors for queries too, but our scheme utilizes trapdoors of size proportional to number of keywords in semantic expansion of query. A detailed comparison of our proposed scheme with existing semantic-aware SSE schemes is given in Table 63.1 along with asymptotic analysis of these schemes.

63.5 Conclusion

In this paper, we proposed a privacy preserving semantic-aware multi-keyword dynamic search scheme which returns documents to the user ranked as per relevance of keywords to documents, based on semantically expanded query keywords. We employ Word2Vec model to ascertain semantically closer terms to query keywords

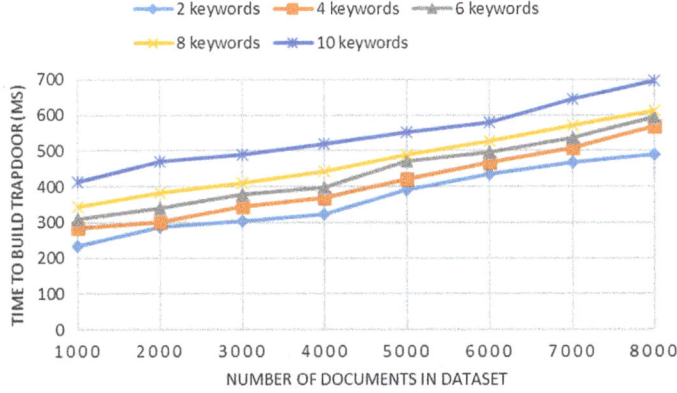

Fig. 63.3 Time to build trapdoors of varying size of queries

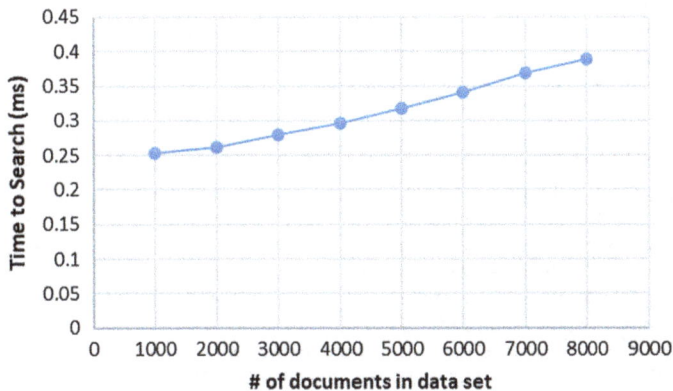

Fig. 63.4 Time to perform semantic search over different size of datasets

to match users' search intention. We use a novel hash-based secure indexing mechanism which prevents most common types of attacks such as file injection attacks. The index stores label-value pairs for each of keyword-document pairs. Upon search for a keyword, the trapdoor received by server is semantically extended to expand query and retrieves matching documents. The proposed scheme is the first scheme which reduces the size of trapdoor and efforts in search process as it does not involve either a vector-based representation of document and queries or a hierarchical indexing structure and also addresses security concerns.

Acknowledgements This work was supported under the Visvesvaraya PhD Scheme (Electronics and IT), with unique awardee number MEITY-PDH-1597. (Visvesvaraya PhD scheme: File Number PhD-MLA-4(63)/2015-16) by Ministry of Electronics and Information Technology, Government of India.

References

1. Song, et al.: Practical techniques for searches on encrypted data. In: Proceeding 2000 IEEE Symposium on Security and Privacy. S&P 2000. IEEE (2000)
2. Goh, E.-J.: Secure indexes. IACR Cryptol. EPrint Arch. **2003**, 216 (2003)
3. Cao, N., et al.: Privacy-preserving multi-keyword ranked search over encrypted cloud data. IEEE Trans. Parallel Distrib. Syst. **25**(1), 222–233 (2013)
4. Wang, B., et al.: Privacy-preserving multi-keyword fuzzy search over encrypted data in the cloud. In: IEEE INFOCOM 2014-IEEE Conference on Computer Communications. IEEE (2014)
5. Mikolov, T., et al.: Efficient estimation of word representations in vector space (2013). arXiv: 1301.3781
6. Etemad, M., et al.: Efficient dynamic searchable encryption with forward privacy. In: Proceedings on Privacy Enhancing Technologies **1**, 5–20 (2018)
7. Moataz, T., et al.: Semantic search over encrypted data. ICT 2013. IEEE (2013)
8. Wang, C., et al.: Achieving usable and privacy-assured similarity search over outsourced cloud data. In: 2012 Proceedings IEEE INFOCOM. IEEE (2012)
9. Fu, Z., et al.: Multi-keyword ranked search supporting synonym query over encrypted data in cloud computing. In: 2013 IEEE 32nd International Performance Computing and Communications Conference (IPCCC). IEEE (2013)
10. Sun, X., et al.: Privacy preserving keyword based semantic search over encrypted cloud data. Int. J. Sec. Appl. **8**(3), 9–20 (2014)
11. Church, K., et al.: Word association norms, mutual information, and lexicography. Comput. Linguist. **16**(1), 22–29 (1990)
12. Xia, Z., et al.: A multi-keyword ranked search over encrypted cloud data supporting semantic extension. Int. J. Multimed. Ubiquitous Eng. **11**, 107–120 (2016)
13. Fu, Z., et al.: Semantic-aware searching over encrypted data for cloud computing. IEEE Trans. Inf. Forens. Sec. **13**(9), 2359–2371 (2018)
14. Dai, H., et al.: Semantic-aware multi-keyword ranked search scheme over encrypted cloud data. J. Netw. Comput. Appl. **147**, 102442 (2019)
15. Hsieh, F.-J, et al.: Semantic multi-keyword search over encrypted cloud data with privacy preservation. In: 2019 IEEE 90th Vehicular Technology Conference. IEEE (2019)
16. Dai, X., et al.: An efficient and dynamic semantic-aware multikeyword ranked search scheme over encrypted cloud data. IEEE Access **7**, 142855–142865 (2019)
17. Zhang, et al.: All your queries are belong to us: the power of file-injection attacks on searchable encryption. In: 25th {USENIX} Security Symposium (2016)

Chapter 64
A Thorough Study on Weblog Files and Its Analysis Tools

N. Vanitha and M. Suriakala

Abstract A web server log is a data file created by a server. Many servers collect data about the information requested for the web server, examples are date, time, client IP address, recommendation, user agent, and so on. This data will help identify specific issues on the web. Log files are stored in the web server, web proxy server, and the client browser. This paper provides a detailed overview of weblog files, analytics tools, and reviews of existing research.

64.1 Introduction

The fast development of communication technology has made the dissemination of information very important. Web Log files are used to track any communication transactions about the multiple user behavior for the specific period, so that it enables as to understand the deep structure of website. The structure of the website can be understood using the Web Log files by recording the users navigation. Weblogs are effectively to clean up certain amounts of useful information. Properly compiled and illustrated log information provides statistics analysis of some important technical information regarding the server's load usage, unusual transactions, or incomplete and failed requests. It also helps with business decisions and website development.

N. Vanitha · M. Suriakala (✉)
Government Arts College, Nandanam, Chennai, India

© The Author(s), under exclusive license to Springer Nature Singapore Pte Ltd. 2021   665
S. C. Satapathy et al. (eds.), *Smart Computing Techniques and Applications*,
Smart Innovation, Systems and Technologies 225,
https://doi.org/10.1007/978-981-16-0878-0_64

64.2   Web Usage Mining

Web application mining refers to the extraction of useful knowledge from web data that includes user information. A weblog files found in three locations:

(i) web servers, (ii) web proxy servers, and (iii) client browsers.

Server Page Logs: These logs usually provide usage data, but there exist two main drawbacks are:

(a)   Logs: These records contain important personal information so server owners can keep them closed.
(b)   It does not record uploaded cached pages.

Proxy Page Logs: A proxy server handles the HTTP requests and sends it to a web server and then returns the results delivered by the web server. Two disadvantages:

(a)   Proxy-server construction is not easy to do.
(b)   Request interception is low.

Client-side logs: Test a website of an existing web browser. HTTP cookies may also be used. Two disadvantages:

(a)   Need special software must install it.
(b)   The operating technique is difficult to achieve compatibility with different operating systems.

64.3   Web Servers

A weblog file is a file created by a web server. Every time a visitor requests any file from the website of their requests, such as a page, image, or video, the current log file is added to the file. It contains information about the user such as time interval, URL, and IP address. There are mainly three types of log file formats used by most servers. They are common log file format, integrated log format, and multiple access logs. Four types of server logs are access log file, agent log file, error log file, and referrer log file.

Access Log: It is one of the major weblog servers; it will record each click, hit, and access of the users.

Agent Log: The file is used to record the information about the user's browser, browser version, and operating system.

Referrer Log: It is used to record the referrer log that a user came from a particular website by using the user's page link.

There is an enormous increase in web users, and everyone wants to reach the right information in a minimum period. The searches performed by the user in the browser are concentrated on the server's weblog data. Valuable information and knowledge live on this data but extracting them is not easy.

64.4 Tools

Weblogging is the process of identifying information on user browsing habits. Web mining is classified into three basic class—web content mining, web architecture mining, and web application mining. There are various tools available that can perform web application mining using web access logs as input and generate reports as output. Some of the web analytics tools are.

Though the tools are more effective, each tool has its limitations and features as follows as the table.

Tools	Primary features	Limitations
Weblog expert	Weblog is a strong log analyzer. It will provide information about the site's visitors, site routes, pages reference, search engines, user agent, and more	Recovering the HTML output can be difficult. Some log file formats are not supported
Logz.io	Logz.io is cloud-based log management. Users can create alerts for log messages and get notified via email	No dashboard with logs and metrics
Sumo logic	It allows analyzing petabyte data and learning to quickly detect patterns	Expensive, missing monitoring
Loggly	Loggly is a cloud based, user-friendly log tracking, and records analysis platform using built-in and integrated collaboration tools It allows us to track events, functions, metrics, and traffic	Rules are restrictive, though they are negotiable via custom plans
Logstash	It is the free open source for processing log data. It parses recordsfrom various sources and used in conjunction with elastic search and Kibana	Lower Performance, high resource usage
Graylog	It provides a quick warning of cyber threats. It has a simple and intuitive UI that enables us to analyze, alert, and report data	Visualization is limited
Paper trail	A paper trail is a log analyzer that automatically scans the records to provide real-time insights. It features real-time searching to quickly identify problems	Mostly output is text-based

(continued)

(continued)

Tools	Primary features	Limitations
Splunk	It is a popular log tracking and analysis site. It also collects, stores, indexes, communicates, displays analyzes, and reports on machine-generated data. Splunk allows real-time alerts	Setup costs and complexity are high
Solar winds log analyzer	It will enable us to do root cause analysis with the help of log monitoring tools and provide a real-time log stream	Limited visualization, configuration, very pricey
Log DNA	It is a cloud service used to analyze log files. It has real-time search, filtering, and debugging capabilities, a robust mechanism to help connect problems to their root cause	Limited visualization
Go access	Go access track web server records in real time using a simple command-line dashboard and analyze traffic metrics	Harder to set up, only basic tracking mechanism
Screaming frog log file analyzer	Allows checking search engine bots, detect crawled URLs, and analyzing the data, etc.	Limited functionality, crawl a limited number of URLs for the free version
Apache log viewer	Apache logs viewer can easily filter and analyze Apache/IIS/Engine log files. The search and filter function is configured. Custom record formats are also supported	Though it is free for unlimited use, some features are locked

Most of the tools are traffic analyzers. Their features are limited to generating reports on website traffic: The user visits and page views, user statistics, operating systems, and web browser and page not found. Some tools analyze user statistics and specific statistics about individual users. There are many business analytics tools available, but most of them have speed limits, which are expensive, inflexible, and difficult to maintain.

64.5 Literature Review

Much research has so far focused to solve this issue on analyzing weblog data. The methods discussed by different researchers used various measures and techniques.

Zhang et al. proposed that the use of association rules analysis in weblog mining is of great importance for achieving a personalized website and personalized service. In this study, the matrix a priori algorithm is used for weblog mining for sogou weblog mining and footwear websites compared to the four association rules algorithms [1].

Koutsoukos et al. show that a new method is proposed for session identification in weblog data based on ambiguous the fuzzy c-means clustering algorithm. The novelty of the approach lies in launching obscure partition matrices using subtractive clustering technology [2].

Cao et al. this paper described an unknown detection system on machine learning for weblog files. The system will be used for analysis log files. This paper first preprocesses log files by using decision tree algorithms based on patterns based on the ruleset, then modeling the normalized data using HMM to create a model that is used to specify the default data state as an abnormal detector [3].

Anuradhab et al. this study is based on a review of 14 college's weblog files. After the estimation, we define the cluster more accurately and present the most accurate results and prediction model based on the Markov chain and Bayesian theorem [4].

Zhu et al. this study proposes an advanced method of F&D to expedite the mining process of WFPS from weblogs, in which weights are allocated by the residence time to find not only frequent visits but also important page collections [5].

Latib et al. the results presented here will facilitate improvements an exploration of weblog files for analysis of weblog files using Hadoop [6].

Agichtein et al. this study explored the use of combining the noisy implicit feedback obtained in the actual web search system to improve search rankings [7].

Kumar et al. this paper is to track visitors' search behavior by using the tool and maintain the website according to this behavior. In model discovery, the association will use different mining techniques such as rule, classification, clustering, and continuous sampling technique to find important information [8].

Patel et al. this research work includes the pre-implementation phase and web application mining, which can be used in industry and application-oriented organizations. Used customized weblogging pre-processing rather than the traditional approach that reduces the size of the original weblog file. The development will show up in time and accuracy [9].

64.6 Conclusion

Weblog files play a significant role in web mining research because it contains user information and user behavior can be extracted from it. As web traffic has already dominated the Internet, massive weblogs are being generated ceaselessly. It is essential and meaningful for operators to mine valuable information and knowledge from the log data. However, the state-less feature of HTTP and increasing dynamics and complexities of web services bring a challenge to web mining in weblogs. Techniques need to be improved to reduce delays in accessing web pages. Weblog mining, however, still has many problems to be solved and in-depth studied. In conclusion, it is hoped that this paper will provide some depth into the tools and review of existing works on log analysis and explained how and why these weblog analyses are important.

References

1. Zhang, H., Song, W., Liu, L., Wang, H.: The application of matrix apriori algorithm in web log mining. In: IEEE 2nd International Conference on Big Data Analysis (2017)
2. Koutsoukos, D., Alexandridist, G., Siolas, G., Stafylopatis, A.: A new approach to session identification by applying fuzzy c-means clustering on web logs. In: IEEE Symposium Series on Computational Intelligence (SSCI) (2016)
3. Cao, Q., Qiao, Y., Lyu, Z.: Machine learning to detect anomalies in web log analysis. In: IEEE International Conference on Computer and Communications (2017)
4. Anuradhab, H.K., Solankic, A.K., Krishna, K.: Progressive machine learning approach with webastro for web usage mining. In: International Conference on Computational Intelligence and Data Science (ICCIDS 2019)
5. Zhu, Z., Xu, Q.: Fast mining frequent page sets from web log by filtering strategy. In: Proceedings of the 2nd International Conference on Computer Science and Application Engineering, October 2018
6. Latib, M.A., Ismail, S.A., Yusop, O.M., Magalingam, P., Azmi, A.: Analysing Log Files For Web Intrusion Investigation Using Hadoop ICSIE '18
7. Agichtein, E., Brill, E., Dumais, S.: Improving web search ranking by incorporating user behavior information, SIGIR (2018)
8. Kumar, M., Meenu: Analysis of visitor's behavior from web log using web log expert tool. In: International Conference on Electronics, Communication and Aerospace Technology ICECA (2017)
9. Patel, R., Panchal, K., Rathod, D.: Efficient log mining from web server using clustering technique. J. Emerg. Technol. Innov. Res. (2016)

Chapter 65
Analysis of Heart Rate Variability Using Different Techniques of Breathing

R. Kalpana, C. R. Sai Krishna, Nihar Jajodia, and M. N. Mamatha

Abstract Heart rate variability plays a major role in the signal processing. The objective of the study is done by considering two important parts of the nervous system of the brain is known as a sympathetic and parasympathetic which are in turn useful for analysis of heart rate variability. These are analyzed using the lower frequency and higher frequency bands. The powers of this ratio will in turn helps to know the quantification of sympathovagal balance. Use of nonlinear methods such as Poincare plot, approximate entropy and correlation dimension has demonstrated in correctly enhancing information from biomedical signals pranayama or breathing exercise leads to better sympathovagal balance from Poincare plot analysis. Sympathetic and parasympathetic nervous system.

65.1 Introduction

Pranayama or yogic breathing has long been reported to spice up health and reduce the implications of stress and strain on the body and modern-day lifestyle has made people to induce mental pressure, especially within the metropolitan cities, due to increasing expectation and thus leading to depression it has been known that health beneficial effects of Pranayama are carried over through by meditation and mainly through improvement of autonomic functions. The bulk of individuals are littered with stress and strain. This could be significantly reduced by practicing certain yoga techniques for our improvement of health benefits, especially for the vital

R. Kalpana (✉) · C. R. Sai Krishna · N. Jajodia
Department of Medical Electronics Engineering, B.M.S College of Engineering, Bangalore 560019, Karnataka, India
e-mail: rk.ml@bmsce.ac.in

M. N. Mamatha
Department of Electronics and Instrumentation Engineering, B.M.S College of Engineering, Bangalore 560019, Karnataka, India
e-mail: mamathamnbms.intn@bmsce.ac.in

© The Author(s), under exclusive license to Springer Nature Singapore Pte Ltd. 2021   671
S. C. Satapathy et al. (eds.), *Smart Computing Techniques and Applications*,
Smart Innovation, Systems and Technologies 225,
https://doi.org/10.1007/978-981-16-0878-0_65

signals of our body. This successively enhances our condition of health. These chronic stresses, affects health and it produces various forms of bodily dysfunction. Therefore, the incidence of stress-induced disorders [1]. That has increased to an oversized within the past few years that particularly increases the cardiovascular abnormalities. It has been studied that several stress and strain-induced disorders occur thanks to autonomic imbalance, the imbalance between sympathetic and parasympathetic functions [1] brain and heart for people that are full of depression and other mental, and chronic diseases. Due to these lifestyle this has a bearing on people that are laid low with diabetics, hypertension obesity, and other stress related syndromes which successively affects the youngsters and young adults at greater measure. However, very less study has been made till date to assess the issues of autonomic balance achieved by practice of pranayama [2] or yogic breathing on reduction of prevalence of such metabolic disorders [1]. There are descriptions from our practice that tells us on how slow pranayama breathing is more practical in improving autonomic functions than the first pranayama breathing. With the use of kapalpathi yoga practice we can reach balance of the sympathetic activity. Though there was an earlier study that reported some extent of dominance in autonomic functions without significant difference in cardio-respiratory functions within the groups practicing left nostril breathing and right nostril breathing exercises. Cardiovascular functions is healthier ensured by strong sympathovagal balance as system is primarily under the control of autonomic system Therefore, effective sympathovagal homeostasis is that the foremost objective of various physical exercises and yogic therapies which are aiming at providing healthy psychological and physical life. Recently spectral analysis of pulse rate variability has been documented as a sensitive measure of autonomic function and dysfunction in health and clinical disorders. Though the results of right and left nostril breathing on sympathetic activity and blood Pressure are known, the implications on variability of heart rate and cardiovascular abnormalities risks don't seem to be yet investigated. Therefore, during this study we've assessed the difference within the results of right nostril breathing and left nostril breathing practice on sympathovagal balance, heart rate variation and cardiovascular abnormalities assessed by spectral analysis of variation of heart rate in young adults with the use of nonlinear analysis (Fig. 65.1).

65.1.1   Data Acquisition

Data was collected from 10 people ageing between 18 and 24 years. This was done in Swami Vivekananda center for Yoga Anusandhana Samsthana or S-VYASA research center. Where the ECG of 10 Male subjects were recorder for four different breathing techniques or 'pranayama'. Data was recorded using an electrocardiogram (ECG).The data acquisition process is done four different days for the four different exercises. The ECG is then segmented into different parts as per the activity in which the subject was involved. These parts contain initial baseline (5 min), three neurocognitive tests before the breathing Exercise (15 mins), three repetitions of the breathing exercise (ANB, BA, LNB, and RNB), and three post breathing neurocognitive tests (Fig. 65.2).

65.1.2   Data Conversion

Data conversion is required to convert ECG waveforms to text and numeric tables for analysis. This is done using a software named Acknowledge software. Acknowledge is a special software that has many user-friendly features in it. User could select any of them for his analysis, few of them are averaging, sample rates, automation and duration's in acquisition modes. Graphical windows are completely customized by allowing multiple views of the identical data. Transformations can be done by rearranging tool bars and by adding custom buttons. There may well be change the color or size of grids, waveforms, backgrounds, text, and more. Acknowledge can output a variety of stimulus paradigms for constant voltage stimulation and constant current. It accepts digital events from computer game environments and stimulus presentation systems like Presentation, E-Prime, and Super Lab and categorizes them for further future analysis and they'll also easily configure arbitrary waveforms, single pulses, and pulse chains for stimulation.

65.1.3   Data Processing

ECG processing involves extraction of parameters like pulse Variability, mean RR interval, LFnu, HFnu, LF/HF, etc. this can be done employing a software called kubios HRV Using the software ECG analysis can be done effectively for finding the suitable parameters used for the analysis such as detection of heart rate, using the amplitudes of R peaks, the software has an ability to accept other file-formats, filtering and trend analysis and automatic detection of QRS components The computation and analysis of Time as well as frequency domain of heart rate variability analysis and also many nonlinear-parameters can be obtained using the proposed software. There are different flexible analysis settings are adopted for optimization of varied datas. In addition to that the respiratory frequency can also computed for our analysis (Fig. 65.3).

65.1.4   Data Analysis

SPSS software was used for Analyzing the Results Obtained from the kubios. There was a comparison done on different breathing exercise through nonlinear methods. Data was processed using SPSS Version 14.0. The parameters suitable for our study is analyzed using analysis of variance and T test is also applied and the value of *p*

< 0.05 was taken for calculation is significant. The related hemodynamic variables such as ratio between low to high frequency and Standard deviations ratio can also calculated

65.2 Linear and Nonlinear Methods

65.2.1 Linear Methods

HRV could be analyzed in time or frequency domain using linear algorithms. This technique assumes. ECG is a stationary signal. Heart rate Variability [1] techniques using linear algorithm [3] are computationally simple. Information content in the RR interval cannot be identified or analyzed through Linear HRV analysis as cardiac functions are complex and nonlinear owing to multiple control loops. Hence there are limitations in the linear system. For the calculation of heart rate variability Time domain features are the most important concern as they calculated from statistical methods and R-R intervals and they can also be correlated with one another. Frequency domain indexes are more elaborated indexes supported spectral analysis, mostly accustomed evaluate the Contribution on HRV of autonomic system (VLF, LF, HF, HF/LF ratio) (Fig. 65.4).

65.2.2 Non-linear Methods

Nonlinear analysis are newly introduced methods to measure rate of variability of heart. They include nonlinear parameters such as, Correlation dimension, Poincare geometry, Hurst exponent, Lypnov exponent, and Approximate Entropy [4, 5]. All complex interactions of electrophysiological, hemodynamic, and humoral variables additionally as by the autonomic and central nervous regulations are understood. These techniques are shown to be powerful tools for characterization of complex biological systems. RR interval contains information about different physiological states thanks to different interacting systems. Hence we get to understand about more complicated physiological data that provides knowledge about vital sign, mental pressure, and respiration. The complex interplay between Parasympathetic and sympathetic [6] stimulation are often detected by Nonlinear pulse Analysis (Fig. 65.5). These nonlinear analysis and methods are applied in Bio-signal Processing and Applications [6–9].

65.3   Results

The following differences in the means of the nine parameters considered in this study to evaluate sympathovagal balance are tabulated below. The nine parameters are (Fig. 65.6): The datas before Pranayama and After Pranayama are tabulted in the following with the subjects taken for Analysis Tables (65.1 and 65.2).

1.   Mean RR interval
2.   Mean Heart Rate
3.   LFnu
4.   HFnu
5.   LF/HF
6.   SD1
7.   SD2
8.   SD2/SD1
9.   Approximate Entropy

Table 65.1   Data before pranayama

Parameters	Sub1	Sub2	Sub3	Sub4	Sub5
Mean HR	71.07	75.59	66.04	83.35	70.47
LF	64.5	48.8	63.9	89.3	65.4
HF	35.5	51	36	10.7	34.4
LF/HF	1.81	0.95	1.74	8.37	1.9
SD1	26	15.6	31.7	59.8	28.1
SD2	71.2	56	107	184	61
SD2/SD1	2.73	3.5	3.37	3.07	2.17

Table 65.2   Data after pranayama

Parameters	Sub1	Sub2	Sub3	Sub4	Sub5
Mean HR	74.55	64.27	68.04	85.35	70.96
LF	45.1	44.8	41.8	60	44.8
HF	54.8	55	58.2	40	54.8
LF/HF	0.823	0.814	0.718	1.499	0.817
SD1	25	31.4	23.2	35.4	30.3
SD2	76.4	60.7	46	69.2	42
SD2/SD1	3.056	1.93	1.98	1.95	1.38

65.4   T Test

T-test is a useful hypothesis applied between the means of two groups which might be related in certain features. To conduct a test with three or more variables, one must use an analysis of variance. To conduct a test with three or more variables, one must use an analysis of variance. T-test was done for various breathing exercise in order to determine the significant nonlinear components. The mean heart rate, SD1, SD2, SD2/SD1, and Approximate Entropy were found to be significant. The mathematical equivalence of LF/HF in the Linear method Correlated with the SD2/SD1 [5]. Thus there was much similarity of LF to SD2 and HF to SD1. The mean heart rate, LF, HF, LFnu, HFnu, SD1, SD2, SD2/SD1, and approximate entropy was calculated for all different exercises. Only few exercises like Alternate breathing and Post Alternate breathing had significance changes with respect to most of the linear and nonlinear parameters. The ratio of standard deviation was seen reduced 1 digit lower from starting data recorded (Figs. 65.7 and 65.8).

65.5   Conclusions

Yogic breathing or pranayama practices were known to influence the neurocognitive abilities, pulmonary and autonomic functions as well as the metabolic and biochemical activities in the body. With the SD1 decreasing, the SD2 increases, and the ratio of SD2/SD1 decreases instantly after the sympathetic stimulation with concurrent change of shape in the Poincare plot. Parasympathetic activity is determined by Standard deviation1 and Sympathetic activity by Standard deviation2. Like the LF/HF the ratio of SD2/SD1 could also determine sympathovagal balance. Sympathovagal changes can be easily detected by the changes in the shape of Poincare plot mainly during the tilt position. We found that the ratio of SD2/SD1 was greater for persons before doing the breathing exercise. After the breathing was done, Ratio gradually decreased and thus maintaining the Sympathovagal balance. The proposed work aims for finding out some more nonlinear parameters such as approximate entropy, Correlation Dimension, Sample entropy, which can also give information about parasympathetic and sympathetic parameters. The above research can enhance the work for people who are suffering from heart-related as well as stress-related for leading them to live a healthy lifestyle and mental peace. The Changing Lifestyle of Man or Woman has brought more stress in them and as a result they undergo through Depression. Yogic breathing not only brings the relax state it is also a powerful Preventive measure for mental peace and Physical Wellness. Overall, we found the practice of yogic breathing safe, when practiced under guidance of a trained teacher.

Fig. 65.1 Block diagram of
the Proposed work

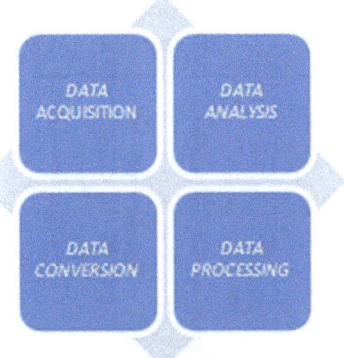

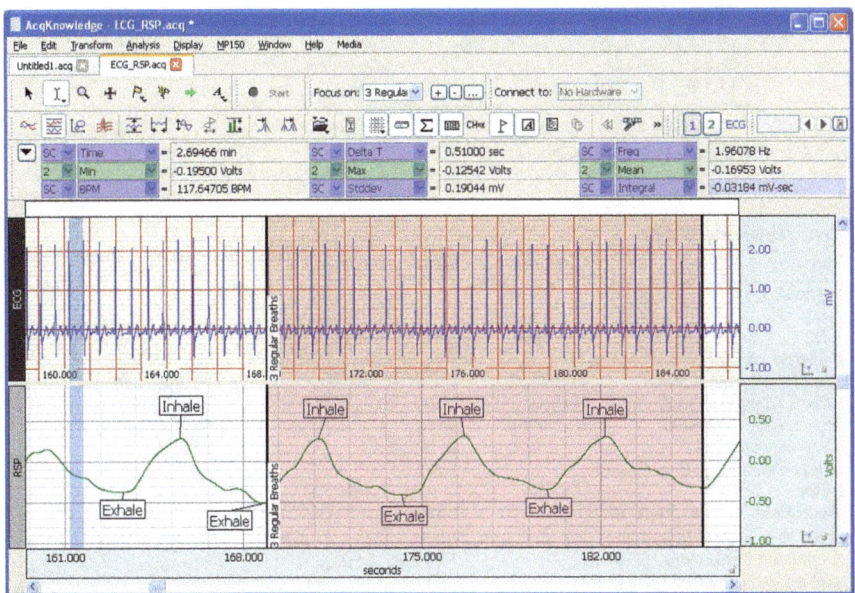

Fig. 65.2 Graphical output of the Proposed work

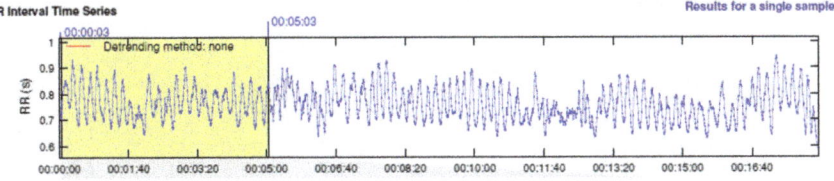

Fig. 65.3 Time Series Analysis

Frequency–Domain Results

FFT spectrum   (Welch's periodogram: 256 s window with 50% overlap)

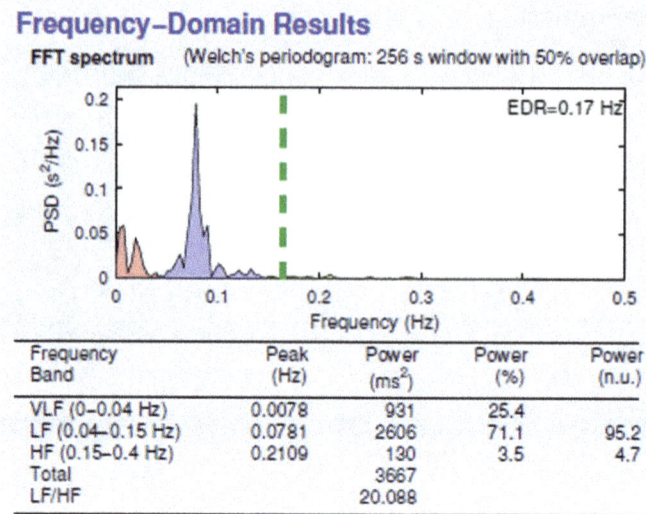

Frequency Band	Peak (Hz)	Power (ms^2)	Power (%)	Power (n.u.)
VLF (0–0.04 Hz)	0.0078	931	25.4	
LF (0.04–0.15 Hz)	0.0781	2606	71.1	95.2
HF (0.15–0.4 Hz)	0.2109	130	3.5	4.7
Total		3667		
LF/HF		20.088		

Fig. 65.4 Frequency Domain Analysis

Nonlinear Results

Variable	Units	Value
Poincare plot		
SD1	(ms)	11.9
SD2	(ms)	39.8
Recurrence plot		
Mean line length (Lmean)	(beats)	11.56
Max line length (Lmax)	(beats)	368
Recurrence rate (REC)	(%)	38.56
Determinism (DET)	(%)	98.82
Shannon Entropy (ShanEn)		3.245
Other		
Approximate entropy (ApEn)		1.169
Sample entropy (SampEn)		1.500
Detrended fluctuations (DFA): $\alpha1$		1.264
Detrended fluctuations (DFA): $\alpha2$		0.844
Correlation dimension (D2)		0.551
Multiscale entropy (MSE)		1.284 – 2.640

Poincare Plot

Fig. 65.5 NonLinear Analysis

B	C	PERSON 1	PERSON 2	PERSON 3	PERSON 4	PERSON 5	PERSON 6	PERSON 7	PERSON 8	PERSON 9	PERSON 10
BASELINE	MEAN RR	847.9	795.9	915.8	741.7	854.1	923.3	736.7	923.9	802.9	1054.1
	MEAN HR	71.07	75.59	66.04	83.35	70.47	65.37	81.7	66.08	75	57.26
	Lfnu	64.5	48.8	63.9	89.3	65.4	53	70.8	77.5	34.6	45
	Hfnu	35.5	51	36	10.7	34.4	47	29.2	22.5	65.2	55
	LF/HF	1.818	0.957	1.744	8.375	1.902	1.126	2.429	3.449	0.531	0.819
	SD1	26	15.6	31.7	59.8	28.1	36.5	17.5	71.7	28.1	51.9
	SD2	71.2	56	107.9	184.7	61.3	93.5	56.2	156.6	61.2	97.7
	SD2/SD1										
	SAMPLE ENTROPY(SampEn)	1.572	1.483	1.324	0.422	1.544	1.524	1.654	1.144	1.855	1.42
	APPROXIMATE ENTROPY(ApEn)	1.142	1.188	1.06	0.625	1.136	1.106	1.287	0.93	1.236	1.022
ANB1	MEAN RR	795.6	681.8	851.7	742.1	808.8	806.7	701.8	754	739.1	919.6
	MEAN HR	76.09	88.44	71.22	83.54	75.1	75.13	86.88	80.44	69.9	66.28
	Lfnu	94.5	96	96.6	89.5	89.9	92.5	97	95	46.7	97.5
	Hfnu	5.5	4	2.1	10.5	10	7.5	3	5	53.3	2.5
	LF/HF	17.165	23.989	28.901	8.51	8.954	12.396	32.738	19.12	0.875	39.521
	SD1	103.9	13.5	22.2	65.7	38.4	35.7	27.2	25.6	31.5	32.9
	SD2	26.1	67.5	124.5	189.6	119.8	110.6	124.9	110.4	93.8	156.6
	SD2/SD1										
	SAMPLE ENTROPY(SampEn)	0.99	0.863	0.861	0.339	0.901	1.118	0.593	0.864	1.149	0.919
	APPROXIMATE ENTROPY(ApEn)	0.928	0.854	0.823	0.55	0.839	0.989	0.682	0.835	0.98	0.835
	MEAN RR	799.5	706.8	862.9	743.6	806.7	825.8	688.3	772.8	720.1	891.5
	MEAN HR	75.64	85.35	70.42	83.2	75.14	73.37	88.3	80.1	83.76	68.48
	Lfnu	94.2	95.8	92.1	90	88.2	92.9	90.8	96.1	26.3	96.9
	Hfnu	5.8	4.2	7.9	10	11.8	7.1	9.2	2.9	72.7	3.1

Fig. 65.6 Nonlinear Parameters

Paired Samples Statistics

		Mean	N	Std. Deviation	Std. Error Mean
Pair 1	BS- MEAN RR	859.630	10	97.5394	30.8447
	ANB1 MEAN RR	780.120	10	71.4596	22.5975
Pair 2	BS- MEAN HR	71.1930	10	7.98113	2.52386
	ANB1 MEAN HR	77.3020	10	7.37001	2.33060
Pair 3	BS- Lfnu	61.280	10	16.2420	5.1362
	ANB1 Lfnu	89.520	10	15.3085	4.8410
Pair 4	BS- Hfnu	38.650	10	16.1960	5.1216
	ANB1 Hfnu	10.340	10	15.3809	4.8639
Pair 5	BS- LF/HF	2.314133955	10	2.286377569	.7230160709
	ANB1 LF/HF	20.82384670	10	14.47099638	4.576130857
Pair 6	BS- SD1	36.690	10	18.5321	5.8604
	ANB1 SD1	39.660	10	26.4082	8.3510
Pair 7	BS- SD2	94.630	10	44.5798	14.0974
	ANB1 SD2	112.380	10	44.8303	14.1766
Pair 8	BS- SD2/SD1	2.701968909	10	.5959161523	.1884452336
	ANB1 SD2/SD1	3.660505478	10	1.541533720	.4874757646
Pair 9	BS ApEn	1.39420	10	.390439	.123468
	ANB1 ApEn	.85970	10	.239189	.075638

Fig. 65.7 Post right nostril breathing

Paired Samples Statistics

		Mean	N	Std. Deviation	Std. Error Mean
Pair 1	BS- MEAN RR	859.630	10	97.5394	30.8447
	Post ANB RR	818.880	10	80.4275	25.4334
Pair 2	BS- MEAN HR	71.1930	10	7.98113	2.52386
	Post ANB HR	74.3370	10	7.46173	2.35961
Pair 3	BS- Lfnu	61.280	10	16.2420	5.1362
	Post ANB LFNU	60.630	10	14.7714	4.6711
Pair 4	BS- Hfnu	38.650	10	16.1960	5.1216
	Post ANB HFNU	39.630	10	14.6075	4.6193
Pair 5	BS- LF/HF	2.314133955	11	2.169048213	.6539926430
	Post ANB LF/HF	1.909300	11	1.0698083	.3225593
Pair 6	BS- SD1	36.690	10	18.5321	5.8604
	Post ANB SD1	31.960	10	12.1728	3.8494
Pair 7	BS- SD2	94.630	10	44.5798	14.0974
	Post ANB SD2	72.150	10	28.6391	9.0565
Pair 8	BS- SD2/SD1	2.701968909	10	.5959161523	.1884452336
	Post ANB SD2/SD1	2.414306247	10	1.084664016	.3430008788
Pair 9	BS ApEn	1.39420	10	.390439	.123468
	Post ANB ApEN	1.12850	10	.091234	.028851

Fig. 65.8 Post alternate nostril breathing

References

1. Aparvinash Saoji, B.R., Raghavendra, N.K., Manjunath: Effects of yogic breath regulation: a narrative review of scientific evidence
2. Schubert, C., Lambertz, M., Nelesen, R.A., Bardwell, W., Choi, J.B., Dimsdale, J.E.: Effects of stress on heart rate complexity—a comparison between short-term and chronic stress. Biol. Psychol. **80**, 325–332 (2009). https://doi.org/10.1016/.j.biopsycho.2008.11.005
3. Shinba, T., Kariya, N., Matsui, Y., Ozawa, N., Matsuda, Y., Yamamoto, K.: Decrease in heart rate variability response to task is related to anxiety and depressiveness in normal subjects. Psychiatry Clin. Neurosci. **62**, 603–609 (2008). https://doi.org/10.1111/j.1440-819.2008.01855.x
4. Rajendra Acharya, U., Paul Joseph, K., Kannathal, N., Lim, C.M., Suri, J.S.: Heart rate variability: a review. Med. Biol. Eng. Comput. **44**, 1031–1051 (2006)
5. Dabire, H., Mestivier, D., Jarnet, J., Safar, M.E., Chau, N.P.: Quantification of sympathetic and parasympathetic tones by nonlinear indexes in normotensive rats. Am. J. Physiol. Heart Circul. Physiol. **275**, H1290–H1297 (1998)
6. Pal, G.K., Agarwal, A., Nanda, N.: Slow yogic breathing through LNB and RNB influences sympathovagal balance, heart rate variability and cardiovascular risks in adults. Department of Physiology, Jawaharlal Institute of Postgraduate Medical Education and Research, Pondicherry, India. PMCID: PMC3978938
7. Joshi, M., Bairagi, S.: Hrv analysis using linear and nonlinear techniques. In diabetic: research article. BMR Med. **2**(1), 1–6
8. Hsu, C.-H., Li, C.-Y.: Poincare plot indexes of HRV detect dynamic autonomic modulation during general anesthesia induction. Department of Anesthesiology, Tri-Service General Hospital, National Defense Medical Center, Taipei, Taiwan
9. Mirescu, S.-C., Harden, S.W.: Nonlinear dynamics methods for assessing HRV in patients with recent myocardial infarction. Burnett School of biomedical Sciences, College of medicine, University of Central Florida, USA

Chapter 66
Minimum Time Search Methods for Unmanned Aerial Vehicles

R. Darsini, Nikita Ganvkar, Karthik Gurunathan, and Ranjita Kumari Dash

Abstract This paper analyzes and compares methods used to evaluate and obtain the best trajectories of multiple Unmanned Aerial Vehicles (UAV) that have to find a lost target as soon as possible. The analysis shows how the high-complexity of this problem, named Minimum Time Search (MTS), makes it a good candidate to solve it with different bio-inspired and probabilistic-based approaches and also elaborates some of the advantages and drawbacks of these approaches. The paper also compares and relates two different mathematical formulations that can be used to evaluate the UAV trajectories based on the mean-time to detect the targets with the sensors on-board the UAVs.

66.1 Introduction

UAVs are aircrafts that do not have a human pilot on board, and they may or may not be remotely controlled. More specifically, UAVs are essential components of Unmanned Aircraft Systems (UAS). The other components of UAS are Ground Control Stations (GCS) and its communication system with the UAVs. The UAV itself comprises of: (i) a frame (the skeleton of the UAV); (ii) a propeller (the physical actuators that hold the UAV in flight); (iii) the engines (makes the propellers rotate); and (iv) the flight controller (generates the actions that ensures the UAV's stability and correct displacement during the flight).

UAVs were originally used for military purposes (e.g. tactical reconnaissance or target tracking), but their applications have been extended to other fields (e.g. inspection, monitoring, surveying, mapping, aerial imaging and video recording). The paper mainly focuses on Search And Rescue (SAR) operations, where the UAVs have to determine the location of one or several lost targets. The main advantage of UAVs over piloted vehicles in these situations is the elimination of the human risk

R. Darsini (✉) · N. Ganvkar
Department of Computer Science and Engineering, PES University, Bengaluru, India

K. Gurunathan · R. K. Dash
Department of Computer Science and Engineering, PES Institute of Technology, Bengaluru, India

© The Author(s), under exclusive license to Springer Nature Singapore Pte Ltd. 2021   681
S. C. Satapathy et al. (eds.), *Smart Computing Techniques and Applications*,
Smart Innovation, Systems and Technologies 225,
https://doi.org/10.1007/978-981-16-0878-0_66

factor in the mission. Moreover, UAVs can offer more endurance and range than their manned counterparts, as the Strengths, Weaknesses, Opportunities, and Threats (SWOT) analysis of UAS-based applications presented in [5] shows.

In terms of SAR operations, time becomes an essential factor in the mission. This is especially true for missions which have a dire aftermath if the target is not found on time. For such scenarios, the target has to be found in the least possible time by a team of UAVs in a SAR mission.

MTS approaches determine UAV's trajectories that cut down the target detection time by considering the available information about the SAR scenario. MTS is often optimized using bio-inspired algorithms and probabilistic-based optimization approaches. A prime bio-inspired algorithm explored in the paper is Ant Colony Optimization (ACO), which can also incorporate a problem-specific heuristic to increment the chances on generating overall good solutions quickly. Other bio-inspired approaches are Evolutionary Algorithms (EA), while probabilistic-based optimization techniques are the Bayesian Optimization Algorithm (BOA) and Cross Entropy Optimization (CEO). Using such approaches, MTS can be used as a framework for single and multi-UAVs SAR mission.

This paper analyses bio-inspired and probabilistic-based optimization algorithms well suited for multi-UAV MTS, highlighting their main advantages and limitations. Besides, after the literature review presented in Sect. 66.2, the optimization approaches are described in Sect. 66.3, the mathematical models for evaluating the solutions are analyzed in Sect. 66.4. The conclusion and future work are presented in Sect. 66.5.

66.2   Literature Overview

This section describes the main bio-inspired and probability-based optimization approaches for MTS. The main advantages and limitations of these approaches are summarized in Table 66.1.

A prime example of the implementation of bio-inspired approaches is [9], where the underlying goal is to locate a lost target as quickly as possible using a fleet of UAVs. This is done by finding the path required to fly quickly to the regions where there are high chances of finding the target (using ACO). The approach in [9] also includes a new heuristic for MTS where the target's spatial and probabilistic properties are exploited while building the UAV trajectory. The performance of this algorithm is tested against various scenarios and is compared with other MTS based algorithms (ad-hoc heuristics in [6], CEO, BOA and Genetic Algorithms- GA). The comparison clearly shows that the approach in [9] provides better solutions than the other algorithms used for MTS.

The feasibility of ACO for MTS is further explored in [11]. This is done by complementing the original MTS heuristic presented in [9] with new terms that account for important constraints such as maintaining the UAV's communication with the GCS and collision avoidance among the UAVs. This algorithm is tested

Table 66.1 Comparison of closely related works based on MTS

Author	Advantages	Limitations
Perez Carabaza et al. [9] MTS-ACO	High level straight segmented trajectories are obtained by exploiting spatial properties of the problem itself	Restricts the UAV mainly towards regions with closer and higher probability of target presence
Perez Carabaza et al. [11] MTS-ACO	Computing UAV trajectories after including the possibility of collision and loss of communication loss with GCS	Synthetic scenarios do not mirror real world situations
Perez Carabaza et al. [7] MTS-ACO	It shortens the detection time of a given target by using a real-coded ACO_R that compares the UAV performance using a complex dynamic model	Artificial ants of ACO_R follow heuristic based methodology similar to patrollers and do not use inputs from the pheromone table
Perez Carabaza et al. [8] MTS-EA	It includes a complex dynamic model for the UAV and evaluates UAV performance with multiple criteria	Computational time increases due to the calculations of four different optimization criteria
Perez Carabaza et al. [10] MTS-EA	It combines intelligence and geographical information to construct target probability models that will be used in real-world missions. It simultaneously optimizes the UAV trajectory and camera movement	Model does not handle probabilistic behavior for multiple targets
Bourgault et al. [1] MTS-Greedy MTS-Control Optimization	It introduces the continuous mathematical formulation of the single UAV MTS problem with arguments that make it applicable to simplified real-world problems	Being a grid based approach (as the others), it is difficult and computationally costly to increase the search area
Lanillos et al. [3] MTS-CEO	It focuses on the discretization of the evaluation of the MTS, applicable to only one UAV at a time instead of including multiple constraints and increasing the uncertainty	Targets with different probability location distributions are out of scope
Lanillos et al. [4] MTS-BOA	Expressions of single-agent MTS problem are extended to accommodate the multi-agent variant	BOA computational cost makes it a slow approach for trajectories with many segments
Meghjani et al. [6] MTS-heuristics	UAVs are driven to overfly highest probability areas by ad-hoc heuristics specially designed for MTS	The heuristics have only been tested with static target scenarios

with pre-defined scenarios and is successfully compared against extended CEO and GA-based MTS algorithms. However, neither the scenarios in [9] nor [11] completely mirror real-world situations as they are theoretically built, to show the superiority of ACO over other algorithms. Besides, both approaches generate trajectories defined by straight-line segments, which can only be followed by rotatory-wing UAVs.

In order to generate curved trajectories that can also be followed by fixed-wing UAVs, the approach in [7] makes use of a complex dynamical model of the UAV, whose reference signals are optimized with ACO_R (an ACO algorithm specifically designed for real-coded domains). Besides, in order to use a novel MTS heuristic in ACO_R, the approach in [7] includes a certain percentage of ants that build their solutions using the proposed heuristic.To deal better with real-world scenarios, the approach in [8] uses an EA for multi-objective constrained problems which includes several criteria: Non-Flying Zones (NFZ) and collision avoidance, fuel consumption, expected detection time, myopia reduction index, and trajectory smoothness. Furthermore, [10] replaces artificial scenarios with real-world ones and it allows to simultaneously optimize both the UAV trajectory and the camera pose. This work also presents three detailed real-world inspired scenarios and includes the impact of different variables (e.g. search area, grid size, and UAV flying height).

To model the uncertainty about the target location in each scenario, all the approaches in Table 66.1 explore the connection between MTS and Bayes theorem. The introductory work in [1] focuses on the condition that a single UAV searches for a non-evading single target, which may be static or in motion. To be able to use Bayes theorem, unknown quantities are cast as random variables, whose information are expressed as probability density functions (PDF). In the MTS problem, an unknown quantity is the target location. Bayes theorem is used to combine new probability information with prior probability information to form a new PDF. In the MTS problem, the Bayes theorem combines the detection measurements provided by the sensors onboard the UAVs with the probability of locating the target just before the observation is obtained to form a new updated PDF. Besides, if the target is in motion, the state of the PDF is also predicted by increasing discrete time intervals of the UAV time of flight. The predicted and updated PDF is used to calculate the *expected time* of target detection, which is optimized through different approaches to obtain the best references for the UAV and their associated trajectories to find the target.

To give a fresh outlook in solving the MTS problem, CEO is applied in [3] to compute the best N number of actions for several UAVs that has to find a single non-evading mobile target in the minimum possible time. This work also proposes to optimize the discounted-time probability of target detection for MTS.

A more extensive use of Bayesian Theory is put to test in [4], which in fact, utilizes the Bayes theorem twice: (i) for evaluating the expected time of target detection in order to compare the routes of the agents and (ii) within an optimization algorithm that exploits Bayesian theorem and probability independence to solve the problem. Besides, it formalizes the MTS problems as a Partially Observable Markov Decision Process (POMDP) using the target initial location belief, and probabilistic models for the target motions and sensors. The approach in [4] also extends the evaluation of the expected time of target detection for multi-UAV MTS problems. This shows that minimising the time taken to find the target increases the probability of finding the target itself, which is successfully tested over five scenarios.

An alternative approach is presented in [6], which introduces several constructive heuristics designed ad-hoc for the MTS problem of a single static target. These heuristics can quickly calculate the UAVs trajectories, avoiding the optimization process of MTS-related criteria that occurs in the other approaches.

Finally, it is worth noting that the previous analyses illustrates how bio-inspired and probability-based approaches can be used to solve different MTS problems successfully. The promising result of ACO, a member of swarm optimization, makes it interesting to analyze the behavior of these types of approaches for MTS. An example is the work in [2], a behaviour-based approach for searching lost targets, with a reduced need of human control. This is done by providing personalization parameters to the UAV, on top of the existing hive mind algorithm (where the UAVs share behaviours and are controlled by a central system). The personalization attributes are defined as: (i) conformity, the importance the UAV gives to the user's suggestions; (ii) sociability, the dispersion among the UAVs; (iii) dedication, the effort put in by the UAV to track the lost target; and (iv) disposition, the UAVs frustration during the negotiation process. These attributes allow the UAVs to behave more efficiently, and not just stick to a centralized control, found in most traditional swarm control approaches.

66.3   Optimization Approaches

This section presents a description of the bio-inspired and probability-based approaches adopted in the previously described MTS techniques.

66.3.1   Ant Colony Optimization (ACO)

Ants possess collective complex social behaviours, which can be implemented to solve different types of combinatorial optimization problems. Specifically, ants tend to find the shortest path from their nest to food by depositing pheromone on the paths, hence creating and following a "pheromone trail'. Since the ants also follow the pheromone trails of other ants, more pheromone on a particular path increases the chance of finding the ideal trail. This ability of ants to find the shortest paths, is the foundation of ACO, a population-based metaheuristic which can be used to estimate solutions for practical optimization problems. This is done by creating a pheromone matrix and then using the matrix to record the paths created by the UAVs. This allows us to find the UAV path with the highest score in the pheromone matrix, which tends to be the suitable trajectory. ACO also allows us to combine the pheromone information with a problem-specific heuristic to guide the UAVs towards better solutions more quickly.

66.3.2 Evolutionary Algorithm (EA)

EAs are a population-based algorithms which incorporate natural selection features into the optimization process. In this algorithm, the population gets better fitness through an iterative process of selection, reproduction and mutation. This algorithm is different from the other approaches, as it randomly samples the solutions that will be crossed. This provides higher chances of obtaining better solutions. It then determines the best solutions that will survive in the following iterations by comparing the new population with the previous one. To handle multiple objective functions and constraints, it contrasts the quality of pairs of solutions based on an overall constraint violation criterion first and on a Pareto comparison of the objective functions second [8]. EA can be applied to maximize the probability of target detection which are searched by the UAVs (whose movement is limited by cardinal directions).

66.3.3 Cross Entropy Approaches (CEO)

CEO uses the cross entropy method, which is a probabilistic-based method used for continuous and combinatorial problems. It is an iterative procedure which can be divided into two phases: (1) using a specified probability function to create random data samples (in our case, trajectories) and (2) updating the parameters of the probability function based on the best data (best current solutions) to construct better samples (solutions) in the succeeding iteration. The CEO approach used in MTS assume the independence among the decision variables of the problem, allowing to generate trajectories that combine the most probable directions to perform at each time step.

66.3.4 Bayesian Optimization Algorithm (BOA)

BOA is a probabilistic optimization algorithm that determines which nodes are the decision variables of the problem, that better fits the group of best solutions of a problem, at each iteration of the Bayesian Network (BN). The BN is then used to sample the set of new solutions that will be evaluated, sorted, and selected as the best set of the following iteration. These two steps are repeated for all iterations as in [1], to determine the relevant relationships among the decision variables, and it returns the best solution found at the end of these iterations.

66.4 Mathematical Evaluation of MTS

This section presents and compares two different models used in the literature to evaluate the expected time of target detection. The two selected models are the ones presented in the introductory work in [1] and the one used in the recent ACO-based work in [11].

In both cases, the unknown target state is modelled with a random variable that has known initial belief and motion probability; the UAV location is known (although not explicitly written in [1]), and the sensor observations can be detection or non-detection measurements, and are related with the target state and UAV location through the sensor likelihood function. All these variables and probability functions used in each paper are summarized in Table 66.2.

66.4.1 Models and MTS Evaluation In [1]

The target initial belief $p(x_0^t)$, target motion model $p(x_k^t|x_{k-1}^t)$ and sensor likelihood $p(z_k|x_{k-1}^t)$ are used in this case to obtain the target belief $p(x_k^t|z_{1:k})$ at any time k given the measurements z_{1_k} up to this time. To do this, it makes use of a recursive bayesian filter that iterates through two steps:

Table 66.2 Variables and probabilities summary

Description	In [1]	In [11]
Time	k	t
UAV indexing	–	u
Target state at given time	x_k^t	v^t
UAV location at given time	–	x_u^t
Sensor observations at given time by given UAV	z_k	z_u^t
Detection observation	D_k	D
Non detection observation	$\overline{D}_k$	$\overline{D}$
Initial belief	$p(x_0^t)$	$P(v_0)$
Target motion probability	$p(x_k^t|x_{k-1}^t)$	$P(v^t|v^{t-1})$
Sensor likelihood	$p(z_k|x_k^t)$	$P(z_u^t|v^t, x_u^t)$
Probability of not detecting the target	Q_k	$P(z_{1:U}^{1:t} = \overline{D}|x_{1:U}^{0:t})$
Expected time to detect the target	MTTD	ET

- *Prediction* phase, which is required when the Probability Distribution Function (PDF) of the target state is time dependent since the target is moving or the ambiguity about its position is increasing.
  This step updates the PDF with the latest observations $p(x_{k-1}^t|z_{1:k-1})$ using the Chapman Kolmogorov expression and the target motion model $p(x_k^t|x_{k-1}^t)$:

$$p(x_k^k|z_{1:k-1}) = \int p(x_k^t|x_{k-1}^t) \cdot p(x_{k-1}^t|z_{1:k-1}) \cdot dx_{k-1}^t \tag{66.1}$$

- *Update* phase, which is performed using the Bayesian theorem and the sensor likelihood $p(z_k|x_k^t)$ after the new observation z_k at time k becomes available:

$$p(x_k^t|z_{1:k}) = \frac{1}{K} \cdot p(z_k|x_k^t) \cdot p(x_k^t|z_{1:k-1}) \tag{66.2}$$

where $K = \int p(z_k|x_k^t) \cdot p(x_k^t|z_{1:k-1}) \cdot dx_k^t$ is a normalization factor to ensure that $p(x_k^t|z_{1:k})$ is a PDF (i.e. $\int p(x_k^t|z_{1:k}) \cdot dx_k^t=1$). Note that for Eq. (2) z_k must equal the value of the observation at time k (i.e. $z_k = D_k$ or $z_k = \overline{D}_k$), and that the likelihood also depends on the UAV location from where the measurement z_k is taken, although it is not implicitly stated in the expression.

To evaluate a given UAV trajectory according to the MTS criterion, the following expressions are used:

- Probability of not detecting $\overline{D}_k$ the target at time step k given previous measurements. $z_{1:k-1}$:

$$r_k = p(\overline{D}_k|z_{1:k-1}) = \int p(\overline{D}_k|x_k^t) \cdot p(x_k^t|z_{1:k-1}) \cdot dx_k^t \tag{66.3}$$

Note that the expression is similar to the one used to evaluate K, when $z_k = \overline{D}_k$.
- Probability of not detecting the target from UAV trajectory.

$$Q_k = P(\overline{D}_{1:k}) = \prod_{i=1}^{k} r_j \tag{66.4}$$

Note that the probability P_k of detecting the target from any point of the UAV trajectory, is the complementary of the previous (i.e. $P_k = 1 - Q_k$).
- Probability of detecting the target at time k, not having it detected at $j < k$

$$p_k = (1 - q_k)Q_{k-1} \tag{66.5}$$

- The Mean Time To Detect (MTTD) the target, which is the evaluation criterion itself:

$$\text{MTTD} = \sum_{k=1}^{\infty} k \cdot p_k \qquad (66.6)$$

Note that previous equation is a discretized version of the usual expression to obtain the mean value of a variable (i.e. $\text{mean}(t) = \int t \cdot p(t) \cdot dt$), with $p(t) = p_k$

66.4.2 Models and MTS Evaluation In [11]

The target initial belief $P(v_0)$, target motion model $p(v_t|v^{t-1})$ and sensor likelihoods $p(z_u^t|v^t, x_u^t)$ are used in this case to obtain directly an "unnormalized" target belief $\tilde{b}(v^t)$ updated up to time k given only non detection measurements up to this time. To obtain this function it uses the following expression:

$$\tilde{b}(v^t) = \prod_{u=1:U} P(z_u^t = \overline{D}|v^t, x_u^t) \quad \cdot \quad \sum_{v^{t-1} \in G_\Omega} P(v^t|v^{t-1}) \cdot \tilde{b}(v^{t-1}) \qquad (66.7)$$

In relation with the previous section's expressions, it is worth to note the following. On one hand, Eq. (66.7) is equivalent to the prediction step stated at Eq. (66.1). Hence, it is used in Eq. (66.7) to redistribute $\tilde{b}(v^{t-1})$ according to the target motion model. On the other hand, Eq. (66.7) is similar to the update step stated at Eq. (66.2). However, Eq. (66.7) includes a product of likelihoods (to account for the measurements of multiple UAVs), lack of the normalization term (required to obtain a proper belief), and assume that the $z_u^t = \overline{D}$ (to be used next to calculate the probability of not detecting the target).

Next, the approach calculates the probability of not detecting the target with Eq. (66.8). This new expression plays the role of Eqs. (66.3) and (66.4) in the previous approach. Moreover, the external summation in Eq. (66.8) stands for the integral in Eq. (66.3), and the product of Eq. (66.4) is not needed since $\tilde{b}(v^t)$ has not been normalized in Eq. (66.7). Hence, this second process is computationally quickly.

$$P(z_{1:U}^{1:t} = \overline{D}|x_{1:U}^{0:t}) = \sum_{v^{t-1} \in G_\Omega} \tilde{b}(v^t) \qquad (66.8)$$

The Expected Time (ET) to detect the target is obtained with Eq. (66.9). Although it plays the same role as Eq. (66.6), the ET is calculated taking advantage that the time is a positive variable and hence its mean value can be obtained as $\text{mean}(t) = \int (1 - P(y \leq t)) \cdot dt$, being $(1 - P(y \leq t))$ for this case, the probability of not detecting the target up to time step t (i.e. $1 - P(y \leq t) = P(z_{1:U}^{1:t})$).

$$ET(x_{1:U}^{0:N}) = \sum_{t=1}^{\infty} P(z_{1:U}^{1:t} = \overline{D}|x_{1:U}^{1:t}) \tag{66.9}$$

Finally, it is worth noting the following. Although the summation in Eq. (66.9) and Eq. (66.6) are up to ∞, it has to be truncated for evaluation purposes. This often makes the ET and MTTD expressions underestimate the mean value of the detection time, causing Eq. (66.9) to have smaller truncation errors than Eq. (66.6).

66.5 Conclusion and Future Work

This paper presents a clear comparison between several bio-inspired and probabilistic optimization algorithms used for UAVs path planning in MTS. Two mathematical models (ACO & Bayesian) have been elaborated as well to emphasize the same. On comparing different literature based on MTS, it is seen that ACO is predominantly used to solve the discrete version of the MTS problem (where the UAVs' flight follows straight lines), while EA is mostly used for its continuous counterpart (where the UAVs' flight follows curved smooth trajectories). Besides, many works use idealistic MTS scenarios and only a handful of them successfully take advantage of ad-hoc constructive heuristics to obtain overall good solutions quickly. Hence, our future work will consist of hybridizing new ad-hoc constructive heuristics for MTS with general optimization meta-heuristics, to obtain new approaches for more realistic MTS scenarios.

References

1. Bourgault, F., Furukawa, T., Durrant-Whyte, H.: Optimal search for a lost target in a Bayesian world. In: Field and Service Robotics, pp. 209–222 (2003)
2. Hexmoor, H., Mclaughlan, B., Baker, M.: Swarm control in unmanned aerial vehicles. In: Proceedings of International Conference on Artificial Intelligence (2005)
3. Lanillos, P., Besada-Portas, E., G-Pajares, Ruz, J.: Minimum time search for lost targets using Cross Entropy Optimization. In: IEEE/RSJ International Conference on Intelligent Robots and Systems, pp. 602–609 (2012)
4. Lanillos, P., Yañez-Zuluaga, J., Ruz, J., Besada-Portas, E.: A Bayesian approach for constrained multi-agent minimum time search in uncertain dynamic domains. In: Proceedings of the 15th Annual Conference on Genetic and Evolutionary Computation, pp. 391–398 (2013)
5. McGuire, M., Rys, M., Rys, A.: A study of how unmanned aircraft systems can support the Kansas Department of Transportation's efforts to improve efficiency, safety, and cost reduction. Kansas State University Transportation Center, Tech. rep. (2016)
6. Meghjani, M., Manjanna, S., Dudek, G.: Multi-target rendezvous search. In: 2016 IEEE/RSJ International Conference on Intelligent Robots and Systems (IROS), pp. 2596–2603. IEEE, New York (2016)
7. Perez-Carabaza, S., Bermudez-Ortega, J., Besada-Portas, E., Lopez-Orozco, J., de la Cruz, J.: A multi-UAV minimum time search planner based on ACO-R. In: Proceedings of the Genetic and Evolutionary Computation Conference, pp. 35–42 (2017)

8. Perez-Carabaza, S., Besada-Portas, E., Lopez-Orozco, J., de la Cruz, J.: A real world multi-UAV evolutionary planner for minimum time target detection. In: Proceedings of the Genetic and Evolutionary Computation Conference, pp. 981–988 (2016)

9. Perez-Carabaza, S., Besada-Portas, E., Lopez-Orozco, J., de la Cruz, J.: Ant colony optimization for multi-UAV minimum time search in uncertain domains. Appl. Soft Comput. **62**, 789–806 (2018)

10. Pérez-Carabaza, S., Besada-Portas, E., López-Orozco, J., Pajares, G.: Minimum time search in real-world scenarios using multiple UAVs with onboard orientable cameras. J. Sens. (2019)

11. Pérez-Carabaza, S., Scherer, J., Rinner, B., López-Orozco, J., Besada-Portas, E.: UAV trajectory optimization for minimum time search with communication constraints and collision avoidance. Eng. Appl. Artif. Intell. **85**, 357–371 (2019)

Chapter 67
A Comparative Study on Breast Cancer Tissues Using Conventional and Modern Machine Learning Models

D. Lakshmi, Srinivas Reddy Gurrela, and Manideep Kuncharam

Abstract Recent advancement in analytical models, availability of hardware resources like GPU, TPU, and cloud infrastructure began to play a pivotal role in healthcare practices and research. It has numerous tools and techniques are available to archive, manage, analyze, and predict large volumes of structured, unstructured, and semi structured data. 'Data Science' plays a vital role in the medical field with the better support for diagnosis, early prediction and cure for the disease. In this research, a comparative study on conventional models and modern methods are carried out using 12 different machine learning models for classification. The response or target variable is the diagnostic observation having two classes viz benign (not cancerous) or malignant (cancerous). The dataset used is The Wisconsin Diagnostic Breast Cancer (WDBC) which is taken from the UCI machine learning repository. The performances of the models are analyzed using the validation matrices, precision, recall, accuracy, and F1 Score. In this research study, we have built the ensemble model using the base estimator as a Random Forest classification model, Principal Component Analysis (PCA), Extra Gradient Boosting (XGBoost) and bagging techniques. Among 11 conventional models, the highest accuracy is obtained for the combined model which is 98.8%. A modern method tSNE is used for dimensionality reduction along with the feature engineering method called k-best method and base model for the classification is Random Forest model is used. This combined approach resulted in the best accuracy 99%. We have obtained a little improved score using PCA and tSNE methods than the reference articles considered in this research.

D. Lakshmi (✉) · M. Kuncharam
B V Raju Institute of Technology, Narsapur, Hyderabad, India
e-mail: lakshmi.d@bvrit.ac.in

S. R. Gurrela
ExcelR, Hyderabad, India

67.1   Introduction

According to World Health Organization (WHO), breast cancer is the second most common cancer in women worldwide especially in developed countries [2]. With the aid of research and development support, the possibility of early detection paves the way to reduce the death rate. Machine learning algorithms are extremely useful in early prediction in order to cure the disease and save the patient.

The main focus of this research study is to classify the patient tumor measurements and medical observations into benign (not a cancer) or malignant (cancerous) using 10 conventional models and one modern method named tSNE. Whereas, 'Benign' is a medical term that signifies that the tumor cells that are not cancerous and will not spread to other areas of the body. 'Malignant' is a medical term that signifies that the tumor is made of cancer cells, and it will affect nearby tissues.

Previous researches using the Wisconsin Diagnostic Breast Cancer (WDBC) [1] dataset have shown high-quality result using machine learning models [2–4, 6–10]. The research study, published in November 2019 by [2] has shown 98.24% accuracy using Genetic programming approach with Adaboost classifier. Similarly, [4] has shown the results with 95.0791% accuracy for the Random Forest classifier. In our research study obtains 98.83% accuracy with convention models and 99% with modern method when compared to the highest accuracy obtained from the research studies [2] and [3].

67.2   Conventional Classification Models

In this section, a comparative study on 11 conventional classification models are discussed based on model validation metrics.

67.2.1   *Data Description*

There are 30 diagnostic measurements or features available in the original dataset. Among these, 29 features are of real number type that are considered as independent variables (X-Variable). The 'Diagnosis' variable is considered as the dependent variable (Y-Variable/Predictive Variable). There are 212 malignant and 357 benign observations are considered for the classification.

67.2.2   Methodology

In this research study, there are 11 classification models which are built to classify the dataset. It follows the various data analytics stages such as data collection, data pre-processing, data transformation, feature selection, and model validation and tuning the hyperparameters in order to increase the accuracy rate. The proposed classification model is used in the WDBC dataset following the sequence of steps like data standardization, dimensionality reduction, hyperparameter tuning, and development of classification model with boosting, bagging, and model validation.

67.2.3   Data Standardization

Data standardization is the basic requirement for most of the machine learning algorithms and also to obtain optimal performance. Data standardization is done on this dataset using 'Min–Max-Normalization'. This method transforms all the features onto the unit scale with the mean as 0 and variance as 1.

67.2.4   Model Development

For all the classification models 70% of the dataset is considered in the training phase and 30% in the testing phase. For the training phase, tenfold cross validation is considered for the classification models.

67.2.5   Model Improvement Techniques

Hyper parameter tuning is the process of optimizing the model parameters to get the best accuracies on train, test, and validation data sets. In this research study, there are three strategies are considered for the model tuning or hyperparameter tuning, such as Grid Search, Random Search, and k-Fold validation techniques.

Grid search is a conventional way to find optimal parameters for a model and it needs search space to be given and model would be built on all the values of the search space and accuracies will be computed. Combinations of hyper parameters will be selected based on best accuracy.

Random search differs from grid search mainly in that it searches the specified subset of hyper parameters randomly instead of exhaustively. The major benefit of this method is with decreased processing time.

In the process of model validation, data set have been divided into 10 folds. There are 9 folds are considered for the training phase and tested on onefold for each iteration. This process was repeated 10 times with different folds in train and test data.

67.3   Results and Discussion

The model validation measures used in classification algorithms are confusion matrix, precision, recall, F1-Score, accuracy shown in Fig. 67.1.

67.3.1   Comparative Study on 10 Classification Models

In the research study [5], there is an imbalance in the predictions with the Support Vector Classifier model (Benign 78.72%, Malignant 0%, and Average 51.10%), whereas, in our research study imbalances have been minimized. And we have obtained 99% for Benign and 93% for Malignant and an average of 95.10%. Although all the 11 classification models are chosen randomly, first model is started with Dummy Classifier as a baseline classifier. All the classification models which are implemented in this study resulted with more than 90% accuracy. However, for the ensemble model Random Forest algorithm is chosen as a base algorithm and combined with PCA, and AdaBoost techniques in order to improve the accuracy. As

Actual

Predicted	Positive	Negative	
Positive	True Positive (TP)	False Negative (FN) **Type II Error**	**Sensitivity/Recall** $\dfrac{TP}{(TP+FN)}$
Negative	False Positive (FP) **Type I Error**	True Negative (TN)	**Specificity** $\dfrac{TN}{(TN+FP)}$
	Precision $\dfrac{TP}{(TP+FP)}$	**Negative Predictive Value** $\dfrac{TN}{(TN+FN)}$	**Accuracy** $\dfrac{TN+TP}{(TP+TN+FP+FN)}$
	F1-Score $\dfrac{2*Recall*Precision}{Recall+Precision}$		

Fig. 67.1 Confusion matrix and validation metrics used in classification models

shown in the Table 67.1, Random Forest classifier resulted with 100% as a highest precision score, Logistic Regression, and Decision Tree Classifier resulted in 98.85% as a highest recall score, Random Forest Classifier using AdaBoost and PCA resulted in 98.74% as a highest F1-Score and 98.83% as a highest accuracy.

Table 67.1 Comparison of 10 classifier models precision, recall, f1-score, and accuracy

S. No.	Model	Precision score (%)	Recall score (%)	F1-score (%)	Accuracy score (%)
1	Logistic regression	93.47	*98.85*	96.08	95.10
2	Linear discriminant analysis (LDA)	*100*	87.35	93.22	95.32
3	Support vector classifier	93.47	98.85	96.08	95.10
4	Gaussian Naïve Bayes	94.25	94.25	94.25	93.00
5	Decision tree classifier	96.20	87.35	91.56	90.20
6	K nearest neighbourhood classifier	98.76	91.95	95.23	94.40
7	Random forest classifier using AdaBoost and PCA	99.16	98.33	*98.74*	*98.83 (0.5)*
8	Gradient boosting using decision tree as a base algorithm	95.40	95.40	95.40	94.40
9	Extra gradient boost using decision tree as a base algorithm	98.30	98.30	98.30	98.00
10	Random forest classifier	96.51	95.40	95.95	95.10
11	Bagging classifier	91.5	97.9	95.8	95.5

67.4 Modern Method Using tSNE

The conventional model named Principal Component Analysis (PCA) is linear transformation technique and this model would lead to poor performance especially when dealing with non-linear manifold structures.

The t-Distributed Stochastic Neighbor Embedding (t-SNE) model has emerged as an admired model for data visualization. It also helps in dimensionality reduction and also for feature extraction. This model has major three steps namely, similarity computation in higher dimensional space, similarity computation in lower dimensional space and minimizes the difference between these two based on conditional probabilities or similarity score in high dimensional and low-dimensional space.

The modern method for dimensionality reduction using t-Distributed Stochastic Neighbour Embedding (t-SNE) and feature engineering method select k-best method, which works based on chi-square method. We have got the best features among all the features ($n = 28$). The base model of the classification is Random Forest Algorithm, which is a bagging method.

SelectKbest method is used for the feature engineering. It is univariate feature selection method. It works by selecting the best features based on univariate statistical tests. This method can be seen as a preprocessing step to an estimator. Scikit-learn exposes feature selection routines as objects that implement the transform method. SelectKBest removes all but the highest scoring features.

Random Forest base classifier is used with 1000 trees and max_features = 5 among 28 features given in the original dataset. The dataset is split with (80:20) for the train and test phase respectively. The tSNE model validation results are shown in the Table 67.2.

A modern method tSNE is used for dimensionality reduction along with the feature engineering method called k-best method and base model for the classification is Random Forest model. This combined approach resulted with the best accuracy 99%.

Among three dimensionality reduction techniques such as Principal Component Analysis (PCA), Linear Discriminant Analysis (LDA), and t-Distributed Stochastic Neighbour Embedding (tSNE), tSNE has resulted with the highest accuracy 99%.

Table 67.2 tSNE model validation matrices

S. No.	tSNE, bagging and random forest classification model	Precision score (%)	Recall score (%)	F1-score (%)	Accuracy score (%)
1	Training phase	98.59	97.61	97.9	99.05
2	Testing phase	95.03	97.22	96.47	99.03

67.5   Conclusion

This research study is conducted in order to find out the best suitable ensemble model on WDBC dataset with the high accuracy of class label prediction. All the 11 classification models are implemented using Python libraries. The ensemble model Random Forest algorithm is chosen as a base algorithm and combined with PCA, AdaBoost, and Bagging are found to be superior among all the 10 conventional models with the highest accuracy of 98.8%. A modern strategy tSNE is utilized for dimensionality reduction along with the feature engineering method called k-best method and base model for the classification is Random Forest model. This ensemble model resulted in the greatest accuracy with 99%. The Python code for all the 11 models are available in the GitHub for the references.

References

1. Wisconsin Diagnostic Breast Cancer (WDBC) dataset: https://archive.ics.uci.edu/ml/datasets/Breast+Cancer+Wisconsin+%28Diagnostic%29
2. Dhahri, H., Maghayreh, E.A., Mahmood, A., Elkilani, W., Nagi, M.F.: Automated breast cancer diagnosis based on machine learning algorithms. J. Healthc. Eng. (2019)
3. Nidhi, M.K., Makkar, S.: Classification of breast cancer tissues using decision tree algorithms. Int. J. Res. Eng. Appl. Manage. **04**(05) (2018). https://doi.org/10.18231/2454-9150.2018.0636
4. Salama, G., Abdelhalim, M.B., Zeid, M.: Experimental comparison of classifiers for breast cancer diagnosis, 180–185 (2012). https://doi.org/10.1109/ICCES.2012.6408508
5. Ferri, C., Hernández-Orallo, J., Modroiu, R.: An experimental comparison of performance measures for classification. Pattern Recogn. Lett. **30**(1), 27–38 (2009). https://doi.org/10.1016/j.patrec.2008.08.010
6. Christobel, A., Sivaprakasam, Y.: An empirical comparison of data mining classification methods. Int. J. Comput. Inf. Syst. **3**(2) (2011)
7. Lavanya, D.: Ensemble decision tree classifier for breast cancer data. Int. J. Inf. Technol. Converge. Serv. **2**(1), 17–24 (2012)
8. Ster, B., Dobnikar, A.: Neural networks in medical diagnosis: comparison with other methods. In: Proceedings of the International Conference on Engineering Applications of Neural Networks, pp. 427–430 (1996)
9. Joachims, T.: Transductive inference for text classification using support vector machines. In: Proceedings of International Conference Machine Learning. Slovenia (1999)
10. Abonyi, J., Szeifert, F.: Supervised fuzzy clustering for the identification of fuzzy classifiers. Pattern Recogn. Lett. **14**(24), 2195–2207 (2003)

Chapter 68
Clinical Decision Support System for Knee Injuries Treatment Using Multi-Agent System

Naveen Dalal and Indu Chhabra

Abstract Diagnosis of sports injuries is a very critical process and its performance depends on the recognition of the relevant symptoms. In this paper, a Multi-Agent-based knee injury detection and diagnosis scheme (MFZS) is introduced that applies fuzzy rules over input symptoms and recommends the relevant treatments. Its performance is compared with traditional fuzzy system (TDFZS) under the constraints of detection accuracy and sensitivity etc. Chi-square and Fisher Exact test is also performed to verify the significance of the outcomes of both schemes.

68.1 Introduction

Clinical decision support systems can be enhanced using the applications of artificial intelligence. Expert systems can improve the overall diagnosis process and sports team can predict the injuries and risks related to a specific sports activity/exercise. Figure 68.1 shows the interaction of different entities with expert system.

Support of medical practitioners is used to design and refine the expert system. Performance and injury risk management may be initiated by sports team whereas injured team member may acquire help regarding diagnosis from expert system. Communication between expert system and these entities suffers from the following barriers and challenges:

Barriers for Medical Practitioners

Designing an expert system requires experience man power as well as highly accurate input data sets to produce rules and response. Sports injury data collection and its

N. Dalal (✉)
Goswami Ganesh Dutta Sanatan Dharam College, Chandigarh, India

I. Chhabra
DCSA, Panjab University, Chandigarh, India

© The Author(s), under exclusive license to Springer Nature Singapore Pte Ltd. 2021
S. C. Satapathy et al. (eds.), *Smart Computing Techniques and Applications*,
Smart Innovation, Systems and Technologies 225,
https://doi.org/10.1007/978-981-16-0878-0_68

Fig. 68.1 Entities interaction with expert system

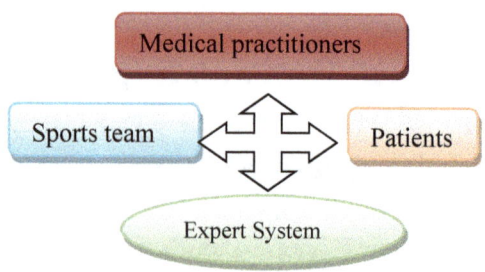

Table 68.1 Challenges for sports team/patients

Challenges for sports team	Challenges for patients
Sports teams may engage in different activities in a specific sports type. Occurrence of an injury depends on the following factors: experience of coach/team member, confidence level, type and frequency of exercises involved in a given sports. All above factors may increase or decrease the risk factor associated with sports injuries. Using an expert system, participants can predict the injuries during an activity as well as medical practitioners can predict the treatment type with respect to injury type	Expert system uses its knowledge base as per the input symptoms provided by current patient and produces a treatment plan. Error prone input may alter the treatment type and patient recovery will suffer due this mismatch. Expert system may produce accurate treatment plan, however supervision of a medical practitioner is still required to crosscheck the treatment recommendations [1–5]

sorting is a very complex task thus may reduce the overall system performance and diagnosis accuracy. Diagnosis prediction may suffer due to obsolete rules and datasets, so medical practitioners should update these input parameters as per current requirements. Large-scale datasets and rules can increase the response time. So there is a need to optimize the rules of an expert system (Table 68.1).

68.2 Requirements of Intelligent Agents for Injury Detection and Diagnosis

Following are the requirements of intelligent agents for injury detection and diagnosis:

In case of single diagnosis system, treatment recommendations are generated as per the input systems and it will not generate no response, if it fails to recognize the input whereas in case of multi agent system, if a diagnosis agent fails to identifies the input symptoms, then request is forwarded to another agent and so on. If all agents fail to resolve the category of treatment, only then treatment for input injury is refused. Data set can be subdivided and assign to each agent, in order to produce quick response. It will also minimize the number of searches to match the input symptoms with available treatments. As per requirements, only specific agent can be invoked for treatment and there is no need to engage all agents in diagnosis process, in this case, resource utilization can be optimized.

68.3   Contribution of Fuzzy Logic in Medical Domain

Kasbe et al. [6] introduced a solution to diagnosis heart disease. Centroid method is used with fuzzy logic for defuzzification process. A dataset is used to produce multiple inputs and ranges. Finally, fuzzy rules are used over output variations (0–5) to determine the disease conditions. Experimental outcomes show its performance in terms of higher accuracy as compared to traditional diagnosis systems. Shaik et al. [7] introduced a fuzzy logic based anemia diagnosis system that takes blood parameters as input and applies fuzzy rules to generate output. Experimental results show that it can identify the various types of disease and can enforce the rules accordingly. Experimental results indicate that it can detect/diagnosis the disease in real-time environment. Thamaraimanalan et al. [8] developed a solution for handheld medical devices that can be used for heart patients. It uses multiple digital signals as input and then fuzzy rules are used to analyze the results. Analysis shows that it can monitor the heart rate of patients with higher accuracy as well as it consumes optimal energy. Farzandipour et al. [9] did the investigation of fuzzy logic based diagnosis solutions and found that their accuracy and outcomes can be refined using the various datasets. Study shows that clinical knowledgebase can be utilized to develop new fuzzy rules and that can be utilized by the healthcare industry for decision making. Akinnuwesi et al. [10] developed a fuzzy logic based expert system for rheumatology. It considers various aspects (symptom and degree of uncertainty/associated risk and causal factors) for diagnosis and decision making. Experimental results show that it can enhance the overall efficiency of healthcare practitioners/diagnosis accuracy etc. as compared to traditional diagnosis solutions. Bautista et al. [11] investigated the relationship between technology, disease, diagnosis and decision makers. Study shows that mapping of all above factors can be used to build highly accurate algorithms (fuzzy logic based) for medical expert system for healthcare industry. Patil et al. [12] explore the various artificial intelligence (AI) based tools (neural networks/fuzzy logic/machine learning) that can be used for healthcare services. Study shows that integration of these tools with BigData can be used to produce a large scale knowledgebase for artificial decision support system for healthcare industry. Singh et al. [13] investigated the contribution of AI-based methods that can be used for the diagnosis of injuries related to neuro system. It considers various factors (neurological pathways/cognition/neurotransmission/disability/behavior disorders) that can be utilized to form the predicates for machine learning algorithms. Analysis shows that AI-based healthcare solutions can maximize the accuracy of diagnosis process. Medeiros et al. [14] presented an interference method for healthcare that uses fuzzy logic to map the risk assessment of patients with decision support system, in order to improve the overall diagnosis process. Analysis shows that mapping of above-discussed parameters can be utilized to develop training datasets for practitioners. Kadhim et al. [15] merged the neural network with fuzzy logic and developed a framework to diagnose the back pain. It uses different health parameters to perform fuzzification/defuzzification. Experimental results show its performance in terms of higher efficiency/accuracy of diagnosis process. Alsmadi [16] used the combination of neutrosophic with fuzzy logic to detect the lesions. A metric is used to define the final output and experimental results indicate that it outperform in terms of accuracy, efficiency as compared to traditional lesions detection schemes. Ahmadi et al. [17] investigated various decision support methods that use fuzzy interference/neural network/decision tree/rational mapping/weighted rules for diagnosis of different diseases (Joint pain/Arthritis/Cancer/Cardiac/Malaria/Hypovolemi). Analysis displays that efficiency and accuracy of diagnosis process can be improved using these methods and the performance of the developed schemes depends over diverse parameters, i.e., sample size/interval/rules etc. Mardani et al. [18] investigated the relationship between technology, domain, healthcare services, environment control, medical waste, risk, patients, practitioners, and diagnosis process, etc. Study shows that there are various AI-based solution that can be used boost the decision making. Analysis identifies some open issues related to selection of fuzzy sets, aggregation operators, rules complexity and standardization communication interface. Mohandes et al. [19] developed a health risk assessment framework for construction industry. It considers the various factors associated with current health, treatment and risks, fuzzy rules, etc. Experimental results show that it can assists the practitioners to track the health issues/diagnosis process and it can be further extended to analyze the relationship between risks and accidental hazards, in order to reduce their probability factor. Lohrmann et al. [20] developed a customized fuzzy feature selection method for diagnosis. It considers various parameters i.e. original features/scale factors/multiple datasets etc. and performs classification to reduce the distance between different features. Similar features are reduced to their minimal value to produce unique feature sets and final outcomes are considered for diagnosis process. Experimental results show its performance in terms of diagnosis accuracy as compared to traditional feature selection/filtering methods.

68.4   Multi-Agent Based Diagnosis

In case of sports, knee injury is a common risk to team that can cause serious health issues. This paper introduces a multi-agent based diagnosis scheme for these injuries. Figure 68.2 shows the involvement of the different agents for injury detection and diagnosis at college/university/hospital level.

User Agent (UA): User Agent gets injury state of the patients and sends it to Diagnoses Agent (DA) for analysis, diagnose, and treatment. Further UA shows the results and treatment plan generated by the DA to the user. It will also respond to the query from the Diagnosis Agent.

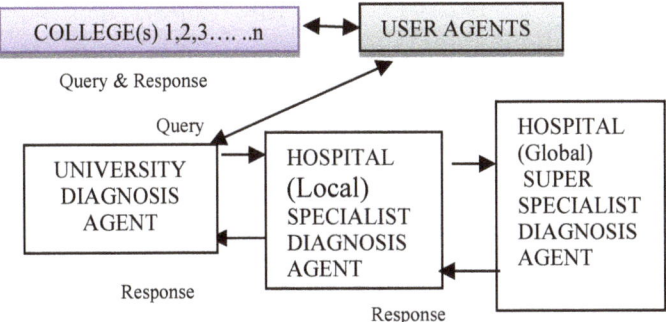

Fig. 68.2 Multi-agents for diagnosis process agents description

UA deals with the data of coaches/doctors, patients, injury details.

Diagnosis Agent (DA): This agent is fully automated and no human intervention is required to diagnosis the sports injury. Its functions are to analysis about patient's state with integration information. The DA will react and ask for more information if necessary and generate treatment plan accordingly. In complicated cases if it needs to consult Specialist Diagnosis Agent (SDA) it can consult. There is one DA at each level i.e. university, hospital in same region, hospital in same national zone (north, south, east, and west). At each level the specialty of DA increases. DA is further categorized as SDA and Super Specialist Diagnosis Agent (SSDA). DA will also save the user data about their personal information, treatment plan for further reference.

Fuzzy Rules for Multi-Agents

Figure 68.3 shows the Fuzzy rules for Multi-agents, in which all agents use their injury and symptom datasets as input and fuzzy rules are applied to produce the final response for all agents are applicable and following is the basic procedure used for query and response. We have used Mamdani Inference method.

Fig. 68.3 Fuzzy rules for multi-agents

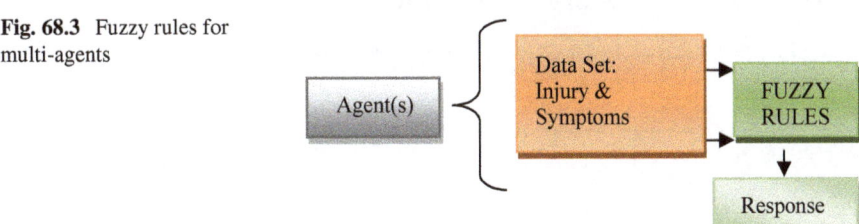

```
RULE 1: IF (kneepain IS low) OR (grade IS gradeone) THEN (treatmentplan IS selfcare);
RULE 4: IF (kneepain IS low) OR (grade IS gradeone) OR (symptoms IS bleeding) THEN (tplan
IS Coldtherapy);
RULE 43: IF (kneepain IS low) OR (grade IS gradeone) OR (symptoms IS instability) THEN
(tplan IS Coldtherapy);
RULE 69: IF (kneepain IS high) AND (grade IS gradethree) AND (symptoms IS
Inabilitytofullystraightenknee) THEN (tplan IS rehabilitation);
```

Step 1: Fuzzification defines membership functions. For instance, we defined knee pain into four linguistic terms, namely symptoms, knee pain, knee strain, and grade

Step 2: Designing the knowledge base by fuzzy rules using an if–then format. These rules are defined using linguistic terms such as:

IF (knee pain IS low) OR (grade IS grade one) THEN (treatment plan IS self-care).

For each linguistic term, a corresponding fuzzy set can either be designed with the help of expert(s) or derived from existing data set.

Step 3: Rule Evaluation is done using input that cause various rules to fire with varying strengths. For instance, many knowledge based rules are fired out for the symptoms, level of knee pain, grade of injury and knee strain. This leads to decisions regarding potential treatment(s), which can then be administered. The outcome of the rules are evaluated to reach to the conclusion.

Step 4: Defuzzification: finding the crisp outcome for output variables is important in this proposed system. As its needs to inform the doctor which treatment plan should be performed, e.g., self-care, cold therapy, surgery, etc.

Algorithm

UA: User Agent, DA: Diagnosis Agent, SDA: Special Diagnosis Agent, SSDA: Super Special Diagnosis Agent, IJS: Injury DataSet, Sym: Symptom DataSet, Usr: Users, AGUsr: User Agent, TP: Treatment Plan.

Initialize (DA, Univ.), IJS {ij1, ij2 ... ijn}, (SDA, Hospital (Local)), (SSDA, Hospital (Global)), (IJS, DA), (IJS, SDA), (IJS, SSDA).

```
//Query (Sender, Receiver), UA sends a query to DA

Query(AGUsr-> Sym, DA)
If FIND(Query-> Sym, IJS, FZR)== TRUE
    Generate_RESPONSE(TP, Query-> Sym)
Else
    Generate_RESPONSE (REFER_TO(SDA),Query-> Sym)
End If

//SDA sends a query to SSDA

Query(SDA, S SDA)
If FIND(SDA.Query-> Sym, IJS, FZR)== TRUE
    Generate_RESPONSE(TP, SDA.Query-> Sym)
Else
    Generate_RESPONSE(NOT_FOUND,SDA.Query-> Sym)
End If
```

```
//DA sends a query to SDA

Query(DA, SDA)
If FIND(DA.Query-> Sym, IJS, FZR)== TRUE
    Generate_RESPONSE(TP, DA.Query-> Sym)
Else
    Generate_RESPONSE(REFER_TO(SSDA), DA.Query-> Sym)
End If

// Response (Sender, Receiver), if invalid symptoms
provided by other agents

If (response == NOT_FOUND)
    ADVISE = "No Rule found for diagnosis for current
    symptom(s), Please provide additional information
    about the injury"
    Generate_ADVISE (ADVISE)
)
End if
```

The above algorithm considers **Injury type**: Muscle injuries **Categories** Quads strain **Symptom**: Sitting Problem, Pain, Climbing Stairs, Sensation, Popping Sound, Bruising, Difficulty Walking. **Treatment**: Rest, Physiotherapy, Surgery. **Symptom are classified into**: LOW/MEDIUM/HIGH **Injury Grade is**: ONE/TWO/THREE.

68.5 Results and Performance Analysis

As per Table 68.2, JADE 4.5 was used as design tool for experimental purposes with jre-7 environment over windows platform with different agents (User Agent/Diagnosis Agent/Special Diagnosis Agent/Super Special Diagnosis Agent).

Registration processes of different diagnosis agent is shown in Fig. 68.4 As per the injury type, different systems can be added, and later on all these are used for by diagnosis process. Figure 68.5 shows that diagnosis agent will generate the response only as per the dataset otherwise request of user agent is refused and user agent will wait for other diagnosis agent for diagnosis. Figure 68.5 shows the message exchange between different agents. It can be observed that request of user agent is sent to a diagnosis agent that can accept or refused the incoming request as per the availability of current injury data set. In case of refusal, another diagnosis agent may accept this request, if it supports the diagnosis for current request.

The Multi-Agent graphical user interface (GUI) has been designed to receive symptoms from the doctor and to display possible treatment plan(s) for knee injury. Following are the result given by Multi-Agent System (MFZS) and Fuzzy System (TDFZS) for the calf strains injury Table 68.3.

Table 68.2 Configuration for analysis

Parameters	Description
Design tool	Java agent and development environment (JADE 4.5)
Environment	Java run time environment-7
Platform	Windows 10 (64 Bit)
Agents	User agent, diagnosis agent, special diagnosis agent, super special diagnosis agent
Scenarios	Performance analysis with: (a) traditional Fuzzy Logic Scheme (TFLS), (b) multi-agent based fuzzy logic scheme (MAFLS)

```
INFO: ----------------------------------------
Agent container Main-Container@192.168.75.1 is ready.
----------------------------------------
ligamentInjuries Diagnosis support available for input Symptom(s)
Hello! User-agent UserAgent1@192.168.75.1:1099/JADE is ready.
Target injury is ligamentInjuries
```

Fig. 68.4 Registration process of Agents (DA/UA)

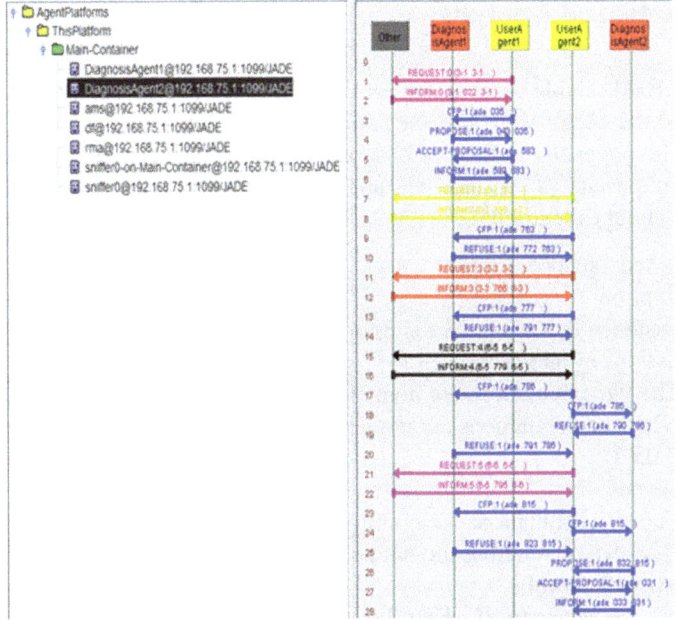

Fig. 68.5 Dynamic response of diagnosis agents

Table 68.3 Treatment recommendations by multi-agent system (MFZS) and fuzzy system (TDFZS)

Muscle injuries

Category	Non surgery					Surgery		
		Elevation	Medication	Rest	Sports massage	Surgery	Unknown	Total
Calf strains	TDFZS	9	10	11	10	21	0	61
	MFZS	10	11	20	10	10	0	61

Table 68.4 Chi-square test: calf strains

Treatment

	Non-surgery	Surgery	Total
TDFZS	40 (45.5) [0.66]	21 (15.5) [1.95]	61
MFZS	51 (45.5) [0.66]	10 (15.5) [1.95]	61
Total	91	31	122 (grand total)

68.6   Statistical Analysis

For about 1000 sample cases tested for different knee injuries the responses of the system and the doctor were collected and tabulated (Table 68.3) for calf strain. To verify the homogeneity of the two observations, a Chi-square test and Fisher Exact test of homogeneity of variance was applied on different knee injuries Table 68.3. show data about Calf strain:

a.   **Data**: See Table 68.3.
b.   Assumption:
c.   It is assumed that we have a simple random sample from each one of the two populations of interest
     H0: The two populations are homogeneous with respect to treatments
     HA: The two populations are not homogeneous with respect to treatments Let
     $\alpha = 0.05$
d.   Hypothesis:
     H0: The two populations are homogeneous with respect to treatments.
     HA: The two populations are not homogeneous with respect to treatments.
e.   **Statistical Formula**:
     Sum all the values for (O–E) 2 /E.
f.   **Calculation of test statistics**: MS-Excel was used for calculating the values.

Table 68.4 Specify the result of Chi-square test for surgery and non-surgery category for Calf Strain. The value calculated for Calf Strain is 5.2329 and p-value is 0.022164 and results are significant at $p < 0.05$. Chi-square test with Yates correction is 4.3247 with p value 0.037563 and results are significant at $p < 0.05$. The Fisher exact test statistic value is 0.0366 and results are significant at $p < 0.05$. In case of chi square and fisher exact test, significance level floats between <0.1 and <1.0 and for most injury sets, it is significant at <0.5 using both schemes.

68.7   Conclusion

In this paper, issues and solutions related to contribution of expert systems in medical domain, role of different stack holders in the development these systems, have been discussed and a new fuzzy logic based multi agent scheme was introduced to guide the diagnosis process of sports knee injuries. Its performance is also compared with the existing fuzzy logic scheme under the constraints of accuracy/sensitivity. Performance analysis shows that each scheme has its own response for different set of knee injuries. And in some cases, both have recommended same treatment and in some cases, both produced different results that affect their accuracy/sensitivity levels which are directly proportional to each other. Four types of knee injury types have been used for experiments and each type has further subdivided into multiple categories. In case of Muscle Injuries, both schemes have 100% accuracy with highest sensitivity level (1) for the treatment recommendation for Torn Quad injury whereas in other categories (Calf Strains/Pulled Quad/Quads Strain), (Figs. 68.6 and 68.7) it is slightly declined for TDFZS as compared to MFZS. In case of Knee cap injury type, both schemes have 100% accuracy with the highest sensitivity level for the category: Dislocated Patella Medial Reefing MPFL Repair MPFL Reconstruction, Tibial Tubercle Transfer, and it is slightly decreased for Patella fractures/Medial Imbrication categories. In case of ligament Injuries, for categories: Anterior cruciate ligament, medial collateral ligament, PCL Knee Injury, Ligament Sprains, etc. accuracy and sensitivity level of TDFZS varies whereas it remains almost constant for MFZS. In case of Cartilage Injuries type, for Category: Meniscus Tears, MFZS has the higher accuracy/sensitivity level of treatment recommendation as compared to TDFZS. It can be observed that in case of each injury category, accuracy and sensitivity level of both schemes varies and in few cases, it is approximately constant. As per the above discussion, it can be concluded that MFZS is more sensitive with respect to input symptoms as compared to TDFZS and it can recommend the accurate treatment types to the patients. Current scope of the work considers the limited sports knee injuries types. In future, MFZS may be utilized to diagnose the other injury types as well. Its performance will be optimized using machine learning algorithms.

Fig. 68.6 Accuracy of treatment recommendations for muscle injuries

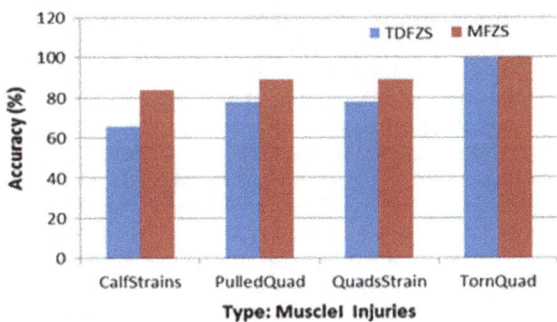

Fig. 68.7 Sensitivity of treatment recommendations for muscle injuries

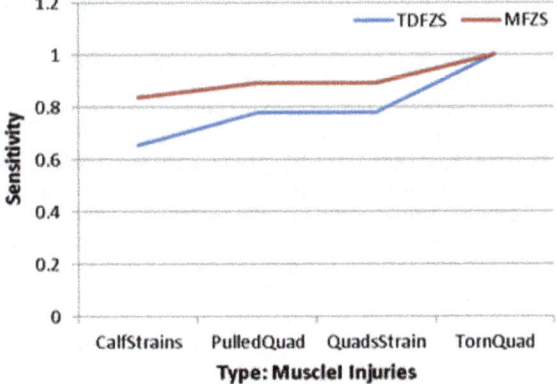

References

1. Sloane, E.B., Silva, R.J.: Clinical Engineering Handbook, Elsevier 2nd Edition Chapter 83, pp. 556–56 (2020)
2. Fister, I., Ljubič, K., Suganthan, P.N., Perc, M., Fistera, F.: Computational intelligence in sports: challenges and opportunities within a new research domain. Appl. Math. Comput. **262**, Elsevier, 178–186 (2015)
3. Aframian, A., Iranpour, F., Cobb, J.: Medical devices and artificial intelligence. Artif. Intell. Healthc. Elsevier, 163–177 (2020)
4. Valle, X., Geli, E.A.L., Tol, J., Hamilton, B., Pruna, R., Til, L., Gutierrez, J.A., Alomar, X., Balius, R., Malliaropoulos, N., Monllau, J.C., Whiteley, R., Witvrouw, E., Samuelsson, K., Rodas, G.: Muscle injuries in sports: a new evidence-informed and expert consensus-based classification with clinical application. Sports Med. **47**, Elsevier, 1241–1253 (2019)
5. Witkowski, E., Ward, T.: Artificial intelligence assisted surgery. Artif. Intell. Healthc. Elsevier, 179–202 (2020)
6. Kasbe, T., Pippal, R.S.: Design of heart disease diagnosis system using fuzzy logic. In: International Conference on Energy, Communication, Data Analytics and Soft Computing (ICECDS). IEEE, pp. 3183–3187 (2017)
7. Shaik, M.F., Subashini, M.M.: Anemia diagnosis by fuzzy logic using LabVIEW. In: International Conference on Intelligent Computing and Control (I2C2), pp. 1–5. Coimbatore, IEEE (2017)
8. Thamaraimanalan, Sampath, T.P.: A low power fuzzy logic based variable resolution ADC for wireless ECG monitoring system. Cognit. Syst. Res. **57**, Elsevier, 236–245 (2019)
9. Farzandipour, M., Nabovati, E., Saeedi, S., Fakharia, E.: Fuzzy decision support systems to diagnose musculoskeletal disorders: a systematic literature review. Comput. Methods Programs Biomed. **163**, Elsevier, 101–109 (2018)
10. Akinnuwesi, B.A., Adegbite, B.A., Adelowo, F., Edomwonyi, U.I., Amumeji, O.T.: Decision support system for diagnosing rheumatic-musculoskeletal disease using fuzzy cognitive map technique. Inform. Med. Unlocked **18**, Elsevier, 1–19 (2020)
11. Bautista, J.A.R., Zavala, A.H., Cárdenas, S.L.C., Ruela, J.A.H.: Review on plantar data analysis for disease diagnosis. Biocybern. Biomed. Eng. **38**(2), Elsevier, 342–361 (2018)
12. Patil, S., Patil, K.R., Patil, C.R., Patil, S.S.: Performance overview of an artificial intelligence in biomedics: a systematic approach. Int. J. Inf. Technol. Springer, 1–11 (2018)
13. Singh, R.O.B., Vishweswaraiah, S., Er, A., Aydas, B., Turkoglu, O., Taskin, B.D., Duman, M., Yilmaz, D., Radhakrishna, U.: Artificial Intelligence and the detection of pediatric concussion using epigenomic analysis. Brain Res. **1726**, Elsevier, 1–22 (2020)
14. Medeiros, I.B.D., Machado, M.A.S., Damasceno, W.J., Caldeira, A.M., Filho, J.B.D.S.: A fuzzy inference system to support medical diagnosis in real time. Proc. Comput. Sci. **122**, Elsevier, 167–173 (2017)
15. Kadhim, M.A.: FNDSB: a fuzzy-neuro decision support system for back pain diagnosis. Cognit. Syst. Res. **52**, Elsevier, 691–700 (2018)
16. Alsmadi, M.K.: A hybrid fuzzy C-means and neutrosophic for jaw lesions segmentation. Ain Shams Eng. J. **9**(4), Elsevier, 697–706 (2018)
17. Ahmadi, H., Gholamzadeh, Shahmoradi, M.L., Nilashi, M., Rashvan, P.: Diseases diagnosis using fuzzy logic methods: a systematic and meta-analysis review. Comput. Methods Programs Biomed. **161**, Elsevier, 145–172 (2018)
18. Mardani, Hooker, R.E., Ozkul, S., Yifan, S., Fei, G.C.: Application of decision making and fuzzy sets theory to evaluate the healthcare and medical problems: a review of three decades of research with recent development. Expert Syst. Appl. **13715**, Elsevier, 202–231 (2019)

19. Mohandes, S.R., Zhang, X.: Towards the development of a comprehensive hybrid fuzzy-based occupational risk assessment model for construction workers. Safety Sci. **115**, Elsevier, 294–309 (2019)
20. Lohrmann, C., Luukka, P., Sabuka, M.J., Kauranne, T.: A combination of fuzzy similarity measures and fuzzy entropy measures for supervised feature selection. Expert Syst. Appl. **11015**, Elsevier, 216–236 (2018)

Chapter 69
Predictive Loan Approval Model Using Logistic Regression

Aiswarya Priyadarsini Behera, Siddharth Swarup Rautaray, Manjusha Pandey, and Mahendra Kumar Gourisaria

Abstract Loan disbursement is one of the fundamental aspects of all banks and financial companies. The profit that comes from loan disbursement is one of the major source of bank's assets. Many applicants apply for a loan in any bank or financial company out of which choosing the appropriate costumer is a herculean task. Though banks nowadays have adopted a more rigorous process for validation and corroboration of a costumer, it still does not guarantee that whichever costumer is chosen has all the right and original required documents. The primary objective of this validation is to choose appropriate costumers, so that their assets can be in good hands and help people who are in need. In this work, we propose a loan recommendation system that will provide an immediate and simple way to choose the right applicant based on validation of features. The aim of this paper is to predict a model for loan disbursement by using regression model. Some weights are given to each feature based on the priority of the bank. This system can predict the feasibility of an applicant for availing the loan. It is beneficial for both employees of the bank and the probable candidates for loan disbursement. Here, we will take the important features from loan dataset. The weight of each appropriate feature can be calculated in loan prediction system. For new test data, same attributes can be processed with their respective weights. A particular ceiling time can be defined to allow the applicants to check the status of their application. A jumping mechanism can be added in the prediction system to allow disbursement of loan on a priority basis. However, one disadvantage of the system being in real-life scenario is that the recommendation system may be biased toward one dominant feature or attribute.

A. P. Behera (✉) · S. S. Rautaray · M. Pandey · M. K. Gourisaria
School of Computer Engineering, KIIT (Deemed to be University), Bhubaneswar, India

S. S. Rautaray
e-mail: siddharthfcs@kiit.ac.in

M. Pandey
e-mail: manjushafcs@kiit.ac.in

© The Author(s), under exclusive license to Springer Nature Singapore Pte Ltd. 2021   715
S. C. Satapathy et al. (eds.), *Smart Computing Techniques and Applications*,
Smart Innovation, Systems and Technologies 225,
https://doi.org/10.1007/978-981-16-0878-0_69

69.1   Introduction

The significance of data analytics in the bank or any financial services sector has been realized at a huge scale. We know that in this technical and advance world, machine is doing various jobs of human and saves more time. Before the AI was introduced, we consumed a lot of time for verification and validation to identify applicant's eligibility. Here, we used logical regression which is also known as logit regression model as machine learning tool for creating a probabilistic and predictive approach to a loan approval prediction model. In regards with this paper, one is the concept in feature selection which is the important attribute to train the model and also apply to a domain of machine learning which is exploited. Advanced analytics assets on a large-scale developing and investing roles are critical to analyze. The applicant revolution enabling clearly defined data ownership and maintenance of high-quality data. A simple linear regression can be replaced by a logistic regression when the input variables are mapped continuously. The reason behind choosing logistic regression is its mathematical flexibility. Now logistic regression is extensively used where a relationship has to be formed among the input variables and a dichotomous output to be obtained.

In Fig. 69.1, mathematically it is represented as:

$$Y = \theta_0 + \theta_1 X \tag{69.1}$$

X is the input used as feature and theta0 and theta1 are parameters of the regression model.

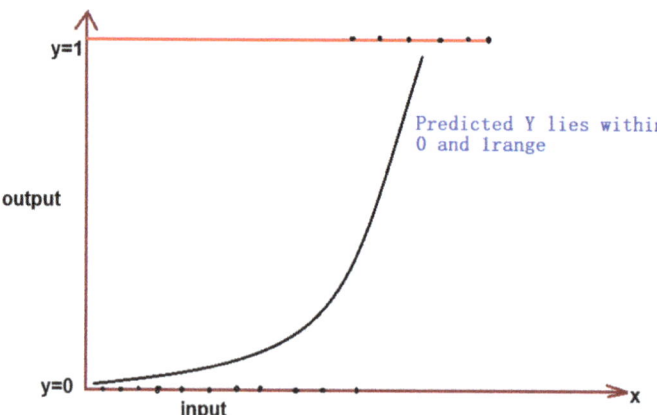

Fig. 69.1 General logistic regression curve

69.2   Literature Survey

Alshouiliy et al. [1] in this paper, by using Azure machine learning (AzureML) platform, introduced a generic IoT. It has been designed by Microsoft to ensure accuracy of the prediction. Iris Calibo et al. [2] in this paper take past nine years data to estimate the contributing features. By using linear regression as visualized on a trend line model, its objective is to focus on credit risk feature selection scoring on loan performance, used for the small and medium enterprise technology upgrading program region. This takes historical data such as debt on asset ratio (DAR), liability ratio ((LR), net profit margin ((NPM), return on investment (ROI), and year. Mashanovich et al. [3] introduced different steps for processing data with multidimensional data of borrowers which is the major step through the data mining and making different group of characteristics. The technique of testing the strength and predictive data throughout the mining view is the objective of the paper which is published at big data zone (DZone) for variable selection and big data analytics in credit score modeling. Vaidya's [4] objective in the paper is to apply machine learning tools and concepts to real-world challenges which is to develop the analytical models. Many industries are now using the capability of AI to grow and get the profit in business. Yadav et al. [5] analyze the credit risk and fraud detection which is really helpful to bank for investors who invest money or give loan to the borrower in the high time; it gives a crystal clear idea. So in Hadoop methodology, it provides different tools for analyzing data in HiveQL and storing data in HDFS.

69.3   Performance of Logistic Regression and Algorithm Used

In our model of logistic regression, we are only focused about the probabilistic value of the outcome-dependent values. Here, we use binary classifier; it will find only success (yes) or failure (no). The value of Y from Eq. (69.1) describes two things: probability of success (p) or failure $(1 - p)$. Also Y must satisfy this condition to come out of logistic regression. Y must be positive and less than equal to 1. Here, we choose our threshold (the probability of success) is more than or equal to 50% (<=0.5), then we can consider the applicant who will get the loan approved. Endmost to come by a better accuracy, we will calculate the model's correctness by using "confusion matrix" which is nothing but the ratio between the actual value and the predicted value. The combination of confusion matrix is true positive (predicted true and actual true), true negative (predicted negative and actual negative), false positive (predicted positive and actual false), and false negative (predicted negative and actual false); usually, it is represented in tabular form to avoid over-fitting (noise of training data) (Fig. 69.2).

Fig. 69.3 describes the algorithm which we have used and tried to getting our simple and best-predicted accuracy model for loan approval status.

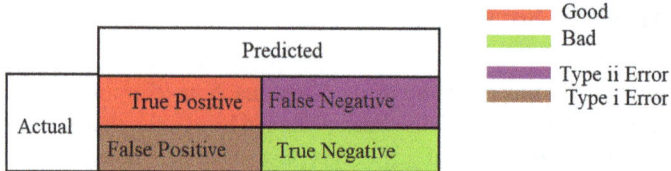

Fig. 69.2 Combination of confusion matrix

69.4   Experimental Setup and Result

In this section, we introduce our experimental setup. In the first step, we download the dataset from Kaggle, then we upload it to our work, and we need to clean and munging it to prepare for our model. There are more steps to be done on the data to be fit with our work and our model. The data contain different attributes that we are choosing attributes for prediction of loan approval to fit into the model. Table 69.1 shows all the attributes which are choosing to fit into the model of loan approval status. Basic information content of the borrowers like the gender, education, relationship status, whether or not self_employed, income status, financially dependent or not, loan amount, and credit_history is considered.

A box plot is shown in Fig. 69.4 for a statistical analysis of the input attributes; box plot or it is known as five-number summary which presents a visualization like max, min, median, lower quartile, and upper quartile point of the data. It is useful in indicating whether a distribution is skewed and whether there are any potential outliers in the dataset, and often used in explanatory data analysis (EDA). This graph is used to show the distribution, its central value, and its variability.

It is pretty tough to predict exactly which people will get the loan, so instead we calculate the probability or likelihood of a person getting it using logistic regression here. For that, we decide a threshold or cut-off value, and then, people with probability higher than that will get the loan or else not. We decided to create a plot (Fig. 69.5) with accuracy and F1 scores for both training and testing set against various threshold values to get the optimal threshold value for our algorithm. From the graph below, we found out that the optimal value for decision threshold is 0.4.

In this Fig. 69.6, it shows the confusion matrix on test data. Here, we obtained the accuracy 86%. To get the F1 measure, we have to calculate the recall (correct prediction out of all positive value) and precision (actual positive and we predict correct out of all positive value); we got 90% F1 measure.

69.5   Conclusion and Future Work

Millions if not billions of people and companies apply for loans in a day all over the world. Bank loans are one of the many parameters that help reveal the performance

Algorithm:

Step1:

input Dataset = Loan_Data

if   Loan_Data=Null or Duplicate

   remove variable

Step2:

Statistical Analysis of data

        Mean,Std,Min,Max,25%,50%,75%
Step3:

Normalize the data
        Min-Max normalization
Step4:

Split the Dataset
    training_data=Loan_Data*T/100   (T is threshold value)
    testing_data=Loan_Data - training_Data
Step5:

Find minimum error average rate R
        R= n   (n is threshold)
    find the best R (minimum avg rate)
Step6:

Select model(Logistic Regression)
    Predict= loan_status class_label

Step7:

 Evaluating the result
    Accuracy=Correct tests / Total tests

Fig. 69.3 Algorithm used in regression model

and credit risks of a financial institution. It not only helps borrowers to take money but also helps investors to invest in order to get higher interest rates. So, credit risk analysis and recommendation system focuses on the identifying the factors that would help the financial institutions to get a credit risk scoring using analytics. In today's day

Table 69.1 Contents of dataset

	Loan_ID	Gender	Married	Dependents	Education	Self_employed	Applicant income	Co-applicant income	Loan amount	Loan_amount_term	Credit_history
0	LP001002	Male	No	0	Graduate	No	5849	0.0	NaN	360.0	1.0
1	LP001003	Male	Yes	1	Graduate	No	4583	1508.0	128.0	360.0	1.0
2	LP001005	Male	Yes	0	Graduate	Yes	3000	0.0	66.0	360.0	1.0
3	LP001006	Male	Yes	0	Not graduate	No	2583	2358.0	120.0	360.0	1.0
4	LP001008	Male	No	0	Graduate	No	6000	0.0	141.0	360.0	1.0
5	LP001011	Male	Yes	2	Graduate	Yes	5417	4196.0	267.0	360.0	1.0
6	LP001012	Male	Yes	0	Not graduate	No	2333	1516.0	95.0	36.0	1.0
7	LP001014	Male	Yes	3+	Graduate	No	3036	2504.0	158.0	360.0	0.0

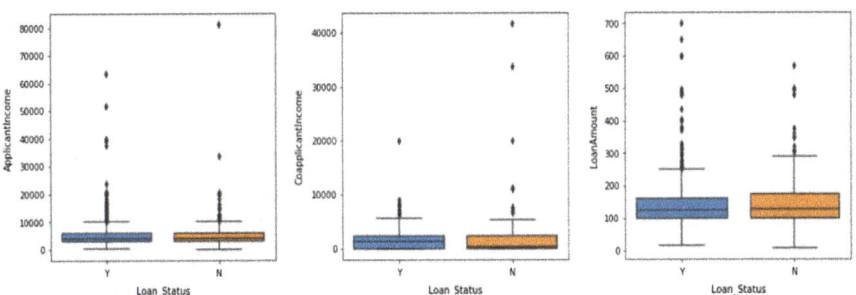

Fig. 69.4 Box plot grouped by applicant_income, co-applicant_income, and loan_amount

Out[22]: <matplotlib.axes._subplots.AxesSubplot at 0x1b2326c4da0>

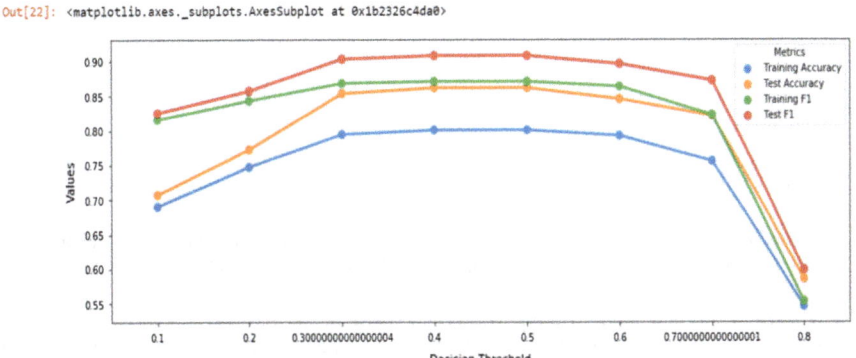

Fig. 69.5 Decision threshold for logistic regression

Fig. 69.6 Confusion matrix
on test data and
mathematical calculation

Out[25]:

Predicted	0	1	All
True			
0	22	16	38
1	1	84	85
All	23	100	123

$$\text{Accuracy} = \frac{\text{True Positive} + \text{True Negative}}{\text{All}} = 0.86$$

$$\text{F1 Measure} = \frac{2*\text{Recall}*\text{Precision}}{\text{Recall} + \text{Precision}} = 0.90$$

and age, both the sales representatives and software developers face increasing challenges in financial services which is complex and requires high involvement. Sales representatives are responsible for identifying tailored product or service recommendations for customers, and software developers are responsible for developing and maintaining that sales support environment. By using regression, we can easily manage data which is taking a less time to identify a particular customer information, and according to the request of customer, we can perform the best prediction model. For logistic regression, we need independent variables for assessment, and otherwise, the regression model tends to over weigh of the dependent attributes. In the future work, we will be doing the predictive analysis by using more techniques, and we will try to have a comparative analysis which will decide which model is the best.

References

1. Kanaujia, P.K.M., Pandey, M., Rautaray, S.S.: A framework for development of recommender system for financial data analysis. Int. J. Inf. Eng. Electron. Business (2017)
2. Kanaujia, P.K.M., Pandey, M., Rautaray, S.S.: Real time financial analysis using big data technologies. In: International Conference on I-SMAC (IoT in Social, Mobile, Analytics and Cloud) (I-SMAC) (2017)
3. Alshouiliy, K., AlGhamdi, A., Agrawal, D.P.: AzureML based analysis and prediction loan borrowers creditworthy. In: 2020 3rd International Conference on Information and Computer Technology (ICICT) (2020)
4. Ferreira, L.E.B., Barddal, J.P., Gomes, H.M., Enembreck, F.: Improving credit risk prediction online peer-to-peer (P2P) lending using imbalanced learning techniques. In: 2017 IEEE 29th International Conference on Tools with Artificial Intelligence (ICTAI) (2017)
5. Bindal, A., Chaurasia, S.: Predictive risk analysis for loan repayment of credit card clients. In: 3rd IEEE International Conference on Recent Trends in Electronics, Information & Communication Technology (RTEICT) (2018)
6. Arun, K., Ishan, G., Sanmeet, K.: Loan approval prediction based on machine learning approach. IOSR J. Comput. Eng. (IOSR-JCE) 18(3), Ver. I, 79–81 (2016). e-ISSN: 2278-0661, p-ISSN: 2278-8727
7. Deng, T.: Study of the prediction of micro-loan default based on Logit model. In: 2019 International Conference on Economic Management and Model Engineering (ICEMME) (2019)
8. Leung, C.K., Kajal, A., Won, Y., Choi, J.M.C.: Big data analytics for personalized recommendation systems. In: IEEE International Conference on Dependable, Autonomic and Secure Computing, International Conference on Pervasive Intelligence and Computing, International Conference on Cloud and Big Data Computing, International Conference on Cyber Science and Technology Congress (2019)
9. Calibo, D.I., Ballera, M.A.: Variable selection for credit risk scoring on loan performance using regression analysis. In: IEEE 4th International Conference on Computer and Communication Systems (2019)
10. Natasha Mashanovich, N.: Variable selection and big data analytics in credit score modeling, DZone Big Data Zone. Available at: https://dzone.com/articles/credit-scoring-part-4-variable-selection (2017)
11. Arutjothi, G., Senthamarai, C.: Prediction of loan status in commercial bank using machine learning classifier. In: Proceedings of the International Conference on Intelligent Sustainable Systems (ICISS 2017) IEEE Xplore Compliant—Part Number: CFP17M 19-ART, ISBN: 978-1-5386-1959-9 (2017)

Chapter 70
Efficient Load Balancing in Distributed Computing Environments with Enhanced User Priority Modeling

Maniza Hijab and Avula Damodaram

Abstract In the distributed computing environments, user-submitted job details such as the expected job completion time estimates are prone to inaccuracies. These inaccurate details force the system to under-perform due to ineffective allocation of processing resources. Addressing the concerns that crop-up due to the presence of these ill-defined user-provided parameters is an important task, which is very much required to get the jobs executed using optimal number of resources. At the same time, studying the ways of deriving accurate job runtimes benefits us in attaining improved load balancing results. The work proposed in the present study tries to apply an enhanced user priority model, taking into account the penalty and aging of job requests. It also combines least load variance method to improve the load balancing across grid resources. Simulation results obtained using realistic workloads bring forward the load balance efficiency of the studied scheme. This also confirms that to consider and infer of improved modeling of these user priority provides gains that outperform the methods void of it. The results attained are equated and contrasted with prevailing algorithms in terms of parameters like response time, wait time, tardiness, and average resource utilization in high-performance computing environments.

70.1 Introduction

High-performance computing (HPC) is unarguably the most widely used computing paradigm to solve complex real-time problems. It uses concurrently executing networked-computing resources in large numbers. It has the capability to bring to life unrelenting performance with simultaneous using thousands of networked-computing servers. Domains that use HPC include healthcare, engineering, space

M. Hijab (✉)
Muffakham Jah College of Engineering & Technology, Hyderabad, India

A. Damodaram
School of Information Technology, JNTUH, Hyderabad, India

© The Author(s), under exclusive license to Springer Nature Singapore Pte Ltd. 2021   723
S. C. Satapathy et al. (eds.), *Smart Computing Techniques and Applications*,
Smart Innovation, Systems and Technologies 225,
https://doi.org/10.1007/978-981-16-0878-0_70

research, urban planning, finance, and business to name some [1]. The intrinsic property of HPC applications is the requirement for soaring processing speeds. This requirement for high computational speeds comes at a cost that is time varying [2].

It is in light of a legitimate concern for HPC clients to tune proficiently to utilize compute resources. Along these lines, the HPC users are centered on the fact to decrease the resource expense while getting optimal execution results.

70.2 Related Work

Various research works have addressed the concerns of modeling the techniques to compensate the inaccurate task execution times submitted by users [3–6]. The main offerings of this article can be summarized as: (i) improving on the existing method of using penalty and aging-based task execution times accuracy modeling [4] and (ii) developing a hybrid task scheduling algorithm for improved load balancing in HPC environments.

Numerous relevant associated works described the user runtime approximations of submitted jobs are frequently imprecise. For instance, as specified in [5] by Tsafrir, more than few of the studies taken up provided startling and counter-intuitive outcomes as well. Despite the fact that some scholars inferred that imprecise estimates are typically desirable over precise ones, other works showed that runtime performance is indifferent to correctness of user's runtime approximations [3, 7].

The work as reported in [1] presents that in four reports of dissimilar high-performance computers, on an average of 55% of submitted jobs ended in using below 20% of their demanded CPU time. Further characteristics were too stated, for illustration, the relationship amid unsuccessful jobs and accurateness, and between job span and correctness. The influence of user job execution time approximations has been a substance of learning in numerous reported studies. The observed study stated in [3] showed that first-come-first-served is insensitive to user runtime approximations.

Nevertheless, using precise runtime approximations definitely will enhance performance on allocated policies that give preference to short jobs, similar to SJF.

In [3], the author examined three realistic task assignments and anticipated a workload generation model in view of reservations. We calculated enhancements of around 30% in processing performance and around 15% in energy consumption.

70.3 Proposed Work

The proposed work studies the implications of using an improved user priority model based on penalty and aging of job requests combined with least load variance method to improve the load balancing across grid resources. The Penalty-based Scheduling

Table 70.1 PSP + Aging Method Priority Assignments

Tag name	Accuracy interval	Priority
C 1	(0.00, 0.05)	1
C 2	[0.05, 0.10)	10
C 3	[0.10, 0.15)	20
C 4	[0.15, 0.20)	25
C 5	[0.20, 0.30)	30
C 6	[0.30, 0.40)	35
C 7	[0.40, 0.52)	40
C 8	[0.52, 0.64)	43
C 9	[0.64, 0.78)	46
C 10	[0.78, 1.00]	49

Policy (PSP) with aging algorithm adjusts the users' job priority over time dynamically to eliminate the job starvation problem. To this end, the initial job priority is assigned according to the model used in [4].

It uses an aging mechanism to manipulate the job submission priority over time. The aim of the aging mechanism is to prevent jobs from starving; see Table 70.1.

For newly scheduled job

$$\text{priority}_{new} = \text{priority}_{calc} \tag{70.1}$$

For already scheduled job

$$\text{priority}_{new} = \text{priority}_{calc} + \text{priority}_{curr} * (\text{job\_waiting time}_{curr} / \text{job\_estimated time}_{curr}) \tag{70.2}$$

We combine the least variance-based load balancing method with the PSP + Aging scheme as follows.

Each grid resource in the distributed computing environment devotes a collection of machines, which are in turn, a combination of processing elements (PEs) when dealing out with user submitted jobs or requests. These processing elements, which are generally labeled as busy PEs, will not be allocated new jobs until they finish existing jobs. The ratio of busy PEs fraction to the overall number of non-busy PEs at a resource at any time contributes to the current load at that resource.

The average load across all the grid resources is computed using the loads at each resource. As and when a grid resource has adequate number of non-busy PEs to execute a user job (i.e., gridlet), it is examined for an additional check. This check is performed by relating the load at a selected grid resource to the average load across all the grid resources. The resource which is found to have the least average load variance is designated for allocated a task. In the following portion, we summarize the significant steps of the realized algorithm.

70.4   Simulation Setup

An extension to the GridSim [8] simulator is ALEA, which implements the features that are required by an innovative job scheduling simulator. The uniqueness of ALEA [9] simulation framework is that the basic structure for various scheduling algorithms, both queue and schedule based, is made readily available for customization. Further, it not only allows the customization of the scheduling algorithms but also has built-in provisions to accept various criteria that are to be used in modeling a distributed computing environment. These parameters are not limited only to common properties like job heterogeneity, resource heterogeneity, job inflow rates, or resource failures. Any customized parameter can also be included.

Algorithm 1: Load Balancing with Enhanced User Priority Model Scheme

Input: (a) A Tasks queue and (b) a set of grid Resources
Output: Task allocation to Resources with improved Load Balancing
Data: R: Resource set, G: Task queue, Ri: Resource instance, Gi: Task instance- and PEs: Processing Elements

```
for each task in G do/* Initialize task priority*/
   Calculate initial priority of Gi;
for each resource in R do/* Calculate average load */
   Find the total fraction of busy PEs in the grid;
Find the average load across the grid;
for each task in G do/*Start resource selection */
   Calculate current priority of Gi;/*update priority */
   if Gi has not enough priority to run then
      break;/* defer scheduling low priority task */
   else
     for each resource in R do
       if required number of PEs is available then
          Calculate the load difference with average load
             across grid;
     Select the resource with least average load variance;
     Update the total Fraction of busy-PEs for selected
        resource;
     Update the average load across the grid;
```

We now describe the organization of the various components of the proposed system which are held together in the computing environment. The anticipated system involves a set of heterogeneous grid resources (R). Each resource is in turn composed of a collection of machines (M). At the next level, the machines put together a collection of processing elements (PEs). Each resource has a variable number of machines (M) accompanying it. The collection of processing elements (PEs) that are run by a machine (M) have varying architectures and speeds.

The ALEA framework accompanies realistic datasets from the Gaia cluster which recorded the log whose details are elaborated in [10]. These are used to test the functionality of the implemented algorithms. One concrete example that we can infer here is the resource with name Gaia [1–60]. It means that it is has 60 machines, each one of it composed of 12 heterogeneous processing elements like the Intel Xeon

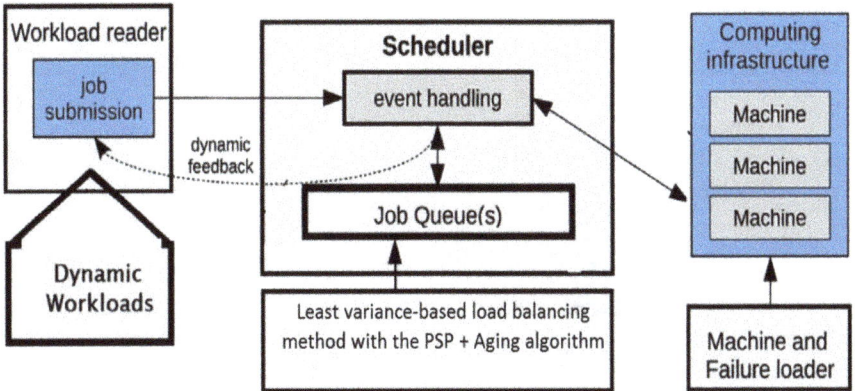

Fig. 70.1 Architecture of ALEA depicting the customized module

L5640 cores and varying speeds in the range of 2 GHz. Tasks, which are otherwise called gridlets, are scheduled on to these grid resources as and when loaded by user applications for processing. The gridlets are allocated on to the resources by means of a unified scheduler [9].

We investigated the schemes discoursed in the previous section using the said realistic data records of tasks offered in the standard workload format (SWF) that were executed on the Gaia cluster [10]. The carefully chosen real-time task logs contain the data for a period of more than a month. It has recorded a count as far as 50 thousand plus jobs allocated on to the grid resources [11].

Below given is Fig. 70.1, which depicts the important modules that constitute the ALEA simulator. The scheduling algorithms module was customized to incorporate the proposed least variance-based load balancing method with the PSP + Aging scheme as follows.

70.5   Results and Discussions

In the figure (Fig. 70.2), after the conduct of experiments with the said data records, a response time comparison is drawn for EDF, P2SDLB [11], and the ALVA (with improved user priority model) algorithm. It brings out that ALVA scheme is 31.8% more efficient than the traditional schemes.

In the figure (Fig. 70.3), with the said data records being used in the experiments, a wait time comparison is drawn for EDF, P2SDLB [11], and the ALVA (with improved user priority model) algorithm. It brings out that ALVA scheme is 89.3% better performing and efficient than the traditional schemes.

In the figure (Fig. 70.4), with the experiments revealing the insights with the said data records, a job tardiness comparison is drawn for EDF, P2SDLB [11], and

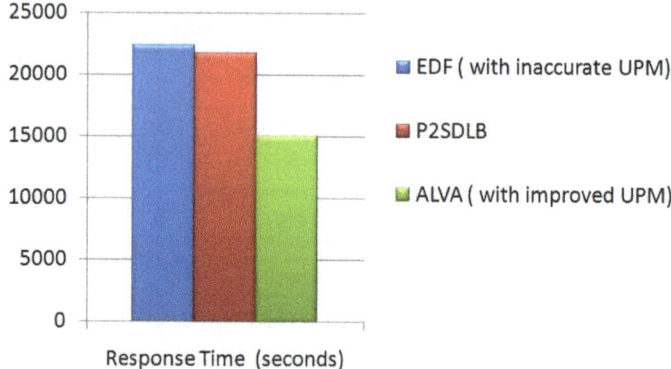

Fig. 70.2 Comparison of response time for the implemented algorithms

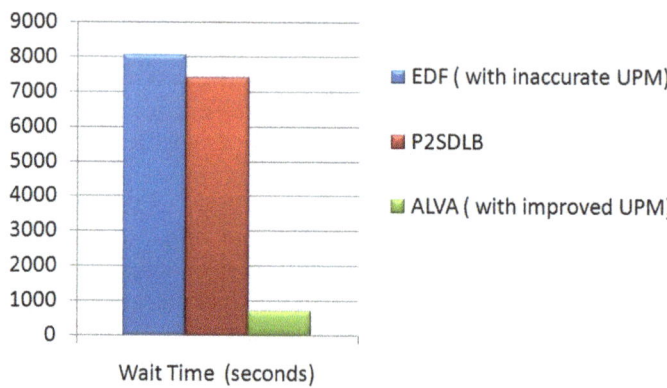

Fig. 70.3 Comparison of job wait time for the implemented algorithms

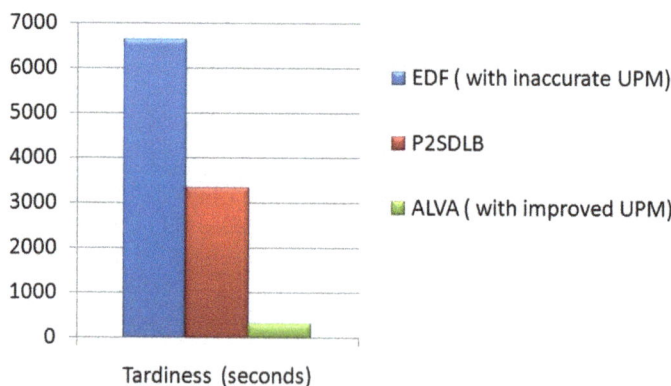

Fig. 70.4 Comparison of job tardiness for the implemented algorithms

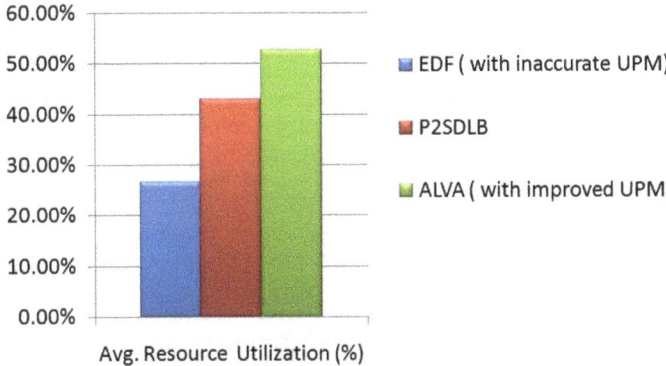

Fig. 70.5 Comparison of average resource utilization for the implemented algorithms

the ALVA (with improved user priority model) algorithm. It brings out that ALVA scheme is 91.1% proficient than the traditional schemes.

In the figure (Fig. 70.5), with the experiments revealing the insights with the said data records, an average resource utilization comparison is drawn for EDF, P2SDLB [11], and the ALVA (with improved user priority model) algorithm. It brings out that ALVA scheme is 32.69% proficient than the traditional schemes.

70.6   Conclusion

In the proposed work, to optimize the response time, wait time, tardiness, and average resource utilization, we use a least average load variance-based algorithm combined with a better user priority model. This algorithm is successful in providing a lower bound on the load across grid resources while ensuring a better response time compared to other methods proposed in literature. With the experiments revealing the insights with the said data records, the ALVA (with improved user priority model) algorithm outclasses traditional solutions in terms of improvement of resource utilization with an optimal response time, wait time, and tardiness.

References

1. Rice, M.: 14 High Performance Computing Applications & Examples. https://builtin.com/har dware/high-performance-computing-applications/. Accessed 19 June 2020
2. Balis, B., et al.: Porting HPC applications to the cloud: a multi-frontal solver case study. J. Comput. Sci. **18**, 106–116 (2017)
3. Rocchetti, N., Iturriaga, S., Nesmachnow, S.: Including accurate user estimates in HPC sched-ulers: ban empirical analysis. In: XXI CongresoArgentino de Ciencias de la Computación (Junín, 2015) 2015

4. Rocchetti, N., et al.: Penalty scheduling policy applying user estimates and aging for super-computing centers. In: Latin American High-Performance Computing Conference. Springer, Cham (2016)
5. Emeras, J., et al.: Evalix: classification and prediction of job resource consumption on HPC platforms. In: Job Scheduling Strategies for Parallel Processing. Springer, Cham (2015)
6. Galleguillos, C., et al.: Constraint programming-based job dispatching for modern HPC applications. In: International Conference on Principles and Practice of Constraint Programming. Springer, Cham (2019)
7. Zotkin, D., Keleher, P.J.: Job-length estimation and performance in backfilling schedulers. In: 8th International Symposium on High Performance Distributed Computing, pp. 236–243 (1999)
8. Buyya, R., Murshed, M.: Gridsim: A toolkit for the modeling and simulation of distribut-edresource management and scheduling for grid computing. Concurr. Comput. Pract. Exp. **14**(13–15), 1175–1220 (2002)
9. Klusáček, D., Tóth, Š., Podolníková, G.: Complex job scheduling simulations with Alea 4. In: Proceedings of the 9th EAI International Conference on Simulation Tools and Techniques (2016)
10. ULHPC: The gaia cluster hpc at university of luxemburg. https://hpc.uni.lu/systems/gaia/. Accessed 26 March 2020
11. Thakor, D., Patel, B.: P2s_dlb: pluggable to scheduler dynamic load balancing algorithm for distributed computing environment. In: Emerging Trends in Expert Applications and Security, pp. 347–355. Springer, Singapore (2019)
12. Guim, F., Corbaĺan, J., Labarta, J.: Prediction f-based models for evaluating backfilling scheduling policies. In: 8th IEEE International Conference on Parallel and Distributed Computing, Applications & Technologies, pp. 9–17 (2007)

Chapter 71
Deep Neural Networks Based Object Detection for Road Safety Using YOLO-V3

Jalaja Tattari, Vineeth Reddy Donthi, Dheeraj Mukirala, and S. Komal Kour

Abstract Traffic safety plays a pivotal role in the society, and hence, there is a grave need for streamlining the protocols and measures taken in our country. This paper gives a comprehensive solution to the situation at hand, a new approach to 'road safety' based on seatbelt and helmet object detection using deep neural networks. The pipeline consists of using the deep learning frameworks and 'You Only Look Once V3 (YOLO-V3)', which is a convolutional neural network-based algorithm, for a hassle-free "real-time object detection." Detection systems in the past re-engineer classifiers or localizers to perform detection. The paper focuses on a single neural network applied to the full image. This network divides the image into regions and predicts bounding boxes and probabilities for each region. Popular image data sources such as the UCI Machine Learning Repository and Google Open Image Datasets were explored, and a dataset was created, which underwent preprocessing. The developed model is very versatile, compact, and was tested on various edge cases after which the model achieved an overall accuracy of 86.3% which was on a higher note compared to the previous works. The program can be readily deployed on CCTV, Webcams, or local computers.

71.1 Introduction

The cutting-edge technology achievable using the deep learning (DL) domain has been giving the mankind glorious results in the recent past. It is in this context that one should think about the possible applications of deep learning with special emphasis on computer vision. The paper revolves around using efficient computer vision algorithms for object detection for the betterment of the current road safety scenario in the country. It was reported by the road transportation authority that over 5 lakh traffic-related accidents occur every year, out of which 1 lakh are deaths. It is a very well-known fact that not wearing helmets and seat belts while driving on the road is a punishable offense under the law, and the offenders are penalized;

J. Tattari (✉) · V. R. Donthi · D. Mukirala · S. Komal Kour
Vasavi College of Engineering, Hyderabad, India

© The Author(s), under exclusive license to Springer Nature Singapore Pte Ltd. 2021   731
S. C. Satapathy et al. (eds.), *Smart Computing Techniques and Applications*,
Smart Innovation, Systems and Technologies 225,
https://doi.org/10.1007/978-981-16-0878-0_71

however, there are many offenders out on the roads, still very reluctant to follow the safety protocols. Governments have been doing a spectacular job raising awareness about the guidelines and also establishing violation detection systems that either use manual detection or algorithms that are less accurate and are not fast.

Technology was static and stagnant earlier, but it is now aggressively dynamic. We are headed toward a future in which machines take over humans with decision-making abilities, hence making models with high output capabilities is of paramount importance. The road safety and transportation domain needs a massive restoration in the pillars of current technologies and techniques. The paper aims at building a real-time, reliable, and a relinquishing model using the "You Only Look Once" (YOLO-V3) algorithm for helmet and seatbelt detection in two wheelers and four wheelers, respectively.

71.2   Related Work

In the paper by Babu et al. [1], they have accomplished object detection using machine learning. The training was carried out with different algorithms like support vector machine (SVM), gradient-boosted trees, etc. The primary aim of the model is to detect the presence of a motorcycle and if there is one, to detect the presence of a helmet on the rider.

In the paper by Nguyen et al. [2], small object detections were implemented using different deep learning algorithms and have proved that YOLO-V3 has faster and accurate predictions. The model was trained on the common objects in context (COCO) dataset.

In the paper by Rohit et al. [3], the videos from a traffic surveillance camera are fed into an object detection model. When the model detects the human class, the output is given to an image classifier which will compare reference image with the new output image and tells if the rider and co-passenger are wearing a helmet or not based on a classification model.

In the paper by Huang et al. [4], they used the YOLO-V3 algorithm to perform traffic flow detections on surveillance videos. The model aims at detecting moving vehicles for improvement of transportation facilities. A dataset containing a wide variety of weather conditions and types of vehicles was used to ensure accurate predictions.

71.3   Proposed Work

A deep learning-based method has been set forth to curb today's alarming increase in road accidents, especially in India. The YOLO-V3 algorithm was used to ensure better accuracy and speed (Fig. 71.1).

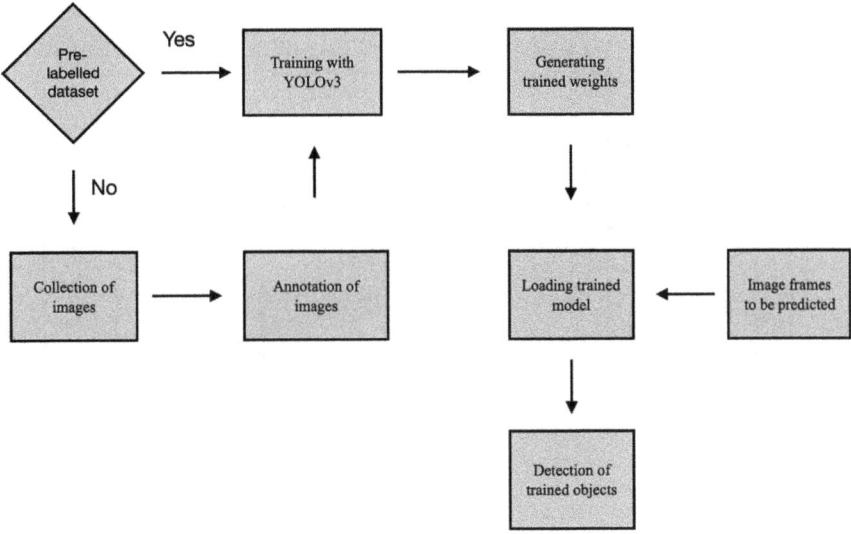

Fig. 71.1 Proposed work flowchart

71.3.1 YOLO-v3

There are many object detection algorithms, YOLO being one of the fastest. Out of all the algorithms, this is opted often because it maintains higher accuracy for real-time detection as well. It detects multiple objects in the feed unlike the other image classification algorithms which only classify the image as a specific class. It finds the objects in an image and draws bounding boxes around them. Exact boundaries of objects can also be drawn using instance segmentation.

The ability of YOLO-V3 to apply a single neural network to the full image rather than apply a model to an image at multiple locations and scales makes it stand out from the other algorithms. This unique ability makes it 1000× faster than R-CNN and 100× faster than Fast R-CNN.

This method divides the input image into small-squared grid cells. A grid cell is held responsible for detecting an object if the center of an object falls in it. Each grid cell predicts the position information of bounding boxes and computes the scores corresponding to the boxes [5].

The feature extractor used in YOLO-V3, i.e., Darknet-53, consists of 53 convolutional layers, while the remaining levels are called resident layers [6].

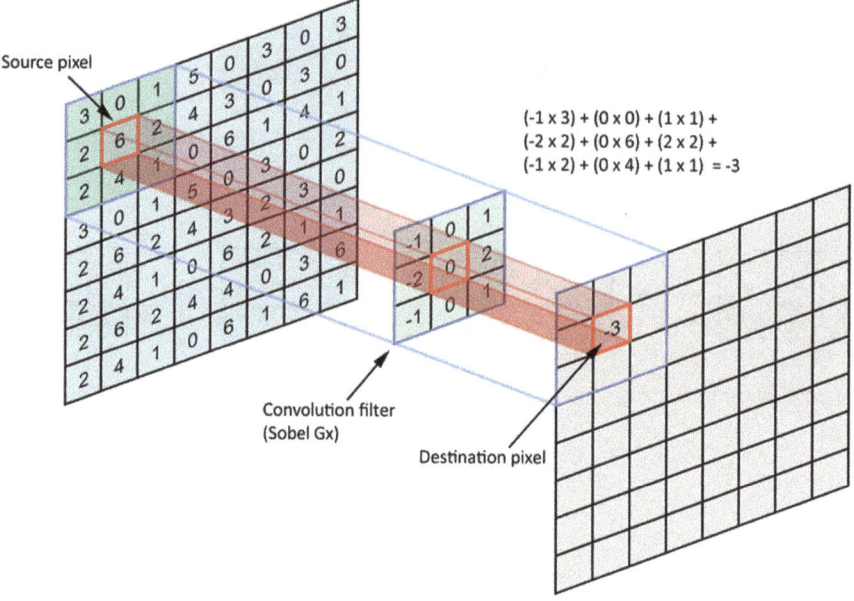

Fig. 71.2 Convolution filter 3D representation

71.3.2 CNN

"A convolutional neural network (CNN) is a deep learning algorithm that can take in an input image, assign importance (learnable weights and biases) to various aspects/objects in the image, and be able to differentiate one from the other" [7].

The employment of convolutional neural networks gives us the ability to automatically learn a large number of filters in parallel specific to a training dataset under the constraints of a specific predictive modeling problem, such as image classification. This ability helps us detect highly specific features anywhere on input images [8]. It converts an image into matrices of numbers. A kernel/filter is a 3 × 3 matrix which is convolved over the entire matrix as shown in Fig. 71.2.[1]

ReLU is a nonlinear activation function like tanh or sigmoid. It takes matrix from a reduced image as input and replaces all the negative values with zero keeping the other values constant.

The softmax activation is normally applied to the very last layer in a neural net. The importance of softmax lies in converting the output of the last layer in a neural network into a probability distribution.

---

[1]Source: https://datascience.stackexchange.com/questions/23183/why-convolutions-always-use-odd-numbers-as-filter-size.

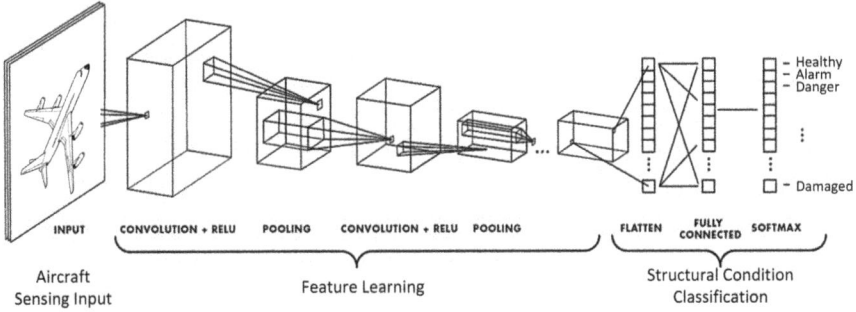

Fig. 71.3 CNN workflow

The entire functioning of CNN's is summarized as shown in Fig. 71.3 [9] [2].

71.3.3 Input

A dataset consisting of different types of helmets and seatbelt altogether 5000 images were taken as input. COCO dataset was used as it is an excellent object detection dataset with 80 classes, 80,000 training images, and 40,000 validation images. A pre-labeled dataset was used to avoid blunders while annotating images, i.e., labeling the images with their respective classes. The dataset was divided into training set and test set in the ratio of 4:1, i.e., for every four images trained, the model will be tested on one image to prevent overfitting of the model.

71.3.4 Training

Google Colab was used for training the model to make use of Tesla K80 GPU (Graphics Processing Units). The model was trained for 3000 iterations to clinch greater accuracy. Though the dataset plays a major role in achieving higher accuracy, preparing configuration files is also equally important. In the data file, the classes must be given the number of classes that you are training the model on, in this case two. Using a piece of code, text files are generated for the training and test sets; these text files must be stored in the variables train and valid in the.data file. This file also must contain another variable "names" which must have the.names file.

---

[2]Tabian, Iuliana; Fu, Hailing; Sharif Khodaei, Zahra. 2019. "A Convolutional Neural Network for Impact Detection and Characterization of Complex Composite Structures." *Sensors* 19, no. 22: 4933.

The .cfg files must be chosen based on the GPU available. Since Google Colab provides GPU, yolo.cfg file was used. The GPU must be set to 1 in the runtime type to fully utilize the hardware. In this file, the following parameters were changed according to our model: batch, subdivisions, filters, and classes.

The filters used in the convolutional layer were calculated using the formula,

$$\text{filters} = (\text{number of classes} + 5) * 3 \tag{71.1}$$

making it 21 in this model. It took 8–9 h approximately to train the network with the above parameters. To save the weights, more often the settings were changed accordingly. The weights generated after 3000 iterations were used to detect the trained objects in random traffic images. [10]

71.4 Experiments and Results

71.4.1 Experiments

The training was carried out on a custom stitched dataset which was examined carefully for imbalance. The dataset consisted around 5000 images which was the backbone of the training process. We have maintained the batch size to be 64 even though batch size 32 would have made our model converge even faster. A backup folder was created in the parent directory to save weights to this folder automatically after every 200 iterations until 1000 iterations and for every 1000 iterations thereafter. We finished training in about 3 days and stopped the training at about 3000 iterations because the model converged to a cogent mAP accuracy of 86.3%.

71.4.2 Results

A few examples of the output object detector are shown in Figs. 71.4 and 71.5.

71.5 Conclusion and Future Work

From the above result and analysis, it is very evident that the generated model is very adaptive in nature. Computer vision algorithms are the need of the hour in the road safety domain in our country. The proposed model is dressed to the nine and has reached higher accuracies than the existing frameworks in the realm of object detection. The role of hyperparameter tuning and optimization is very essential for clinching higher performances. This paper is capable of paving a tenacious path in

Fig. 71.4 Helmet detection

Fig. 71.5 Seatbelt detection

the future of deep learning in several areas like self-driving vehicles, monitoring crops for disease detection, medical diagnostics, etc., one of the major applications being penalizing traffic offenders to ensure public safety.

References

1. Ramesh Babu, D.R., Rathee, A., Kalita, K., Deo, M.S.: Helmet detection on two-wheeler riders using machine learning. In: ARSSS International Conference (2018)
2. Nguyen, N.-D., Do, T., Ngo, T.D., Le, D.-D.: An Evaluation of Deep Learning Methods for Small Object Detection, Hindawi (2020)
3. Rohith, C.A., Nair, S.A., Nair, P.S., Alphonsa, S.: An efficient helmet detection for MVD using deep learning. In: Proceedings of the Third International Conference on Trends in Electronics and Informatics (2019)
4. Huang, Y.-Q., Zheng, J.-C., Sun, S.-D., Yang, C.-F., Liu, J.: Optimized YOLOv3 algorithm and its application in traffic flow detections. Appl. Sci. (2020)
5. Zhao, L., Li, S.: Object detection algorithm based on improved YOLOv3, Computer Science and Engineering (2020)
6. Vidyavani, A., Dheeraj, K., Rama Mohan Reddy, M., Naveen Kumar, K.H.: Object detection method based on YOLOv3 using deep learning networks. IJITEE (2019)
7. Saha, S.: A Comprehensive Guide to Convolutional Neural Networks—The ELI5 way, TowardsDataScience (2018)
8. Brownlee, J.: How do convolutional layers work in deep learning neural networks? Deep Learn. Comput. Vision (2019)
9. Tabian, I., Fu, H., Sharif Khodaei, Z.: A convolutional neural network for impact detection and characterization of complex composite structures. Sensors **19**(22), 4933 (2019)
10. Masurekar, O., Jadhav, O., Kulkarni, P., Patil, S.: Real time object detection using YOLOv3. IRJET (2020)

Chapter 72
Log-Based Authentication for Cloud Environments

Nalini Subramanian and G. Shobana

Abstract In the world of cloud era, several organizations achieve benefits in terms of capital expenditure and operational expenditure. Security in adopting cloud is one of the important challenges that need to be addressed for several organizations. In order to increase the security, cloud should provide an advanced authentication framework for accessing its resources. This paper introduces the need for advanced authentication, based on the security issues encountered in cloud computing at various levels, since advanced authentication provides the logging functionality. When a local account is used to log on to a computer, that computer performs both logon and authentication. When a domain account is used to log on to a computer, the computer being accessed makes the connection, but a domain controller in the domain that owns the user account authenticates. So, this paper explores log-based advanced authentication, and it also depicts how the usage of dimensionality reduction technique and support vector machine (SVM) reduces the complexity of an authentication framework without sacrificing the accuracy.

72.1 Introduction

Data is at the heart of the crypto-cloud, and data breach is identified as the number one threat by the cloud security alliance (CSA). The life cycle of data from source to destruction must be secure, as storing data in a remote location and the concept of multi-tenancy can lead to data leaks. Data security is classified into data in transmission and data at rest. Inactive data attracts hackers more than transmitting data. Data security can be broadly classified into three basic levels, namely the communication level, the processing level, and the service level agreement (SLA) level.

N. Subramanian (✉)
Prathyusha Engineering College, Tiruvallur, India

G. Shobana
Department of Computer Applications, Madras Christian College (Affiliated to University of Madras), Chennai, India

© The Author(s), under exclusive license to Springer Nature Singapore Pte Ltd. 2021   739
S. C. Satapathy et al. (eds.), *Smart Computing Techniques and Applications*,
Smart Innovation, Systems and Technologies 225,
https://doi.org/10.1007/978-981-16-0878-0_72

Typically, authorized users can access the resource from the server once they are authenticated by the server. The goal is to provide a strong authentication framework for cloud environments. The authenticators used and their combination play a very important role. When a unique authenticator is used, it can easily be forged. This leads to a single point of failure. A combination of these is used because a one-factor authentication system often fails to provide high security. Multi-factor authentication is required to improve security depending on the type of application. Simply increasing the number of authentication levels alone does not increase security. On the contrary, it involves costs, time, space, etc., considerable to provide strong authentication. Three fundamental elements must be combined: what the user knows, what he has, and what he is. This multi-layered authentication framework enables secure and strong authentication for cloud environments.

Log details kept on the server are used to prevent system performance degradation and to monitor critical events. Logs are local files saved when an event occurs in the system. Events can be the insertion of a file, the deletion of a file, and its access or its modification. Security events are data or system security-related events, and the corresponding log file is called the security log. Event logs can be classified as application log, system log, and security logs, among which this work focuses on the security log, which is related to system security. An invalid connection or disconnection falls within these criteria.

The number of failed login attempts prevents or reduces data theft, giving it the highest priority for authentication. Logging of all events is not recommended as it exposes the data to hackers. Depending on the type of application, only critical events require regular monitoring. The user details provided as input along with specific information such as username, date and time, IP address, status, URL, and number of login attempts are used to improve security and authentication. As the total number of parameters increases, a dimensionality reduction technique such as PCA is required to reduce system complexity. Otherwise, it leads to false negatives.

72.2 Related Work

In the article [1], the authors presented an approach called log-based anomaly detection (LAD). The proposed approach finds the communication rules used in mobile ad-hoc networks using the logs of router. Reasoning rules are used to detect both active and passive attacks triggered during the phase of routing. Using Merkle hash tree, the files are secured, and confidentiality of the nodes is improved. The methodology detects malicious nodes without the usage of special nodes nominated for verification. This approach is implemented in NS3 using router environment to prove that mobile ad-hoc security is increased. It is proved that the accuracy rate is increased from 2.28 to 29.22%. From the results, it is confirmed that the proposed approach consumes little memory and time.

Allani and Ghannouchi [2], authors proposed an approach to evaluate the business process models as well as analysis of event logs based on the previous execution.

The models evaluated are written in BPMN2.0 (Business Process Model Notation) and executed using activity explorer online service. In order to verify the results, the experiments have been done on two different business processes.

The process of implementing the PKI improvements assured on logs has been proposed by [3] in two ways. First, the PKI influence model should encourage each other to implement a PKI enhancement and determine that potential PKI enhancements should focus their initial efforts on the browser vendor implementation. Two things, as a promising vendor-based solution, the authors implemented status filters, here use a Bloom filter to monitor the status of the distribution and effectively defend against demotion attacks. Protocols have been improved for the current PKI TLS. The findings have led to promising implementation strategies for the log-based improvements in PKI and raise further questions for further fruitful research.

A log-based analysis tool to evaluate the computer system of Web applications was proposed by [4]. This tool integrates software registry with an infrastructure. Even though software and infrastructure issues are complicated, software engineers can overcome only system approach errors. The system approach comprises of five phases: software and infrastructure preparation, log collection, log review data, and log data discovery. To a simple local area network of a small-scale Web application system, the tool was applied. The usefulness of the proposed tool was confirmed, when a software engineer detects file system crash errors such as "404" and errors of type "no response." The tool was also partially applied to a large-scale computer system with many Web applications and a large network. Using the proofreading and tracing tool, the causes of a true authentication failure were found. The causes enlisted both infrastructure problem with a software problem which are combined. It is confirmed that errors are not caused only by an issue in software. Software engineers are only responsible to distinguish a software problem when compared to an infrastructure problem using the proposed tool.

In [5], we presented real-world examples that would lead to broken instances of the idealized AuthA protocol and the One-Encryption-Key-Exchange (OEKE) protocol. The results show that the AuthA protocol can be instantiated in an insecure manner and does not exist to distinguish between safe and insecure instances. Therefore, without a strict metric for ideal numbers, the value of demonstrable security in the ideal encryption model is limited.

The log-based intrusion detection system uses user logs from Web servers, where login information is recorded on every request. The downside to log-based Web intrusion detection is that it can only be performed after transactions have been completed; therefore, prevention of attacks is impossible [6]. Focused on user behavior, to discover the difference between a common user and a malicious user and to try to prevent the attack before it happens.

72.3   Proposed Framework

When the user logs in, they are authenticated by their name and password. At the
second level of authentication, the fingerprint is used. Fingerprints can be forged
using jelly wax. Therefore, the network time protocol (NTP) timestamp is used for
authentication. An authentication system is strong when a combination of credentials,
such as a password, a biometric key, and a session key (token) is used.

To increase the accuracy of the strong authentication framework (SAF), the log
details kept on the server are used with the parameters provided by the user to
provide high security as in Fig. 72.1. In addition to client details, log details can
be accessed to provide an advanced authentication infrastructure. Log details can
be viewed according to application requirements. Typically, they will be queued at
regular intervals and used only when needed to resolve a problem. The details can be
viewed when needed, in order to increase the security of the application. Using this
information increases the security of the cloud server that requires authentication.
The information requested can be collected from the various logs available.

There are different types of logs such as audit log, secure log, and access log.
The details kept in the log file vary depending on the needs of the application.
Starting with the single autonomous system and moving to a network of distributed
systems, logs are kept to track and identify attackers. Any user movement, such
as the file they are accessing, URL details, system usage date and time, system IP
address, action/status, for example, connect/failed if parameter action or code is used
to a state like 404, 200, etc., is stored. This list of parameters varies depending on
the application requirements. Some secure log settings are identified as containing
specific user details required for authentication. The number and type of parameters

Fig. 72.1 Log-based
authentication framework

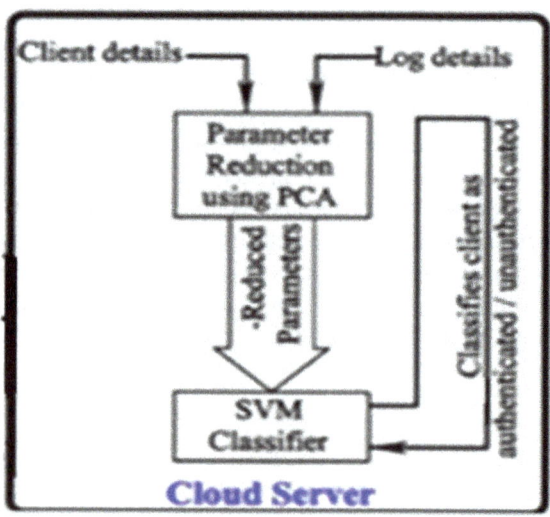

vary dynamically depending on the type of user login. Variations and increases in the number of parameters lead to the need for a technique to help reduce them.

PCA is used to decrease the number of features prior to the classifier, which classifies the users as authenticated/unauthenticated.

72.3.1 Support Vector Machine

Support vector machine is a supervised machine learning model. When machine learning is implemented on a cloud server with suitable feature compression, it is certainly possible to obtain a very accurate model for predicting authenticated users. In general, the number of parameters taken should be larger to ensure accurate results. But this increases the complexity of the model along with the increase in false negatives and sometimes even false positives. This framework reduces the overall complexity and enhances the cloud environment, to permit only authenticated (true negatives) users to access resources when compared to unauthenticated (true positives) users.

The following are the overview of implementing SVM:

Step 1:   Dataset to be loaded
Step 2:   The loaded dataset is summarized
Step 3:   It is visualized
Step 4:   The model is evaluated, and predictions are made.

According to [7], the occurrence of an attack is represented by 1, and the non-occurrence of an attack is represented by 0. When attack and alarm are 0, it is identified as true negative. With attack as 0 and alarm as 1, it is false positive. If both are 1, then it is true positive, and if only attack is 1 and alarm is 0, it is identified as false negative.

72.4 Classification Phase Algorithm

The classifier algorithm is stated as follows:

Input: web_log data set

Output: Classified as authenticated/Unauthenticated users

1. Using pandas, numpy, matplotlib, and sklearn packages, SVM classification phase is implemented.
2. Predefined libraries like classificiation_report and confusion_matrix are imported.
3. Import the preprocessed dataset in.csv format (https://www.kaggle.com/sha won10/web_log_dataset#weblog.csv)

4.   Dividing the data into attributes and labels. X = weblog.drop('action', axis = 1) and y = weblog['action']
5.   Dividing training and testing sets, X_train, X_test, y_train, y_test = train_test_split (X, y, test_size = 0.20).
6.   Model is created with SVC () of sklearn.svm, then fit () is used to train the model. Kernel parameter is set to linear, and svclassifier.fit (X_train, y_train) is invoked.
7.   Predictions are performed using predict () of svclassifier as, y_pred = svclassifier.predict (X_test)
8.   Model is evaluated by computing the confusion matrix and classification report, using y_test and y_pred.

Among the available models, SVM yields the highest percentage of accuracy for small datasets. The model is executed on the test data, and the results are summarized in a final accuracy score, as a confusion matrix and a classification report.

72.5   Experimental Evaluation

Once the confusion matrix is obtained, the precision, recall, accuracy, and error can be computed as in Eqs. (72.1)–(72.4). They are the parameters for the evaluation of the classifier.

$$\text{precision} = \text{true} - \text{positive}/(\text{true} - \text{positive} + \text{false} - \text{positive}) \qquad (72.1)$$

$$\text{recall} = \text{true} - \text{positive}/(\text{true} - \text{positive} + \text{false} - \text{negative}) \qquad (72.2)$$

$$\text{accuracy} = \text{true} - \text{positive} + \text{true} - \text{negative}$$
$$/(\text{true} - \text{negative} + \text{true} - \text{positive} + \text{false} - \text{negative} + \text{false} - \text{positive}). \qquad (72.3)$$

$$\text{error} = 1 - \text{accuracy}. \qquad (72.4)$$

The confusion matrix and classification report for the sample of 100 records are tested with the model. Accuracy, precision, and recall are computed for 100 users. The number of false negatives is reduced when the number of record size increases as in Table 72.1. This is done from the two classes of input, namely authenticated and unauthenticated. TN = 13, FP = 4, FN = 5, and TP = 18 form the list of data, the model interpreted accurately, and accuracy is 77.5%. The precision will be 72%, and recall is 76%

Increase in accuracy is seen with increase in the number of users. Accuracy is 0.77 initially for a record size of 100, as the record size increases the accuracy also

Table 72.1 Computation of the parameters

Records	Precision	Recall	Accuracy
100	0.72	0.76	0.775
200	0.89	0.75	0.81
300	0.89	0.8	0.85
400	0.95	0.84	0.9
500	0.97	1	0.96

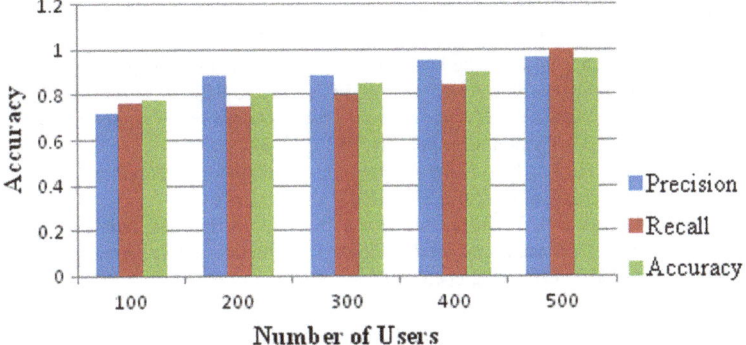

Fig. 72.2 Precision, recall, and accuracy

increases. It is 0.96 which is 96% when the record size is 500, precision is 0.97, and recall is 1 as shown in Fig. 72.2.

72.6 Conclusion

Log security and reliability are important for effective system maintenance and monitoring. This leads to the conclusion that logs are an important source for the security control system, as hackers primarily modify log data to hide their invasive behavior. The log file includes the person's personal information, so it should be protected. The security and authenticity of the log are guaranteed to protect against attackers. With the help of using dimensionality reduction technique, large numbers of redundant and irrelevant features are reduced. 'Support vector machines' work on smaller datasets, compared to other models. They can be stronger and more powerful as building models. This proposed work accurately classifies users as authenticated and unauthenticated. It includes SVM for classification and PCA for feature selection to improve accuracy and reduces the complexity. The false positive rates have been eliminated when compared to other models. The security has increased, once unauthenticated users are limited by a framework.

References

1. Li, T., Ma, J., Pei, Q., Shen, Y., Sun, C.: Log-based anomalies detection of MANETs routing with reasoning and verification. In: Proceedings, APSIPA Annual Summit and Conference, pp. 240–246 (2018)
2. Allani, O., Ghannouchi, S.A.: Verification of BPMN 2.0 process models: an event log-based approach. In: Conference on ENTERprise Information Systems/International conference on Project MANagement/Conference on Health and Social Care Information Systems and Technologies,CENTERIS/ProjMAN/HCist 2016; Proc. Comput. Sci. **100**, pp. 1064–1070 (2016)
3. Matsumoto, S., Szalachowski, P., Perrig, A.: Deployment challenges in log-based PKI enhancements, pp. 1–7 (2015)
4. Hanakawa, N., Obana, M.: A log-based trace and replay tool integrating software and infrastructure. Int. J. Softw. Eng. Appl. (IJSEA) **9**(4), 1–19 (2018)
5. Zhao, Z., Dong, Z., Wang, Y.: Security analysis of a password-based authentication protocol proposed to IEEE 1363. Theoret. Comput. Sci. **352**, 280–287 (2006)
6. Chanthini, S., Latha, K.: Log based internal intrusion detection for web applications. Int. J. Adv. Res. Ideas Innov. Technol. **5**(3), 350–353 (2019)
7. Subramanian, N., Andrews, J.: Strong authentication framework using statistical approach for cloud environments. In: Concurrency and computation: Practice and Experience (2018). ISSN: 1532-0634. https://doi.org/10.1002/cpe.4870

Chapter 73
Development of a Wireless Sensor Node for Electrical Power Distribution Monitoring System

Sanjay Chaturvedi, Arun Parakh, and H. K. Verma

Abstract Today, systems around us are expanding rapidly every day, whether it be communication systems, power distribution systems or security systems, or any other interwoven system in our daily lives. With the increasing size, these systems are becoming complex to manage as they are interdependent on several parameters. This problem of managing systems is handled very well by incorporating smartness into these systems. A system can be made smart by using current technologies like IoT and WSN. Where in IoT, the sensors are Internet-enabled and send data directly over the Internet, and the case with WSN is different in the sense that WSN does not need to have any connection to the Internet and can be used independently. The use of WSN-based systems requires wireless sensor nodes to sense the parameters of the system under observation. The available works present different ways of developing the sensor nodes. Still, the common problem with them is the cost of development and the complexity and steep learning curve to reproduce those sensor nodes within the academia for research purposes. In this work, we present a simple approach to design a wireless sensor node for application in a power monitoring system. We built the sensor node around an open-source development board, Arduino Uno, using simple circuits and readily available sensors and used ZigBee technology, an open standard, for wireless communication. We implemented the developed sensor node in our work 'wireless sensor network-based power monitoring system'. Here, we measure three critical parameters at a power distribution point viz voltage, current, and frequency. The measured values are transmitted to the central server from where the distribution system is monitored. The operating voltage of power lines varies between 230 and 245 V, and the current varies to a maximum of 30 A in our case. The frequency ranges from 49.8 to 50.25 Hz. The results obtained in implementation have a marginal error with an approximate mean square error of less than 0.5% of the parameter's maximum operating value. These results have validated our developed sensor node and the power monitoring system.

S. Chaturvedi (✉) · A. Parakh · H. K. Verma
Shri G. S. Institute of Technology & Science, Indore, India

© The Author(s), under exclusive license to Springer Nature Singapore Pte Ltd. 2021   747
S. C. Satapathy et al. (eds.), *Smart Computing Techniques and Applications*,
Smart Innovation, Systems and Technologies 225,
https://doi.org/10.1007/978-981-16-0878-0_73

73.1   Introduction

A wireless sensor network is a network of many sensing nodes, which are spatially distributed. These nodes continuously monitor the system's parameters under observation and relay the measured data to a central server over a wireless communication channel. These sensor networks are used in the monitoring of health [1], monitoring of power grids [2], detecting fire [3], in energy management systems [4], and for factory automation [5], to name few.

A wireless sensor network's three main components include sensor nodes, wireless communication link, and a central unit. The sensor node is based on some processing element like a microcontroller, which, when integrated with sensors and a wireless transceiver, forms the wireless sensor node. Wireless communication is done either by the use of available wireless network services or using an ad-hoc network. And the central unit is generally a computer responsible for coordinating with all the nodes in the sensor network and managing the data received from the sensor nodes [6]. This paper presents the development work of one of the main components of a wireless sensor network, the wireless sensor node. We find a fair amount of work in the literature where wireless sensor nodes have been built using different development boards and other approaches. Few works with different methods include the work of Prasad et al. [7], which presents a power monitoring system for smart grids. In this work, they used a power sensing IC ADE7757 to develop their sensor node. Suryadevara et al. [8] in their work have developed a monitoring and control system for household appliances. They used the processor available in XBee radio modules for the development of sensor nodes. The work done by Arora et al. [9] has presented two systems for measuring voltage, one based on PIC microcontroller PIC16F877A and the other based on the AT89S52 microcontroller of the 8051 families. This work also validates the use of 8-bit microcontrollers for AC signal sensing applications. Ahamed et al. [10] have made an energy meter using a dsPIC33F microcontroller and an external ADC. Several other works have used Rasberry Pi for developing sensor nodes.

The problem in the available works was the absence of a smooth transition from the theory being taught to the hardware implementation. We find a lack of simple digital and analog circuits that can be realized easily. We see that many of the work emphasized making the sensor node efficient in terms of power consumption, etc., rather than on the application of those nodes itself, which leads them to follow complex approaches for sensor node development, making development costly and time taking. Also, we find a minimal number of works utilizing the open-source tools and hardware for system development.

This paper presents the steps involved in developing a wireless sensor node using an open-source development board, Arduino Uno. We have designed units used for signal conditioning to make signals suitable for use as an input to the microcontroller. We have focused our approach to maximize the use of simple circuits like the Schmitt trigger in frequency measurements, RC filters, resistive voltage dividers, and also readily available components like potential and current transformer and the open-source development board. With this approach, we reduce the cost and time of development.

The paper proceeds with a basic introduction to the wireless sensor node and the process of its development in the next section. The following section presents the results obtained on the implementation of the developed node. We have used this sensor node in a wireless sensor network prototype system developed to measure electrical parameters at a power distribution point. The last section concludes our work.

73.2   Development of Wireless Sensor Node

In this section, we present the steps followed in the development of the wireless sensor node. A wireless sensor node is a device that has sensors interfaced to a processing unit and a wireless transceiver connected to the processing unit for wireless communication, as shown in Fig. 73.1. This section starts with a discussion on selecting a microcontroller for the development of a sensor node. It then proceeds with the design and implementation of different sensor node units: the voltage measurement, current measurement, frequency measurement, and wireless communication units.

73.2.1   Microcontroller Selection

Going through the literature, we see that a wireless sensor node's general structure includes some sensors for measurement and a radio device for wireless communication and data transmission. This gives us an idea of the minimum requirements

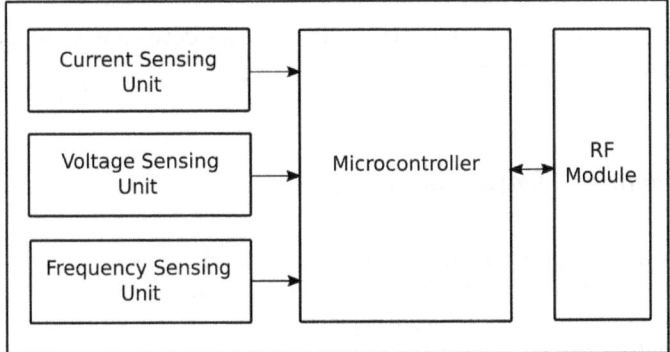

Fig. 73.1   Block diagram of proposed wireless sensor node

Table 73.1 Comparison of AT89c51 and ATmega328P microcontrollers

Features	AT89c51 (8051 Family)	ATmega328P (Arduino uno)
GPIO Pins	32	14 + 6 (Analog Input)
Built-In ADC	NA	6 channel ADC
Flash memory	4 kB	32 kB
On-Chip RAM	128 Bytes	2 kB
Timer/counters	2 16 bit	2 8-bit, 1 16-bit
Communication Ports	1 UART	1 UART
		1 I2C
		1 SPI
Clock speed	12 MHz	16 MHz
1 machine cycle	12 clock cycle	1 clock cycle

associated with the microcontroller like availability of a sufficient number of digital input/output pins, analog to digital converters, timers, and counters, communication ports, processing speed, etc. So, when we evaluate different microcontrollers and microprocessors available for embedded system development, we see that using a 16-bit or 32-bit processing unit in our application would be an overkill. Since we do not have to deal with large number's computations, there is no need for the high processing power of high bit depth microprocessors. This boils down our search to 8-bit processing units.

Based on the availability of development resources like development environments and hardware, development community base, and the microcontroller's learning curve, we select two 8 bit microcontrollers best suited for our purpose. A comparison of these two microcontrollers is made, as shown in Table 73.1, and ATmega328P is chosen for our work. The fact that Arduino Uno is an open-source development board based on our chosen microcontroller helps us to develop our sensor node fast utilizing the vast amount of technical documents already available. Also, it brings a universal appeal to our design as the work can be reproduced easily.

After selecting an 8-bit microcontroller for our node development, we proceed to develop other units of our sensor node.

73.2.2 Voltage Measurement Unit

To measure the voltage using a microcontroller, we need to condition the signal under observation, suitable for use as an input to the microcontroller [11–13]. The input range of the microcontroller we are using is 0–5 V DC, and the signal to be measured is around 230 V AC. So, we need to step down the voltage level using a step-down transformer to a safe working level. We use a 230 V/12-0-12 V potential transformer for this purpose. The reduced voltage obtained from the transformer is then rectified

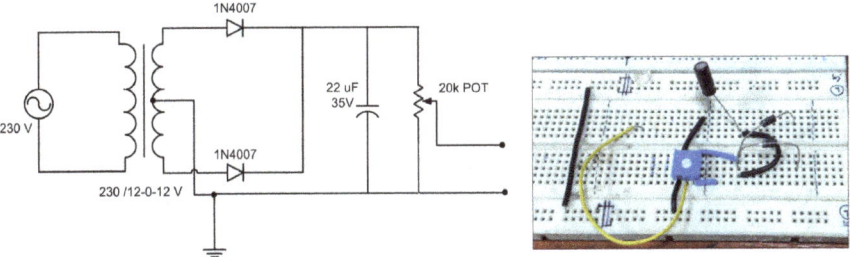

Fig. 73.2 Circuit diagram and breadboard implementation of voltage unit

using a full-wave rectifier. The rectified signal is then passed through an RC filter to get a smooth DC input signal, which we feed to the microcontroller's analog input, as shown in Fig. 73.2.

73.2.3   Current Measurement Unit

For the measurement of current, we will use a widely used Hall effect current sensor ACS712. It is available in three different ranges, and we select the lowest value sensor, i.e., 5 A range. With this sensor, the measurement of current up to 5 A can be easily done by using the current sensor directly. But that is not possible for high values of current. The current level is reduced using a current transformer similar to the approach adopted in the voltage measurement case to measure high current value. Here, we use a 30/5 A Tape Wound current transformer. The current obtained from the secondary is fed to the selected current sensor, and the sensor produces a voltage signal proportional to the applied current. This voltage signal is provided to the analog input of the microcontroller, as shown in Fig. 73.3.

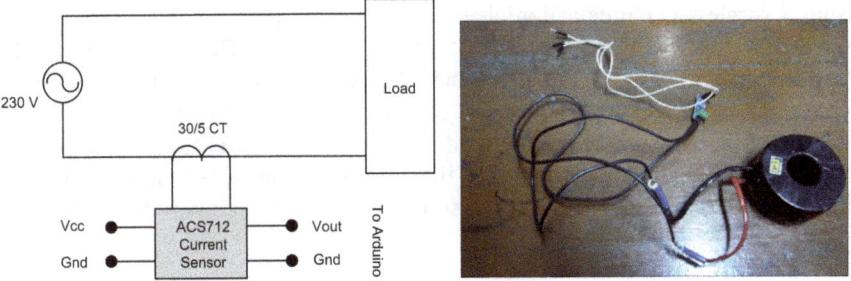

Fig. 73.3 Circuit diagram and hardware setup used in current unit

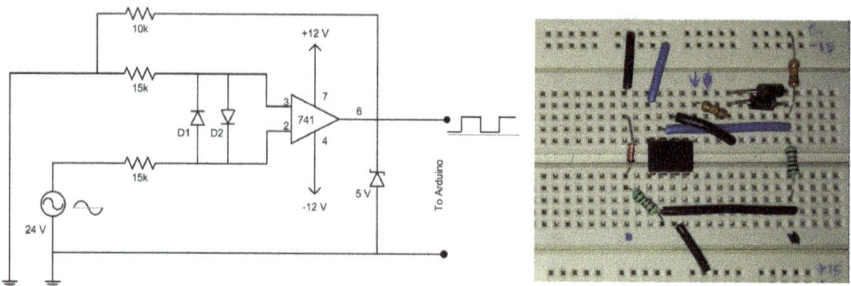

Fig. 73.4 Schmitt trigger circuit and its breadboard implementation

73.2.4 Frequency Measurement Unit

We have many methods available for the measurement of the frequency of alternating signals. The most commonly used techniques include using a zero-crossing detector circuit and a microcontroller that monitors and counts the events of zero-crossing in an alternating signal. The other method is converting the sinusoidal signal into square pulses and then measuring the pulse period using the microcontroller to determine the signal's frequency.

We have used the second method in our work, where we convert the sinusoid signal to a square wave using a Schmitt trigger, as shown in Fig. 73.4. This square wave is clipped using a Zener diode to discard the negative half of the square pulses, and the positive half is fed to the PWM pin of the microcontroller to measure the period.

73.2.5 Communication Unit

The wireless sensor node is called wireless because it is enabled to communicate using a wireless communication channel. For this purpose, we have selected the XBee radio module, which operates at a frequency of 2.4 GHz and works on the ZigBee standard, which is an open standard. These modules work in two different modes, AT mode and API mode, thus enabling us to have either a wireless serial link or a mesh network as per the need. We have used the AT mode of operation. The necessary configuration setting of the module is done as per the work [6], and then, the modules are connected to the serial port of the microcontroller as shown in Fig. 73.5.

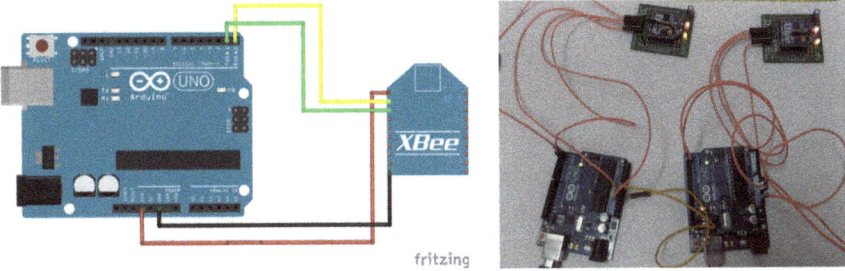

Fig. 73.5 Circuit diagram and implementation of Arduino Uno and XBee connection

73.3 Implementation of Proposed System

We have used the above developed wireless sensor node in our work, where we have implemented a power monitoring system using wireless sensor network. The power monitoring system is developed to monitor the basic electrical parameters at any distribution point in the distribution network within our institute campus. The system includes the above created wireless sensor node and the communication channel and data logger developed in our earlier work of WSN-based data logger [6].

The initial prototype is tested in the lab to validate the working of the complete system, as shown in Fig. 73.6. In this test, the sensor node takes electrical voltage measurements across the lamp load, the current drawn by the load, and the frequency of the applied voltage. On request from the central node, all these measurements are sent back to it using the ZigBee-based wireless serial link.

The existing monitoring system of our institute uses Elite 440 multi-line three phase meters made by Secure Meters Limited. These meters are good for operation in high tension feeder lines. But when it comes to their use in distribution lines, they are an overkill as we need to measure only few electrical parameters. Also, these

Fig. 73.6 Final setup of the power monitoring system prototype

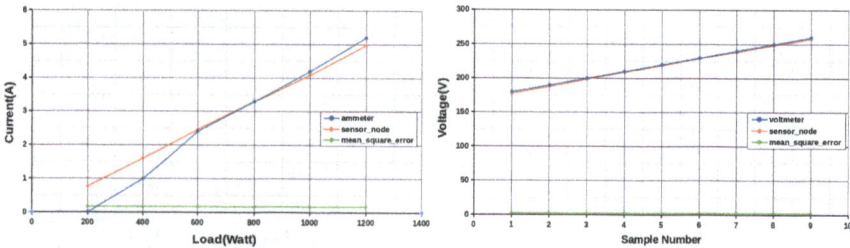

Fig. 73.7 Results of current measurement and voltage measurement

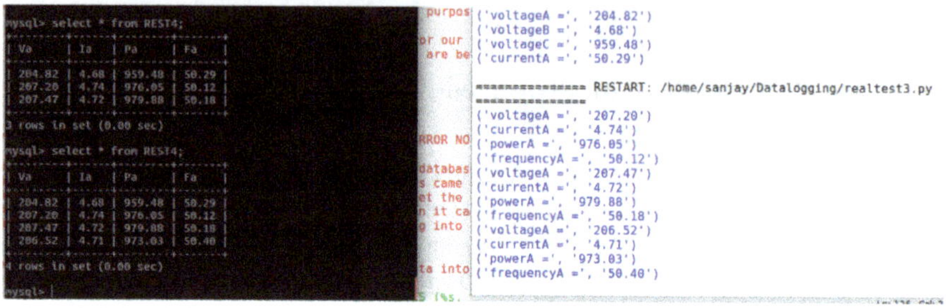

Fig. 73.8 Data loaded into the database and data received from wireless sensor node

meters are costly and are proprietary item leaving us with no scope of customization. Our proposed node was not only able to do accurate measurements with precision but was also very cheap compared to meters that are used. Also, the modular design of our wireless sensor node gives us the flexibility to customize them for specific application. The results obtained in our prototype testing were as expected, as shown in Figs. 73.7 and 73.8, and have validated the working of the wireless sensor node and the complete power monitoring system. The test outcome also validates our approach of using open-source hardware and software tools to develop a wireless sensor network. It also shows that the basic circuits being taught in theory can be realized in house using locally available components at low prices.

73.4 Conclusion and Future Scope

We made a wireless sensor node for our monitoring application. This sensor node measures three electrical parameters, voltage, current, and frequency of AC power. With this work, we have presented that a primary sensor node with good performance in terms of precision of measured values can be developed with limited resources and cost. The use of open-source resources like the Arduino Uno development board and basic circuits helps us focus more on the system's application than the system itself. In the future, we can optimize the work of microcontrollers to bring down the unnecessary delays in processes running on the microcontroller. Also, we will work more on improving the signal conditioning units to minimize the noise levels in the signal. By doing this, we will have optimum performance from our sensor nodes.

References

1. Lee, D.S., Lee, Y.D., Chung, W.Y., Myllyla, R.: Vital sign monitoring system with life emergency event detection using wireless sensor network. In: SENSORS, 2006 IEEE, pp. 518–521. IEEE, New York (2006, October)
2. Yerra, R.V.P., Bharathi, A.K., Rajalakshmi, P., Desai, U.B.: WSN based power monitoring in smart grids. In: 2011 Seventh International Conference on Intelligent Sensors, Sensor Networks and Information Processing, pp. 401–406. IEEE, New York (2011, December)
3. Cheong, P., Chang, K.F., Lai, Y.H., Ho, S.K., Sou, I.K., Tam, K.W.: A ZigBee-based wireless sensor network node for ultraviolet detection of flame. IEEE Trans. Ind. Electron. **58**(11), 5271–5277 (2011)
4. Han, D.M., Lim, J.H.: Smart home energy management system using IEEE 802.15. 4 and Zigbee. IEEE Trans. Consumer Electron. **56**(3), 1403–1410 (2010)
5. Korber, H.J., Wattar, H., Scholl, G.: Modular wireless real-time sensor/actuator network for factory automation applications. IEEE Trans. Ind. Inform. **3**(2), 111–119 (2007)
6. Chaturvedi, S., Parakh, A., Verma, H.K.: Development of a WSN based data logger for electrical power system. In: Gani, A., Das, P., Kharb, L., Chahal, D. (eds.) Information, Communication and Computing Technology. ICICCT 2019. Communications in Computer and Information Science, vol. 1025. Springer, Singapore (2019). https://doi.org/10.1007/978-981-15-1384-8_7
7. Yerra, R.V.P., Bharathi, A.K., Rajalakshmi, P., Desai, U.B.: WSN based power monitoring in smart grids. In: 2011 Seventh International Conference on Intelligent Sensors. Sensor Networks and Information Processing (ISSNIP), pp. 401–406. IEEE, New York (2011)
8. Suryadevara, N.K., Mukhopadhyay, S.C., Kelly, S.D.T., Gill, S.P.S.: WSN-based smart sensors and actuator for power management in intelligent buildings. IEEE/ASME Trans. Mechatron. **20**(2), 564–571 (2015)
9. Arora, J., Rawat, S.S.S., Srinivasan, K., Puri, V. et al.: Design and development of digital voltmeter using different techniques. In: 2014 International Conference on Green Computing Communication and Electrical Engineering (ICGCCEE), pp. 1–5. IEEE, New York (2014)
10. Ahamed, T., Sreedevi, A.: Design and development of pic microcontroller based 3 phase energy meter. Int. J. Innov. Res. Sci. Eng. Technol. **3** (2014)
11. John Errington's Experiments with an Arduino. http://www.skillbank.co.uk/arduino/measure.htm
12. Circuits4you. http://circuits4you.com/2016/05/13/arduino-ac-voltage/
13. All About Circuits. https://forum.allaboutcircuits.com/

Chapter 74
CommonKADS and Ontology Reasoner: Bulky-Baggage Case Study

**Edwin Fabricio Lozada T., Wladimir L. Tenecota,
Paul Santiago Pullas Tapia, and Livio Danilo Miniguano Miniguano**

Abstract This paper presents the development of an agent based on an ontology reasoner based on the CommonKADS methodology. An ontology is built for the representation of the shoplifting trick "bulky-baggage" in the Protégé tool and in Java source code on the Jena library. And to build the agent, the Pellet reasoner is used both for Protégé and for its implementation in Java. As a result, the products that were stolen from the supermarket are taken by inference.

74.1 Introduction

Shoplifting or shoplifting refers to the theft of merchandise in stores during working hours. It is one of the most prevalent non-violent crimes in society, and the one that has increased the most in recent years. [1], this being a problem that costs customers and stores entrepreneurs billions of dollars per year since products tend to increase since entrepreneurs have to pay for lost merchandise, and their security and customers are obliged to purchase the products at higher costs. Shoplifters (shoplifters) use various tricks to steal merchandise from stores; one of them is the bulky-baggage. This trick

E. F. Lozada T. (✉)
Carrera de Software, Universidad Regional Autónoma de Los Andes, Ambato, Ecuador
e-mail: ua.edwinlozada@uniandes.edu.ec

W. L. Tenecota
Facultad de Ciencias Humanas de la Educación y Desarrollo Social, Universidad Tecnológica, Indoamérica, Ambato, Ecuador
e-mail: wladimirtenecota@uti.edu.ec

P. S. Pullas Tapia
Facultad de Ciencias Humanas y de la Educación, Universidad Técnica, Ambato, Ecuador
e-mail: paulspullas@uta.edu.ec

L. D. Miniguano Miniguano
Facultad de Ciencias Humanas de la Educación y Desarrollo Social, Instituto Superior
Tecnológico Sucre, Quito Ecuado, Universidad Tecnológica, Indoamérica, Ambato, Ecuador
e-mail: dminiguano@tecnologicosucre.edu.ec

© The Author(s), under exclusive license to Springer Nature Singapore Pte Ltd. 2021   757
S. C. Satapathy et al. (eds.), *Smart Computing Techniques and Applications*,
Smart Innovation, Systems and Technologies 225,
https://doi.org/10.1007/978-981-16-0878-0_74

consists of the camouflage of the merchandise in larger packages. Packages that include gloves (gloves), newspapers (newspapers), packages (packages), and any item that can be carried in the hands and can camouflage the objects to be stolen. The packages are known as booster boxes and can even be gift-wrapped [2].

Video surveillance cameras provide a large amount of knowledge that can be used by knowledge engineering. The monitoring and detection of activities are very important in the security of a business. At present, it is possible for any user to connect surveillance cameras of their business and receive alarms on their mobile or email. In addition to giving them a greater sense of security, the automatic recognition tasks of human activities in video sequences are in high demand both by society and by industry and security agencies [3]. Therefore, to include ontologies in the procedures implemented by the agents, indirectly some reasoner is needed. Its main reason is that ontologies only explicitly describe a domain vocabulary, while reasoners obtain the implicit vocabulary hidden in them. [4, 5]. In this way, reasoners provide agents with a greater flow of information in order to improve the results of the objectives they pursue [6].

CommonKADS It is a methodology applied to the analysis and construction of knowledge-based systems (CBS). It consists of a set of models that allow various perspectives to be expressed from a given point that result in a set of activities for modeling. The models that comprise it are: Organization, Task, Agent, Knowledge, Communication, and Design. The CommonKADS methodology was developed and proposed by a group of researchers from various countries in the European community through a program called ESPRIT. And it is used in IC for the development of Systems Based on Knowledge (SBC) in a similar way to the methods used by Software Engineering. CommonKADS covers the complete software development cycle through a set of interrelated models that capture the main features of a system and its environment. The SBC development process is based on completing a set of templates of the models that are associated with states that represent the milestones in the development of each of them [7, 8]. The domain of this work is shoplifting under the buggy-baggage trick and aims to represent this knowledge in an ontology based on the CommonKADS methodology and to finally create an agent based on an ontology reasoner. The agent is built in Java with the Jena 2.6.4 libraries[1] with a Pellet reasoner 2.3.0[2]. This paper has been organized as follows in Sect. 74.2; the work methodology is presented. In Sect. 74.3, the experimentation is presented, and in Sect. 74.4 the conclusions, product of the experiment.

---

[1]https://jena.apache.org/

[2]https://mail.google.com/mail/u/0/?ui=2&view=bsp&ver=ohhl4rw8mbn4

74.2 Methodology

In accordance with the CommonKADS methodology for knowledge-based systems, the following models were developed.

74.2.1 Organization Model

See Tables 74.1 and 74.2.

74.3 Experimentation

To represent the knowledge of the buggy-baggage trick, an ontology called buggy-baggage was built in Protégé 3.4.5 for graphic display. For which the classes, properties and individuals of this profile are defined with their respective rules. In order to obtain the objects that were stolen, the following classes are defined: "Hidden_Objects" with the rule of obtaining all the products that disappeared in the second phase of a two-time period (on_Period_1, on_Period 2), "Shoplifter" which obtains the IDs of the customers who took one of the hidden objects and "Shoplifter-BuggyBaggage" which acquires the type of bulky package that the shoplifter used to hide the product.

Then Protégé's Pellet reasoner is used to see the inferences that occur in the buggy-baggage ontology, which allows to visualize the individuals that were inferred from the rules that were defined in the ontology.

As a result, we have in the class "Hidden_Object", "Shoplifter" and "Shoplifter-Buggy Baggage" the data of the objects that were hidden by one of the customers who entered the supermarket. Likewise, the Pellet reasoner is applied at the Java code level. As a result, the products that were stolen from the Supermarket by some of the customers are also presented. And what was the bulky bundle that the shoplifter used to hide the products. The following figure shows as results that customer_2 hid the products clothing_2 and shoes_2 in the bulky package buggy_baggage_2.

Table 74.1 Template OM-1: identification in the organization of problems and opportunities oriented to knowledge

Organization model	Problems and opportunities OM-1 worksheet	
Organizational context	Supermarkets are urban commercial establishments that sell consumer goods, including food, clothing, hygiene items, perfumery and cleaning They are stores that can be part of a chain, generally in the form of a franchise and can have more locations in the same city, state or country Their main characteristics are: They have their items on display so they are more prone to theft; they generally offer products at a low price. Also, their main objective is to generate profits so they try to counteract the low profit margin with a high sales volume	
SWOT matrix	*Weaknesses* The security techniques it has are not sufficient to counteract the high degree of theft they face. Clarke (2002) also tells us that people seem to have less inhibitions about shoplifting than they do about people. For they know that there is little probability that they will be discovered; And if they do, in most cases they can give some credible excuse like forgetting to pay *Threats* The articles and products they offer are on display so they are more prone to theft. Additionally, these sales methods are so com- mon today that they provide many opportunities for shoppers to manipulate merchandise and hide it in their clothing or in bags. (Clarke, 2002) *Strengths* They sell perishable and non-perishable consumer products so they are the favorites for customers *Opportunities* A high volume of sales generates greater profits	
Knowledge-intensive problems with the time factor as a determining factor in carrying out the process	One of the techniques that shoplifters (shoplifters) use to steal objects is bulky-baggage (bulky packages) that consists of camouflaging the articles or products in bulky packages So one of the main ones has to do with the fact of recording the specific knowledge of the tasks that shoplifters perform with the bulky-baggage trick. This will avoid the loss of large volumes of products due to theft and will allow the people who comment on it to be identified through the construction of an agent An effective response to the bulky-baggage trick of shoplifting means: fewer repeat offenders, decreased loss, increased sales, and increased profits	
List of problems with their associated priority	Taking into account the factor of time and economic resources as a determining factor, the following problems have been defined	
	Problems	*Priority*
	The constant theft of articles and products	High
	Theft of articles with the bulky-baggage trick	High
	Offer all your products and items on display	Low
	High economic investment for security personnel	High

Table 74.2 Worksheet CM-1: description of the transaction SEND ALERT

Agent: motion sensor	
Model agent	Agent leaf AM-1
Name	Motion sensor
	[Cognex Corporation (2011), mentions the PatMax technology that allows the location of objects to detect their movement]
Location in the organization	The motion sensor will be activated the moment a product begins to move, two sensors will be located in each supermarket stand
Type agent	It is a software agent that performs the process of detecting the movement of a product that will be being mobilized by a person
Restrictions temporary	The motion sensor agent should always be perceiving the movements of all supermarket products according to its stand
Communicates with	It will communicate with the task MPRO (movement of a product)
Knowledge	The sensing agent has knowledge of what can be recognized as a product and a person [López (2011) mentions some of the techniques used to carry out this activity, with the temporal derivative method]
Responsibilities	The motion sensing agent must detect the movements made by the product, which will be made by a person
Agent: ontology	
Model agent	Agent leaf AM-1
Name	Ontology
Location in the organization	The ontology has as entities the actions that a product can carry out, such as moving or hiding and by which person it will be carried out in order to infer if a product disappears in a scene and thereby determine who is the possible person who stole the product
Type agent	It is a software-type agent where all the entities involved in a product movement scene can be established
Restrictions temporary	The ontology must have the possible actions that a product can perform such as moving or hiding
Communicates with	The ontology agent will communicate with the task MPRO: movement of a product and LALR: issue alert
Knowledge	The ontology will have knowledge of the products that exist in the store, with the possible actions (move, hide) that it can take and the people who have the product in a certain period of time of the determined stand
Responsibilities	Track the movement of a product in a scene to be able to launch an alert when that product disappears from the scene
Communications model	Description of transaction worksheet CM-1
Transaction name/identifier	SEND ALERT
Information object	Broadcast an alert signal to the guards
Agents involved	Motion sensor + security personnel
Communication plan	See Fig. 6
Restrictions	In the event of an alert being generated, an alert signal is sent to the human agent
Information exchange specification	This operation is of the report type

```
Type: Hidden_Products
          Individual: clothing_2
          Individual: shoes_2

Type: Shoplifter
          Individual: customer_2

Type: ShoplifterBuggyBaggage
          Individual: buggy_baggage_2
BUILD SUCCESSFUL (total time: 6 seconds)
```

Code 1. The results of the hidden products of the "buggy-baggage" ontology inferred with the reasoning agent Pellet (Static Agent).

A dynamic agent is also built from the dynamic agent (Fig. 18) that allows us to know if a client is a shoplifter or not. Its operation is based on having the customer's data, the bulky package and the type of product that it takes as inputs; to output if this is a shoplifter; if the type of bulky package that I enter is a bulky-baggage for theft; and if the product is hidden or not.

74.4 Conclusions

CommomKADS is a comprehensive methodology, encompassing corporate knowledge management, engineering knowledge and analysis, the path to knowledge-intensive and application systems design. It offers theories, methods and scientific techniques to facilitate the representation of knowledge, we can model and understand the mental reasoning processes used by knowledge experts and reproduce them in a knowledge-based system, with a high focus at the engineering level. Protégé is a framework that has a set of tools for the construction of domain models and applications based on knowledge through ontologies, in a visual way, where the classes are organized in a hierarchy to represent the outstanding concepts of a domain, a set of spaces associated with the classes to describe their properties and relationships, and a set of instances of said individual classes of the concepts that contain specific values for their properties; You can also make inferences with reasoners that are installed as application plugins. It is Java-based, extensible, and provides a plug-and-play environment that makes it a flexible foundation for rapid prototyping and application development. Apache Jena is a Java framework to facilitate the creation of applications for the Semantic Web. It offers a collection of Java tools and libraries for development and linking with data from applications, tools and servers. It also includes an API for managing OWL ontologies and an inference rule based on a reasoning engine with RDF and OWL data sources. The library has Java classes to help developers write code to handle RDF, WL, and SPARQL with W3C recommendations. The inference engine is rule based to carry out reasoning based on OWL ontologies and storage strategies for saving RDF triples in memory or on disk.

Fig. 74.1 Ontology
"buggy-baggage" in Protégé

Fig. 74.2 Ontology
"buggy-baggage" Inferred
with the reasoner Pellet in
Protégé

References

1. Aloysius, J., Arora, A., Venkatesh, V.: Shoplifting in mobile checkout settings: cybercrime in retail stores. Inf. Technol. People (2019)
2. Elnahla, N.; From the Panopticon to Amazon go: an overview of surveillance in retailing. In: Proceedings of the Conference on Historical Analysis and Research in Marketing (2019)
3. L. Belova, «Experience of Artificial Intellience Implementation in Japan,» de *E3S Web of Conferences*, 2020.
4. Greco, L., Ritrovarto, P., Vento, M.: On the use of semantic technologies for video analysis. J. Ambient Intell. Human. Comput. (2020)
5. Gomez, H.F., Martínez-Tomás, R., Arias, S.A., Zamorano, M.R.: Using semantic technologies and the OSU ontology for modelling context and activities in multi-sensory surveillance systems. Int. J. Syst. Sci. **45**(4), 798–809 (2014)
6. Quan, Z., Haarslev, V.: A framework for parallelizing OWL classification in description logic reasoners (2019)
7. Zhigalov, K., Avetisyan, K., Markova, S.: Training video surveillance system for the purpose of object identification with the help of neural networks. In: IOP Conference Series: Materials Science and Engineering, vol. 537, no. 5 (2019)
8. Lulic, L., Stipanovic, B., Smoljic, H.: The role and representation of expert systems in making decisions on grating credit in banks in the republic of Croatia. In: Economic and Social Development: Book of Proceedings, pp. 289–303 (2019)

Chapter 75
Accelerated Sorting of Apples Based on Machine Learning

Kiron Deb, Shobhin Basu, Vamshi Krishna Palakurthy, and Nabarun Bhattacharyya

Abstract Accurate classification and grading are important for the quality assurance of apples since it is necessary to ensure that they retain their freshness and texture until they reach the end user following transportation and storage. We demonstrate that machine learning makes it feasible to separate high-quality apples from the poor-quality ones, which have signs of rotting, insect damage, and other forms of degradation. An alternative to the prevalent manual process of sorting and classification of apples is proposed. We focus on apples grown in orchards in Kashmir and USA. We analyze images taken following harvest using two machine learning models—an open source model as well as a well-known commercial tool. We also propose a hybrid approach for sorting perishable inventory that streamlines the process by using a high-throughput sorter based on image classification. We weed out the apples with surface blemishes in the sorting stage to identify potentially premium grade apples. The apples sorted by our approach are further screened by existing industry-standard procedures such as manual sorting or automated processes like acoustic analysis for classification into different grades. Using this two-step approach, we reduce the time for isolating the higher quality apples compared to a process that relies on a single-step sorting process only. This also achieves greater consistency and economies of scale.

75.1 Introduction

Automated classification and sorting of fruits and vegetables following their harvest have attracted a good amount of research interest in the recent past [1–3]. Most of the

K. Deb (✉)
Calcutta International School, Kolkata, India

S. Basu
Fremont High School, Sunnyvale, CA, USA

V. K. Palakurthy · N. Bhattacharyya
Centre for Development of Advanced Computing, Kolkata, India

© The Author(s), under exclusive license to Springer Nature Singapore Pte Ltd. 2021
S. C. Satapathy et al. (eds.), *Smart Computing Techniques and Applications*,
Smart Innovation, Systems and Technologies 225,
https://doi.org/10.1007/978-981-16-0878-0_75

sorting and grading, however, are still carried out manually. An efficient automated process would lead to greater consistency and economies of scale, and avoid problems with subjectivity. Food and groceries are low-margin, competitive businesses. Any improvement in determining fruit quality will reduce wastage during distribution and storage, which in turn will lead to higher profitability and customer satisfaction. For fruits like apples, which are considered a premium fruit for domestic consumption as well as export, a data-driven, consistent, and reliable sorting method is essential to gain a competitive advantage.

The criteria used to classify fruits vary from fruit to fruit. In the case of apples, both the appearance and the physical firmness are considered the two important criteria. Image analysis could serve as the first layer of quality assurance, which would discard apples that have poor appearance such as spots and blemishes which would turn off a prospective buyer. This is to be followed by a second round of inspection by manually feeling the apples. Alternatively, an automated process using acoustic sensors [4] could be used for this second stage. However, the initial screening should eliminate the need for second level of sorting for a large percentage of apples, if the goal is to select the premium grade of fruits.

For analyzing the apple images using machine learning, tools such as Amazon Sagemaker [5], Microsoft Azure [6] or Google Cloud AutoML Vision [7] could be used. However, for the purpose of this research, we wanted to share code using which other researchers can reproduce similar results with their own datasets. Hence, we decided to use ImageAI, an open source Python-based library for this purpose to train and validate our model. Google AutoML Vision is then used to validate our approach.

75.2 Description of Method

75.2.1 Training Data

We collected an image dataset consisting of 422 images of apples of different varieties, such as golden delicious, fuji, Washington red delicious, honeycrisp, as well as Kashmir variety. Each apple in this dataset has three images taken 120 degrees apart. This results in three images per apple. Many of these images were acquired using a cell phone camera with a resolution of 3024 × 2730. These images were annotated as "healthy" or "damaged." As the attached image shows, the imperfections on the apple's surface are easily visible to the naked eye.

75.2.2 Training Program

We adopted supervised learning as the process for producing a machine learning model. In the process of supervised learning, we feed a computer several data inputs along with the correct corresponding labels, so that it can infer patterns or correlations and generate a model that can predict the label of a previously unseen input with sufficiently high accuracy. In our case, the data inputs are the apple images, and the classification labels are "healthy" and "damaged."

We used ImageAI, a Python-based library that facilitates the training and implementation of computer vision models to help generate the model [8]. ImageAI uses as its backbone other libraries (dependencies): TensorFlow, OpenCV, Keras, and NumPy.

The underlying algorithm in the Python program for detecting the location of the apple in each image is YOLOv3 [9]. The YOLOv3 object detection algorithm is leveraged through transfer learning, wherein the knowledge gained in a pretrained YOLOv3 model is used as a starting point for training a new customized model in order to reduce training time and improve accuracy. By transfer learning through ImageAI, we produced a model that not only identified where each apple is in the image but also gave each detected apple a classification indicating its grade.

75.2.3 Testing Program

We feed the three images representing one apple into the image analysis software. The output is a message stating that the apple is damaged if any of the three images appears to have a spot or a blemish. In addition, the confidence score, which gives as a percentage the likelihood of the prediction being correct, is given as an output for each image. Representative outputs are given in Table 75.1. If the apple passes the image quality test, it proceeds for further testing to determine its firmness. Figure 75.2 shows the block diagram of the entire classification and sorting process.

75.3 Results and Analysis

We analyzed images of 32 apples from orchards of Kashmir obtained by Centre for Development of Advanced Computing (C-DAC), Kolkata using the above machine learning technique. Of these, we found 14 to be with some surface blemish, and the remaining 18 are eligible for further inspection, so that these can be certified as premium grade. Thus, 44% of apples could be eliminated right away by adding this image classification process. We expect this to lead to improved efficiency and sorting time.

Table 75.1 Apple images and corresponding model prediction

Image	Model prediction	Confidence score (%)
	healthy_apple	85.54
	healthy_apple	83.43
	healthy_apple	71.21
	damaged_apple	90.86
	damaged_apple	98.04

Fig. 75.1 Example image from our dataset consisting of 422 images

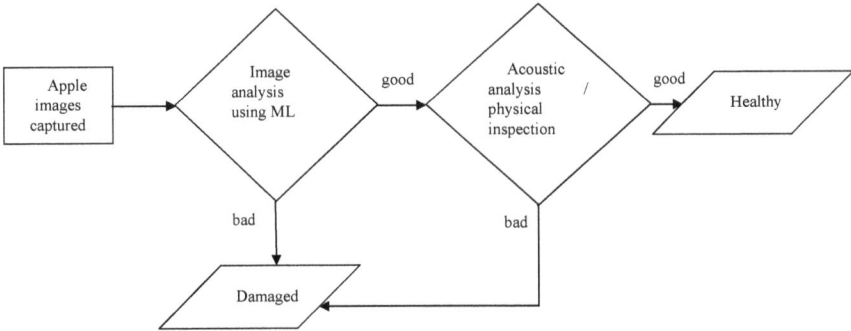

Fig. 75.2 Block diagram of the entire classification and sorting process

75.4 Validation Using Google Cloud AutoML Vision

As a validation of our approach, we used a commercially available machine learning product to predict apple quality. The product chosen for this purpose was Google Cloud AutoML Vision. There was no need to write code in this approach. By uploading the same dataset of apple images in the product's user interface, we were able to train two AutoML models. The first one is an object detection model that is capable of running in the cloud only. It was trained on a subset of 181 images where we drew the bounding box around the apples using the labeling tool in this product. The second one is a classification model that used the full dataset consisting of 422 images and is capable of running on edge devices. It can be exported from the product. Since YOLOv3 also has object detection as its model objective, a batch prediction request was sent for the test dataset of Kashmir apples to the AutoML object detection model hosted in Google Cloud. The predictions were similar to those shown in s with confidence scores typically over 95%, and lowest at 79%. Figure 75.3 shows the AutoML classification model that we trained on 422 images.

75.5 Further Work

Future enhancements could include developing a conveyor belt system to capture the apple images, and an automated analysis program using the machine learning algorithm. The conveyor belt system could be further integrated with a second level of automated system to test the firmness of the apples using non-destructive means such as acoustic sensing. Analysis of sound signals generated when the apple is physically tapped by a small hammer would identify any internal damage masked by a blemish-free surface appearance.

The apples that do not pass the image analysis test can be further screened for the degree of blemish, so that some of these apples can be sold, perhaps at a discount, if the magnitude of the blemish is reasonable. This will require further training of

ICN_V3_422Images   ⋮

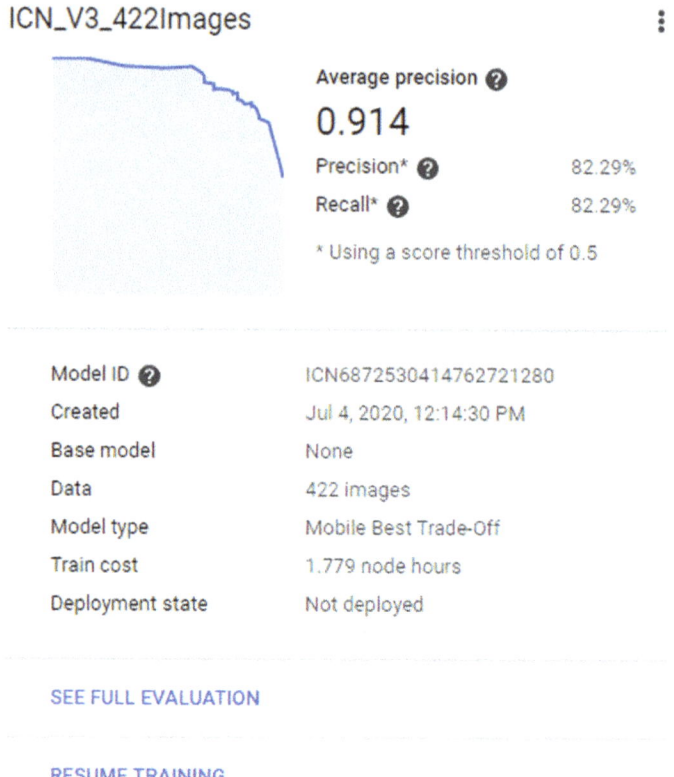

Average precision ❓

0.914

Precision* ❓   82.29%

Recall* ❓   82.29%

* Using a score threshold of 0.5

Model ID ❓	ICN6872530414762721280
Created	Jul 4, 2020, 12:14:30 PM
Base model	None
Data	422 images
Model type	Mobile Best Trade-Off
Train cost	1.779 node hours
Deployment state	Not deployed

SEE FULL EVALUATION

RESUME TRAINING

Fig. 75.3 AutoML Vision screenshot, adapted for style, shows the evaluation metrics from the classification model

the machine learning model since a high level of precision would be needed. For example, even though the third image given in Table 75.1 has a small spot, the model decided that the image indicated a healthy apple. However, the confidence score is relatively low at 71%, suggesting that though the apple crosses the threshold for healthiness, it is on the lower end and can hence be assigned a slightly lower grade. We intend to use the exported version of the AutoML classification edge model to build Web and mobile apps, so that interesting cases like these can be collected for training enhanced models.

75.6 Conclusion

We present a mechanism of non-destructive testing of freshly picked apples using machine learning. The proposed technique would serve as a first-level screen for separating the highest quality of apples based on their images alone. Initial tests on a typical dataset of freshly picked apples indicate that up to 40–45% could be screened out if the goal is to identify superior apples free from any surface blemish. The tools used are open source software available at no cost. This is expected to lead to easier uptake at the beginning as no investment for software is required. Alternatively, commercial products like Google AutoML vision could be used as these maybe faster to implement and would require no coding effort.

Acknowledgements We have taken permission from competent authorities to use the images/data as given in the paper. In case of any dispute in the future, we shall be wholly responsible. Figure 3 is a section of the Google AutoML Vision UI. Google and the Google logo are registered trademarks of Google LLC, used with permission.

References

1. García-Ramos, F.J., Valero, C., Homer, I., Ortiz-Cañavate, J., Ruiz-Altisent, M.: Nondestructive fruit firmness sensors: a review. Span. J. Agric. Res. **3**(1), 61–73 (2005). https://doi.org/10.5424/sjar/2005031-125
2. Nourain, J.: Application of acoustic properties in non-destructive quality evaluation of agricultural products. Int. J. Eng. Technol. **2**(4) (2012). ISSN: 2049-3444
3. Abbott, J.A.: Quality measurement of fruits and vegetable, Elsevier. Postharvest Biol. Technol. **15**, 207–225 (1999). https://doi.org/10.1016/S0925-5214(98)00086-6
4. Abbott, J.A., Massie, D.R., Upchurch, B.L., Hruschka, W.R.: Nondestructive sonic firmness measurement of apples. Trans. Am. Soc. Agric. Eng. **38**(5), 1461–1466 (1995)
5. Amazon Sagemaker Image Classification. https://docs.aws.amazon.com/sagemaker/latest/dg/image-classification.html. Accessed 01 Sept 2020
6. Microsoft Azure Custom Vision. https://docs.microsoft.com/en-us/azure/cognitive-services/custom-vision-service/. Accessed 01 Sept 2020
7. AutoML Vision Beginner's Guide. https://cloud.google.com/vision/automl/docs/beginners-guide. Accessed 01 Sept 2020
8. Moses, O.J.: ImageAI, an open source python library built to empower developers to build applications and systems with self-contained. Comput. Vis. Capabilities (2018)
9. Redmon, J., Farhadi, A.: YOLOv3: an incremental improvement, 8 Apr 2018. arxiv.org/abs/1804.02767

Chapter 76
Execution of Smart Contacts Concurrently Using Fine-Grain Locking and Asynchronous Functions

Hemalatha Eedi and Pattan Asif Khan

Abstract Smart contracts are widely popularized to transfer an asset securely from one device to another. Smart contracts contain the blockchain's business logic that revolutionized the concept of self-enforcing contracts without the need for a middle man. Many use cases thus are being generated day by day. Smart contracts allow us to build decentralized applications through blockchains. The transactions in the smart contract are interrelated; they define the overall flow of the application. To maintain integrity between the transactions, the execution of the smart contracts is sequential and has no conflicts. These contracts' execution is slow and time-consuming. Thus, we cannot leverage the current systems' complete resource utilization. To overcome this, we propose a safer multi-threading model using fine-grain locking and thread executor service mechanism to concurrently execute the smart contract which also provides integrity among the transactions. Our results prove the faster and safer execution of smart contracts.

76.1 Introduction

Blockchain is essentially a decentralized distributed ledger, shared among a set of peers who communicate using a shared protocol. Each blockchain contains data stored in a format called "blocks," and each block contains a list of transactions, timestamps, and additional data required to execute the transactions. All the blocks are cryptographically linked to each other, i.e., each block is linked to the previous block using the previous block hash [1].

A smart contract is a code or a piece of a program, that can be used to transfer entities between two parties with strict monitoring, and these details cannot be manipulated by any means. Further, after the transfer, all the details are logged into the

H. Eedi · P. A. Khan (✉)
JNTUH College of Engineering Hyderabad, Hyderabad, Telangana, India

H. Eedi
e-mail: hemamorarjee@jntuh.ac.in

© The Author(s), under exclusive license to Springer Nature Singapore Pte Ltd. 2021   773
S. C. Satapathy et al. (eds.), *Smart Computing Techniques and Applications*,
Smart Innovation, Systems and Technologies 225,
https://doi.org/10.1007/978-981-16-0878-0_76

blockchain to prevent any entity's illegal authorization. Two entities, "the miner" and "the validator," execute the smart contract. Miners are the nodes that own a block in the blockchain, and they execute all the transactions present in the block sequentially. Validators are the nodes that re-execute the transactions to verify the miner's identity, and on successful verification, they update their ledgers accordingly. Thus, all of the nodes in the chain maintain integrity in the ledger state. To ensure integrity among different nodes or to have the same state of the ledger in all the nodes, all the transactions are executed sequentially [2], resulting in the same state. This mechanism ensures the integrity and consistency of the state of the ledger across the system [3]. However, it fails to leverage the absolute power of multi-core machines today; also, the mechanism works slower to parallel mechanisms used by current database systems.

In this paper, we discuss mechanisms to ensure the state's integrity in all the systems, propose a concurrent system leveraging the power of multi-threading, improve the utilization of current computational systems, and achieve better results than the existing sequential mechanism using locking techniques [4]. Fine-grain locking, in this type of locking mechanism, rather than blocking the whole object and non-related processes by a thread [5] it concentrates on locking what is required, thus ensuring a higher level of concurrency, although with more number of locks, we have an over the head of maintenance, and it ensures security and a higher level of parallelism. Asynchronous functions refer to the routines, where there is no block scoping, i.e., one block of code does not depend on the other function or wait for other functions to execute the code [6]. All the tasks run parallelly in the background without interrupting the flow of each other. Fine-grain systems most implementations are based on the synchronous mechanism to ensure there is no race condition, but there is a possibility of relaxation for few functions to run asynchronously. We leverage this advantage and follow safe thread programming to achieve better results [7].

76.2 Related Work

In [8], the authors propose a method using lockless mechanisms and concurrent graphs; they use a software transactional memory (STM)-based methodology to ensure conflict serializability and integrity of all the transactions. STMs take care of most of the problems caused by concurrent access to data, crash failure, and a mechanism for recovery. In [9], the author proposed the idea of parallel blockchains, i.e., mining and validation on parallel blockchains. There can be discussed about updating, maintaining single history when operations performed on parallel blockchains; he discusses techniques such as transaction sharding, where transactions are distributed to pools of miners, and each pool can access a single shard, and a pool can only do execute transactions present in the shard. A reduced wait between block addition with a minimum number of conflicts is also added. In [10], in this paper, author proposed a locking-based mechanism, where he uses traditional locking methods to lock on a resource, then execute the transaction, and record the locking procedure in the mining phase. Later in the validation phase, based on the record, a Happens Before graph is generated; all the transactions are executed through the graph and using a fork-join approach. The author proposes a parallel mechanism and raises issues about the data conflicts that could occur through the given mechanism. In [11], the author proposed the ACE (asynchronous and concurrent) execution mechanism, and he uses his infrastructure to have separate roles for users, miners, and service providers. Service providers are responsible for ordering and executing transactions. In [12], the author proposed a method where the state managements and smart contract executions are separated from the original blockchain insertions and updations; the author also mentions a separate pipeline model for verification and creation of blocks in parallel. In [13], the author proposed a transaction splitting algorithm for the execution of smart contracts. This helps in resolving the issues of synchronization while executing the smart contracts in parallel.

76.3 Proposed Methodology

The paper proposes the following methodology, to ensure concurrent execution of smart contract transactions. In the first step, a miner does parallel guessing of the nonce, using the brute force approach; nonce can be guessed parallel as the operation is self-isolated, and there is no dependency in guessing the nonce. At the end of this step, a miner would have successfully gained a block in the blockchain. After finding the nonce successfully, the miner who has won the block executes the smart contract; for example, we take a simple transfer and receive a smart contract. Each of the thread acquires a vertex lock on an entity present in the smart contract; here, an entity is referred to as the transaction's participants or an address present in the transaction.

During mining, all the instructions present in the node are executed atomically; atomic instructions are referred to the instructions that are executed fully, and only once, i.e., no two threads can acquire a lock simultaneously on the same instruction or resource. For each acquired lock on a transaction, we create a task and push it into the executor pool, and the threads acquire tasks from the executor pool through the blocking queue and execute the task. When all the transactions are executed, the program stops execution. Using the order of transaction execution, we generate a Happens Before Graph; a Happens Before graph is a directed acyclic graph (DAG) where nodes are the participating entities and edges consisting of relations between them. During validation, for each parent vertex in the Happens Before graph on which a lock is obtained. We try to acquire the child nodes or the entities to which the transaction must happen; we push it into a fork-join pool. All the statements are then executed by threads present in the pool.

The current paper works on permissionless blockchain like ethereum [14], where smart contracts are written in a language called solidity [15]. Solidity is a turing complete language and does not allow random execution of the transactions. Ethereum virtual machine (EVM) on which the code is executed runs on a single core

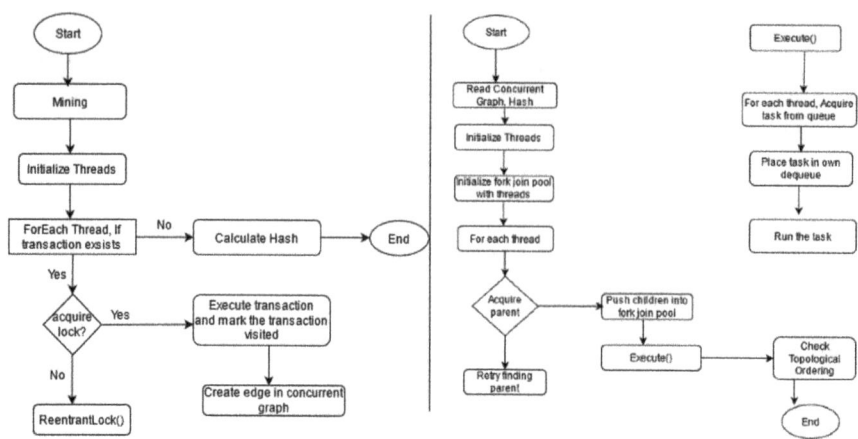

Fig. 76.1 Complete flow of mining and validation

system; this does not allow multi-threading or execution of transactions concurrently or in parallel, reducing the speed of executing the transactions.

To overcome this problem, the paper proposes to use Java, which is currently being used in permission blockchains, to write smart contracts and demonstrate the results using the same. Also, we do a comparative study based on the results obtained from both the systems. JVM has robust support for multi-threading and can handle threads well (Fig. 76.1).

In the mining process, a miner tries to guess the nonce to obtain a block in the chain to execute the smart contracts. Once a miner guesses the nonce correctly, he initializes a set of threads to execute the smart contract. Each thread in the transaction tries to acquire the lock on the objects present in a single transaction; if a lock is acquired, the transaction is executed successfully. Otherwise, a reentrant lock, i.e., the thread, tries to keep trying for a lock. As most transactions are distinct, the execution is done quickly. But in case there is a race condition in the reentrant lock, we can set a timeout for a thread after the thread retries exceed a certain threshold. Once the transaction is executed, the transaction is added as an edge into the concurrent graph. In the end, we generate the final hash of all executed transactions, which can be used for verification by the other peers present in the network.

In the validation process, the validators get a concurrent graph as input; the validator creates a set of threads; each thread finds a parent node and pushes all its children into a fork-join pool [16]. Each thread in the fork-join pool has its independent flow and acquires tasks from the queue; each thread pushes the tasks into its dequeue, using which all the tasks are executed. The threads which do not acquire tasks can barge in other tasks or acquire tasks from other threads. Thus, we must ensure no thread executes a transaction twice. Finally, after all the transactions are executed, a topological sort can be performed to verify whether all transactions are executed successfully, and the final hash generated matches the hash provided by the miner; if it does, block is accepted, else it is rejected.

Algorithm 1 MineParallel(T):

Input: Transactions present in the given smart contract

Output:- A Happens Before Graph(G)

　　　Final Hash of Transactions(H) and order of Execution

```
function MineParallel(T):
  forEach Transaction A in T:
    Obtain the lock on entities present in the Transaction
      if(lock):
        Task.submit() //Submit Task into the thread pool
          ExecuteTransaction(A)
            if(vertex in G):
              addEdge(i,j)//CreateEdge in ConcurrentGraph
            Else:
              addVertex(A) //Createvertex if not found
              addEdge(i,j)
              hash()
              record()
      Else:
        ReentrantLock()
  Based on Locking log generate a Happens Before Graph(G)
  return (G,H)
```

```
Algorithm 2 ExecuteTransaction(A)

Input: Happens Before Graph

Output: Validation result

function ExecuteTransaction(A):

  For Transactions in Blocking queue:

    Obtain a transaction T from the queue

      If Transaction is Atomic:

        Execute The Transaction

          Update the state of atomic variables

            Else:

              Obtain lock on entities in the Transaction

                if(lock):

                      Complete the task execution

                      Release the lock

                else:  tryLock()
```

The concurrent miner uses the following technique to execute the transactions. For each transaction, say A in the transactions (T) list. A thread tries to obtain a lock on the entities present in the transaction. If it obtains a lock successfully on all the entities, it submits the transaction into the thread pool to execute the task. To make sure we do not execute a transaction multiple times, we can store each transaction data inside a ConcurrentHashMap() (H1), containing the list of already executed transactions. ConcurrentHashMap in Java is a thread-safe and also makes sure safe accessibility of data by the threads.

In the execution algorithm, whenever a task is submitted into the queue, the threads responsible for the execution of the task fetch a task. The blocking queue ensures that no two threads access the same resource. Hence, we do not have to worry about conflicts in such cases. For each task, we would need to acquire a lock. To know whether a particular entity or node is locked or it is not, we maintain a ConcurrentHashMap() which would contain the list of entities, and whether the entity is locked or not. After fetching the lock successfully, the task is executed by the thread. In case lock is not obtained, we do not do a thread.sleep() operation. If all threads are busy, then we might leave the core unused. To overcome that we use tryLock(), the concept of tryLock() in Java is that a thread tries continuously to obtain a lock on an object for a certain period and does not leave the core. Even after the mentioned time frame if a lock is not obtained, then the thread goes to sleep.

As in-memory operations are quicker and obtaining locks, checking for locks does not take significant time and is mostly O(1) operations. An equivalent of this, we could use a "mapping()" construct similar to a hash map for locking in smart contracts of ethereum. The two operations, acquiring the locks and executing the transactions, can run in parallel, without conflicts, but, while we use locking techniques, the program flow restricts to either of the blocks, and the other threads do not perform any task, so while we submit tasks to the thread pool, they are all stored in a queue and wait for the execution.

Algorithm 3 Validator-Validator Function

Input - A Happens before Graph(G) and the result of serial transactions

Output - The same Hash generated after the execution of smart contract

```
function Validator(G):

     /*The graph execution starts from a parent nodes,
Parent node is a node with zero incoming edges, is non
dependant on any transactions*/

  For each parent node in G:

   if(getLock(Node parent):

     parent.[adjList].fork()

     parent..[adjList].join()

  Each Thread(Ti) does the following:

     Try to acquire a lock on a parent node,

     Push all the transactions list into Fork-Join pool

     Each thread in fork-join pool

         acquires a set of tasks Using Per-Queue threads

     Each Thread load tasks in local Queue and execute

     In case a thread does not acquire a task,
     It barges-in into local queues of other threads.

     After executing all the transactions call join()

     Return result
```

The advantage of the fork-join procedure in this process is that fork-join pools mostly will not allow tasks that cause conflicts to execute concurrently, and fork-join pools work well in the case of smart contracts because all the statements are mostly dependent and are short, thus providing us with a safer state at the end of the all join operations. In the end, we can compare both states of validator and miner if both reach the same state. Thus, both of them are having a good history, and we can update the history locally.

76.3.1   Implementation in Java

We use the concurrent package present in Java to generate the required number of threads using the FixedThreadPool class to avoid concurrency-based issues. The number of threads running will always be equal to the number of cores present in the system since Java implements OS-level threads that take a core of the system to run the thread. All the threads that do not acquire a task will be waiting until they are assigned with the task. All the threads which have acquired a task work along with the CompletableFuture class to execute the tasks asynchronously, i.e., task execution does not restrict the flow in a single block and gives us the facility to keep the main thread running. CompletableFuture implements a call-back-based mechanism to execute the task and can run in the background without disturbing the overall flow. We use AtomicInteger class to store data as it restricts access to the same variable by threads simultaneously and provides safe access to variables.

In the validation process, we create a fork-join pool, each thread in the pool has its dequeue where it stores the tasks, and to acquire a task, fork() is called. The threads then execute the tasks which are acquired via fork and are present in their dequeue. Here, some of the threads might get fewer tasks and can complete their execution fastly. In this case, the barge-in operation occurs; each thread tries to acquire tasks from the other queue and execute the tasks. This operation can happen because all of them use a similar shared memory. The barge-in operation improves the speed significantly. But, it comes with the overhead to check whether a task is previously executed, this problem of lookup can be solved using the concurrent hash map where we can keep a record of transactions which are already executed. At the end of this phase, we can perform a topological ordering of the graph to check whether all the transactions are executed or not.

76.3.2   Smart Contracts Used for Results

We use the Ballot.sol, SimpleAuction.sol, and TransferContract.sol smart contracts to test our methodology. A ballot is a smart contract that represents an electronic voting system. It consists of functions as a vote() and WinningProposal() functions similar to that of regular voting systems. It also has added functions such as delegate() and RighttoVote(), which can be used to transfer one's vote and to check whether a voter is a valid voter or not.

It is a smart contract where multiple bidders try to bid for an item, and there is a single owner of the contract. The owner initiates the bidding process. Each user who bids has his specific address given along with the bid amount; the bid takes place using the bid() function of the contract; the bidding amount and the bidder's address are stored in a "mapping." Finally, except for the winner, all the amounts of other bidders are reimbursed using the withdraw() function.

Operation	Ballot	Simple auction	Transfer
Mining	1.71x	1.32x	1.59x
Validation	1.83x	1.6x	1.9x

Table 76.1 Average speedup times of the contracts compared to as of serial execution

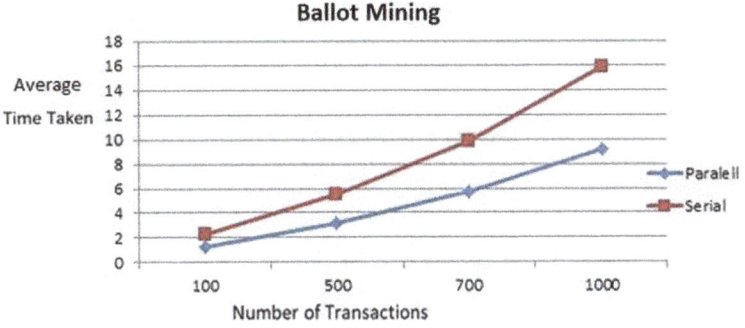

Fig. 76.2 Ballot mining

It is a smart contract used to transfer an amount or an entity from one address to another. It consists of method transfer(), using which we can transfer an asset from one address to another receive (); this method's functionality is to acquire an asset. Check() function to check whether an asset is received or not. A "mapping" consists of details of users between whom the transfer is going to happen. After the execution of this smart contract, the successful transfer of assets is completed.

76.4 Experimental Results

We run our simulations on an Intel(R) Core(TM) i7-8550U CPU @ 1.80 GHz with octave cores and 16 GB RAM. The codes run on the windows 64-bit operating system using Java libraries. We carried out executions using a randomly generated graph with varied nodes. Each smart contract consists of a minimum of 700 transactions. The transactions go up to 2000 per contract. The input taken is in the format that suits the need of the contract. The average speed up the time of the contract execution is given in Table 76.1. When compare to serial execution, our method improves the speed as shown in Figs. 76.2, 76.3, 76.4, 76.5, 76.6, and 76.7.

76.5 Conclusion and Future Scope

Finally, we can conclude that the methods proposed above bring a faster execution of the smart contracts than the serial execution; we can select a set of functions

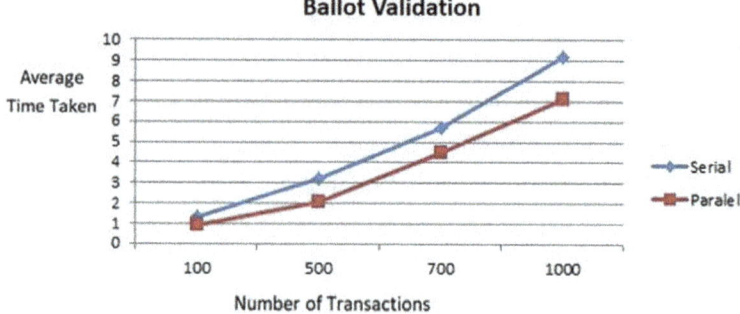

Fig. 76.3 Ballot validation

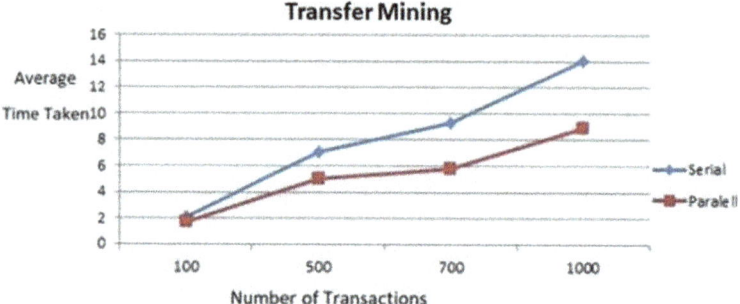

Fig. 76.4 Transfer mining

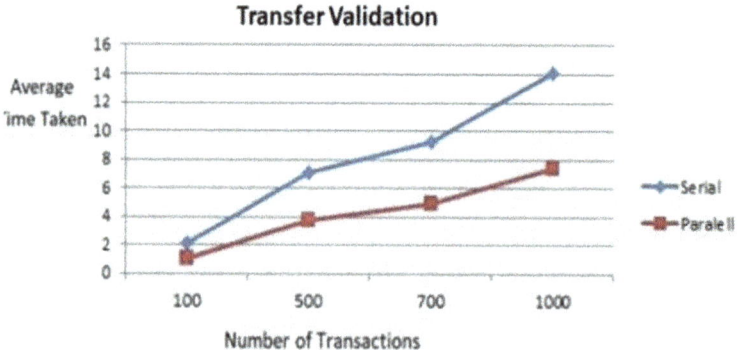

Fig. 76.5 Transfer validation

that can run asynchronously or tasks that can be allowed to run asynchronously
and improve the performance significantly. Although the method comes up with the
advantage of faster execution, maintaining the locks is an overhead and can increase
the program's gas cost. There is also a problem that can optimize the execution further

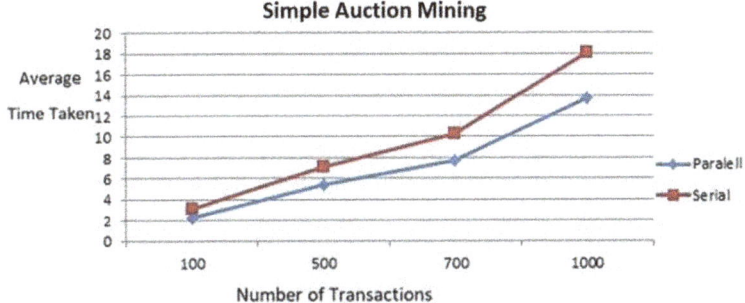

Fig. 76.6   Auction mining

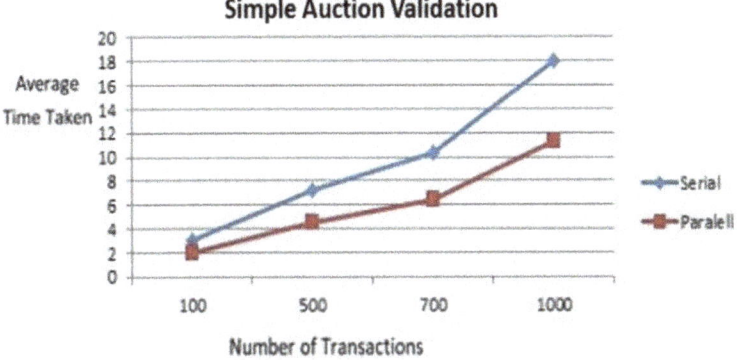

Fig. 76.7   Auction validation

which is to select which function to run asynchronously, and selecting these functions is sometimes hard and can require a complete understanding of the smart contract code. In the future, advanced optimization techniques can improve task assignments to the threads in the fork-join pool.

References

1. Nakamoto, S.: Bitcoin: a peer-to-peer electronic cash system May 2009
2. Herlihy, M.P., Wing, J.M.: Linearizability: a correctness condition for concurrent objects. ACM Trans. Program. Lang. Syst. (TOPLAS) **12**(3), 463–492 (1990)
3. Flanagan, C., Abadi, M.: Types for safe locking. In: European Symposium on Programming, pp. 91–108. Springer, Berlin, Heidelberg (1999)
4. Szabo, N.: Formalizing and securing relationships on public networks. First Monday (1997)
5. Dice, D., Shalev, O., Shavit, N.: Transactional locking II. In: International Symposium on Distributed Computing, pp. 194–208. Springer, Berlin, Heidelberg (2006)
6. Kleiman, S., Shah, D., Smaalders, B.: Programming with threads. Sun Soft Press, Mountain View (1996)

7. Oaks, S., Wong, H.: Java Threads: Understanding and Mastering Concurrent Programming. O'Reilly Media, Inc. (2004)

8. Anjana, P.S., Kumari, S., Peri, S., Rathor, S., Somani, A.: An efficient framework for optimistic concurrent execution of smart contracts. In: 2019 27th Euromicro International Conference on Parallel, Distributed and Network-Based Processing (PDP), pp. 83–92. IEEE (2019)

9. Fitzi, M., Gazi, P., Kiayias, A., Russell, A.: Parallel chains: improving throughput and latency of blockchain protocols via parallel composition. IACR Cryptol. ePrint Arch, 1119. (2018). https://eprint.iacr.org/2018/1119

10. Dickerson, T., Gazzillo, P., Herlihy, M., Koskinen, E.: Adding concurrency to smart contracts. Distrib. Comput. 1–17 (2019). https://doi.org/10.1145/3087801.3087835

11. Wüst, K., Matetic, S., Egli, S., Kostiainen, K., Capkun, S.: ACE: asynchronous and concurrent execution of complex smart contracts. IACR Cryptol. EPrint Arch. **2019**, 835 (2019)

12. Yu, L., Tsai, W.T., Li, G., Yao, Y., Hu, C., Deng, E.: Smart-contract execution with concurrent block building. In: 2017 IEEE Symposium on Service-Oriented System Engineering (SOSE), pp. 160–167. IEEE (2017)

13. Yu, W., Luo, K., Ding, Y., You, G., Hu, K.: A parallel smart contract model. In: Proceedings of the 2018 International Conference on Machine Learning and Machine Intelligence, pp. 72–77 (2018). https://doi.org/10.1145/3278312.3278321

14. Ethereum. https://github.com/ethereum/

15. Solidity https://solidity.readthedocs.io/

16. Naftalin, M.: Mastering Lambdas: Java Programming in a Multicore World. McGraw-Hill Education Group (2014)

Chapter 77
Artificial Intelligence-Based Automatic Evaluation Engine

Sai Gayatri Vadali and Gali Suresh Reddy

Abstract Automatic evaluation of transcripts deals with comparing texts and evaluating them using the best available methods in NLP. The core part of this project is to enable the machine to evaluate descriptive transcripts all by itself. This reduces the human effort correcting huge answers in educational institutions and crowd examination platforms. Finding syntax and semantics out of text has been of great effort since years. As a part of this research, many NLP models like entity recognition, POS tagging, and topic modeling like LSA and LDA have come into existence. We are using a combination of these to get the semantics out of a transcript to eventually evaluate it.

77.1 Introduction

Natural language processing in answer evaluation has been a great research domain since years. Various NLP methods have been proposed in the literature to reduce blurring effects. Many of these methods are based on preprocessing and post-processing data. These techniques help us build machines which can understand natural language. In other words, natural language processing enhances the analysis on text data by removing the noise and the inconsistencies of the text. Natural language processing consists of different techniques, i.e., tokenizing [1], POS tagging, parsing, semantic similarity [2], and language model. NLTK [3] provides tool for above method. Its methods can be broadly classified into two that is preprocessing and modeling stage. POS tagging, parsing, topic modeling [4], WordNet-based similarity [5, 6], and Word2vec [7] are special domain methods.

S. G. Vadali · G. Suresh Reddy (✉)
VNR Vignana Jyothi Institute of Engineering and Technology, Hyderabad, Telangana, India

© The Author(s), under exclusive license to Springer Nature Singapore Pte Ltd. 2021   785
S. C. Satapathy et al. (eds.), *Smart Computing Techniques and Applications*,
Smart Innovation, Systems and Technologies 225,
https://doi.org/10.1007/978-981-16-0878-0_77

77.1.1 Motivation

Auto-evaluation engine helps in evaluating the descriptive transcripts with high speed, accuracy, and transparency. This kind of a tool can help evaluators, teachers, instructors, and interviewers by reducing the time needed for evaluation and by giving them the basic understanding of student's capabilities even before they start evaluating the transcripts.

There has been a lot of research in this area since the beginning of natural language processing and text processing. Using the conventional technologies together with recent advancements in deep learning, a novel approach has been developed to build the engine.

77.1.2 Relevant Work

Nevertheless, automating descriptive answers can be particularly difficult because they cannot be restricted to certain rules. Also, getting semantics out of text can be very difficult. Although there are parsers like that of by Stanford dependency parsing [8] for grammar checking, certain problems like dealing complex sentences were not solved. Getting semantics using Word2vec [9] also did not solve the problem of correcting answers accurately.

As our approach is one such tries evaluate open-ended answers for generic essays, we have come up with a rule-based parser and WordNet-based and topic modeling-based semantic extraction methods.

We use a novel approach of making syntactic check of answer followed by semantic check. These two together give an overall analysis of understanding and expressive ability of the candidate in writing essays.

Following research works have already been made in NLP relevant to grammar check and semantic extraction which have been used in our present work. Ng et al., describe latent Dirichlet allocation (LDA) [10], a generative probabilistic model for collections of discrete data such as text corpora. Wenli [11], the basic principle of classic traditional information retrieval model, is the machine matching of the key word, namely retrieval based on keywords.

Also, VADER sentiment analysis [12] has been used for sentiment analysis at the end of semantic extraction. The idea of text summarization has also been included when converting bigger paragraphs of suggested answers to smaller answers. Character-level text classification has been studied for punctuation analysis. LSTM based question and answering model has been analyzed [13]. Grammar checking tool is based on deterministic grammar generation.

77.2 Background Work

77.2.1 Spell Check

A passage full of spelling mistakes may become very difficult to read and understand. A single letter change in the word may lead to change in the meaning of whole of the sentence, for example, consider the words bear and bare. With a small change in placement of letters, the meaning has changed totally.

77.2.2 Grammar Check

In natural languages like English, just like programming languages, following grammar rules are very essential to build understandable sentences. These rules form Syntax of the language. Syntax can be built on sentence structure like noun, verb phrases (rule-based parsers) or with the dependency of the rest of a sentence on a word (Stanford Parser). Techniques like CFG have been used for this purpose. Context-free grammar known as CFG helps in building a rule-based grammar parser. Context-free grammar-based tree can be parsed either top-down or bottom-up approaches. CFG is based on building tree with non-terminals and terminals. Non-terminals in natural language constitute NP, VP, and N which are finite. But terminals which are nouns like Joy and verbs like come are infinite. This poses challenge to rule-based parsers.

77.2.3 Semantic Extraction

Semantic extraction helps in extracting the meaning and understanding from a given paragraph. Techniques like topic modeling can help in extracting the semantics.

77.2.3.1 Topic Modeling

Topic modeling is the methodology in which topic of the paragraph is extracted. It extracts not only a single topic but then topics which constitute important part of the passage. Topic modeling has become famous by the techniques like LDA, LSI. LDA is based on the most usage of the word, and LSI is based on semantic indexing.

77.2.3.2 Sentiment Analysis

Sentiment analysis can help us understand the opinion of a person with respect to a topic. This helps us understand the personality and emotional ability of the person. Particularly there can be questions like

Do you think growth in investment banking—a sign of growth of nation?

Questions like this give a clear understanding of person's opinions, weightage to his/her points, analytical, and persuasive capabilities, and sentiment analysis helps us just do that.

77.3 Problems Posed by Existing Systems

77.3.1 No Generic Grammar Checker

English is a diversified language with many different grammar rules. The syntactic engines present right now cannot handle all the different types of sentences effectively. For each sentence, a separate grammar must be defined.

77.3.2 Extracting Context or Semantics of the Paragraph

The existing methods cannot correctly find the context of the sentences. Semantics has mostly been restricted to getting meaning from dictionary. The present method is based on vector embeddings and hence not totally reliable. Synonyms cannot be handled properly.

- I went fishing for some sea bass.
- The bass line of the song is too weak.
- Here 'bass' has different meanings in different sentences.

77.3.3 Dealing with Variety of Sentences

Complex sentences have different sentences in same sentence. Present parsers cannot identify the work put in making good complex sentences. Handling complex sentences itself is a difficult thing.

77.3.4 Humor and Sentiment Analysis

Humor and sentiment analysis can be difficult for a machine to understand. Often, understanding semantics is a very difficult task. A fascinating series of experiments sheds light on how humor affects the brain's functioning. When horizontal moving lines are presented to the left visual field (processed by the right hemisphere), and vertical moving lines are presented to the right visual field (processed by the left hemisphere) subjects report seeing either vertical stripes or horizontal stripes, but only very rarely both (a "crosshatch" of intersecting lines). This is called "binocular rivalry"—meaning only one image makes it to our awareness at a time, and the other is not consciously noticed. This is missing with the machines. Hence, humor becomes difficult.

77.3.5 Time and Space Complexity

The present system has difficulties handling time and space complexities. When the sentences are large that is, if the sentence is complex or have many numbers of words, the parser generally takes lot of time to generate parse tree. If that is the case, the system may go into halt state. If it is recursive descent parser, it goes into infinite recursion.

77.4 Proposed Work

The answer written by the candidate undergoes syntax check, semantic check followed by sentiment analysis. Here is a case study of it that occurs.

Let us try to see how the engine works for answer written to the given question "Write about Corruption." The answer written by the candidate in the text box undergoes following steps.

77.4.1 Preprocessing of Text

Text processing occurs at two times—before syntax check and before semantic check. Before syntax check, the passage is divided into sentences. and POS tagging is applied. Before semantic check, techniques like removal of stop words and punctuation are applied.

- Parts of speech (POS) tagging helps us to classify words of sentence into nouns, pronouns, and such English grammar parts of speech. We used Penn Treebank

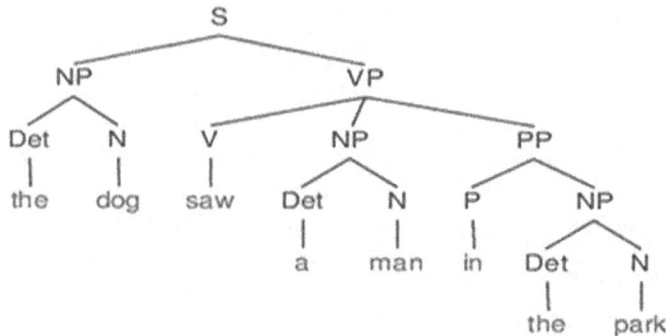

Fig. 77.1 Parse tree generated syntax rules

POS tag set for tagging. It has 36 tags in total. It includes different kinds of nouns, verbs, prepositions, punctuation, adverbs, etc.

- Chunking is the process of joining the words into phrases like noun and verb phrases. Chunking helps to join words and parse them through the rules. It is one of the crucial steps in text preprocessing.

77.4.2 Syntax Analysis

Syntax analysis is a process where the preprocessed sentence is parsed through a parse tree. The parse tree is generated based on syntax rules that have been formulated earlier. If a sentence is completely parsed through the syntax tree, we can say that the sentence is grammatically correct. If not, it has some grammar issues. Following is one example of the same (Fig. 77.1).

77.4.3 Semantic Analysis

Topic modeling is a process of extracting important topics from text. We find the topic from the question as well as from the answer written by the candidate. If they match, then the answer is with in context of the question. LSI and LDA are the topic modeling techniques used in our model.

77.4.4 Sentiment Analysis

The next step is to apply sentiment analysis to find out the sentiment of the text. This acts as an additional feature to find out the characteristics of the text.

77.4.5 Final Score

Final score is obtained by taking scores from all the above steps. For the answer in "corruption," following score is obtained. The scores obtained for the answer are as follows.

Syntactic score—75
Semantic score—41
Sentiment—0.89

We observe the scores are very satisfactory.

77.5 Experimentation and Results

The parser has been used to parse different types of answers written by the candidate as well as questions that can be asked.

77.5.1 Different Types of Answers that Can Be Written by Candidate

Following are the various ways in which a person can write the answer:

1. Grammatically correct answer with good semantic match.
2. Answer written by the candidate is gibberish and irrelevant to question.
3. Grammatically correct sentence but no semantic match.

One example is illustrated below with pictures. Question based on corruption was asked, and suggested answer was given. Then, the answer written by a good candidate is given. Refer Fig. 77.2 for answer by a good candidate and Fig. 77.3 for suggested answer.

A good syntactic and semantic score of 100 and 65 have been given, respectively, by the machine for above answer.

Corruption is misuse of authority to earn monetary gains in wrong means. Corruption is highly prevalent in many developed and developing nations. It can be controlled only with right governance and attitude of people. Politicians, government officials should not yield to corrupted sources of income. Huge penalties should be imposed on people who resort to unethical corrupted practices.

Fig. 77.2 Answer by a good candidate

As we all know that corruption is very bad thing. It inhibits the individual growth as well as society and country growth and development. It is social evil which is playing humans body and mind socially, economically and intellectually. It is continuously making its roots so deeply because of the increasing human greediness towards money, power and position. Corruption is the misuse of authority, public position, natural or public resources, power, etc. by someone to gain his/her personal gratifications. According to the sources, it has been identified that India ranks three in the highly corrupted countries. Corruption is highly spread in the field of civil service, politics, business and other illegal fields. India is a famous country for its democracy but it is corruption which disturbs its democratic system. Politicians are highly responsible for all type of corruption in the country. We chose our leaders by having lots of expectations to them to lead our country in the right direction. In the starting they make us lots of promises however, just after the voting they forget all that and involve in corruption. We are sure that our India would be corruption free a day when our political leaders would be free of greediness and use their power, money, status and position in right direction to lead the country, not their own luxury and personal wishes.

We should select very honest and trustworthy leaders to lead our India just like our earlier Indian leaders such as Lal Bahadur Shastri, Sardar Vallabh Bhai Patel, etc. Only such political leaders can reduce and finally end the corruption from India. Youths of the country should also need to be aware of all the reasons of corruption and get together to solve it in group. Increasing level of the corruption needs to take some heavy steps to get control over

Fig. 77.3 Suggested answer

77.6   Conclusions and Future Work

Writing has been the source of our knowledge as well as a powerful way of expressing our ideas in a way which stays forever. Also, a beautifully written essay can depict the personality as well as creative power of the individual. Auto-evaluation engine with these features of grammar checker, semantic analysis, and sentiment extraction can be a very good tool to assess as well as preserve the written wealth of the people. The present tool can be extended into an advanced version by using deep learning and generative language modeling for semantic analysis. Probabilistic methods like PCFG and generative methods like language modeling can help in revolutionizing syntactic checking of the text. Generative methods enhance semantic check to a great extent.

References

1. Vijayarani, S.: Adv. Comput. Intell. Int. J. (ACII) **3**(1) (2016)
2. Meng, L., Huang, R., Gu, J.: A review of semantic similarity measures in wordnet. Int. J. Hybrid Inf. Technol. **6**(1) (2013)
3. Manning, C.D., Surdeanu, M., Bauer, J.: The stanford core NLP natural language processing toolkit. In: Proceedings of 52nd Annual Meeting of the Association for Computational Linguistics: System Demonstrations, pp. 55–60. Baltimore, Maryland USA, June 23–24 (2014)
4. Jelodhar, H., Wagi, Y., Yuan, C.: Latent Dirichlet allocation (lda) and topic modelling, applications, survey. Nanjing University of Science and Technology (2018)
5. Pedersen, T.: WordNet: similarity—measuring the relatedness of concepts. In: Proceedings of the 19th national conference on Artificial intelligence, pp. 1024–1025 (2004)
6. Miller, G.A.: WordNet: a lexical database for English. Commun. ACM **38**(11) (1995)
7. Zhang, Y., Ma, L.: UsingWord2Vec to process big text data. In: Big Data (Big Data), 2015 IEEE International Conference on 29 Oct–1 Nov (2015)
8. Chen, D.: Christopher D Manning. Stanford University, A fast and accurate Dependency Parser using neural networks (2014)
9. Víta, V.K.: Word2vecbased system for recognizing partial textual entailment. In: Computer Science and Information Systems (FedCSIS), 2016 Federated Conference on 14 Sept (2016)
10. Blei, D.M., Andrew, Jordan, M.I.: Latent Dirichlet allocation. University of California, Berkeley. Berkeley, C A 94720. J. Mach. Learn. Res. **3**, 993–1022 (2003)
11. Deshmukh, A., Hegde, G.: A literature survey on latent semantic indexing. Int. J. Eng. Invent. **1**(4), 01–05 (2012). ISSN: 2278-461
12. Hutto, C.J., Gilbert, E.: VADER: A parsimonious rule-based model for sentiment analysis of social media text. In: Proceedings of the 8th International Conference on Weblogs and Social Media, ICWSM (2014)
13. Wang, D., Nyberg, E.: A long short-term memory model for answer sentence selection in question answering. In: 7th International Joint Conference on Natural Language Processing, Association for Computational Linguistics, pp. 707–712, July (1995)

Chapter 78
SmartRPL: Secure RPL Routing Protocol for Internet of Things Using Smart Contracts

P. Subhash, P. Navya Sree, and M. Pratyusha

Abstract Numerous nodes are connected to Internet of Things (IoT), nodes need to communicate and synchronize with each other, and these nodes are resource constrained with limited processing power, memory, energy with unstable links, and low bandwidth. There is a standard routing protocol for IoT-LLN (low-power and Lossy networks) known as RPL (routing protocol for low-power and Lossy networks); this protocol is used for communication between low power devices and exchange data. For secure communication among IoT nodes, the routing protocols require central authority which results in various security attacks and can severely degrades the network performance. In the proposed system, the blockchain-based framework with smart contracts is used to offer secure route establishment between IoT nodes in the network and protects the control packets information from being modified by an external attacker. Thus, it can prevent the formation of variety of attacks and to ensure secure route formation between nodes in the network.

78.1 Introduction

Routing protocol defines how nodes communicate and with one another and distribute data which allows them to select the route between the nodes [1]. Routing in IoT nodes is performed using RPL which is a proactive routing protocol for LLN such as IoT; these devices in the network are arranged in the form of directed acyclic graphs (DAGs) which has multiple destination-oriented directed acyclic graphs (DODAGs); each DODAG is in the tree-based structure which is created by the root node; these nodes are arranged such that no loops are formed in the network. RPL protocol is effective and secure routing protocol. These messages are carried via ICMPV6 messages and use the DODAG information object (DIO), DODAG information solicitation (DIS), and destination advertisement object (DAO) control messages used in RPL protocol [2]. In addition to this, offering security during the path establishment

P. Subhash (✉) · P. Navya Sree · M. Pratyusha
VNR Vignana Jyothi Institute of Engineering and Technology, Hyderabad, India

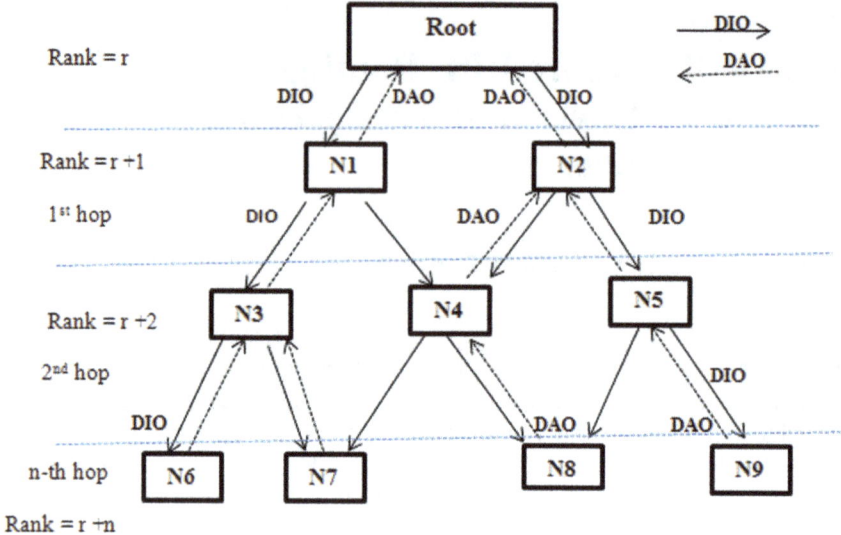

Fig. 78.1 Control messages in RPL protocol

is also utmost important to prevent most of the hazardous vulnerabilities and to safe-guard the network. This in turn gives the significant improvements in the network performance.

In this paper, we introduce a decentralized blockchain-based RPL routing using smart contracts (smartRPL) which allows IoT network nodes from distinct vendors to trust one another and cooperate during the data communication.

Figure 78.1 shows the broadcast of RPL protocol control messages during the path establishment

78.1.1 Routing Protocol for Low Power and Lossy Networks

The process starts with building the DODAG root. The root acts as a gateway which sends the DIO messages to its neighboring nodes which has information like rank, RPL instance ID, version number, and DODAG ID; these DIO message are multi-casted downwards depending on the received messages; the neighboring nodes decide if to join the root node or not. If the nodes want to join the DODAG root, then it has upward path, DAO messages are sent toward the root node, and at this point, the nodes compute their rank and select the parent based on the rank. The node selects the parent which has the lowest rank. Any new node can join the DODAG by sending the DIS message.

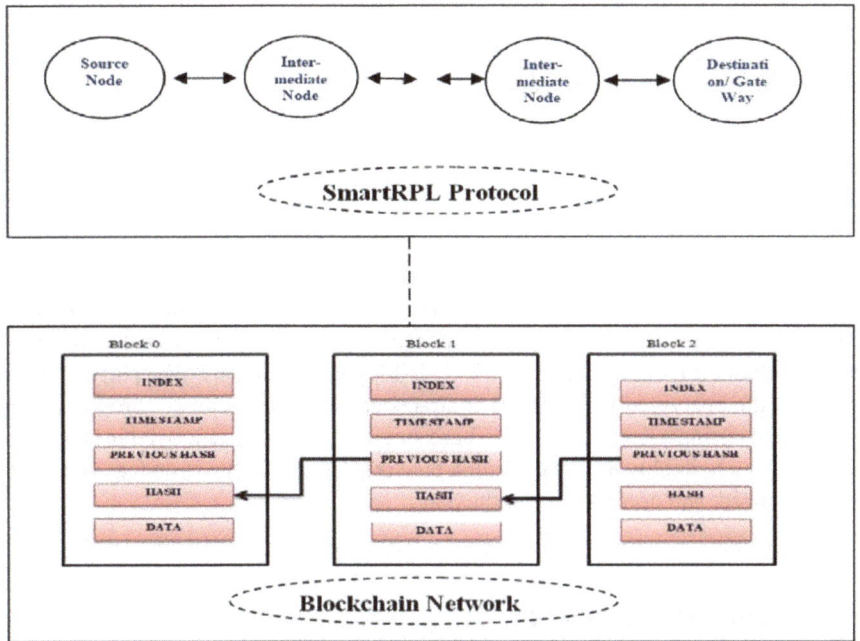

Fig. 78.2 Architecture of the proposed system

78.1.2   Blockchain Technology

Blockchain [3] was invented by Satoshi Nakamoto. It is chain of blocks which are interlinked to one another with the help of previous hash in the backward direction, every block contains details of the transactions, and this data which is stored in the blockchain network cannot be altered by a single node because of its distributed nature. Each block consists of index, timestamp, previous hash, hash, and data; there is unique ID for each and every block, and these are present in the sequential pattern. Figure 78.2 gives the structure of blockchain network with the number of elements that are present in each block in the network and shows the link among blocks.

78.1.3   Ethereum

Ethereum was discovered by Buterin [4]; it is an open source and decentralized platform that executes the programs called smart contracts. Ethereum is a platform specifically designed to build decentralized applications which provides transparency, availability, and auditability and removes intermediaries and allows users to interact with each other directly through a global network. Ether is the main cryptocurrency which is used to pay transaction fee, and the average blocktime is 10–15 s. Ethereum can store transaction and documents.

78.1.4 Smart Contract

A smart contract is an agreement between two parties inside the smart contracts; there are set of conditions. All the contracts are based on terms, and conditions are recorded explicitly once the data written in the contract cannot be changed, updated, or deleted; they maintain the local copy in each block; it does not require any third party, and the data is in the encrypted form. Smart contracts are written in strength programming language [5] using a Remix IDE [6]. These smart contracts are deployed in ethereum platform, and to deploy the smart contracts, it requires the transaction fee. These are secure, trusted, and correct to store the data.

78.2 Problem Statement

IoT-LLNs are connected to various nodes in the network and exchange the data. These nodes should support parametric- constrained routing where nodes should select the best path to pass the data within the network. RPL routing protocol is used for LLNs, and it is vulnerable to different attacks [7] because of its resource-restraint nature to prevent these attacks; there are cryptographic algorithms, but these are managed using central authority, and these are difficult and expensive to implement. So, to support the trust between the nodes without any central authority, we propose the decentralized approach to find the best route to pass the data from source to destination, and it ends the single-point failure.

1. We propose a decentralized approach to set up a secure path from source to destination.
2. Performance metrics are considered to evaluate the performance.
3. Smart contracts are used to secure the data, and these are deployed in ethereum platform.

78.3 Literature Review

Few works that have been found in the literature focuses RPL security using various approaches. However, to the best of our knowledge, none of the work has been found focusing RPL routing security using blockchain-based smart contracts mechanism.

Dvir et al. proposed (VeRA) to protect the RPL protocol from the rank attack and version number attack; it prevents the attack node from changing version number and rank; it is implemented using hash chain method [8], but this approach is vulnerable to rank attack.

Glissa et al. [9] proposed an approach to prevent rank modifications using threshold and hash chain authentication mechanism; it limits the rate of increase or decrease of rank values; it uses one-way hashed function such as secure hash algorithm 1 (SHA1). However, this approach causes the added overheads in the start to use hashing techniques.

Perrey et al. [10] proposed an approach TRAIL to solve the topology inconsistencies and discovers the fake nodes in the network. However, there are message overheads and increase the network size.

Ramezan et al. proposed a protocol [11] to prevent black hole and gray hole attacks using blockchain technology; it is used to find the trusted path between the source to destination. However, there are security issues in this model.

Conti et al. proposed Split [12] to secure the network from rank and Sybil attacks; a lightweight remote confirmation technique is used to prevent the network, but performance metrics are not considered in this approach.

Zhong et al. proposed an approach named Sprite [13] which is for mobile ad-hoc network. This approach needs the central authority to send the proof of message that data packet is received; it is vulnerable if the malicious node mounts the data to get there rewards.

78.4   Proposed System

In the proposed model, each IoT node creates the smart contracts to store the data; it has various parameters to find the trusted route and pass the data from source to destination. It uses RPL control messages to setup the path. It has different states to verify the route and pass the data. Figure 78.2 shows the structure of the proposed method which has a communication link (dotted lines) between the IoT network running RPL protocol to check the smart contracts present in the blockchain network for the correctness of the fields of control packets received via regular operations of routing protocol. If the fields of the control packets received by neighboring nodes are correct after checking with the information present in the smart contracts, then the intermediate node offers route via node through which it receives unmodified control packets. This approach continues till the secure path is offered from source to destination.

1. **Route Request**: To establish the secure route to the destination, we have to find the best path from source node to destination, and the source IoT node sends the DIO message which contains the information of contract address of the source node, rank, hop count, tokens, route request bond, and route request validity. As a bond, source IoT node transfers some of its tokens to the intermediate nodes to find the optimal route to destination and earn tokens. Route request bond also specifies the time until which the request is valid.

Table 78.1 Protocol parameters

S. no.	Parameter	Description
1	Contract address	Address of node
2	Rank	It calculates the individual node position from the root node
3	Hop count	It is count of hops from source node
4	State	Indicates the state
5	Route request bond	No. of tokens paid by source node to intermediate node as body
6	Route request expiry	It is route requested expiry time
7	Timestamp	Time at which the block is created in blockchain

2. **Route Offered**: The neighboring nodes send the valid route to the destination by DAO message and sends its information to its parent node maximum of three nodes which are stored in the smart contract. If the neighboring nodes are unaware of the routes and wants to join the network, it can pass the DIS message.
3. **Route Accepted**: The source IoT nodes verify the conditions to accept the route which is offered it selects based on its internal policies such as rank, hop count, and low-cost route from the offered routes to increase the throughput and security.
4. **Route Passed**: It selects the valid route based on the conditions, and the state of the contract is changed to data passed; if any node is unable to relay the packets, then the node's address is listed as blacklist address.
5. **Stop**: Anytime the source node can stop the routing process (Table 78.1).

78.5 Performance Evaluation

To evaluate smartRPL method, simulation study is performed using Cooja simulator with Contiki OS 2.6 [14] setup with 64 nodes in a grid structure and IEEE (Institute of Electronics and Electrical Engineers) 802.15.4 MAC (media access control). This implements the RPL routing protocol with an attack scenario and SRPL operations. Since our proposed work is blockchain based and do not allow the attackers to change the contents present in the RPL control packets, it ensures secure and reliable path establishment. This allow us to analyze the performance of the proposed method in comparison with the other two mentioned approaches. During the performance analysis study, it has been clearly seen that the proposed method has shown the significant improvements in terms of the key metrics such as packet delivery ratio (PDR) and throughput. The proposed model (smartRPL) has been evaluated and compared with RPL protocol with rank modification attack and SRPL [9] protocol. The results are significant and outperforms the existing two methodologies.

78.5.1 Packet Delivery Ratio (PDR)

Ensuring the secure path establishment among nodes in the network guarantee to offer higher packet delivery ratio. This can be defined as:

$$\text{Packet Delivery Ratio} = \frac{\text{No. of packets received}}{\text{No. of packets sent}}$$

As the number of attacker nodes increases, the proposed model shows higher packet delivery rate when it is compared with the RPL with rank attack and SRPL methodologies.

78.5.2 Throughput

Since the proposed mechanism offers blockchain-based secure routing approach, it do not allow the attacker nodes to change the control packets information. Thus, it can offer higher overall network throughput performance. Throughput can be defined as an average number of data packets received by destination per unit amount of time slot.

$$\text{Throughput} = \frac{\text{Data packets successfully received by destination}}{\text{Time taken}}$$

Simulation results show that the proposed model is highly effective in protecting the network against modification attacks as the blockchain has a public ledger system and undeniable properties. This reports better performance outcomes as shown in Figs. 78.3, 78.4, and 78.5. These simulations are performed in Contiki operating system [14].

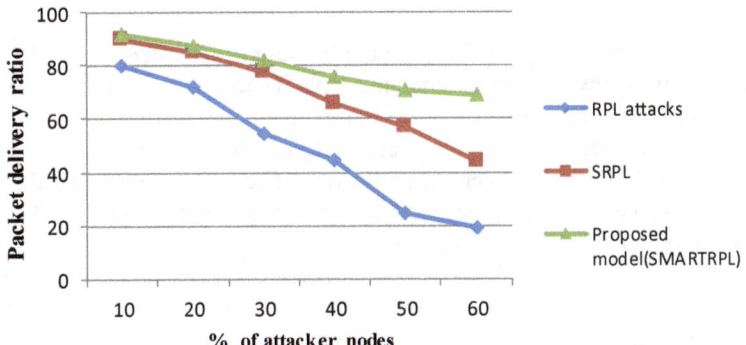

Fig. 78.3 Packet delivery ratio versus no. of attacker nodes

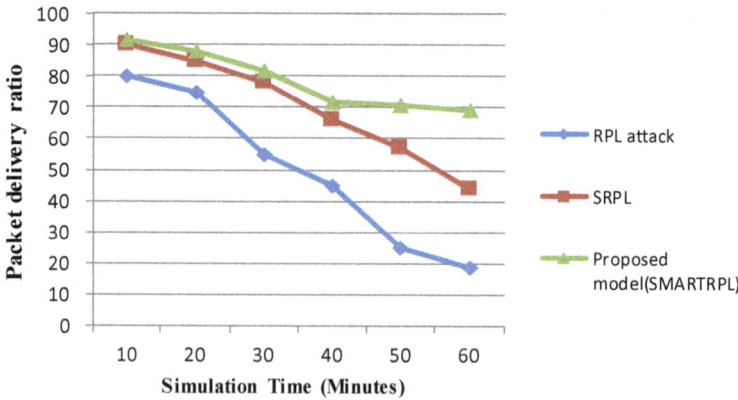

Fig. 78.4 Packet delivery ratio versus simulation time

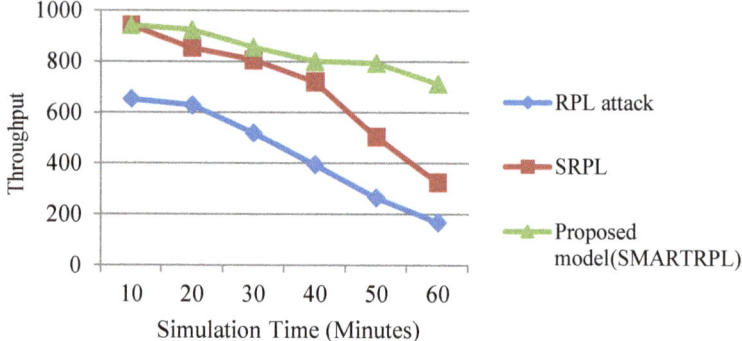

Fig. 78.5 Packet delivery ratio versus simulation time

78.6 Conclusion

RPL is the most promising routing protocol because of its efficiency for IoT-LLN devices describes how the nodes communicate with each other from source to desti-nation. In the proposed system, a novel blockchain-based smart contracts method is used which is a decentralized, distributed ledger, and it creates a trustworthy IoT ecosystem which is capable to eradicate the risk of central authority and the nodes in the network; it establishes a secure route to pass the data from source IoT node to the destination node. This approach does not require a node to be equipped with a specialized hardware and avoid the RPL control packets from being changed by the attacker nodes during transmission without using more complex cryptographic algo-rithms. It uses smart contracts between node to node and checks for the correctness of control packets fields to set up a valid secure route between IoT nodes to gateway or root and vice versa to pass the data securely during the data transmission phase in the network.

References

1. Airehrour, D., Gutierrez, J., Ray, S.K.: Secure routing for internet of things: a survey. J. Netw. Comput. Appl. **66**, 198–213 (2016)
2. Winter, T., Thubert, P., Brandt, A., Hui, J., Kelsey, R., Levis, P., Pister, K., Struik, R., Vasseur, J., Alexander, R.: RPL: IPv6 routing protocol for low-power and lossy networks (rfc 6550) (2012)
3. Zheng, Z., et al.: An overview of blockchain technology: architecture, consensus, and future trends. In: 2017 IEEE International Congress on Big Data (Big Data congress). IEEE (2017)
4. Buterin, V.: Ethereum: A Next-generation Smart Contract and Decentralized Application Platform (2014). https://github.com/ethereum/wiki/wiki/%5BEnglish%5D-White-Paper
5. https://solidity.readthedocs.io/en/v0.4.24/introduction-to-smart-contracts.html
6. https://remix.ethereum.org/
7. Mayzaud, A., Badonnel, R., Chrisment, I.: A Taxonomy of Attacks in RPL-Based Internet of Things (2016)
8. Dvir A., Buttyan, L.: VeRA-version number and rank authentication in rpl. In: 2011 IEEE Eighth International Conference on Mobile Ad-Hoc and Sensor Systems. IEEE (2011)
9. Glissa, G., Rachedi, A., Meddeb, A.: A secure routing protocol based on RPL for internet of things. In: 2016 IEEE Global Communications Conference (GLOBECOM). IEEE (2016)
10. Perrey, H., et al.: TRAIL: topology authentication in RPL(2013). arXiv:1312.0984
11. Ramezan, G., Leung, C.: A blockchain-based contractual routing protocol for the internet of things using smart contracts. Wirel. Commun. Mob. Comput. (2018)
12. Conti, M. et al.: SPLIT: a secure and scalable RPL routing protocol for internet of things. In: 2018 14th International Conference on Wireless and Mobile Computing, Networking and Communications (WiMob). IEEE (2018)
13. Zhong, S., Chen, J., Yang, Y.R.: Sprite: a simple, cheat-proof, credit-based system for mobile ad-hoc networks. In: IEEE INFOCOM 2003. Twenty-Second Annual Joint Conference of the IEEE Computer and Communications Societies (IEEE Cat. No. 03CH37428), vol. 3. IEEE (2003)
14. Get Started with Contiki. https://www.contiki-os.org/start.html

Chapter 79
Customized Music Classification and Recommendation System Based on Classifiers of Neural Networks and Sensor Embedded on Smart Devices

V. Roopa, A. ChristyJeba Malar, R. Rekanivetha, R. Thanga Pradeep Kumar, R. Sarveshwaran, and A. Prithiksha Parameshwari

Abstract Advancements in the process of embedding sensors in vehicles are helpful in recognizing the emotions of the person for annotating large music data set. The existing automated music playlist consists of collaborative filtering or content-based approaches in providing a fitting extension to a given playlist. This paper proposes a contextual-based music recommendation system that recommends songs based on the learning from user emotion. The user emotion is classified by the sensor embedded in the vehicle or the wearable computing device which is attached to the photo plethysmography (PPG) physiological sensors. This will recommend the songs to the user based on the context from the classified dataset. Prior to recommendation, songs are classified into genres considering the similarity in temporal structure and emotion classification is based on sensor values. This paper further proposes genre classification process with two steps comprising of feature extraction and music genre classification. The emotion classification from sensor values are may be based on decision tree, and Support vector and KNN algorithms of machine learning. Music

V. Roopa · A. ChristyJeba Malar (✉) · R. Rekanivetha · R. Thanga Pradeep Kumar · R. Sarveshwaran · A. Prithiksha Parameshwari
Department of Information Technology, Sri Krishna College of Technology, Coimbatore, Tamilnadu, India
e-mail: a.christyjebamalar@skct.edu.in

V. Roopa
e-mail: v.roopa@skct.edu.in

R. Rekanivetha
e-mail: rekharajkumar2000@gmail.com

R. Thanga Pradeep Kumar
e-mail: thangapradeep007@gmail.com

R. Sarveshwaran
e-mail: sarvesh.rajendran@gmail.com

A. Prithiksha Parameshwari
e-mail: prithiksha99@gmail.com

© The Author(s), under exclusive license to Springer Nature Singapore Pte Ltd. 2021   805
S. C. Satapathy et al. (eds.), *Smart Computing Techniques and Applications*,
Smart Innovation, Systems and Technologies 225,
https://doi.org/10.1007/978-981-16-0878-0_79

genre classification and recommendation with emotional and stress management
provides a relaxation for the users while driving.

79.1 Introduction

Any song lasts for minutes, which is longer in duration. So, audio signal which is
given as input for each and every song contains millions of samples with a sample
frequency which is higher of about 10 khz. We prefer splitting a song into many
frames so as to enable subsequent timbre extraction at frame level to signal analysis.
This may lead to an efficient extraction process since we may require the application
of spectral analysis only to signals of short time. Next, we have to choose an effective method for modeling the timbre features with the duration of 10–100 ms. The
statistical properties of non-statistical series depend on time. Therefore on applying
feature analysis at different frame level with the same audio signal. Thus timbre
information is captured in a reliable manner. The wearable computing devices have
sensors to measure heart rate, pulse rate and also provide the basis for physiological
interpretations.

A feature content which is rhythmic includes the information about the exploration of beat tracking and tempo. Therefore beat spectrum is used for representation of rhythm. Frequency information is dealt by pitch content features obtained by
using techniques of pitch detection. Nowadays deep learning techniques are explored
regarding music genre classification. Thus the system provides the set of features for
music which is direct and enables modeling. It explores the usage of Support vector
in classification of music genre through K-Nearest Neighbour and Gaussian models.
This music recommendation model extracts the correlation between the user emotion
and the music genre.

Smart cars can have automatic play system enables such feature by extensively
reading the sensor inputs and play the songs automatically with these predicted
values. This customizes the genre of music being played and oozes out the imbalance
which is the need of the hour. The proposed Machine learning algorithms emphasizes
greater accuracy. The genre of music played in the proposed system used Convolution
Neural Network (CNN) with classifiers as hidden convolution layers.

79.2 Literature Survey

Any song lasts for minutes, which is longer in duration. So, audio signal which is
given as input for each and every song contains millions of samples with a sample
frequency which is higher of about 10 khz. We prefer splitting a song into many frames
so as to enable subsequent timbre extraction at frame level to signal analysis. This may
lead to an efficient extraction process since we may require the application of spectral
analysis only to signals of short time. Next we have to choose an effective method

for modeling the timbre features with the duration of 10–100 ms. The statistical properties of non-statistical series depends on time. Therefore on applying feature analysis at different frame level, we assume audio signals to be stationary. Thus timbre information is captured in a reliable manner. A feature content which is rhythmic includes the information about the exploration of beat tracking and tempo. Therefore beat spectrum is used for representation of rhythm. Frequency information is dealt by pitch content features obtained by using techniques of pitch detection. Not much work is explored regarding music genre classification. Thus the system provides the set of features for music which is direct and enables modeling. It explores their usage for classification of music genre through K-Nearest Neighbour and Gaussian models.

Fluctuation patterns from "focus" and "gravity". The combination of similarity and fluctuation patterns is mainly done for the addition of complementary information. Since it is sensitive to the effects that were produced, there is no usage of descriptors for overall loudness. There is no usage of spectral descriptors too as they are covered by the measure of spectral similarity.

A small audio segment is mainly used as it contains the information for characterizing the contents of a song because repetitions may be observed inherently in a musical structure. This idea is preferred as the computation required can be reduced. Proposed algorithms mostly use a small segment from audio for every title. To avoid introductions, 30s segment starts 30s after the piece is started.

Music can be categorized into genres in different ways. Music genre classification is useful to organize albums and songs which can be able to place music recommendations to the customers while driving or in downhearted. There are many genres of music classifications like pop, classical, and folk are available. There are different tones available for each genre. Around world 40 million songs are available in the digital library. It is time-consuming job to classify the music manually by listening the music. The person must have patience to listen the music and knowledge about genres. Some companies use music classification to place recommendation systems to the users (ex. Spotify, Soundcloud, Apple Music). Song recommendation system estimates the correlation between the users past listening history and songs.

Biofeedback-based music recommendation system [1] which consists of sensor-equipped headphones Septimu that records the heartbeat rate at the server. The recorded contextual information of the person was analyzed to suggest the dynamic music to maintain the target heart rate. A drowsiness detection system was used which records electroencephalogram (EEG) reading recorded by a sensor [2]. Artificial Neural Network (ANN), Support vector machine (SVM), and k-nearest neighbor (kNN) are integrated into the classification system to classify the drowsy state of the driver. Genetic algorithms also used to adjust weights in the classifiers [3]. Brainwaves classification and music recommendation are highly proved by experiments. Abdul et al. [4] generate implicit user rating for the music by calculating the correlation between user psychological data and music feature. Deep convolutional neural networks extract user behavior whereas weighted feature extraction (WFE) generates implicit user rating for music to get the correlation.

Different recommendation algorithms are evaluated using different types of measurement [5] which did not care about user emotion. An automatic stress reliving system was designed for music listeners [6]. It has a wireless module with finger type sensor, which converts heartbeat rate to stress index level. Liu et al. [7] recommends a music playlist for the person which dynamically changes based on user heartbeat measurement. Yoon et al. [8] have developed a music recommendation system by feature selection, contextual information and past playlist history. Rosa et al. [9] implemented a music recommendation framework comprises of user profile with music preferences. The user's sentiment is analyzed by extracting the user's social media post. The framework recommends songs based on the metric.

Music recommendation system [10] involves the tasks of genre classification, song identification, feature extraction, and chord recognition. Shazam Entertainment is a music recommendation service available over internet, identifying artist and entire song receiving an audio sample (www.shazam.com). By Example (QBE) is a music search service available in Shazam. Many classification types are possible in music genre listed as Music Classification, Artist Identification, Mood Classification, and Music Annotation. Classifier learning is important to map the feature vector to the output label. Many audio features are considered in learning. Short-term, long-term features, semantic features are the considered categories in audio. Spectral Bandwidth, Amplitude Spectrum Envelop, Spectral Flux, Spectral centroid are some of the timbre features considered in music classification. Different timbres are possible by different sound sources of music. Markov et al. [11] have used Gaussian Process in music classification and emotion estimation which performed consistently better than support vector machine. Some of the features considered in music signal processing are,

Mel Frequency Cepstral Coefficients (MFCCs): Describe the shape of the spectral envelop used in speech recognition as well as audio similarity measures.

Line spectral Pairs (LSP): It is used to represent linear prediction coefficient to represent spectral envelop of a song in compressed form. It is the main parameter used in speech coding or synthesis.

Timbre (TMBR): Ability to distinguish different instruments. It is a group of four scalar features which include spectral centroid, spectral rolloff, spectral flux, and zero crossings.

Spectral Flatness Measure (SFM): SFM measures the differentiation between tone line and noise-like sounds.

Chromagram (CHR): Related to pitch is a complement of tone height. In music era, Two notes are separated one or more octaves and produce the same chorma.

79.3 Proposed System

The system classifies the genre of the songs in the repository of a person's music application through a predefined dataset and from that Convolution Neural Networks

(CNN) that helps in the classification of the music through the music features like MFCC, Mel's spectrum. GTZan and Million Songs Data set (MSD) are the mainly used datasets. GTZan is preferred as it has been widely used by many researchers. The Million Song Dataset (MSD) contains the meta data and derived features of the track such as artist, terms of the artist, similar artists, danceability, energy, duration, beats, tempo, loudness, and time signature and release year. There are seven genres such as blues, electronic, jazz, metal, rap, reggae, and rock are available with 1000 tracks in MSD. Mel frequency cepstral coefficients are mainly used in preprocessing technique without performing result. The features of classification include the percentage of training and test samples, the linear separation features for algorithms like SVM (Support Vector Machines), PCA (Principal Component Analysis), as in Fig. 79.1 independent target features, the sample overfitting percentage, overall accuracy, and performance measures. These results had training and modified testing data and the results are based on some pre-requisite labels used for classification. The enhanced networking model for classification with hidden layers represented the artificial neural networks. The proposed system has the classification part of fully connected layers that provides the classified recommendations as the output. In this module the features of a song will be selected and evaluated. Rhythm of the song has unique content features are movement of music signals over time, the rhythm and its regular interval, cycle of the beat, the rhythmic tempo and the lapse time signature. The rhythm structure is considered as feature set which is extracted from beat histogram. The melody and harmony information are also considered as features set and are extracted using pitch detection techniques. The music genres consists of the following features including:

- Tempo: The tempo of a piece of music measured in beats per minute which determines the speed of playing. In this experiment, the tempo of the entire song is taken into account. Any audio editor software can be used to calculate the tempo value of a song.

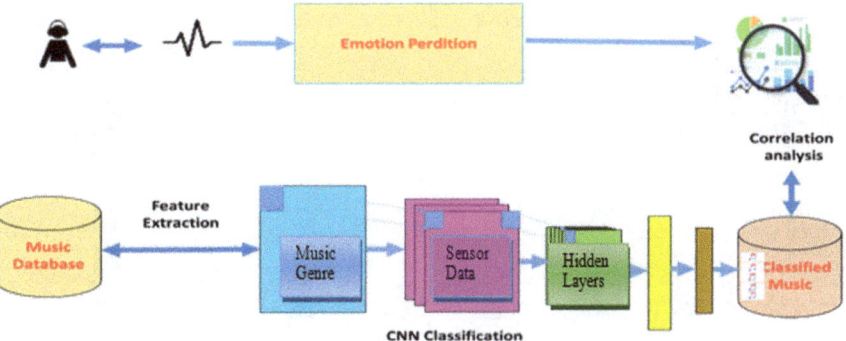

Fig. 79.1 Architecture diagram of proposed system using convolution neural networks and smart devices for music genre classification

- Beats: The basic unit of rhythm is beat. The beat value of the entire song if converted into a single digital value and is used as a feature for training the model.
- Chroma_feature: Chroma-based features represent the entire music in a condensed form. It is used in chord recognition, harmonic melody similarity analysis.
- RMSE: The Root Mean Square Error (RMSE) is used as an error metric. Residual measures the difference between the value predicted by model and the observed value.
- Spectral centroid: The spectral centroid is associated with brightness of the sound.
- Spectral Bandwidth: The spectral bandwidth refers to the spectral magnitude at one-half the peak maximum.
- Rolloff: Roll-off is the steepness of a transfer function with frequency.
- Zero_crossing_rate: The zero-crossing rate is the rate of sign-changes along a signal. This feature is highly used in MIR (Music Information Retrival).
- MFCC: Mel Frequency Cepstral Coefficients (MFCCs) The mel frequency cepstral coefficients (MFCCs) of a signal are a small set of features represents the short-term power spectrum of sound (usually about 10–20).

Sample music data set is fed into the classifier for training and it is classified according to the classifier results obtained. Once the classifier completes the classification, the sensor data from physical devices are fed to convolution layers for CNN algorithm implementation. The physical devices send the output value of simulation such as heart rate, pulse rate, physical activity, and stress rates at regular intervals to the trained neural network. These values are considered as the hidden layer of convolution pooling. The popular in store classifiers for songs and the emotion attached classified as in Fig. 79.2 shows the genre classification with sample music data set of 1000 songs. Techno—For party hard, for celebrating accomplishments. Listened usually with high volume. Euphoric Trance—To chill out after a long day or bad day at work. Mild volume with a sip of wine will be perfect. Uplifting music—Listened

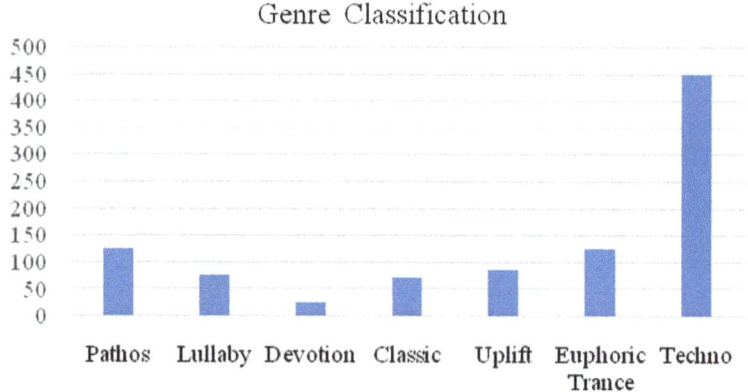

Fig. 79.2 Genre classification of sample data

during start of the day to derive needed energy to go after and accomplish the tasks planned for the day. Classical music—Music to be listened when in the mood of refreshing an orderly mind by enjoying the defined protocols in music. Western Classical and Carnatic music though they seem extreme they follow rigorous protocols which refreshes mind. Devotional music—This music is to direct mind and stay within the core belief system of the listener music. Lullaby—This music is for babies and toddlers. To make the child happy. They are very expressive and are considered impactful for child's growth. Folk music—The native music of the land. The rawness seen in this music is the beauty of this music. Brings out the natural characteristics of the listener in a more rustic way. It is also sung to ease the task at hand. Yelelo in India is sung by fishermen and farmers to ease their work and bring about bonding. Pathos—This is sad music. In other words, soothing a sad heart. This is music for the heart and not mind. Bereavement, broken heart, lost relationship are examples of pathos music situations.

Fig. 79.3 Genre classification on liveliness from spotify features

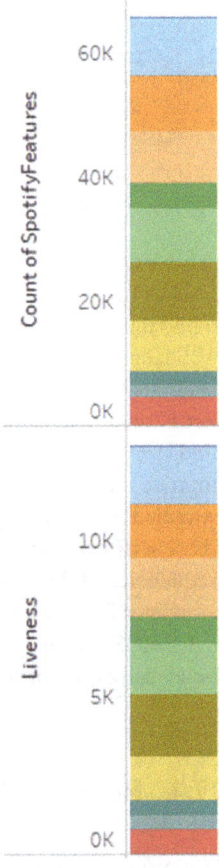

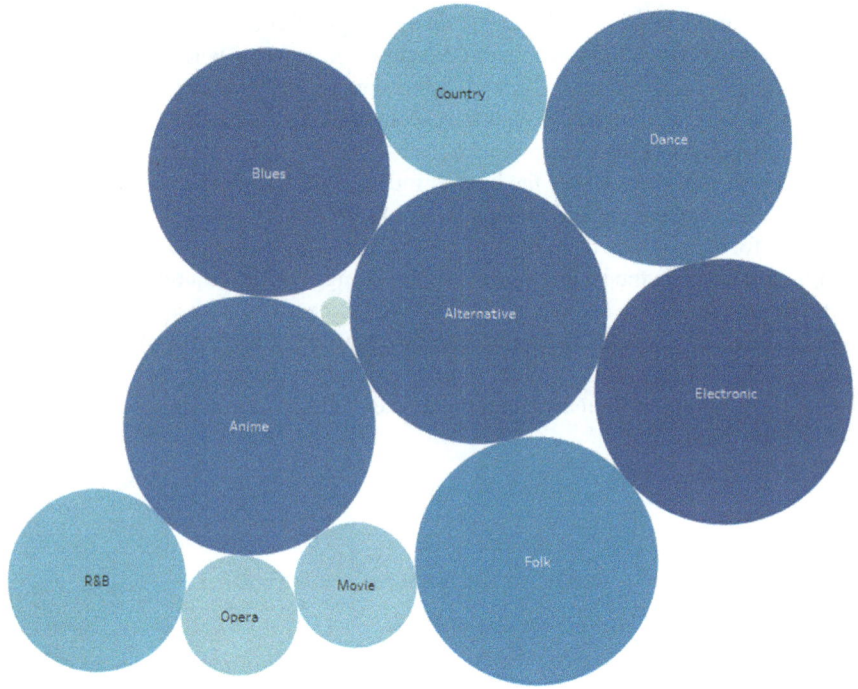

Fig. 79.4 Genre classification of spotify data

The existing model is implemented using Convolution Neural Network algorithm to find the genre of a given music with its feature data. CNN is used for classification [12]. Convolutional Neural Network is a type of deep learning algorithm consists of single input and single output layer with multiple hidden layers. Each hidden layer consists of series of convolution layer and followed by pooling layers. The commonly used activation function is RELU. The final convolution has back propagation to improve the final output. Mel Frequency Cepstral Coefficients (MFCC) features are extracted for each song. Fourier transform is applied to convert music into frequency. These extracted features are act as an input to neurons. Visual representation of frequencies of the music signal is represented by spectrogram. Table 79.1 shows the parameter for Training.

LibROSA is the python library needed to work on music data. Figure 79.5 shows the visualization of music spectrum.

79.4   Result and Discussion

The proposed system model with Convolution Neural Network (CNN) algorithm uses the feature and target values of the dataset are set as in Fig. 79.6, test and train

Table 79.1 Training parameters

Parameter	Value
Input layer	128 × 660 neurons
Convolutional layer	16 different 3 × 3 filters
Max pooling layer	2 × 4
Convolutional layer	32 different 3 × 3 filters
Max pooling layer	2 × 4
Convolutional layer	32 different 3 × 3 filters
Max pooling layer	2 × 4
Dense layer	64 neurons
Output layer	10 neurons for the 10 different genres

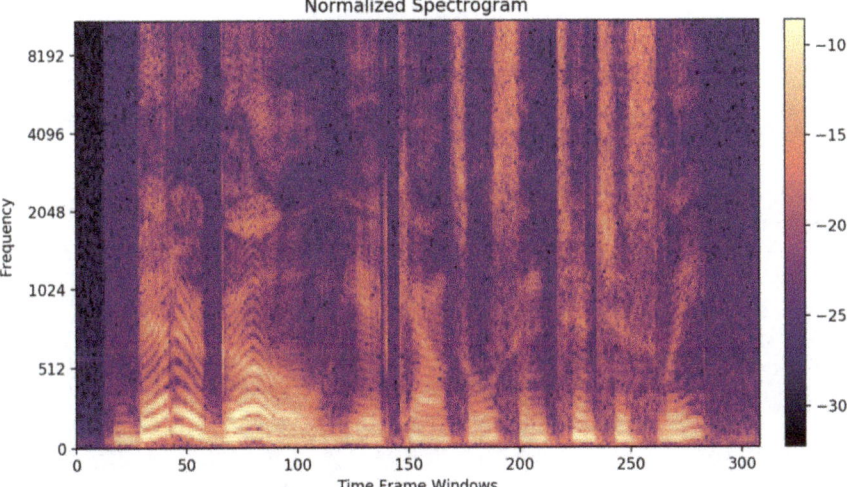

Fig. 79.5 Spectogram

data set are split, and in addition that a test size and random states that provides an optimal result.

The classified music data is further fed to these hidden layers. Thus an accuracy score is calculated. The music applications like JioSaavn, Wynk Music, Gaana, BigFM and many other applications have the online streamline of songs with more than 50 million songs in store. The accuracy and performance measures are the ease out of emotional stress and the involvement of music and mind to play together.

From the designed model the accuracy of the model obtained was 0.87 i.e., 87%. The input given to predict the genre was predicted successfully for all the inputs during the first trial and in the second trial for the second set of values the prediction for the 75% of the input data. The KNN with 75% accuracy was raised to the performance of 87% with CNN algorithm on convolution neural network as hidden

Fig. 79.6 Feature-based
CNN Classifier

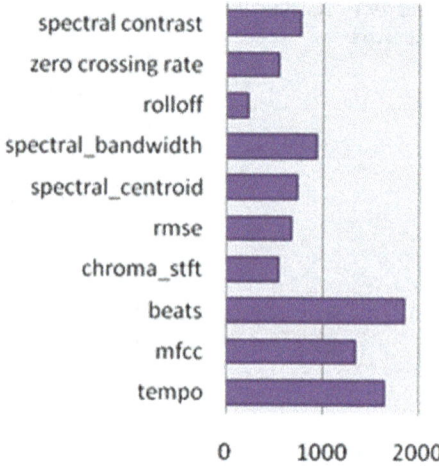

layer as in Fig. 79.7 has the accuracy performance measures. Table 79.2 shows the

Fig. 79.7 Performance
measures

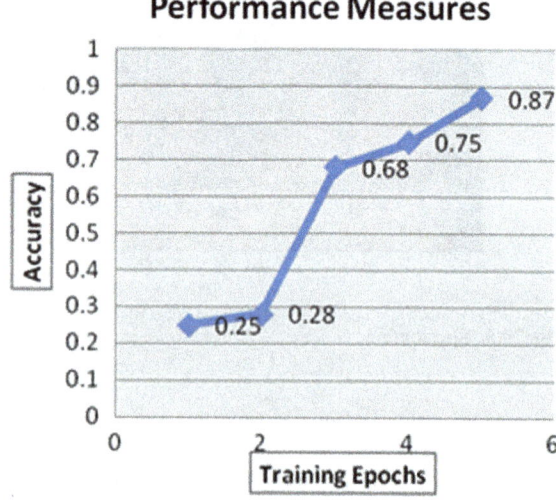

Table 79.2 Comparative
study on different classifiers

Classifier	Accuracy
Decision tree	0.25
PCA	0.28
SVM	0.68
KNN	0.75
CNN	0.87

comparative study of classification algorithms is given.

79.5 Conclusion

This paper proposed a music genre classification and recommendation system using Convolution Neural networks. The feature sets are extracted using Mel Spectrum. The recent innovations in wearable devices are much needed for emotional predictions. Smart cars, smart homes, and many innovations are possible with the predictive values from the physical devices. The proposed system thus classifies music based on these sensor-enabled physical devices with music data set from many online stream applications. The system with features discussed above have many hidden layers as pooling over the convolution hidden layers. The classified music is thus played with the emotional simulation inputs across these hidden layers. The future scope is to build smart applications embedded with cognitive therapy for addressing emotional imbalance via music therapy. The increasing need of emotional intelligence with cognitive study can be developed using brain simulations over neural networks.

References

1. Nirjon, S., Dickerson, R.F., Li, Q., Asare, P., Stankovic, J.A., Hong, D., Zhao, F.: Musicalheart: a hearty way of listening to music. In: 10th ACM Conference on Embedded Network Sensor Systems, pp. 43–56. ACM (2012)
2. Liu, N.H., Hsieh, S.J., Tsai, C.F.: An intelligent music playlist generator based on the time parameter with artificial neural networks. Expert Syst. Appl. **37**(4), 2815–2825 (2010)
3. Liu, N.H., Chiang, C.Y., Hsu, H.M.: Improving driver alertness through music selection using a mobile EEG to detect brainwaves. Sensors **13**(7), 8199–8221 (2013)
4. Abdul, A., Chen, J., Liao, H.Y., Chang, S.H.: An emotion-aware personalized music recommendation system using a convolutional neural networks approach. Appl. Sci. **8**(7), 1103 (2018)
5. Isinkaye, F.O., Folajimi, Y.O., Ojokoh, B.A.: Recommendation systems: principles, methods and evaluation. Egypt. Inform. J. **16**(3), 261–273 (2015)
6. Shin, I.H., Cha, J., Cheon, G.W., Lee, C., Lee, S.Y., Yoon, H.J., Kim, H.C.: Automatic stress-relieving music recommendation system based on photoplethysmography-derived heart rate variability analysis. In: 36th Annual International Conference of the IEEE Engineering in Medicine and Biology Society, pp. 6402–6405. IEEE (2014)
7. Liu, H., Hu, J., Rauterberg, M.: Music playlist recommendation based on user heartbeat and music preference. In: 2009 International Conference on Computer Technology and Development, pp. 545–549. IEEE (2009)
8. Yoon, K., Lee, J., Kim, M.U.: Music recommendation system using emotion triggering low-level features. IEEE Trans. Consum. Electron. **58**(2), 612–618 (2012)
9. Rosa, R.L., Rodriguez, D.Z., Bressan, G.: Music recommendation system based on user's sentiments extracted from social networks. IEEE Trans. Consum. Electron. **61**(3), 359–367 (2015)
10. Deger, A., Yusuf, Y., Kamasak, M.E.: Emotion based music recommendation system using wearable physiological sensors. IEEE Trans. Consumer Electron. **14**(8) (2018)

11. Markov, K., Matsui, T.: Music genre classification using Gaussian process models. In: 2013 IEEE International Workshop on Machine Learning for Signal Processing (MLSP), pp. 1–6. IEEE (2013)
12. Costa, Y.M., Oliveira, L.S., SillaJr, C.N.: An evaluation of convolutional neural networks for music classification using spectrograms. Appl. Soft Comput. **52**, 28–38 (2017)

Chapter 80
Localization of Optic Cup-Disc in Retinal Images Using Morphological Filters for Glaucoma Detection

Deepika Pal, Vikrant Bhateja, Archita Johri, and Babita Pal

Abstract Glaucoma is one of the main diseases causing permanent blindness. It is marked by an increase in Intraocular Pressure (IOP) which results in changing the shape of Optic disc and Optic cup. Segmentation and Localization of Optic Disc and Optic Cup for the automated detection and diagnosis of Glaucoma is necessary. Morphological Filters with the method of thresholding of the retinal fundus images have been adapted for segmentation of Region of Interest (ROI). The quality measurement of the segmentation of the ROI can be determined by the use of Similarity Coefficients. The resultant segmented binary images have been obtained with the value of Dice and Jaccard representing overlapping and similarity check between the segmented image and the ground truth. Cup to Disc Ratio has been calculated for the temporary analysis of Glaucoma.

80.1 Introduction

Glaucoma is the second largest cause of permanent blindness all over the world. Glaucoma patients or suspects develop symptoms like in any other ophthalmic disease. Hence it is difficult for the doctor to diagnose it at an early stage to start treatment at the earliest. Glaucoma patient's optic cup and disc start changing shape and size, eye of the suspect swells, redness is also developed, etc. In order to analyze Glaucoma in any patient some important measures have to be considered such as inheritance of the disease from his ancestors [1], any other history of ophthalmic diseases, cup and disc measurements, etc. There are many techniques adapted for the segmentation and localization of ROI. Empirical Wavelet Transform and B spline coefficient had been deployed by Maheshwari et al. [2]. EWT method dealt with decomposition of

D. Pal · V. Bhateja (✉) · A. Johri · B. Pal
Department of Electronics and Communication Engineering, Shri Ramswaroop Memorial Group of Professional Colleges (SRMGPC), Faizabad, Road, Lucknow, U.P. 226028, India

Dr. APJ Abdul Kalam Technical University, Lucknow, U.P., India

S. C. Satapathy et al. (eds.), *Smart Computing Techniques and Applications*,
Smart Innovation, Systems and Technologies 225,
https://doi.org/10.1007/978-981-16-0878-0_80

signals for both public and private databases. Three-Fold and Ten-Fold Cross Validation had been considered for it. The decomposed components obtained are arranged in an order of low to high-frequency ranges. Superpixel based segmentation by using Simple Linear Iterative Clustering (SLIC) algorithm had been used by Cheng et al. and Zhou et al. [3, 4]. SLIC approach dealt with tuning of optic disc boundary. Generation of super pixels for retinal fundus images had been considered for Glaucoma detection. Combination of Hough Transform, Polar Transform, and Morphological Operations had been used by Zahoor and Fraz [5]. Optic Disc localization had been performed by Circular Hough Transform (CHT), adaptive thresholding had been used for the segmentation of Optic Disc. CHT was used for the analysis of circular objects or components of the retinal image. Local image fitting model with oval-shaped constraint (LIFO) has been introduced by Gao [6]. LIFO had been used for localization of optic disc. Contour was formed using thresholding methods in combination with saliency map detection. Other OD segmentation for the COSFIRE filter was narrated by Guo et al. [7]. Vasculature selective and Disk selective COSFIRE filters were adopted for localization of OD in Glaucoma detection. Contrast Limited Adaptive Histogram Equalization (CLAHE) algorithm had been used for extraction of ROI by Sevastopolsky [8]. In order to combat for demerits of previous work on CAD of Glaucoma, an improved method of segmentation and localization of ROI based on combination of thresholding and morphological operation of retinal fundus images has been proposed in this paper. The rest of the paper is organized in the following sections: Sect. 2 discusses the proposed approach for segmentation and localization using morphological filter, thresholding method, ellipse fitting IQA, and CDR evaluation, Sect. 3 discusses the outcome of the proposed technique, Sect. 4 discusses the conclusion of the proposed work.

80.2 Proposefd Segmentation Methodology Using Morphological Filters and Thresholding

A. Morphology

Morphological filtering is the branch of science that deals with various operations related to different shapes and structures in any analysis. Morphology is tool for extracting various objects of different shape and size. Changing the size of ROI according to the need of our study or analysis [9]. It is applied on the images with a particular shape of structuring element which is desired at the output. Structuring element are different shapes matrices used for extraction of any object, component or ROI from any image or signal for its further processing. They are broadly classified as flat and non-flat structuring element. These includes structuring element of various shapes such as square, rectangle, disk and ball, etc. [10]. The selection criteria for the shape and size of structuring element depends on the need for analysis in which it is being used. Often used morphological operations include Erosion, Dilation

[11], Opening and Closing, etc. Morphology has various applications such as Image Processing, Signal Processing, etc. [12].

B.   Thresholding and Ellipse Fitting Method

Thresholding is the method of extraction of significant information from the images automatically. Thresholding is the method of segmentation of various objects or parts of the image containing relevant information for further analysis [13]. In the proposed methodology Global Thresholding has been used. It is considered when any components or classes of image have different information. Global thresholding is considered when differences between two classes or parameters are distinct enough to be divided into foreground and background pixels in a binary image [14]. In this method, only one threshold is considered over the entire image to depict difference in parameters, etc. Global Thresholding is represented by Eq. 80.1 [15]:

$$g(x, y) = \begin{Bmatrix} 0 \text{ if } f(x, y) < T \\ 1 \text{ if } f(x, y) \geq T \end{Bmatrix} \tag{80.1}$$

Ellipse fitting algorithm is used for smoothening the ROI based on a least square fitting algorithm. It is based on the best-fit curve that has a minimal sum of deviations squared from a given data. It allows fitting the ellipse on a certain data points in a particular ROI [16].

C.   Proposed Methodology for Segmentation and Localization

The proposed approach for OD and OC segmentation for the diagnosis of glaucoma using combination of Morphological filter. Simple operations of Morphology such as dilation and binary area filtering have been adapted for the segmentation module. Dilation is the morphological operation for adding the pixels at the region of interest. Binary area filtering is used for the extraction of the considered region or the region of interest by removing other unnecessary components of the image [17]. OD segmentation has been done on the enhanced red channel images. Standard Deviation has been calculated from the enhanced red channel images then thresholding has been applied to it. Binary area filtering has been used for removal of objects other than the region of interest i.e., OD. The segmented binary image of OD has been subjected to ellipse fitting for the smoothing of the disc. Finally, the segmented OD has been obtained from the retinal fundus images. OC segmentation has been carried out in the same way by computing its standard deviation of the green channel enhanced image. Then thresholding has been carried out then the resultant has been subjected to the binary area filtering. The filtered image is subjected to dilation by considering the 'disk' type structuring element of size 3 [17]. Ellipse Fitting has been applied to the binary image obtained. Finally, the segmented OC has been extracted. The extraction of ROI has been shown in Fig. 80.1.

Fig. 80.1 Morphological
segmentation module

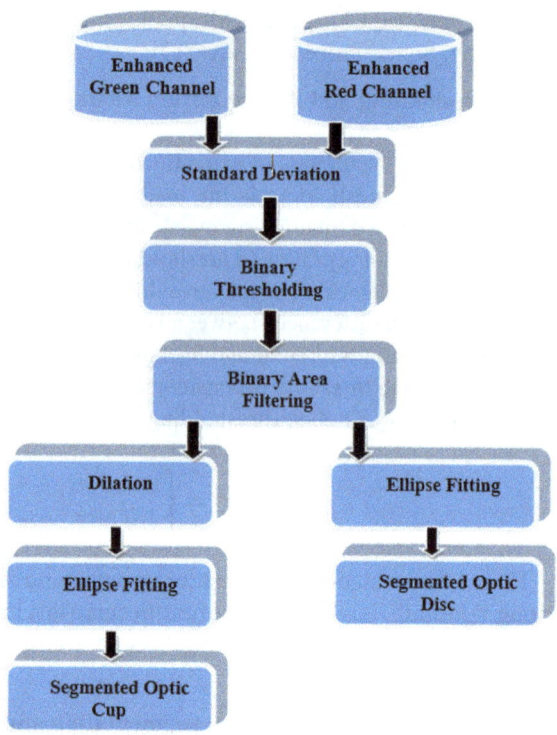

D.   Proposed Evaluation of Cup to Disc Ratio

The differentiation between glaucomatous and non-glaucomatous retinal fundus images has been done by computing Cup to Disc Ratio (CDR). CDR is calculated as the ratio of vertical diameter of cup to the vertical diameter of the disc. The value of CDR for retinal fundus image greater than 0.5 then the retinal fundus image is of glaucomatous eye else it is non-glaucomatous eye [18, 19]. The database considered for the CAD of Glaucoma is publicly available DRISHTI-GS. The dataset has been collected from Arvind Hospital, Madurai. It consists of total 101 images out of which 69 are Glaucomatous and 32 are normal retinal fundus images [20]. Figure 80.2 depicts calculation of OD and OC in a retinal fundus image.

The red circle in the image represents OD and green circle represents OC in the fundus image. For the computation of CDR vertical diameter of both disc and cup have been considered.

E.   Image Quality Assessment using Similarity Coefficients

A similarity coefficient represents the similarity between two documents, two comparative quantities. A similarity coefficient is a function that computes the degree of similarity between a pair of text objects [21]. Similarity Coefficients for the assessment of segmentation have been proposed in this paper. In proposed method, Dice and

Fig. 80.2 OD and OC in the
retinal fundus image [20]

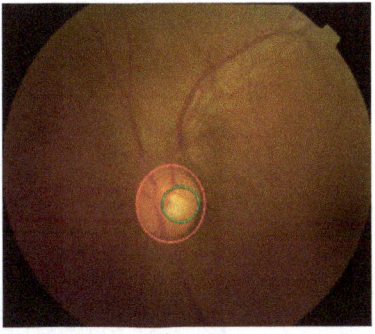

Jaccard are the similarity coefficients considered for the assessment of the segmented
ROI. The equations for Dice [22] and Jaccard [22] are as follows

$$\text{Dice} = 2\frac{A \cap G}{|A| + |G|} \tag{80.2}$$

$$\text{Jaccard} = \frac{|A \cap G|}{|A \cup G|} \tag{80.3}$$

80.3   Results and Discussions

The database consists of publicly available Drishti-GS dataset. It contains a total
of 101 images. The dataset is divided into 50 training and 51 testing image sets.
The images in the dataset are either glaucomatous or non-glaucomatous. All the
images were considered for the analysis. Segmentation is the process of obtaining
relevant information for the analysis of any problem. The segmentation results of
the proposed methodology have been shown in Figs. 80.3 and 80.4. Figure 80.3
shows segmentation of optic cup whereas Fig. 80.4 shows segmentation of optic disc.
Qualitative analysis of segmentation of ROI is performed using Sorenson similarity
coefficients that are Dice and Jaccard. Segmentation results for Test_Image #A and
Test_Image #B have been shown in Table 80.1.

IQA for the segmented ROI has been given in the Table 80.1. Various values of
Dice and Jaccard represent that segmentation of the retinal fundus image has been
carried out. Values of CDR depict that image is glaucomatous or non-glaucomatous.

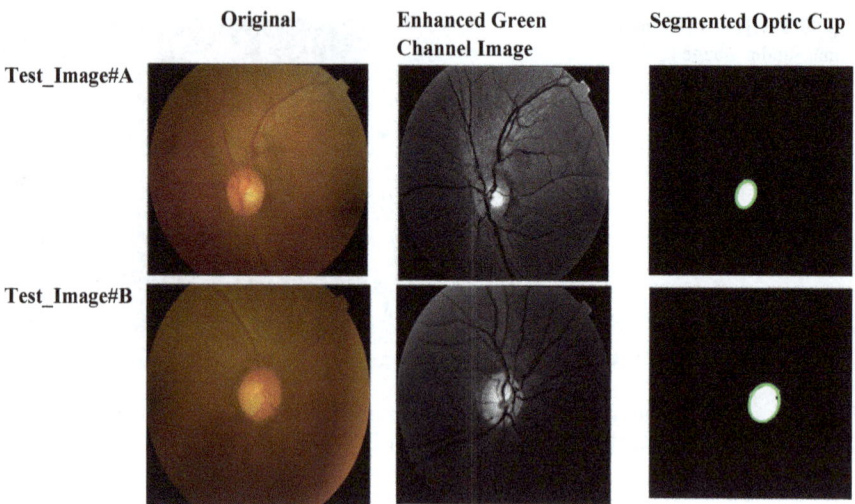

Fig. 80.3 Segmentation of optic cup

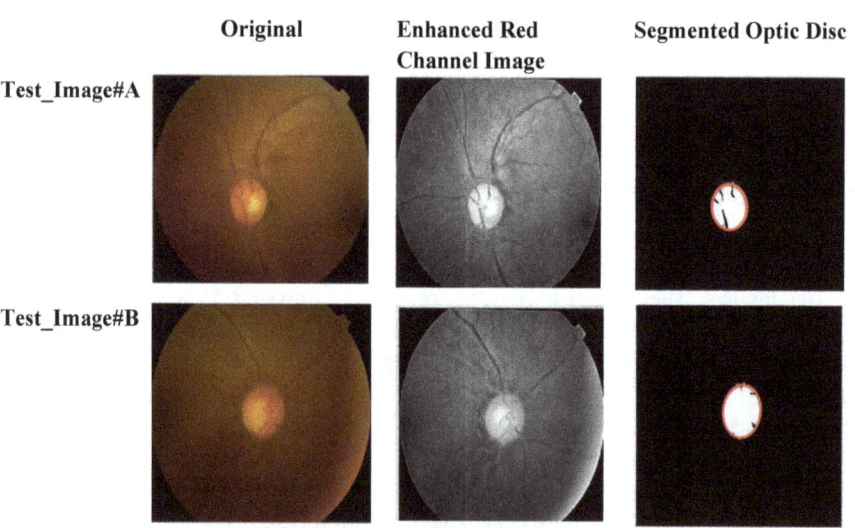

Fig. 80.4 Segmentation of optic disc

Table 80.1 IQA of Segmented ROI and CDR for glaucoma diagnosis

	Dice of OC	Jaccard of OC	Dice of OD	Jaccard of OD	CDR
Test_Image#A	0.0022	0.0011	0.0207	0.0104	0.6145
Test_Image#B	0.2793	0.1623	0.27292	0.15802	0.7594

80.4 Conclusion

CAD approach is a crucial step in diagnosis of various diseases thereby reducing the probability of error due to manual procedure or interference. The values of Dice and Jaccard are in the range between 0 and 1 indicating that segmentation has been done. Values of CDR is more than 0.5 in both the case indicating eye with less probability of affecting from glaucoma whereas the other one is glaucomatous with the value of CDR 0.7594. The proposed algorithm is simple for the segmentation of relevant objects or components of image. The proposed methodology can be improved by incorporating various features such as ISNT ratio, blood vessels removal, etc. Convolutional Neural Network (CNN) could also be used in order to increase the accuracy of the system. Various optimization algorithms could also be added to avoid tuning of the parameters.

References

1. Kavitha, S., Zebardast, N., Palaniswamy, K., Wojciechowski, R., Chan, E.S., Friedman, D.S., Venkatesh, R., Ramulu, P.Y.: Family history is a strong risk factor for prevalent angle closure in a south Indian population. Ophthalmology **121**(11), 2091–2097 (2014)
2. Maheshwari, S., Pachori, R.B., Acharya, U.R.: Automated diagnosis of glaucoma using empirical wavelet transform and correntropy features extracted from fundus images. IEEE J. Biomed. Health Inform. **21**(3), 803–813 (2016)
3. Cheng, J., Liu, J., Xu, Y., Yin, F., Kee Wong, D.W., Tan, N.-M., Tao, D., Cheng, C.Y., Aung, T., Wong, T.Y.: Superpixel classification based optic disc and optic cup segmentation for glaucoma screening. IEEE Trans. Med. Imag. **10**(10), 1–15 (2013)
4. Zhou, W., Wu, C., Yi, Y., Duet, W.: Automatic detection of exudates in digital images using superpixel multi-feature classification. IEEE Access **5**(9), 17077–17088 (2017)
5. Zahoor, M.N., Fraz, M.M.: Fast optic disc segmentation in retina using polar transform. IEEE Access **5**(7), 12293–12300 (2017)
6. Gao, Y.: Automatic optic disc segmentation based on a modified local image fitting model with shape prior information. J. Healthc. Eng. Shenyang Liaoning China **3**, 1–10 (2019)
7. Guo, J., Azzopardi, G., Shi, C., Jansonius, N.M., PetkoV, N.: Automatic determination of vertical cup to disc ratio in retinal fundus images for glaucoma screening. IEEE Access **7**(1), 8527–8541 (2019)
8. Sevastopolsky, A.: Optic disc and cup segmentation methods for glaucoma detection with modification of U-net convolutional neural network. Pattern Recogn. Image Anal. Adv. Math. Theory Appl. 1–13 (2017)
9. Bhadauria, A.S., Nigam, M., Arya, A., Bhateja, V.: Morphological filtering based enhancement of MRI. In: Proceedings of 2nd International Conference on Computing, Communication and Control Technology (IC^4T), Lucknow, (U. P.), India, pp. 54–56, October, 2018
10. Structuring Element is available at: https://in.mathworks.com/help/images/structuring-elements.html
11. Bhateja, V., Devi, S.: A novel framework for edge detection of microcalcifications using a nonlinear enhancement operator and morphological filter. In: IEEE 3rd International Conference on Electronics Computer Technology (ICECT), Kanyakumari, India, vol. 5, pp. 425–430, April, 2011

12. Srivastava, A., Alankrita, A., Raj, Bhateja, V.: Combination of wavelet transform and morphological filtering for enhancement of magnetic resonance images. In: International Conference on Digital Information Processing and Communications, Ostrava, Czech Republic, pp. 460–474, 7–9 July 2011

13. Verma, R., Mehrotra, R., Bhateja, V.: A new morphological filtering algorithm for preprocessing of electrocardiographic signals. In: Proceedings of 4th International Conference on Signal and Image Processing, pp. 193–201, January, 2013

14. Bhargavi, K., Jyothi, S.: A survey on threshold based segmentation technique in image processing. Int. J. Innov. Res. Devel. 3(12), 234–239 (2014)

15. Castelman, K.R.: Digital Image Processing, Tsinghua Univ Press (2003)

16. Nayak, J., Acharya, R., Bhat, P.S., Shetty, N., Lim, T.C.: Automated diagnosis of glaucoma using digital fundus images. J. Med. Syst. 33, 337–346, August, 2008

17. Gonzalez, R.C., Woods, R.E.: Digital image processing, 3rd ed. Pearson Education, Prentice Hall, Chap. 10, pp. 711–819 (2009)

18. Anusorn, C.B., Kongprawechnon, W., Kondo, T., Sintuwong, S., Tungpimolrut, K.: Image processing technique using cup to disc ratio. Thammasat Int. J. Sci. Technol. 18(1), 22–34 (2013)

19. Singh, M., Virk, J.K., Singh, M.: A simple approach to cup-to-disk ratio determination for glaucoma screening, vol. 6, No. 2, pp. 177–182, September, 2015

20. Sivaswamy, J., Krishnadas, S.R., Joshi, G.D., Jain, M., Tabish, A.: DRISHTI-GS: retinal image dataset for optic nerve head (ONH) segmentation. In: IEEE 11th International Symposium on Biomedical Imaging, Beijing, China, pp. 1–4, July, 2014

21. Thada, V., Jaglan, V.: Comparison of Jaccard, Dice, cosine similarity coefficient to find best fitness value for web retrieved documents using genetic algorithm. Int. J. Innov. Eng. Technol. (IJIET) 2(4), 202–205 (2013)

22. Shrivastava, N., Bharti, J.: Empirical analysis of image segmentation techniques. In: International Conference on Smart Trends for Information Technology and Computer Communications, Jaipur, India, pp. 143–153, August, 2016

Chapter 81
Region Labeling Based Brain Tumor Segmentation from MR Images

Vikrant Bhateja, Mansi Nigam, and Anuj Singh Bhadauria

Abstract Brain tumor due to their increasing rate and high uncertainty has become a curse for mankind. For their effective diagnosis, automatic systems called Computed Aided Diagnosis (CAD) systems have developed that help in tumor analysis without manual interference. However, due to high variability in tumors, their segmentation from MR images is a challenging task. This paper proposes an improved tumor segmentation methodology that is an extension to simple thresholding technique. In this method, different regions of the binary images are labeled and are segregated on the basis of solidity and area. Then the region having solidity around 50% and maximum area is extracted as tumor. This segmented region is further dilated to include edema tissues surrounding the tumor. The performance of the methodology is justified by the results obtained in which only the tumorous region has been extracted indicating successful segmentation without inclusion of other brain tissues.

81.1 Introduction

Brain tumor is one of the most common causes for the increasing mortality among children and adults in the world. It is a group of abnormal cells that grows inside the brain creating pressure on the other tissues and the skull. With the increase in the number of people who develop brain tumor there is an immediate need to develop an automated system for early detection of brain tumor. Therefore, Computer Aided Diagnosis (CAD) systems are developed that are used to assist radiologists in interpreting the medical images with the help of automatic assisted tools. CAD technologies are employed to improve the diagnostic accuracy of the radiologists [1] by providing a second opinion about tumor diagnosis [2]. Magnetic Resonance Imaging (MRI) is one of the most reliable technique for brain tumor analysis because

V. Bhateja (✉) · M. Nigam · A. S. Bhadauria
Department of Electronics and Communication Engineering, Shri Ramswaroop Memorial Group of Professional Colleges (SRMGPC), Faizabad, Road, Lucknow, U.P. 226028, India

Dr. APJ Abdul Kalam Technical University, Lucknow, U.P., India

it provides various images with different contrasting features thereby providing means for detailed analysis of tumorous tissues. One of the most crucial step of CAD systems is tumor segmentation from the brain image [3]. Segmentation is the process in which tumor region is separated from the entire brain image to detect its presence of an abnormal area. It is an important step and needs to be performed with utmost care because diagnosis at higher levels depends on the region extracted here. Incorrect segmentation may lead to false diagnosis of tumor which can be affected the patient severely [4, 5]. There have been a number of techniques use for tumor segmentation amongst which some of the existing approaches includes the concept of Hybrid and Variational Expectation Maximization, Tian et al. [6] utilized this technique to segregate the tumor region from other objects in the image. The proposed approach was easy to implement but did not consider the spatial relationship between neighboring pixels and was complicated to implement. Overcoming this problem Zhang et al. [7] presented a framework for medical image analysis which used Support Vector Machine (SVM) for tumor segmentation. Although this technique reduced time consumption but could not be used for inhomogeneous images which was generally a problem with medical images. To increase the reliability of the system Ji et al. [8] proposed fuzzy logic technique for segregation of infected tissues. Further, to extract the edge and texture information along with the infected region, Mohsen et al. [9] presented Feedback Pulse Coupled Neural Network based technique. However, CAD of brain tumor through MRI required more efficient and simple technique, abiding to which Somasundaram et al. [10] proposed a segmentation technique based on the adaptive intensity Thresholding. The above approach worked with dataset of any orientation and contrast but the selection of a particular threshold value to extract the exact tumor region excluding the brain tissues was a manual task and required clear knowledge about brain anatomy. Thresholding technique cannot be solely used to segregate the tumor region as it separates the infected area along with the objects of similar intensity. To brush-off the above challenges, this paper proposes a segmentation technique which employs thresholding and followed by region labeling techniques where the thresholded tissues are further subjected to another extraction on the basis of tumor solidity. This results into a segmented image containing only the abnormal tumor tissue. The remaining paper has been arranged into the following sections: Sect. 81.2 discusses the concept of region labeling and describes the proposed methodology; Sect. 81.3 explains and summarizes the results obtained and discusses them; Sect. 81.4 summarizes the conclusion of the work.

81.2 Proposed Methodology for Tumor Segmentation

81.2.1 Region Labelling

Image segmentation plays a vital role in medical image processing. It is the process of extraction of the region of interest by dividing the image into different regions and the level up to which this division occurs depends on the problem being dealt. The segmentation is carried in order to extract the important details from which information can be perceived. For tumor diagnosis the segmentation is done to an extent that will yield the tumor area without the inclusion of brain tissues. The proposed segmentation methodology is based on Region Labeling. It is a technique in which different regions of a binary image are labeled so that different regions can be dealt separately and each region has its own identity. This helps in examining each region independently on the basis of certain properties and categorizing them. This distinction aids extraction of region that has properties resembling the tumor and therefore, tumor segmentation is done. In the proposed methodology, the property of tumor being exploited is solidity. Tumor solidity is 50% [11] which is very high compared to other brain tissues and hence region having this solidity level is identified as tumorous region using the process as illustrated in Fig. 81.1.

Fig. 81.1 Block diagram of proposed region labeling based segmentation technique

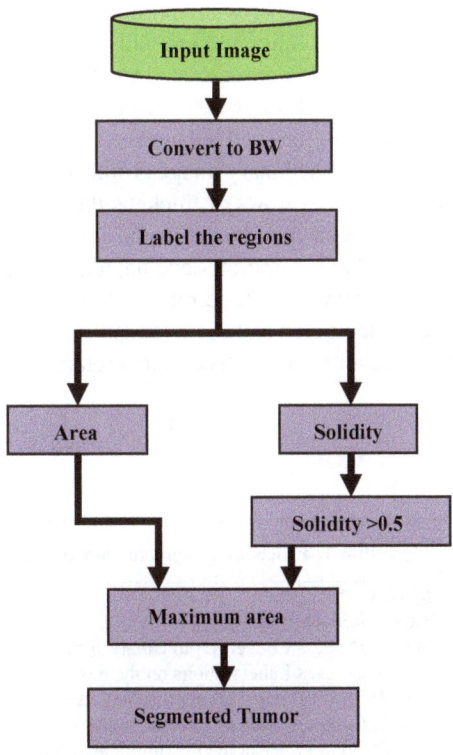

81.2.2 Proposed Region Labeling Based Tumor Segmentation

This paper suggests an improved method for segmentation of tumors from brain MR images using region labeling. In the first step, before labeling the regions, the image is converted into binary color map using thresholding which is a process of finding a specific intensity in a grayscale image that separates two distinct regions of an image, foreground, and background [12]. The proposed segmentation technique as shown in Algorithm 1 uses a threshold of value 0.3 for binarization because Region of Interest (ROI) i.e. Tumor is of very high intensity and is therefore always higher than this level. The pixel values greater than 0.3 are scaled to 1 which is white color (foreground) and those less than it are scaled down to 0 which is black color (background). Now the image consists of distinct regions of white color one amongst which is tumor region which is a dense region of uniform intensity and this image is subjected to region labeling. The white portion of the image is labeled into two classes, firstly on the basis of solidity and secondly on the basis of area. The regions having solidity greater than 50% are extracted and then amongst them the object of maximum area is obtained. This portion extracted is the segmented tumor region. In order to obtain the segregated tissues and balanced edges, the image is subjected to morphology-based operations. The binary image is dilated using a structuring element of size 5. Dilation [13] operation grows or thickens the objects in a binary image. It also bridges the gaps between the objects in the image [14] and is operated using Eq. 81.1. This process also incorporates edema, region of active cells surrounding the tumor region which otherwise will be removed due to its low-intensity value. The dilated image is complemented and the resultant image is subtracted from the input image which removes all the other tissues and retains the segmented tumor region.

$$X \oplus se = \{z|(se)_z \cap X \neq \emptyset\} \qquad (81.1)$$

Algorithm 1: Procedural Steps for Segmentation using Region Labelling.

BEGIN
Step 1: *Input* MR image [I].
Step 2: *Process* Convert [I] to Binary format [I_1] using 0.3 threshold.
Step 3: *Process* Label regions on the basis of area and solidity.
Step 4: *Process* Select the regions having solidity greater than 50%.
Step 5: *Compute* Area of selected regions.
Step 6: *Process* Select the maximum area region [I_2].
Step 7: *Define* Structuring element (*se*) of size 5.
Step 8: *Compute* Dilation [I_3] of image [I_2] using Eq. (1).
Step9: *Compute* Complement [I_4] of [I_3].
Step 10: *Output* Subtract Step 9 and Step 2.
END

81.3   Results and Discussions

81.3.1   Experimental Results

The brain MR images used for simulation of the proposed work are taken from the Whole Brain Atlas [15]. Segmentation of tumor region cannot be done only by thresholding because it extracts not only the tumor region but also other regions and tissues of similar intensity. To overcome this problem thresholding is followed by solidity-based segmentation. This method segregates the region of maximum area having solidity more than 50%. This leaves the resultant image with only a single region which is the region of interest i.e. tumor as shown in Fig. 81.2. The segmentation stage is the end of the brain tumor detection and first step in diagnosis. From the results, it is evident that the segmented region helps to identify whether the image is of tumor infected brain or not.

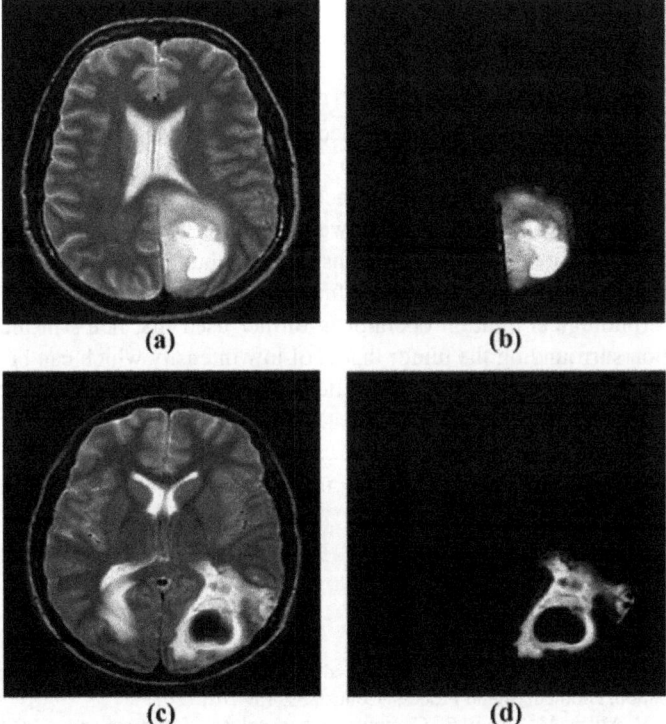

Fig. 81.2 a Original MRI image Test_Image#A. **b** Segmented Test_Image#A. **c** Original Test_Image#B. **d** Segmented Test_Image#B

81.3.2   Discussions

The above figure shows two MR images Test_Image#A and Test_Image#B and their respective segmented output images. The original Test_Image#A has a high, uniform intensity region at the lower half of the left occipital lobe and a slightly greyish edema region surrounding the tumor region. Along with this, there is a bright ventricular structure in the centre on the image which is a normal brain tissue. Upon segmentation the tumorous region is segregated from the other ventricular details with edema intact. In the Test_Image#B there are three main structures of similar intensities of which one is in the center, one at the lower half of the right occipital lobe and the other at lower half of the left occipital lobe. After segmentation, the homogeneous, dense region of maximum area present at the left occipital lobe along with the edema region is separated from the other objects. Therefore, the proposed methodology for tumor segmentation helps to obtain a separate tumor region but also preserve its original shape and structure.

81.4   Conclusion

A detailed analysis of the tumor region is necessary for proper detection and diagnosis through CAD systems which is possible by correct identification of tumor in the MR Images. In this paper, an extended approach of brain tumor segmentation from MR Images is suggested. Thresholding followed by region labeling is deployed which not only extract tumor solely by its intensity levels but also uses an important its property which is solidity that helps in further removal of similarly leveled brain tissues. Morphological dilation operator is further used that helps in inclusion of edema region surrounding the tumor that is of low intensity which can be observed in the results simulated and is usually omitted by thresholding. This inclusion is very important as it indicates the grade and exact dimensions of the tumor. Since in this technique precise details of tumor are also captured this is very helpful for diagnostic stage where features of tumor play a very important role to determine its grade.

References

1. Mohana, G., Subashini, M.M.: MRI based medical image analysis: survey on brain tumor grade classification. Biomed. Signal Process. Control **39**, 139–161 (2019)
2. Bhateja, V., Misra, M., Urooj, S.: Computer-aided analysis of mammograms. In: Non-Linear Filters for Mammogram Enhancement, pp. 21–27. Springer, Singapore (2020)
3. Basavaraju, H.T., et al.: Arbitrary oriented multilingual text detection and segmentation using level set and Gaussian mixture model. Evolut. Intell. 1–14 (2020)
4. Bhadauria, A.S., Bhateja, V., Nigam, M., Arya, A.: Skull stripping of brain MRI using mathematical morphology. In: Smart Intelligent Computing and Applications, pp. 775–780. Springer, Singapore (2020)

5. Bhateja, V., et al.: Two-stage multi-modal MR images fusion method based on parametric logarithmic image processing (PLIP) model. Pattern Recogn. Lett. (2020)
6. Tian, G., Xia, Y., Zhang, Y., Feng, D.: Hybrid genetic and variational expectation-maximization algorithm for Gaussian-mixture-model-based brain MR image segmentation. IEEE Trans. Inf. Technol. Biomed. **15**(3), 373–380 (2011)
7. Zhang, N., Ruan, S., Lebonvallet, S., Liao, Q., Zhu, Y.: Kernel feature selection to fuse multi-spectral MRI images for brain tumor segmentation. Comput. Vis. Image Underst. **115**, 256–269 (2011)
8. Ji, Z., Suna, Q., Xiab, Y., Chena, Q., Xiaa, D., Feng, D.: Generalized rough fuzzy C-means algorithm for brain MR image segmentation. Comput. Methods Programs Biomed. **108**, 644–655 (2011)
9. Mohsen, H., El-Dahshan, E.A., Salem, A.M.: A machine learning technique for MRI brain images. In: Proceedings of 8th IEEE Conference on Informatics and Systems, pp. 161–16. Cairo, Egypt (2012)
10. Somasundaram, K., Kalaiselvi, T.: Automatic brain extraction methods for T1 magnetic resonance images using region labeling and morphological operations. Comput. Biol. Med. **41**(8), 716–725 (2011)
11. Radhi, A.A.: Efficient algorithm for the detection of a brain tumor from an MRI images. Int. J. Comput. Appl. **170**(10), 38–42 (2017)
12. Laddha, R.R., Ladhake, S.A.: A review on brain tumor detection using segmentation and threshold operations. Int. J. Comput. Sci. Inf. Technol. **5**(1), 607–611 (2014)
13. Arya, A., Bhateja, V., Nigam, M., Bhadauria, A.S.: Enhancement of brain MRT1/T2 images using mathematical morphology. In: Proceedings of 3rd International Conference on ICT, vol. 933, pp. 833–840. Springer Singapore (2019)
14. Gonzalez, R.C., Woods, R.E.: Digital Image Processing, pp. 689–794. Pearson Education, Chap. 10 (2009)
15. The Whole Brain Atlas, https://www.med.harvard.edu/aanlib/home.html

Chapter 82
A Comparative Evaluation of Decomposition Methods Based on Pitch Estimation of Piano Notes

U. Vamsi Krishna, R. Priyamvada, G. Jyothish Lal, V. Sowmya, and K. P. Soman

Abstract Wavelet decomposition, variational mode decomposition, and dynamic mode decomposition are the latest signal processing tools that are recently being utilized in the music domain. Most of the work on these algorithms in music domain shows results based on pitch contour. None of the work mentions the effect of different frequency ranges (octaves) on these algorithms. In this paper, the evaluation is performed to understand the effect of different octaves on these decomposition algorithms based on pitch estimation of piano notes. Wavelet decomposition, variational mode decomposition, and dynamic mode decomposition methods are evaluated based on pitch estimation for different octaves. The purpose of this evaluation is to identify the most suitable method for pitch estimation. A comparative evaluation is performed on piano recordings taken from the database of the electronic music studio, University of Iowa. Absolute mean logarithmic error is used as the metric for evaluating the algorithms. The evaluation of algorithms is performed on seven different octaves ranging from 1 to 7. Variational mode decomposition performed better throughout the 7 octaves. Wavelet decomposition also performed well but was less accurate than variational mode decomposition. Dynamic mode decomposition was the least accurate among all the methods.

82.1 Introduction

In the music context, a note is a symbol denoting a musical sound. In western music, notes are ordered on a logarithmic scale [1]. The frequency of all the notes is standardized based 440 Hz which is the frequency of note A4. The frequency of note n is given as $440 \times 2^{\frac{n-49}{12}}$ Hz where n ranges from 1 to 88 on a standard piano. Pitch and frequency are related concepts, the pitch is considered to be the fundamental frequency of the note [2]. An octave is an interval between a musical pitch and another

U. Vamsi Krishna · R. Priyamvada · G. Jyothish Lal (✉) · V. Sowmya · K. P. Soman
Center for Computational Engineering and Networking (CEN), Amrita School of Engineering, Amrita Vishwa Vidyapeetham, Coimbatore, India
e-mail: g_jyothishlal@cb.amrita.edu

S. C. Satapathy et al. (eds.), *Smart Computing Techniques and Applications*,
Smart Innovation, Systems and Technologies 225,
https://doi.org/10.1007/978-981-16-0878-0_82

Fig. 82.1 Piano notes with corresponding notations

pitch which is double its frequency [3]. There are 12 notes in an octave (denoted as C, C#, D...) as shown in Fig. 82.1 [1]. There are 7 complete octaves and each octave contains 12 notes. The 0th octave contains 3 notes that are A0, Bb0, B0, and the 8th octave contains 1 note, i.e., C8. More information regarding pitch and octaves can be found in [1]. The frequency of the piano notes increases exponentially, so the gap between the adjacent frequency of the note increases. For example, the frequency of A3 note 220 Hz which is in 3rd octave. The frequency next to A3 note, i.e., Bb3 is 233.1 Hz. The difference between the notes is 13.1 Hz. In contrast, for the 7th octave, A7 frequency 3520 Hz and the frequency of the next note, i.e., Bb7 3729 Hz. The frequency difference between the notes 209 Hz which is greater than in 3rd octave. If the +10 Hz error is added to both the notes, A3 frequency value will change 230 Hz which is close to the next note, i.e., Bb3. But for A7, if the same amount of error is added then it 2530 Hz which is still close to A7. This observation indicates that the effect of constant error decreases with respect to the octave. Typically used metric like RMSE might not be able to capture the error dependency. Logarithmic metrics can be used to deal with such dependencies. Section 82.4.1 talks about how the metric is selected for comparing the algorithms over octaves. There are many works based on pitch estimation algorithms. YIN is one of the classical pitch estimation algorithms [4] used for fundamental frequency estimation. The work presented in [5] shows how it fails for higher octaves of the piano. The parametric method in [5] also faces problems in detecting pitch for higher octaves. A genetic algorithm strategy is used to estimate the pitch for piano notes [6]. It is compared with three methods that are YIN, parametric F_0 estimation, and nonparametric F_0 estimation. Comparison results show that the genetic algorithm performs almost equal to the compared methods but parametric method performed better. Furthermore, pitch estimation is also required in pitch styling in speech signals [7]. Pitch styling is done to obtain a smooth pitch contour. Wavelet decomposition (WD) is utilized to estimate the pitches for pitch styling. Recent work on wavelet transform shows a modified Morlet wavelet filter to estimate the notes [8]. This work compares the proposed model with a fast Fourier transform and original Morlet wavelet for single pitch as well as multi-pitch estimation. The estimation of notes is done only for a few selected octaves. In one of the works, variational mode decomposition (VMD) is used to find the pitch contour in monophonic Turkish music [9]. The estimated frequencies are compared with the spectrogram of the signal. There is no metric evaluation mentioned for the estimated frequencies. Dynamic mode decomposition (DMD) is widely used for frequency estimation in many fields [10]. Recent work shows a windowed DMD approach which is a replacement for short-term Fourier transform to estimate the pitch contour [11]. As it is just a conceptual idea, there are no results that show concrete evidence of using the DMD algorithm for pitch estimation. A brief explanation of the three mentioned algorithms is given as follows.

82.1.1 Wavelet Decomposition

In wavelet decomposition, scaling and shifting are done to capture the time and frequency information present in the signal. There are many kinds of wavelets like Morlet, Haar, Daubechies, etc. [12]. The selection of these wavelets is dependent on the type of signal. Different level of decomposition leads to a different type of scaling. The generic criterion of scaling is $\frac{1}{2^n}$ where n is the level of decomposition. High scaling value will give higher frequencies accurately and lower scaling value for lower frequency (i.e., bigger window size will give details on lower frequencies and vice-versa).

82.1.2 Variational Mode Decomposition

VMD is a signal processing algorithm used to decompose an input signal into different band-limited sub-components which are called modes. These components are closely packed around the center frequency which is found in the decomposition process [13]. It uses optimization in frequency and time domain to find the optimal modes present in the signal. Here, the optimization is used to find the optimal functions. The mathematics of optimal function finding are known as variational calculus, hence, the name variational mode decomposition. It works on two constraints; the total sum of decomposed signals must be equal to the original and frequency variation from the central frequency (bandwidth) is minimum. The analytical form is obtained by using the Hilbert transform. Frequency mixing is done on the analytical function to find the central frequency present in the corresponding mode. Then, frequency spectrum shifting is done to the baseband by mixing complex exponential which is adjusted with the early estimated central frequency.

Mathematically, we can express as,

$$\min_{m_k(t), \omega_k} \sum_k \left\| \frac{\partial}{\partial t} \left[\left(\delta(t) + \frac{j}{\pi t} \right) * m_k(t) \right) e^{-j\omega_k t} \right] \right\|_2^2 \tag{82.1}$$

$$\text{s.t.} \sum_k m_k(t) = f(t)$$

where $m_k(t)$ is the kth mode of original signal $m(t)$ with center frequency w_k and k is the integer index of the mode. Augmented Lagrangian multiplier is used to convert the constrained to unconstrained problem.

$$L(m_k, w_k, \lambda) = \alpha \sum_k \left\| \frac{\partial}{\partial t} \left[\left(\partial(t) + \frac{j}{\pi t} \right) * m_k(t) \right) e^{-j\psi_k t} \right] \right\|_2^2$$
$$+ \left\| f - \sum_k m_k \right\|_2^2 + \langle x, f - \sum_k m_k \rangle \qquad (82.2)$$

where f is the original signal.

82.1.3 Dynamic Mode Decomposition

DMD is a data-driven method used to predict dynamics of the stationary system. It uses singular value decomposition (SVD) as its core algorithm and is used to estimate the low-rank dynamics [14]. Data snapshots from the signal are stored in the matrices X_1 and X_2. Where X_1 is the matrix containing 1 to n snapshots and the X_2 is the delayed version containing snapshots from 2 to $n + 1$.

$$X_1 = \begin{bmatrix} \vdots & \vdots & & \vdots \\ x_1 & x_2 & \dots & x_n \\ \vdots & \vdots & & \vdots \end{bmatrix}$$

And a delayed version of the snapshot is taken

$$X_2 = \begin{bmatrix} \vdots & \vdots & & \vdots \\ x_2 & x_3 & \dots & x_{n+1} \\ \vdots & \vdots & & \vdots \end{bmatrix}$$

The core part of DMD is to find a matrix A such a way that

$$X_2 \approx A X_1 \qquad (82.3)$$

The eigenvalues of matrix A are used to estimate the frequencies f present in the signal.

$$f = \frac{imag(\log(\lambda))}{2\pi \Delta t} \qquad (82.4)$$

where λ is the eigenvalues of the A matrix which is computed using SVD. Δt is the time-step of the signal. Further, details are given in the [15].

82.2 Dataset Description

Dataset is taken from the database of electronic music studios, the University of Iowa [16]. It contains .aiff stereo (left and right information) format 88 audio files. Each file contains an individual note of Steinway & Sons Model B piano and is recorded using Neumann KM 84 microphone. The dataset is noise-free. The sampling and bit rates of audio files are 44.1 KHz and 16-bit, respectively.

82.3   Methodology

All audio files in the dataset contained more silence regions after and before the original piano note, as shown in Fig. 82.2a. Different lengths of the window are applied on the signal to find out the region where the energy of the signal is high as shown in Fig. 82.2b. After finding the appropriate region, decomposition algorithms are applied. After analyzing the signal, the frequency with the highest energy is taken as the fundamental frequency. Figure 82.3 shows the flow diagram of the method.

82.4   Results and Discussions

82.4.1   Selection of Metrics

Root mean square logarithmic error (RMSLE), RMSE and mean absolute logarithmic error (MALE) are compared to show which metrics can capture the error dependency mentioned in Sect. 82.1. Equations 82.5, 82.6 and 82.7 are RMSLE, RMSE and MALE, respectively.

$$\sqrt{\sum \frac{(\log(y_{\text{ref}} + 1) - \log(y_{\text{pred}} + 1))^2}{N}} \tag{82.5}$$

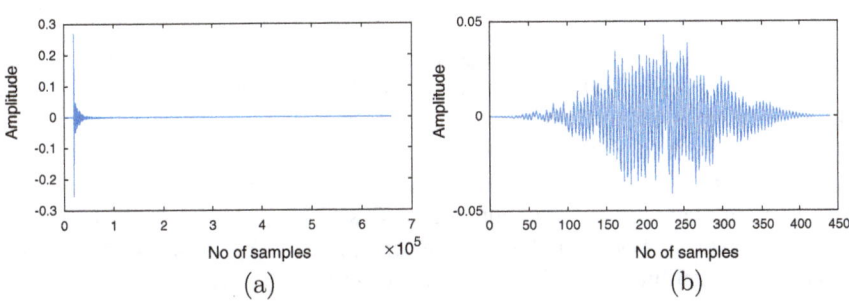

Fig. 82.2   a Original signal and **b** windowed signal

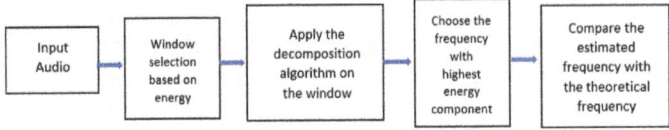

Fig. 82.3   Methodology

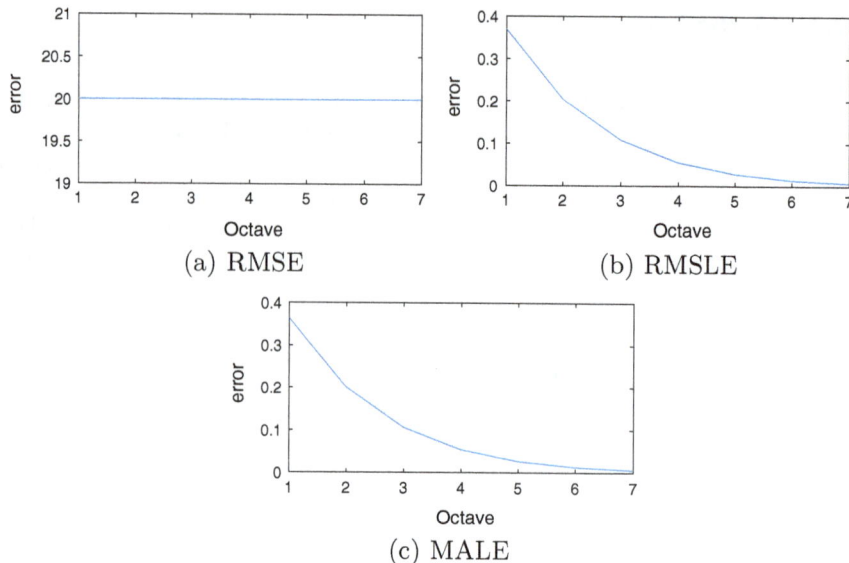

Fig. 82.4 Comparison of metrics over 7 octaves

$$\sqrt{\sum \frac{(y_{\text{ref}} - y_{\text{pred}})^2}{N}} \qquad (82.6)$$

$$\sum \frac{abs(\log(y_{\text{ref}}) - \log(y_{\text{pred}}))}{N} \qquad (82.7)$$

where y_{ref} is reference value, y_{pred} is predicted value and N total number of values. A constant error of ± 20 Hz is added to all the 88 piano frequencies. As shown in Fig. 82.4a, RMSE remains constant throughout the 8 octaves. On the other hand, as shown in Fig. 82.4b, c, RMSLE and MALE are decreasing with respect to the octaves, respectively. The mathematical reason for RMSLE and MALE to work better is due the logarithm function utilized in both the metrics. The expression for the frequency of notes is given as $f(n) = 440 \times 2^{\frac{n-49}{12}}$, where n is the integer index of piano note. By applying logarithmic function on $f(n)$, we get

$$\log(f(n)) = \log(440) + \frac{n - 49}{12} log(2)$$

where $\log(f(n))$ is linear in frequency and the difference between frequencies becomes equal for all octaves. MALE and RMSLE perform identically, but RMSLE uses square root, mean and square operations. On the other hand, MALE uses mean and logarithmic difference. This makes MALE computationally effective.

Table 82.1 Tunable parameters for WD

Octave	Fs (Hz)	Window length (s)	Decomposition level
1	441	0.5	1
2	4410	0.02	2
3	4410	0.5	3
4	4410	0.25	2
5	4410	2	1
6	22,050	0.4	2
7	44,100	0.02	2

Table 82.2 Tunable parameters for VMD

Octave	Fs (Hz)	Window length(s)	K	α
1	220	1	4	20,000
2	2205	0.2	1	15,000
3	2940	0.15	1	10,000
4	4410	0.1	1	10,000
5	4410	0.1	3	10,000
6	8820	0.05	5	10,000
7	8820	0.05	6	5000

Table 82.3 Tunable parameters for DMD

Octave	Fs (Hz)	Window length(s)	Data snapshot length (N)	D
1	221	4.9	100	1
2	294	3.74	100	1
3	441	2.49	180	1
4	882	1.13	180	1
5	2940	1.5	180	1
6	8820	0.12	100	2
7	44,100	0.02	180	1

82.4.2 Parameter Tuning

Tables 82.1, 82.2 and 82.3 show the tunable parameters for WD, VMD and DMD algorithms, respectively. Each algorithm is tuned based on the least MALE. Hanning, Hamming and Blackman windows are used but there is no significant change in the MALE. Applying any of the mentioned window will produce almost the same result. In this paper, Blackman window is used. Sampling rate F_s and window length are the common parameters for all the three algorithms. Table 82.1 contains the parameters

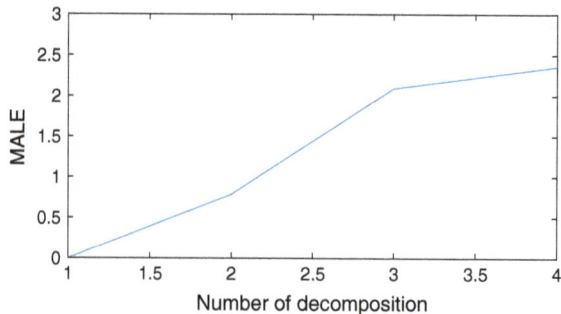

Fig. 82.5 WD parameter tuning for 5th octave

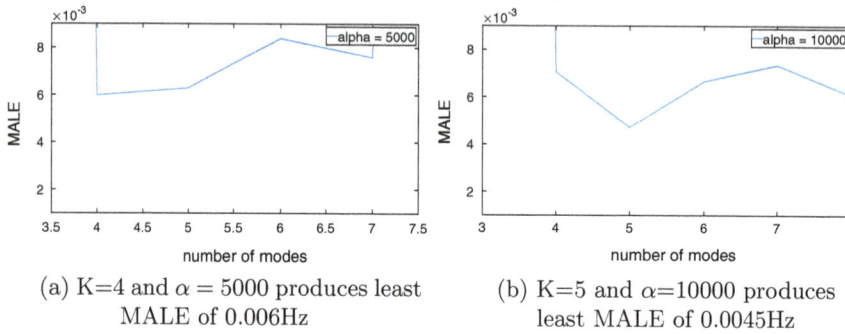

Fig. 82.6 VMD parameter tuning for 6th octave

selected for the WD algorithm. Figure 82.5 shows the selection of the decomposition level with respect to the MALE for octave 5. First level decomposition is selected because it provides the least 0.003 Hz MALE, as shown in Fig. 82.5. Same criterion is used for all the other octaves. Second level decomposition is applied on 2nd, 4th, 6th and 7th octave. First level decomposition is applied on the 1st and 5th octave. The third level of decomposition is applied on the 3rd octave. VMD parameters are given in Table 82.2. K is the number of modes, and α is the bandwidth parameter. Figure 82.6 shows the tuning of parameters based on MALE for octave 6. Figure 82.6(a) shows that $K = 4$ produces the least 0.006Hz MALE for $\alpha = 5000$. Figure 82.6(b) shows that $k = 5$ and $\alpha = 10,000$ gives 0.0045Hz of MALE which is lower than the previous scenario. $K = 5$ and $\alpha = 10,000$ are selected for octave 6. Similarly, for all the octaves, same criterion is applied. α is selected as 5000 for the 7th octave. For 3rd, 4th, 5th and 6th octave, α is 10000 and for 1st and 2nd octave, α is taken as 20000 and 15000, respectively. DMD parameters are given in Table 82.3. Here, D is the number of modes and N is the length of snapshots. Figure 82.7 shows the parameters tuned for octave 6 based on MALE with snapshots 100 and 180 for comparison. For $D = 1$ and $N = 100$, DMD gives 1.005 Hz MALE, as shown in Fig. 82.7a. For $D = 1$ and $N = 180$ DMD gives 1.0065 Hz MALE, as shown in Fig. 82.7b. Comparing both the error,

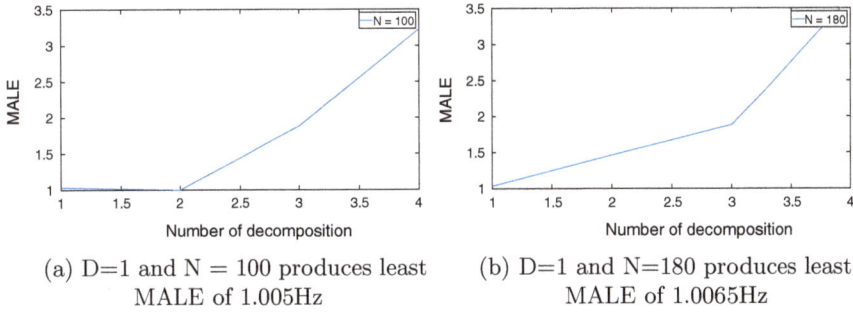

(a) D=1 and N = 100 produces least MALE of 1.005Hz

(b) D=1 and N=180 produces least MALE of 1.0065Hz

Fig. 82.7 DMD parameter tuning for 3th octave

DMD performs better for $D = 2$ and $N = 100$ for 6th octave. Similarly, parameter for all the octaves is selected. For octave 6, the second decomposition is selected, and for the rest of the octaves, the first decomposition is selected. For octave 3rd, 4th, 5th and 7th, the N is taken as 180, and for octave 1st, 2nd, 6th, it is taken as 100. Table 82.4 shows that VMD performed well for all the octaves and overall accuracy is also high. The expression for accuracy is given as

$$\text{Accuracy} = \frac{N_{\text{est}}}{N_{\text{total}}} \times 100$$

where N_{est} is number of notes correctly estimated and N_{total} is total number of notes. As shown in Table 82.2, the bandwidth parameter α keeps on decreasing with an increase in octave. The higher the frequency, the lower the α value. The error for lower octaves is very high, this is because VMD cannot detect frequencies which are close to each other. From octave 1st to 4th, WD did not perform well with respect to VMD. From 5th to 7th octave, WD is able to estimate all the notes correctly.

82.5   Conclusions and Future Scope

The evaluation of all three algorithms shows that VMD performed better for all the octaves. WD also performed well but not accurate as VMD. DMD did not perform well for any of the octave, which makes DMD the least choice for pitch estimation application. While decomposing the signal, VMD has a frequency mixing criteria which can capture the fundamental frequency in conditions where harmonics are as strong as fundamental frequency. WD uses high pass and low pass filters for decomposition which can capture fundamental frequency given the harmonics are not too close. Unlike VMD and WD, DMD does not have any such property with respect to frequency filtering. This could be one of the reasons why DMD was not able to perform for any octave. The evaluation also shows how the algorithms are dependent

Table 82.4 Evaluation of the algorithms

Octave	VMD		WD		DMD	
	MALE	Correct notes	MALE	Correct notes	MALE	Correct Notes
1	0.3536	2	0.385	1	1.66	2
2	0.061	11	0.396	2	1.57	1
3	0.046	11	0.24	5	1.66	1
4	0.0043	12	0.23	8	1.21	3
5	0.0033	12	0.003	12	1.0003	2
6	0.0045	12	0.006	12	1.0005	8
7	0.0133	12	0.015	12	0.74	3
Total notes	72		52		20	
Accuracy (%)	85.71		61.9		23.81	

on sampling frequency and window length for different octaves. Further, experiments can be conducted to identify the optimal sampling frequency and window length to accurately estimate pitches for all the octaves. Multi-pitch estimation can also be evaluated on these algorithms based on MALE. Furthermore, noise can be added to evaluate the algorithms in different conditions. VMD's performance is not expected to reduce due to its wiener filter properties [13]. DMD is very sensitive to noise which can further lower its performance [17]. Evaluating the modified versions of these algorithms like multi-resolution DMD [18], elastic regression VMD [9], etc., can give better insights.

References

1. Klapuri, A., Davy, M.: Signal Processing Methods for Music Transcription. Springer Science & Business Media, Berlin (2007)
2. Goto, M.: A real-time music-scene-description system: predominant-F0 estimation for detecting melody and bass lines in real-world audio signals. Speech Commun. **43**, 11 (2004). https://doi.org/10.1016/j.specom.2004.07.001
3. Smith, W., Cheetham, S.: A Dictionary of Christian Antiquities. John Murray, London (1875). Archived from the original on 30 Apr 2016
4. Cheveigné, A., Kawahara, H.: YIN, a fundamental frequency estimator for speech and music. J. Acoust. Soc. Am. **111**, 1917–30 (2002). https://doi.org/10.1121/1.1458024
5. Emiya, V., David, B., Badeau, R.: A parametric method for pitch estimation of piano tones. In: IEEE International Conference on Acoustics, Speech, and Signal Processing, 1. I-249 (2007). https://doi.org/10.1109/ICASSP.2007.366663
6. Inácio, T., Miragaia, R., Reis, G., Grilo, C., Fernandéz, F.: Cartesian genetic programming applied to pitch estimation of piano notes. In: IEEE Symposium series on computacional inteligence (SSCI) (2016)
7. Wang, D., Narayanan, S.: Piecewise linear stylization of pitch via wavelet analysis, pp. 3277–3280 (2005)

8. Kumar, N., Kumar, R.: Wavelet transform-based multipitch estimation in polyphonic music. Heliyon **6**(1) (2020)
9. Öztürk, B., Akan, A., Bozkurt, B.: Fundamental frequency estimation for monophonical Turkish music by using VMD. In: 2015 23rd Signal Processing and Communications Applications Conference, SIU 2015—Proceedings, pp. 1022–1025 (2015). https://doi.org/10.1109/SIU.2015.7130006
10. Mohan, N., Soman, K.P., Sachin, K.S.: A data-driven approach for estimating power system frequency and amplitude using dynamic mode decomposition. In: International Conference and Utility Exhibition on Green Energy for Sustainable Development (ICUE), vol. 2018, pp. 1–9. Phuket, Thailand (2018). https://doi.org/10.23919/ICUE-GESD.2018.8635792
11. Pogorelyuk, L., Rowley, C.: Melody Extraction For Mirex 2016 Using Dynamic Mode Decomposition. Mirix **2017**, (March 2016)
12. Soman, K.P., Ramachandran, K.I.: Insight Into Wavelets: From Theory to Practice. PHI Learning Pvt. Ltd. (2010)
13. Dragomiretskiy, K., Zosso, D.: Variational mode decomposition. IEEE Trans. Sign. Process. **62**(3), 531–544 (2014)
14. Schmid, P.J.: Dynamic mode decomposition of numerical and experimental data. J. Fluid Mech. **656** (2008). https://doi.org/10.1017/S0022112010001217
15. Kutz, J.N., et al.: Dynamic mode decomposition: data-driven modeling of complex systems. Soc. Ind. Appl. Math. (2017)
16. Mazunik, E., Cash, M.: University of Iowa Electronic Music Studios piano dataset. Available at http://theremin.music.uiowa.edu/MISpiano.html (2001)
17. Hemati, M.S., Rowley, C.W., Deem, E.A., et al.: De-biasing the dynamic mode decomposition for applied Koopman spectral analysis of noisy datasets. Theor. Comput. Fluid Dyn. **31**, 349–368 (2017). https://doi.org/10.1007/s00162-017-0432-2
18. Kutz, J., Fu, X., Brunton, S., Erichson, N.: Multi-resolution dynamic mode decomposition for foreground/background separation and object tracking, pp. 921–929 (2015). https://doi.org/10.1109/ICCVW.2015.122
19. Salamon, J., Gomez, E.: Melody extraction from polyphonic music signals using pitch contour characteristics. IEEE Trans. Audio Speech Language Process. **20**(6), 1759–1770 (2012)
20. Lal, G.J., Gopalakrishnan, E.A., Govind, D.: Epoch estimation from emotional speech signals using variational mode decomposition. Circuits Syst. Sign. Process. **37**, 3245–3274 (2018)

Correction to: Periocular Segmentation Using K-Means Clustering Algorithm and Masking

V. Sandhya and Nagaratna P. Hegde

Correction to:
Chapter 31 in: S. C. Satapathy et al. (eds.),
Smart Computing Techniques and Applications, **Smart**
Innovation, Systems and Technologies 225,
https://doi.org/10.1007/978-981-16-0878-0_31

In the original version of the chapter 31, the following belated correction has been incorporated: Reference 19 has been removed from both reference list and citation. The chapter has been updated with this change.

The updated version of this chapter can be found at
https://doi.org/10.1007/978-981-16-0878-0_31

Author Index

S. C. Satapathy et al. (eds.), *Smart Computing Techniques and Applications*,
Smart Innovation, Systems and Technologies 225,
https://doi.org/10.1007/978-981-16-0878-0

Lightning Source UK Ltd.
Milton Keynes UK
UKHW020607110722
405674UK00001B/24